ANNUAL REVIEW OF
ECOLOGY AND
SYSTEMATICS

ANNUAL REVIEW OF ECOLOGY AND SYSTEMATICS

VOLUME 29, 1998

DAPHNE GAIL FAUTIN, *Editor*
University of Kansas

DOUGLAS J. FUTUYMA, *Associate Editor*
State University of New York at Stony Brook

FRANCES C. JAMES, *Associate Editor*
Florida State University

http://www.AnnualReviews.org   science@annurev.org   650-493-4400

ANNUAL REVIEWS   4139 EL CAMINO WAY   P.O. BOX 10139   PALO ALTO, CALIFORNIA 94303-0139

ANNUAL REVIEWS
Palo Alto, California, USA

International Standard Serial Number: 0066-4167
International Standard Book Number: 0-8243-1429-8
Library of Congress Catalog Card Number: 71-135616

⊗ The paper used in this publication meets the minimum requirements of American
National Standard for Information Sciences—Permanence of Paper for Printed Library
Materials, ANSI Z39.48-1992.

TYPESETTING BY TECHBOOKS, FAIRFAX, VA
PRINTED AND BOUND IN THE UNITED STATES OF AMERICA

Annual Review of Ecology and Systematics
Volume 29 (1998)

CONTENTS

SOME RELATED ARTICLES IN OTHER *ANNUAL REVIEWS*

From the *Annual Review of Earth and Planetary Sciences*, Volume 26, 1998:

Early History of Arthropod and Vascular Plant Associations, Conrad C. Labandeira

Ecological Aspects of the Cretaceous Flowering Plant Radiation, Scott L. Wing and Lisa D. Boucher

Isotopic Reconstruction of Past Continental Environments, Paul L. Koch

From the *Annual Review of Energy and Environment*, Volume 22, 1997:

The Development of the Science of Aquatic Ecosystems, Ruth Patrick

The Role of Moisture Transport Between Ground and Atmosphere in Global Change, D. Rind, C. Rosenzweig, and M. Stieglitz

Terrestrial Ecosystem Feedbacks to Global Climate Change, Daniel A. Lashof, Benjamin J. DeAngelo, Scott R. Saleska, and John Harte

Codes on Environmental Management Practice: Assessing Their Potential as a Tool for Change, Jennifer Nash and John Ehrenfeld

From the *Annual Review of Entomology*, Volume 43, 1998:

Phylogeny and Evolution of Host-Parasitoid Interactions in Hymenopter, J. B. Whitfield

Biodiversity of Stream Insects: Variation at Local, Basin, and Regional Scales, Mark R. Vinson and Charles P. Hawkins

Biological Control of Weeds, Rachel E. Cruttwell McFadyen

Higher-Order Predators and the Regulation of Insect Herbivore Populations, Jay A. Rosenheim

The Ecology and Behavior of Burying Beetles, Michelle Pellissier Scott

Evolution and Ecology of Spider Coloration, G. S. Oxford and R. G. Gillespie

From the *Annual Review of Microbiology*, Volume 52, 1998:

Thinking about Bacterial Populations as Multicellular Organism, James A. Shapiro

Bacteria as Modular Organisms, John H. Andrews

Cooperative Organization of Bacterial Colonies: From Genotype to Morphotype, Eshel Ben-Jacob, Inon Cohen, and David L. Gutnick

(*continued*)

From the *Annual Review of Phytopathology*, Volume 36, 1998:

From the *Annual Review of Plant Physiology and Plant Molecular Biology*, Volume 49, 1998:

ANNUAL REVIEWS is a nonprofit scientific publisher established to promote the advancement of the sciences. Beginning in 1932 with the *Annual Review of Biochemistry*, the Company has pursued as its principal function the publication of high-quality, reasonably priced *Annual Review* volumes. The volumes are organized by Editors and Editorial Committees who invite qualified authors to contribute critical articles reviewing significant developments within each major discipline. The Editor-in-Chief invites those interested in serving as future Editorial Committee members to communicate directly with him. Annual Reviews is administered by a Board of Directors, whose members serve without compensation.

For the convenience of readers, a detachable order form/envelope is bound into the back of this volume.

Annu. Rev. Ecol. Syst. 1998. 29:1–21

MOLECULAR TRANS-SPECIES POLYMORPHISM

Jan Klein, Akie Sato, Sandra Nagl, and Colm O'hUigín
Max-Planck Institut für Biologie, Abteilung Immungenetik, Corrensstrasse 42,
D-72076 Tübingen, Germany; e-mail: jan.klein@tuebingen.mpg.de

KEY WORDS: trans-species polymorphism, *Mhc*, self-incompatibility

ABSTRACT

Trans-species polymorphism (TSP) is the occurrence of similar alleles in related species. Excluding instances in which the similarity arose by convergent evolution, TSP is generated by the passage of alleles from ancestral to descendant species. Closely related, recently diverged species, such as those of the Lake Victoria cichlid flock, may share neutral alleles, but long-lasting TSPs occur only in genetic systems evolving under balancing selection. Two such systems have been studied extensively, the major histocompatibility complex (*Mhc*) of jawed vertebrates and the self incompatibility (*SI*) system of flowering plants. Allelic lineages that diverged many millions of years ago and passed through numerous speciation events have been described in both systems. The lineages may differ at up to 50% of their coding sites, both synonymous and nonsynonymous. The differences arise by the process of incorporation of mutations, which is different from the process of fixation. TSP, on the one hand, complicates phylogenetic analysis, but on the other, it is a useful tool for the study of speciation.

BACKGROUND

In a population of randomly breeding individuals, a newly arisen mutation will suffer one of three fates. The great majority of mutations will be eliminated from the population shortly after their emergence— they will become extinct. A small minority will spread through the population and replace their wild-type counterparts—they will become fixed. Mutations in a third category will achieve a certain equilibrium frequency and then persist in the population for very long periods of time; ultimately, however, they too will either be lost or

1

0066-4162/98/0001$08.00

fixed. Which of these three fates a given mutation will suffer depends mainly on two factors—the selective value of the mutation and chance and their interplay being influenced by the effective size of the population, N_e. Mutations can be selectively neutral, deleterious, or advantageous. The fate of all three types of mutation is influenced by chance (random genetic drift), the effect of which is inversely proportional to the population size. A deleterious mutation under negative selection becomes lost under most circumstances. An advantageous mutation under positive selection can either become fixed (in the case of directional selection) or established at an equilibrium frequency (in the case of balancing selection). A neutral mutation becomes either extinct or fixed, unless it occurs in the vicinity of a site under balancing selection, in which case it, too, may persist for a long time at an equilibrium frequency.

A neutral mutation destined to become fixed needs an average $4N_e$ generations to achieve the fixation frequency of 1 (33). [The average fixation time of a mutation under directional selection is shorter than $4N_e$ generations; its length depends on the selection coefficient, s, in the following manner: $t = (2/s)ln(N_e)$ generations; see 33.] If during this period the species diverges into two or more new forms, the transient neutral polymorphism may be passed on from the ancestral to the descendent species and thus be shared by the latter. One then refers to an *ancestral, shared,* or *trans-species polymorphism* (Figure 1, see color insert) (37, 39). A mutation under balancing selection may persist in a population much longer than the lifetime of a species and hence become part of a long-lasting trans-species polymorphism (TSP). In the case of a neutral (transient) TSP, it is improbable that another mutation in the same gene will achieve appreciable frequency during the time the first mutation is on its way to fixation. Here, therefore, the alleles rarely differ at more than one site, except in very large populations. By contrast, balanced polymorphisms often persist for such a long time that differences at multiple sites may accumulate. The neutral and balanced TSPs have different idiosyncracies and are therefore treated separately.

NEUTRAL TRANS-SPECIES POLYMORPHISMS

It is notoriously difficult to determine whether a mutation is neutral, semi-deleterious, or mildly advantageous. However, mutations in certain regions of the genome, such as intergenic segments, introns, and parts of the 3′ untranslated region (3′UTR), have a high probability of being neutral. Species undergoing rapid diversification, such as those of the haplochromine species flock in Lake Victoria, East Africa, are particularly suitable for studying neutral TSPs. The estimates of the number of haplochromine species in Lake Victoria range from 200 to 500, although many species have disappeared recently (24). Because

the Lake apparently dried out approximately 12,600 years ago (32), it is assumed that most, if not all, of its haplochromines arose from a single ancestral
species within this period (23, 24, 60, 62). The young age of the flock is also
indicated by an extreme difficulty in finding interspecies differences in either
nuclear (48, 74) or mitochondrial (83) genes, even though the species are well
differentiated morphologically and behaviorally, particularly in terms of trophic
specializations (24).

Polymorphism in both the mitochondrial and nuclear DNA is, however, by
no means rare in the Lake Victoria haplochromines. Both neutral and balanced polymorphisms have been detected. The balanced polymorphisms are
described later; the apparently neutral polymorphisms have been detected by
sequencing 3'UTRs or introns of randomly selected nuclear genes. Of the
several polymorphisms uncovered, we discuss one example only, that in the
gene for glucose-6-phosphatase (G6P), an enzyme involved in the terminal steps
of gluconeogenesis and glycogenolysis. The comparison of G6P cDNA from
two species, *Haplochromis nubilis* and *H. xenognathus*, revealed two forms of
the G6P gene differing in a three-base pair (bp) insertion/deletion (indel) and a
single-base pair substitution in the 3'UTR (S Nagl, W Mayer, H Tichy &
J Klein, unpublished data). Further study showed that the two forms occur
in both species and hence represent a shared polymorphism that, because of
its nature and location, is most likely neutral. The same polymorphisms occur
in all nine other species of *Haplochromis* screened. These polymorphisms are
also present in *Haplochromis* species endemic to the satellite lakes (e.g. in Lake
Nabugabo), as well as in species inhabiting rivers of the Lake Victoria basin
(e.g. *H. bloyeti*, *H. burtoni*, *H. sparsidens*, and *H. katavi*).

Cichlid fishes of Lakes Malawi and Tanganyika possess only one of the two
forms, and species of the genus *Tropheus* possess different forms. (The species
flocks of Lakes Victoria and Malawi are believed to have been founded by ancestors that migrated via river systems from Lake Tanganyika, the oldest of the
three great East African lakes; see Reference 60). Both G6P polymorphisms
must have therefore been present in the ancestral riverine species (possibly
H. bloyeti-like; see 25) that founded the Lake Victoria species flock. It was
then passed on to the descendant species as they emerged during the explosive
diversification. Hence the polymorphisms must be >12,000 years old and, although neutral, must have survived all speciation events that created the extant
flock. They thus represent genuine neutral TSPs.

A similar conclusion follows from the study of several other neutral polymorphisms found in a variety of genes of the Lake Victoria haplochromine
fishes. No neutral polymorphism has, however, been found to be shared by
cichlid fishes of Lake Victoria and those of the other two great lakes (S Nagl,
W Mayer, H Tichy, & J Klein, unpublished data). Since the average persistence

time of neutral mutations in a population is $4N_e$ generations, since the observed polymorphisms have persisted for at least 12,000 years, and since the generation time of cichlid fish is 3 years, from the relationship $4N_e \geq 6,000$ generations we get $N_e \geq 1000$ generations.

We are not aware of any shared ancestral polymorphism involving an allele under directional selection; all the nonneutral TSPs involve alleles under balancing selection. The two best-characterized balanced TSPs are those at the class I and class II major histocompatibility complex (*Mhc*) loci and loci controlling self-incompatibility responses in flowering plants. We therefore focus on these two genetic systems and describe other systems only briefly.

BALANCED TRANS-SPECIES POLYMORPHISM OF *Mhc* CLASS I AND CLASS II LOCI

All gnathostomes possess two groups of genes that code for proteins capable of binding short peptides and displaying them on the cell surface. The two groups are referred to as the class I and class II genes of the *Mhc* (38). The peptides are ordinarily derived from the body's own (self) proteins, but during an infection they may also come from the proteins of the pathogens (non-self). The displayed non-self peptides are recognized, together with the presenting class I or class II molecules, by a family of specialized receptors on the surfaces of the thymus-derived lymphocytes, and the recognition initiates the anticipatory (adaptive) form of immune response. The proteins encoded in the two groups of *Mhc* genes differ in structure, in the manner in which they acquire peptides, and in the type of immune response they preferentially initiate. The class I and II genes are, however, clearly related; they apparently were derived from a common ancestor. In most tetrapods the two classes of loci are closely linked in a single *Mhc* cluster, albeit intermingled with other loci (44). In bony fish, the two groups are on different chromosomes. Although the structure of the class I and class II genes themselves and the tertiary structure of the encoded proteins are conserved, the composition of the *Mhc* clusters is highly unstable, having been subjected to repeated cycles of expansion and contraction during their evolution in the various taxa (42, 43). As a result of these upheavals, the clusters often contain large numbers of pseudogenes, truncated genes, and gene fragments. The functional loci (i.e. those that code for molecules capable of presenting peptides to T lymphocytes) are of two kinds, especially in the class I category. Some are highly polymorphic and their encoded proteins are capable of presenting a wide spectrum of peptides, whereas others are oligomorphic or monomorphic and their encoded proteins are specialized in presenting a narrow range of peptides and sometimes also non-peptidic fragments. We discuss the former, frequently referred to as classical genes.

Polymorphism of the classical *Mhc* loci is concentrated in the second of the 5–6 class II gene exons and in the second and third of the 7–8 class I gene exons (40). Within these exons, the polymorphism is largely restricted to sites that specify the amino acids of the peptide-binding region (PBR). The polymorphism is very high in terms of both numbers of alleles and genetic distances between alleles. In the human *Mhc* (the *HLA* complex) at the most polymorphic class I (*HLA-B*) and II (*HLA-DRB1*) loci, 150 and 156 alleles have thus far been registered, respectively (4). Even more spectacular than the high number of alleles, however, are the large genetic distances between the alleles. A pair-wise comparison of, for example, *HLA-DRB1* alleles reveals distribution of genetic distances around a mean of 14 nucleotide differences with a range of 1 to 35 differences in exon 2 alone, which is 270 bp long.

How, then, has this large polymorphism arisen? Originally, immunologists and population geneticists (see, for example, Reference 69) thought it to be the result of rapid intraspecific accumulation of mutations assisted by a gene conversion–like process. An alternative to this interpretation was proposed by Klein (37, 39), who argued that the large genetic distances between alleles reflected allele divergence times in excess of species divergence times. He was led to this TSP hypothesis by the occurrence of serologically indistinguishable Mhc allomorphs (products of allelic genes) in populations separated geographically for long periods of time (20). To test the possibility that the serological identification was imprecise and that in reality the allomorphs might have been similar but not identical, Arden & Klein (1) applied the method of tryptic peptide mapping to the serologically indistinguishable allomorphs isolated from mouse species believed to have diverged more than 1 million years ago (MYA). The allomorphs were indistinguishable also by this method, leaving little doubt that they were in fact identical. This report was thus the first experimental demonstration of TSP at the *Mhc* loci. When methods for studying *Mhc* variability directly at the DNA level came into widespread use, interspecies sharing of *Mhc* allelic lineages could be demonstrated by three groups of investigators. We showed that two *H2-Ab* allelic lineages distinguishable by a two-codon indel were shared by different species of the genus *Mus* and even by the rat, *Rattus norvegicus* (15, 16). Wakeland and colleagues (59, 90) provided evidence for the sharing within the genus *Mus* of two other *H2-Ab* lineages differing in restriction fragment polymorphism and the absence of a short interspersed element (SINE). Lawlor and colleagues (51), as well as Mayer and coworkers (58), demonstrated that certain genes at the *HLA-A* and -*B* loci are more similar in their sequence to certain chimpanzee *Mhc* alleles than to other *HLA* alleles. Trans-species *Mhc* polymorphisms have since been reported, not only in mammals but also in other vertebrates, including bony fishes (22).

BALANCED TRANS-SPECIES POLYMORPHISM
OF SELF-INCOMPATIBILITY LOCI

Angiosperms and fungi have evolved genetic systems that prevent self-fertilization and thus promote outcrossing (27). In flowering plants, the self-incompatibility (*SI*) systems prevent or inhibit the development of an incompatible pollen tube upon the deposition of a pollen grain on the stigma of the pistil. Here, an interaction between identical *S*-gene products leads to the rejection of the self pollen, whereas lack of interaction between nonidentical *S*-gene products allows the non-self pollen tube to develop. In fungi, the *SI* systems prevent either the fusion of gametes or the development of the diploid phase after the fusion. In this case, identical gene products of the mating system fail to interact so that no fertilization occurs, whereas complementarity between nonidentical products allows fertilization to proceed. The molecular mechanisms underlying the SI responses have not been elucidated fully in any of the systems; it is known, however, that they operate on different principles in angiosperms and fungi. Moreover, in both angiosperms and fungi several systems have evolved that achieve the same phenotypic effect (self-incompatibility) by different means.

In angiosperms, an *S*-gene product of the pollen can be derived either from the diploid tissue of the anther tapetum so that the phenotype is determined by two *S* alleles of the parent plant (= sporophytic SI), or it is derived from a single *S* allele of the pollen's haploid genome (= gametophytic SI). An example of the sporophytic SI is provided by the cabbage, *Brassica oleracea*, of the family Cruciferae. In this species, two loci—the S-receptor kinase or *SRK*, and the *S* locus glycoprotein or *SLG*—have been identified in a 200-kb segment of the *S*-region, both expressed primarily in the stigma. The SRK consists of an extracellular S-domain, a transmembrane region, and an intracytoplasmic serine/threonine protein kinase domain. The secreted form of SLG consists of the S-domain, which shows sequence similarity with the S-domain of the SRK; membrane-anchored forms of SLG have, however, also been described. The *SRK* and *SLG* genes presumably arose from a common ancestor by gene duplication. The two *S*-genes seem to coevolve since the S-region of *SLG* in a given haplotype is more similar to the S-region of SRK of that haplotype than to that of SLG in another haplotype. The S-regions of both genes are highly polymorphic. The *SLG* and *SRK* gene products are believed to be somehow involved in the recognition of self-incompatibility (27). The pollen presumably produces an unidentified ligand that influences the interaction between the SLG and the SRK S-domains. The interaction is thought to induce dimerization of SRK molecules, which in turn leads to autophosphorylation of serine and threonine residues in the cytoplasmic part of the SRK (27). The resulting signal-transduction pathway then culminates in localized inhibition of pollen-tube development.

Of the gametophytic SIs, three systems have been partially characterized at the molecular level. In Solanaceae, as represented by *Petunia* or the tobacco plant *Nicotiana*, the product of the *S*-gene in the pistil is a glycoprotein with RNase activity. The protein part of the secreted molecule contains two hypervariable and five conserved regions. The former account for the high polymorphism of the *S*-gene; two of the latter (C2 and C3) have high sequence similarity with various RNases. The hypervariable regions presumably interact with an as-yet-unidentified pollen *S*-gene product. It is believed that in the case of self-pollination the S-RNase produced by the stylar cells is recognized and internalized by receptors (translocators) on the pollen-tube surface (27). Once inside the cell, the enzyme degrades messenger and ribosomal RNA and thus inhibits further pollen tube development. In Papaveraceae, represented by the field poppy, *Papaver rhoeas*, the *S*-gene product of the stigma neither is an RNase nor does it show significant sequence similarity to any other known protein. Following self-pollination, the secreted product presumably binds to a receptor on the pollen-tube surface (possibly the product of the pollen *S*-gene) and the interaction leads to the generation of a second messenger that stimulates the release of Ca^{2+} from intracellular stores. The increased level of intracellular Ca^{2+} then initiates biochemical reactions that inhibit pollen-tube development. In the grasses of the Poaceae family, represented by *Phalaris coerulescens*, the *S*-gene of the pollen encodes thioredoxin H protein, which has also been implicated in the stigmatic response of *Brassica*. Rejection of self-pollen occurs by a direct effect on the synthesis of the pollen-tube wall.

The fact that some of the S loci are highly polymorphic has been known for some time. For example, Emerson (13) identified 37 *S*-alleles in *Oenothera organensis* in a population of 1000 plants. In Solanaceae, 30–50 *S*-alleles are commonly found per population (52), and in *Brassica campestris*, the total number of *S*-alleles has been estimated to be >100 (66). At the *S*-loci, as at the *Mhc* loci, the polymorphism is concentrated in certain hypervariable regions, and in these regions it occurs mainly at specific sites (9). The *S*-locus polymorphism also resembles the *Mhc* polymorphism in that the *S*-alleles may differ from each other at >50% of the sites. The trans-species character of the *S*-locus polymorphism has been demonstrated for both the gametophytic (31, 73) and the sporophytic (12) *SI* systems; it has also been shown in *SI* systems of fungi (55). In the gametophytic *SI* system of Solanaceae, Ioerger and coworkers (31) found that some of the *Nicotiana alata S*-alleles were genetically closer to certain ones of *Petunia inflata* than to other *N. alata* alleles. Simulation studies revealed that the observed pattern of shared polymorphism must have arisen before the divergence of the two species (31).

TRANS-SPECIES POLYMORPHISM IN OTHER GENETIC SYSTEMS

Long-lasting and hence presumably balanced polymorphisms have also been described for other systems. Seemingly the first description of TSP was provided by von Dungern & Hirschfeld (11), who observed that a human antiserum specific for human antigen A of the ABO system agglutinated erythrocytes of a chimpanzee. This finding was extended to other ape species and other blood group antigens by Landsteiner & Miller (50) and by several other investigators (reviewed in 3). An A-like antigen occurs in 90% of common chimpanzee (*Pan troglodytes*) individuals, 19% of gibbons (*Hylobates* sp.), 56% of orangutans (*Pongo pygmaeus*), and most species of Old World monkeys (4–100%), as well as New World monkeys (55–100%). Similarly, a B-like antigen exists in the gorilla (100%), gibbon (42%), siamang (*Symphalangus*, 100%), Old World monkeys (8–100%), and New World monkeys (27–100%).

Other examples of purported TSP include other blood group antigens [e.g. the primate antigens of the MN system (reviewed in 81)], serum allotypes [e.g. the primate Gm system (92)], the ability of humans and other primates to taste phenyl-thiocarbamide (PTC) (7, 19), and mollusc shell color and pattern (21, 72). In all these studies, however, the description of the TSP was limited to the phenotype, and the possibility was not excluded that the polymorphism evolved in each species independently. In at least one of these systems convergence proved, indeed, to be partially responsible for the shared polymorphism. In the ABO system, the antigenic differences are generated by the action of glycosyltransferases, which attach sugar residues to a basic oligosaccharide chain (3). The human A antigen is produced by the attachment of N-acetyl-D-galactosamine (GalNAc) to the ultimate residue of the oligosaccharide—a reaction catalyzed by the enzyme N-acetylgalactosaminyltransferase. The B antigen arises when the galactosyltransferase specified by the *B* gene catalyzes the attachment of D-galactose (D-Gal) to the ultimate residue. Sequence comparisons of the genes coding for these two enzymes in humans and apes have revealed sharing of polymorphic differences in the region specifying the active site (57). Intron sequences, however, supported trans-species evolution within the genus *Homo* and possibly *Australopithecus*, the two genes having diverged 3 MYA, but not between *Homo* and *Pan* or the other ape species tested (68). Thus, the *Mhc* and *Sl* remain the only genetic systems for which long-lasting TSPs have been demonstrated convincingly at the DNA level.

PROPERTIES OF BALANCED TRANS-SPECIES POLYMORPHISM

The Concept of Polymorphism and of Allelic Lineages

To qualify as polymorphism, an allele must attain a certain frequency in the gene pool. The definitions of polymorphism and allelism are thus tied to the notion of a gene pool. Since the boundaries of a gene pool are those of a species, strictly speaking polymorphism cannot be trans-specific and genes from different species are not alleles. The meaning of trans-species polymorphism is therefore different from intraspecies polymorphism. We define TSP as the sharing of allelic lineages between species that are either in an ancestor-descendant relationship or are extant species derived from a common ancestor. Here, then, frequency is not a parameter of the definition, although it of course underlies intraspecific polymorphism on which the concept of allelic lineage is based. This brings us to the question of what an allelic lineage is.

An allelic lineage can best be defined in terms of the coalescent theory (25, 29, 34, 85) or, more precisely, the lines-of-descent theory (26) as a group of alleles at a given locus in a particular species whose coalescence time (i.e. the time of origination of a new allele) antedates the species' divergence time. This definition requires specification of the time point in the history of the alleles at which coalescences are to be scored, for the number of lineages will depend on it. The point can lie anywhere within an interval delineated by the species' divergence time at one end and the coalescence time of all the alleles at the other end. The choice of the time point will be dictated by the specific aim of a study, but two points are particularly relevant: one given by the emergence time of a species and the other by the divergence time of the two most closely related extant species from a common ancestor. The former is of interest because of its relevance to the nature of the speciation process, the latter because a comparison of two related species normally defines TSP. The choice of either of the two is not without problems, however, some of which we discuss.

To determine how many allelic lineages at a given locus have been passed to an extant species from its immediate ancestor, we first make a phylogenetic tree of all the extant alleles at this locus, using an algorithm such as the unweighted pair-group method with arithmetic mean (UPGMA) that assumes constancy of evolutionary rate. We then convert genetic distances to time intervals and on the resulting time scale identify the point representing the species' emergence time. A line drawn from this point perpendicular to the branches of the tree identifies the allelic lineages: All alleles that diverged from a given branch-time line intersection constitute a single allelic lineage. There are three major problems with this method. First, since most of the interallelic differences are at sites that are under balancing selection, the assumption of rate constancy may not be warranted. This problem can partially be alleviated by focusing the comparisons on presumably neutrally evolving sites—synonymous codon positions and introns. (It should be pointed out, however, that neither for the *Mhc* nor for the *SI* genes has any evidence been put forward for inconstancy of evolutionary rates at nonsynonymous sites.) Second, the usual method of obtaining a time scale for a UPGMA tree is not applicable to trans-specifically evolving gene systems, since it assumes congruence of gene and species trees. This problem can, however, be solved by using the minimum-minimum method of divergence time estimation (78). In this method, a large collection of allelic sequences at orthologous loci is assembled from related species, the alleles are arranged into all possible interspecies combinations, the genetic distances are calculated for each pair, and the pairs with the smallest distance are identified. These pairs are assumed to have diverged at approximately the same time as the two species and are therefore used to calculate the evolutionary rate and thus to calibrate the time scale. Third, the emergence time of a species is not easy to determine precisely. The direct ancestor of an extant species is normally never available because, according to the prevailing theory of speciation (cladogenesis; see 71), it ceases to exist at the moment when the new species are born. Fossils of the direct ancestor are rarely available, and if they are, their relationship to the extant species is difficult to establish. As for molecular estimates, the coalescence time of a neutral gene not only has a large standard error, it may also yield misleading information in a different way. As was mentioned earlier, neutral sequences can and do evolve trans-specifically, even though their coalescence patterns are relatively shallow. The estimated coalescence time of neutrally evolving genes therefore does not necessarily mark the beginning of a species. It is therefore best to present the species' emergence time not as a point on a time line but as an interval defined by the standard error of the estimate and the probability distribution of the coalescence time. The number of allelic lineages is then a range specified by this interval.

Age of Allelic Lineages

At the moment of its creation, whether by gene duplication or some other means, a new locus is present as a single copy in a gene pool. It is therefore monomorphic and so polymorphism obviously cannot be older than a locus at which it occurs. The emergence time of a locus thus sets an upper limit to the age of an allelic lineage. The lower limit is set, according to our definition of an allelic lineage, by the emergence time of an extant species sampled for polymorphism. During the interval from the birth of a locus to the birth of an extant species, some or all of the allelic lineages can be lost and new lineages then begin to diverge. The loss of allelic lineages can result from random genetic drift, especially during a phase of reduced population size (bottleneck) or from a shift in selection pressure.

The *Mhc* and *Sl* loci have different evolutionary histories and this fact has influenced the TSP in these two systems. The evolution of the *Mhc* is punctuated by cycles of gene expansion and contraction effected by gene duplications and deletions, respectively (43, 64). Each contraction cleans the slate in terms of polymorphism and marks the beginning of divergence of new alleles and new allelic lineages. Because the expansions-contractions occur irregularly, the *Mhc* loci and polymorphisms are of different ages in the various taxa. To illustrate these age variations, we use the example of the best-characterized *Mhc*, the human leukocyte antigen or *HLA* complex. Its main polymorphic class I loci are *HLA-A*, *-B*, and *-C*; its main polymorphic class II loci are *HLA-DRB1*, *-DQA1*, *-DQB1*, and *-DPB1*. Each of these loci has a distinct evolutionary history and idiosyncracies in terms of polymorphism. Orthologs of the three class I loci have been found only in primate mammals; each of the non-primate orders has its own set of functional, polymorphic class I loci (40). Even in primates, however, the *HLA-A*, *-B*, and *-C* orthologs are apparently functional and polymorphic only in Catarrhini, whereas recent Platyrrhini and presumably also prosimians use different polymorphic sets (92). From sequence comparisons and using the evolutionary rate of 1.4×10^{-9} /synonymous site/year, we estimate that the *A* and *B* loci diverged approximately 50 MYA and *C* diverged from *B* about 30 MYA. The *A* locus may have therefore diverged from an ancestral *B*-like locus after or close to the time of the Platyrrhini-Catarrhini split, which is currently estimated to have occurred 55 MYA (56). Similarly, the divergence of the *C* locus may have postdated or occurred close to the separation of Old World monkeys from hominids, an event currently dated to 36 MYA (56). By a similar approach, the divergence times of the oldest allelic lineages at the *HLA-A*, *-B*, and *-C* loci can be estimated as being 18, 22, and 20 MYA. Thus, although the *B* locus is older than the *A* and *C* loci, its oldest lineages diverged later than the oldest *A*- and *C*-locus lineages.

In contrast to the class I loci, the same class II gene families are found not only in primates, but in other orders of eutherian mammals (40) and some of them possibly also in marsupials (80). They are, however, absent in nonmammalian vertebrates, which have evolved their own class II gene families. In teleost fishes, for example, every order thus far tested has a different set of class II (and also class I) loci derived from distinct ancestor genes (42). The presence of the same class II gene families in the various mammalian taxa does not mean, however, that they are equally utilized. Even a cursory survey has revealed significant differences in the functional permanence and hence polymorphism of the different genes. The mouse, for example, uses only the *DR* and *DQ* homologs (40), whereas the mole rat, *Spalax ehrenbergi*, relies primarily on the *DP* homologs (65).

Even though the human *DRB1*, *DQA1*, *DQB1*, and *DBP1* loci are all very old, their polymorphisms are of different ages. At the *DRB1* locus, the oldest allelic lineages diverged approximately 55 MYA, presumably after the separation of Platyrrhini and Catarrhini, as the lineages are found only in the latter (75, 76; but see section on Convergence). The deep divergence time of these lineages is indicated not only by large intron sequence differences but also by the presence of dateable insertion elements (75, 76). Compared with *DRB1*, the polymorphisms at the other *HLA* class II loci are considerably younger. We estimate the deepest divergences of allelic lineages at the *DQA1*, *DQB1*, and *DPB1* loci to have occurred 25–30 MYA, 28 MYA, and 15 MYA, respectively. The *DPB1* polymorphism is thus the youngest of all the *HLA* loci, although it, too, began to unfold before the emergence of hominids. All these estimates are based on differences at synonymous sites of full-length cDNA sequences, so they are not influenced by the processes specifically affecting the nonsynonymous sites specifying the peptide-binding region.

The angiosperm *SI* loci apparently undergo less frequent expansions and contractions compared with the *Mhc* loci. In some *SI* systems, a single pair of loci appears to be orthologous among the various species of an angiosperm family (27). In other systems, remnants of multiple pairs are indicative of an expansion followed by deletion of some of the copies, and haplotype-specific rearrangements have been described (5). But on the whole, the *SI* system appears to be more stable than the *Mhc*, perhaps because of the need to coevolve the two interactive components. Each *S* haplotype encodes at least two gene products that are coadapted structurally; frequent rearrangements in the *SI* region would disrupt the tight linkage that underlies the coadaptation.

Because of the relative stability of the region, the *SI* polymorphism can persist for a very long time. Indeed, in some angiosperm *SI* systems, the deepest divergence of allelic lineages has been estimated to have occurred more than 70 MYA (31). It is generally believed that the ancestors of the *SI* loci were involved in functions other than self-incompatibility, perhaps defense against parasites (28, 53), and that only when the angiosperms began to radiate explosively during the Cretaceous did a change in function occur. This probably happened approximately 120 MYA (89), and it is conceivable that the oldest allelic lineages may have already begun to diverge at that time.

Number of Allelic Lineages

Allelic lineages can be counted either by comparing polymorphisms of related extant species (i.e. shared polymorphism) or by extrapolating from a descendant to an antecedent species (i.e. ancestral polymorphism). The former way of counting can be illustrated by three examples of species that diverged at different times. The first example is provided by the haplochromine species flock of Lake Victoria. Although the flock is comprised of several hundred species (23) and although most of the species are reproductively isolated in their natural habitats, the flock may be fewer than 12,000 years old (32). This young age of the flock is reflected in the *Mhc* polymorphism of the species constituting it (35, 70; H Sültmann, N Takahata, Y Satta, H Tichy & J Klein, in preparation). The various species share not only allelic lineages but also alleles (i.e. not just similar but identical alleles are common occurrences among the species). Although the species may differ in the frequencies of the individual alleles and in that not all of them have all the alleles, the overall profiles of the polymorphism at both the class I (B Murray, P Nielsson, H Sültmann & J Klein, unpublished data) and class II (35, 70; H Sültmann, N Takahata, Y Satta, H Tichy & J Klein, in preparation) are quite similar. Obviously, during the short interval since their divergence the species have not had enough time to generate their own noticeable species-specific polymorphism. They have largely maintained the polymorphism they presumably inherited from the riverine species that was their ancestor.

The second example is provided by the comparison of Lake Victoria and Lake Malawi haplochromine cichlid flocks. Morphological and molecular evidence suggests that the flocks in these two lakes originated more than 2 MYA from a common ancestor, presumably living in the third great East African lake, Lake Tanganyika (60). The longer separation time is clearly reflected in the *Mhc* polymorphism. Identical alleles shared between species of the two flocks are either absent or rare, but lineages within which genes in the two flocks differ at only a few sites are common (35, 70). Although some of the species may lack some of the lineages, most lineages are well represented among the various species in the two lakes. In this example, therefore, the species of the two flocks are beginning to differentiate from one another in their *Mhc* polymorphism by diversification and sorting of allelic lineages.

The third example illustrates a later stage still in the diversification process. *Danio rerio* and *D. albolineatus*, two species of fish family Cyprinidae, diverged from their last common ancestor approximately 13 MYA (61). During this time their *Mhc* genes diverged to the extent that no identical alleles at the class II loci are shared by the two species (22). Closely related genes are still found, however, within the same allelic lineages, some of them differing at as few as two or three sites in the most polymorphic part (exon 2).

The number of allelic lineages scored at the time of birth of a species will depend on the age of the species, the size of the founding population, and the evolutionary history of the locus. In very young species, such as the Lake Victoria haplochromines, which have been founded by large populations (H Sültmann, N Takahata, Y Satta, H Tichy & J Klein, in preparation), most of the alleles presumably present in the ancestral species persist in the descendant forms. Here, then, virtually all alleles are scored as allelic lineages that made it through the speciation phase. An example of a moderately young species is *Homo sapiens sapiens*, estimated to have emerged 150,000 to 200,000 YA (82). In this species, the number of *Mhc* allelic lineages is lower than the number of alleles because of polymorphism generated in the postspeciation phase. The estimated numbers of allelic lineages for the *HLA-A, -B, -C, -DRB1, -DQA1, -DQB1,* and *-DPB1* loci are 36, 75, 35, 64, 8, 19, and 14, respectively (4).

The effect of population size on the number of allelic lineages is illustrated by the comparison of the *SI* polymorphism between the horsenettle, *Solanum carolinense*, and the ground cherry, *Physalis crassiflora* (73). Although the two species are closely related, *P. crassiflora* possesses only 2 of the 28 allelic lineages found in *S. carolinense*. The difference is attributed to loss of allelic lineages during a postulated bottleneck phase that may have occurred at the time of speciation or at any time subsequently.

Diversification of Allelic Lineages

Allelic lineages evolve by a process that differs from fixation of mutations in a population. Instead of spreading through a gene pool until it reaches the frequency of 1, a mutation under balancing selection or a neutral mutation closely linked to a site that is under balancing selection spreads only until it reaches its equilibrium frequency and then fluctuates around it. We call the process by which allelic lineages diversify *incorporation* (45). The equilibrium frequencies of different mutations or of a combination of mutations may differ within an allelic lineage so that a pairwise comparison of alleles then gives a distribution of differences like the one shown in Figure 2, and the tree of the lineage appears bushy. Neutral mutations remain linked to the selected mutation in whose vicinity they occurred until recombination separates them. Recombination is therefore the only way by which a mutation could be transferred from one lineage to another. The existence of allelic lineages and their long persistence times indicate that recombinations between lineages are rare. Incorporation of mutations under selection is accelerated in comparison to neutral mutations. The latter, even when linked to a site maintained by balancing selection, incorporate by random genetic drift, but their probability of persistence in the gene pool is increased by the association with the selected site. Neutral sites closely linked to sites under selection can therefore be expected to show higher polymorphism than other sites. This effect decreases with distance from the selected sites owing to an increase in the frequency of recombination (C O'hUigin, R Dawkins & J Klein, in preparation).

Incorporation rates have been estimated for the human *Mhc* loci (77). Their average values of 1.2×10^{-9} per synonymous site per year and 5.9×10^{-9} per

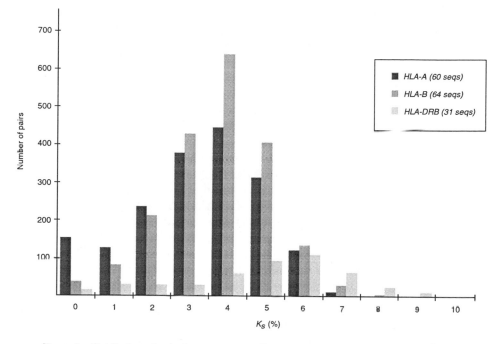

Figure 2 Distribution of pairwise synonymous distances (K_s) for 60 *HLA-A*, 64 *HLA-B*, and 31 *HLA-DRB1* alleles. Full-length sequences only were chosen, and distances were calculated for all synonymous sites using the method of Li (54).

nonsynonymous PBR site per year are no greater the substitution rates of most non-*Mhc* loci (46). Although rapid evolution of *HLA-B* genes, purportedly by gene conversion, has been reported (2, 93), the evidence remains anecdotal. By contrast, a considerable body of data obtained in diverse vertebrate taxa supports the notion that *Mhc* genes evolve slowly, even at nonsynonymous PBR sites that are under balancing selection (46). Average evolutionary rates apparently also characterize the *SI* loci (8, 26, 49, 73, 88). Differences in the degree of diversification exist between allelic lineages at the *Mhc* loci. For example, in contrast to other *HLA-DRB1* lineages, which are all very bushy in appearance on phylogenetic trees, the *DRB1*07* and *DRB1*09* lineages appear trimmed, consisting only of the main branch and one or two side branches. This appearance cannot be attributed to a young age of the lineage because, in terms of sequence differences, it is highly divergent from other lineages, and furthermore, it is shared, in the same trimmed form, between humans and chimpanzees. In this case, presumably, selection favors specific sets of mutations, and once these are incorporated, no further diversification of nonsynonymous PBR sites is tolerated.

A pairwise comparison of synonymous diversity among *Mhc* alleles reveals a difference between class I and class II sequences. The mean of the diversity distribution is higher, the spread is larger, and the tail is longer in the class II alleles at most of the class II loci in comparison to the class I loci (Figure 2). This difference is probably a reflection of the greater age of class II loci and polymorphisms relative to the class I loci and polymorphisms.

The coalescent theory predicts a certain pattern of the coalescence process that, if appropriately scaled, is the same for neutral genes (29, 34, 84) and for genes under balancing selection (85). In this pattern, for example, the time required for all the extant alleles save two to coalesce is the same as that required for the last two alleles to coalesce. Alleles at some of the *Mhc* and *SI* loci, however, do not behave according to this expectation. The oldest *HLA-DRB1* allelic lineages, for example, diverge in short succession. A similar phenomenon is seen also at some of the *SI* loci (26, 31, 49). Deviations of this sort might happen by chance alone when one or several of the assumptions on which the coalescent theory is based (demographically stable population, symmetric balancing selection, infinite allele model of mutation, absence of recombination) are violated. In the case of the *DRB1*, the disturbing factor might have been the instability of the population (N Takahata, personal communication).

Agents of Selection

Explaining the long persistence time of allelic lineages without invoking selection would require prohibitively large population sizes. That selection is responsible for the TSP of the *Mhc* and *SI* loci is indicated by several observations (46), most notably by the excess of nonsynonymous compared with synonymous changes at specific sites. At the *Mhc* loci, these are the sites that code for the PBR (30); at the *SI* loci, the function of the selected sites has not been established with certainty, but it is widely assumed that they specify amino acid residues involved in the interaction between the products of two genes, one expressed in the pollen and the other in the pistil (9). In both systems the selection is of the balancing type, but its precise nature is not well understood. In the *SI* system, the polymorphism is believed to be maintained by frequency-dependent selection (63), in the *Mhc* by overdominant selection (86), although frequency-dependent selection has not been excluded. In the case of the *Mhc*, the selection pressure is believed to be exerted by parasites (10, 36), but direct evidence for this supposition is limited and alternative or complementing agents have also been proposed (79). In the *SI* systems, the pressure is presumably exacted by inbreeding depression resulting from self-fertilization (18). The selection pressures in the two systems are thus of a different nature, although there are certain superficial similarities between them.

Ancestral Polymorphism Versus Convergence

The presence of similar alleles in different species does not always signal persistence of ancient polymorphisms. In some cases, the similarity can result from convergent evolution (67). Convergence can be distinguished from ancestral polymorphism by comparison of synonymous sites. An example of convergence mimicking ancestral polymorphism between distant taxa is provided by the primate *Mhc-DRB1* alleles. Comparisons of this gene's exon 2 sequences revealed the presence in the Platyrrhini of alleles that were more closely related to certain Catarrhini than to other Platyrrhini alleles (17). Yet sequencing of the introns of these genes, even those flanking exon 2, failed to support this relationship (K Kriener, C O'hUigin & J Klein, in preparation). The similarity in exon 2, which specifies the PBR, apparently arose by selection-driven parallel evolution.

SIGNIFICANCE OF TRANS-SPECIES POLYMORPHISM

To an evolutionary biologist, the phenomenon of TSP presents two sides, one a nuisance and the other beneficial. The former aspect reveals itself as a complicating factor in phylogenetic analyses by the observation that gene trees do not necessarily match species trees. This problem, which can potentially occur in analysis of any locus, selected or neutral, has recently been discussed (6) and so need not be detailed. The beneficial side of trans-species polymorphism, which is only now being appreciated, is that a polymorphism that passes through the speciation phase has the potential of revealing a great deal about the phase itself. For example, effective population size is an important parameter influencing trans-species persistence of a polymorphism. By the same token, therefore, TSP is an important source of information about the size of the population before, during, and after speciation (41). At the time of the mitochondrial Eve frenzy, we applied a simple computer simulation algorithm to the polymorphism of the *HLA* loci and estimated that the founding effective population size of the human species was large, on the order of 10^4 individuals (41). Subsequently, more refined estimates based on an analytic approach (47, 87) confirmed and extended the initial analysis. It has been argued that the *HLA* loci are evolving very fast at their PBR-encoding sites and that the *Mhc*-based estimates of founding population size are therefore flawed (14). This argument is, however, invalidated by estimates based exclusively on synonymous sites; even when the PBR-encoding sites are excluded from the analysis, the range of the number of allelic lineages existing 150,000 to 200,000 YA remains large, and their passage therefore requires a large founding population. The human species is not an isolated case in this regard: Large founding population sizes are required in all other species studied thus far, including the cichlid fishes of the Great East African Lakes (35, 70; H Sültmann, N Takahata, Y Satta, H Tichy & J Klein, in preparation) and Darwin's finches of the Galapagos Islands (89).

Founding population size is one aspect of speciation that can be investigated with the help of TSP. Another is the pattern of species divergence. In diagrams, speciation is usually depicted as the splitting of one ancestral into two descendant species (i.e. as bifurcation). There is, however, very little evidence that this is, indeed, how species arise, and alternatives, such as speciation by multifurcation or star-shaped phylogeny, cannot be excluded in most cases. TSP provides an opportunity to investigate experimentally how the splitting process proceeds. Speciation happens so fast that phylogenetic trees based on fixation of mutations cannot be expected to have the power necessary to resolve it into steps. It should, however, be possible to use differential segregation of allelic lineages into emerging species and differences in allelic frequencies to reveal the phylogenetic relationships among the emerging species. Other aspects of speciation can be investigated in similar ways.

ACKNOWLEDGMENTS

We thank Niamh Ni Bhleithin for editorial assistance, Beth Coffey for the preparation of computer graphics, Yoko Satta for SI locus sequence files, and Naoyuki Takahata for helpful comments on the manuscript.

Visit the *Annual Reviews home page* at
http://www.AnnualReviews.org

Literature Cited

1. Arden B, Klein J. 1982. Biochemical comparison of major histocompatibility complex molecules from different subspecies of *Mus musculus*: evidence for trans-specific evolution of alleles. *Proc. Natl. Acad. Sci. USA* 79:2342–46
2. Belich MP, Madrigal JA, Hildebrand WH, Zemmour J, Williams RC, et al. 1992. Unusual *HLA-B* alleles in two tribes of Brazilian Indians. *Nature* 357:326–29
2a. Blancher A, Klein J, Socha WW, eds. 1997. *Molecular Biology and Evolution of Blood Group and MHC Antigens in Primates*. Berlin: Springer-Verlag. 570 pp.
3. Blancher A, Socha WW. 1997. The ABO, Hh and Lewis blood group in human and nonhuman primates. See Ref. 2a, pp. 30–92.
4. Bodmer J, Marsh SGE, Albert ED, Bodmer WF, Bontrop RE, et al. 1997. Nomenclature for factors of the HLA system, 1996. *Hum. Immunol.* 53:98–128
5. Boyes DC, Nasrallah ME, Vrebalov J, Nasrallah JB. 1997. The self-incompatibility (S) haplotypes of *Brassica* contain highly divergent and rearranged sequences of ancient origin. *Plant Cell* 9:237–47
6. Brower AVZ, DeSalle R, Vogler A. 1996. Gene trees, species trees, and systematics: a cladistic perspective. *Annu. Rev. Ecol. Syst.* 27:423–50
7. Chiarelli B. 1963. Sensitivity to P.T.C. (phenol-thiocarbamide) in primates. *Folia Primatol.* 1:88–94
8. Clark AG. 1997. Neutral behaviour of shared polymorphism. *Proc. Natl. Acad. Sci. USA* 94:7730–34
9. Clark AG, Kao T-h. 1991. Excess nonsyn-

onymous substitution at shared polymorphic sites among self-incompatibility alleles of Solanaceae. *Proc. Natl. Acad. Sci. USA* 88:9823–27

9a. David CS, ed. 1987. *H-2 Antigens, Genes, Molecules, Function.* New York: Plenum

10. Doherty PC, Zinkernagel RM. 1975. Enhanced immunological surveillance in mice heterozygous at the *H-2* gene complex. *Nature* 256:50–52

11. Dungern E von, Hirschfeld L. 1911. Über Vererbung gruppen-spezifischer Strukturen des Blutes (III). *Immunitätsforschung* 8:526–30

12. Dwyer KG, Balent MA, Nasrallah JB, Nasrallah MG. 1991. DNA sequences of self-incompatibility genes from *Brassica campestris* and *Brassica oleracea*: polymorphism predating speciation. *Plant Mol. Biol.* 16:481–86

13. Emerson S. 1938. The genetics of self-incompatibility in *Oenothera organensis. Genetics* 23:190–202

14. Erlich IIA, Bergstrom TF, Stoneking M, Gyllensten U. 1996. HLA sequence polymorphism and the origin of humans. *Science* 274:1552–54

15. Figueroa F, Günther E, Klein J. 1988. MHC polymorphism pre-dating speciation. *Nature* 335:265–67

16. Figueroa F, Klein J. 1987. Origin of H-2 polymorphism. See Ref. 9a, pp. 61–76.

17. Figueroa F, O'hUigin C, Tichy H, Klein J. 1994. The origin of primate *Mhc-DRB* genes and allelic lineages as deduced from the study of prosimians. *J. Immunol.* 152:4455–65

18. Fisher R. 1961. A model for the generation of self-sterility alleles. *J. Theor. Biol.* 1:411–14

19. Fisher RA, Ford EB, Huxley J. 1939. Taste testing the anthropoid apes. *Nature* 144:750

20. Götze D, Nadeau J, Wakeland EK, Berry RJ, Bonhomme F, et al. 1980. Histocompatibility-2 system in wild mice. X. Frequencies of H-2 and Ia antigens in wild mice from Europe and Africa. *J. Immunol.* 124:2675–81

21. Gould SJ. 1988. Prolonged stability in local populations of *Cerion agassizi* (Pleistocene-Recent) in Great Bahama Bank. *Paleobiology* 14:1–18

22. Graser R, O'hUigin C, Vincek V, Meyer A, Klein J. 1996. Trans-species polymorphism of class II Mhc loci in danio fishes. *Immunogenetics* 44:36–48

23. Greenwood PH. 1974. The cichlid fishes of Lake Victoria, East Africa: the biology and evolution of a species flock. *Bull. Br. Mus. Nat. Hist. Zool. Suppl.* 6:1–134

24. Greenwood PH. 1981. *The Haplochromine Fishes of the East African Lakes.* Ithaca, NY: Cornell Univ. Press

25. Griffiths RC, Tavaré S. 1994. Sampling theory for neutral alleles in a varying environment. *Philos. Trans. R. Soc. Ser. B* 344:403–10

26. Hinata K, Watanabe M, Yamakawa S, Satta Y, Isogai A. 1995. Evolutionary aspects of the *S*-related genes of the *Brassica* self-incompatibility system: synonymous and nonsynonymous base substitutions. *Genetics* 140:1099–104

27. Hiscock SJ, Kües U, Dickinson HK. 1996. Molecular mechanisms of self-incompatibility in flowering plants and fungi—different means to the same end. *Trends Cell Biol.* 6:421–28

28. Hodgkin T, Lyon GD, Dickinson HG. 1988. Recognition in flowering plants: a comparison of the *Brassica* self-incompatibility system and plant pathogen interactions. *New Phytol.* 110:557–69

29. Hudson RR. 1982. Estimating genetic variability with restriction endonucleases. *Genetics* 100:711–19

30. Hughes AL, Nei M. 1988. Pattern of nucleotide substitution at major histocompatibility complex class I loci reveals overdominant selection. *Nature* 355:167–70

31. Ioerger TR, Clark AG, Kao T-H. 1990. Polymorphism at the self-incompatibility locus in Solanaceae predates speciation. *Proc. Natl. Acad. Sci. USA* 87:9732–35

32. Johnson TC, Scholz CA, Talbot MR, Kelts K, Ricketts RD, et al. 1996. Late Pleistocene desiccation of Lake Victoria and rapid evolution of cichlid fishes. *Science* 273:1091–93

33. Kimura M, Ohta T. 1969. The average number of generations until fixation of a mutant gene in a finite population. *Genetics* 61:763–71

34. Kingman JFC. 1982. The coalescent. *Stoch. Process Appl.* 13:235–48

35. Klein D, Ono H, O'hUigin C, Vincek V, Goldschmidt T, et al. 1993. Extensive MHC variability in cichlid fishes of Lake Malawi. *Nature* 364:330–34

36. Klein J. 1977. Evolution and function of the major histocompatibility complex: facts and speculations. In *The Major Histocompatibility System in Man and Animals,* ed. D. Götze, pp. 339–78. New York: Springer-Verlag

37. Klein J. 1980. Generation of diversity at MHC loci: Implications for T-cell receptor repertoires. In *Immunology 80,* ed. M Fougereau, J Dausset, pp. 239–53. London: Academic

38. Klein J. 1986. *Natural History of the Major*

Histocompatibility Complex. New York: Wiley

39. Klein J. 1987. Origin of major histocompatibility complex polymorphism: the trans-species hypothesis. *Hum. Immunol.* 19:155–62

40. Klein J, Figueroa F. 1986. Evolution of the major histocompatibility complex. *CRC Crit. Rev. Immunol.* 6:295–386

41. Klein J, Gutknecht J, Fischer N. 1990. The major histocompatibility complex and human evolution. *Trends Genet.* 6:7–11

42. Klein J, Klein D, Figueroa F, Sato A, O'hUigin C. 1997. Major histocompatibility complex genes in the study of fish phylogeny. In *Molecular Systematics of Fishes*, ed. TD Kocher, CA Stepien, pp. 271–83. New York: Academic

43. Klein J, Ono H, Klein D, O'hUigin C. 1993. The accordion model of Mhc evolution. In *Progress in Immunology*, ed. J Gergely, G Petranyi, pp. 137–43. Heidelberg: Springer-Verlag

44. Klein J, Sato A. 1998. Birth of the major histocompatibility complex. *Scand. J. Immunol.* 47:199–209

45. Klein J, Satta Y, O'hUigin C, Mayer WE, Takahata N. 1992. Evolution of the primate DRB region. In *HLA 1991, Proc. Int. Histocompat. Works., 11th, Yokohama, Jpn.*, ed. T Kimiyoshi, M Aizawa, pp. 45–56. Oxford: Oxford Univ. Press

46. Klein J, Satta Y, O'hUigin C, Takahata N. 1993. The molecular descent of the major histocompatibility complex. *Annu. Rev. Immunol.* 11:269–95

47. Klein J, Satta Y, Takahata N, O'hUigin C. 1993. Trans-specific Mhc polymorphism and the origin of species in primates. *J. Med. Primatol.* 22:57–64

48. Kornfield IL. 1991. *Genetics*. In *Cichlid Fishes. Behavior, Ecology and Evolution*, ed. MHA Keenleyside, pp. 103–28. London: Chapman & Hall

49. Kusaba M, Nishio T, Satta Y, Hinata K, Ockendon D. 1997. Striking sequence similarity in inter- and intra-specific comparisons of class I SLG alleles from *Brassica oleracea* and *Brassica campestris*: implications for the evolution and recognition mechanism. *Proc. Natl. Acad. Sci. USA* 94: 7673–78

50. Landsteiner K, Miller CP Jr. 1925. Serological studies on the blood of the primates, II. The blood groups in anthropoid apes. *J. Exp. Med.* 43:853–62

51. Lawlor DA, Ward FE, Ennis PD, Jackson AP, Parham P. 1988. HLA-A and B polymorphism predate the divergence of humans and chimpanzees. *Nature* 335:268–71

52. Lawrence MJ, Lane MD, O'Donnell S, Franklin-Tong VE. 1993. The population genetics of the self-incompatibility polymorphism in *Papaver rhoeas*. V. Cross-classification of the S-alleles of samples from three natural populations. *Heredity* 71:581–90

53. Lee H-S, Singh A, Kao T-h. 1992. RNase X2, a pistil-specific ribonuclease from *Petunia inflata*, shares sequence similarity with solanaceous *S* proteins. *Plant Mol. Biol.* 20:1131–41

54. Li W. 1993. Unbiased estimation of the rates of synonymous and nonsynonymous substitution. *J. Mol. Evol.* 36:96–99

55. Lukens L, Yicun H , May G. 1996. Correlation of genetic and physical maps at the *A* mating-type locus of *Coprinus cinerus*. *Genetics* 144:1471–77

56. Martin RD. 1993. Primate origins: plugging the gaps. *Nature* 363:223–34

57. Martinko JM, Vincek V, Klein D, Klein J. 1993. Primate ABO glycosyl-transferases: evidence for trans-species evolution. *Immunogenetics* 37:274–78

58. Mayer WE, Jonker M, Klein D, Ivanyi P, van Seventer G, Klein J. 1988. Nucleotide sequences of chimpanzee Mhc class I alleles: evidence for trans-species mode of evolution. *EMBO J.* 7:2765–74

59. McConnell TJ, Talbot WS, McIndoe RA, Wakeland EK. 1988. The origin of Mhc class II gene polymorphism within the genus *Mus*. *Nature* 332:651–54

60. Meyer A. 1993. Phylogenetic relationships and evolutionary processes in East African cichlid fishes. *Trends Ecol. Evol.* 8:279–84

61. Meyer A, Biermann CH, Ort' G. 1993. The phylogenetic position of the zebrafish (*Danio rerio*), a model system in developmental biology: an invitation to the comparative method. *Proc. R. Soc. London Ser. B* 252:231–36

62. Meyer A, Kocher TD, Basasibwaki P, Wilson AC. 1990. Monophyletic origin of Lake Victoria cichlid fishes suggested by mitochondrial DNA sequences. *Nature* 347:550–53

63. Nasrallah JB. 1997. Evolution of the *Brassica* self-incompatibility locus: a look into *S*-locus gene polymorphisms. *Proc. Natl. Acad. Sci. USA* 94:9516–19

64. Nei M, Gu X. Sitnikova T. 1997. Evolution by the birth-and-death process in multigene families of the vertebrate immune system. *Proc. Natl. Acad. Sci. USA* 94:7799–806

65. Nizetic D, Figueroa F, Dembic Z, Nevo E, Klein J. 1987. Major histocompatibility complex gene organization in the mole rat, *Spalax ehrenbergi*. Evidence for transfer of

function between class II genes. *Proc. Natl. Acad. Sci. USA* 84:5828–32

66. Nou IS, Watanabe M, Isogai A, Hinata K. 1993. Comparison of S-alleles and S-glycoproteins between two wild populations of *Brassica campestris* in Turkey and Japan. *Sex. Plant Reprod.* 6:79–86

67. O'hUigin C. 1995. Quantifying the degree of convergence in primate *Mhc-DRB* genes. *Immunol. Rev.* 143:123–40

68. O'hUigin C, Sato A , Klein J. 1997. Evidence for convergent evolution of A and B blood group antigens in primates. *Hum. Genet.* 101:141–48

69. Ohta T. 1991. Role of diversifying selection and gene conversion in evolution of major histocompatibility complex loci. *Proc. Natl. Acad. Sci. USA* 88:6716–20

70. Ono H, O'hUigin C, Tichy H, Klein J. 1993. Major-histocompatibility-complex variation in two species of cichlid fishes from Lake Malawi. *Mol. Biol. Evol.* 10: 1060–72

71. Otte D, Endler JA. 1989. *Speciation and Its Consequences*. Sunderland, MA: Sinauer

72. Owen DF. 1966. Polymorphism in Pleistocene land snails. *Science* 152:71–72

73. Richman AD, Uyenoyama MK, Kohn JR. 1996. Allelic diversity and gene genealogy at the self-incompatibility locus in the Solanaceae. *Science* 273:1212–16

74. Sage RD, Loiselle PV, Basasibwaki P, Wilson AC. 1984. Molecular versus morphological change among cichlid fishes of Lake Victoria. In *Evolution of Fish Species Flocks*, ed. AA Echelle, I Kornfield, pp. 185–201. Orono, ME: Univ. Maine Press

75. Satta Y, Mayer WE, Klein J. 1996. Evolutionary relationships of *HLA-DRB* genes inferred from intron sequences. *J. Mol. Evol.* 42:648–57

76. Satta Y, Mayer WE, Klein J. 1996. HLA-DRB intron 1 sequences: implications for the evolution of HLA-DRB genes and haplotypes. *Hum. Immunol.* 51:1–12

77. Satta Y, O'hUigin C, Takahata N, Klein J. 1993. The synonymous substitution rate of the major histocompatibility complex loci in primates. *Proc. Natl. Acad. Sci. USA* 90: 7480–84

78. Satta Y, Takahata N, Schšnbach C, Gutknecht J, Klein J. 1991. Calibrating evolutionary rates at major histocompatibility complex loci. In *Molecular Biology of the Major Histocompatibility Complex*, ed. J Klein, D Klein, pp. 51–62. Heidelberg: Springer-Verlag

79. Serjeantson SW. 1989. The reasons for MHC polymorphism in man. *Transplant. Proc.* 21:598–601

80. Slade RW, Mayer WE. 1995. The expressed class II a-chain genes of the marsupial major histocompatibility complex belong to eutherian mammal gene families. *Mol. Biol. Evol.* 12:441–50

81. Socha WW, Blancher A. 1997. The MNSs blood group system. See Ref. 2a, pp. 93–112

82. Stringer C, McKie R. 1996. *African Exodus. The Origins of Modern Humanity*. London: Jonathon Cape

83. Sturmbauer C, Meyer A. 1992. Genetic divergence, speciation and morphological stasis in a lineage of African cichlid fishes. *Nature* 358:578–81

84. Tajima F. 1983. Evolutionary relationship of DNA sequences in finite populations. *Genetics* 105:437–60

85. Takahata N. 1990. A simple genealogical structure of strongly balanced allelic lines and trans-species evolution of polymorphism. *Proc. Natl. Acad. Sci. USA* 87: 2419–23

86. Takahata N, Satta Y, Klein J. 1992. Polymorphism and balancing selection at major histocompatibility complex loci. *Genetics* 130:925–38

87. Takahata N, Satta Y, Klein J. 1995. Divergence time and population size in the lineage leading to modern humans. *Theor. Popul. Biol.* 48:198–218

88. Uyenoyama MK. 1995. A generalized least-squares estimate for the origin of sporophytic self-incompatibility. *Genetics* 139:975–92

89. Vincek V, O'hUigin C, Satta Y, Takahata N, Boag PT, et al. 1997. How large was the founding population of Darwin's finches? *Proc. R. Soc. London Ser. B* 264:111–18

90. Wakeland EK, Tarruzzee RW, Lu C C, Potts W, McIndoe RA, et al. 1987. The evolution of MHC class II genes within the genus *Mus*. See Ref. 9a, pp. 139–53

91. Wang AC, Shuster J, Fudenberg HH. 1969. Evolutionary origin of the Gm peptide of immunoglobulins. *J. Mol. Biol.* 41:83–86

92. Watkins DI. 1995. The evolution of major histocompatibility class I gene in primates. *Crit. Rev. Immunol.* 15:1–19

93. Watkins DI, McAdam SN, Liu X, Strang CR, Milford EL, et al. 1992. New recombinant HLA-B alleles in a tribe of South Amerindians indicate rapid evolution of MHC class I loci. *Nature* 357:329–33

Annu. Rev. Ecol. Syst. 1998. 29:23–58

PRINCIPLES OF PHYLOGEOGRAPHY AS ILLUSTRATED BY FRESHWATER AND TERRESTRIAL TURTLES IN THE SOUTHEASTERN UNITED STATES

DeEtte Walker and John C. Avise

Department of Genetics, University of Georgia, Athens, Georgia 30602;
e-mail: walker@bscr.uga.edu, avise@bscr.uga.edu

KEY WORDS: historical biogeography, mitochondrial DNA, molecular evolution, conservation
genetics

ABSTRACT

Geographic patterns in mtDNA variation are compiled for 22 species of freshwater and terrestrial turtles in the southeastern United States, and the results are employed to evaluate phylogeographic hypotheses and principles of genealogical concordance derived previously from similar analyses of other vertebrates in the region. The comparative molecular findings are interpreted in the context of intraspecific systematics for these taxa, the historical geology of the area, traditional nonmolecular zoogeographic information, and conservation significance. A considerable degree of phylogeographic concordance is registered with respect to (*a*) the configuration of intraspecific mtDNA subdivisions across turtle species, (*b*) the principal molecular partitions and traditional morphology-based taxonomic boundaries, (*c*) genetic patterns in turtles versus those described previously for freshwater fishes and terrestrial vertebrates in the region, and (*d*) intraspecific molecular subdivisions versus the boundaries between major zoogeographic provinces as identified by composite ranges of species in the Testudines. Findings demonstrate shared elements in the biogeographic histories of a diverse regional biota. Such phylogeographic concordances (and discordances) have ramifications for evolutionary theory as well as for the pragmatic efforts of taxonomy and conservation biology.

23

0066-4162/98/1120-0023$08.00

In the study of dispersal and distribution of animals, it is important to see that the physical conditions lead, and that in a more or less definite succession the flora and fauna follow; thus the fauna comes to fit the habitat as a flexible material does a mold. The time is passed when faunal lists should be the aim of faunal studies. The study must not only be comparative, but genetic, and much stress must be laid on the study of the habitat, not in a static, rigid sense, but as a fluctuating or periodical medium.

Charles Adams, 1901

INTRODUCTION

The spatial genetic architecture of any species is likely to be a complex outcome of contemporary demographic and ecological forces acting upon a preexisting population structure that was molded by biogeographic factors operative throughout the evolutionary history of a species. Molecular methods are well suited for (*a*) describing current population genetic structures and (*b*) recovering historical components of those structures. A particularly useful molecule is animal mitochondrial (mt) DNA, which, by virtue of a rapid rate of change and a nonrecombining mode of asexual transmission through female lineages, permits powerful phylogeographic inferences at the levels of conspecific populations and closely related species (9, 39, 67). Given the great diversity of ecological and evolutionary factors that can influence genealogical structures, an idiosyncratic phylogeographic outcome might be expected for each species.

Nonetheless, comparative molecular assessments of many freshwater fishes, terrestrial vertebrates, and maritime species in the southeastern United States have revealed repeated patterns at several levels (reviews in 6, 8). These studies prompted the original formulation of phylogeographic hypotheses (9) and principles of genealogical concordance (8, 10), which are summarized in Table 1. Because these concepts were motivated (rather than independently tested) by the regional biogeographic data available at the time, they were considered provisional ideas pending further empirical evaluation. Thus, an important question is whether these phylogeographic hypotheses and concordance trends may prove generalizable to other taxonomic groups and to other regional biotas. Here we provide a summary of the results of one such extended set of independent tests: comparative evaluations of intraspecific phylogeographic patterns in the Testudines (turtles and tortoises) distributed across the southeastern United States.

BACKGROUND

This research on the Testudines was motivated by comparative phylogeographic patterns reported for conspecific populations within each of several freshwater fish species in the southeastern United States (11, 14, 40, 54): bowfin (*Amia*

Table 1 Phylogeographic hypotheses and principles of genealogical concordance

---

Phylogeographic Hypotheses (from Reference 9)

  I. Most species are composed of geographic populations whose members occupy recognizable genealogical branches of an extended intraspecific pedigree.

 II. Species with limited or "shallow" phylogeographic population structure have life histories conducive to dispersal and have occupied ranges free of long-standing impediments to gene flow.

III. Intraspecific monophyletic groups distinguished by large genealogical gaps usually arise from long-term extrinsic (biogeographic) barriers to gene flow.

Aspects of Genealogical Concordance (theoretical corollaries of Phylogeographic Hypothesis III; after Reference 8)

A. Concordance across sequence characters within a gene (yields statistical significance for putative gene-tree clades).

B. Concordance in significant genealogical partitions across multiple genes within a species (establishes that gene-tree partitions register population-level phylogenetic partitions).

C. Concordance in the geography of gene-tree partitions across multiple codistributed species (implicates shared historical biogeographic factors in shaping intraspecific phylogenies).

D. Concordance of gene-tree partitions with spatial boundaries between traditionally recognized biogeographic provinces (implicates shared historical biogeographic factors in shaping intraspecific phylogenies and organismal distributions).

---

calva); mosquitofish (*Gambusia affinis/Gambusia holbrooki*); largemouth bass (*Micropterus salmoides*); and four species of sunfish (*Lepomis punctatus, Lepomis microlophus, Lepomis gulosus*, and *Lepomis macrochirus*). Within each of these species, deep and geographically oriented phylogenetic "breaks" typically have distinguished populations in the eastern portion of the species' range (river drainages primarily along the Atlantic coast and in peninsular Florida) from those to the west (most drainages entering the Gulf of Mexico from western Georgia or Alabama to Louisiana). An example involving the spotted sunfish (*Lepomis punctatus*) is presented in Figure 1.

Additional genetic substructure is evident within some of these east-west phylogeographic units, but these differences typically were "shallow" (low mtDNA sequence divergence) relative to the matrilineal separations between regions. In some of these fishes, notably *M. salmoides* (44), *L. macrochirus* (11), and the *Gambusia* complex (54), contact zones of introgressive hybridization also have been documented by allozymes (in conjunction with mtDNA) in geographically intermediate populations primarily in western Georgia and eastern Alabama.

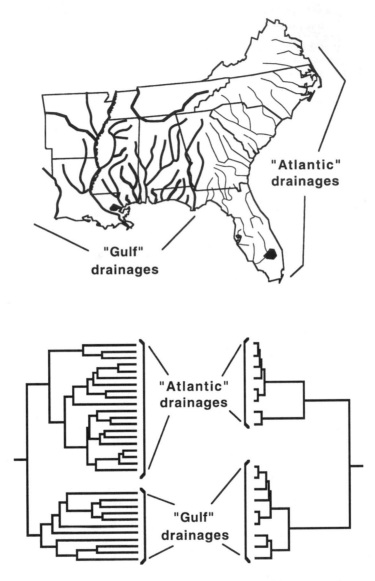

faunal similarity
(241 fish species)

mtDNA genealogy
(*Lepomis punctatus*)

The major mtDNA phylogeographic subdivisions in freshwater fishes are consistent with historical biogeographic scenarios that invoke long-term (Pleistocene or earlier) separations of drainages and the fishes they contain into the Atlantic coast region and the Gulf coast region. The subvisions are also consistent with relatively recent physical connections permitting interdrainage exchange of fishes within each area. Such patterns are concordant with a traditional class of biogeographic information: the distributional boundaries of species. In a compilation of ranges for the 241 freshwater fish species native to the southeastern region, the most fundamental break was identified at the Apalachicola River it that forms part of the Alabama-Georgia boundary (58). Thus, a phenetic clustering of 31 drainages based on a presence-absence matrix of species revealed two basal assemblages (Figure 1): (*a*) an eastern (mostly Atlantic coast) group composed of all rivers from the Savannah to the Suwannee and (*b*) a western (Gulf coast) group composed of the Apalachicola and all drainages westward to Louisiana (Figure 1). Most of the intraspecific breaks in fish mtDNA reported previously (11, 14, 54) fall in this same general area as well, thus yielding concordance aspects C and D as defined in Table 1.

Several studies of terrestrial vertebrates in the southeastern United States also have revealed strong intraspecific phylogeographic partitioning in mtDNA and a tendency (though not as consistent or well documented as in fishes) for concordant spatial patterns (8). Populations in peninsular Florida (and in some cases the adjoining Atlantic coast) tend to be strongly differentiated from those to the north and west. These results have been attributed to the insularization effects of a Floridian peninsula that was relatively isolated periodically during the Pleistocene and earlier (18, 23). Such isolating effects on faunal distributions were predicted many years ago: Remington (48) first emphasized the large number of species whose populations display a "suture zone" of secondary contact situated at the boundary between the Floridian peninsula and the continental mainland. Remington afforded this contact region a status equal to that of only five other major suture zones in North America.

←——————————————————————————————

Figure 1 (*Top*) Map of the 10 southeastern states that are of primary interest in the current analysis. *Heavy lines* depict most of the river drainages in the Gulf of Mexico freshwater biotic province, and *light lines* indicate river drainages in the Atlantic province. (*Bottom left*) Phenogram summarizing faunal similarities among 31 southeastern river drainages, based on a presence-absence matrix of 241 fish species in the area (after Reference 58). Note the basal split between the Atlantic and Gulf regions. (*Bottom right*) Phenogram summarizing relationships among mitochondrial (mt) DNA haplotypes observed in the spotted sunfish (*Lepomis punctatus*) sampled across more than a dozen major drainages in the region (after Reference 14). Note that the fundamental split in the mtDNA gene tree distinguishes conspecific specimens of the Atlantic coast drainages from those of the Gulf coast drainages.

Thus, in addition to tests of the general phylogeographic hypotheses and concordance aspects listed in Table 1, comparative studies of the Testudines in the southeastern United States permit independent evaluations of more specific biogeographic predictions. With respect to life history, aptitude for dispersal, and potential response to vicariant biogeographic effects, aquatic turtles as a group should be intermediate between freshwater fishes and low-vagility terrestrial mammals. Most turtles are associated with aquatic environments but are able to move on land; tortoises are slow moving and have no aquatic affiliations. Thus, if previous genetic findings for other freshwater and terrestrial vertebrate taxa in the southeastern United States are a guide, any deep phylogenetic partitions uncovered within species of Testudines might also tend to distinguish populations (*a*) in peninsular Florida from those on the main body of the continent and/or (*b*) in the Atlantic coastal region from those to the west and along the Gulf coast.

PHYLOGEOGRAPHIC OUTCOMES BY SPECIES

A total of 35 species of Testudines are native to the southeastern United States. Twenty-two of these (63%) have been the subject of molecular analyses based on mtDNA restriction sites or sequences (Table 2). Typically, a genetic study involved the assay of about 240–500 base pairs per individual either as an accumulation of data from multiple-restriction enzyme assays or as sequences obtained directly from particular mitochondrial genes (e.g. the control region). Many studies involved assays of scores of specimens, often sampled from populations scattered throughout the region. Table 2 is also a compilation of information on genetic variability either as reported directly in the original publications or as calculated by us from the data provided.

Following are brief species-by-species descriptions of major mtDNA phylogeographic studies. The original papers should be consulted for details. Typically, the authors used (*a*) parsimony and/or maximum likelihood (59) as applied to DNA sequence data or to presence-absence restriction-site matrices and (*b*) neighbor-joining (53) and/or phenetic clustering (56) as applied to genetic distance matrices. In no case did these alternative phylogenetic procedures yield inconsistent or conflicting outcomes with respect to the major mtDNA intraspecific phylogroups that are the focus of this review. Furthermore, these phylogroups invariably received strong statistical support by criteria such as bootstrapping (20) in phylogenetic appraisals presented in the original publications.

Thus, for simplicity and for ease of visual comparison across studies, results summarized below are presented in the form of cluster phenograms (and associated geographic maps) as plotted on common scales of estimated nucleotide sequence divergence. For the most part, we are not concerned with genetic structure of local populations within the principal phylogroups, although (small sample sizes notwithstanding) such shallower genealogical structure was pronounced for most species in terms of spatial heterogeneity in haplotype frequencies (e.g. Figure 2).

Freshwater Turtles

STERNOTHERUS MINOR (MUSK TURTLE)   In *Sternotherus minor*, a small-bodied turtle confined to the southeastern United States, two morphological subspecies have been recognized: *S.m. minor* to the east and *S.m. peltifer* to the west (Figure 3). In both restriction-site and direct-sequencing assays, Walker et al (61) observed numerous mtDNA haplotypes, all of which were spatially localized (Figure 2). These haplotypes cluster into two distinct phylogroups whose geographic orientations align well with these subspecies as traditionally recognized (Figure 3). Because the mean genetic distance between these intraspecific phylogroups is considerably greater than observed haplotype distances within either assemblage, net sequence divergence is large: $p \cong 0.032$ in the restriction-site assays (Table 2).

To explain the geographic distributions of the two subspecies, Iverson (25) suggested that an ancestral *S. minor* stock invaded southeastern North America during the Miocene and subsequently became vicariantly subdivided into two units—one in peninsular Florida and the other in what is now north-central Alabama. The current distributions were suggested to be a result of post Miocene-Pliocene dispersal from these refugial areas, probably facilitated for ancestral *S.m. peltifer* by a well-known historical connection of the current Tennessee River system to rivers draining southward through Alabama into Mobile Bay (37, 57).

STERNOTHERUS DEPRESSUS (FLATTENED MUSK TURTLE)   The range of *Sternotherus depressus*, confined to the Black Warrior River in northern Alabama (Figure 3), is completely encircled by that of *S. minor peltifer*. The flattened musk turtle has been of questionable taxonomic status (see discussion in 65), but it is thought to be related closely to *S. minor* and is currently on the federal list of threatened and endangered species. Notwithstanding its odd distribution and an uncertain genetic etiology for a characteristic flattened carapace, this form is phylogenetically distinct in mtDNA composition from both *S. m. peltifer* and *S. m. minor* as well as from all other kinosternid turtle species in the southern United States (Figure 4).

STERNOTHERUS ODORATUS (STINKPOT)   *Sternotherus odoratus* is traditionally considered monotypic: It is relatively uniform in morphology and life history features throughout its range, so no taxonomic subspecies have been recognized (49, 55, 60). However, striking geographic differentiation was uncovered

Table 2 Summary information for terrestrial and freshwater turtles genetically surveyed from the southeastern United States

Assay method	Number of base pairs[a]	Number of subspecies[b]	Phylogeographic unit (mtDNA)[c]	Number of individuals	Number of mtDNA haplotypes	Genotypic diversity[d]	Nucleotide diversity[e]	Net sequence divergence between units[f]	Reference
Whole-mtDNA restriction fragment length polymorphisms (RFLP)									
Sternotherus minor	367	2	A	18	3	0.451	0.001		
			B	34	7	0.817	0.004		
Total				52	10	0.859	0.017	A vs B = 0.032	61
Sternotherus odoratus	395	0	A	41	9	0.836	0.007		
			B	41	5	0.712	0.002		
			C	16	2	0.125	0.000		
Total				98	16	0.899	0.016	A vs B = 0.014	64
								A vs C = 0.028	
								B vs C = 0.024	
Kinosternon subrubrum	434	3	A	16	1	0.000	0.000		
			B	24	7	0.798	0.003		
			C	19	8	0.868	0.005		
			D	5	4	0.900	0.008		
Total				64	20	0.902	0.043	A vs B = 0.070	62
								A vs C = 0.067	
								A vs D = 0.053	
								B vs C = 0.048	
								B vs D = 0.027	
								C vs D = 0.030	
Kinosternon baurii			A	5	2	0.400	0.001		
			B	13	6	0.718	0.003		
Total	359	0		18	8	0.817	0.006	A vs B = 0.010	62
Trachemys scripta			A	40	1	0.000	0.000		
			B	25	1	0.000	0.000		
Total	240	2		65	2	0.482	0.003	A vs B = 0.006	12, C[g]
Graptemys pseudogeographica (A)		7		39	3	0.527	0.002	A vs B = 0.010	

Taxon	Mean no. base pairs[a]	No. subspecies[b]	n	No. haplotypes	$1-\Sigma f_i^2$[d]	Mean pairwise sequence divergence[e]	Net sequence divergence[f]	Reference
Graptemys pulchra (B)		4	16	4	0.800	0.005		
Graptemys geographica (C)		1	12	1	0.000	0.000	A vs C = 0.028 B vs C = 0.026	
Total	400	12 species	57	8	0.803	0.013		31
Control-region sequence								
Gopherus polyphemus[h]								
A			1	1	0.000	0.000		
B			1	1	0.000	0.000	A vs B = 0.021	
Total	363–429	0	2	2	0.000	0.000		42
Sternotherus minor								
A			18	4	0.661	0.003		
B			34	13	0.915	0.014	A vs B = 0.013	
Total	430	2	52	17	0.925	0.017		61
Sternotherus depressus	402	0	6	4	0.866	0.007		65
Chelydra serpentina	409	2	66	3	0.172	0.001		63
Macroclemys temminckii								
A			18	1	0.000	0.000	A vs B = 0.026	
B			47	4	0.420	0.002	A vs C = 0.028	
C			95	6	0.540	0.005	B vs C = 0.017	
Total	420	0	160	11	0.837	0.015		51, 52
Deirochelys reticularia								
A			3	2	0.675	0.002		
(B + C)			45	7	0.471	0.002	A vs (B + C) = 0.043	
Total	436	3	50	9	0.533	0.006		C[g]

[a]Mean number of base pairs assayed per individual (sequenced directly, or contained within the surveyed recognition sites).

[b]Number of subspecies conventionally recognized based primarily on morphology (from Reference 17).

[c]Number of moderately or highly distinct phylogeographic units as identified in the mtDNA assays (see text).

[d]$1 - \Sigma f_i^2$, where f_i is the frequency of the ith mtDNA haplotype.

[e]Mean pairwise sequence divergence between individuals within a mtDNA phylogeographic unit.

[f]Net sequence divergence $= p_{xy} - 0.5(p_x + p_y)$ where p_{xy} is the mean pairwise mtDNA sequence divergence between phylogeographic units x and y, and p_x and p_y are the mean pairwise sequence distances within x and y.

[g]C, current study.

[h]This species also was surveyed through polymerase chain reaction–based RFLP assays of mtDNA sequences (see Reference 42 for details).

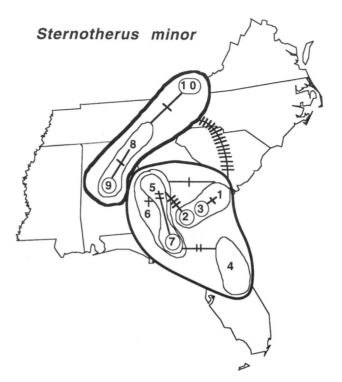

Figure 2   Parsimony network for the 10 different mitochondrial (mt) DNA haplotypes observed in restriction-site assays of the musk turtle *Sternotherus minor*, superimposed over the geographic source of the collections (after Reference 61). *Circles* and *other lines* encompass the ranges within which particular haplotypes were observed among the total of 52 specimens assayed, and *slashes* across branches of the network indicate numbers of restriction-site changes along each path.

in the mtDNA assays, with all haplotypes spatially localized. Three major phylogroups with regional distributions are evident (Figure 5): (*a*) group C in Florida and south Georgia, (*b*) group B along the Atlantic seaboard from Georgia to Virginia, and (*c*) group A in all locales to the west. Within the latter assemblage, two phylogeographic subgroups occur (64): one from northern Alabama to western Virginia and the other in western sites from southern Missouri through Mississippi and Louisiana.

KINOSTERNON SPECIES (MUD TURTLES)   For reasons that will become apparent, two traditionally recognized congeners in the southeastern United States are considered together. The range of *Kinosternon subrubrum* encompasses most of the region, where three parapatric morphological subspecies typically

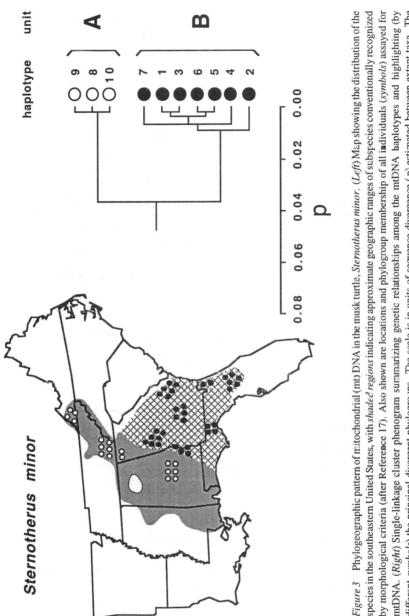

Figure 3 Phylogeographic pattern of mitochondrial (mt) DNA in the musk turtle, *Sternotherus minor*. (*Left*) Map showing the distribution of the species in the southeastern United States, with *shaded regions* indicating approximate geographic ranges of subspecies conventionally recognized by morphological criteria (after Reference 17). Also shown are locations and phylogroup membership of all individuals (*symbols*) assayed for mtDNA. (*Right*) Single-linkage cluster phenogram summarizing genetic relationships among the mtDNA haplotypes and highlighting (by different *symbols*) the principal divergent phylogroups. The scale is in units of sequence divergence (*p*) estimated between extant taxa. The open area in northern Alabama describes the range of a related threatened species, the flattened musk turtle, *Sternotherus depressus* (see text).

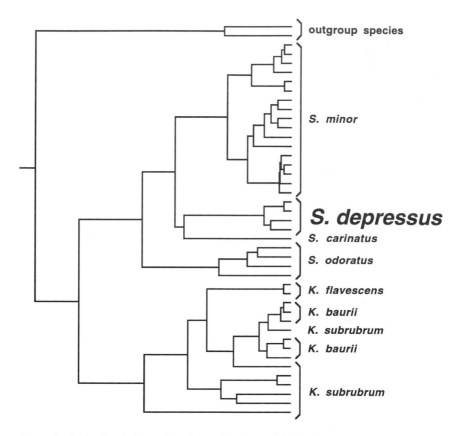

outgroup species

S. minor

S. depressus

S. carinatus

S. odoratus

K. flavescens

K. baurii

K. subrubrum

K. baurii

K. subrubrum

Figure 4 Distinctive phylogenetic position of the threatened flattened musk turtle, *Sternotherus depressus*, within a broader array of some presumed relatives in the genera *Sternotherus* and *Kinosternon* (Kinosternidae) (after Reference 65). These particular assays involved control region sequences of the mitochondrial mtDNA molecule. They were based on samples chosen to represent the major phylogroups identified by restriction fragment length polymorphisms (RFLPs) in more extensive geographic surveys.

have been recognized (Figure 6). In terms of mtDNA, four major phylogroups are evident (Figure 6): (*a*) group D, confined to the Florida peninsula (consistent with the traditionally described range of *K. subrubrum steindachneri*); (*b*) group C, along the Atlantic seaboard from south Georgia to Virginia (part of the traditional *K. subrubrum subrubrum*); (*c*) group B, in a central region from the Florida Panhandle and western Georgia to Mississippi (also *K. subrubrum subrubrum*); and (*d*) group A, from the west (in the traditional range of *K. subrubrum hippocrepis*).

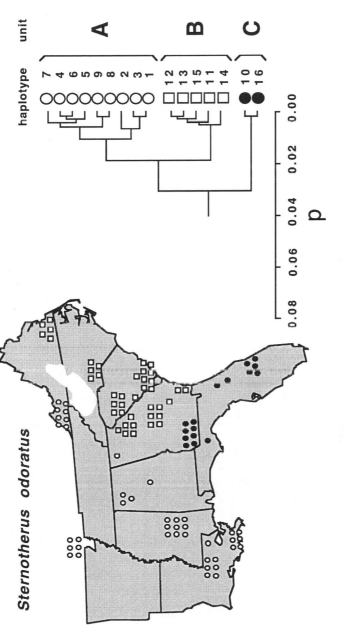

Figure 5 Phylogeographic pattern of mitochondrial (mt) DNA in the stinkpot, *Sternotherus odoratus*. (*Left*) Map showing the distribution of the species in the southeastern United States (after Reference 17). Also shown are locations and phylogroup membership of all individuals (*symbols*) assayed for mtDNA. (*Right*) Single-linkage cluster phenogram summarizing genetic relationships among the mtDNA haplotypes and highlighting (by different *symbols*) the principal divergent phylogroups. The scale is in units of sequence divergence (*p*) estimated between extant taxa. See text for further explanation.

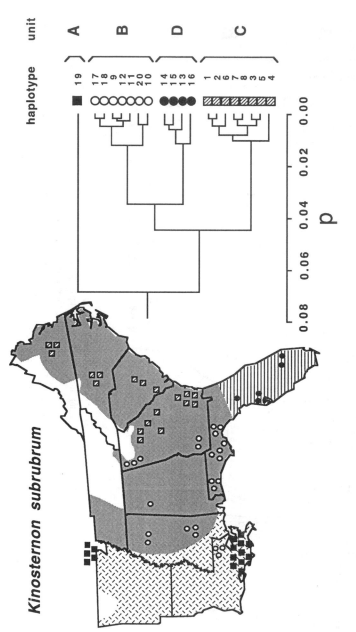

Figure 6 Phylogeographic pattern of mitochondrial (mt) DNA in the mud turtle, *Kinosternon subrubrum*. (*Left*) Map showing the distribution of the species in the southeastern United States, with *shaded regions* indicating approximate geographic ranges of subspecies conventionally recognized by morphological criteria (after Reference 17). Also shown are locations and phylogroup membership of all individuals (*symbols*) assayed for mtDNA. (*Right*) Single-linkage cluster phenogram summarizing genetic relationships among the mtDNA haplotypes and highlighting (by different *symbols*) the principal divergent phylogroups. The scale is in units of sequence divergence (*p*) estimated between extant taxa. See text for further explanation.

However, the genetic situation becomes more complicated when a close taxonomic relative, *Kinosternon baurii*, is included in the comparison (62). This taxon has two moderately different mtDNA phylogroups (Figure 7): One is confined to the Florida peninsula, and the other occurs along the Atlantic coast. Both of the *K. baurii* mtDNA units are related more closely to haplogroup C in *K. subrubrum* (the Atlantic coast assemblage) than is this C assemblage to other haplogroups within *K. subrubrum*. Indeed, all haplotypes in *K. baurii* are embedded phylogenetically within the broader diversity of the C clade of *K. subrubrum* (Figure 4).

Several explanations are possible (62). Perhaps hybridization has led to an introgressive transfer of mtDNA lineages between two otherwise long-separated biological species of mud turtles along the Atlantic seaboard. Alternatively, *K. baurii* may be a recent phylogenetic derivative of *K. subrubrum* in this area and has not yet evolved noticeable differences from its ancestor in mtDNA composition. Under this hypothesis, *K. subrubrum* and *K. baurii* might be good biological species with the former being paraphyletic to the latter in matrilineal genealogy. Under either of these scenarios, a secondary invasion of the Floridian peninsula by *K. baurii* could account for the sympatric occurrence there of highly divergent mtDNA lineages (the C and D phylogroups) within the *Kinosternon* complex. Another possibility is that *K. baurii* is confined to the Floridian peninsula and that turtles along the Atlantic seaboard represent *K. subrubrum* exclusively. Indeed, because of morphological similarity between the two species, particularly along the Atlantic seaboard, there has been much discussion in the literature as to whether the range of *K. baurii* truly extends into this area (21, 26, 27, 30). In the absence of direct evidence from nuclear genes, we cannot resolve these possibilities.

In any event, at least four major mtDNA phylogroups are present within this complex of *Kinosternon* mud turtles in the southeastern United States, and their distributions overall bear considerable likeness to those discussed above for *S. odoratus*.

GRAPTEMYS SPECIES (MAP TURTLES)   Many turtle groups tend to be relatively conservative morphologically (relative to birds, for example). However, the carapaces and heads of *Graptemys* species display varied and strikingly beautiful designs, from which the moniker map turtles derives. About a dozen forms in the southeastern United States traditionally are recognized at the taxonomic level of species, and most are endemic to particular drainages along the Gulf coast (Figure 8). No species occur in Atlantic coastal drainages or in peninsular Florida.

Lamb et al (31) examined all of these forms for mtDNA restriction sites as well as nucleotide sequences from the *cyt b* gene and control region. About

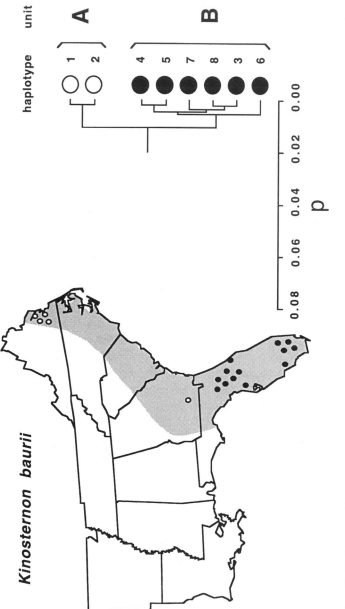

Figure 7   Phylogeographic pattern of mitochondrial (mt) DNA in the striped mud turtle, *Kinosternon baurii*. (*Left*) Map showing the distribution of the species in the southeastern United States (after Reference 17). Also shown are locations and phylogroup membership of all individuals (*symbols*) assayed for mtDNA. (*Right*) Single-linkage cluster phenogram summarizing genetic relationships among the mtDNA haplotypes and highlighting (by different *symbols*) the principal divergent phylogroups. The scale is in units of sequence divergence (*p*) estimated between extant taxa. See text for further explanation.

five of the taxa assayed could not be distinguished by restriction sites or *cyt b* sequences, and depths of genetic separations in the *Graptemys* complex overall were small by the standards summarized above for intraspecific phylogroups within other turtle species (Table 2). Nonetheless, three monophyletic lineages within *Graptemys* could be discerned in assays of mtDNA restriction fragment length polymorphisms (RFLPs) (Figure 8): (*a*) a *G. pulchra* group of four taxonomic species in central coastal rivers of the Gulf states; (*b*) a *G. pseudogeographica* group of five species mostly to the west but overlapping spatially with members of the *G. pulchra* group to some extent; and (*c*) *Graptemys geographica*, a widespread species in the central and northern United States.

The genealogical data taken at face value suggest that the *Graptemys* complex has been taxonomically oversplit at the species level. Perhaps this is because shell characters have been available to distinguish adjacent populations, many of which now appear from the molecular data not to have been long separated historically. Indeed, the *Graptemys* complex appears to display less mtDNA lineage diversification overall than do conspecific populations of several kinosternid species (compare Figure 8 with Figures 3, 5, and 6).

TRACHEMYS SCRIPTA (SLIDER)   *Trachemys scripta* is widespread and abundant, with two named subspecies in the current coverage area: *T. s. scripta*, mostly along the Atlantic coast, and *T. s. elegans*, to the west. In field guides (e.g. 17), these forms usually are depicted as "intergrading" in western Georgia, Alabama, and the Florida Panhandle.

With respect to mtDNA, Avise et al (12) first surveyed a small number of specimens of this species, but additional specimens in the current study bring the total sample size to $N = 65$. Mitochondrial variation in this species is extremely limited: Two haplotypes were observed, and these differed by only three assayed restriction sites ($p \cong 0.006$). Nonetheless, the two lineages show a strong geographic orientation generally consistent with the described subspecies ranges (Figure 9). However, two individuals with the western haplotype A were observed in the Atlantic coastal plain.

DEIROCHELYS RETICULARIA (CHICKEN TURTLE)   *Deirochelys reticularia* occurs primarily in coastal plain regions throughout the study area (Figure 10). In control region sequences, the mtDNA of some specimens display a variant feature of potential cladistic import that has not been reported in other turtle species: a relatively large (10 bp) deletion. This deletion, which appears to be a derived condition by reference to outgroup species, is present in all chicken turtles in the eastern portion of the species' range (Atlantic coast and peninsular Florida) but is absent to the west (Figure 10). However, within the western region, large sequence divergence estimates (mean $p \cong 0.045$) distinguish

Graptemys species (map turtles)

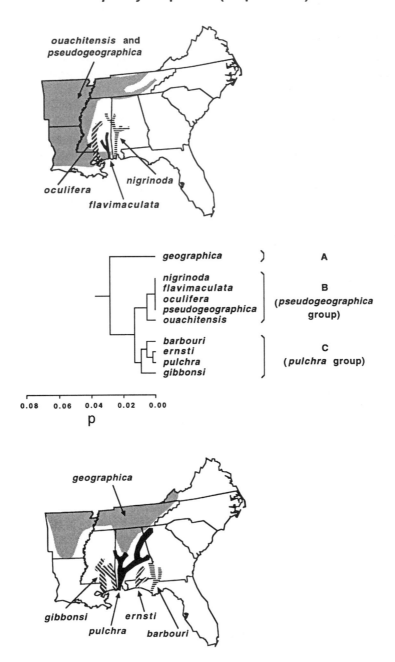

samples in Missouri from those along the Gulf coastal plain. The pattern suggests an ancient separation in the mtDNA gene tree plus a more recent derivation for the distinctive matrilines along the Atlantic coast and peninsular Florida (Figure 10).

MACROCLEMYS TEMMINCKII (ALLIGATOR SNAPPING TURTLE)   Striking phylogeographic structure in mtDNA is evident at two spatial scales within *Macroclemys temminckii*, a highly aquatic species (51, 52). First, samples from each of several river drainages entering the Gulf of Mexico display fixed differences in haplotype frequencies, suggesting severe restrictions on contemporary interdrainage gene flow. Second, much deeper phylogenetic separations in the mtDNA gene tree distinguish three regional population assemblages (Figure 11) that the authors (52) refer to as evolutionarily significant units. The most distinctive of these units is confined to the Suwannee River, the only major drainage in peninsular Florida currently inhabited by the species. The two other principal mtDNA units characterize populations in all drainages from the Pensacola River in western Florida to the Trinity River in Texas and all drainages in the Floridian Panhandle between the Pensacola and the Suwannee.

CHELYDRA SERPENTINA (COMMON SNAPPING TURTLE)   *Chelydra serpentina* occurs throughout eastern and central North America. Two subspecies are recognized: *C. s. osceola*, in the Florida peninsula, and *C. s. serpentina*, elsewhere on the continent. A survey of mtDNA control-region sequences from samples across the southeastern United States revealed almost no variation within or between populations (63). A single haplotype characterized 60 of the 66 specimens surveyed, and two variant haplotypes differed from it by one and two mutational changes (Figure 12). This paucity of mtDNA variation could be attributed to some unknown molecular mechanism or selective peculiarity that has arrested mtDNA evolution in snapping turtles, but these reasons seem unlikely because a broader geographic survey of the species uncovered moderate mtDNA sequence differences between North American specimens and those in Central and South America (45). Also, in similar molecular assays, the snapping turtle proved to be highly distinct from its closest living relative, the alligator snapping turtle (51, 52).

←——

Figure 8   Phylogeographic patterns of mtDNA in map turtles of the genus *Graptemys*. Maps show the distribution of the species in the southeastern United States, with *shaded regions* indicating approximate geographic ranges of species conventionally recognized by morphological criteria (after Reference 17). See text for further explanation. (Center) Single-linkage cluster phenogram summarizing genetic relationships among the mtDNA haplotypes of these species. The scale is in units of sequence divergence (f) between species.

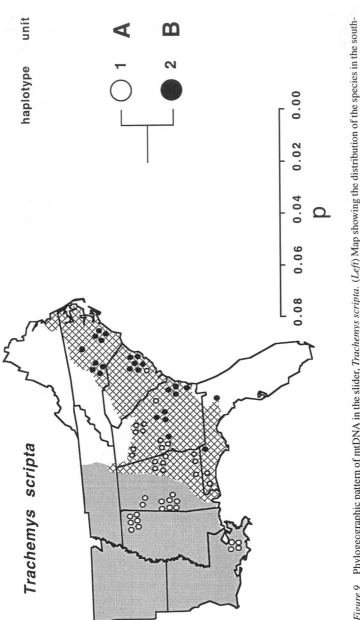

Figure 9   Phylogeographic pattern of mtDNA in the slider, *Trachemys scripta*. (*Left*) Map showing the distribution of the species in the south-eastern United States, with *shaded regions* indicating approximate geographic ranges of subspecies conventionally recognized by morphological criteria (after Reference 17). Also shown are locations and phylogroup membership of all individuals (*symbols*) assayed for mtDNA. (*Right*) Single-linkage cluster phenogram summarizing genetic relationships among the mtDNA haplotypes and highlighting (by different *symbols*) the principal divergent phylogroups. The scale is in units of sequence divergence (*p*) estimated between extant taxa. See text for further explanation.

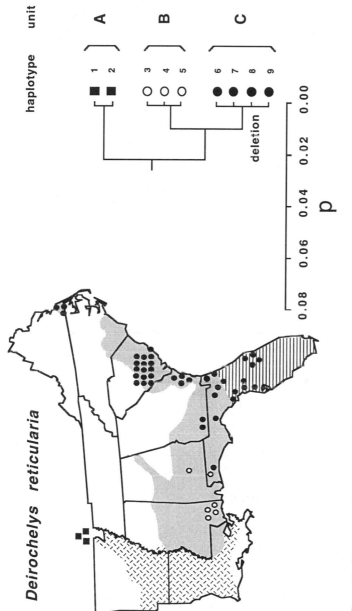

Figure 10 Phylogeographic pattern of mtDNA in the chicken turtle, *Deirochelys reticularia*. (*Left*) Map showing the distribution of the species in the southeastern United States, with *shaded regions* indicating approximate geographic ranges of subspecies conventionally recognized by morphological criteria (after Reference 17). Also shown are locations and phylogroup membership of all individuals (*symbols*) assayed for mtDNA. (*Right*) Single-linkage cluster phenogram summarizing genetic relationships among the mtDNA haplotypes and highlighting (by different *symbols*) the principal divergent phylogroups. The scale is in units of sequence divergence (*p*) estimated between extant taxa. See text for further explanation.

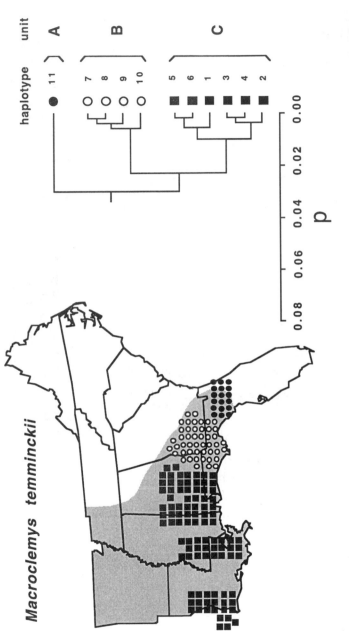

Figure 11 Phylogeographic pattern of mtDNA in the alligator snapping turtle, *Macroclemys temminckii*. (*Left*) Map showing the distribution of the species in the southeastern United States (after Reference 17). Also shown are locations and phylogroup membership of all individuals (*symbols*) assayed for mtDNA. (*Right*) Single-linkage cluster phenogram summarizing genetic relationships among the mtDNA haplotypes and highlighting (by different *symbols*) the principal divergent phylogroups. The scale is in units of sequence divergence (*p*) estimated between extant taxa. See text for further explanation.

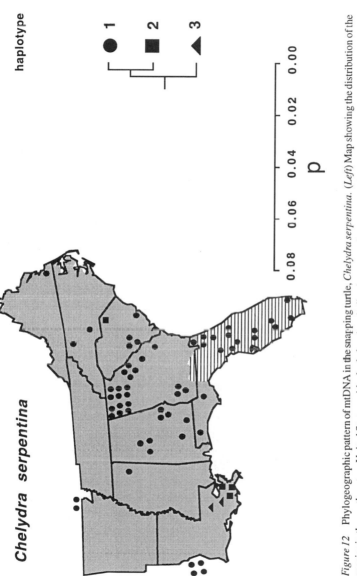

Figure 12   Phylogeographic pattern of mtDNA in the snapping turtle, *Chelydra serpentina*. (*Left*) Map showing the distribution of the species in the southeastern United States, with *shaded regions* indicating approximate geographic ranges of subspecies conventionally recognized by morphological criteria (after Reference 17. Also shown are locations and phylogroup membership of all individuals (*symbols*) assayed for mtDNA. (*Right*) Single-linkage cluster phenogram summarizing genetic relationships among the mtDNA haplotypes and highlighting (by different *symbols*) the principal divergent phylogroups. The scale is in units of sequence divergence (*p*) estimated between extant taxa. See text for further explanation.

Terrestrial Turtle

GOPHERUS POLYPHEMUS (GOPHER TORTOISE)   The gopher tortoise is the sole land-confined turtle in the southeastern United States. This threatened species occurs in sand-scrub habitats in Florida and in the coastal plain from eastern South Carolina to Louisiana. A genetic survey (42) revealed numerous mtDNA haplotypes that are grouped into two major phylogenetic lineages with a strong geographic configuration (Figure 13). One phylogroup characterizes samples from western Georgia and the Floridian Panhandle to Louisiana, and the other is confined to peninsular Florida, southern Georgia, and South Carolina. The latter assemblage also consists of two recognizable subgroups, one of which is present only in mid-Florida.

PHYLOGEOGRAPHIC HYPOTHESES AND GENEALOGICAL CONCORDANCE

These comparative data on the intraspecific matrilineal histories of several species of Testudines in the southeastern United States permit independent tests of phylogeographic hypotheses and principles of genealogical concordance previously derived from similar genetic studies of other freshwater and terrestrial vertebrates in the region. These phylogeographic concepts and their corollaries are appraised in order of their appearance in Table 1.

Hypothesis I: Populations of Most Species Display Significant Phylogeographic Structure

Perhaps not surprisingly, the hypothesis that populations of most species display significant phylogeographic structure is supported abundantly by mtDNA data for the Testudines. With the exception of the snapping turtle, all broadly distributed species surveyed across the southeastern region show striking matrilineal population structure at various spatial scales and inferred temporal depths. Given the limited mobility of individuals in most turtle species, perhaps this local structure is to be expected, as is a window of opportunity for the evolution of deeper interregional separations in response to longer-term biogeographic barriers. This latter opportunity appears to have been realized, as evidenced by the major phylogeographic breaks identified within nearly all of the broadly distributed turtle species surveyed.

Hypothesis II: Nonsubdivided, High-Dispersal Species May Have Limited Phylogeographic Structure

The common snapping turtle, C. serpentina, is the only surveyed species without pronounced mtDNA phylogeographic structure. Although shallow or modest matrilineal structure might yet be detected in more sensitive molecular assays,

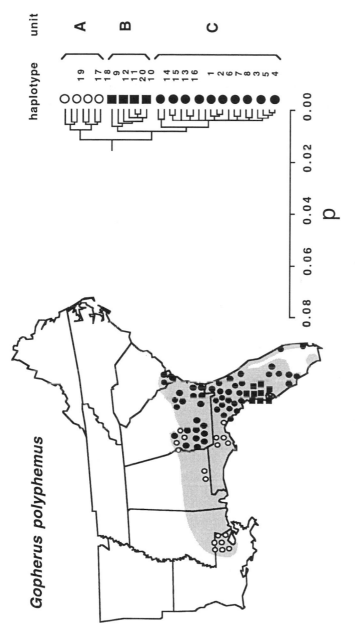

Figure 13 Phylogeographic pattern of mtDNA in the gopher tortoise, *Gopherus polyphemus*. (*Left*) Map showing the distribution of the species in the southeastern United States (after Reference 17). Also shown are locations and phylogroup membership of all individuals (*symbols*) assayed for mtDNA. (*Right*) Single-linkage cluster phenogram summarizing genetic relationships among the mtDNA haplotypes and highlighting (by different *symbols*) the principal divergent phylogroups. The scale is in units of sequence divergence (*p*) estimated between extant taxa. See text for further explanation. Because of large sample sizes, not all individuals are shown.

the available data strongly suggest that long-standing evolutionary separations have not been a part of the phylogeographic history of contemporary populations of the common snapping turtle in the southeastern United States. Perhaps individuals of this species are unusual among the surveyed turtles with respect to high dispersal capabilities (19) and attendant imperviousness to historical biogeographic barriers that appear to have affected other species.

A related possibility has to do with the observation that among all North American turtles, *C. serpentina* (and *Chrysemys picta*) are tolerant to cold and were "always among the first to invade formerly glaciated areas at the end of the Wisconsin" (24, p. 45). The southeastern United States was never covered by Pleistocene glaciers, but its climate was far cooler at times of glacial maxima, and some drainages traversing the South received frigid meltwaters from northern glaciers during warming episodes (37). The unusual cold tolerance and perhaps the high dispersal capability of *C. serpentina* may indicate that the species was not confined to isolated pockets of warm water habitat during the late Tertiary and Quaternary and thus was less subject to historical population subdividing by climatic events and/or shifting watersheds.

Hypothesis III: Major Phylogeographic Units Within a Species Reflect Long-Term Historical Barriers to Gene Flow

Most of the widely distributed turtle species assayed display deep matrilineal separations on a regional geographic scale. However, major splits in a gene tree (such as that for mtDNA) cannot automatically be equated with deep separations in a population tree (5, 22, 32, 41, 43). Thus, it has been argued (10), additional evidence in the form of genealogical concordance is required to establish by hard criteria that major phylogenetic branches in a gene tree register major branches in an organismal phylogeny. Four aspects of genealogical concordance (Table 1) can be distinguished, all of which represent logical corollaries of phylogeographic hypothesis III.

CONCORDANCE ACROSS CHARACTERS WITHIN A GENE   Every deep phylogenetic split in the intraspecific gene tree of a turtle species is, by definition, supported concordantly by multiple restriction-site or sequence characters in mtDNA. Thus, this category of genealogical concordance merely identifies the major gene-tree phylogroups worthy of further biogeographic consideration.

CONCORDANCE OF GENEALOGICAL PARTITIONS ACROSS MULTIPLE GENES   The concordance of genealogical partitions across multiple genes cannot be evaluated critically in the turtles studied because comparable genealogical evidence from multiple nuclear genes is unavailable for comparison. In the absence of such direct information, a potential surrogate can be employed: traditional subspecies definitions. To the extent that morphology-based intraspecific taxonomy reflects substantial population-level differences in nuclear genomes, this category of genealogical concordance can be addressed, at least in part.

Agreement exists between mtDNA phylogeography and taxonomic definitions for several of the turtles assayed: within *Sternotherus minor* (Figure 3) and *Trachemys scripta* (Figure 9), to a partial extent within *Kinosternon subrubrum* (Figure 6), and with regard to the phylogenetic distinctiveness of *Sternotherus depressus* from the other species of Kinosternidae (Figure 4). However, in other cases such agreement is lacking. Thus, the mtDNA data provide no evidence for a special phylogenetic distinctiveness of the Floridian subspecies of *Deirochelys reticularia* (Figure 10) or *Chelydra serpentina* (Figure 12) or for long-standing evolutionary separations among several of the recognized species of *Graptemys* turtles (Figure 8). Conversely, relatively deep phylogenetic separations in mtDNA are apparent within the taxonomically monotypic *Sternotherus odoratus* (Figure 5), *Macroclemys temminckii* (Figure 11), and, to a lesser extent, *Gopherus polyphemus* (Figure 13).

For these cases of mtDNA discordance with traditional taxonomy, two primary possibilities remain: (*a*) The existing taxonomy does not reflect significant phylogeographic partitions or (*b*) the mtDNA phylogenies are misleading in this regard. Described next are two aspects of genealogical concordance that suggest that the mtDNA gene trees are meaningful registers of phylogeographic population histories and hence that current taxonomy in several cases may need revision.

CONCORDANCE OF GENEALOGICAL PARTITIONS ACROSS MULTIPLE CODISTRI-BUTED SPECIES   As was true in earlier studies of freshwater fishes, a remarkable result of the current review is the level of general agreement across species of Testudines in the spatial positions of major mtDNA phylogeographic units across the southeastern United States. In *Sternotherus minor* (Figure 3), *Sternotherus odoratus* (Figure 5), *Kinosternon subrubrum* (Figure 6), *Trachemys scripta* (Figure 9), *Deirochelys reticularia* (Figure 10), *Macroclemys temminckii* (Figure 11), and *Gopherus polyphemus* (Figure 13), recognizable phylogenetic separations in the mtDNA gene tree distinguish populations in peninsular Florida and/or those along the Atlantic coast from populations in western (Gulf coast) areas. These regions also bear striking resemblance to the areas inhabited by major mtDNA phylogroups within several southeastern US fish species (Figure 1).

Furthermore, as was the case for the freshwater fishes surveyed, in only a few cases of the turtles are additional deep mtDNA subdivisions evident within the surveyed region. Exceptions to this statement involve far western forms in

K. subrubrum (Figure 6) and *D. reticularia* (Figure 10) and a central Alabama form of *Sternotherus* (*S. depressus*; Figures 3, 4). These cases embellish but do not contradict the tendency for the above-mentioned phylogeographic distinctions between eastern and western regions.

CONCORDANCE OF GENEALOGICAL PARTITIONS WITH BIOGEOGRAPHIC PROVINCES IDENTIFIED BY INDEPENDENT EVIDENCE   Faunal lists are the traditional data by which regional biotic provinces are identified. For example, as described earlier, an analysis of geographic ranges for the 241 species of freshwater fishes in the southeastern United States identified Atlantic (including peninsular Florida) and Gulf drainages as the two most distinctive regions faunistically (Figure 1). These two regions also show general agreement with the geographic distributions of major intraspecific mtDNA phylogroups in several fish species (e.g. Figure 1). Does a similar concordance between composite faunal distributions and intraspecific mtDNA phylogroups exist for the Testudines?

Following the general procedures employed by Swift et al (58) in analyses of fish faunal provinces, we compiled range information (17, 19) for all 35 native species of freshwater and terrestrial turtles that inhabit the southeastern United States. Presence or absence of each species was determined for each of 48 grids on a southeastern map (Figure 14), and similarity coefficients (Jaccard coefficients of association; see 56) between grids were calculated. The resulting similarity matrix was clustered phenetically (by UPGMA; see 50), with results shown in Figure 14.

The geographic picture of the Testudines bears strong resemblance to that of the freshwater fishes: A basal split distinguishes the Atlantic coast and Floridian region from locales to the north and west across most of the remainder of the survey area. As already described, this pattern agrees well with geographic trends in the distributions of major mtDNA phylogroups within several of the turtle species surveyed. Where deep phylogeographic splitting was observed, almost invariably the major phylogroups were oriented in an eastern (Atlantic)

Figure 14   Faunal assessments of Testudines in the southeastern United States based on composite species distributions. The map shows the two basal faunal assemblages (*shaded* and *unshaded*) for all native turtle species in the region as identified in a cluster analysis (phenogram shown in the *lower half of the figure*) of a matrix of faunal similarity coefficients for the grid squares. Note the basal distinction between "Atlantic" and "Gulf" areas. A somewhat different grouping method based on turtle "species richness" gives nearly identical results (1). *Numbered areas* on the map depict freshwater faunal ecoregions or aquatic ecological units as identified by differences in the assemblages and subassemblages of fish species (36).

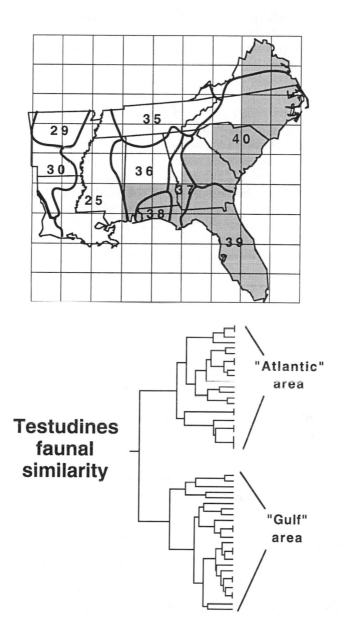

versus western (Gulf) pattern generally consistent with the major break in the overall faunal distributions.

We suspect that these two seemingly different sources of information on zoogeographic provinces—composite faunal lists and intraspecific phylogeographic partitions—may have similar evolutionary etiologies reflecting longstanding ecological or biogeographic impediments to interregional dispersal. The ranges of species clearly reflect dispersal barriers. Similarly, we suggest that phylogeographic breaks within species that are widely distributed across biogeographic provinces are due to evolutionary subdivision, with populations in both biotic provinces having survived for current observation. If populations in only one of the historical regions had survived, that species as a whole would be merely another contributor to the concordant distributional faunal lists upon which biotic provinces traditionally have been recognized.

On a finer geographic scale, Maxwell et al (36) identified several additional "ecoregions" or "aquatic ecological units" in the southeastern United States (Figure 14) that might be considered biogeographic "subprovinces" from traditional evidence. One of these (ecoregion 39) encompasses the Floridian peninsula and southern Georgia and is approximately coincident with Remington's (48) demarcation of the unique Floridian biome. Another of Maxwell et al's units (ecoregion 40) encompasses all of the Atlantic coastal plain from Georgia to Virginia. This distinction between peninsular Florida and the northern coastal plain was mirrored closely by phylogenetic partitions in mtDNA within *S. odoratus* (Figure 5), *K. subrubrum* (Figure 6), and *K. baurii* (Figure 7). To the west, a more complicated array of ecoregions appear in Maxwell et al's analysis, and in general these bear a less clear alignment to phylogeographic patterns observed in molecular studies of the Testudines.

PHYLOGEOGRAPHIC SEPARATIONS

The comparative phylogeographic patterns summarized here for the Testudines are similar in yet another regard to those reported previously for the freshwater fishes in the southeastern United States. In mtDNA studies of these fishes, the geographic distributions of the principal phylogroups were similar across the assayed species, but the absolute magnitudes of the estimated net sequence divergences between these mtDNA clades varied by more than an order of magnitude: e.g. from $p \cong 0.006$ in the bowfin (*Amia calva*) to $p \cong 0.082$ in the redear sunfish (*Lepomis microlophus*) (14). Under a conventional molecular clock calibration for vertebrate mtDNA ($p \cong 0.02$ between a pair of lineages per million years; Reference 16), these estimates at face value imply times of intraspecific phylogroup separation ranging from roughly 300,000 years ago (bowfin) to 4,100,000 years before the present (redear sunfish).

Similarly, net sequence divergence estimates between the major intraspecific phylogroups in the turtles differ by more than an order of magnitude (Table 2): from $p \cong 0.006$ in *Trachemys scripta* to $p \cong 0.070$ in *Kinosternon subrubrum* (phylogroups A and B). Under the conventional clock noted above, these translate into population divergence estimates ranging from 300,000 to 3,500,000 years ago. From mtDNA studies of other Testudine taxa, a fivefold to tenfold slower pace of mtDNA sequence evolution has been suggested (12, 15, 29; see also 33–35, 46, 47). As applied to the phylogroups under current review, these values imply intraspecific separation times that could range from 1.5 Mya to as much as 35 Mya.

Microevolutionary rate assessments are particularly challenging because de novo sequence change postdating population separations must be distinguished from sequence differences attributable to lineage sorting from a polymorphic common ancestor (7). In principle, the former is independent of effective population size whereas the latter is critically dependent on historical population demographic factors (that typically remain unknown from independent evidence). Our use of "net" sequence divergence (as defined in Table 2) is an attempt to correct for ancestral polymorphism, but it assumes that levels of variation in extant taxa are representative of that in an ancestral population.

Apart from these and additional concerns about estimation errors, at least two ad hoc biological explanations can be advanced for the large variance in the sequence divergence values. First, perhaps extant turtle populations were subdivided at widely different times. During the Pliocene and especially the Pleistocene, repeated episodes of climatic alteration promoted ecophysiographic changes in the southeastern landscape and forced either range shifts or extinctions on its biotic elements. Thus, lineage separations within various species might trace to different climatic cycles depending in part on the patterns of extinction of former regional isolates. Furthermore, some of the phylogeographic footprints might trace to earlier Tertiary times, when (for example) the central highlands of Florida probably existed periodically as one or more islands relatively isolated from the continent.

Second, perhaps a mtDNA "clock" ticks at widely varying paces in different turtle lineages. Considerable controversy (beyond the scope of this paper) exists about the calibration and reliability of molecular clocks (38), and considerable empirical evidence exists for severalfold mtDNA rate variation among taxonomic groups (e.g. 2, 4, 33–35, 46, 47, 66) and sometimes even among closely related lineages (68).

Our recent molecular findings on several kinosternid species (in *Sternotherus* and *Kinosternon*) run counter to our prior experience with several other marine, estuarine, freshwater, and terrestrial species of Testudines, in which unusually low (by vertebrate standards) levels of intraspecific and interspecific mtDNA sequence divergence had been the norm (12, 15, 28, 29). The highly variable mtDNA genotypes in the Kinosternidae do not by themselves speak directly to molecular clock calibrations, but they do provide different impressions of the magnitudes of intraspecific population variation and differentiation as compared with several other turtle groups. One extinct kinosternid turtle (*Xenochelys formosa*) has been described from the Oligocene, and the extant species *Sternotherus odoratus* and *Kinosternon subrubrum* have been described from Pliocene deposits (19). This considerable antiquity for these kinosternid turtles provides an opportunity for ancient intraspecific lineage separations. On the other hand, *Chelydra serpentina* also has been reported from the Pliocene (19), yet no deep mtDNA lineage separations were evident.

RELEVANCE TO CONSERVATION BIOLOGY

Regardless of the particular historical scenarios invoked to account for the phylogeographic patterns in these terrestrial and freshwater turtles, the molecular data on matrilineal separations carry ramifications for management and conservation efforts in two principal regards (13). First, the data are germane to taxonomic and systematic issues for particular endangered (or other) species considered individually. For example, the federally threatened *Sternotherus depressus* was a problematic taxon because of its peculiar range and because of uncertainties about the genetic basis of its oddly flattened carapace. The mtDNA data indicate that the matrilines of this population had a relatively ancient evolutionary separation from those of other kinosternids in the region. To this extent, the basis for the existing taxonomic recognition is bolstered, as are special conservation efforts that have stemmed from it.

Second, the intraspecific genetic architectures of multiple species add to knowledge of the faunal provinces that, we argue, should be appreciated as major centers of biogeographic diversity in biome-based conservation efforts. With recognition of the relative integrity and tendency toward uniqueness of major historical biotic assemblages, conservation planning at the ecosystem level might be instituted in conjunction with traditional species-focused management efforts to enhance the effectiveness and impact of conservation programs on a regional scale (8).

SUMMARY

We have identified four aspects of genealogical concordance that apply empirically to several species of turtles (as well as to other freshwater and terrestrial vertebrates) in the southeastern United States. Many species-idiosyncratic phylogeographic outcomes also are apparent in the mtDNA analyses. Depending on the context, both the idiosyncrasies and the generalized trends can be relevant to taxonomic and conservation efforts.

In the future, it will be of interest to evaluate phylogeographic hypotheses and principles of genealogical concordance by conducting similar comparative analyses on other regional biotas. Outcomes may differ. For example, perhaps high-latitude ecoregions will tend to lack genealogical concordance because no long-term in situ evolution can have taken place in recently glaciated areas. A more cosmopolitan view of phylogeography would recognize such trends as well. Regardless of the outcomes of such future studies, the comparative genetic analyses of faunas in the southeastern United States already have contributed insights into the historical nature of particular biotic provinces and of the evolutionary factors that can contribute to the composite architectures of species on a regional scale.

ACKNOWLEDGMENTS

Work has been supported by the National Science Foundation. DW was supported by a National Institutes of Health Training Grant in Genetics. We thank members of the Avise laboratory for comments on the manuscript, G. Ortí for computer assistance, and KA Buhlmann and PE Moler for help with collections.

> Visit the *Annual Reviews home page* at
> http://www.AnnualReviews.org

Literature Cited

1. Abell R, Olson D, Dinerstein E, Walters S, Hurley P, et al. 1998. *A Conservation Assessment of the Freshwater Ecoregions of North America.* Washington, DC: World Wildlife Fund
2. Adachi J, Cao Y, Hasegawa M. 1993. Tempo and mode of mitochondrial DNA evolution in vertebrates at the amino acid level: rapid evolution in warm blooded vertebrates. *J. Mol. Evol.* 36:270–81
3. Adams CC. 1901. Baseleveling and its faunal significance, with illustrations from southeastern United States. *Am. Nat.* 35:839–52
4. Aufrray JC, Vanlerberghe F, Britton-Davidian J. 1990. The house mouse progression in Eurasia: A paleontological and archeaozoological approach. *Biol. J. Linn. Soc.* 41:13–25
5. Avise JC. 1989. Gene trees and organismal histories: a phylogenetic approach to population biology. *Evolution* 43:1192–208
6. Avise JC. 1992. Molecular population structure and the biogeographic history of a regional fauna: a case history with lessons for conservation biology. *Oikos* 63: 62–76
7. Avise JC. 1994. *Molecular Markers, Natural History and Evolution.* New York: Chapman & Hall. 511 pp.
8. Avise JC. 1996. Toward a regional conservation genetics perspective: phylogeography of faunas in the southeastern United States. In *Conservation Genetics: Case Histories from Nature,* ed. JC Avise, JL Hamrick, pp. 431–70. New York: Chapman & Hall. 512 pp.
9. Avise JC, Arnold J, Ball RM, Bermingham E, Lamb T, et al. 1987. Intraspecific phylogeography: the mitochondrial DNA bridge between population genetics and systematics. *Annu. Rev. Ecol. Syst.* 18:489–522
10. Avise JC, Ball RM Jr. 1990. Principles of genealogical concordance in species concepts and biological taxonomy. *Oxford Surv. Evol. Biol.* 7:45–67
11. Avise JC, Bermingham E, Kessler LG, Saunders NC. 1984. Characterization of

mitochondrial DNA variability in a hybrid swarm between subspecies of bluegill sunfish (*Lepomis macrochirus*). *Evolution* 38:931–41

12. Avise JC, Bowen BW, Lamb T, Meylan AB, Bermingham E. 1992. Mitochondrial DNA evolution at a turtle's pace: evidence for low genetic variability and reduced microevolutionary rate in the Testudines. *Mol. Biol. Evol.* 9:457–73

13. Avise JC, Hamrick JL, eds. 1996. *Conservation Genetics: Case Histories from Nature*. New York: Chapman & Hall. 512 pp.

14. Bermingham E, Avise JC. 1986. Molecular zoogeography of freshwater fishes in the southeastern United States. *Genetics* 113:939–65

15. Bowen BW, Meylan AB, Ross JP, Limpus CJ, Balazs GH, Avise JC. 1992. Global population structure and natural history of the green turtle (*Chelonia mydas*) in terms of matriarchal phylogeny. *Evolution* 46:865–81

16. Brown WM, George M Jr, Wilson AC. 1979. Rapid evolution of animal mitochondrial DNA. *Proc. Natl. Acad. Sci. USA* 76:1976–81

17. Conant R, Collins JT. 1991. *A Field Guide to Reptiles and Amphibians*. Boston, MA: Houghton Mifflin. 450 pp.

18. Ellsworth DL, Honeycutt RL, Silvy NJ, Bickham JW, Klimstra WD. 1994. Historical biogeography and contemporary patterns of mitochondrial DNA variation in white-tailed deer from the southeastern United States. *Evolution* 48:122–36

19. Ernst CH, Lovich JE, Barbour RW. 1994. *Turtles of the United States and Canada*. Washington, DC: Smithsonian Inst. Press. 578 pp.

20. Felsenstein J. 1985. Confidence limits on phylogenies: an approach using the bootstrap. *Evolution* 39:783–91

21. Gibbons JW, Nelson DH, Patterson KK, Greene JL. 1979. The reptiles and amphibians of the Savannah River Plant in west-central South Carolina. In *Proc. 1st South Carolina Endangered Species Symp.*, ed. DM Forsythe, WB Ezell Jr., pp. 133–42. Columbia, SC: South Carolina Wildlife and Marine Resources Department, and Charleston, SC: The Citadel

22. Harvey PH, Leigh Brown AJ, Maynard Smith J, Nee S, eds. 1996. *New Uses for New Phylogenies*. Oxford, UK: Oxford Univ. Press. 349 pp.

23. Hayes JP, Harrison RG. 1992. Variation in mitochondrial DNA and the biogeographic history of woodrats (*Neotoma*) of the eastern United States. *Syst. Biol.* 41:331–44

24. Holman JA, Andrews KD. 1994. North American Quaternary cold-tolerant turtles: distributional adaptations and constraints. *Boreas* 23:44–52

25. Iverson JB. 1977. Geographic variation in the musk turtle, *Sternotherus minor*. *Copeia*, pp. 502–17

26. Lamb T. 1983a. On the problematic identification of *Kinosternon* (Testudines: Kinosternidae) in Georgia, with new state localities for *Kinosternon baurii*. *Georgia J. Sci.* 41:115–20

27. Lamb T. 1983b. The striped mud turtle (*Kinosternon baurii*) in South Carolina, a confirmation through multivariate character analysis. *Herpetologica* 39:383–90

28. Lamb T, Avise JC. 1992. Molecular and population genetic aspects of mitochondrial DNA variability in the diamondback terrapin, *Malaclemys terrapin*. *J. Hered.* 83:262–69

29. Lamb T, Avise JC, Gibbons JW. 1989. Phylogeographic patterns in mitochondrial DNA of the desert tortoise (*Xerobates agassizi*), and evolutionary relationships among the North American gopher tortoises. *Evolution* 43:76–87

30. Lamb T, Lovich J. 1990. Morphometric validation of the striped mud turtle (*Kinosternon baurii*) in the Carolinas and Virginia. *Copeia*, pp. 613–18

31. Lamb T, Lydeard C, Walker RB, Gibbons JW. 1994. Molecular systematics of map turtles (*Graptemys*): a comparison of mitochondrial restriction site versus sequence data. *Syst. Biol.* 43:543–59

32. Maddison WP. 1995. Phylogenetic histories within and among species. In *Experimental and Molecular Approaches to Plant Biosystematics*, ed. PC Hoch, AG Stephenson, pp. 273–87. St. Louis, MO: Monogr. Syst. Missouri Botan. Garden 53

33. Martin AP. 1995. Metabolic rate and directional nucleotide substitution in animal mitochondrial DNA. *Mol. Biol. Evol.* 12:1124–31

34. Martin AP, Naylor GJP, Palumbi SR. 1992. Rates of mitochondrial DNA evolution in sharks are low compared with mammals. *Nature* 357:153–55

35. Martin AP, Palumbi SR. 1993. Body size, metabolic rate, generation time, and the molecular clock. *Proc. Natl. Acad. Sci. USA* 90:4087–91

36. Maxwell JR, Edwards CJ, Jensen ME, Paustian SJ, Parrot H, Hill DM. 1995. *A Hierarchical Framework of Aquatic Ecological Units in North America (Nearctic Zone)*. USDA Forest Service General Tech. Rep. NC–176, St. Paul, MN

37. Mayden RL. 1988. Vicariance biogeography, parsimony, and evolution in North American freshwater fishes. *Syst. Zool.* 37:329–55

38. Mindell DP, Thacker CE. 1996. Rates of molecular evolution: phylogenetic issues and applications. *Annu. Rev. Ecol. Syst.* 27:279–303

39. Moritz CC, Dowling TE, Brown WM. 1987. Evolution of animal mitochondrial DNA: relevance for population biology and systematics. *Annu. Rev. Ecol. Syst.* 18:269–92

40. Nedbal MA, Philipp DP. 1994. Differentiation of mitochondrial DNA in largemouth bass. *Trans. Am. Fish. Soc.* 123:460–68

41. Neigel JE, Avise JC. 1986. Phylogenetic relationships of mitochondrial DNA under various demographic models of speciation. In *Evolutionary Processes and Theory*, ed. E Nevo, S Karlin, pp. 515–34. New York: Academic

42. Osentoski MF, Lamb T. 1995. Intraspecific phylogeography of the gopher tortoise, *Gopherus polyphemus*: RFLP analysis of amplified mtDNA segments. *Mol. Ecol.* 4:709–18

43. Pamilo P, Nei M. 1988. Relationships between gene trees and species trees. *Mol. Biol. Evol.* 5:568–83

44. Philipp DP, Childers WF, Whitt GS. 1983. A biochemical genetic evaluation of the northern and Florida subspecies of largemouth bass. *Trans. Am. Fish. Soc.* 112:1–20

45. Phillips CA, Dimmick WW, Carr JL. 1996. Conservation genetics of the common snapping turtle (*Chelydra serpentina*). *Conserv. Biol.* 10:397–405

46. Rand DM. 1993. Endotherms, ectotherms, and mitochondrial genome-size variation. *J. Mol. Evol.* 37:281–95

47. Rand DM. 1994. Thermal habit, metabolic rate and the evolution of mitochondrial DNA. *Trends Ecol. Evol.* 9:125–31

48. Remington CL. 1968. Suture-zones of hybrid interaction between recently joined biotas. *Evol. Biol.* 2:231–428

49. Reynolds SL, Seidel ME. 1983. Morphological homogeneity in the turtle *Sternotherus odoratus* (Kinosternidae) throughout its range. *J. Herpetol.* 12:113–20

50. Rohlf FJ. 1990. *NTSYS—PC Version 1.6.* Setauket, NY: Exeter

51. Roman J. 1997. Cryptic evolution and population structure in the alligator snapping turtle (*Macroclemys temminckii*). MS Thesis. Univ. Fla., Gainesville. 38 pp.

52. Roman J, Santhuff S, Moler P, Bowen BW. 1998. Cryptic evolution and population structure in the alligator snapping turtle (*Macroclemys temminckii*). *Conserv. Biol.* In press

53. Saitou N, Nei M. 1987. The neighbor-joining method: a new method for reconstructing phylogenetic trees. *Mol. Biol. Evol.* 4:406–25

54. Scribner KT, Avise JC. 1993. Cytonuclear genetic architecture in mosquitofish populations and the possible roles of introgressive hybridization. *Mol. Ecol.* 2:139–49

55. Seidel ME, Reynolds SL, Lucchinno RV. 1981. Phylogenetic relationships among musk turtles (genus *Sternotherus*) and genic variation in *Sternotherus odoratus*. *Herpetologica* 37:161–65

56. Sneath PHA, Sokal RR. 1973. *Numerical Taxonomy.* San Francisco: Freeman. 573 pp.

57. Stejneger L. 1923. Rehabilitation of a hitherto overlooked species of musk turtle of the southern states. *Proc. US Natl. Mus.* 62:1–3

58. Swift CC, Gilbert CR, Bortone AS, Burgess GH, Yerger RW. 1986. Zoogeography of the freshwater fishes of the southeastern United States: Savannah River to Lake Ponchartrain. In *Zoogeography of North American Freshwater Fishes*, ed. CH Hocutt, EO Wiley, pp. 213–65. New York: Wiley. 866 pp.

59. Swofford DL, Olsen GJ, Waddell PJ, Hillis DM. 1996. Phylogenetic inference. In *Molecular Systematics*, ed. DM Hillis, C Moritz, BK Mable, pp. 407–514. Sunderland, MA: Sinauer. 2nd ed.

60. Tinkle DW 1961. Geographic variation in reproduction, size, sex ratio, and maturity of *Sternotherus odoratus* (Testudinata: Chelydridae). *Ecology* 42:68–76

61. Walker D, Burke VJ, Barák I, Avise JC. 1995. A comparison of mtDNA restriction sites vs. control region sequences in phylogeographic assessment of the musk turtle (*Sternotherus minor*). *Mol. Ecol.* 4:365–73

62. Walker D, Moler PE, Buhlmann KA, Avise JC. 1998a. Phylogenetic patterns in *Kinosternon subrubrum* and *K. baurii* based on mitochondrial DNA restriction analyses. *Herpetologica* 54:174–84

63. Walker D, Moler PE, Buhlmann KA, Avise JC. 1998b. Phylogeographic uniformity in mitochondrial DNA of the snapping turtle (*Chelydra serpentina*). *Anim. Conserv.* 1:55–60

64. Walker D, Nelson WS, Buhlmann KA, Avise JC. 1997. Mitochondrial DNA phylogeography and subspecies issues in the monotypic freshwater turtle *Sternotherus odoratus. Copeia*, pp. 16–21

65. Walker D, Ortí G, Avise JC. 1998c. Phylogenetic distinctiveness of a threatened aquatic turtle (*Sternotherus depressus*). *Conserv. Biol.* In press

66. Wallace GP, Arntzen JW. 1989. Mitochondrial-DNA variation in the crested newt superspecies: limited cytoplasmic gene flow among species. *Evolution* 43:88–104

67. Wilson AC, Cann RL, Carr SM, Palumbi SR, Prager EM, et al. 1985. Mitochondrial DNA and two perspectives on evolutionary genetics. *Biol. J. Linn. Soc.* 26:375–400

68. Zhang Y, Ryder OA. 1995. Different rates of mitochondrial DNA sequence evolution in Kirk's dik-dik (*Madoqua kirkii*) populations. *Mol. Phylogeny Evol.* 4:291–97

Annu. Rev. Ecol. Syst. 1998. 29:59–81

THE FUNCTIONAL SIGNIFICANCE OF THE HYPORHEIC ZONE IN STREAMS AND RIVERS

Andrew J. Boulton,[1] *Stuart Findlay,*[2] *Pierre Marmonier,*[3] *Emily H. Stanley,*[4] *and H. Maurice Valett*[5]

[1]Division of Ecosystem Management, University of New England, Armidale, 2351 New South Wales, Australia, e-mail: aboulton@metz.une.edu.au; [2]Institute of Ecosystem Studies, Millbrook, New York 12545; [3]University of Savoie, G.R.E.T.I., ESA-CNRS #5023, 73376 Le Bourget du Lac, France; [4]Department of Zoology, Oklahoma State University, Stillwater, Oklahoma 74078-3052; [5]Department of Biology, Virginia Polytechnic Institute and State University, Blacksburg, Virginia 24061

KEY WORDS: aquatic ecosystems, hydrology, scale, ecotone, model

ABSTRACT

The hyporheic zone is an active ecotone between the surface stream and groundwater. Exchanges of water, nutrients, and organic matter occur in response to variations in discharge and bed topography and porosity. Upwelling subsurface water supplies stream organisms with nutrients while downwelling stream water provides dissolved oxygen and organic matter to microbes and invertebrates in the hyporheic zone. Dynamic gradients exist at all scales and vary temporally. At the microscale, gradients in redox potential control chemical and microbially mediated nutrient transformations occurring on particle surfaces. At the stream-reach scale, hydrological exchange and water residence time are reflected in gradients in hyporheic faunal composition, uptake of dissolved organic carbon, and nitrification. The hyporheic corridor concept describes gradients at the catchment scale, extending to alluvial aquifers kilometers from the main channel. Across all scales, the functional significance of the hyporheic zone relates to its activity and connection with the surface stream.

59

INTRODUCTION

Traditionally, most ecological research on groundwater and rivers has treated groundwater and rivers as distinct entities and has focused on within-system issues (16). One reason for this distinction reflects historical perspectives in disciplinary focus: Most groundwater studies are by hydrologists, whereas ecologists have been more interested in rivers (145). Another reason may lie in the marked differences between these two environments. Rivers typically have currents generating turbulence, short water residence time, variable discharge and physicochemical conditions, unidirectional transport of nutrients, sediments and biota, and a dynamic channel morphology, and they are well lit. In contrast, alluvial groundwater environments are more stable, have longer water residence times, exhibit laminar flow, are permanently dark, and change little in sediment bed structure (16, 20, 57, 110, 158).

Recently, attention has turned to the ecology of the *interface* between these two environments because we now recognize the connections via exchange of water, nutrients, other materials, and biota between the surface stream and alluvial groundwater. This intervening zone is the hyporheic zone (HZ) (104). Although a rich lexicon of definitions now exists (see reviews in 16, 57), the most functional emphasizes the dynamic ecotone model, where exchange of river and groundwater occurs (54, 149). Key aspects of this definition include the difficulty of defining the boundaries of this zone because they vary in time and space (12, 155, 157, 158), the shared features of the surface stream and underlying groundwater (often existing as gradients at a range of scales), and the importance of the permeability of this ecotone to the functions that occur within (54, 147, 149).

Therefore, in general terms, the hyporheic zone can be defined as a spatially fluctuating ecotone between the surface stream and the deep groundwater where important ecological processes and their requirements and products are influenced at a number of scales by water movement, permeability, substrate particle size, resident biota, and the physiochemical features of the overlying stream and adjacent aquifers.

This review focuses on the functional significance of the HZ as an ecotone viewed at several scales. Scale provides a useful framework for organizing the wealth of information we have on the HZ and assessing the hierarchy of processes (e.g. 52, 82). Where we have more information and ability to predict processes at certain scales (such as the reach scale), this review examines our ability to extrapolate among scales. We describe regulatory factors at each scale and specify potential impediments to extrapolating across scales or stream ecosystems. Rather than exhaustively review the literature on the HZ (see 16, 53, 77, 109, 157), we critique the current status of research on the functional significance of the HZ, addressing the following questions:

1. What hydrological, chemical, and biological processes occur in the HZ, and how are they interrelated at a range of scales?

2. How do these processes and their interactions influence ecological processes occurring in the surface stream?

3. What features determine the functional significance of the HZ to stream and river ecosystems?

4. What are the promising future research directions in this field, and how do they relate to river management?

We identify relevant temporal and spatial scales of these issues, emphasizing the roles of natural disturbance (e.g. floods) and human activities (e.g. catchment land use, flow regulation) on the functional significance of the HZ in streams and rivers. The relationship of the HZ to compartments other than the surface stream (Figure 1) is reviewed fully elsewhere (16, 53, 102). We contend that the significance of the HZ to the surface stream relates to its activity (e.g. nutrient transformations, respiration rates) and connection (e.g. via hydrological exchange). Both of these features are influenced by sediment characteristics and hydrology at a range of scales.

The explicit recognition of scale for describing hierarchies in patterns and processes and generating hypotheses in ecology has proved valuable (1, 82). Scale issues have been central to some conceptual models in stream ecology (e.g. 52, 97, 112) and have been used as a framework from studies of individuals (e.g. 108) to entire ecosystems (e.g. 130). However, few efforts have been made to explicitly put hyporheic research into a scale context (e.g. Figure 2 in 57) and link the relationships between physical and biological processes. We have adopted this scale-based approach, and although the relationships and gradients

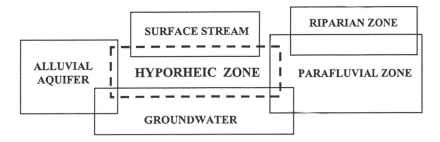

Figure 1 Simplified schematic diagram of the hydrological compartments that can interact with the hyporheic zone. Alluvial aquifers typify floodplain rivers with coarse alluvium and are often considered synonymous with groundwater. The parafluvial zone lies under the active channel, which lacks surface water, and it can interact with subsurface water of the riparian zone.

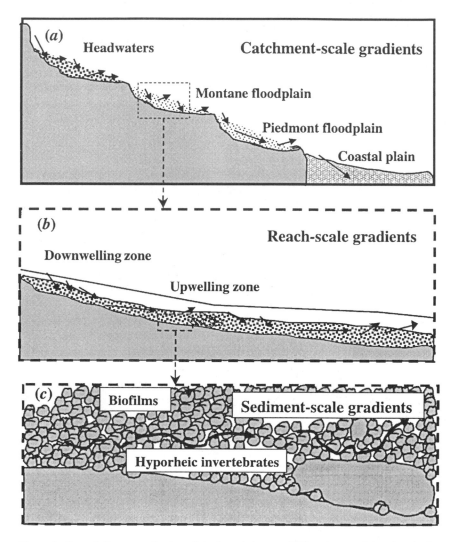

Figure 2 Lateral diagrammatic view of the hyporheic zone (HZ) at three spatial scales. At the catchment scale (*a*), the hyporheic corridor concept predicts gradients in relative size of the HZ, hydrologic retention, and sediment size (126). At the reach scale (*b*), upwelling and downwelling zones alternate, generating gradients in nutrients, dissolved gases, and subsurface fauna. At the sediment scale (*c*), microbial and chemical processes occur on particle surfaces, creating microscale gradients. *Arrows* indicate water flow paths.

occur along a continuum, we focus on three discrete spatial scales: sediment, reach, and catchment (Figure 2).

SEDIMENT-SCALE PROCESSES

Particle Size, Interstitial Flow Pathways, and Microbial Activity

Fine-scale granulometric features (size, shape, and composition of sediments) derive from catchment scale geological processes and determine most physical and chemical processes in the HZ (16, 36, 89, 133). Interstitial flow patterns are a product of hydraulic gradient (direction and strength of flows, Figure 2) and streambed porosity. These flows are turbulent and irregular, creating zones of rapid, slow, and no flow (dead zones). Even where flows are brisk, dead zones exist in sheltered regions, and anaerobic processes can occur. Hence, a seemingly well-oxygenated HZ contains anoxic and hypoxic pockets associated with irregularities in sediment surfaces, small pore spaces, or local deposits of organic matter (31, 84). This heterogeneity enables biologically and chemically disparate microzones to co-occur, facilitating diverse ecological processes in a small volume. Gradients or microzones may exist because there is no hydrological exchange to break them down.

Ammonification, nitrification, and denitrification often all occur as soon as water enters the HZ (72, 73, 75, 76). These sediment-scale transformations of dissolved nitrogen (N), controlled by oxygen availability, influence the nutrient status of upwelling water with concomitant effects on surface stream processes (7, 59, 60, 141–143). Phosphorus (P) concentration in interstitial water is also affected by oxygen distribution; loss of oxygen, change in redox status, and subsequent release of P from bound iron or manganese play a role in P availability (31). Bacterial alkaline phosphatase activity has been documented in the HZ, and breakdown of organic P may be an important source of this nutrient for surface and subsurface biota (93). Less explored at the microscale is the significance of lithological and geochemical processes that can regulate availability of N and P. Sediments with high cation exchange capacity (resulting from their chemical composition and particle size) will readily sorb ammonium (139) and inorganic P (80, 131).

Key processes at this fine spatial scale include those that alter the size or amount of interstitial space or oxygen availability, including clogging of interstices by fine particles (114, 119, 162) or the translocation of fine particulate organic matter (11, 85). Distribution of particulate organic matter among sediments is particularly important in its role as a surface and substrate for microbes and a structure that alters porosity and hydraulic conductivity (15, 16). It also acts either as a source of inorganic nutrients after mineralization or as a potential sink for ions such as phosphate through colloidal interactions. Biofilms predominate on small particles because of their large surface area (5, 81), resulting in a negative correlation between mean sediment particle size and bacterial abundances in river sediments (22). Their low porosity and influence on water velocity allow smaller sediments to trap fine organic matter, so they are generally associated with high organic matter content (84) that stimulates biofilm growth (22). Intense bacterial activity may so reduce oxygen concentrations that anaerobic processes such as denitrification predominate, fueled by labile organic matter from the surface (73) or parafluvial zones (23).

Interstitial Fauna

The interstitial spaces among sediment particles in the HZ of many streams and rivers are occupied by a diverse array of aquatic invertebrates, termed the "hyporheos" (104, 159). The hyporheos includes many types of crustaceans, segmented worms, flatworms, rotifers, water mites, and juvenile stages of aquatic insects (8, 34, 64, 109, 125, 157). Biofilms provide an important food source for the hyporheos (2, 8, 16, 158). Therefore, variables that affect the extent, composition, and food quality of the biofilms probably influence the distribution of grazing invertebrates (7, 8, 36, 56, 57, 85, 132, 133). Their feeding activities may enhance biofilm productivity (e.g. 8, 98) and break down coarse particulate detritus trapped in sediments, increasing their surface area for microbial attack. There has been little research on the dietary requirements of the hyporheos (30, 35, 158), but it seems that predatory subsurface invertebrates are particularly diverse (8, 12, 27). There is still much we need to learn about what fuels the hyporheic food web and how these energy sources vary in streams with different bed porosities, discharges, and organic matter inputs from their catchments.

Large numbers of hyporheic invertebrates may be collected [e.g. up to 10,711 in 3L (133)]. Most of these are meiofauna, less than 1 mm long when adult (64, 107). Their small size and high reproductive rate imply that they are important at the sediment scale, regulating microbial productivity and providing food for larger hyporheic invertebrates and even fish in the surface stream (107). Invertebrate activity (burrowing, formation of fecal pellets) can alter interstitial flow paths (8, 33, 34, 40), influencing the physical and chemical processes described above. The influx of fine sediments can render the HZ uninhabitable either directly by clogging the spaces or indirectly by reducing interstitial flows and flushing of nutrients, gases, and wastes (16, 99, 114). In keeping with the concept of the HZ as a dynamic ecotone, we can summarize these processes and their products as inputs and outputs whose compartments and residence times vary temporally and spatially. The magnitude and directions of inputs and outputs become relevant at higher scales (i.e. reach and catchment), as they relate to adjacent habitats such as the surface stream (Figure 1). For example, at a local scale, fine silt may enter trout spawning gravels due to sediment runoff from a cleared catchment upstream, reducing the interstitial flow of water and hence the supply of dissolved oxygen and other requirements of developing fish eggs (25, 96). Until the next flushing flow (58), these impaired spawning gravels reduce trout recruitment and fish densities in the overlying streams. Thus, an input of silt at the local scale of a gravel bed may have ramifications at the stream scale by altering the food web.

Future Research at the Sediment Scale

Research is needed to determine the extent to which relationships observed at the sediment scale can be extrapolated to the reach scale (10–100 m). For instance, many small-scale studies (e.g. 45, 68, 92) show that microbial processes, including respiration and growth, are tightly related to sediment organic content. The relationship between hyporheic respiration and organic matter (OM) matches that found in surface sediments (45), which implies that information derived from surface sediments can help explain factors controlling microbial processes in the HZ. Although we can predict hyporheic bacterial production in a reach because we know the distribution of sediments with various OM contents, to understand the functional significance of the HZ in that reach, we must also know the magnitude of the hydrologic exchange between hyporheic and surface habitats because this exchange provides the actual link (e.g., 7, 43, 60, 76, 142, 143, 145, 151, 154). How well do reach-scale hydrological models approximate sediment-scale water movements? This question poses a major research challenge (see also 158).

Most workers acknowledge the importance of sediment-scale processes [e.g. redox-sensitive chemical gradients (31, 101, 138, 139)], but technological and sampling limitations still hamper advances at this scale. These limitations also apply to sampling fauna at fine scales. There is a wide range of collecting methods, such as freeze-coring (13, 86, 87), pumping interstitial water (6, 9, 10, 12, 39, 153), digging pits in exposed sediments (21, 134), hand-coring (107), standpipe coring (159), and hyporheic pots (42, 94), but comparative research is needed to reveal the differences in efficiency of extraction and selectivity of these methods as well as the choice of appropriate mesh size under different conditions (63, 64, 158). Some pumping methods, for example, may sample interstitial water from regions distant from the end of the sampling tube; this method precludes replicate sampling (9, 27) and is selective (49). Until reliable, quantitative data can be collected, ecological studies such as complete food web analyses are probably impossible (30, 57).

REACH-SCALE PROCESSES

Flow Paths and Hydrologic Retention

Our best perception of the functional significance of the HZ may be at the reach scale because this has been the scale at which most workers have explored the connection of the surface stream with the HZ. The most obvious linkage is via hydrological exchange in upwelling and downwelling regions that form in response to reach-scale geomorphological features such as discontinuities in slope and depth of riffle-pool sequences, the shape of the channel and its bars, the roughness and permeability of the streambed, and obstacles (e.g. macrophytes, boulders) that extend into the channel and alter surface flow paths (16, 118, 154, 158). Commonly, decreasing stream depth at the end of a pool forces surface water down into the sediments (downwelling), displacing interstitial water that may travel for some distance before upwelling into the surface stream (Figure 2b). Tracer experiments (e.g. 60, 67, 72, 76, 140, 160) indicate that flow paths are usually more complex than this and can respond to other factors such as flooding and riparian transpiration. Geomorphological features such as depth to bedrock are also relevant, especially in rivers with shallow HZ; for some of these, the ecological role of the HZ may be less important to the total stream ecosystem (9, 43).

Horizontal flows entering and leaving stream banks (56) and gravel bars (72, 91, 150, 160, 161) are functionally equivalent to downwelling and upwelling through the streambed (76). Together, these flow paths contribute to hydrologic retention (*sensu* 100), a delay in transport that occurs when water enters flow paths moving more slowly than the surface stream. Hydrologic retention is strongly influenced by granulometric features. For example, among three catchments in New Mexico differing in geologic composition, retention was least in fine-grained sedimentary sandstone and highest in the bed of poorly sorted cobbles and boulders of a granitic catchment (100). Similarly, storage zone residence times increased with increasing particle size, indicating not only that more water was exchanged between the stream and aquifer, but also that water remained in the subsurface longer before it returned to the stream (100).

Within any reach, there is a maze of flow paths of different lengths, directions, and velocities. Because streams and aquifers exchange water horizontally and vertically, flow dynamics are inherently three dimensional. However, most hydrologic studies have used single-dimensional models (review in 135), and only recently have two-dimensional models been used (67, 160, 161, 163). Preliminary results from two-dimensional models have been encouraging. For example, a hydrological model for a lowland stream-floodplain system showed that although the magnitude of fluxes changed with season and water table conditions, the general shape of the flow net connecting the stream, HZ, and floodplain remained constant, suggesting geomorphic control over the direction of exchange (160). Three-dimensional models will contribute more explicit information but require geophysical data that are difficult to obtain. Nevertheless, this information is crucial for our knowledge of the significance of the HZ to the surface stream and adjacent habitats.

Longitudinal Gradients

Longer hyporheic retention time promotes interaction between the biofilms on sediment particles and the nutrients and carbon entrained in subsurface flow paths (contact time *sensu* 43). Patterns in variables such as temperature (27, 156), alkalinity (131), nutrients (23, 60, 72, 141, 142, 146, 158), dissolved organic carbon (44, 46), and dissolved oxygen (7, 27, 46, 70, 91, 129, 144, 154, 158) within the HZ reflect the influx of surface water or the movement of water along a hyporheic flow path. Movement of water through porous sediments has been likened to an ion chromatograph (50), with differential separation and retention of solutes as water travels down the gradient (3). Several researchers have demonstrated hyporheic nitrification by showing the accumulation of nitrate along a flow path (72, 138). These gradients are typically coupled with oxygen depletion because of the mineralization of organic matter (23, 75), thus highlighting the role of the HZ in regenerating inorganic N, which may later become available to nutrient-limited surface biota (142, 143).

Longitudinal trends in nutrients, dissolved oxygen, and the hyporheos matched the direction and magnitude of hydrological exchange and varied in response to flooding and drying in a desert stream reach in Arizona (10, 129). Similar trends are evident in mesic rivers (128). In a regulated channel of the Rhône River, longitudinal changes in dissolved oxygen, particulate organic matter, and hyporheic fauna correlated with flow paths through a 1200 m gravel bar (91). Furthermore, these patterns varied with changes in contact time and interstitial flow rate as a result of variation in stream discharge (91; reviewed in 39), although there was also some spatial variation in response to granulometric features (36).

The Significance of the HZ to Surface Stream Biota

In streams where hydrological exchange with the HZ is active, ecological patterns that are correlated with locations of upwelling zones (Figure 2) are evident. Upwelling hyporheic water rich in nutrients can promote "hot spots" of productivity in the surface stream (7, 26, 151). For example, in some desert streams, the metabolically active HZ generates nitrate that normally limits primary production (61, 62). Upwelling water thus promotes algal activity, resulting in longitudinal gradients of nitrogen uptake in the surface stream (59) and altering benthic algal composition (142, 144). Furthermore, after floods scour the algae, succession is more rapid at these upwelling zones (142, 143).

Aquatic macrophyte distribution may also be influenced by subsurface nutrient concentrations and water movement (48). Convection patterns below *Chara* hummocks apparently benefit the plants by drawing nutrient-rich water toward the rhizomes (69). Few studies have been made on the direct effects of vertical exchange of hyporheic water on surface invertebrates (111). Effects are probably trivial because of dilution in most rivers and the stronger influences of other variables on stream benthos such as substrate and current velocity. However, in floodplain habitats where flushing effects are low, the amount of upwelling groundwater has been found to correlate with benthic faunal composition (e.g. 18, 47) and macrophyte distribution (41).

It has been proposed that the HZ provides an important refuge for surface invertebrates from floods and droughts, predation, and deterioration in surface water quality (reviewed in 7, 8, 16, 38, 90, 106). These invertebrates range in life history strategies from those that spend most of their life in the stream and enter the HZ only briefly (occasional hyporheos, *sensu* 159) to those with a hyporheic larval stage but with subadult and adult stages that leave the HZ (amphibites; 57, 126). Individuals from virtually every insect family and most other groups found at the surface have been collected from the HZ, although few of these collections have been from depths exceeding 50 cm (8, 14).

For many small instars, the HZ is a refuge from the shear stress of strong currents and the more variable conditions that occur in the surface stream (e.g. extreme water temperatures). This more stable environment generates relatively protected and predictable conditions for eggs, pupae, and diapausing stages of invertebrates (113), and the development of fish embryos of several species (66, 99). Success of the development of salmon embryos in spawning gravels is correlated with interstitial dissolved oxygen (25, 96), and human activities leading to siltation are of concern to fisheries managers (162).

Future Research at the Reach Scale

AN APPROACH TO ASSESSING THE RELATIVE IMPORTANCE OF VARIABLES Numerous variables influence the significance of the HZ to the surface stream. Fundamentally, each variable affects the activity of the HZ, its connection with the surface stream, or both. However, the relative importance of these variables at sediment and reach scales and over time is unclear. At the reach scale, physical features such as granulometric characteristics, permeability and porosity, stream morphology (riffle/pool transitions, channel constrictions, lateral deposits), and topography (stream size, stage, slope) are relevant because they influence, among other things, hydrological exchanges. The first research challenge is to rank the controlling factors or to provide a predictive framework for this approach. The second is to expand our hydrological models to incorporate three dimensions and to explore the extent to which sediment-scale flow paths can be extrapolated to reach-scale hydrological processes. Without reliable hydrological models, it is difficult to identify the importance of exchanges at various scales of space and time and predict changes in features of the surface stream that reflect hyporheic processes.

One simple starting place involves estimation of the ratio of water moving through hyporheic sediments to surface stream flow. Surface velocities can be measured or estimated using Manning's equation (58), while Darcy's Law allows estimation of subsurface velocity from hydraulic conductivity (ease with which subsurface water flows) and slope (51). Discharges in the channel and the HZ are obtained by multiplying the respective velocities by their relative cross-sectional areas (A_s = cross-sectional area of the HZ, A = cross-sectional area of the stream) to obtain a rough estimate of the proportion of water moving down the channel relative to subsurface flow. By varying factors such as slope, hydraulic conductivity, and A_s/A, we can generate values for surface/channel velocities and discharges that span natural stream conditions. For example, A_s/A can range from roughly 5 (i.e. the hyporheic cross section is five times the channel cross section) (146) to almost zero in bedrock streams (101). Similarly, although the calculations only approximate actual velocities, they match realistic surface water (0.1–2 m/s) and hyporheic velocities (0.00001–0.01 m/s). The resultant proportions of hyporheic versus channel flow vary over 4–5 orders of magnitude when plotted as a function of A_s/A with channel discharge 100–1000 times greater than hyporheic discharge except at high values of A_s/A (Figure 3).

Based on these approximations, the contribution of the HZ to the entire stream ecosystem is likely to be greatest when a relatively high proportion of the total discharge flows at intermediate velocities (allowing time for transformation processes, etc) through a relatively large HZ. This model attempts to integrate reach-scale variables such as the relative flows through given cross-sections of the HZ and the overlying surface stream with sediment-scale variables such as nutrient transformations and diffusion from the biofilms, resulting in predictions that may be extrapolated to a catchment scale.

As the proportion of surface water passing through the HZ will normally be less than 100%, the relevant question becomes, "How big a difference in biogeochemical processing is necessary for the HZ to be functionally significant?" Future research could use this model as a springboard to answer this question and to relate activity in the HZ to the degree of connection between the HZ and the surface stream. This model also allows at least first-order ranking of the controlling variables to generate testable predictions and to compare different stream reaches. Such a simple approach may suffice until more tractable hydrological models are readily available to ecologists.

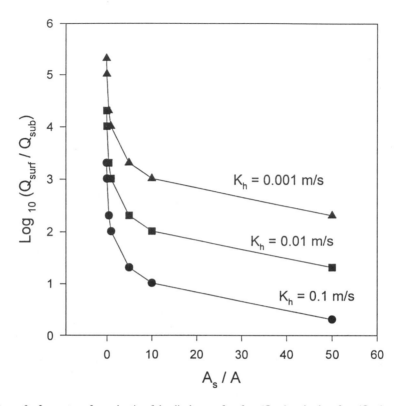

Figure 3 Log$_{10}$-transformed ratio of the discharge of surface (Q$_{surf}$) and subsurface (Q$_{sub}$) water derived from simple empirical models of channel and subsurface flow. K$_h$ is hydraulic conductivity, and A$_s$/A is the cross-sectional area of the subsurface storage zone relative to the open channel. With this model, the hyporheic zone (HZ) is hypothesized to be most significant to stream ecosystem function when a relatively high proportion of total discharge travels at intermediate velocities through a relatively large HZ (see text).

EXPERIMENTAL APPROACHES More experimental work is needed to explore the causal mechanisms and hypotheses generated by reach-scale descriptive surveys and to test hypotheses generated by modeling approaches such as that described above. There are methodological limitations to overcome (see 7, 63, 64, 105, 158), but we urgently need to test hypotheses about the factors that influence the rate of nutrient regeneration, that control microbial processes, and that determine patterns in composition and abundance of the hyporheos in the HZ. For example, interesting patterns in distribution of invertebrates in the HZ have often been noted, but the *causes* of this patchiness are not obvious and are usually ascribed to physical, chemical, hydrological, and sedimentary features based on correlative data collected at rather broad scales (e.g. 9, 10, 12, 16, 27–29, 33, 34, 36, 86, 88, 89, 94, 113, 115, 128, 132, 133, 151–153, 157–159). Furthermore, these distributional patterns are modified by drying (10, 24), changes in surface water quality (74), and flooding (10, 37, 88, 90, 106), indicating a need for experiments at the reach scale that manipulate flows and hydrological exchange patterns. Although small-scale experiments (e.g. 11, 106) provide limited insights into ecological aspects such as local migration rates, they are probably at the wrong scale for assessing reach-scale responses to manipulations of hydrology or other variables.

Technologically, there is scope for innovative approaches to these larger-scale manipulations, and we may be able to design suitable experiments to take advantage of river and riparian zone restoration measures (148). Given the lack of information about the role of the HZ as a "storage area" for recolonization after natural and human-induced disturbances, there is a need to obtain more reliable data before making generalizations about population resilience and resistance at the stream-reach scale (83, 136).

CATCHMENT-SCALE PROCESSES

The Hyporheic Corridor Concept

Few studies have been conducted at this broadest scale, and theoretical models predicting how the HZ varies within a catchment are in their infancy (28, 57, 126, 155). Stanford & Ward (126) proposed the hyporheic corridor concept (HCC), which emphasizes the connections and interactions between the HZ and the catchment. Alluvial flow paths and residence time are suggested to control hyporheic biodiversity and ecosystem metabolism. The subsurface continuum formed by the "hyporheic corridor" has a lateral component connecting riparian zones, anabranches, paleochannels, and floodplain aquifers (up to 3 km from the main channel; 127) that generates a wide array of landscape features whose temporal variability relates to their degree of connection and the discharge regime in the river. Along the river continuum, vertical hydrological exchange between the HZ and the surface stream occurs at a series of points (Figure 2a). These points correlate with reaches with limited HZs interspersed with unconstrained alluvial reaches, like beads on a string (28). Thus, the HCC identifies catchment-scale processes whereby (a) production in the main channel is strongly influenced by upwelling nutrient-rich water, (b) riparian zone structure and dynamics reflect hyporheic flow patterns, and (c) the spatial and temporal variability in hydrological exchange processes and linkages promote exceptional biodiversity within the landscape.

Hyporheic development is predicted to be least in headwater streams (126, 155), peak in the intermediate reaches, and then decline in lowland rivers, where lower alluvial hydraulic conductivities inhibit hydrologic exchange relative to lateral linkages mediated by flooding (78). While many of these predictions remain untested, in a study of solute transport in headwater streams to lowland rivers (32), A_s/A values were found to be highest for low-order streams whereas absolute storage zone size was greatest in unconstrained lowland rivers. On the other hand, studies of hyporheic flow paths in montane upland streams (100, 163) reported hyporheic flow paths much higher than predicted.

At the landscape level, variation among basins may relate to geologic control of sediment alluvial characteristics (71, 100) and patterns of runoff and catchment drainage (79). Associated with contrasting parent lithology, differing sediment hydraulic conductivity dictated the size of the HZ and the rate of exchange between the stream and the aquifer in three headwater streams (100, 146). The size of the HZ increased nearly 300-fold from a first-order stream draining sandstone and volcanic tuff compared with one originating from the cobbles and boulders of a granitic basin.

Catchment-Scale Ecological Studies of the HZ

Although biogeographic patterns of distribution and evolutionary pathways of several invertebrate groups occurring in the HZ and associated groundwater habitats are well studied (e.g. 4, 35, 132), catchment-scale studies of the ecology of invertebrates in the HZ are rare (116, 152). Broad-scale patterns do not seem as obvious as reach-scale patterns. For example, despite substantial variation in elevation (ca 2000 m) along its length, longitudinal patterns in the composition of fauna associated with the surface gravel of alluvial aquifers at nine sites along the South Platte River, Colorado, were only weakly associated with altitude (153). There was no correlation between altitude and the interstitial faunal composition deeper in the sediments, which suggested site-specific geomorphic features may be more important (152). Conversely, a survey of 14 sites across the eastern United States using comparable sampling methods indicated that correlations of faunal composition with sediment size, oxygen concentration, and organic matter were weak (133), which implied that other factors regulated these hyporheic communities.

The two best-known catchment-level studies of the HZ and associated groundwater environments are those of the Rhône River (France) and the Flathead River (Montana, United States). The interstitial fauna of these environments has been studied for almost two decades. Seminal work by Gibert et al (55) first drew attention to the faunal richness of the alluvial aquifers of the Rhône, and the role of hydrology and geomorphology in structuring hyporheic assemblages in space and over time is now well established (see review in 39). In 1974, many invertebrates (including stonefly nymphs) were reported from deep in the alluvial sediments of the Tobacco River, northwestern Montana (124), and this reporting heralded the discovery of diverse (>70 taxa) invertebrate assemblages from the larger Flathead River system and further evidence for hydrological and geomorphological control of interstitial faunal composition (125; see review in 127).

Future Research and River Management at the Catchment Scale

Parallels between the conclusions from these two catchment-wide studies are striking. At a broad scale, the complexity of habitats and resultant biodiversity support predictions of the HCC. The vast areas of interfaces at a range of scales (sediment scale to riparian and aquifer ecotones) produce physicochemical gradients across which substantial fluxes and transformations of organic matter, nutrients, and other materials occur. These observations have important ramifications for river management and thus are an obvious research priority (56, 103). Both of these catchments are occupied by humans whose activities affect ecological processes occurring in the associated groundwater and hyporheic components of these rivers (e.g. cultural eutrophication, sedimentation, flow regulation; see 39, 127).

Hydrologic fluxes between various compartments and the HZ (Figure 1) mean that this zone both receives and contributes anthropogenic pollutants. Sewage discharges to surface water can significantly increase interstitial and sediment-associated nutrient concentrations, depleting hyporheic oxygen (19, 117) and fundamentally altering hyporheic biogeochemical structure and function. Similarly, chemicals in agricultural runoff can move from surface water into groundwater with little change in concentration (122, 123). If degradation occurs, it is likely to happen within the HZ rather than in deeper groundwater zones (120). Conversely, heavy metals (137), pesticides, and anthropogenic nutrients (95) can move from groundwater into surface water through the hyporheic interface. To manage these pollutants properly, we must learn more about the ecological ramifications of organic chemical, nutrient, or heavy metal loading to the HZ.

Regulation of the Rhône River, France, altered the bed geomorphology (aggradation and degradation of 4–5 m), reversing the direction of aquifer/river interaction by changing the relative elevation of the riverbed and alluvial aquifer and substantially altering the composition of the hyporheos (29). A similar uncoupling of the riparian, river, and aquifer subsystems occurred along the Rhine River and was associated with increasing geomorphic and hydrologic manipulation (17). Upstream portions of the Rhine have become entrenched, isolating the river from lateral interactions with the floodplain, drying out spring brooks, and restricting aquifer recharge to areas of the riverbed within the constraints of the hydroelectric canals.

Given the significance of hydrological connections to the role of the HZ, human activities that alter quantity and quality of sediment transport are important. These include dam construction, which reduces long-term sediment loading, and road building, farming, housing construction, suburban development, and logging, which can increase sediment loads (162). Such land uses enhance transport of fine sediment into the streambed, clogging sediments (16, 99). The impacts on fish have been studied (e.g. 121, 162), but little is known about the physical and chemical changes that are likely to occur in the HZ (114). However, negative impacts on the HZ are not always evident. Despite large amounts of fine sediment generated during construction of the Thomson Dam, Victoria, there was little silt deposition in the HZ (87). Consequences of clearing such as bank slump and removal of riparian vegetation have been postulated to influence the HZ of several small streams in New Zealand (9), although river restoration measures may prove costly.

CONCLUSIONS

The importance of bed permeability and hydrological flow patterns has been a recurring theme at a range of scales in this review and in the many attempts to synthesize and identify the key features controlling ecosystem processes in the HZ (reviewed in 7, 16, 36, 43, 57, 65, 132, 151, 157–159). At the catchment and reach scales, permeability and flow patterns determine the proportion of discharge through the HZ, which has been hypothesized to influence how biogeochemical processes in the HZ affect stream ecosystem metabolism (43).

Ultimately, the significance of the HZ to the surface stream is a function of its activity and extent of connection. Although some fine-scale measurements of hyporheic activity have been obtained (e.g. rates of respiration, nitrification), it has not yet been demonstrated that these measurements can be extrapolated to reach and catchment levels. We know the principal variables controlling hyporheic metabolic activity and the connection of the HZ to the surface stream, especially at the reach scale, but we lack a framework for assessing the relative importance of these variables across systems. Research on the HZ awaits some technological advances in hydrological modeling, reach-scale experimental procedures, and sampling methods. Although it may be argued that the HZ is important only in a limited subset of streams (i.e. relatively large HZ and A_s/A, metabolically active, substantial hydrological exchange), its role in these streams can underpin understanding of how they function, exemplified by research on desert streams and lowland gravel-bed rivers. Further, a simple budget approach undoubtedly overlooks some of the special properties and processes (e.g. nitrification and upwelling "hot spots" of productivity) that render the HZ functionally significant to the surface stream at a range of spatial scales.

ACKNOWLEDGMENTS

We are grateful to Drs Stuart Fisher and Judy Meyer for their encouragement to write this review. For constructive comments on earlier drafts, we thank Prof. Janine Gibert and Drs Richard Marchant, Marie-José Olivier, Dave Strayer, Kerry Trayler, Philippe Vervier, and Steve Wondzell. We also thank the French Ministry for the Environment (Grant #94049) and the French Ministry of Research (Grant #96N60/0014) for support to Pierre Marmonier, and the Australian Research Council for supporting Andrew Boulton.

> **Visit the *Annual Reviews* home page at**
> **http://www.AnnualReviews.org**

Literature Cited

1. Allen TFH, Starr TB. 1982. *Hierarchy: Perspectives for Ecological Complexity.* Chicago: Univ. Chicago Press
2. Bärlocher F, Murdoch LH. 1989. Hyporheic biofilms—a potential food source for interstitial animals. *Hydrobiologia* 184:61–67
3. Bencala KE, Kennedy VC, Zellweger GW, Jackman AP, Avanzino RJ. 1984. Interactions of solutes and streambed sediments. 1. An experimental analysis of cation and anion transport in a mountain stream. *Water Resour. Res.* 20:1797–803
4. Botosaneanu L, Holsinger JR. 1991. Some aspects concerning the colonisation of the subterranean waters: a response to Rouch and Danielopol. *Stygologia* 6:11–39
5. Bott TL, Kaplan LA. 1985. Bacterial biomass, metabolic state, and activity in stream sediments: relations to environmental variables and multiple assay comparisons. *Appl. Environ. Microbiol.* 50:508–22
6. Bou C, Rouch R. 1967. Un nouveau champ de recherches sur la faune aquatique souterraine. *C. R. Acad. Sci. Paris* 265:369–70
7. Boulton AJ. 1993. Stream ecology and surface-hyporheic exchange: implications, techniques and limitations. *Aust. J. Mar. Freshwater Res.* 44:553–64
8. Boulton AJ. 1999. The ecology of subsurface macrofauna. See Ref. 77, In press
9. Boulton AJ, Scarsbrook MR, Quinn JM, Burrell GP. 1997. Land-use effects on the hyporheic ecology of five small streams near Hamilton, New Zealand. *N.*

Z. J. Mar. Freshwat. Res. 31:609–22
10. Boulton AJ, Stanley EH. 1995. Hyporheic processes during flooding and drying in a Sonoran Desert stream. II. Faunal dynamics. *Arch. Hydrobiol.* 134:27–52
11. Boulton AJ, Stibbe SE, Grimm NB, Fisher SG. 1991. Invertebrate recolonization of small patches of defaunated hyporheic sediments in a Sonoran Desert stream. *Fresh. Biol.* 26:267–77
12. Boulton AJ, Valett HM, Fisher SG. 1992. Spatial distribution and taxonomic composition of the hyporheos of several Sonoran Desert streams. *Arch. Hydrobiol.* 125:37–61
13. Bretschko G. 1990. The effect of escape reactions on the quantitative sampling of gravel stream fauna. *Arch. Hydrobiol.* 120:41–49
14. Bretschko G. 1991. The limnology of a low order alpine gravel stream (Ritrodat-Lunz study area, Austria). *Verh. Int. Ver. Limnol.* 24:1908–12
15. Bretschko G, Leichtfried M. 1989. Distribution of organic matter and fauna in a second order alpine gravel stream (Ritrodat-Lunz area, Austria). *Verh. Int. Ver. Limnol.* 23:1333–39
16. Brunke M, Gonser T. 1997. The ecological significance of exchange processes between rivers and groundwater. *Freshw. Biol.* 37:1–33
17. Carbiener R, Trémolière M. 1990. The Rhine Rift Valley groundwater-river interactions: evolution of their susceptibility to pollution. *Reg. Rivers* 5:375–89
18. Castella E, Amoros C. 1988. Freshwater macroinvertebrates as functional describers of the dynamics of former river

beds. *Verh. Int. Ver. Limnol.* 23:1299–305

19. Chambers PA, Prepas EE. 1994. Nutrient dynamics in riverbeds—the impact of sewage effluent and aquatic macrophytes. *Water Res.* 28:453–64

20. Chappel FH. 1993. *Groundwater Microbiology and Geochemistry.* New York: Wiley

21. Chappuis PA. 1942. Eine neue Methode zur Untersuchung der Grundwasserfauna. *Acta Sci. Math. Nat. Kolozsvar* 6:3–7

22. Claret C, Fontvieille D. 1997. Characteristics of biofilm assemblages in two contrasted hydrodynamic and trophic contexts. *Microb. Ecol.* 34:49–57

23. Claret C, Marmonier P, Boissier J-M, Fontvieille D, Blanc P. 1997. Nutrient transfer between parafluvial interstitial water and river water: influence of gravel bar heterogeneity. *Freshw. Biol.* 37:657–70

24. Clinton SM, Grimm NB, Fisher SG. 1996. Response of a hyporheic invertebrate assemblage to drying disturbance in a desert stream. *J. N. Am. Benthol. Soc.* 15:700–12

25. Coble DW. 1961. Influence of water exchange and dissolved oxygen in redds on survival of steelhead trout embyos. *Trans. Am. Fish. Soc.* 90:469–74

26. Coleman RL, Dahm CN. 1990. Stream geomorphology: effects on periphyton standing crop and primary production. *J. N. Am. Benthol. Soc.* 9:293–302

27. Cooling MP, Boulton AJ. 1993. Aspects of the hyporheic zone below the terminus of a South Australian arid-zone stream. *Aust. J. Mar. Freshwater Res.* 44:411–26

28. Creuzé des Châtelliers M. 1991. Geomorphological processes and discontinuities in the macrodistribution of the interstitial fauna. A working hypothesis. *Verh. Int. Ver. Limnol.* 24:1609–12

29. Creuzé des Châtelliers M, Reygrobellet JL. 1990. Interactions between geomorphological processes, benthic and hyporheic communities: first results on a by-passed canal of the French Upper Rhône River. *Reg. Rivers* 5:139–58

30. Culver DC. 1994. Species interactions. See Ref. 53, pp. 271–85

31. Dahm CN, Trotter EH, Sedell JR. 1987. Role of anaerobic zones and processes in stream ecosystem productivity. In *Chemical Quality of Water and the Hydrological Cycle*, ed. RC Averett, DM McKnight, pp. 157–78. Chelsea, MI: Lewis

32. D'Angelo DJ, Webster JR, Gregory SV, Meyer JL. 1993. Transient storage in Appalachian and Cascade mountain streams as related to hydraulic characteristics. *J. N. Am. Benthol. Soc.* 12:223–35

33. Danielopol DL. 1984. Ecological investigations on the alluvial sediments of the Danube in the Vienna area—a phreatobiological project. *Verh. Int. Ver. Limnol.* 22:1755–61

34. Danielopol DL. 1989. Groundwater fauna associated with riverine aquifers. *J. N. Am. Benthol. Soc.* 8:18–35

35. Danielopol DL, Creuzé des Châtelliers M, Moeszlacher F, Pospisil P, Popa R. 1994. Adaptation of Crustacea to interstitial habitats: a practical agenda for ecological studies. See Ref. 53, pp. 217–43

36. Dole-Olivier M-J, Marmonier P. 1992. Patch distribution of interstitial communities: prevailing factors. *Freshw. Biol.* 27:177–91

37. Dole-Olivier M-J, Marmonier P. 1992. Effects of spates on the vertical distribution of the interstitial community. *Hydrobiologia* 230:49–61

38. Dole-Olivier M-J, Marmonier P, Beffy J-L. 1997. Response of invertebrates to lotic disturbance: Is the hyporheic zone a patchy refugium? *Freshw. Biol.* 37:257–76

39. Dole-Olivier M-J, Marmonier P, Creuzé des Châtelliers M, Martin D. 1994. Interstitial fauna associated with the alluvial floodplains of the Rhône River (France). See Ref. 53, pp. 313–46

40. Edler C, Dodds WK. 1992. Characterization of a groundwater community dominated by *Caecidotea tridentata* (Isopoda). See Ref. 124a, pp. 91–109

41. Eglin I, Roeck U, Robach F, Trémolière M. 1997. Macrophyte biological methods used in the study of the exchange between the Rhine River and the groundwater. *Water Res.* 31:503–14

42. Eisenmann H, Traunspurger W, Meyer EI. 1997. A new device to extract sediment cages colonized by microfauna from coarse gravel river sediments. *Arch. Hydrobiol.* 139:547–61

43. Findlay S. 1995. Importance of surface-subsurface exchange in stream ecosystems: the hyporheic zone. *Limnol. Oceanogr.* 40:159–64

44. Findlay S, Sobczak WV. 1996. Variability in removal of dissolved organic carbon in hyporheic sediments. *J. N. Am. Benthol. Soc.* 15:35–41

45. Findlay S, Sobczak WV. 1999. Microbial communities in hyporheic sediments. See Ref. 77, In press
46. Findlay S, Strayer DL, Goumbala C, Gould K. 1993. Metabolism of streamwater dissolved organic carbon in the shallow hyporheic zone. *Limnol. Oceanogr.* 38:1493–99
47. Foeckler F. 1990. Die Bewertung von Lebensräum auf der Basis ihrer biozönotischen Charakterisierung—am Beispiel von Wassermolluskengesellschaften in Donau-Augewässern. *Schr. R. F. Landschaft. Nat.* 32:143–63
48. Fortner SL, White DS. 1988. Interstitial water patterns: a factor influencing the distribution of some lotic aquatic vascular macrophytes. *Aquat. Bot.* 31:1–12
49. Fraser BG, Williams DD. 1997. Accuracy and precision in sampling hyporheic fauna. *Can. J. Fish. Aquat. Sci.* 54:1135–41
50. Freeman C, Chapman PJ, Gilman K, Lock MA, Reynolds B, et al. 1995. Ion exchange mechanisms and the entrapment of nutrients by river biofilms. *Hydrobiologia* 297:61–65
51. Freeze RA, Cherry JA. 1979. *Groundwater.* Englewood Cliffs, NJ: Prentice-Hall
52. Frissell CA, Liss WJ, Warren CE, Hurley MD. 1986. A hierarchical framework for stream habitat classification: viewing streams in a watershed context. *Environ. Manage.* 10:199–214
53. Gibert J, Danielopol DL, Stanford JA, eds. 1994. *Groundwater Ecology.* San Diego: Academic
54. Gibert J, Dole-Olivier M-J, Marmonier P, Vervier P. 1990. Surface water-groundwater ecotones. In *The Ecology and Management of Aquatic-Terrestrial Ecotones*, ed. RJ Naiman, H Décamps, pp. 199–225. London: Parthenon
55. Gibert J, Ginet R, Mathieu J, Reygrobellet J-L, Seyed-Reihani A. 1977. Structure et fonctionnement des écosystèmes du Haut Rhône francais. IV. Le peuplement des eaux phréatiques, premiers résultats. *Ann. Limnol.* 13:83–97
56. Gibert J, Marmonier P, Vanek V, Plénet S. 1995. Hydrological exchange and sediment characteristics in a riverbank—relationship between heavy metals and invertebrate community structure. *Can. J. Fish. Aquat. Sci.* 52:2084–97
57. Gibert J, Stanford JA, Dole-Olivier M-J, Ward JV. 1994. Basic attributes of groundwater ecosystems and prospects for research. See Ref. 53, pp. 7–40
58. Gordon ND, McMahon TA, Finlayson BL. 1992. *Stream Hydrology: An Introduction for Ecologists.* Chichester, UK: Wiley
59. Grimm NB. 1987. Nitrogen dynamics during succession in a desert stream. *Ecology* 68:1157–70
60. Grimm NB, Fisher SG. 1984. Exchange between interstitial and surface water: implications for stream metabolism and nutrient cycling. *Hydrobiologia* 111:219–28
61. Grimm NB, Fisher SG. 1986. Nitrogen limitation in a Sonoran Desert stream. *J. N. Am. Benthol. Soc.* 5.2–15
62. Grimm NB, Fisher SG, Minckley WL. 1981. Nitrogen and phosphorus dynamics in hot desert streams of the southwestern USA. *Hydrobiologia* 83:303–12
63. Hakenkamp CC, Palmer MA. 1992. Problems associated with quantitative sampling of shallow groundwater invertebrates. See Ref. 124a, pp. 101–10
64. Hakenkamp CC, Palmer MA. 1999. The ecology of hyporheic meiofauna. See Ref. 77, In press
65. Hakenkamp CC, Valett HM, Boulton AJ. 1993. Perspectives on the hyporheic zone: integrating hydrology and biology. Concluding remarks. *J. N. Am. Benthol. Soc.* 12:94–99
66. Hansen EA. 1975. Some effects of groundwater on brown trout redds. *Trans. Am. Fish. Soc.* 1975:100–10
67. Harvey JW, Bencala KE. 1993. The effect of streambed topography on surface-subsurface water exchange in mountain catchments. *Wat. Resour. Res.* 29:89–98
68. Hedin LO. 1990. Factors controlling sediment community respiration in woodland stream ecosystems. *Oikos* 57:94–105
69. Hendricks SP, White DS. 1988. Hummocking by lotic *Chara*: observations on alterations of hyporheic temperature patterns. *Aquat. Bot.* 31:13–22
70. Hendricks SP, White DS. 1991. Physicochemical patterns within a hyporheic zone of a northern Michigan river, with comments on surface water patterns. *Can. J. Fish. Aquat. Sci.* 48:1645–54
71. Hinton MJ, Schiff SL, English MC. 1993. Physical properties governing groundwater flow in a glacial till catchment. *J. Hydrol.* 142:229–49
72. Holmes RM, Fisher SG, Grimm NB. 1994. Parafluvial nitrogen dynamics in a desert stream ecosystem. *J. N. Am. Benthol. Soc.* 13:468–78
73. Holmes RM, Jones JB, Fisher SG, Grimm NB. 1996. Denitrification in a nitrogen-limited ecosystem. *Biogeochemistry* 33:125–46

74. Jeffrey KA, Beamish FW, Ferguson SC, Kolton RJ, MacMahon PD. 1986. Effects of the lampricide, 3-trifluoromethyl–4-nitrophenol (TFM) on the macroinvertebrates within the hyporheic region of a small stream. *Hydrobiologia* 134:43–51

75. Jones JB, Fisher SG, Grimm NB. 1995. Nitrification in the hyporheic zone of a desert stream ecosystem. *J. N. Am. Benthol. Soc.* 14:249–58

76. Jones JB, Holmes RM. 1996. Surface-subsurface interactions in stream ecosystems. *Trends Ecol. Evol.* 11:239–42

77. Jones JB, Mulholland P, eds. 1999. *Surface-Subsurface Interactions in Streams.* New York: Academic/Landes Biosci. In press

78. Junk WJ, Bayley PB, Sparks RE. 1989. The flood pulse concept in river-floodplain systems. *Spec. Pub. Can. Fish. Aquat. Sci.* 106:110–27

79. Kelson KI, Wells SG. 1989. Geologic influences on fluvial hydrology and bedload transport in small mountainous watersheds, northern New Mexico, U.S.A. *Earth Surf. Process. Landforms* 14:671–90

80. Klotz RL. 1988. Sediment control of soluble reactive phosphorus in Hoxie Gorge Creek, New York. *Can. J. Fish. Aquat. Sci.* 45:2026–34

81. Kondratieff PF, Simons GM. 1985. Microbial colonization of seston and free bacteria in an impounded river. *Hydrobiologia* 128:127–33

82. Kotliar NB, Wiens JA. 1990. Multiple scales of patchiness and patch structure: a hierarchical framework for the study of heterogeneity. *Oikos* 59:253–60

83. Lancaster J, Hildrew AG. 1993. Flow refugia and the microdistribution of lotic macroinvertebrates. *J. N. Am. Benthol. Soc.* 12:385–93

84. Leichtfried M. 1991. POM in bed sediments of a gravel stream (Ritrodat-Lunz study area, Austria). *Verh. Int. Ver. Limnol.* 24:1921–25

85. Lenting N, Williams DD, Fraser BG. 1997. Qualitative differences in interstitial organic matter and their effect on hyporheic colonisation. *Hydrobiologia* 344:19–26

86. Marchant R. 1988. Vertical distribution of benthic invertebrates in the bed of the Thomson River, Victoria. *Aust. J. Mar. Freshw. Res.* 39:775–84

87. Marchant R. 1989. Changes in the benthic invertebrate communities of the Thomson River, southeastern Australia, after dam construction. *Reg. Rivers* 4:71–89

88. Marchant R. 1995. Seasonal variation in the vertical distribution of hyporheic invertebrates in an Australian upland river. *Arch. Hydrobiol.* 134:441–57

89. Maridet L, Philippe M, Wasson JG, Mathieu J. 1996. Spatial and temporal distribution of macroinvertebrates and trophic variables within the bed sediment of three streams differing by their morphology and riparian vegetation. *Arch. Hydrobiol.* 136:41–64

90. Marmonier P. 1991. Effect of alluvial shift on the spatial distribution of interstitial fauna. *Verh. Int. Ver. Limnol.* 24:1613–16

91. Marmonier P, Dole M-J. 1986. Les Amphipodes des sédiments d'un bras court-circuité du Rhône: logique de répartition et réaction aux crues. *Rev. Fr. Sci. Eau* 5:461–86

92. Marxsen J. 1996. Measurement of bacterial production in stream-bed sediments via leucine incorporation. *FEMS Microbiol. Ecol.* 21:313–25

93. Marxsen J, Schmidt H-H. 1993. Extracellular phosphatase activity in sediments of the Breitenbach, a central European mountain stream. *Hydrobiologia* 253:207–16

94. McElravy EP, Resh VH. 1991. Distribution and seasonal occurrence of the hyporheic fauna in a northern California stream. *Hydrobiologia* 220:233–46

95. McMahon PB, Litke DW, Paschal JE, Dennehy KF. 1994. Ground water as a source of nutrients and atrazine to streams in the South Platte River Basin. *Wat. Resour. Bull.* 30:521–30

96. McNeil WJ. 1962. Variations in the dissolved oxygen content of intragravel water in four spawning streams of southeastern Alaska. *US Fish Wildl. Serv. Spec. Sci. Rep. Fish.* 402:1–32

97. Minshall GW. 1988. Stream ecosystem theory: a global perspective. *J. N. Am. Benthol. Soc.* 7:263–88

98. Montagna PA. 1995. Rates of metazoan meiofaunal microbivory: a review. *Vie Milieu* 45:1–9

99. Moring JR. 1982. Decrease in stream gravel permeability after clear-cut logging: an indication of intergravel conditions for developing salmonid eggs and alevin. *Hydrobiologia* 88:295–98

100. Morrice JA, Valett HM, Dahm CN, Campana ME. 1997. Alluvial characteristics, groundwater-surface water exchange and hydrologic retention in headwater streams. *Hydrol. Process.* 11:253–67

101. Mulholland PJ, Marzolf ER, Webster

JR, Hart DR. 1997. Evidence that hyporheic zones increase heterotrophic metabolism and phosphorus uptake in forest streams. *Limnol. Oceanogr.* 42: 443–51

102. Naiman RJ, Décamps H. 1997. The ecology of riparian zones. *Annu. Rev. Ecol. Syst.* 28:621–58

103. Notenboom J, Plénet S, Turquin M-J. 1994. Groundwater contamination and its impact on groundwater animals and ecosystems. See Ref. 53, pp. 477–504

104. Orghidan T. 1959. Ein neuer Lebensraum des unterirdischen Wassers: der hyporheische Biotop. *Arch. Hydrobiol.* 55:392–414

105. Palmer MA. 1993. Experimentation in the hyporheic zone: challenges and prospectus. *J. N. Am. Benthol. Soc.* 12:84–93

106. Palmer M, Bely AE, Berg KE. 1992. Response of invertebrates to lotic disturbance: a test of the hyporheic refuge hypothesis. *Oecologia* 89:182–94

107. Palmer MA, Strayer DL. 1996. Meiofauna. In *Methods in Stream Ecology*, ed. FR Hauer, GA Lamberti, pp. 315–37. San Diego: Academic

108. Peckarsky BL, Cooper SD, McIntosh AR. 1997. Extrapolating from individual behavior to populations and communities in streams. *J. N. Am. Benthol. Soc.* 16:375–90

109. Pennak RW. 1988. Ecology of freshwater meiofauna. In *Introduction to the Study of Meiofauna*, ed. RP Higgins, H Thiel, pp. 39–60. Washington, DC: Smithsonian

110. Petts GE, Calow P. 1996. Fluvial hydrosystems: the physical basis. In *River Flows and Channel Forms*, ed. GE Petts, P Calow, pp. 1–5. Oxford: Blackwell

111. Plénet S, Gibert J, Marmonier P. 1995. Biotic and abiotic interactions between surface and interstitial systems in rivers. *Ecography* 18:296–309

112. Poff NL. 1997. Landscape filters and species traits: towards mechanistic understanding and prediction in stream ecology. *J. N. Am. Benthol. Soc.* 16:391–409

113. Pugsley CW, Hynes HBN. 1986. The three dimensional distribution of winter stonefly nymphs, *Allocapnia pygmaea*, within the substrate of a southern Ontario river. *Can. J. Fish. Aquat. Sci.* 43:1812–17

114. Richards C, Bacon KL. 1994. Influence of fine sediment on macroinvertebrate colonization of surface and hyporheic stream substrates. *Great Basin Nat.* 54:106–13

115. Rouch R. 1988. Sur la répartition spatiale des Crustacés dans le sous-écoulement d'un ruisseau des Pyrénées. *Ann. Limnol.* 24:213–34

116. Rouch R, Danielopol DL. 1997. Species richness of microcrustacea in subterranean freshwater habitats. Comparative analysis and approximate evaluation. *Int. Rev. Ges. Hydrobiol.* 82:121–45

117. Rutherford JC, Wilcock RJ, Hickey CW. 1991. Deoxygenation in a mobile-bed river. I. Field studies. *Wat. Res.* 25:1487–97

118. Savant SA, Reible DD, Thibodeaux LJ. 1987. Convective transport within stable river sediments. *Wat. Resour. Res.* 23:1763–68

119. Schälchli U. 1992. The clogging of coarse gravel river beds by fine sediment. *Hydrobiologia* 235/236:189–97

120. Schwarzenbach RP, Giger W, Hoehn E, Schneider JK. 1983. Behavior of organic compounds during infiltration of river water to groundwater. Field Stud. *Environ. Sci. Technol.* 17:472–79

121. Scrivener JC, Andersen BC. 1984. Logging impacts and some mechanisms that determine the size of spring and summer populations of coho salmon fry (*Oncorhynchus kisutch*) in Carnation Creek, British Columbia. *Can. J. Fish. Aquat. Sci.* 41:1097–105

122. Squillace PJ, Caldwell JP, Schulmeyer PM, Harvey CA. 1996. Movement of agricultural chemicals between surface water and groundwater, Lower Cedar River Basin, Iowa. *USGS Wat. Supply Pap. No.* 2448

123. Squillace PJ, Liszewski MJ, Thurman EM. 1992. Agricultural chemical interchange between ground and surface water, Cedar River Basin, Iowa and Minnesota—A study description. *USGS Open File Rep.* 92–85

124. Stanford JA, Gaufin AR. 1974. Hyporheic communities of two Montana rivers. *Science* 185:700–2

124a. Stanford JA, Simons JJ, eds. *Proc. First Int. Conf. Ground Water Ecol.* Bethesda, MD: Am. Water Res. Assoc.

125. Stanford JA, Ward JV. 1988. The hyporheic habitat of river ecosystems. *Nature* 335:64–66

126. Stanford JA, Ward JV. 1993. An ecosystem perspective of alluvial rivers: connectivity and the hyporheic corridor. *J. N. Am. Benthol. Soc.* 12:48–60

127. Stanford JA, Ward JV, Ellis BK. 1994. Ecology of the alluvial aquifers of the Flathead River, Montana. See Ref. 53, pp. 367–90

128. Stanley EH, Boulton AJ. 1993. Hydrology and the distribution of hyporheos: perspectives from a mesic river and a desert stream. *J. N. Am. Benthol. Soc.* 12:79–83

129. Stanley EH, Boulton AJ. 1995. Hyporheic processes during flooding and drying in a Sonoran Desert stream. I. Hydrologic and chemical dynamics. *Arch. Hydrobiol.* 134:1–26

130. Stanley EH, Fisher SG, Grimm NB. 1997. Ecosystem expansion and contraction in streams. *BioScience* 47:427–36

131. Stewart AJ. 1988. Alkalinity dynamics in a hard-water prairie margin stream. *Arch. Hydrobiol.* 112:335–50

132. Strayer DL. 1994. Limits to biological distributions in groundwater. See Ref. 53, pp. 287–310

133. Strayer DL, May SE, Nielsen P, Wolheim W, Hausam S. 1997. Oxygen, organic matter, and sediment granulometry as controls on hyporheic animal communities. *Arch. Hydrobiol.* 140:131–44

134. Strayer DL, O'Donnell EB. 1988. Aquatic microannelids (Oligochaeta and Aphanoneura) of underground waters of southeastern New York. *Am. Midl. Nat.* 119:327–35

135. Stream Solute Workshop. 1990. Concepts and methods for assessing solute dynamics in stream ecosystems. *J. N. Am. Benthol. Soc.* 9:95–119

136. Townsend CR, Hildrew AG. 1994. Species traits in relation to a habitat templet for river systems. *Freshw. Biol.* 31:265–75

137. Trémolières M, Eglin I, Roeck U, Carbiener R. 1993. The exchange process between river and groundwater on the Central Alsace floodplain (Eastern France). 1. The case of the canalised River Rhine. *Hydrobiologia* 254:133–48

138. Triska FJ, Duff JH, Avanzino RJ. 1990. Influence of exchange flow between the channel and hyporheic zone on nitrate production in a small mountain stream. *Can. J. Fish. Aquat. Sci.* 47:2099–111

139. Triska FJ, Jackman AP, Duff JH, Avanzino RJ. 1994. Ammonium sorption to channel and riparian sediments: a transient storage pool for dissolved inorganic nitrogen. *Biogeochemistry* 26:67–83

140. Triska FJ, Kennedy VC, Avanzino RJ, Zellweger GW, Bencala KE. 1989. Retention and transport of nutrients in a third-order stream in northwestern California: hyporheic processes. *Ecology* 70:1893–905

141. Valett HM, Dahm CN, Campana ME, Morrice JA, Baker MA, et al. 1997. Hydrologic influences on groundwater-surface water ecotones: heterogeneity in nutrient composition and retention. *J. N. Am. Benthol. Soc.* 16:239–47

142. Valett HM, Fisher SG, Grimm NB, Camill P. 1994. Vertical hydrologic exchange and ecological stability of a desert stream ecosystem. *Ecology* 75:548–60

143. Valett HM, Fisher SG, Grimm NB, Stanley EH, Boulton AJ. 1992. Hyporheic-surface water exchange: implications for the structure and functioning of desert stream ecosystems. See Ref. 124a, pp. 395–405

144. Valett HM, Fisher SG, Stanley EH. 1990. Physical and chemical characteristics of the hyporheic zone of a Sonoran Desert stream. *J. N. Am. Benthol. Soc.* 9:201–15

145. Valett HM, Hakenkamp CC, Boulton AJ. 1993. Perspectives on the hyporheic zone: integrating hydrology and biology. Introduction. *J. N. Am. Benthol. Soc.* 12:40–43

146. Valett HM, Morrice JA, Dahm CN. 1996. Parent lithology, surface-groundwater exchange, and nitrate retention in headwater streams. *Limnol. Oceanogr.* 41:333–45

147. Vanek V. 1997. Heterogeneity of groundwater-surface water ecotones. In *Groundwater/Surface Water Ecotones: Biological and Hydrological Interactions and Management Options*, ed. J. Gibert, J Mathieu, F Fourier, pp. 151–61. Cambridge: Cambridge Univ. Press

148. Vervier P, Dobson M, Pinay G. 1993. Role of interaction zones between surface and ground waters in DOC transport and processing: considerations for river restoration. *Freshw. Biol.* 29:275–84

149. Vervier P, Gibert J, Marmonier P, Dole-Olivier M-J. 1992. A perspective on the permeability of the surface freshwater-groundwater ecotone. *J. N. Am. Benthol. Soc.* 11:93–102

150. Vervier P, Naiman RJ. 1992. Spatial and temporal fluctuations of dissolved organic carbon in subsurface flow of the Stillaguamish River (Washington, USA). *Arch. Hydrobiol.* 123:401–12

151. Ward JV. 1989. The four-dimensional nature of lotic ecosystems. *J. N. Am. Benthol. Soc.* 8:2–8

152. Ward JV, Voelz NJ. 1990. Gradient analysis of interstitial meiofauna along a longitudinal stream profile. *Stygologia* 5:93–99

153. Ward JV, Voelz NJ. 1994. Groundwater fauna of the South Platte River system, Colorado. See Ref. 53, pp. 391–423
154. White DS. 1990. Biological relationships to convective flow patterns within stream beds. *Hydrobiologia* 196:149–58
155. White DS. 1993. Perspectives on defining and delineating hyporheic zones. *J. N. Am. Benthol. Soc.* 12:61–69
156. White DS, Elzinga CH, Hendricks SP. 1987. Temperature patterns within the hyporheic zone of a northern Michigan river. *J. N. Am. Benthol. Soc.* 6:85–91
157. Williams DD. 1984. The hyporheic zone as a habitat for aquatic insects and associated arthropods. In *The Ecology of Aquatic Insects*, ed. VH Resh, DM Rosenberg, pp. 403–55. New York: Praeger Scientific
158. Williams DD. 1993. Nutrient and flow vector dynamics at the hyporheic/ groundwater interface and their effects on the interstitial fauna. *Hydrobiologia* 251:185–98

159. Williams DD, Hynes HBN. 1974. The occurrence of benthos deep in the substratum of a stream. *Freshw. Biol.* 4:233–56
160. Wondzell SM, Swanson FJ. 1996. Seasonal and storm dynamics of the hyporheic zone of a 4th-order mountain stream. I: Hydrological Processes. *J. N. Am. Benthol. Soc.* 15:3–19
161. Wondzell SM, Swanson FJ. 1996. Seasonal and storm dynamics of the hyporheic zone of a 4th-order mountain stream. II: Nutrient cycling. *J. N. Am. Benthol. Soc.* 15:20–34
162. Wood PJ, Armitage PD. 1997. Biological effects of fine sediment in the lotic environment. *Environ. Manage.* 21:203–17
163. Wroblicky GJ, Campana ME, Valett HM, Dahm CN. 1998. Seasonal variation in surface-subsurface water exchange and lateral hyporheic area of two stream-aquifer systems. *Water Resour. Res.* 34:317–28

Annu. Rev. Ecol. Syst. 1998. 29:83–112

ENDANGERED MUTUALISMS: The Conservation of Plant-Pollinator Interactions

Carol A. Kearns,[1,4] *David W. Inouye,*[2,4] *and Nickolas M. Waser*[3,4]

[1]EPO Biology, Environmental Residential Academic Program, C.B.176, University of Colorado, Boulder, Colorado 80309; [2]Department of Biology, University of Maryland, College Park, Maryland 20742; [3]Department of Biology, University of California, Riverside California 92521; [4]Rocky Mountain Biological Laboratory, P.O. Box 519, Crested Butte Colorado 81224; e-mail: kearnsca@rtt.colorado.edu

KEY WORDS: agriculture, ecosystem services, fragmentation, habitat alteration, species invasions

ABSTRACT

The pollination of flowering plants by animals represents a critical ecosystem service of great value to humanity, both monetary and otherwise. However, the need for active conservation of pollination interactions is only now being appreciated. Pollination systems are under increasing threat from anthropogenic sources, including fragmentation of habitat, changes in land use, modern agricultural practices, use of chemicals such as pesticides and herbicides, and invasions of non-native plants and animals. Honeybees, which themselves are non-native pollinators on most continents, and which may harm native bees and other pollinators, are nonetheless critically important for crop pollination. Recent declines in honeybee numbers in the United States and Europe bring home the importance of healthy pollination systems, and the need to further develop native bees and other animals as crop pollinators. The "pollination crisis" that is evident in declines of honeybees and native bees, and in damage to webs of plant-pollinator interaction, may be ameliorated not only by cultivation of a diversity of crop pollinators, but also by changes in habitat use and agricultural practices, species reintroductions and removals, and other means. In addition, ecologists must redouble efforts to study basic aspects of plant-pollinator interactions if optimal

0066-4162/98/1120-0083$08.00

management decisions are to be made for conservation of these interactions in natural and agricultural ecosystems.

INTRODUCTION

To persist on planet Earth, humans depend on "life-support services" provided by biological, geological, and chemical processes in healthy ecosystems. Services such as the cycling of nutrients and regulation of climate are widely recognized. Other such services are less well known, among them biological processes arising from interactions among species, including enhancement of other species' populations by beneficial biotic agents. The pollination of flowering plants is a prime example: Without pollination by animals, most flowering plants would not reproduce sexually, and humans would lose food and other plant products (22).

One measure of the immense value of ecosystem services is monetary value. A recent estimate places a conservative overall mean value per annum of 33 trillion American dollars on all ecosystem services (40); the component due to pollination services is $112 billion. Independent estimates placed the annual value of pollination for crop systems at $20 billion (102) to $40 billion in the United States alone (159); for global agriculture, the estimated value is $200 billion (172). Of pollinators other than honeybees, the value to US crop yields may be as high as $6.7 billion per year (141).

The economic importance of pollination, and its esthetic and ethical values, makes it clear that the conservation of pollination systems is an important priority. In this paper, we describe the ecological and evolutionary nature of plant-pollinator interactions and review evidence that they are increasingly threatened by human activities. We then discuss potential management solutions to ameliorate the "pollination crisis" and highlight areas that call for further research.

THE NATURE OF PLANT-POLLINATOR INTERACTIONS

Modern angiosperms comprise an estimated 250,000 species (81), and most of these—by some estimates over 90% (22, p. 274)—are pollinated by animals, especially insects. Bees alone comprise an estimated 25,000–30,000 species worldwide, all obligate flower visitors (22, 206, 215, 237). The ranks of flies, butterflies and moths, beetles, and other obligate or facultative insect flower visitors surely are several times as large, to which must be added species of birds in several families (35), bats, and small mammals. The number of flower-visiting species worldwide may total nearly 300,000 (141).

Relatively few plant-pollinator interactions are absolutely obligate. Most are more generalized on the part of both plants and animals, and they also vary through time and space (61, 62, 78, 79, 181, 232). For example, the shrub *Lavandula latifolia* in southern Spain is visited by 54 insect taxa from 3 orders, with insects varying substantially in their quality as pollinators (75–77). If added into that is the number of plant species each pollinator visits, the "connectance" of plant and pollinator species in a food web can be high. Jordano (93) reported an average connectance C of about 0.3 for fragments of 36 pollination webs, where C is the realized fraction of the product of n pollinator species and m plant species in the web. C should decline with size of a web, but perhaps not as strongly as previously thought (130, 162).

Recognizing most pollination interactions as being far from obligate fundamentally changes the perception of their conservation. We must abandon the perspective that to lose one plant species is to lose one or more animal species via linked extinction, and vice versa. If the fundamental ecological nature of pollination "interaction webs" is that they are relatively richly connected and shift in time and space, depending in part on the landscape context (20), then the job of conservation biologists is made more subtle and complex.

One major root of generalized interactions is opportunism on the part of both plants and pollinators. To understand this, consider what might be called the fundamental evolutionary nature of pollination. Plants and animal pollinators are mutualists, each benefiting from the other's presence (13; see also 19, 222). But the mutualism is neither symmetrical nor cooperative. Indeed, pollination derives evolutionarily from relationships that were fully antagonistic (44, 167). The goals of plants and animal pollinators remain distinct—in most cases reproduction on the one hand and food gathering on the other—and this leads to conflict of interest rather than cooperation (83, 233, 239, 240). One place to see this conflict is in the behavior of animals such as bees that "rob" flowers for nectar (87).

The conflict of interest dictates that natural selection will act in divergent ways on plants and pollinators. Pollinators are agents of selection and gene flow from the perspective of plants (30) and are involved in evolutionary events ranging from plant speciation to molding floral phenotype. But floral phenotypes are not simply those that are optimal for the animals (84). Conversely, plants select for features of the animal phenotype (200), but the result is not optimal for the plants. The most basic evolutionary outcome that is common across both plants and pollinators is efficiency of each in exploiting what for each is a valuable or critical resource. One common manifestation is opportunism and flexibility on the part of pollinators toward plants, and vice versa.

To devise the best possible strategies for management, conservation, or restoration of pollination systems, it is essential to have several elements in place. We need excellent knowledge of the natural history of plants and pollinators. And we need an appreciation for interaction webs and a "Darwinian perspective" on how natural selection is likely to have shaped behavior, morphology, and other aspects of the phenotype of plants and pollinators.

THE POLLINATION CRISIS

Endangered Pollinators and Plants

Disruption of pollination systems, and declines of certain types of pollinators, have been reported on every continent except Antarctica. Although large regions of each continent have not been evaluated, we can assume that disruption is widespread because the causes are widespread phenomena associated with human activities. The overall picture is of a major pollination crisis (22). The causes include habitat fragmentation and other changes in land use, agriculture and grazing, pesticide and herbicide use, and the introduction of non-native species.

Biological Effects of Fragmentation

Many threats to pollination systems stem from fragmentation of once-continuous habitat. Fragmentation creates small populations from larger ones, with attendant problems that include increased genetic drift, inbreeding depression, and (for very small populations) increased risk of extinction from demographic stochasticity (7, 58, 191). Furthermore, fragmentation increases spatial isolation and the amount of edge between undisturbed and disturbed habitat, both of which can harm pollination (133).

If the isolation of fragmented populations becomes greater than the foraging range of pollinators, if the local pollinator population becomes small enough, or if wide-ranging pollinators avoid small populations, the outcome may be reduced pollination services. Limitation of pollen receipt occurs in many plant species. Burd (23) found evidence for pollen limitation for 62% of 258 species surveyed. The degree of limitation typically varies among years, within a season, among sites within a season, and among plants flowering synchronously within a site (54 and references therein).

Population size contributes to pollen limitation. For example, both male function (pollen removal) and female function (fruit set) are functions of population size for three Swedish orchid species (67; see also 110; see 195 for pollinator visitation rates). Some studies of endangered plants have specifically implicated a lack of effective pollinators. Pavlik et al (152) found that seed set of *Oenothera deltoides* ssp. *howellii* was 26% and 37% of maximum in 2 years and suggested scarcity of hawkmoths as a cause; a related species growing in an unfragmented habitat had seed set that was 65% of maximum. Spatial isolation of plants or populations can also play a role. For example, isolated plants of *Cynoglossum officinale* receive fewer approaches by bumblebees than patches of these plants (108). Percy & Cronk (155) studied an endemic of the island of St. Helena with a total population of 132 adult trees. Pollination is accomplished by small syrphid flies, and pollen delivery declines beyond 50 m; thus, isolated trees are effectively left without pollination.

Pollen limitation does not always imply a dangerous conservation situation. It is often the natural condition due, among other things, to stochasticity in flower visitation (24). Johnson & Bond (92) found widespread pollen limitation in wildflowers in the mountains near Cape Town, South Africa. They attributed this pattern to a general scarcity of pollinators and, in some cases, to lack of floral rewards.

Population size can affect aspects of pollination other than pollen limitation. For example, the composition of the pollinator fauna often differs in flower patches of different size (195, 202). In some cases, such a faunal change may result in higher per-flower visitation rates in small populations (202).

Pollination services are also likely to be affected by density of a plant population, which will sometimes, but not always, covary with population size (114). Thomson (219) and Schmitt (192) reported declines in pollination services at low density for several species in the Asteraceae. Seed set in the desert annual plant *Lesquerella fendleri* depends on the number of conspecifics flowering within 1 m, but not farther away, and behavior of small insect pollinators appears to be the cause (175). Density-related declines in the quality of each pollinator visit (the proportion of conspecific versus foreign pollen delivered) can be more important than parallel declines in the quantity of visits (112, 113, 116).

Interactions of population size, density, and spatial isolation are likely to have even more complex effects on pollination, and these interactions require further study. For example, outcrossing rate is unrelated to population size of an endangered *Salvia* species, but high plant density (in combination with low frequencies of male steriles) promotes outcrossing in hermaphrodites (227). Groom (73) reported that pollen limitation depends on both population size and isolation in a species of *Clarkia*. Of particular concern is an Allee effect— a threshold density, population size, or combination thereof—below which pollinators no longer visit flowers. In a species of *Banksia*, populations below a threshold size produce few or no seeds, presumably in part because of pollen limitation (121; see also 156; for a theoretical approach see 86).

Small plant populations resulting from fragmentation tend to suffer from increased genetic drift and inbreeding depression (58, 228). This may be due to increased geitonogamy, as pollinators may visit a higher proportion of flowers on individual plants, resulting in more self-fertilization (108). Inbreeding depression may explain why small populations of *Ipomopsis aggregata* are more susceptible to environmental stress and have reduced germination success (80). In general, knowledge of the mating systems of plants often is important for conservation. Self-incompatibility may further compound the dangers of small population size by reducing the availability of suitable mates (27, 49, 122).

Studies of Pollination in Fragments

Recent studies illustrate some of the range of fragmentation-related effects on pollination systems. Most of these effects are clearly deleterious. Aizen & Feinsinger (2, 3), who studied habitat fragmentation in dry thorn forest in Argentina, found fragmentation-related declines in pollination, fruit set, and seed set for most of the 16 plant species examined. For at least two species, frequency and taxon richness of native flower-visitors declined with decreasing fragment size, but visitation by introduced Africanized honey bees tended to compensate for loss of visits by natives in small fragments. Honeybees can be successful in disturbed and fragmented habitats (2, 3, 90, 183), and fragmentation may hasten the spread of Africanized bees (2, 3) and the demise of native pollinators (179, 180).

Spears (203) found that pollen dispersal to neighboring plants is significantly reduced in island populations relative to mainland populations of the same species. Pollinator limitation on islands separated by fewer than 10 km from the mainland may foreshadow the fate of many increasingly isolated mainland plant species. For example, seed set in *Dianthus deltoides* declined in habitat islands even though nectar availability was equivalent to that in an undisturbed "mainland" (90).

A few studies have addressed fragmentation and pollination in tropical areas. Powell & Powell (164) used fragrance baits to determine that male euglossine bees, which are pollinators of many neotropical orchids, would not cross cleared areas as small as 100 m between forest habitats. Allozyme heterozygosity, polymorphism, and effective number of alleles decline in small and isolated populations of the tropical tree *Pithecellobium elegans* (74). In seeming contradiction to this apparent genetic erosion in fragments, pollen dispersal by hawkmoths appears to be substantial for this species and seems easily capable of connecting isolated trees and those in fragments to the rest of the population (32).

The generation of new edges as forests are fragmented will change both abiotic and biotic components of the environment. Murcia (138, 139) divided biotic effects into (*a*) direct effects that involve changes in the abundance and distribution of species and (*b*) indirect effects that involve changes in species interactions, including pollination. She detected no consistent changes in pollination levels at a forest edge in Columbia, which suggests that the primary influence of fragmentation is through the creation of smaller populations and the isolation they experience.

The response of insects to fragmentation is poorly understood (50). Bowers (17) studied bumble bee colonization, extinction, and reproduction in subalpine meadows of different sizes. The number and diversity of queens that colonize meadows at the beginning of the summer are positive functions of meadow area, although by mid- to late summer the flower composition of meadows govern species composition and the subsequent reproduction of colonies.

Not all studies have detected negative effects. Stouffer & Bierregaard (209) sampled understory hummingbirds in Amazonian forest before and for nine years after fragmentation. Two species present before isolation did not change in abundance, but one became nearly twice as common, and five were captured only after fragmentation. In contrast to insectivorous birds, the hummingbirds appeared to be plastic in habitat preferences.

Olesen & Jain (144) described how fragmentation can harm not only pollination, but also interactions that plants have with seed dispersers and other mutualists. Loss of these interactions could lead to an extinction vortex with potentially catastrophic consequences for biodiversity. An improved understanding of such effects is critical for conservation (169).

Effects of Agricultural Practices on Wild Pollinators

Humans depend on animal pollination directly or indirectly for about one third of the food they eat (147, 172). Pollination is required for seed production (e.g. alfalfa, clover), to increase seed quality (e.g. sunflower) and number (e.g. caraway), for fruit production and quality (e.g. orchard fruits, melons, tomatoes), to create hybrid seed (e.g. hybrid sunflower), and to increase uniformity in crop ripening (e.g. oilseed rape) (39).

Several features associated with modern agriculture make farms poor habitat for wild bees and other pollinators. Crop monocultures sacrifice floral diversity, and consequently diversity of pollinating insects, over large areas (6, 147, 246). For example, cultivated orchards surrounded by other orchards have significantly fewer bees than orchards surrounded by uncultivated land (193), and the number of bumblebees on crops increases with proximity to natural habitats (246). Chemical fertilizers, pesticides, and herbicides harm pollinators. In addition, marginal land is increasingly cultivated (52, 101, 103, 147, 225 and references therein), resulting in (a) loss of wild vegetation to support pollinators, (b) fewer areas where bees can nest, (c) fewer larval host plants for butterflies, and (d) less-varied microhabitats for egg laying and larval development (52, 246). For example, since 1938, Britain has lost 30% of its hedgerow habitats, which provide floral resources and nesting sites for wild bees at the margins of cultivated fields (146).

Elimination of many native pollinators is an unappreciated price that has been paid for increased food production over the last 50 years (172, 224, 225). These pollinators are lost to adjacent natural ecosystems and to crop pollination as well. Although honeybees have long been considered the most important crop pollinators (references in 10, 147), wild pollinators are also important (165) and can be managed to provide "free" services (10, 39, 165).

Shortages of bees to pollinate crops have now been predicted in both Europe and the United States (146, 224). At least 264 crop species from 60 families are grown in the European Union, 84% of which are believed to be dependent on insect pollination (244). The best evidence for declines in bee populations comes from Europe (38, 143, 147, 172, 246), although similar losses have occurred elsewhere.

Damage is not restricted to agricultural situations in industrialized countries. Vinson et al (229) documented disruption in pollination systems following the clearing of tropical dry forest in Costa Rica to provide land for grazing and agriculture. Where livestock are raised, native grasses are commonly replaced with introduced forage grass, which burns more readily and hotly than native grasses. Fires from private lands spread to adjacent forest reserves, threatening native plants and the insects and bats that pollinate most of them. The direct effect of fire is not the only problem. Some specific relationships exist between anthophorid bees such as *Centris* and oil plants of the family Malphigiaceae. Several species of *Centris* depend on finding dead wood with holes formed by wood-boring insects, a resource that disappears when forests are cleared. Many oil-producing plants burn, and those that survive produce less oil. Bees in the dry forests appear to be decreasing in both numbers and diversity, and trees that historically provided bee resources, and depended on bees for outcrossing, are disappearing.

Grazing

Grazing threatens pollinators through removal of food resources, destruction of underground nests and potential nesting sites, and other more subtle mechanisms (70, 96, 211).

Sugden (211) studied sheep grazing practices in California and the effects on pollinators of an endemic vetch (*Astragalus monoensis*) and found evidence of nest destruction, pollinator food removal by sheep, and direct trampling of bees. Bees at risk included *Anthidium, Anthophora, Bombus, Callanthidium, Colletes, Hoplitis*, and *Osmia*. Another example of removal of food resources by grazing is the loss of willow shrubs (*Salix* spp.) due to cattle along riparian areas. These willows are important browse for livestock (186) and provide nectar and pollen for spring-emerging bumblebee queens and other pollinators; their loss may harm the pollinators and, in turn, other species of plants that flower later in the summer.

Pesticides

Pesticides pose a major threat to pollinators (9). Ironically, the greatest use of pesticides is on crop plants where pollinators are most often limited. Pollinators also are harmed by pesticide application in grasslands (18, 154, 215), forests, (101), urban areas (103), and even tourist resorts (47). An increasing awareness of environmental risks has helped reduce pollinator poisonings in industrialized nations (103), but pesticide-induced declines in bee abundance are still being reported from developing countries (43).

Bee poisoning from insecticides first became a problem in the United States in the 1870s (91), but advances in agricultural technology and elaboration of new chemicals exacerbated the problem after World War II (5, 91). Poisoning of honeybees (on which most attention has been focused) can result in direct mortality, abnormal communication dances, inability to fly, and displacement of queens (91). Foraging honeybees can contaminate the hive with pesticides or other pollutants. Pesticides, arsenic, cadmium, PCBs, fluorides, heavy metals, and radionucleotides (after the 1986 Chernobyl accident) have all been reported in contaminated honey or pollen (103).

In the 1970s, Kevan (98–100) cautioned about the disruptive effects of pesticides on native pollinators, and his predictions have been borne out. The best example is a long-term study conducted in Eastern Canada (99, 101, 103, 104, 106, 161). From 1969 until 1978, spruce budworm was controlled by aerial spraying of Fenitrothion, an organophosphate that is highly toxic to bees. Commercial blueberry production in the region largely depended on pollination by as many as 70 species of native insects. Blueberry crops failed in 1970 and subsequent years (102). Populations of bumblebees and andrenid and halictid bees declined in blueberry fields near sprayed forests (99), and reproduction of native plants was depressed (218, 221). Native bees showed steady signs of recovery after Fenitrothion was replaced with a less harmful insecticide (101).

In the western United States, broad-spectrum insecticides are used to control grasshoppers on rangelands (215). Spraying occurs during the flowering of a number of threatened or endangered endemic plants (18) and coincides with the foraging period of most native bees (154). Spraying is prohibited in a 3-mile radius around points where listed plants are known to occur, but the 3-mile figure is arbitrary because little is known about flight distances of the pollinators (154). Some of these listed species appear to have pollinator-limited seed production (63), and their persistance will be related to successful pollination (194, 215).

Herbicides

Herbicide use affects pollinators by reducing the availability of nectar plants (47, 100). In some circumstances, herbicides appear to have a greater effect than insecticides on wild bee populations (11,47). Herbicide spraying and mechanical weed control in alfalfa fields can reduce nectar sources for wild bees. The magnitude of the effect for each species is related to the length of its seasonal flight period. Many bees have a flight period that extends beyond the availability of alfalfa flowers. Some of these bee populations show massive declines due to the lack of suitable nesting sites and alternative food plants (11).

Honeybee Declines

More than 9000 years ago, humans realized they could harvest honey from the stores of some bees (69). Humans have taken honeybees with them as they settled new regions of the world (21). Honeybees have been domesticated and naturalized in temperate areas of Australia, North America, and South America for centuries (before 1641 in North America) (196), whereas extensive naturalization in tropical regions is much more recent (183). Although *Apis mellifera* is native to western Asia, it is not widely naturalized in other parts of Asia, where five other species of *Apis* naturally occur (37, 183).

Today, bee products are still valuable, but the value of crop pollination is far greater (22; references in 10). Honeybees, which are generalists and will pollinate many crops, are easily managed and transported (147). Some suggest the annual value of honeybee pollinated crops in the United States alone is as high as $10 billion (235; see also 201, 224).

Recently, honeybees have been declining. More than 20% of the cultivated honeybee colonies in the United States have been lost since 1990 (85,235), along with most feral honeybees (235). The number of commercially managed colonies has declined from a peak of 5.9 million in the 1940s to 4.3 million in 1985 and 2.7 million in 1995 (85). Declines are severe in some regions. For example, in 1994, California almond growers had to import honeybees from as far away as Florida (235). The European community supports an estimated 7.5 million managed honeybee colonies (244, 245), and these are believed to have been declining since 1985 (245).

Two parasitic mites, *Varroa jacobsoni* and *Acarapsis woodi*, have been particularly damaging to honeybees. *Varroa* spread from its original host, the Asiatic honeybee (*Apis cerana*), when *A. mellifera* was introduced to Asia (57). The mites had spread from Asia to Europe by 1950, to North Africa by 1970, to South America by 1971, and to North America by 1987 (136). A bee infected by *Varroa* loses protein to the parasite, resulting in lowered life expectancy. Also, bacteria penetrate holes in the exoskeleton formed by the mites (174). Existence of *A. woodi*, the tracheal mite, was first documented in England in 1921; subsequently it spread to continental Europe, Asia, Africa, South America, and North America (42, 57). Entire bee colonies become infected, resulting in decreased brood production, decreased honey production, and high winter mortality (48). Beekeepers can attempt to control both mites with chemicals, but *Varroa* mites are beginning to exhibit resistance (236). Treatment can be costly, and chemical residues may appear in honey. New control techniques are being developed, but the difficulty of mite control is causing a decline in beekeeping, particularly among hobbyists (103, 235).

Africanized honeybees also are implicated in honeybee decline in the Americas. The term Africanized has been used to describe hybrids between honeybees of European descent and African subspecies *A. mellifera scutellata* (173). Taylor (214) suggested that the term neotropical African bees be used for the feral colonies in South and Central America that still retain the African phenotype distinguishable by morphology, behavior, and genetics, and that the term Africanized bees be used to refer to bees found primarily in apiaries that show clear evidence of hybridization. The failure to make these distinctions has led to differing predictions about the spread of the bees (214). African queens were released accidently in Brazil in 1956 (136) and rapidly dominated colonies of European descent. The bees became established in the United States in 1990 (22). The predominately African phenotype may be restricted to the warmer climate of the southern United States, but the variable hybrid Africanized phenotypes may be able to survive farther north (214). Neotropical African bees display several features that make them undesirable for apiculture. They swarm when colonies are relatively small and have little honey, and they leave an area when environmental conditions become unfavorable (64). Furthermore, their reputation for aggressive behavior is responsible for negative public attitudes and a decline in beekeeping (22, 34, 201).

Non-Native Pollinators

The introduction of non-native pollinators has the potential to harm native pollination systems. For example, fig wasps were introduced to California in 1899, at which point non-native trees that had been grown there for decades began to produce fruits (51). Because of the introduction of their wasp pollinators, some fig species are now weedy pests in parts of the continental United States, Hawaii, and New Zealand (68, 132). The introduction of bumblebees into areas sometimes have negative results. Non-native *Bombus terrestris* were brought to Japan to pollinate greenhouse tomatoes but soon escaped and became naturalized (I Washitani, personal communication). Because of their aggressive nature, queens are able to take over the hives of native bumblebees by killing the queen, and ecologists fear serious declines in native bumblebee species. Queens of the native Japanese *Bombus diversus* are important pollinators of at least one endangered plant, *Primula sieboldii* (234), and cannot be replaced by *B. terrestris*. *B. terrestris* has also invaded parts of Israel in recent decades, expropriating nectar resources to the apparent detriment of native bees

(46a). *Xylocopa* carpenter bees are pollinators of some plants but are also well known as nectar robbers (87). Little is known about the impact of this bee on native species of flowers or pollinators in Hawaii, where it was introduced (82).

By far the most significant introduction of non-native pollinators involves honeybees, whose movement by humans to all areas of the globe can be considered a major, uncontrolled ecological experiment. Honeybees in some cases might benefit wildflowers by excluding native pollinators from crops (245), but they are often poor pollinators of crops and native flowers compared with native insects (10, 115, 147, 148, 165, 172, 184, 224, 239). Furthermore, honeybee colonies require prodigious amounts of pollen and nectar, and worker bees fly long distances and recruit to rich floral resources (21, 183). Thus, honeybees may compete with native pollinators for resources, leading to reduced species diversity of pollinators. Honeybees also are likely to affect the reproduction of native plants, perhaps even facilitating the spread of weedy non-native plants (4, 8, 82, 128, 188; see also 26). Whether or not honeybees aid in the spread of introduced plants, the presence of these plants may disrupt natural pollination systems because native pollinators sometimes prefer them at the expense of native plants (230).

Competition with honeybees has been implicated in the decline of buprestid beetles in western Australia (109). These jewel beetles are important pollinators in arid mallee scrub vegetation. Sugden & Pyke (212) demonstrated competition by introducing honeybees into an alpine area of Australia and examining the nesting and reproductive success of a generalist native bee. Honeybees remove as much as half of all the available nectar from flowers of the Australian bottlebrush, *Callistemon rugulosus*, and New Holland honeyeaters respond by visiting individual flowers less frequently and expanding their feeding territories (149, 150). Honeybees visit many other Australian plants and on some species remove over 90% of the available resources (151). Roubik et al (176–178, 182, 183, 185) studied competition between African honeybees and native pollinators in South and Central America. In French Guiana, African honey bees are common visitors to *Mimosa pudica* (183). Patches dominated by honeybees had the lowest levels of seed and fruit production, whereas highest levels occurred in patches visited by native *Melipona* bees. Honeybees have been increasing in moderately disturbed, mixed forest-savanna habitats (from 20% of visitors in 1977 to 99% of visitors in 1994), which suggests that they are displacing native insects. Honeybees were introduced onto Santa Cruz Island, off the coast of California, in the 1880s and can now be found foraging on more than one third of the island's plant species (223, 238). Removal of honeybee colonies from the eastern half of the island over the past few years suggests an inverse relationship between honeybee abundance and native bee abundance. Experiments in old field in New York state show that the native megachilid bee *Osmia pumila* suffers reduced brood cell production and pupal mass, and increased brood parasitism in the presence of honeybees (K. Goodall, unpublished).

The hypothesis of competition is not supported by all studies. Sugden et al (213) reviewed 24 studies conducted on four continents and three islands; 16 detected competition under some conditions whereas 8 produced ambiguous results. Although Africanized honeybees reached the neotropics two decades ago and the foraging behavior of native bees changes when honeybees are present, there is no strong evidence of declines in native bee populations (25). Perhaps this is unsurprising: Where honeybees monopolize a rich resource, native species may shift to other flowers and there may be no effect on their population size (150, 183, 190). Also, effects of competition are difficult to detect, if they occur, against the background of natural variation in pollination systems (25, 183). The idea that honeybees automatically compete with natives is probably naïve (183), and more studies, including ones of longer duration, are needed.

POTENTIAL MANAGEMENT SOLUTIONS

Conservation of Habitats and Pollinators

Conservation biology is undergoing a paradigm shift away from single-species conservation efforts and toward habitat, ecosystem, and regional efforts. Pollinators should benefit from this change, because the pollinators of many plant species are not yet identified and stand to gain protection from blanket conservation efforts. Also, it is difficult to convince the public to devote resources to protecting small insect pollinators whose aesthetic beauty is not obvious to the unaided eye. The broad context of habitat- or ecosystem-level conservation efforts is especially appropriate for pollination systems because of the web of interactions that links plant species via pollinators (216, 232).

Studies of several systems demonstrate why an ecosystem-based conservation strategy is valuable. A rare orchid in the western United States, *Spiranthes diluvialis*, requires pollinators, so management plans must encompass the maintenance of bumblebees, which may be at risk from insecticide spraying on public rangeland (197). The habitat must also be managed for appropriate nest sites for bumblebees, and for floral diversity to provide nectar (the orchid produces none) and pollen for the whole flight season of bumblebees (199). Petit & Pors (158) calculated the carrying capacity for nectar-feeding bats on the island of Curaçao by using the daily availability of flowers on three species of columnar cacti. They estimated the carrying capacity for one bat species at 1200, about 300 more than the actual population, and suggested that removal of native vegetation on the island should be strictly regulated to prevent further decline. Cropper & Calder (45) attributed the lack of seed set of the rare and endangered Australian orchid *Thelymitra epipactoides* to the absence of pollinators and suggested elimination of natural fire as the root cause. Burning stimulates flowering in many coastal heathland species, which helps to maintain high pollinator species diversity. Kwak et al (118) pointed out the value of other plant species in attracting bumblebees to small populations of the rare Dutch plant *Phyteuma nigrum*.

The dependence of wild pollinator populations on appropriate habitat is increasingly recognized. A study of margins of agricultural fields (119) pointed out that small areas with flowering plants can be very effective at attracting and maintaining pollinator populations, including Syrphidae and other Diptera. Habitat could be managed to encourage bumblebee and honeybee populations by providing a seasonal succession of suitable forage plants, protecting them from pesticides and herbicides and providing for long-term set-aside of fields (38, 145; see also 242). The last recommendation makes sense because butterflies and bumblebees tend to prefer flowers of perennials and because ground-nesting bees avoid recently disturbed areas (38). Such a policy could also benefit insect species that are not crop pollinators, e.g., satyrid butterflies (53).

Conservation of bee habitat may be the best means of reversing declines in pollinator populations (172). In many parts of the world this may mean conservation of human-made habitats, some of which prove to be good substitutes for threatened or destroyed natural habitats (47, 107, 242). Many bee species have colonized restored areas along the Rhine River. Levees can provide prime bee habitat, especially when built of sand and gravel and managed for high floral diversity (107). Day (47) argued that as technology becomes more important and farming starts to decline in Europe, hedgerows, pastures, and woodlands should be regenerated. Disturbed urban areas may also be favorable for some bee communities (189), although multiple types of habitat may be required to satisfy both foraging and nesting requirements (241).

Some pollinators only need a relatively small patch of habitat near their host plants, but others require large areas. In Santa Rosa National Park in Costa Rica, there are at least 40 species of sphingid moths, which pollinate at least 50 plant species as adults and which live for one generation in the park during the beginning of the rainy season before moving to other parts of the country for the rest of the year (89). For these and other migratory pollinators, conservation efforts can require large geographical areas and even international cooperation. Perhaps the most extreme examples are migrants such as hummingbirds, butterflies, and moths, which may be important pollinators along migratory routes extending for thousands of kilometers (12, 28, 71, 231).

Maintenance of Populations and Species in the Absence of Pollinators

Relatively few examples exist of the absolute loss of pollinators, but this may reflect only our ignorance. Steiner (204) reported the loss of a specialized oil-collecting pollinator of a rare South African shrub, although subsequent work (205) led to discovery of a population where the predicted specialist pollinator was still present. Sipes & Tepedino (198, p. 164) suggested that one interpretation of the low visitation and fruit set to a rare plant from the western United States is that the original pollinator "is no longer consistently found within the plants' distribution." Lord (125) described a New Zealand liane that has lost its bat pollinator.

One effort that may, at least in the short term, prove fruitful for conservation is hand pollination of plants that have lost natural pollinators. For example, *Trifolium reflexum*, a prairie species threatened by loss of habitat, was brought into cultivation at the Chicago Botanic Garden, where hand pollination yielded thousands of seeds for additional restoration efforts (208). Hand pollination has also been used for two Hawaiian species of *Brighamia*, whose few remaining individuals have apparently lost their native pollinators (22), and for an endangered orchid in Illinois (168).

Biosphere 2, an experiment in which a small human population was sealed in a (mostly) closed environment for 2 years, included a diversity of plants. All pollinators quickly went extinct so that most plant species "had no future beyond the lifetime of individuals already present" (33). One conclusion is that maintenance of normal plant-pollinator relationships is difficult and that people in such circumstances in the future should be prepared for hand-pollinating.

Another possible solution is the intentional introduction of exotic pollinators, although there are risks (10, 51, 101, 105). The first known example was the introduction of bumblebees to pollinate red clover in New Zealand (51, 65). More recently, weevils were introduced to pollinate oil palms in Malaysia (101), providing services valued at $3 million per year (72, 187).

Changed Agricultural Practices and Uses of Pesticides and Herbicides

In the United States alone, crop production is reduced by about 8000 species of insects, 2000 species of weeds, 160 types of bacteria, 250 types of viruses, and 8000 species of pathogenic fungi (9). Pesticides and herbicides seem an attractive solution because they can rapidly reduce numbers of problem organisms. However, new chemicals must be continually developed as pests evolve resistance and for other reasons (9). One alternative is to move to more labor-intensive control methods that are more "friendly" to pollinators. For example, some USDA studies comparing organic farms and nearby farms using pesticides showed similar crop yields (9). The organic farms controlled pests in ways that encouraged natural predators of pests and created more favorable habitats for pollinators.

There is an increasing emphasis on preventing pollinator loss due to application of crop pesticides. Toxicity levels of pesticides to honeybees are generally known (103), but this has not been useful in determining the effects on other bees (142). Toxicity is in part related to surface-to-volume ratio (91), so that bumblebees may be more tolerant, and small solitary bees more susceptible, than honeybees. In addition, details of pesticide use (such as timing, method of application, and formulation) can affect toxicity (43, 142). Crops can be sprayed before or after flowering to minimize the chances of harming pollinators (66). However, leafcutter bees may collect contaminated leaf tissue for nest construction even when crops are not in flower (142). Timing of application within the day can also be critical. Although honeybees are not active at night, some bees, such as *Nomia*, rest in crop fields at night where they would be susceptible to night spraying (142). Bees such as *Apis* and *Nomia* forage as far as 13 km from the nest (142), so spraying may affect bees that nest far from fields. Honeybee apiaries can be either moved or closed-up during pesticide application, but native bees are not as fortunate. Compounds such as benzaldehyde, propionic anhydride, and some amines may prove useful in repelling bees from fields during pesticide application (142). Bran-baits instead of pesticide spraying could be used to kill grasshoppers in rangelands, thereby potentially reducing pollinator mortality (153).

Few studies have systematically documented declines of bees other than honeybees (but see 99, 103 and references therein). Documentation can be difficult because baseline data are generally unavailable and often the importance of non-*Apis* bees is poorly understood (142). However, enough is known about pesticide problems that much can be done to reduce pollinator losses (103). Kevan (103) suggested regulation and certification for pesticide users. In many countries, regulations are in place but violations carry minimal penalties (103).

Reintroductions of Plants and Pollinators

Reintroduction of endangered plants is still relatively uncommon (60). No plant reintroduction to date appears to have been stimulated by the need to support pollinator populations, although existing pollinators may have benefited. Maunder's (131) paper on plant reintroduction does not mention pollinators, nor does that by Falk et al (60).

One potential problem of reintroducing a plant species into an area is that during its absence some native pollinators may have vanished. This loss would be most serious if the plant had a single pollinator species, but such species appear to be in the minority, and it is common for a plant to have multiple, sometimes very numerous, pollinators (232). Given the variability among years that can be observed in pollinator populations (e.g. 31, 79, 95, 127, 157), multiple pollinators may often be necessary for plant persistence (232).

Hawaii provides one example of an introduction that inadvertently filled the role of a recently extinct pollinator. Cox (41) described the pollination of *Freycinetia arborea*, the indigenous ieie vine, by *Zosterops japonica* (Japanese White-eye, introduced in 1929). Museum specimens of three native birds, two extinct and one endangered, carried pollen grains from the plant, indicating that they were among the original pollinators. Lammers et al (120) reported that White-eyes also visit flowers of an endemic lobelioid, *Clermontia arborescen*. Not all Hawaiian plants have been so lucky; some have gone extinct whereas others are very rare.

Removal of Alien Pollinators

Animals have been intentionally introduced because of their role as pollinators (e.g. honeybees, the alfalfa leafcutter bee). Some intentional introductions involve animals that pollinate but were not introduced for that reason (e.g. *Zosterops* in Hawaii, possums in New Zealand) and some unintentional introductions involve pollinators (e.g. cabbage butterflies, fig wasps). In only a few cases have there been calls for the removal of introduced pollinators. The European bumblebees that were introduced to Japan as pollinators of greenhouse crops escaped to establish feral populations. An effort to eradicate them is underway (M Ono, personal communication). *B. terrestris* was also introduced in about 1992 to Tasmania, where an attempt to eradicate it has had little success (163).

Domestication of Wild Bees and Other Pollinators

Research on non-*Apis* bees as crop pollinators has a long history (15, 224), but it recently has achieved new significance (220, 243). As early as the 1980s, concerns were raised about the need for an increased diversity of pollinators for agriculture in North America (148, 172). At least 50 native bee species have been cultivated experimentally or commercially (43, 172, 224, 225). Parker et al (148) also discussed the use of dipterans as possible crop pollinators.

A few success stories illustrate the potential for non-*Apis* bees as pollinators. The leafcutter bee (*Megachile rotundata*) was introduced from Asia into North America and is the primary pollinator of plants grown to produce alfalfa seed (171, 220). In 1977, *Osmia cornifrons* was introduced from Japan as a pollinator of apples; it has now been distributed to 23 states and 2 Canadian provinces (148; see also 172, 225). In the tomato industry, bumblebees can replace humans equipped with electric vibrators (the flowers require "buzz pollination" to release pollen) or sprayers with synthetic plant hormones to induce fruit production (124, 148, 172). The bumblebee business originated in The Netherlands about a decade ago and has now spread as far as North America and Japan (124, 148, 172).

Legal Protection

Nearly 25% of the planet's vascular plant species may become extinct within the next 50 years (170), and 22% of the species in the United States is currently of conservation concern (59). The situation for most pollinators appears less bleak because the numbers are smaller, but this may only reflect poorer knowledge of them. Both plants and pollinators can be afforded legal protection through the Endangered Species Act in the United States and internationally via listing in the Convention on International Trade in Endangered Species of Wild Fauna and Flora (CITES). In the United States, only 390 of the 639 species of flowering plants afforded protection by the Endangered Species Act had recovery plans as of December 1997 (http://www.fws.gov/~r9endspp/plt1data.html), and only 16 species of butterflies, 1 species of fly, 1 species of moth, and 2 species of skippers (Lepidoptera) were included in the list as endangered or threatened (http://www.fws.gov/~r9endspp/invdata.html#Insects) as of that date. Three vertebrate pollinators, two flying fox species and the lesser long-nosed bat (*Leptonycteris curasoae*), were also listed as of December 1997. The International Union for Conservation of Nature and Natural Resources (IUCN) now lists 165 genera of vertebrate pollinators (including 186 species) of conservation concern (140), which suggests a need for legal protection for many more.

Public Education

Education efforts have helped bring publicity to bee conservation efforts in Europe, particularly for bumblebees. The Watch Trust for Environmental Education engaged thousands of volunteers, mostly children, to document the abundance and distribution of *Bombus* species and to provide information on preferred plant species (117, 146). The success of this survey inspired a similar program in The Netherlands.

In 1995, The Arizona-Sonora Desert Museum launched the Forgotten Pollinators Campaign. The focus was to draw attention to the impending pollination crisis. The campaign included publication of a book (22), media campaigns, a research program conducted by volunteers, development of pollinator gardens at the museum, and other efforts to increase public awareness of the importance of pollinator conservation.

PRIORITIES FOR FURTHER RESEARCH

In virtually all cases, biologists must provide scientific information for conservation decisions based on less-than-perfect knowledge. The best approach is to base scientific input on the consensus of experts; this is vastly preferable to no scientific input at all or to that of a small minority (55). At the same time, it is important for pollination biologists to map out a research program for filling major gaps in our knowledge, as we attempt to do here.

The Ecology of Animal Pollinators

Typical ecosystems at intermediate latitudes harbor as many as several hundred pollinating insect species, most belonging to Hymenoptera, Lepidoptera, Diptera, and Coleoptera (79, 111, 134, 157, 210, 247). The vast majority of hymenopteran pollinators are solitary bees (237). Compared with our understanding of social bees, we still have much to learn about the nesting biology, demography, and trophic ecology of most solitary bees and about the composition of local species assemblages (137, 215). Relative abundances of given species of solitary bees fluctuate spatially and temporally (31, 157), and we need to understand how this relates to floral resources (215). We also need to learn more about the degree of specialization of individual bee species and the degree to which even specialists may use other plant species (31, 46). The picture for other insect orders is further complicated by the fact that larvae may require food plants that differ from those of adults. We need to learn how to manage landscapes that will support the entire life cycle of such species (22, 14). Our knowledge of larval ecology is best for the Lepidoptera because of the intense interest of naturalists in butterflies (56). More effort needs to be expended in learning comparable information about dipteran and coleopteran life cycles and larval diets. The role of flies as pollinators in many ecosystems seems to have been underestimated until recently (94, 157, 226, 247).

Links Between Pollination and Plant Population Dynamics

The diversity of pollinators is matched by local diversity of plant species and temporal and spatial variation in species composition. For example, Tepedino & Stanton (217) reported substantial year-to-year variation in relative abundances and phenologies of different flowers in a shortgrass prairie in Wyoming (see also 88, 166). Thus, a pollinator foraging for floral reward experiences a complex and fluctuating marketplace. It is important to characterize variation in floral abundance more carefully and to study how pollination contributes to it. Ecologists have assumed that pollination plays an important role in plant population dynamics, but there is virtually no empirical evidence for it. We do know that pollination is often limiting to seed production (23), although resources (207) or both pollination and resources simultaneously (29, 135) can also be limiting. However, we need more experimental manipulations of seed input, seedling establishment, and other stages of the life cycle with measurement of subsequent changes (if any) in plant population size and structure (1, 126, 129).

In particular, it would be useful to design such studies so they help us to predict how reduction in pollination services will influence the demography of plant species that are threatened because of fragmentation of other anthropogenic insults.

The Nature of Interaction Webs

Pimm (160) distinguished four aspects of ecological stability, one of which is resilience—the degree to which an ecosystem resists further change following initial change. Pollination webs are threatened with the loss of component species and addition of non-natives. The substantial connectance of pollination webs makes us suspect that such changes will elicit additional ones, perhaps even cascades of extinction. To our knowledge, nobody has modeled resilience (or other aspects of stability) specifically for mutualistic interactions such as those of plants and pollinators, much less studied resilience of such systems empirically.

CONCLUDING THOUGHTS

The natural history knowledge of pollination gained over the last several centuries shows that animal-mediated pollination is essential for the sexual reproduction of most higher plants. Although many plants are iteroparous, with multiple opportunities for sexual reproduction, spread by clonal propagation or other asexual means or having a dormant seed stage, these life-history features cannot compensate in the long term for a chronic loss of pollination services (16). A reduction in plant fecundity is of clear concern for agroecosystems but equally problematical for natural ecosystems. There is indeed a strong argument to be made that pollination interactions are keystones in both human-managed and natural terrestrial ecosystems (102).

In spite of centuries of study, our understanding of interactions between plants and animal pollinators is far from complete. Appreciating this was our motivation for stressing that continued research is essential to the long-term conservation of pollination systems. At the same time, we agree with others in political and scientific circles who urge ecologists to become more active in educating those around them about issues in conservation biology. The evidence on multiple fronts is sufficiently alarming to conclude that there is an ongoing and pending ecological crisis in pollination systems. Although there are dangers in sounding the alarm for a pollination crisis, and hurdles to be overcome in explaining the issues to a wider audience, the alternatives hold far greater risks.

Our understanding of the keystone role that pollinators can play in ecosystems around the world, and the risks faced by both pollinators and the plants they visit, has increased greatly during the past few decades. Research on endangered plants, including rehabilitation and reintroduction programs, is more likely now than in the past to include consideration of breeding systems and the potential need for pollinators in management plans (97, 123). The conservation of insects and their habitats is now a topic for discussion in the scientific literature (36). A decade ago, Feinsinger (62) found only two papers that clearly related conservation and animal-flower interactions; now these topics are written about frequently, as our review shows. Much progress has been made since Kevan's plea arising from concern about the damage to pollinators from pesticide and herbicide use in Canada (100). The most encouraging progress is that we now recognize much more clearly what problems exist and what we need to know to solve them.

At the same time, many challenges lie ahead. We must redouble our research efforts on basic aspects of pollination systems at a time when it is difficult to obtain financial support for work that lacks immediate management applications. The pace of change in ecosystems and growth of threats to pollination systems promise to increase in the future. We face accelerated alteration of habitat by a growing human population, linked with accelerated invasion of non-native species, and the prospect of global climate change, which threatens to decouple plants and pollinators phenologically and ecologically (166). Although the challenges are daunting, they must be met with our most determined efforts as ecologists and citizens.

Visit the *Annual Reviews* home page at
http://www.AnnualReviews.org

Literature Cited

1. Ackerman JD, Sabat A, Zimmerman JK. 1996. Seedling establishment in an epiphytic orchid: an experimental study of seed limitation. *Oecologia* 106:192–98
2. Aizen MA, Feinsinger P. 1994. Forest fragmentation, pollination, and plant reproduction in a Chaco dry forest, Argentina. *Ecology* 75:330–51
3. Aizen MA, Feinsinger P. 1994. Habitat fragmentation, native insect pollinators, and feral honey bees in Argentine "Chaco Serrano." *Ecol. Appl.* 4:378–92
4. Allen RB, Wilson JB. 1992. Fruit and seed production in *Berberis darwinii* Hook., a shrub recently naturalized in New Zealand. *NZ J. Bot.* 30:45–55
4a. Allen-Wardell G, Bernhardt P, Bitner R, Burques A, Buchmann S, Cane J, et al. 1998. The potential consequences of pollinator declines on the conservation of biodiversity and stability of food crop yields. *Conserv. Biol.* 12:8–17
5. Anderson LD, Atkins EL. 1968. Pesticide usage in relation to beekeeping. *Annu. Rev. Entomol.* 13:213–38
6. Banaszak J. 1996. Ecological bases of conservation of wild bees. See Ref. 130a, pp. 55–62
7. Barrett SCH, Kohn JR. 1991. Genetic and evolutionary consequences of small population size in plants: implications for conservation. In *Genetics and Conservation of Rare Plants*, ed. DA Falk, KE Holsinger, pp. 3–30. New York: Oxford Univ. Press
8. Barthell JF, Randall JM, Thorp RW, Wenner AM. 1994. Invader assisted invasion: pollination of star-thistle by

feral honey bees in island and mainland ecosystems. *Bull. Ecol. Soc. Am.* 75:10 (Abstr.)

9. Batra SWT. 1981. Biological control in agroecosystems. *Science* 215:134–39

10. Batra SWT. 1995. Bees and pollination in our changing environment. *Apidologie* 26:361–70

11. Benedek P. 1972. Possible indirect effect of weed control on population changes of wild bees pollinating lucerne. *Acta Phytopathol. Acad. Sci. Hung.* 7:267–78

12. Bertin RI. 1982. The ruby-throated hummingbird and its major food plants: ranges, flowering phenology, and migration. *Can. J. Zool.* 60:210–19

13. Bertin RI. 1989. Pollination biology. In *Plant-Animal Interaction*, ed. WG Abrahamson, pp. 23–86. New York: McGraw-Hill

14. Blair RB, Launer AE. 1997. Butterfly species diversity and human land use: species assemblages along an urban gradient. *Biol. Conserv.* 80:113–25

15. Bohart GE. 1972. Management of wild bees for the pollination of crops. *Annu. Rev. Entomol.* 17:287–312

16. Bond WJ. 1994. Do mutualisms matter? Assessing the impact of pollinator and disperser disruption on plant extinction. *Philos. Trans. R. Soc. London Ser. B* 344:83–90

17. Bowers MA. 1985. Bumble bee colonization, extinction, and reproduction in subalpine meadows in northeastern Utah. *Ecology* 66:914–27

18. Bowlin RW, Tepedino VJ, Griswold TL. 1993. The reproductive biology of *Eriogonum pelinophilum* (Polygonaceae). See Ref. 199a, pp. 296–302

19. Bronstein JL. 1994. Conditional outcomes in mutualistic interactions. *Trends Ecol. Evol.* 9:214–17

20. Bronstein JL. 1995. The plant-pollinator landscape. In *Mosaic Landscapes and Ecological Processes*, ed. L Hansson, L Fahrig, G Merriam, pp. 257–88. London: Chapman & Hall

21. Buchmann SL. 1996. Competition between honey bees and native bees in the Sonoran Desert and global bee conservation issues. See Ref. 130a, pp. 125–42

22. Buchmann SL, Nabhan GP. 1996. *The Forgotten Pollinators*. Washington, DC: Island. 292 pp.

23. Burd M. 1994. Bateman's principle and plant reproduction: the role of pollen limitation in fruit and seed set. *Bot. Rev.* 60:83–139

24. Burd M. 1995. Ovule packaging in stochastic pollination and fertilization environments. *Evolution* 49:100–9

25. Butz Huryn VM. 1997. Ecological impacts of introduced honeybees. *Q. Rev. Biol.* 72:275–97

26. Butz Huryn VM, Moller H. 1995. An assessment of the contribution of honey bees (*Apis mellifera*) to weed reproduction in New Zealand protected natural areas. *NZ J. Ecol.* 19:111–22

27. Byers DL. 1995. Pollen quantity and quality as explanations for low seed set in small populations exemplified by *Eupatorium* (Asteraceae). *Am. J. Bot.* 82:1000–6

28. Calder WA. 1987. Southbound through Colorado: migration of rufous hummingbirds. *Natl. Geogr. Res.* 3:40–51

29. Campbell DR, Halama KJ. 1993. Resource and pollen limitations to lifetime seed production in a natural plant population. *Ecology* 74:1043–51

30. Campbell DR, Waser NM, Meléndez-Ackerman EJ. 1997. Analyzing pollinator-mediated selection in a plant hybrid zone: hummingbird visitation patterns on three spatial scales. *Am. Nat.* 149:295–315

31. Cane JH, Payne JA. 1993. Regional, annual, and seasonal variation in pollinator guilds: Intrinsic traits of bees (Hymenoptera: Apoidea) underlie their patterns of abundance at *Vaccinium ashei* (Ericaceae). *Ann. Entomol. Soc. Am.* 86: 577–88

32. Chase MR, Moller C, Kessell R, Bawa KS. 1996. Distant gene flow in tropical trees. *Nature* 383:398–99

33. Cohen JE, Tilman D. 1996. Biosphere 2 and biodiversity: the lessons so far. *Science* 274:1150–51

34. Collins AM. 1988. Genetics of honeybee colony defense. See Ref. 143a, pp. 110–17

35. Collins BG, Paton DC. 1989. Consequences of differences in body mass, wing length and leg morphology for nectar-feeding birds. *Aust. J. Ecol.* 14: 269–89

36. Collins NM, Thomas JA, eds. 1989. *The Conservation of Insects and Their Habitats*. London: Academic

37. Connor LJ, Rinderer T, Sylvester HA, Wongsiri S, eds. 1993. *Asian apiculture. Proc. First Int. Conf. Asian Honey Bees and Bee Mites*. Cheshire, CT: Wicwas

38. Corbet SA. 1995. Insects, plants and succession: advantages of long-term set-aside. *Agric. Ecosyst. Environ.* 53:201–17

39. Corbet SA, Williams IH, Osborne JL. 1991. Bees and the pollination of crops

and wild flowers in the European Community. *Bee World* 72:47–59

40. Costanza R, d'Arge R, de Groot R, Farber S, Grasso M, et al. 1997. The value of the world's ecosystem services and natural capital. *Nature* 387:253–60

41. Cox PA. 1983. Extinction of the Hawaiian avifauna resulted in a change of pollinators for the ieie, *Freycinetia arborea*. *Oikos* 41:195–99

42. Crane E. 1988. Africanized bees, and mites parasitic on bees, in relation to world beekeeping. See Ref. 143a, pp. 1–12

43. Crane E, Walker P. 1983. *The Impact of Pest Management of Bees and Pollination*. London: Int. Bee Res. Assoc.

44. Crepet WL. 1983. The role of insect pollination in the evolution of the angiosperms. In *Pollination Biology*, ed. L Real, pp. 29–50. New York: Academic

45. Cropper SC, Calder DM. 1990. The floral biology of *Thelymitra epipactoides* (Orchidaceae), and the implications of pollination by deceit on the survival of this rare orchid. *Plant Syst. Evol.* 170:11–27

46. Cruden RW. 1972. Pollination biology of *Nemophila menziesii* (Hydrophyllaceae) with comments on the evolution of oligolectic bees. *Evolution* 26:373–89

46a. Dafni A, Schmida A. 1996. The possible ecological implications of the invasion of *Bombus terrestria* (L.) (Apidae) at Mt. Carmel, Israel. In *The Conservation of Bees*, ed. A Matheson, SL Buchman, C O'Toole, P Westrich, IH Williams, pp. 183–200. London: Academic

47. Day MC. 1991. *Towards the Conservation of Aculeate Hymenoptera in Europe*. Strasbourg, France: Counc. Eur.

48. Delfinado-Baker M. 1988. The tracheal mite of honey bees: a crisis in beekeeping. See Ref. 143a, pp. 493–97

49. DeMauro MM. 1993. Relationship of breeding system to rarity in the lakeside daisy (*Hymenoxys acaulis* var *glabra*). *Conserv. Biol.* 7:542–50

50. Didham RK, Ghazoul J, Stork NE, Davis AJ. 1996. Insects in fragmented forests: a functional approach. *Trends Ecol. Evol.* 11:255–60

51. Donovan BJ. 1990. Selection and importation of new pollinators to New Zealand. *NZ Entomol.* 13:26–32

52. Dover J, Sotherton N, Gobbett K. 1990. Reduced pesticide inputs on cereal field margins: the effects on butterfly abundance. *Ecol. Entomol.* 15:17–24

53. Dover JW. 1996. Factors affecting the distribution of satyrid butterflies on arable farmland. *J. Appl. Ecol.* 33:723–34

54. Dudash MR. 1993. Variation in pollen limitation among individuals of *Sabatia angularis* (Gentianaceae). *Ecology* 74:959–62

55. Ehrlich PR, Ehrlich AH. 1996. *Betrayal of Science and Reason*. Washington, DC: Island. 335 + xiii pp.

56. Ehrlich PR, Raven PH. 1964. Butterflies and plants: a study in coevolution. *Evolution* 18:586–608

57. Eickwort GC. 1988. The origins of mites associated with honeybees. See Ref. 143a, pp. 327–38

58. Ellstrand NC, Elam DR. 1993. Population genetic consequences of small population size: implications for plant conservation. *Annu. Rev. Ecol. Syst.* 24:217–42

59. Falk DA. 1992. From conservation biology to conservation practice: strategies for protecting plant diversity. In *Conservation Biology: the Theory and Practice of Nature Conservation, Preservation and Management*, ed. PL Fiedler, SK Jain, pp. 397–431. New York: Chapman & Hall

60. Falk DA, Millar CI, Olwell M, eds. 1996. *Restoring Diversity. Strategies for Reintroduction of Endanger. Plants*. Washington, DC: Island. 505 pp.

61. Feinsinger P. 1983. Coevolution and pollination. In *Coevolution*, ed. DJ Futuyma, M Slatkin, pp. 282–310. Sunderland, MA: Sinauer Assoc.

62. Feinsinger P. 1987. Approaches to nectarivore-plant interactions in the new world. *Rev. Chil. Hist. Nat.* 60:285–319

63. Fitts RD, Tepedino VJ, Griswold TL. 1993. The pollination biology of Arizona cliffrose (*Purshia subintegra*), including a report on experimental hybridization with its sympatric congener *P. stansburiana* (Rosaceae). See Ref. 199a, pp. 359–68

64. Fletcher DJC. 1988. Relevance of the behavioral ecology of African bees to a solution to the Africanized-bee problem. See Ref. 143a, pp. 55–61

65. Free JB, Butler CG. 1959. *Bumblebees*. New York: Macmillan

66. Free JB, Ferguson AW. 1980. Foraging of bees on oil-seed rape (*Brassica napus* L.) in relation to the stage of flowering of the crop and pest control. *J. Agric. Sci.* 94:151–54

67. Fritz AL, Nilsson LA. 1994. How pollinator-mediated mating varies with population size in plants. *Oecologia* 100:451–62

68. Gardner RO, Early JW. 1996. The naturalisation of banyan figs (*Ficus* spp, Moraceae) and their pollinating wasps (Hymenoptera: Agaonidae) in New Zealand. *NZ J. Bot.* 34:103–10

69. Gauld ID, Collins NM, Fitton MG. 1990. *The Biological Significance and Conservation of Hymenoptera in Europe.* Strasbourg, France: Counc. Eur. 47 pp.

70. Gess FW, Gess SK. 1993. Effects of increasing land utilization on species representation and diversity of aculeate wasps and bees in the semi-arid areas of southern Africa. See Ref. 121a, pp. 83–113

71. Grant KA, Grant V. 1967. Effects of hummingbird migration on plant speciation in the California flora. *Evolution* 21:457–65

72. Greathead DJ. 1983. The multi-million dollar weevil that pollinates oil palms. *Antenna* 7:105–7

73. Groom MJ. 1998. Allee effects limit population viability of an annual plant. *Am. Nat.* 151:487–96

74. Hall P, Walker S, Bawa K. 1996. Effect of forest fragmentation on genetic diversity and mating system in a tropical tree, *Pithecellobium elegans. Conserv. Biol.* 10:757–68

75. Herrera CM. 1987. Components of pollinator "quality": comparative analysis of a diverse insect assemblage. *Oikos* 50:79–90

76. Herrera CM. 1988. Variation in mutualisms: the spatio-temporal mosaic of a pollinator assemblage. *Biol. J. Linn. Soc.* 35:95–125

77. Herrera CM. 1989. Pollinator abundance, morphology, and flower visitation rate: analysis of the "quantity" component in a plant-pollinator system. *Oecologia* 80:241–48

78. Herrera CM. 1996. Floral traits and plant adaptation to insect pollinators: a devil's advocate approach. In *Floral Biology: Studies on Floral Evolution in Animal-Pollinated Plants,* ed. DG Lloyd, SCH Barrett, pp. 65–87. New York: Chapman & Hall

79. Herrera J. 1988. Pollination relationships in southern Spanish Mediterranean shrublands. *J. Ecol.* 76:274–87

80. Heschel MS, Paige KN. 1995. Inbreeding depression, environmental stress, and population size variation in scarlet gilia (*Ipomopsis aggregata*). *Conserv. Biol.* 9:126–33

81. Heywood VH, ed. 1993. *Flowering Plants of the World.* New York: Oxford Univ. Press

82. Howarth FG. 1985. Impacts of alien land arthropods and mollusks on native plants and animals in Hawai'i. In *Hawai'i's Terrestrial Ecosystems: Preservation and Management,* ed. CP Stone, JM Scott, pp. 149–79. Honolulu: Univ. Hawaii Coop. Natl. Park Resour. Study Unit

83. Howe HF. 1984. Constraints on the evolution of mutualism. *Am. Nat.* 123:764–77

84. Hurlbert AH, Hosoi SA, Temeles EJ, Ewald PW. 1996. Mobility of *Impatiens capensis* flowers: effect on pollen deposition and hummingbird foraging. *Oecologia* 105:243–46

85. Ingram M, Nabhan GP, Buchmann S. 1996. Impending pollination crisis threatens biodiversity and agriculture. *Tropinet* 7:1

86. Ingvarsson PK, Lundberg S. 1995. Pollinator functional response and plant population dynamics: pollinators as a limiting resource. *Evol. Ecol.* 9:421–28

87. Inouye DW. 1983. The ecology of nectar robbing. In *The Biology of Nectaries,* ed. TS Elias, BL Bentley, pp. 153–73. New York: Columbia Univ. Press

88. Inouye DW, Calder WA, Waser NM. 1991. The effect of floral abundance on feeder censuses of hummingbird abundance. *Condor* 93:279–85

89. Janzen DH. 1987. Insect diversity of a Costa Rican dry forest: Why keep it, and how? *Biol. J. Linn. Soc.* 30:343–56

90. Jennersten O. 1988. Pollination in *Dianthus deltoides* (Caryophyllaceae): effects of habitat fragmentation on visitation and seed set. *Conserv. Biol.* 2:359–66

91. Johansen CA. 1977. Pesticides and pollinators. *Annu. Rev. Entomol.* 22:177–92

92. Johnson SD, Bond WJ. 1997. Evidence for widespread pollen limitation of fruiting success in Cape wildflowers. *Oecologia* 109:530–34

93. Jordano P. 1987. Patterns of mutualistic interactions in pollination and seed dispersal: connectance, dependence asymmetries, and coevolution. *Am. Nat.* 129:657–77

94. Kearns CA. 1992. Anthophilous fly distribution across an elevation gradient. *Am. Midl. Nat.* 127:172–82

95. Kearns CA, Inouye DW. 1994. Fly pollination of *Linum lewisii* (Linaceae). *Am. J. Bot.* 81:1091–95

96. Kearns CA, Inouye DW. 1997. Pollinators, flowering plants, and conservation biology. *BioScience* 47:297–307

97. Kesseli RV. 1992. Population biology and conservation of rare plants. In *Applied Population Biology*, ed. SK Jain, LW Botsford, pp. 69–90. Dordrecht, The Netherlands: Kluwer Acad.

98. Kevan PG. 1974. Pollination, pesticides and environmental quality. *BioScience* 24:198–99

99. Kevan PG. 1975. Forest application of the insecticide Fenitrothion and its effect on wild bee pollinators (Hymenoptera: Apoidea) of lowbush blueberries (*Vaccinium* spp.) in southern New Brunswick, Canada. *Biol. Conserv.* 7:301–9

100. Kevan PG. 1975. Pollination and environmental conservation. *Environ. Conserv.* 2:293–98

101. Kevan PG. 1986. Pollinating and flower visiting insects and the management of beneficial and harmful insects and plants. In *Biological Control in the Tropics: Proc. First Reg. Symp. Biol. Control, Univ. Pertanian Malaysia, Serdang, 4–6 Sept, 1985*, ed. MY Hussein, AG Ibrahim, pp. 439–52. Serdang, Selangor, Malaysia: Penerbit Univ. Pertanian

102. Kevan PG. 1991. Pollination: keystone process in sustainable global productivity. *Acta Hortic.* 288:103–9

103. Kevan PG. 1998. Pollinators in agroecosystems: their keystone role in sustainable productivity and biodindication. In *Biodiversity in Agroecosystems Role of Sustainability and Biodindication*, ed. M Paoletti. New York: Elsevier. In press

104. Kevan PG, Clark EA, Thomas FG. 1990. Pollination: a crucial ecologial and mutualistic link in agroforestry and sustainable agriculture. *Proc. Entomol. Soc. Ontario* 121:43–48

105. Kevan PG, Laverty TM. 1990. A brief survey and caution about importing alternative pollinators into Canada. *Can. Beekeep.* 15:176–77

106. Kevan PG, Plowright RC. 1989. Fenitrothion and insect pollination. In *Environmental Effects of Fenitrothion Use in Forestry: Impacts on Insect Pollinators, Songbirds, and Aquatic Organisms*, ed. WR Ernst, PA Pearce, TL Pollock, pp. 13–42. Dartmouth, Nova Scotia: Environ. Canada

107. Klemm M. 1996. Man-made bee habitats in the anthropogenous landscape of central Europe—substitutes for threatened or destroyed riverine habitats? See Ref. 130a, pp. 17–34

108. Klinkhamer PGL, de Jong TJ, de Bruyn G-J. 1989. Plant size and pollinator visitation in *Cynoglossum officinale*. *Oikos* 54:201–4

109. Knowles D. 1983/4. Flying jewels. *Geo* 5:46–57

110. Krannitz PG, Maun MA. 1991. An experimental study of floral display size and reproductive success in *Viburnum opulus*: importance of grouping. *Can. J. Bot.* 69:394–99

111. Kratochwil A. 1988. Co-phenology of plants and anthophilous insects: a historical area-geographical interpretation. *Entomol. Gen.* 13:67–80

112. Kunin WE. 1992. Density and reproductive success in wild populations of *Diplotaxis erucoides* (Brassicaceae). *Oecologia* 91:129–33

113. Kunin WE. 1993. Sex and the single mustard: population density and pollinator behavior effects on seed-set. *Ecology* 74:2145–60

114. Kunin WE. 1997. Population size and density effects in pollination: pollinator foraging and plant reproductive success in experimental arrays of *Brassica kaber*. *J. Ecol.* 85:225–34

115. Kwak MM. 1987. Pollination and pollen flow disturbed by honeybees in bumblebee-pollinated *Rhinanthus* populations? In *Disturbance in Grasslands*, ed. J van Andel, JP Bakker, RW Snaydon, pp. 273–83. Dordrecht, The Netherlands: Dr. W. Junk

116. Kwak MM. 1995. Pollination ecology and endangered plant species. In *Proc. Second CONNECT Workshop on Landscape Ecol.*, ed. F Skov, J Komdeur, G Fry, J Knudsen, pp. 54–57. Kalo, Denmark: Environ. Res. Inst.

117. Kwak MM. 1996. Bumble bees at home and at school. In *Bumble Bees for Pleasure and Profit*, ed. A Matheson, pp. 12–23. Cardiff, UK: Int. Bee Res. Assoc.

118. Kwak MM, Kremer P, Boerrichter E, van den Brand C. 1991. Pollination of the rare species *Phyteuma nigrum* (Campanulaceae): flight distances of bumblebees. *Proc. Exp. Appl. Entomol.* 2:131–36

119. Lagerlöf J, Stark J, Svensson B. 1992. Margins of agricultural fields as habitats for pollinating insects. *Agric. Ecosyst. Environ.* 40:117–24

120. Lammers TG, Weller SG, Sakai AK. 1987. Japanese White-eye, an introduced passerine, visits the flowers of *Clermontia arborescens*, an endemic Hawaiian Lobelioid. *Pac. Sci.* 41:74–77

121. Lamont BB, Klinkhamer PGL, Witkowski ETF. 1993. Population fragmentation may reduce fertility to zero in

Banksia goodii—a demonstration of the Allee effect. *Oecologia* 94:446–50

121a. LaSalle J, Gauld ID, ed. 1993. *Hymenoptera and Biodiversity.* Wallingford, UK: CAB Int.

122. Les DH, Reinhartz JA, Esselman EJ. 1991. Genetic consequences of rarity in *Aster furcatus* (Asteraceae), a threatened, self-incompatible plant. *Evolution* 45:1641–50

123. Lesica P. 1993. Loss of fitness resulting from pollinator exclusion in *Silene spaldingii* (Caryophyllaceae). *Madroño* 40:193–201

124. Light N. 1994. Abuzz about bumblebees. *Am. Fruit Grow.* 114:20–21

125. Lord JM. 1991. Pollination and seed dispersal in *Freycinetia baueriana*, a dioecious liane that has lost its bat pollinator. *NZ J. Bot.* 29:83–86

126. Louda SM, Potvin MA. 1995. Effect of inflorescence-feeding insects on the demography and lifetime fitness of a native plant. *Ecology* 76:229–45

127. Lubbers AE, Lechowicz MJ. 1989. Effects of leaf removal on reproduction vs. belowground storage in *Trillium grandiflorum. Ecology* 70:85–96

128. Mal TK, Lovett-Doust J, Lovett-Doust L, Mulligan GA. 1992. The biology of Canadian weeds. 100. *Lythrum salicaria. Can. J. Plant Sci.* 72:1305–30

129. Maron JL, Simms EL. 1997. Effect of seed predation on seed bank size and seedling recruitment of bush lupine (*Lupinus arboreus*). *Oecologia* 111:76–83

130. Martinez ND. 1992. Constant connectance in community food webs. *Am. Nat.* 139:1208–18

130a. Matheson A, Buchmann SL, O'Toole C, Westrich P, Williams IH, ed. 1996. *The Conservation of Bees.* New York: Academic

131. Maunder M. 1992. Plant reintroduction—an overview. *Biodivers. Conserv.* 1:51–61

132. McKey D. 1989. Population biology of figs: applications for conservation. *Experientia* 45:661–73

133. Menges ES. 1995. Factors limiting fecundity and germination in small populations of *Silene regia* (Caryophyllaceae), a rare hummingbird-pollinated prairie forb. *Am. Midl. Nat.* 133:242–55

134. Moldenke AR, Lincoln PG. 1979. Pollination ecology in montane Colorado: a community analysis. *Phytologia* 42:349–79

135. Montalvo AM, Ackerman JD. 1987.

Limitation to fruit production in *Ionopsis utricularioides* (Orchidae). *Biotropica* 19:24–31

136. Morse RA. 1988. Preface. See Ref. 143a, pp. xvii

137. Morse RA. 1991. Honeybees forever. *Trends Ecol. Evol.* 6:337–38

138. Murcia C. 1995. Edge effects in fragmented forests: implications for conservation. *Trends Ecol. Evol.* 10:58–62

139. Murcia C. 1996. Forest fragmentation and the pollination of neotropical plants. In *Forest Patches in Tropical Landscapes*, ed. J Schelhas, R Greenberg, pp. 19–36. Washington, DC: Island

140. Nabhan GP. 1996. *Pollinator Redbook.* Vol. 1: *Global List of Threatened Vertebrate Wildlife Species Serving as Pollinators for Crops and Wild Plants.* Tucson, AZ: Arizona-Sonora Desert Mus. & Forgot. Pollinat. Camp. Monogr.

141. Nabhan GP, Buchmann SL. 1997. Services provided by pollinators. In *Nature's Services. Societal Dependence on Natural Ecosystems*, ed. GC Daily, pp. 133–50. Washington, DC: Island

142. Natl. Res. Counc. Can. 1981. *Pesticide-Pollinator Interactions. NRC Assoc. Comm. Sci. Criteria Environ. Qual. Publ. NRCC No. 18471.* Ottawa, Canada: Natl. Res. Counc. Can. Environ. Secr.

143. Nat. Conserv. Counc. 1991. *A Review of the Scarce and Threatened Bees, Wasps and Ants of Great Britain.* Peterborough, UK: Nat. Conserv. Counc.

143a. Needham GR, Page RE Jr, Delfinado-Baker M, Bowman CE, eds. 1988. *Africanized Honey Bees and Bee Mites.* New York: Wiley

144. Olesen JM, Jain SK. 1994. Fragmented plant populations and their lost interactions. In *Conservation Genetics*, ed. V Loeschcke, J Tomiuk, SK Jain, pp. 417–26. Basel: Birkhäuser

145. Osborne JL, Corbet SA. 1994. Managing habitats for pollinators in farmland. *Aspects Appl. Biol.* 40:207–15

146. Osborne JL, Williams IH, Corbet SA. 1991. Bees, pollination and habitat change in the European Community. *Bee World* 72:99–116

147. O'Toole C. 1993. Diversity of native bees and agroecosystems. See Ref. 121a, pp. 169–96

148. Parker FD, Batra SWT, Tepedino VJ. 1987. New pollinators for our crops. *Agric. Zool. Rev.* 2:279–304

149. Paton DC. 1985. Food supply, population structure, and behaviour of New Holland Honeyeaters *Phylidonyris novaehollandiae* in woodlands near

Horsham, Victoria. In *Birds of Eucalypt Forests and Woodlands: Ecology, Conservation, and Management*, ed. A Keast, HF Recher, H Ford, D Saunders, pp. 222–30. Sydney, Aust.: R. Aust. Ornithol. Union and Surry & Beatty

150. Paton DC. 1993. Honeybees in the Australian environment. *BioScience* 43:95–103

151. Paton DC. 1996. *Overview of Feral and Managed Honeybees in Australia: Distribution, Abundance, Extent of Interactions with Native Biota, Evidence of Impacts and Future Research.* Canberra, Aust.: Aust. Nat. Conserv. 71 pp.

152. Pavlik BM, Ferguson N, Nelson M. 1993. Assessing limitations on the growth of endangered plant populations: 2. Seed production and seed bank dynamics of *Erysimum capitatum* ssp. *angustatum* and *Oenothera deltoides* ssp. *howellii*. *Biol. Conserv.* 65:267–78

153. Peach ML, Alston DG, Tepedino VJ. 1994. Bees and bran bait: Is carbaryl bran bait lethal to alfalfa leafcutting bee (Hymenoptera: Megachilidae) adults or larvae? *J. Econ. Entomol.* 87:311–17

154. Peach ML, Tepedino VJ, Alston DG, Griswold TL. 1993. Insecticide treatments for rangeland grasshoppers: potential effects on the reproduction of *Pediocactus sileri* (Englem.) Benson (Cactaceae). See Ref. 199a, pp. 309–19

155. Percy DM, Cronk QCB. 1997. Conservation in relation to mating system in *Nesohedyotis arborea* (Rubiaceae), a rare endemic tree from St Helena. *Biol. Conserv.* 80:135–45

156. Petanidou T, den Nijs HCM, Ellis-Adam AC. 1991. Comparative pollination ecology of two rare Dutch *Gentiana* species, in relation to population size. *Acta Hortic.* 288:308–12

157. Petanidou T, Ellis WE. 1993. Pollinating fauna of a phryganic ecosystem: composition and diversity. *Biodivers. Lett.* 1:9–22

158. Petit S, Pors L. 1996. Survey of columnar cacti and carrying capacity for nectar-feeding bats on Curaçao. *Conserv. Biol.* 10:769–75

159. Pimentel D, Wilson C, McCullum C, Huang R, Dwen P, et al. 1997. Economic and environmental benefits of biodiversity. *BioScience* 47:747–57

160. Pimm SL. 1991. *The Balance of Nature? Ecological Issues in the Conservation of Species and Communities.* Chicago: Univ. Chicago Press

161. Plowright RC, Rodd FH. 1980. The effect of aerial spraying on hymenopterous pollinators in New Brunswick. *Can. Entomol.* 112:259–69

162. Polis GA. 1991. Complex trophic interactions in deserts: an empirical critique of food-web theory. *Am. Nat.* 138:123–55

163. Pomeroy N. 1997. *Message from BOMBUS-L listserv list, 27 March.*

164. Powell AH, Powell GVN. 1987. Population dynamics of male euglossine bees in Amazonian forest fragments. *Biotropica* 19:176–79

165. Prescott-Allen C, Prescott-Allen R. 1986. *The First Resource: Wild Species in the North American Economy.* New Haven, CT: Yale Univ. Press

166. Price MV, Waser NM. 1998. Effects of experimental warming on plant reproductive phenology in a subalpine meadow. *Ecology* 79:1261–71

167. Proctor M, Yeo P, Lack A. 1996. *The Natural History of Pollination.* Portland, OR: Timber

168. Pyle RM. 1997. Burning bridges. *Wings* 21:22–23

169. Rathcke BJ, Jules ES. 1993. Habitat fragmentation and plant-pollinator interactions. *Curr. Sci.* 65:273–77

170. Raven PH. 1987. The scope of the plant conservation problem world-wide. In *Botanic Gardens and the World Conservation Strategy*, ed. D Bramwell, O Hamann, V Heywood, H Synge, pp. 19–20. London: Academic

171. Richards KW. 1984. *Alfalfa Leafcutter Bee Management in Western Canada. Publication 1495/E.* Ottawa, Can.: Agric. Can.

172. Richards KW. 1993. Non-*Apis* bees as crop pollinators. *Rev. Suisse Zool.* 100:807–22

173. Rinderer TE. 1988. Evolutionary aspects of the Africanization of honey-bee populations in the Americas. See Ref. 143a, pp. 13–28

174. Ritter W. 1988. *Varroa jacobsoni* in Europe, the tropics, and subtropics. See Ref. 143a, pp. 349–69

175. Roll J, Mitchell RJ, Cabin RJ, Marshall DL. 1997. Reproductive success increases with local density of conspecifics in the desert mustard *Lesquerella fendleri*. *Conserv. Biol.* 11:738–46

176. Roubik DW. 1978. Competitive interactions between neotropical pollinators and Africanized honey bees. *Science* 201:1030–32

177. Roubik DW. 1980. Foraging behavior of competing Africanized honeybees and stingless bees. *Ecology* 61:836–45

178. Roubik DW. 1983. Experimental community studies: time-series tests of competition between African and neotropical bees. *Ecology* 64:971–78

179. Roubik DW. 1989. *Ecology and Natural History of Tropical Bees*. Cambridge, UK: Cambridge Univ. Press

180. Roubik DW. 1991. Aspects of Africanized honey bee ecology in tropical America. In *The "African" Honey Bee*, ed. M Spivak, DJC Fletcher, MD Breed, pp. 259–81. Boulder, CO: Westview

181. Roubik DW. 1992. Loose niches in tropical communities: Why are there so few bees and so many trees? In *Effects of Resource Distribution on Animal-Plant Interactions*, ed. MD Hunter, T Ohgushi, P Price, pp. 327–54. New York: Academic

182. Roubik DW. 1993. Tropical pollinators in the canopy and understory: field data and theory for stratum preferences. *J. Insect Behav.* 6:659–73

183. Roubik DW. 1996. African honey bees as exotic pollinators in French Guiana. See Ref. 130a, pp. 173–82

184. Roubik DW. 1996. Measuring the meaning of honey bees. See Ref. 130a, pp. 163–72

185. Roubik DW, Moreno JE, Vergara C, Wittmann D. 1986. Sporadic food competition with the African honey bee: projected impact on neotropical social bees. *J. Trop. Ecol.* 2:97–111

186. Sampson AW. 1952. *Range Management, Principles and Practices*. New York: Wiley

187. Samways MJ. 1994. *Insect Conservation Biology*. London: Chapman & Hall

188. Sanford MT. 1996. A pollination crisis? *APIS Apic. Issues Answ. Fla. Coop. Ext. Newsl.* 14:1

189. Saure C. 1996. Urban habitats for bees: the example of the city of Berlin. See Ref. 130a, pp. 47–53

190. Schaffer WM, Zeh DW, Buchmann SL, Kleinhans S, Schaffer MV, Antrim J. 1983. Competition for nectar between introduced honey bees and native North American bees and ants. *Ecology* 64:564–77

191. Schemske DW, Husband BC, Ruckelshaus MH, Goodwillie C, Parker IM, Bishop JG. 1994. Evaluating approaches to the conservation of rare and endangered plants. *Ecology* 75:584–606

192. Schmitt J. 1983. Flowering plant density and pollinator visitation in *Senecio*. *Oecologia* 60:97–102

193. Scott-Dupree CD, Winston ML. 1987. Wild bee pollinator diversity and abundance in orchard and uncultivated habitats in the Okanagan Valley, British Columbia. *Can. Entomol.* 119:735–45

194. Senft D. 1990. Protecting endangered plants. *Agric. Res.* 38:16–18

195. Sih A, Baltus M. 1987. Patch size, pollinator behavior, and pollinator limitation in catnip. *Ecology* 68:1679–90

196. Simmons CH Jr, ed. 1996. *Plymouth Colony Records. Wills and Inventories, 1633–1669*. Camden, ME: Picton

197. Sipes SD, Tepedino VJ. 1995. Reproductive biology of the rare orchid, *Spiranthes diluvialis*: breeding system, pollination, and implications for conservation. *Conserv. Biol.* 9:929–38

198. Sipes SD, Tepedino VJ. 1996. Pollinator lost? Reproduction by the enigmatic Jones Cycladenia, *Cycladenia humilis* var. *jonesii* (Apocynaceae). In *Southwestern Rare and Endanger. Plants: Proc. 2nd Conf., 1995 Sept. 11–14. Flagstaff, AZ. General Technical Report RM-GTR-283*, ed. J Maschinski, HD Hammond, L Holter, pp. 158–66. Fort Collins, CO: USDA For. Serv. Rocky Mt. For. Range Exp. Stn.

199. Sipes SD, Tepedino VJ, Bowlin WR. 1993. The pollination and reproductive ecology of *Spiranthes diluvialis* Sheviak (Orchidaceae). See Ref. 199a, pp. 320–33

199a. Sivinski R, Lightfood K, eds. 1993. *Proc. Southwest. Rare and Endanger. Plant Conf*. Santa Fe, NM: NM For. Resour. Conserv. Div.

200. Smith TB, Freed LA, Lepson JK, Crothers JH. 1995. Evolutionary consequences of extinctions in populations of a Hawaiian honeycreeper. *Conserv. Biol.* 9:107–13

201. Southwick EE, Southwick L Jr. 1992. Estimating the economic value of honey bees (Hymenoptera: Apidae) as agricultural pollinators in the United States. *J. Econ. Entomol.* 85:621–33

202. Sowig P. 1989. Effects of flowering plant's patch size on species composition of pollinator communities, foraging strategies, and resource partitioning in bumblebees (Hymenoptera: Apidae). *Oecologia* 78:550–58

203. Spears EE Jr. 1987. Island and mainland pollination ecology of *Centrosoma virginianum* and *Opuntia stricta*. *J. Ecol.* 75:351–62

204. Steiner KE. 1993. Has *Ixianthes* (Scrophulariaceae) lost its special bee? *Plant Syst. Evol.* 185:7–16

205. Steiner KE, Whitehead VB. 1996. The consequences of specialization for pollination in a rare South African shrub,

Ixianthes retzioides (Scrophulariaceae). *Plant Syst. Evol.* 201:131–38

206. Stephen WP, Bohart GE, Torchio PF. 1969. *The Biology and External Morphology of Bees, with a Synopsis of the Genera of Northwestern America.* Corvallis, OR: Oreg. State Univ. Agric. Exp. Stn.

207. Stephenson AG. 1981. Flower and fruit abortion: proximate causes and ultimate functions. *Annu. Rev. Ecol. Syst.* 12:253–79

208. Stolzenburg W. 1993. Lucky clovers. *Nat. Conserv.* 43:6

209. Stouffer PC, Bierregaard RO Jr. 1995. Effects of forest fragmentation on understory hummingbirds in Amazonian Brazil. *Conserv. Biol.* 9:1085–94

210. Struck M. 1994. Flowers and their insect visitors in the arid winter rainfall region of southern Africa: observations on permanent plots. Composition of the anthophilous insect fauna. *J. Arid. Env.* 28:45–50

211. Sugden EA. 1985. Pollinators of *Astragalus monoensis* Barneby (Fabaceae): new host records; potential impact of sheep grazing. *Gt. Basin Natur.* 45:299–312

212. Sugden EA, Pyke GH. 1991. Effects of honey bees on colonies of *Exoneura asimillimia*, an Australian native bee. *Aust. J. Ecol.* 16:171–81

213. Sugden EA, Thorp RW, Buchmann SL. 1996. Honey bee-native bee competition: focal point for environmental change and apicultural response in Australia. *Bee World* 77:26–44

214. Taylor OR. 1988. Ecology and economic impact of African and Africanized honey bees. See Ref. 143a, pp. 29–44

215. Tepedino VJ. 1979. The importance of bees and other insect pollinators in maintaining floral species composition. *Gt. Basin Nat. Mem.* 3:139–50

216. Tepedino VJ, Sipes SD, Barnes JL, Hickerson LL. 1997. The need for "extended care" in conservation: examples from studies of rare plants in the western United States. *Acta Hortic.* 437:245–48

217. Tepedino VJ, Stanton NL. 1980. Spatiotemporal variation in phenology and abundance of floral resources on shortgrass prairie. *Gt. Basin Natur.* 40:197–215

218. Thaler GR, Plowright RC. 1980. The effect of aerial insecticide spraying for spruce budworm control on the fecundity of entomophilous plants in New Brunswick. *Can. J. Bot.* 58:2022–27

219. Thomson JD. 1981. Spatial and temporal components of resource assessment by flower-feeding insects. *J. Anim. Ecol.* 50:49–59

220. Thomson JD. 1993. The queen of forage and the bumblebee revolution: a conference with an attitude. *Trends Ecol. Evol.* 8:41–42

221. Thomson JD, Plowright RC, Thaler GR. 1985. Matacil insecticide spraying, pollinator mortality, and plant fecundity in New Brunswick forests. *Can. J. Bot.* 63:2056–61

222. Thomson JN, Pellmyr O. 1992. Mutualism with pollinating seed parasites amid co-pollinators: constraints on specialization. *Ecology* 73:1780–91

223. Thorp RW. 1996. Resource overlap among native and introduced bees in California. See Ref. 130a, pp. 143–51

224. Torchio PF. 1990. Diversification of pollination strategies for U.S. crops. *Environ. Entomol.* 19:1649–56

225. Torchio PF. 1991. Bees as crop pollinators and the role of solitary species in changing environments. *Acta Hortic.* 288:49–61

226. Totland O. 1993. Pollination in alpine Norway: flowering phenology, insect visitors, and visitation rates in two plant communities. *Can. J. Bot.* 71:1072–79

227. van Treuren R, Bijlsma R, Ouborg NJ, Van Delden W. 1993. The effects of population size and plant density on outcrossing rates in locally endangered *Salvia pratensis. Evolution* 47:1094–104

228. van Treuren R, Bijlsma R, Van Delden W, Ouborg NJ. 1991. The significance of genetic erosion in the process of extinction. I. Genetic differentiation in *Salvia pratensis* and *Scabiosa columbaria* in relation to population size. *Heredity* 66:181–89

229. Vinson SB, Frankie GW, Barthell J. 1993. Threats to the diversity of solitary bees in a neotropical dry forest in Central America. See Ref. 121a, pp. 53–82

230. Waring GH, Loope LL, Medeiros AC. 1993. Study on the use of alien versus native plants by nectarivorous forest birds on Maui, Hawaii. *Auk* 110:917–20

231. Waser NM. 1979. Pollinator availability as a determinant of flowering time in ocotillo (*Fouquieria splendens*). *Oecologia* 39:107–21

232. Waser NM, Chittka L, Price MV, Williams N, Ollerton J. 1996. Generalization in pollination systems, and why it matters. *Ecology* 77:279–96

233. Waser NM, Price MV. 1983. Optimal and actual outcrossing in plants, and the

nature of plant-pollinator interaction. In *Handbook of Experimental Pollination Biology*, ed. CE Jones, RJ Little, pp. 341–59. New York: Van Nostrand Reinhold

234. Washitani I. 1996. Predicted genetic consequences of strong fertility selection due to pollinator loss in an isolated population of *Primula sieboldii*. *Conserv. Biol.* 10:59–64

235. Watanabe ME. 1994. Pollination worries rise as honey bees decline. *Science* 265:1170

236. Watkins M. 1997. Resistance and its relevance to beekeeping. *Bee World* 78:15–22

237. Wcislo WT, Cane JH. 1996. Floral resouce utilization by solitary bees (Hymenoptera: Apoidea) and exploitation of their stored foods by natural enemies. *Annu. Rev. Entomol.* 41:257–86

238. Wenner AM. 1993. The honey bees of Santa Cruz. *Bee Cult.* 121:272–75

239. Westerkamp C. 1996. Pollen in bee-flower relations: some considerations on melittophily. *Bot. Acta* 109:325–32

240. Westerkamp C. 1997. Keel blossoms: bee flowers with adaptations against bees. *Flora* 192:125–32

241. Westrich P. 1996. Habitat requirements of central European bees and the problems of partial habitats. See Ref. 130a, pp. 1–16

242. Westrich P, Schwenninger HR. 1997. Habitatwahl, Blütennutzung und Bestandsentwicklung der Zweizelligen Sandbiene (*Andrena lagopus* Latr.) in Südwest-Deutschland (Hymenoptera, Apidae). *Z. Ökol. Nat.schutz* 6:33–42

243. Williams CS. 1995. Conserving Europe's bees: why all the buzz. *Trends Ecol. Evol.* 10:309–10

244. Williams IH. 1996. Aspects of bee diversity and crop pollination in the European Union. See Ref. 130a, pp. 63–80

245. Williams IH, Simpkins JR, Martin AP. 1991. Effect of insect pollination on seed production in linseed (*Linum usitatissimum*). *J. Agric. Sci.* 117:75–79

246. Williams PH. 1986. Environmental change and the distributions of British bumble bees (*Bombus* Latr.). *Bee World* 67:50–61

247. Wolf M, Shmida A. 1995. Association of flower and pollinator activity in the Negev Desert, Israel. *Adv. Geoecol.* 28:173–92

Annu. Rev. Ecol. Syst. 1998. 29:113–40

THE ROLE OF INTRODUCED SPECIES IN THE DEGRADATION OF ISLAND ECOSYSTEMS: A Case History of Guam[1]

Thomas H. Fritts

US Geological Survey, Patuxent Wildlife Research Center, National Museum of Natural History, MRC 111, Washington, DC 20560; e-mail: Thomas_Fritts@usgs.gov

Gordon H. Rodda

US Geological Survey, Patuxent Wildlife Research Center, Fort Collins, Colorado 80525-3400; e-mail: gordon_rodda@compuserve.com

KEY WORDS: Guam Island, *Boiga irregularis*, brown treesnake, extinction

ABSTRACT

The accidental introduction of the brown treesnake (*Boiga irregularis*) on Guam around 1950 induced a cascade of extirpations that may be unprecedented among historical extinction events in taxonomic scope and severity. Birds, bats, and reptiles were affected, and by 1990 most forested areas on Guam retained only three native vertebrates, all of which were small lizards. Of the hypotheses to account for the severity of this extinction event, we find some support for the importance of lack of coevolution between introduced predator and prey, availability of alternate prey, extraordinary predatory capabilities of the snake, and vulnerabilities of the Guam ecosystem. In addition, there were important interactions among these factors, especially the presence of introduced prey (possessing coevolutionary experience) that were thus able to maintain their populations and provide alternate prey to the introduced predator while it was driving the native prey species to extinction. This complex of vulnerabilities is common on oceanic islands.

113

INTRODUCTION

Guam, the largest island of Micronesia [54,100 hectares (ha)], has a remarkable ecological history. In the latter half of the 20th century, Guam lost virtually all its native bird species (36, 62, 73, 81). By early 1998, only three of Guam's 13 native forest bird populations retained even a slender hold on survival (Table 1). The largest population of the Micronesian starling (*Aplonis opaca*) was restricted to an urban area and numbered about 50 birds; the cave-roosting island swiftlet (*Aerodramus vanikorensis bartschi*) occupied only a single site, where it numbered in the low hundreds; and the most endangered population (94, 98) was that of the Marianas crow (*Corvus kubaryi*), which had one known pair and fewer than 20 individuals (C Aguon, personal communication).

Less well known is the loss of other vertebrate taxa. Of three native mammal species, the Mariana fruit bat (*Pteropus mariannus* subsp. *mariannus*) survives, but its long-term prospects are very much in doubt, with failure of recruitment extending more than a decade and residual adults now numbering just over 100 individuals (G Wiles, personal communication).

All other native vertebrates in the forests of Guam are reptiles. Of the 10 to 12 native species, six survive somewhere on the island, but only three lizards are found throughout: a native blue-tailed skink (*Emoia caeruleocauda*), a native mourning gecko (*Lepidodactylus lugubris*), and a prehistoric introduction or native (61) house gecko (*Hemidactylus frenatus*).

While other recent extinction events have involved greater numbers of species (32), we are unaware of any that have involved such a diversity of major vertebrate taxa and have had an impact on such a large percentage of the indigenous species. True to Western tradition, the villain is believed to be a serpent, a previously obscure nocturnal arboreal colubrid from Australasia, the brown treesnake (*Boiga irregularis*). We review the Guam biodiversity crisis with the objective of answering three questions: To what extent were the extirpations due to the introduction of the snake? What ecological features led to so extreme an outcome? What is projected to happen to the snake population and the ecosystem in the absence of most native prey species?

EVALUATION OF ECOLOGICAL IMPACTS

Evaluating the influence of an added species involves not only determining the effect of the species, but also considering other potential causes of the same outcome. In the case of Guam, for example, the introduction of the snake was followed shortly by independent introductions of the musk shrew (*Suncus murinus*), an arboreal lizard—the green anole (*Anolis carolinensis*), a bird—the black drongo (*Dicrurus macrocercus*), and a terrestrial lizard—the

Table 1   Status of native terrestrial vertebrates breeding on the island of Guam in 1998

Vertebrate	Surviving species	Exterminated by brown treesnake (*Boiga irregularis*)
Birds		
Pelagic	0 of 4	3
Brown booby *Sula leucogaster*		
(breeding stopped before snake)		
Fairy tern *Gygis alba*		
(breeding ended—snake)		
Brown noddy *Anous stolidus*		
(breeding ended—snake)		
White-tailed tropic bird *Phaethon lepturus*		
(breeding ended—snake)		
Near-shore	1 of 1	0
Pacific reef heron *Egretta sacra* (present)		
Wetland/grassland	2 of 4	0?
Common moorhen *Gallinula chloropus*		
subsp. *guami* (Endangered—habitat loss)		
Yellow bittern *Ixobrychus sinensis* (present)		
Marianas mallard *Anas platyrhynchos*		
subsp. *oustaleti* (extirpated—habitat loss)		
White-browed crake *Poliolimnas cinereus*		
(extirpated—not snake)		
Forest	3 of 13	9?
Island swiftlet *Aerodramus vanikorensis*		
subsp. *bartschi* (one colony surviving)		
Marianas crow *Corvus kubaryi*		
(< 20 individuals—snake)		
Marianas starling *Aplonis opaca* (remnant		
populations in urban areas—snake)		
Bridled white-eye *Zosterops conspicillata*		
subsp. *conspicillata* (extirpated—snake)		
Guam flycatcher *Myiagra freycineti*		
(extinct—snake)		
Guam rail *Rallus owstoni*		
(extirpated—snake; captive)		
Mariana fruit-dove *Ptilinopus roseicapilla*		
(extirpated—snake)		
Micronesian honeyeater *Myzomela rubratra*		
(extirpated—snake)		
Micronesian kingfisher *Halcyon*		
cinnamomina subsp. *cinnamomina*		
(extirpated—snake; captive)		

(Continued)

Table 1   (*Continued*)

Vertebrate	Surviving species	Exterminated by brown treesnake (*Boiga irregularis*)
Micronesian megapode *Megapodius laperouse* subsp. *laperouse* (extirpated—not snake)		
Nightingale reed-warbler *Acrocephalus luscinia* subsp. *luscinia* (extirpated—snake?)		
Rufous fantail *Rhipidura rufifrons* (extirpated—snake)		
White-throated ground-dove *Gallicolumba xanthonura* (extirpated—snake)		
Mammals	1 of 3	?
Mariana fruit bat *Pteropus m.* subsp. *mariannus* (one colony—snake)		
Little Mariana fruit bat *Pteropus tokudae* (extinct—loss not attributable)		
Sheath-tailed bat *Emballonura semicaudata* (extirpated—loss not attributable)		
Reptiles	6 of 10–12	3–5
Blue-tailed skink *Emoia caeruleocauda* (present)		
House gecko *Hemidactylus frenatus* (present)		
Mourning gecko *Lepidodactylus lugubris* (present)		
Moth skink *Lipinia noctua* (localized)		
Pelagic gecko *Nactus pelagicus* (localized)		
Brahminy blind snake *Ramphotyphlops braminus* (not definitely native—present)		
Snake-eyed skink *Cryptoblepharus poecilopleurus* (extirpated—snake?)		
Azure-tailed skink *Emoia cyanura* (no recent records—shrew?)		
Blue-tailed copper-striped skink *Emoia impar* (no recent records—shrew?)		
Mariana skink *Emoia slevini* (extirpated—snake? shrew?)		
Spotted-belly gecko *Perochirus ateles* (extirpated—snake)		
Tide-pool skink *Emoia atrocostata* (no definite records)		

curious skink (*Carlia* cf *fusca*). In addition, it is necessary to consider ecological alterations brought on by other changes in the ecosystem, for example,
deforestation. Fortunately for biodiversity conservation, but unfortunately for
ecological understanding, species introductions cannot be replicated. In most
cases they cannot even be tested directly with introductions into or exclusions
from naturalistic enclosures. We are forced to rely on the plausibility of competing scenarios. We evaluated proposed scenarios for the effects of introduced
species on Guam in light of nine inquiries:

1. Is the scenario consistent with what is known about ecological interactions in
   similar ecosystems? For example, have other island ecosystems been found
   more vulnerable to loss of primary forest or introductions of generalist predators where there were none before?

2. Is the proposed ecological interaction plausible on trophic grounds? Was
   the putative prey found in the diet of the putative predator? Did all members
   of the same prey guild show similar declines?

3. Did the putative prey attract the putative predator in substantial numbers
   in naturalistic field trials? For example, if brown treesnakes are alleged to
   decimate bird nests, are snakes drawn to traps baited with bird eggs?

4. Is the proposed scenario plausible on numerical grounds? For example,
   were shrews numerous enough to have been responsible for the observed
   decline in the pelagic gecko, *Nactus pelagicus*, given the shrew's normal
   diet?

5. Is the recorded expansion of the predator population temporally and geographically consistent with the observed declines in putative prey? This
   comparison must take into account the longevity of the species if the hypothesized predator is believed to interrupt reproduction but not harm adults.

6. Is the size distribution of proposed prey consistent with predatory capabilities
   of the proposed predator? In the Guam example, species losses in the 1980s
   affected only those species with adult sizes in the range 4–125 g. Was this
   consistent with the known dietary habits of the proposed predator?

7. Are proposed predator and prey syntopic? For example, the decline of the
   pelagic gecko, a terrestrial species, occurred at a time when brown treesnakes
   were not known to forage terrestrially. We infer that the snake was not a
   probable cause of the gecko's decline.

8. Did observed losses also occur in localities from which the putative predator was absent? For example, the scenario that brown treesnakes were responsible for the demise of Guam's population of pelagic geckos is greatly

weakened by evidence that the gecko concurrently disappeared from the islands of Saipan and Tinian but not Rota (all three of which lack the snake). However, the hypothesis that shrews were involved is supported by this test (i.e. shrews colonized Guam, Saipan, and Tinian but not Rota).

9. Did experimental removal of the putative predator result in a population rebound by the proposed prey species? This type of data is available only for the interaction between the snake and several lizards (6) but positive results constitute strong evidence. Interpretation of negative evidence is limited by the possibility that the experimental removal was carried out for too short a period (60) or was otherwise unnatural.

Among the alternate hypotheses that we considered for each proposed species interaction were those involving habitat deterioration (24, 66), environmental contaminants (34, 79), introductions of disease organisms (80, 82), alternate predator effects (5), direct human exploitation (96, 97), and competition (17).

To evaluate ecological interactions, it is desirable to have periodic measures of abundance for the constituent species. For birds, a number of published and unpublished records exist that give relative abundance (e.g. annual reports of Guam's Division of Aquatic and Wildlife Resources), as well as several estimates of absolute abundance (16, 25, 26, 64). Some absolute population densities for introduced small mammals also exist, especially after World War II (2, 4, 5, 35). In recent years, there have been more-or-less complete counts of the one surviving bat colony on Guam (98).

Replicable population estimates are generally lacking for reptiles. We partially filled this void by intensively sampling representative plots (10 × 10 m) of forest land (70). We separated the canopy of each plot from adjoining vegetation, and we blocked ground-level movement of lizards by erecting a small fence of greased aluminum flashing. These barriers were installed during the lizards' inactive period. We then removed all vegetation in small pieces and counted the number of each lizard species trapped within an isolated plot. The average number of lizard individuals captured per plot on Guam was 130, indicating that a reasonable sample was obtained. These counts have demonstrated that the biomass of reptiles in Guam forests exceeds that of all other vertebrate taxa. Reptiles provide a significant food resource to both native (Micronesian kingfisher, *Halcyon cinnamomina* subsp. *cinnamomina*) and introduced (snake) predators. By comparing lizard densities among habitat-stratified plots possessing and lacking the snake (6), and between islands possessing the snake (Guam) and lacking the snake (Saipan), and integrating this information with published counts of lizards (31, 47, 48, 68, 69, 77, 99, 101), we can roughly estimate the probable lizard densities that occurred on Guam prior and subsequent to the arrival of the snake.

Our analysis focuses on Guam's primary natural terrestrial habitat—forest. Species that have never occurred in significant portions of Guam forests are omitted (e.g. brown noddies, *Anous stolidus*, which once roosted in isolated colonies but did not occur throughout the forest), as are feral and introduced species that did not play a significant role in the vertebrate food web of most localities: sambar deer (*Cervus mariannus*), Asiatic water buffalo (*Bubalus bubalis*), feral pig (*Sus scrofa*), feral dog (*Canis familiaris*), feral cat (*Felis catus*), marine toads (*Bufo marinus*), eastern dwarf tree frog (*Litoria fallax*), feral chicken (*Gallus gallus*), black francolin (*Francolinus francolinus*), blue-breasted quail (*Coturnix chinensis*), pigeon (*Columba livia*), Eurasian tree sparrow (*Passer montanus*), and chestnut mannikin (*Lonchura malacca*).

A SHORT ECOLOGICAL HISTORY OF GUAM'S FOREST VERTEBRATES

Prehistoric Extirpations

It is likely that Guam experienced the same pattern of anthropogenic extinctions suffered by other Pacific islands (21, 85, 87). Research on Rota (61, 86), the island nearest Guam, as well as the islands just north of Rota (61, 88), indicates that a significant portion of the native fauna disappeared about the time of human colonization (ca. 1500 BC). For example, Rota lost 13 of 22 (59%) avian species, including one shearwater, one tern, one duck, one megapode, three rails, two pigeons, one parrot, one swift, one monarch flycatcher, and one parrotfinch. Reptile and mammal losses are less well documented but probably include at least one large gecko (61). Thus the historical fauna of Guam includes only part of the native fauna.

Humans not only caused extinctions, they added species, especially mammals and lizards. In addition to the usual assortment of domestic livestock (dogs, pigs), prehistoric humans were probably responsible for the establishment of Asian black rats (*Rattus tanezumi*), mutilating geckos (*Gehyra mutilata*), oceanic geckos (*Gehyra oceanica*), and possibly monitors. The exotic geckos may have induced population declines in the native geckos but are not known to have eliminated any native species.

Historic Losses

From the time of Magellan (ca. 1520) until the 20th century, few additional species were lost, and the species additions (deer, Philippine turtle-dove *Streptopelia bitorquaata*, Polynesian rat *Rattus exulans*, house mouse *Mus domesticus*) were relatively inconsequential for the vertebrate food webs discussed below. The Micronesian megapode (*Megapodius laperouse*), which was not common on Guam during the historical period (probably related to a shortage of suitable soils for oviposition) and was subject to fairly intense human exploitation (3, 93), may have survived into the 20th century. It is difficult to know what role it would have played in the food web had its numbers not been so limited by human predation. The turtle-dove was also widely hunted, but it persisted and today is a food source for brown treesnakes (11, 12).

Recent Perturbations

The modern era in Guam began with the American Navy's effort to wrest control of the island from the Japanese during World War II. The Guam assault was part of a coordinated invasion of the Marianas. The conquest of Saipan took longer than American planners had anticipated, causing the preinvasion bombardment of Guam to be extended for several weeks (51). As a result, about 80% of the island's structures were destroyed. No one seems to have quantified the damage to the island's natural habitats.

During the subsequent wartime buildup for the planned invasion of Japan, Guam's civilian population of 21,838 was augmented by more than 200,000 soldiers (54). Quarters, warehouses, and airfields were built largely on previously undeveloped land. This buildup was further expanded in some localities for the Korean War (early 1950s) and later (1960s, 1970s) cold-war activities. After the 1950s, many of the clearings reverted to forest, although much of the regrowth was tangantangan (*Leucaena leucocephala*), an introduced leguminous tree (15, 27).

Heavy but unrecorded doses of DDT and allied pesticides were broadcast on the island for several decades after the war (1). The aggregate impacts of the postwar habitat destruction and pesticide contamination are not easy to discern, because the levels of use were not recorded and populations of the possibly affected species were not monitored. Nonetheless, no species disappeared from the forests of Guam at that time (24). The insectivorous nightingale reed-warbler (*Acrocephalus luscinia* subsp. *luscinia*) disappeared from central Guam around 1968, possibly as a delayed result of pesticide contamination but more likely as a result of a combination of brown treesnake predation and wetland habitat destruction (37). Several other insectivorous bird species were judged to be inexplicably rare in the 1960s, perhaps as a result of bioamplification of contaminants or snake predation. The insectivorous sheath-tailed bat (*Emballonura semicaudata*) was lost in the mid 1970s (41), possibly as a result of pesticide contamination. However, the bat's numbers were so poorly documented that inferences about the date or cause of its extirpation can neither be supported nor refuted. In addition, several poorly documented lizard species disappeared at an undetermined time in the postwar period, possibly because of pesticide contamination or habitat loss. However, none are known to have required pristine habitat, and when pesticide residues were first sampled in Guam wildlife in the 1980s, lizards were not found to harbor high levels (34).

THE IRRUPTION OF THE SNAKE   Brown treesnakes probably arrived on Guam shortly after World War II as an unintended consequence of the salvage of derelict war materials from the New Guinea area (74). In particular, the huge American naval base at Manus (an island in the Admiralty Group, north of Madang, Papua New Guinea) was used as a staging point for vehicles, aircraft, and other supplies that had been sitting in the jungle since the battles. Undoubtedly, some of the items had snakes in them when they were transported from surrounding areas to Manus and from Manus to Guam.

Snake colonization was first evident in the southern part of Guam nearest Apra Harbor (74, 81). Spread was not well documented, but it was relatively slow in comparison to that of the other postwar irruptions of vertebrates on Guam (below). In contrast to the shrew irruption, there is evidence for only one locus of snake colonization, the harbor area and adjoining naval supply depots.

Quantification of snake abundances did not begin until 1985. In the early 1980s, Savidge (81) polled local residents throughout Guam to determine the date on which they first became aware of the snake. Residents at the far northern part of the island (35 km from the harbor) were not aware of the snakes until the 1980s. The dates Savidge compiled for local awareness of the snake should probably be viewed as when the snake became relatively abundant. For example, although Savidge found that residents of Ritidian Point at the northern tip of the island became aware of the snake around 1982, a visiting herpetologist captured a snake there in 1968. Thus, the snake had reached all parts of Guam by about 1970, but the crest of the irruption moved away from the port as a concentric spreading wave. This slow buildup is consistent with the relatively low vagility and modest fecundity of the species (77).

The buildup of snakes in the southern part of the Guam in the 1950s and 1960s was concurrent with disappearance of other vertebrates in southern Guam. There are no data for lizards, but residents were well aware of the disappearance of noisy birds and edible fruit bats. Bird and bat surveys by Guam's Division of Aquatic and Wildlife Resources were discontinued in the 1970s, as there were no more native endotherms to count. The three species that persisted longest were the Mariana fruit bat, Guam rail (*Rallus owstoni*), and island swiftlet (25, 37). The long-lived rail and bat persisted for about a decade beyond the dates when small birds disappeared from a locality. The swiftlet persists.

Island swiftlets differ from the other two species by roosting in caves. They are capable of echolocation and rarely, if ever, perch except in the roost cave (3). During the 1960s and 1970s, when the other bird and bat species were being extirpated from southern Guam, the swiftlet was disappearing from caves throughout Guam, except for the single roost that remains (91). In this cave, the birds perch and raise young in saliva-mud nests glued to the ceiling. Snakes forage at the entrance to and along the walls of the cave (G Wiles, personal communication; J Morton, personal communication) and are capable of gripping and bridging among ceiling stalactites in swiftlet caves elsewhere on the island (since abandoned by the birds), but snakes do not seem capable of scaling the smooth ceiling of the surviving roost cave. Brown treesnakes can accurately strike prey in total darkness (38), and a snake that had reached a cup nest in total darkness would seem to have no problem detecting immobile prey by olfaction or vomerolfaction alone. Thus, the snake could harm swiftlet populations but only those living in caves in which it could scale the walls.

The pattern of range contraction of island swiftlets on Guam was geographically the opposite of that of the other birds: Swiftlets vanished first from the northern part of the island. This has led some observers to search for a unique cause of endangerment for this bird. Swiftlets persisted in northern Guam until about 1980, however (37), a time when snakes were well established there, probably exceeding 50/ha by 1980 (74). Thus, the snake population was already about an order of magnitude higher than that recorded for other relatively large snake species away from water or dens (55, 77).

Brown treesnake population enumeration began in 1985, at which time the density was about 100 snakes/ha at a site on the northern end of the island (28, 74). Subsequent estimates have all been lower, suggesting that the peak of the irruption in northern Guam occurred around 1985. From 1985 to 1990, the snake population declined in northern Guam and exhibited signs of food stress (very high mortality among adult females, little recruitment, high proportion of emaciated individuals, and other signs). In the period 1992–1996, the condition of adult snakes improved and prey abundances rebounded from the extraordinary lows around 1989.

THE IRRUPTION OF THE MUSK SHREW   The musk shrew (*Suncus murinus*) irruption occurred much more quickly than did that of the snake. The shrew was first detected in Guam at several sites around Apra Harbor in 1953 (57). The introduction is assumed to have been accidental; a likely source is the Philippines (5, 57). For the first year of the Guam colonization, the shrew was found only near the harbor, but by the end of the second year the shrew was found over most of the northern three quarters of the island (57). Shrew populations covered the entire island by 1958, including remote forested areas that are not the preferred habitat of this species (4). Mammalogists have speculated that unintentional vehicular transport must have spread the shrew (4, 57). It does not seem likely that the shrew was able to naturally expand its population over the 35-km expanse it occupied in one year. Furthermore, if it was capable of expanding so rapidly, one would have expected it to have first colonized the southern end of the island, which was closer to the point of introduction. It is possible that the south experienced undocumented colonizations. It is also possible that the shrew had greater difficulty colonizing those parts of the island, such as the south, that were well populated with brown treesnakes. In contrast, shrews reaching the northern end of the island would not have been subject to snake predation in 1955. Although both species entered Guam via the port, the snake was the first colonist of the port and southern areas of the island, whereas the shrew was the first colonist of the northern end of Guam.

Shrew trap lines on Guam in 1962–1964 had capture rates of around 15%, about 129% of the comparable value for 1958 (5). Capture rates fell from 1960 to 1981 to 1994 (K King, C Grue, C Fecko, unpublished information). In most forested areas of northern Guam the shrew is now too rare to detect. The peak of the irruption probably occurred before 1981, and the decline of shrews extended over more than a decade. Data are insufficient to determine the year of the crest of the shrew population irruption, but anecdotes of its abundance in the early 1960s have not been matched in recent years on Guam.

In contrast to Guam, where shrew populations rose and fell sharply, the nearby and, until recently, snake-free island of Saipan has maintained high shrew populations since its colonization in 1962. A 1997 mark-recapture study of shrews indicated a density of about 55/ha (95% CL 51–71; S Vogt, unpublished information). Barbehenn (5) estimated that shrews on Guam in 1962 (near the time of the crest of their irruption) numbered about 15.5/ha. If catch rates are proportional to abundance, the current shrew density in appropriate habitat on Guam would be about 0.6/ha.

Barbehenn (5) suggested that shrews were responsible for an order of magnitude decline in house mice (*Mus domesticus*) on Guam from 1958 to 1969 and that shrews might be impacting terrestrial lizards, of which he saw none on Guam "during hundreds of hours tending trap lines in the fields during 1962 to 1964." In 1998, this statement seems incredible, for both native and introduced skinks are present in surprising numbers in suburban areas as well as disturbed and undisturbed forests. For example, in snake-free and shrew-free plots of tangantangan forest on Guam, we found an average density of 13,200 skinks/ha, a density that declines in the presence of either snakes (to 8850/ha on Guam) or shrews (to 2200/ha on Saipan). The paucity of skinks throughout Saipan (70) is consistent with an adverse effect of introduced shrews on terrestrial lizards.

Two of Guam's terrestrial lizards have disappeared since the introduction of the shrew: the pelagic gecko (*Nactus pelagicus*) and the Mariana skink (*Emoia slevini*). There is no unequivocal evidence that the shrew affected the Mariana skink (48) but the pattern of extirpations of the gecko is consistent with shrew involvement. For example, the gecko is relatively common on Rota (shrews absent), but is gone from (in the case of Tinian and Saipan) or highly localized in (in the case of Guam) the shrew-occupied parts of its historical range in the southern Mariana Islands. The situation on Guam is complex because of the highly localized present distribution of the gecko (69). On Guam, the gecko occurs in relatively undisturbed forest in several portions of the island but is widespread only in southern forests (69). The area occupied by geckos in the southern portion of the island was first colonized by the snake, whereas the area of the island first colonized by the shrew is generally devoid of pelagic geckos. Furthermore, the surviving populations of pelagic geckos on Guam exhibit an unexpectedly high level of arboreality, which would be consistent with behavior that prevents predation by terrestrial predators.

THE IRRUPTION OF THE GREEN ANOLE   The green anole (*Anolis carolinensis*) was purposely released on Guam around 1955 by a citizen who judged it beneficial for insect control (23). The chronology of the spread of this diurnal arboreal lizard is not known, but it increased in density and spread over much of the island over the next 20 years (50). After that time, however, its abundance waned and it became rare in most nonurban areas (76). Both its rarity in nonurban areas and its primary effect on Guam's ecology were illustrated by sampling conducted by B Smith and T Fritts in 1985. They collected snakes and lizards (snake prey) in the Northwest Field area of northern Guam to determine if the snakes were preying on lizards in the same relative proportions that the lizards were discovered by herpetologists. Over half (52%) of 91 lizards preyed upon by the 168 snakes collected were anoles, whereas only 4.3% of the 494 lizards collected by herpetologists were anoles, even though anoles are conspicuous to humans. These data suggest that anoles are unusually vulnerable to snake predation. Anoles have not been found on Northwest Field since 1985, presumably reflecting a population decline in the lizard brought on by extremely effective snake predation.

The discrepancy between utilization and apparent availability of Guam anoles is mirrored by the exceptional ability of brown treesnakes to find agamid lizards in Australia (84) at night. In both cases the prey are taken at night. Brown treesnakes cruise slowly through the twig ends of foliage at night; thus, they are able to discover and capture lizards while the lizards sleep on the ends of branches. Guam's native lizards do not sleep in such locations. Thus, one potentially major effect of the anole's introduction to Guam was that it provided a prey item that was uniquely suited to the snake's manner of foraging. Because of its rarity and unique niche, it is unlikely that the anole currently has a direct impact on the welfare of Guam's native lizards.

THE IRRUPTION OF THE BLACK DRONGO   The black drongo (*Dicrurus macrocercus*) was translocated from Taiwan to Rota for insect control shortly before World War II (37). It is believed to have colonized northern Guam through overwater dispersal from Rota in the 1950s (R Ryder, personal communication) or

1960s (37, 43). By 1967, it was reported to be the fourth most common bird in Guam roadside counts (R Ryder, personal communication). It had reached the southern end of the island by 1970, when it was judged common in central Guam (40). By 1981, it was fourth in population density, having about one sixth the abundance of the Micronesian starling (25). In the mid-1980s, drongo counts along roadsides began a steep decline, especially in northern Guam (according to annual reports of Guam's Division of Aquatic and Wildlife Resources).

Drongos are strongly territorial, aggressive birds that are believed to displace smaller birds that might otherwise nest within their territories. Although this has been proposed to account for declines in species of smaller birds, the only attempt to document a demographic impact failed to show an effect (43). Because drongo abundances on Guam have declined since these experiments, it is likely that present effects are relatively small.

THE IRRUPTION OF THE CURIOUS SKINK   In vertebrate biomass, the most significant of the postwar irruptions was that of the curious skink (*Carlia* cf *fusca*). The cause and origin of this colonization is unknown, as is the exact species (45). The early years of the irruption are undocumented, although it appears to have first colonized central Guam (23).

The skink arrived on Guam in the midst of the shrew irruption, which probably limited the skink's initial population growth and spread. [Unfortunately, the only surviving evidence is of relative abundance of this skink in comparison to the native skinks (31, 68).] Those data show that the curious skink rapidly expanded its populations and soon came to dominate collections of skinks (31, 74) from Guam (75%), Tinian (>90%), and Saipan (>95%). The growing dominance of *Carlia* in collections might have meant that the introduced skink was displacing the native skinks, but that inference is only partially supported by direct assessments of abundance. *Carlia* were not as dominant as is suggested by the museum collections. Because most herpetological collections are made along roadsides, in habitats that are highly disturbed, it is probable that skinks in disturbed habitats are overrepresented in museum collections. Curious skinks may also be more readily detected because they are bold and inquisitive. Our total removal samples indicate that it constitutes only a small fraction of the skink fauna in native forest (8.3%) and highly disturbed ravine forest (10.2%). Even in highly disturbed tangantangan forest it barely constitutes a majority (55%). Therefore, it seems likely that the rise of the curious skink from 0% to about 75% of skink collections in the interval 1960 to 1990 represents primarily the increasing dominance of curious skinks in disturbed habitats. Severe typhoons, which were unusually frequent in Guam in the 1990s, may have increased the amount of disturbed habitat; thus the irruption of the curious skink may not yet have crested.

By the 1990s, the curious skink constituted the primary prey item for most brown treesnakes (75). The smallest snakes are relatively arboreal in their habits and consume substantial numbers of geckos along with skinks; intermediate-sized snakes eat almost exclusively *Carlia*; and the largest snakes shift to a diet including endotherms, especially rats, but also including skinks. Therefore, the skinks make a major contribution to sustaining populations of the introduced snake.

The Consequences of Five Irruptions

To illustrate trophic interactions, we prepared food webs for northern Guam in 1945 (Figure 1), 1965 (Figure 2), and 1995 (Figure 3). We obtained mean masses of each species from our collections or from the literature (35, 37, 41, 56) and multiplied these by estimated absolute population densities to obtain crude estimates of biomass density for each species.

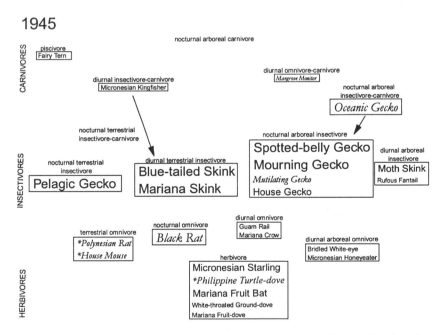

Figure 1 Typical vertebrate food web for northern Guam in 1945. *Italic*, introduced species; *asterisks*, historic introductions; *type size*, relative biomass abundance by order of magnitude from 0.01 to >10 kg/hectare (ha). Biomass densities were grouped by order of magnitude into four classes (0.01–0.099 kg/ha; 0.1–0.99 kg/ha; 1.0–9.9 kg/ha; and >10 kg/ha). Species represented by <0.01 kg/ha were considered trophically insignificant and were omitted from the figures. The figures show major trophic interactions within the vertebrates, and the niche box labels indicate the major trophic interactions between vertebrate and nonvertebrate species (Figures 1–3). See text and Table 1 for additional information.

1965

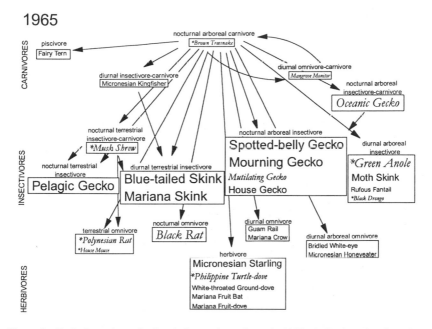

Figure 2   Typical vertebrate food web for northern Guam in 1965. *Italic*, introduced species; *asterisks*, historic introductions; *type size*, relative biomass abundance by order of magnitude from 0.01 to >10 kg/hectare (ha). Biomass densities were grouped by order of magnitude into four classes (0.01–0.099 kg/ha; 0.1–0.99 kg/ha; 1.0–9.9 kg/ha; and >10 kg/ha). Species represented by <0.01 kg/ha were considered trophically insignificant and were omitted from the figures. The figures show major trophic interactions within the vertebrates, and the niche box labels indicate the major trophic interactions between vertebrate and nonvertebrate species (Figures 1–3). See text and Table 1 for additional information.

The vertebrate forest food web on Guam today bears little resemblance to that prior to the postwar introductions (Figures 1–3). The most striking change from 1945 to 1995 is the reorganization of the food web from one in which virtually all components were native (indicated by plain typeface) and wherein vertebrates fed on nonvertebrates (plants and invertebrates) to one in which almost all major components are introduced vertebrates (italic font) that prey on other introduced species.

In 1945 (Figure 1), carnivory (consumption of vertebrates) was limited to kingfishers eating skinks and very large geckos occasionally eating smaller geckos. Mangrove monitors consumed eggs and small vertebrates opportunistically, but this interaction does not seem to have been demographically significant for the prey species. Thus, the soon-to-be-filled niches of nocturnal arboreal carnivore and nocturnal terrestrial insectivore-carnivore were vacant in 1945. It is not clear if the monitor is a native species, a prehistoric introduction,

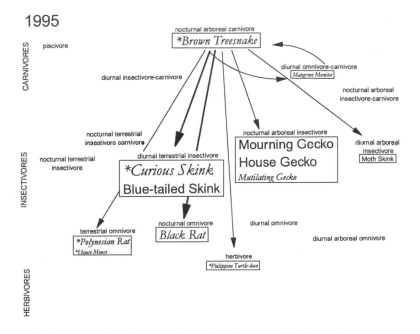

Figure 3 Typical vertebrate food web for northern Guam in 1995. *Italic*, introduced species; *asterisks*, historic introductions; *type size*, relative biomass abundance by order of magnitude from 0.01 to >10 kg/hectare (ha). Biomass densities were grouped by order of magnitude into four classes (0.01–0.099 kg/ha; 0.1–0.99 kg/ha; 1.0–9.9 kg/ha; and >10 kg/ha). Species represented by <0.01 kg/ha were considered trophically insignificant and were omitted from the figures. The figures show major trophic interactions within the vertebrates, and the niche box labels indicate the major trophic interactions between vertebrate and nonvertebrate species (Figures 1–3). See text and Table 1 for additional information.

or a recent introduction [which it is on more northern islands, although data from Guam are unresolved (G Pregill, personal communication)]. In addition to the monitor, there were only six consequential introduced species in northern Guam forests in 1945: three rodents, one game bird (Philippine turtle-dove), and two geckos. The native species dominated the food web, including twelve birds, three bats, and seven lizards.

By 1965 (Figure 2), the number of native species had not changed, but the addition of several predators, especially the shrew and the snake, had radically altered the number of predatory links within the vertebrate community (Figure 2). This increase in carnivory did not initially result in a great diminution in the abundances of the prey species. In part this was because the predators had not reached their peak abundance, and in part it was due to their eating each other. For example, the shrew diminished the abundances of several terrestrial lizards, but the snake was not only eating the shrew (thereby relieving pressure on the lizards), it was also reducing lizard predation by eating kingfishers. Thus energy utilization was shifted up the food chain, but the energetic consequence of this consumption was diffused among several trophic levels and among many prey species. A few key prey species, many of them introduced, help sustain rapid growth in predator populations. This phenomenon was greatly reinforced by the irruption of the curious skink in the 1970s and 1980s.

In the mid-1980s, however, pressure by introduced predators overwhelmed many of the native endotherms, putting additional pressure on the surviving prey species. By 1995 (Figure 3), native vertebrate prey species were so diminished that the food web was becoming comparatively top-heavy and holey: piscivores, nocturnal terrestrial insectivores, diurnal insectivore-carnivores, nocturnal arboreal insectivore-carnivores, and diurnal arboreal omnivores were entirely missing (Figure 3). Several additional niches were unfilled over most of northern Guam because the surviving species were localized and rare: nocturnal terrestrial insectivore-carnivores, diurnal omnivores, herbivores, and diurnal arboreal insectivores. With the exception of blue-tailed skinks, the vertebrate forest food web consisted of snakes preying on introduced rats, introduced skinks, and introduced geckos (the house gecko may be a prehistoric introduction).

NONVERTEBRATE IMPACTS   Whatever the cause of this radical rearrangement of the food web, it seems likely that repercussions will be felt outside the vertebrate community. Although there may have been some compensatory increases in insectivory by the surviving vertebrate and invertebrate insectivores, one mammalian and many avian insectivores were lost, presumably increasing insect abundances at some cost to crop production. Newly introduced insects may also find it easier to colonize Guam. Pollination and seed dispersal services of native birds to native plants were also lost (18, 53). Subjectively, Guam seems to have a much higher density of web-building spiders than nearby islands; this is consistent with experiments in the West Indies on the importance of predation and competition between spiders and lizards on islands (19, 83) but has not been investigated on Guam. Spiders that place conspicuous filaments (stabilimenta) in their webs, presumably to avoid bird damage to their webs, do so less often on Guam than on nearby islands that have forest birds (39).

As a result of the loss of avian and mammalian herbivores on Guam, one would expect to see some reduction in herbivory, especially frugivory, which may be only partially offset by compensatory increases by invertebrate frugivores. Invertebrate frugivory can be a substantial economic burden on agriculture, so it is perhaps surprising that this phenomenon has not been investigated on Guam. The impact on native plant species of reduced vertebrate herbivory is harder to judge, and to our knowledge no one has attempted to separate the effects of this from the concurrent impacts of introduced ungulates (especially pig and deer) that we do not consider here.

TO WHAT EXTENT ARE FACTORS OTHER THAN INTRODUCED SPECIES RESPONSIBLE?

Of the factors that have been suggested as contributing to the Guam biodiversity crisis (introduced predators, habitat loss, contaminants, introduced diseases, competition from introduced species, and direct human exploitation), only introduced predators and habitat loss are believed to be generally important. Pesticide contamination may have played a role in depressing insectivore populations prior to about 1970, but it has not been linked to the loss of any species (34, 79), and pesticide contamination should have waned in importance prior to the bulk of extinctions in the 1980s. Significant problems with introduced diseases have not been discovered (80, 82). Competition between native and introduced lizards (9, 10) and birds (43) has been discounted but will be difficult to dismiss unequivocally. Direct human exploitation has been a concern primarily for the Marianas fruit bat (97), but this species continued to decline for a decade after elimination of direct exploitation pressure. Furthermore, the early bat losses in southern Guam (56) are so strikingly similar to those observed concurrently among birds that they seem likely to have been due to snake predation, which would have been more frequent but less obvious to human observers.

The role played by habitat loss is more subtle because of the strong interaction between the effects of introduced predators and habitat loss (i.e. the introduced species thrive in disturbed habitats), and the long evolutionary history native vertebrates in the Marianas have with catastrophic habitat disturbance following typhoons. Because the average time between typhoons on Guam is shorter than the interval needed to restore primary forest after a typhoon, few or no native species are rigidly dependent on old growth (24). Still, the issue is complex enough to warrant explicit analysis.

Are any of the native vertebrates restricted to old native forest? Habits are unknown for a few species (e.g. little Mariana fruit bat, *Pteropus tokudae*), but the others were known to occur in disturbed habitats on either Guam or a comparable nearby island. The spotted-belly gecko (*Perochirus ateles*) may now be restricted to native forest on Saipan, perhaps owing to competition with recently introduced geckos. On Cocos Island, however, this gecko is frequently seen on buildings, a highly disturbed habitat (47). The remaining species are well documented to occur in both native forest and disturbed habitats, although they may reach higher densities in native forest (13, 14). A few species, such as the Micronesian honeyeater (*Myzomela rubrata*), do better in disturbed habitats (16).

Could postwar species losses on Guam have occurred as a result of habitat destruction? This seems improbable. No species were lost as a result of habitats having been impacted by the war. During the following period of habitat regrowth, nearly every species contracted in distribution and most were ultimately extirpated. Three caveats are appropriate: (*a*) The lack of importance of forest habitat deterioration does not apply to wetland species, which suffered severe and potentially extinction-causing loss of habitat in the postwar period (89, 92); (*b*) The presence of so much disturbed habitat facilitated the irruptions of all five of the postwar introductions (63); (*c*) Guam's rapidly expanding human population is now converting swaths of former forest into suburban landscapes, which will place a severe constraint on future species restoration. Nonetheless, we have not seen compelling evidence that the loss of any of Guam's forest vertebrates was primarily attributable to habitat degradation.

Considering all nine criteria for appraising the impacts of an introduced species, we believe that the brown treesnake was responsible for the extirpation or decline of virtually all of the native forest birds (Table 1). This excludes the wetland birds but includes sea birds such as fairy terns (*Gygis alba*) that nested in forested areas. Pesticides and habitat deterioration may have played a role in population declines, especially in the immediate postwar period, but outright extirpation is consistent only with the high predatory impact of the snake. The record for mammals is much less clear, but the current demographic strains being experienced by the Mariana fruit bat do seem to be a result of snake predation. Insufficient information exists to evaluate the causes of the extirpation of the other two bats.

The situation for lizards is more complex. The musk shrew irruption probably greatly diminished the pelagic gecko and may have impacted the Mariana skink. It may have played a role in the elimination of several poorly known skinks from southern Guam not discussed in this paper (the tide-pool skink *Emoia atrocostata*, the azure-tailed skink *Emoia cyanura*, and the blue-tailed copper-striped skink *Emoia impar*). *Boiga irregularis* was probably a key player in, if not the sole cause of, the reduction or extirpation of the snake-eyed skink (*Cryptoblepharus poecilopleurus*, which is restricted to southern Guam and is too poorly known for comprehensive evaluation), the spotted-belly gecko, and the introduced mutilating gecko and oceanic gecko. The oceanic gecko probably reduced population densities of the spotted-belly gecko before the brown treesnake arrived (61).

WHY SUCH AN EXTREME EFFECT?

Four explanations have been suggested to account for the unusual number and taxonomic breadth of extinctions on Guam. The first is that the resident species, having evolved in an essentially predator-free environment (Figure 1), lacked the predator defenses that would have spared mainland species from extirpation (42, 81). Based on comparison with mainland areas that have experienced introductions of snakes, and on differences between Guam's native and introduced prey species in vulnerability to snake predation, this argument is supported (73). No introduced species is known to have been extirpated during the postwar extinctions; in contrast, almost all of the native species either declined to near the limit of population viability or were lost.

With one exception, however, predator defenses of Guam species have not been tested. Campbell (6) compared the defensive behaviors of two Guam geckos, the house gecko (probably a prehistoric introduction) and the mourning gecko (native). The native species rarely (37%) fled from the scene of a simulated predator approach, whereas the introduced species routinely (76%) avoided the sites of nighttimes disturbances ($P < 0.001$ by Fisher test). This is an example of the well-known phenomenon of island tameness (62).

Pimm (58, 59) and others (46, 75) emphasized the trophic role of less vulnerable prey species. They argued that the abundance of prey species that were capable of sustaining brown treesnake predation (primarily introduced species) increased the abundance of the snake to a level at which the native prey species, taken opportunistically, could not persist. In this regard it is noteworthy that the various introduced species that irrupted in the postwar period (Figures 2 and 3) thrived, and in some cases increased their abundance, in the face of heavy predation by the brown treesnake. Had the introduced species not been present, it seems probable that the snake could not have attained nearly as high a density as it did. Indeed, had the snake been limited to feeding on native species, it would have had virtually nothing to eat in 1985 in northern Guam. Instead the snake reached maximum densities of around 100/ha.

The brown treesnake possesses certain characteristics that make it a particularly attractive candidate for disrupting island ecosystems. It is nocturnal, a generalist feeder, and an exceptional climber (29). Generalist predators are known to be especially problematic (22, 58). Nocturnal predators exploit the inability of most passerines to fly safely after dark. And the exceptional climbing skills of this snake give it access to virtually all refugia and nest sites except for the swiftlet nests glued to the ceilings of smooth-walled caves. Ebenhard (22) tied arboreality to greater impacts of introduced species, especially on islands.

The snake also possesses certain characteristics that limit its impact on the native fauna (90). For example, it is a relatively slow reproducer and has a longer generation time than most of the endangered species; thus it should not greatly overshoot the carrying capacity of the prey. This particular snake does possess certain attributes that make extinctions of island prey more likely (nocturnality, arboreality), but these attributes are not unique among snakes (73). At least 20 species within the genus *Boiga* share the problem-causing attributes (33), as do many other nocturnal arboreal snakes. On balance, it is difficult to ascribe the Guam crisis solely to the attributes of the invader snake.

A fourth hypothesis is that Guam has a uniquely vulnerable ecology. Being an oceanic island, Guam has a moderate climate, one that would not preclude the colonization of most invading species (73). It is also likely to have high densities of lizard prey. Our review of studies on lizard assemblage density strongly supports the generality of MacArthur's "excess density compensation" for island lizards (44), as tentatively indicated for islands in the Gulf of California by Case (8). For example, among 15 relatively simple assemblages (10 or fewer lizard and frog species) on oceanic islands, the mean biomass density of lizard assemblages averaged 16.30 kg/ha, whereas the comparable value for 23 tropical mainland locations was only 0.63 kg/ha. It is not clear, however, that oceanic islands generally have high biomasses of avian or mammalian prey (20, 44, 49), and in any event, neither of these attributes is unique to the island of Guam [Guam originally had a lower density of birds than comparable mainland areas (81)]. Moderate climates and excess density compensation should apply equally to all tropical oceanic islands.

In summary, the lack of coevolution between predator and prey was probably a major contributing factor to the severity of the Guam biodiversity crisis. In addition, the brown treesnake would not have reached such high densities if it had not had at its disposal a large and resilient prey base of introduced species. The introduced species did better in disturbed habitats; thus the native species suffered both from habitat loss and greater pressure from exotic predators in the disturbed habitat. Furthermore, the snake and the Guam environment were well suited to introduced species problems, but neither the snake nor the environment was unique; these problems could arise with any number of predators on any number of ecosystems in which coevolutionary experience between predator and prey is lacking.

WHAT WILL HAPPEN NOW?

Having heard that the native vertebrates of Guam are largely extirpated, many nonbiologists assume that the snake will run out of food. Our indications are that the food supply in northern Guam was more plentiful in 1995—well after the crest of the irruption—than it was in 1990 (closer in time to the crest), and we estimate that snake populations now averaging about 3 kg/ha can be easily supported by the remaining prey, even though important prey types are now limited to introduced species (Figure 3). We estimate that only seven prey species retain biomass densities of greater than 0.1 kg/ha. Both significant endotherm prey species in forested areas of northern Guam are introduced rats: the Polynesian rat (ca. 0.2 kg/ha) and black rat (ca. 3 kg/ha). In addition, geckos abound: the mutilating gecko (ca. 0.5 kg/ha), the house gecko (ca. 2.5 kg/ha), and the mourning gecko (ca. 2.5 kg/ha, but this species may be too small a prey for any but the youngest snakes). Finally, and most abundantly, there are leaf-litter skinks: the blue-tailed skink (ca. 6 kg/ha) and the vital curious skink (ca. 16 kg/ha). The adequate predator:prey ratio of 10:1 does not tell the whole story, of course. The recruitment rate of prey species is much greater than is that of the snake. Thus, there is no reason to predict that present snake populations cannot be sustained indefinitely on introduced prey. Furthermore, the pace of new species introductions has been accelerating over recent decades and it is probable that new food items will become available for the snake.

A new problem confronting the brown treesnake is that the food items upon which it now depends are primarily diurnal, terrestrial species of skinks that sleep in relatively sheltered locations during the night (M McCoid, personal communication). The curious skink becomes active at first light, which may increase its contact with the historically nocturnal brown treesnake (45). This may account for at least part of the greater representation of the curious skink in brown treesnake diets, in comparison to the later-arising blue-tailed skink. Whichever skink is taken, however, requires the snake to forage during the day on the ground. Perhaps this is the reason that brown treesnakes on Guam became significantly more diurnal and more terrestrial in the late 1980s when native endothermic prey vanished. While the growing diurnality of the snake has not been quantitatively verified except by the pattern of electrical power outages shown in Figure 4, anecdotal measures of activity (e.g. time when snakes are seen being caught in traps) are consistent with a profound change since the early 1980s. Snakes do not appear to be very active during the early afternoon (when skinks are also relatively quiescent), but total daytime activity approaches 50% of all movements.

In addition, significant ground-level activity was not observed in brown treesnakes on Guam prior to 1988 but is now the modal condition observed in some localities, especially of snakes large enough to depend on skink prey (65). It is difficult to know if other snake species would have had the behavioral flexibility shown by the brown treesnake to change in a few years from nocturnal arboreality to habits including diurnal terrestriality, but this foraging flexibility undoubtedly contributed to the severity of the impact the snake had on Guam wildlife.

Although skinks and geckos are abundant, they are small. Thus, the dietary shift of adult snakes from native endotherms (1980) to lizards (1990) resulted in a shift from food appropriate for adult snakes to food for juveniles. One would expect this to transform survivorship curves from high adult/low juvenile survivorship to low adult/high juvenile survivorship. Although age-stratified survivorship in brown treesnakes has not been measured, we have observed

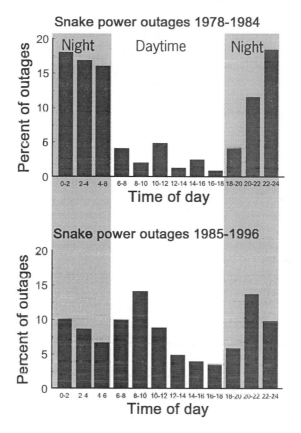

Figure 4 Change in diurnality of snake activity as measured by time of snake-induced power outages (1978–1996). Outages are attributed to a snake if an electrocuted snake was found at the site of the fault. Outages along lines not operated 24 h/day are omitted.

dramatic shifts in size distributions (71). Most notably, the percentage of the detected population that is mature has dropped from about 60% in the snake's native range, to 48% on Guam in the early 1980s, to about 25% on Guam since 1986 (77). The Guam values are surprisingly low for a snake with such a small clutch size (3–12 eggs). During 1985–1995, clutch sizes of brown treesnakes on Guam averaged about five, for which a typical snake would have a mature fraction of about 80% (55). There are problems with undersampling juvenile brown treesnakes, but this undersampling should produce a conservative error in this comparison with other snake species (67). Thus, although the brown treesnake has been able to maintain high population densities on Guam following the extirpation of native prey, it has done so at the cost of high adult mortality, probably caused by a shortage of large prey.

An additional potential source of change for the Guam population is human management. At the present time, management activities are local, influencing only the abundance and size distribution of snakes in the vicinity of port, airport, and cargo-handling facilities. However, future development of biological control agents or broadcast toxicants could result in population depletion islandwide (7). In addition, the technology for fencing snakes out of high priority wildlife management sites is maturing (72), and may soon be in place to eliminate or greatly reduce snake densities over significant areas of Guam forests.

Until islandwide snake control is operational, Guam will continue to be a source of propagules for accidental transport to other islands (28). Based on the climatic tolerances of the snake, the risk of colonization would be highest for tropical islands, but would include any locally mesic location that does not undergo a hard freeze (7). Based on the amount of commerce conducted through Guam ports, the areas of highest risk are Micronesian islands and Hawaii, followed by US West Coast ports and all US Gulf Coast locations with military facilities (30). The Caribbean possessions of the United States also receive goods, particularly military supplies and emergency equipment transferred to assist in hurricane recovery efforts. Based on the ecological interactions observed on Guam, risk factors for recipient locations include prey that lack coevolutionary experience with comparable predators, abundant vulnerable prey, and prey that share habitats with relatively invulnerable prey (e.g. species possessing coevolutionary experience with nocturnal arboreal snakes).

Taken together, these risk factors point to the Mariana, Hawaiian, and Caroline islands as being most at risk. The Mariana and Caroline islands are so similar to Guam that a brown treesnake colonization of those islands may be expected to play out in a manner very similar to that observed on Guam (albeit on a time scale commensurate with the size of the island). Hawaii is less predictable because its biotic community differs from Guam's in several major ways. First, there are introduced species for which there is no analog in the Marianas, such as mongoose (52). Mongoose now depress lizard populations on Hawaii, but would provide a high-value prey for large snakes. If their populations were to decline appreciably in the face of a brown treesnake infestation, the mongoose decline might release previously depressed populations of ground lizards and thereby provide juvenile snakes with an additional rich food source. Second, Hawaii is unique in that the native endotherms have already been eliminated from most low elevation sites by diseases and introduced species (95). The replacement of the native avifauna by several exceedingly abundant exotic birds (e.g. zebra doves, *Geopelia striata*) may accelerate snake population buildup.

Thus the pace and impacts of colonization are less predictable. The rate of colony expansion may be sufficiently slow to permit effective control actions to be implemented. Initial efforts to control an apparent colonization of Saipan have not yielded hoped-for successes, however. Additional research on the control of the brown treesnake may bring new tools to this very urgent task.

ACKNOWLEDGMENTS

We thank the US Departments of Defense and Interior for supporting this work. A host of volunteers and technicians helped us carry out the field work; we appreciate the labors of all. Earl Campbell was a major source of information and inspiration for much of our work. Kathy Dean-Bradley and Linda Wolfe provided indispensable computational and administrative assistance. We thank C Aguon, C Grue, K King, MJ McCoid, J Morton, G Pregill, R Ryder, S Vogt, and G Wiles for graciously sharing unpublished findings. Earl Campbell, Kathy Dean-Bradley, Gad Perry, and Renée Rondeau suggested improvements to the manuscript.

> **Visit the *Annual Reviews* home page at**
> **http://www.AnnualReviews.org**

Literature Cited

1. Baker RH. 1946. Some effects of the war on the wildlife of Micronesia. *Trans. N. Am. Wildl. Conf.* 11:207–13
2. Baker RH. 1946. A study of rodent populations on Guam, Mariana Islands. *Ecol. Monogr.* 16:393–408
3. Baker RH. 1951. The avifauna of Micronesia, its origin, evolution, and distribution. *Univ. Kans. Publ. Mus. Nat. Hist.* 3:1–359
4. Barbehenn KR. 1962. The house shrew on Guam. In *Pacific Island Rat Ecology*, ed. TI Storer, pp. 247–56. Honolulu: Bull. Bishop Mus. 225
5. Barbehenn KR. 1974. Recent invasions of Micronesia by small mammals. *Micronesica* 10:41–50
6. Campbell EW III. 1996. *The effect of brown tree snake* (Boiga irregularis) *predation on the island of Guam's extant lizard assemblages*. PhD thesis. Ohio State Univ., Columbus, OH. 83 pp.
7. Campbell EW III, Rodda GH, Fritts TH, Bruggers RL. 1998. An integrated management plan for the brown treesnake (*Boiga irregularis*) on Pacific islands. See Ref. 78, In press
8. Case TJ. 1975. Species numbers, density compensation and colonizing ability of lizards on islands in the Gulf of California. *Ecology* 56:3–18
9. Case TJ, Bolger DT. 1991. The role of introduced species in shaping the distribution and abundance of island reptiles. *Evol. Ecol.* 5:272–90
10. Case TJ, Bolger DT, Richman AD. 1992. Reptilian extinctions: the last ten thousand years. In *Conservation Biology: the Theory and Practice of Nature Conservation Preservation and Management*, ed. P Fiedler, S Jain, pp. 91–125. New York: Chapman & Hall
11. Conry PJ. 1987. Ecology of the Philippine turtle-dove on Guam. *Guam Div. Aquat. Wildl. Res. Tech. Rep. 6.* 69 pp.
12. Conry PJ. 1988. High nest predation by brown tree snakes on Guam. *Condor* 90: 478–82
13. Craig RJ. 1990. Foraging behavior and microhabitat use of two species of white-eyes (*Zosteropidae*) on Saipan, Micronesia. *Auk* 107:500–5
14. Craig RJ. 1992. Territoriality, habitat use and ecological distinctness of an endangered Pacific Island reed-warbler. *J. Field Ornithol.* 63:436–44

15. Craig RJ. 1994. Regeneration of native Mariana Island forest in disturbed habitats. *Micronesica* 26:97–106
16. Craig RJ. 1996. Seasonal population surveys and natural history of a Micronesian bird community. *Wilson Bull.* 108:246–67
17. Craig RJ, Taisacan E. 1994. Notes on the ecology and population decline of the Rota bridled white-eye. *Wilson Bull.* 106:165–69
18. Denton GRW, Muniappan R, Marutani M. 1991. The distribution and biological control of *Lantana camara* in Micronesia. *Micronesica Suppl.* 3:71–81
19. Dial R, Roughgarden J. 1995. Experimental removal of insectivores from rain forest canopy: direct and indirect effects. *Ecology* 76:1821–34
20. Diamond JM. 1970. Ecological consequences of island colonization by southwest Pacific birds. II. The effect of species diversity on total population density. *Proc. Natl. Acad. Sci. USA* 67:1715–21
21. Dye T, Steadman DW. 1990. Polynesian ancestors and their animal world. *Am. Sci.* 78:207–15
22. Ebenhard T. 1988. Introduced birds and mammals and their ecological effects. *Swed. Wildl. Res. Vilrevy* 13:1–107
23. Eldredge LG. 1988. Case studies of the impacts of introduced animal species on renewable resources in the U.S.-affiliated Pacific islands. In *OTA Commissioned Papers, Integrated Renewable Resource Management for U.S. Insular Areas ("Islands Study" 1987)*, Vol. 1, ed. CK Imamura, E Towle, pp. A.26–46. Washington, DC: US Cong. Off. Technol. Assess.
24. Engbring J, Pratt HD. 1985. Endangered birds in Micronesia: their history, status, and future prospects. In *Bird Conservation*, ed. SA Temple, pp. 71–105. Madison: Univ. Wis. Press
25. Engbring J, Ramsey FL. 1984. Distribution and abundance of the forest birds of Guam: results of a 1981 survey. *US Fish Wildl. Serv., FWS/OBS-84/20.* 54 pp.
26. Engbring J, Ramsey FL, Wildman VJ. 1986. *Micronesian forest bird survey, 1982. Saipan, Tinian, Aguiguan, and Rota.* Honolulu: Rep. US Fish Wildl. Serv.
27. Fosberg FR. 1960. The vegetation of Micronesia. *Bull. Am. Mus. Nat. Hist.* 119:1–75
28. Fritts TH. 1988. The brown tree snake, *Boiga irregularis*, a threat to Pacific Islands. *Biol. Rep.* 88(31). Washington, DC: US Fish Wildl. Serv. 36 pp.
29. Fritts TH, Chiszar D. 1998. Snakes on electrical transmission lines: patterns, causes, and strategies for reducing electrical outages due to snakes. See Ref. 78, In press
30. Fritts TH, McCoid MJ, Gomez DM. 1998. Dispersal of snakes to extralimital islands: incidents of the brown treesnake, *Boiga irregularis*, dispersing to islands in ships and aircraft. See Ref. 78, In press
31. Fritts TH, Rodda GH. 1995. Invasions of the brown tree snake. In *Our Living Resources: a Report to the Nation on the Distribution, Abundance, and Health of U.S. Plants, Animals, and Ecosystems*, ed. E LaRoe, G Farris, C Puckett, P Doran, M Mac, pp. 454–56. Washington, DC: US Natl. Biol. Serv.
32. Goldschmidt T, Witte F, Wanink J. 1993. Cascading effects of the introduced Nile perch on the detritivorous/phytoplanktivorous species in the sublittoral areas of Lake Victoria. *Conserv. Biol.* 7:686–700
33. Greene HW. 1989. Ecological, evolutionary, and conservation implications of feeding biology in Old World cat snakes, genus *Boiga* (*Colubridae*). *Proc. Calif. Acad. Sci. Ser. 4* 46:193–207
34. Grue CE. 1985. Pesticides and the decline of Guam's native birds. *Nature* 316:301
35. Jackson WB. 1967. Productivity in high and low islands with special emphasis to rodent populations. *Micronesica* 3:5–15
36. Jaffe M. 1994. *And No Birds Sing.* New York: Simon and Schuster. 283 pp.
37. Jenkins JM. 1983. The native forest birds of Guam. *Am. Ornith. Union Ornithol. Monogr. 31.* 61 pp.
38. Kardong KV, Smith PR. 1991. The role of sensory receptors in the predatory behavior of the brown tree snake, *Boiga irregularis* (Squamata: Colubridae). *J. Herpetol.* 25:229–31
39. Kerr AM. 1993. Low frequency of stabilimenta in orb webs of *Argiope appensa* (Araneae: Araneidae) from Guam: an indirect effect of an introduced avian predator? *Pac. Sci.* 47:328–37
40. Kobayashi K. 1970. Observation of the birds of the Mariana Islands. *Tori* 20(88):24–29
41. Lemke TO. 1986. Distribution and status of the Sheath-tailed Bat (*Emballonura semicaudata*) in the Mariana Islands. *J. Mammal.* 67:745–46
42. Loope LL, Mueller-Dombois D. 1989. Characteristics of invaded islands, with special reference to Hawaii. In *Biological Invasions: a Global Perspective*, ed. JA Drake, HA Mooney, F diCastri, RH

Groves, FJ Kruger, et al, pp. 257–80. Chichester, UK: Wiley

43. Maben AF. 1982. *The feeding ecology of the black drongo* Dicrurus macrocercus *on Guam*. MS thesis. Calif. State Univ., Long Beach. 87 pp.

44. MacArthur RH, Diamond JM, Karr JR. 1972. Density compensation in island faunas. *Ecology* 53:330–42

45. McCoid MJ. 1997. *Interactions of* Carlia *cf.* fusca (Scincidae) *with the herpetofauna of Guam*. MS thesis. Texas A&M, Kingsville, Tex. 98 pp.

46. McCoid MJ. 1998. Established exotic reptiles and amphibians of the Mariana Islands. See Ref. 78, In press

47. McCoid MJ, Hensley RA. 1994. Distribution and abundance of *Perochirus ateles (Gekkonidae)* in the Mariana Islands. *Herpetol. Rev.* 25:97–98

48. McCoid MJ, Rodda GH, Fritts TH. 1995. Distribution and abundance of *Emoia slevini (Scincidae)* in the Mariana Islands. *Herpetol. Rev.* 26:70–72

49. Moller H, Craig JL. 1987. The population ecology of *Rattus exulans* on Tiritiri Matangi Island, and a model of comparative population dynamics in New Zealand. *N. Z. J. Zool.* 14:305–28

50. Moore PH. 1977. An ecological survey of pristine terrestrial communities on Guam. In *Guam Coastal Manag. Prog. Tech. Rep.*, Vol. 1, ed. D Bonvouloir, P McMakin, pp. 1 45. Agana, Guam: Govt. Guam Bur. Plan.

51. Morison SE. 1953. *New Guinea and the Marianas, March 1944–Aug. 1944*. Boston: Little, Brown. 420 pp.

52. Moulton MP, Pimm SL. 1986. Species introduction to Hawaii. In *Ecology of Biological Invasions of North America and Hawaii*, ed. HA Mooney, JA Drake, pp. 231–49. New York: Springer-Verlag

53. Muniappan R. 1988. Biological control of the weed, *Lantana camara* in Guam. *J. Pl. Prot. Trop.* 5:99–101

54. Palomo T. 1984. *An Island in Agony*. Guam: Bank of Guam. 261 pp.

55. Parker WS, Plummer MV. 1987. Population ecology. In *Snakes: Ecology and Evolutionary Biology*, ed. RA Siegel, JT Collins, SS Novak, pp. 253–301. New York: Macmillan

56. Perez GSA. 1972. Observations of Guam bats. *Micronesica* 8:141–49

57. Peterson GD Jr. 1956. *Suncus murinus*, a recent introduction to Guam. *J. Mammal.* 37:278–79

58. Pimm SL. 1987. Determining the effects of introduced species. *Trends Ecol. Evol.* 2:106–8

59. Pimm SL. 1987. The snake that ate Guam. *Trends Ecol. Evol.* 2:293–95

60. Pimm SL. 1992. *The Balance of Nature: Ecological Issues in the Conservation of Species and Communities*. Chicago: Univ. Chicago Press. 434 pp.

61. Pregill GK. 1998. Squamate reptiles from prehistoric sites in the Mariana Islands, Micronesia. *Copeia* 1998:64–75

62. Quammen D. 1996. *The Song of the Dodo*. New York: Scribner's

63. Ramakrishnan PS, Vitousek PM. 1989. Ecosystem-level processes and the consequences of biological invasions. In *Biological Invasions: a Global Perspective*, ed. JA Drake, HA Mooney, F diCastri, RH Groves, FJ Kruger, et al, pp. 281–300. Chichester, UK: Wiley

64. Ramsey FL, Harrod LA. 1995. *Results from avian surveys of Rota and Tinian Islands, Northern Marianas, 1982 and 1994*. Honolulu: Rep. US Fish Wildl. Serv.

65. Rodda GH. 1992. Foraging behaviour of the brown tree snake, *Boiga irregularis*. *Herpetol. J.* 2:110–14

66. Rodda GH, Campbell EW III, Derrickson SR. 1998. Avian conservation research needs in the Mariana Islands, western Pacific Ocean. In *Avian Conservation: Research and Management*, ed. J Marzluff, R Sallabanks. Washington, DC: Island Press

67. Rodda GH, Fritts TH. 1992. Sampling techniques for an arboreal snake, *Boiga irregularis*. *Micronesica* 25:23–40

68. Rodda GH, Fritts TH. 1992. The impact of the introduction of the brown tree snake, *Boiga irregularis*, on Guam's lizards. *J. Herpetol.* 26:166–74

69. Rodda GH, Fritts TH. 1996. *Reptiles and Amphibians: Faunal Survey for the Ordnance Annex, Naval Activities, Guam*. Honolulu: US Fish Wildl. Serv.

70. Rodda GH, Fritts TH. 1997. *Absolute population densities of Saipan lizards*. Saipan, Mariana Islands: Rep. Div. Fish Wildl.

71. Rodda GH, Fritts TH. 1997. *Modeling of brown tree snake size distributions highlights sampling problems and high adult mortality*. Presented at Annu. Meet. Am. Soc. Ichthyol. Herpetol., 77th, Seattle, p. 254. Seattle: Am. Soc. Ichthyol. Herpetol.

72. Rodda GH, Fritts TH, Campbell EW III. 1998. The feasibility of controlling the brown treesnake in small plots. See Ref. 78, In press

73. Rodda GH, Fritts TH, Chiszar D. 1997. The disappearance of Guam's wildlife;

new insights for herpetology, evolutionary ecology, and conservation. *Bio-Science* 47:565–74

74. Rodda GH, Fritts TH, Conry PJ. 1992. Origin and population growth of the brown tree snake, *Boiga irregularis*, on Guam. *Pac. Sci.* 46:46–57

75. Rodda GH, Fritts TH, McCoid MJ, Campbell EW III. 1998. An overview of the biology of the brown treesnake, *Boiga irregularis*, a costly introduced pest on Pacific islands. See Ref. 78, In press

76. Rodda GH, Fritts TH, Reichel JD. 1991. The distributional patterns of reptiles and amphibians in the Mariana Islands. *Micronesica* 24:195–210

77. Rodda GH, McCoid MJ, Fritts TH, Campbell EW III. 1998. Population trends and limiting factors in *Boiga irregularis*. See Ref. 78, In press

78. Rodda GH, Sawai Y, Chiszar D, Tanaka H, eds. 1998. *Problem Snake Management: Habu and Brown Treesnake Examples*. Ithaca, NY: Cornell Univ. Press. In press

79. Savidge JA. 1985. Pesticides and the decline of Guam's native birds. *Nature* 316:301

80. Savidge JA. 1986. *The role of disease and predation in the decline of Guam's avifauna*. PhD thesis. Univ. Ill., Champaign-Urbana. 79 pp.

81. Savidge JA. 1987. Extinction of an island forest avifauna by an introduced snake. *Ecology* 68:660–68

82. Savidge JA, Sileo L, Siegfried LM. 1992. Was disease involved in the decimation of Guam's avifauna? *J. Wildl. Dis.* 28:206–14

83. Schoener TW, Spiller DA. 1987. Effects of lizards on spider populations: manipulative reconstruction of a natural experiment. *Science* 236:949–52

84. Shine R. 1991. Strangers in a strange land: ecology of Australian colubrid snakes. *Copeia* 1991, pp. 120–31

85. Steadman DW. 1989. Extinction of birds in eastern Polynesia: a review of the record and comparison with other Pacific island groups. *J. Archaeol. Sci.* 16:177–205

86. Steadman DW. 1992. Extinct and extirpated birds from Rota, Mariana Islands. *Micronesica* 25:71–84

87. Steadman DW. 1995. Prehistoric extinctions of Pacific Island birds: biodiversity meets zooarchaeology. *Science* 267:1123–31

88. Steadman DW. 1995. *Determining the Natural Distribution of Resident Birds in the Mariana Islands*, phase 1. *Prelim. Rep. (22 Apr. 1995) Results Field Work Conduct. Jun.–Jul. 1994*, US Fish Wildl. Serv., Honolulu, Hawaii

89. Stinson DW, Ritter MW, Reichel JD. 1991. The Mariana common moorhen: decline of an island endemic. *Condor* 93:38–43

90. Townsend CR. 1991. Exotic species management and the need for a theory of invasion ecology. *N. Z. J. Ecol.* 15:1–3

91. US Fish and Wildlife Service. 1992. *Recovery Plan for the Mariana Islands Population of the Vanikoro Swiftlet*, Aerodramus vanikorensis bartschi. Portland, OR: US Fish Wildl. Serv. 49 pp.

92. US Fish and Wildlife Service. 1992. *Recovery Plan for the Mariana Common Moorhen (Gallinule)*, Gallinula chloropus guami. Portland, OR: US Fish Wildl. Serv. 55 pp.

93. US Fish and Wildlife Service. 1997. *Technical/Agency Draft Recovery Plan for the Micronesian Megapode*. Portland, OR: US Fish Wildl. Serv. 73 pp.

94. US National Research Council. 1997. *The Scientific Bases for Preservation of the Mariana Crow*. Washington, DC: Natl. Acad. Press. 91 pp.

95. van Riper C III, van Riper SG, Goff ML, Laird M. 1986. The epizootiology and ecological significance of malaria in Hawaiian land birds. *Ecol. Monogr.* 56:327–44

96. Wheeler ME. 1979. The Marianas fruit bat: management history, current status and future plans. *Cal.-Nev. Wildl. Trans. 1979*, pp. 149–65

97. Wiles GJ. 1987. The status of fruit bats on Guam. *Pac. Sci.* 41:148–57

98. Wiles GJ, Aguon CF, Davis GW, Grout DJ. 1995. The status and distribution of endangered animals and plants in northern Guam. *Micronesica* 28:31–49

99. Wiles GJ, Amerson AB Jr, Beck RE Jr. 1989. Notes on the herpetofauna of Tinian, Mariana Islands. *Micronesica* 22:107–18

100. Wiles GJ, Guerrero JP. 1996. Relative abundance of lizards and marine toads of Saipan, Mariana Islands. *Pac. Sci.* 50:274–84

101. Wiles GJ, Rodda GH, Fritts TH, Taisacan EM. 1990. Abundance and habitat use of reptiles on Rota, Mariana Islands. *Micronesica* 23:153–66

Annu. Rev. Ecol. Syst. 1998. 29:141–77

EVOLUTION OF HELPING BEHAVIOR IN COOPERATIVELY BREEDING BIRDS

Andrew Cockburn

Evolutionary Ecology Group, Division of Botany and Zoology, Australian National University, Canberra ACT 0200 Australia; e-mail: andrew.cockburn@anu.edu.au

KEY WORDS: inclusive fitness, cooperative breeding, polyandry, kin selection, avian mating systems

ABSTRACT

It has recently been argued that the paradox of helping behavior in birds has been solved (73). This optimism may be premature. I argue that there is no obvious dichotomy between cooperative societies based on natal philopatry and the formation of extended families, and those formed via recruitment of unrelated individuals into coalitions. Tests of the effect of helping behavior suggest that kinship may have been overemphasized for male helpers but underestimated for females. The first studies applying molecular techniques to resolve genealogy in these societies suggest that reproductive sharing occurs commonly across all types of social organization. Incest avoidance may be an important constraint on sharing in families, but molecular techniques have thus far been inappropriate to assess its importance. The interests of males and female helpers may be quite different because females often have less opportunity to inherit a territory vacancy on the death of the breeder, less opportunity to court mates by helping them, and less opportunity to share reproduction without perturbing the size of the brood. We still have only a weak understanding of sex biases in helping behavior.

INTRODUCTION

Cooperative breeding in birds describes social systems wherein more than two individuals combine to rear a single brood of young. Such systems generate the paradox that some individuals, called helpers, are caring for young that are not their own. Studies of cooperative breeders include some of the most

141

0066-4162/98/1120-0141$08.00

impressive long-term field research ever undertaken (19, 137, 223, 240). Most studies are guided by a widely accepted framework involving dissociation of two questions: (*a*) Why do birds remain philopatric instead of dispersing to breed independently? (*b*) Why do these philopatric birds provide care to the offspring raised on their territory, instead of ignoring or seeking to harm them? In a recent review, Emlen (73) argued that these questions are largely answered. He suggested (73:229) that the original paradox disappeared with "the widespread confirmation that: (i) helpers frequently *do* improve their chances of becoming breeders by staying at home and helping temporarily; and (ii) they frequently *do* obtain large indirect genetic benefits by helping to rear collateral kin."

Emlen (71, 73, 80) expanded these arguments to develop a "theory of the family," which attempts to predict how mutual interests and conflicts between family members will be resolved. One limitation to this interesting approach is that it could potentially divert us from recognizing parallels between families and societies formed through coalitions of unrelated individuals, of which the best known example is the dunnock, *Prunella modularis* (44).

In the following review, I suggest that these parallels are of great importance and that we are poised for a reinvigoration of the study of cooperation through a number of conceptual, empirical, and technical advances. Advances include (*a*) greater general recognition of the separate interests of males and females in avian mating systems (43) and the complexity of female reproductive tactics (90); (*b*) the use of the new techniques of molecular ecology that resolve genealogical relationships among individuals (23) and allow the gender of young birds to be determined (97); and (*c*) new studies of cooperative societies that are not organized as families of more or less related individuals. My review assesses the evidence for the hypotheses that have been used to explain helping behavior and why help is sometimes offered predominantly by only one sex.

IS THERE A DICHOTOMY BETWEEN COOPERATIVE POLYANDRY AND COOPERATIVE FAMILIES?

Hartley & Davies (102) suggested that it is heuristically useful to distinguish between cooperative polyandry (wherein several males compete for matings with a single female) from help based on collateral kinship (wherein the helpers are full or partial sibs, or uncles or aunts, of the brood that they help rear).

These authors concede the existence of cases that blur this distinction, particularly where dominant males and birds that have dispersed to a helper role compete to mate with the female they help (e.g. 202). However, I suggest that in addition to older reports of societies that fit uncomfortably in either of these frameworks, e.g. groove-billed anis, *Crotophaga sulcirostris* (229), new reports of cooperative societies reveal extraordinary combinations of helping by unrelated and related individuals, and they suggest that any attempt to erect a dichotomy is an artificial and ultimately unhelpful one. Three case studies illustrate this continuum.

White-winged trumpeters, *Psophia leucoptera*, are denizens of wet forest in Peru and have taxonomic affinities to cranes. They breed in groups, which typically comprise three unrelated males, two unrelated females, and birds less than two years of age that are the progeny of the dominant female and possibly several of the adult males, as all adult males copulate with the dominant female (65, 219, 220). There is a distinct dominance hierarchy among both adult males and females. All birds except the subordinate adult female contribute extensively to feeding and defense of chicks. Subordinate adult males frequently change groups, may help on several territories, and may return to a territory that they have previously abandoned. Subordinate females sometimes arrive on a territory in groups and are met with violence from the dominant female, hindering any attempt to approach the nest. The dominant male will also attack the subordinate female if she solicits copulations from him. However, once accepted, the subordinate female is unlikely to leave and inevitably inherits dominant status on the death of the senior female.

Riflemen, *Acanthisitta chloris*, are small suboscine passerines from New Zealand. Pairs typically produce two broods each season. Once the first brood has been hatched for four days, the pairs are assisted by two sorts of male helpers, to which they are generally unrelated (217, 218). Casual helpers visit the nests of a number of pairs but provision at each nest only occasionally. Regular helpers assist only a single pair and feed at a rate comparable to that of the adults that produced the brood. These helpers do not generally assist with the second brood, which instead are provisioned by the young from the first brood. Regular helpers have a very high chance of later pairing with the first brood progeny that they have helped, while casual helpers appear well placed to replace any males that die during the breeding season.

White-browed scrubwrens, *Sericornis frontalis*, are common small oscine passerines that live in a variety of habitats in Australia. They live in year-round territories and breed in pairs or groups comprising a single hen and between two and four males (162, 163, 238). There is a clear dominance hierarchy among males. Most males are group members because of natal philopatry, but group rearrangements and natal dispersal mean that male groups can also comprise coalitions of unrelated males. Subordinate males related to the female generally do not provision the young and gain no access to paternity. However, subordinates unrelated to the female are likely both to provision her offspring and to obtain fertilizations.

These case studies not only illustrate the difficulties in deciding whether a society is a family or a coalition of unrelated individuals but also illustrate some important issues that are not always addressed in analyses of help. (*a*) Members of the same social group may obtain very different sorts of benefits from helping; (*b*) some group members in avian societies decline to help yet are apparently tolerated by adults; (*c*) help is sometimes declined by adult birds, despite the apparent willingness of potential helpers to provision or defend young; (*d*) the question of why birds join groups in order to help must be considered in parallel with the question of why some birds remain philopatric in their natal territory; (*e*) some individuals help at several nests; (*f*) birds do not always restrict care to individuals to which they are most closely related and may do the opposite.

WHY DO HELPERS HELP?

Costs and benefits have been reviewed previously on a number of occasions (19, 70, 75, 87, 138). The question "why help?" can be approached by means of several levels of analysis, which have been thoughtfully reviewed by Koenig & Mumme (138).

Help, as we have already seen, is not inevitable. Non-helping individuals occur in societies where subordinates are typically unrelated males (222), related males (162), and unrelated females (220), and where both males and females provision (76). More remarkably, group membership is not necessary for the provision of help. For example, in the moustached warbler, *Acrocephalus melanopogon*, non-territorial floaters often intrude on the territory when the home male is away from the nest and assist the females in incubation, feeding of nestlings, and defense of chicks (83). Help and philopatry are clearly not synonymous.

Costs of Help

One of the most stimulating recent contributions to the debate on helping behavior has been the suggestion by Jamieson (122, 123) that helping is essentially neutral and is a reflection of a general behavioral response in adult birds to the presence of begging young. According to this argument, helping per se is unselected but is a reflection of a selection for provisioning behavior by reproductive birds. Helping therefore emerges as a direct consequence of the presence of supernumeraries near a nest. There is little doubt that birds can be induced to provision unrelated young, and this forms the selective pressure for interspecific brood parasitism. Jamieson (123) suggests that, before proceeding to test adaptive explanations of helping, all studies should first test the null hypothesis that provisioning by helpers is undifferentiated from provisioning by adults. Some studies have attempted tests of this sort and failed to demonstrate any difference (62). However, other studies show that helping behavior differs from dominant behavior in a way that is not simply explained by age (75). Most importantly, many societies contain birds that decline to help despite the presence of begging young, and often the context of refusal to help is well understood (78).

The costs of providing care to offspring have been extensively investigated in the broader literature on parental care (31, 90, 133). Costs include reduced body condition, reduced survival, and reduced future fecundity. Birds are sensitive to their own ability to provide care. They adjust their level of investment accordingly to mitigate these costs (e.g. 185). Apart from obvious differences such as egg-laying, there are costs to care that affect the sexes differently. For example, male great tits suffer increased parasite burdens when brood sizes are increased, while females do not (209).

If these costs are also experienced by helpers, they should provide one restraint on the expression of unselected helping. The costs have been little investigated. In the most compelling analysis, stripe-backed wren helpers matched for sex and from the same cooperative group in the same year showed lower survival if they had provisioned at a high rate (201). One promising approach to these questions is to attempt to induce help in birds that do not normally provision, by reducing the costs they suffer. The first successful induction of helping resulted from enhancement of food availability to juvenile moorhen, *Gallinula chloropus* (66). Perhaps the most comprehensive set of evidence that helping can be costly comes from white-winged choughs (107). Helpers contribute to incubation only when group size is small, and they lose weight in proportion to the amount of time they spend incubating. Low weight compromises survival over winter (112). Fledglings are helped for up to six months but provisioning ceases in winter, presumably in response to a decline in food availability. However, provisioning can then be switched back on by food supplementation (39).

ADAPTIVE EXPLANATIONS FOR HELP

Adaptive reasons for help fall into six classes (Table 1): (*a*) enhanced production of nondescendant kin; (*b*) payment of rent, which allows access to other benefits of living on a territory or in a group; (*c*) direct access to parentage; (*d*) enhancement of the territory or group size in a way that improves later opportunity for direct reproduction; (*e*) enhancement of social circumstances via formation of alliances that improve the prospect of reproduction; or (*f*) acquisition of skills or prolonged maturation that facilitates later reproduction.

Table 1 Benefits of helping behavior for helpers or subordinates in cooperatively breeding birds

Benefit	Gender of helper receiving the benefit?	Family structure (relatedness required)?	Effects on the dominant or senior member of the group
Increased production of collateral kin			
a. Immediate benefit	Both	Yes	Positive
b. Deferred benefit	More likely for gender that does not inherit the territory (often female)	Yes	Positive
Payment-of-rent or mutualism			
a. Access to the benefits of group living (e.g. efficient foraging, coordinated vigilance)	Both	No	Positive, though there can be conflicts of interest over optimum group size
b. Inheritance of the territory or part of the territory, or obtaining a platform for dispersal to nearby breeding vacancies	Inheritance often only for males, both benefit from short-distance dispersal	If due to parental facilitation	Can be beneficial if the subordinates are kin
Access to mating opportunities			
a. Shared reproduction within the group	Both	No	Can be negative for the gender of the helper; but often positive for the opposite gender to the helper
b. Access to reproduction outside the group	Both	No	Neutral or negative
c. Courtship of future mates	Only known for males	No	Could be useful to females, but harmful for males
Improvement of local conditions			
a. Establishment and enlargement of groups enhances territory size or defense, enhancing reproduction or making it possible	Both	Yes	Can be obligatory but negative, have mixed benefits, or be highly beneficial
b. Helping increases the chances that the helper will gain the services of the helped young as a helper	Both	Yes	Beneficial
Establishment of strategic alliances and enhancement of "privilege"	Both	No	Could be neutral, but can also be harmful to adults and senior subordinates
Improved skills			
a. Learning skills for reproduction by copying others or practice	Both	No	May be neutral, or beneficial through enhanced productivity and "cheap" parental investment
b. Acquisition of skills during prolonged parental investment	Both	Yes	As above

Enhanced Production of Nondescendant Kin

IMMEDIATE BENEFITS   The central theme of Emlen's theory of the family is the importance of production of nondescendant kin through assisting the reproduction of relatives (71). Testing this hypothesis has two necessary elements. First, it should be possible to measure a benefit of the effect of help on the productivity of offspring by the recipients of help. Second, helpers should be related to the individuals they assist.

Early attempts to demonstrate that help enhanced the production of nondescendant kin revolved around measuring the difference between the productivity of pairs with and without helpers. These attempts became well entrenched as textbook examples of the application of Hamilton's rule [$rB > C$ (16, 68, 93)].

Very strong correlations between helper number and productivity have sometimes been reported, e.g. in white-fronted bee-eaters, *Merops bullockoides* (77); *Campylorhynchus* wrens (200); white-winged choughs, *Corcorax melanorhamphos* (110). In many other highly social species, such correlations are absent, e.g. *Turdoides* babblers (87, 171, 248). However, positive correlations are of negligible use in assessing the benefits of help to either the donor or recipient (17). This is because dissecting the direction of causation is difficult, particularly for territorial species, wherein it is impossible to distinguish direct effects of helpers on productivity from an effect of territory or parental quality on parental output, in turn leading to increased numbers of helpers. Even without the complexity of helpers, statistical attempts to disentangle the relative importance of territory quality and parental ability in non-cooperative species have proved difficult and have sometimes generated contradictory results (e.g. 89, 119).

A variety of attempts have been made to exclude statistically the effect of territory or parental quality. The most convincing of these involve comparing the performance of the same individuals on the same territory with and without helpers, with due care to control for improvement of ability with age (Table 2). Multivariate modelling has probably controlled adequately for territory quality in some cases (118) but has certainly not in others (81). More seriously, I do not feel that parental ability has ever been modelled using appropriate statistical procedures, which would involve treating the identity of the pair as a random variable in a mixed model (4) and controlling for the age of both participants.

It is also unclear whether strong conclusions can be drawn from a lack of correlation between help and productivity. Where help is facultative, it may be provided only where it is needed, leading to equivalent success for groups in both classes (163). For example, Reyer & Westerterp (208) showed that they could induce pied kingfishers, *Ceryle rudis*, to accept unrelated helpers by experimentally increasing brood size, minimizing costs associated with the enlarged brood.

Table 2   Attempts to demonstrate an effect of helping on reproductive success of offspring, which do not rely on a correlation or multiple regression approach

Treatment (Species)	Effect	Sex of helper	Source
Experimental removal of helpers			
Grey-crowned babbler *Pomatostomus temporalis*	Positive	Both	21
Florida scrub jay *Aphelocoma coerulescens*	Positive	Both	178
Seychelles warbler *Acrocephalus sechellensis*	Positive in good habitats, negative in other habitats	Female	145, 147
Moorhen *Gallinula chloropus*	None	Unsexed juveniles	153
Long-tailed tit *Aegithalos caudatus*	None; provisioning only measured but adults compensated fully	Male	104
Experimental manipulation of a major source of mortality			
White-winged chough *Corcorax melanorhamphos*	Small groups did as well as large groups if provided with food	Both	7
Stripe-backed wren *Campylorhynchus nuchalis*	Small groups did as well as large groups if protected from predation	Both	1
Pied kingfisher *Ceryle rudis*	Helpers were recruited when brood size was enlarged, minimizing brood reduction	Male	208
Comparison of the same pair on the same territory with and without help			
Superb fairy-wren *Malurus cyaneus*	None	Male	94
Pinyon jay *Gynmorhinus cyanocephalus*	None	Male	164
Variation in the amount of help in groups of the same size			
White-browed scrubwren *Sericornis frontalis*	None	Male	163
White-throated magpie-jay *Calocitta formosa*	Positive	Female bias	121
Splendid fairy-wren *Malurus splendens*	Positive effect of females, but not of males or group size	Both	13
Experimental manipulation and variation in the amount of help in groups of the same size			
Dunnock *Prunella modularis*	Positive	Male	46

One solution to the weak evidence produced by correlations is to turn to experiment. Removal of helpers depressed productivity in four species but had no effect in two others (Table 2). There are three reasons to hesitate before performing experiments of this sort. First, the long-term studies that have helped us understand the demographic basis of cooperative breeding involve immense effort, which either fully preoccupies investigators or leads to a justifiable reluctance to interrupt such studies with experimental intervention. Second, removals do not test explicitly for an effect of help because they do not distinguish the effect of group size from the effects of supplementary provisioning (138, 163). Third, even the simplest intervention (temporary removal of a helper) can profoundly disrupt group dynamics, affecting many parameters other than the amount of food received by offspring (59, 125, 177).

The other experimental approach has been to try to change the success of pairs or small groups by moderating (1, 7) or exacerbating (208) the effects of the apparent cause of their poor performance (Table 2). Although these experiments are illustrative, they often manipulate more parameters than help and require careful interpretation.

An alternative, very powerful approach is to examine groups of equivalent size that differ mainly in the amount of assistance provide by helpers (138). For example, in white-browed scrubwrens, auxiliaries either provision at a rate comparable to the breeding pair or fail to provision at all. An effect of help can thus be tested independent of an effect of group size, but in exhaustive analysis, help conferred no obvious benefits to the adults (163). By contrast, additional help appeared to be beneficial in white-throated magpie-jays, *Calocitta formosa*, and dunnocks, controlling for group size (Table 2).

Females may help more than was thought   Compilation of the tests that do not rely on correlations reveals a hitherto unnoticed pattern (Table 2). Where helping is male-biased, the benefits of helping for the offspring of the recipient are often difficult to document or absent (two out of six studies show an effect). The helpers providing the benefit are not related to the dominants in the two studies where there is an effect (dunnocks and pied kingfishers). By contrast, help is beneficial where females are the predominant helpers (2/2) or where both sexes provide care (5/5). In the latter case, the gender of the helpers is not always known, but in splendid fairy-wrens, *Malurus splendens*, an increase in the number of female helpers increases group productivity, whereas an increase in the number of males or the group size does not (13). In one of the best designed experimental studies, Mumme (178) showed that the cause of enhanced productivity in Florida scrub jays, *Aphelocoma coerulescens*, was effective vigilance against nest predation, which other studies have shown is facilitated by coordinated sentinel duty by both adults and helpers (167). However, while male helpers provide more food and are more vigilant than females during most of the year, female helpers are most vigilant during the nesting period (100). Hailman et al (100) speculate that this is because females stay close to the nest to facilitate participation in incubation and brooding, but they provide little evidence in support of this argument.

It is possible that because female help has attracted attention only recently, studies may have been designed with greater sensitivity to the difficulties of demonstrating an effect of help. This highlights the importance of better tests in some of the studies that have hitherto relied on correlational evidence. Nonetheless, current data support the general conclusion that female helpers, when present, may help adults more than males do, and suggest that in many cases of male help it is necessary to look to direct benefits to the helper.

Do birds usually help kin?   Although the absence of an effect on productivity in many species weakens arguments based on inclusive fitness, it is necessary in the cases where there is a discernible benefit to proceed to the second question and ask whether the recipients of help are close relatives. In some species there are very strong correlations between relatedness and whether birds provision (71). For example, Seychelles warblers (*Acrocephalus sechellensis*) in polygynous territories have a choice of nests to provision, and they help only at the nests containing the most closely related nestlings, aiding both parents in preference to nests where the father is paired to an unrelated female and declining to help where they are unrelated to both adults (144). Relatedness is less likely to influence the amount of provisioning (40, 78).

However, in a suite of other species, helping behavior is not associated with relatedness (51, 60, 62, 186). Absence of this correlation is not sufficient to reject an inclusive fitness argument, as indiscriminate helping could still return an indirect benefit provided that young are commonly related to the helpers, as certainly occurs in some of these species. A stronger case for rejection occurs where helpers are often unrelated to the adults they help (e.g. 44), or, more remarkably, where helping is preferentially directed to broods to which the helpers are unrelated (162, 238).

Taken collectively with the failure to demonstrate consistent productivity increases (Table 2), it is clear that immediate gains in production of nondescendant kin are not universal in many avian societies, necessitating investigation of alternative hypotheses.

DEFERRED PRODUCTION OF NONDESCENDANT KIN   An alternative benefit of help to the recipient could arise through increased survival of the recipient adults because of reduced reproductive effort (38). Immediate effects such as rapid renesting leading to higher seasonal productivity are taken into account in most of the productivity analyses reported in Table 2 and are not discussed further. Deferred inclusive fitness benefits could also be important for the helpers, as their parents may have more young through higher subsequent fecundity and/or higher survival. Adults might also be more likely to have a helper in future years as a consequence of help in previous years (181) or have a better helper in future years as a consequence of improved foraging and provisioning skills (18). In eusocial insect societies, many helpers are sterile, and reproductives enjoy a massive benefit through enhanced lifespan (130). One study suggests that these effects can also be significant in birds (181). However, I believe the analysis used suffers from two deficiencies. First, the method of assessing inclusive fitness involved double counting (92, 197). Second, there has been no attempt to discount any gains by taking into account the delay experienced by the helper in receiving the direct benefits that accrue from acquisition of a breeding vacancy through queuing (see below). Future benefits may be most important when the helper cannot take over the territory, which is often true for females (see below).

Evidence for load-lightening   Load-lightening hypotheses predict that the acquisition of help should reduce parental workload and hence increase survival or future fecundity. Unfortunately, correlational evidence for an effect on parental survival and reproduction is even more fraught with circularity than are measurements of productivity. If territories are of high quality, it is probable that survival of both adults and young will be enhanced, generating an indirect correlation between the number of helpers and survival of the adult.

This hypothesis is amenable to direct experiment following the methods developed to test the cost of parental care in non-cooperative species. The impact of changes in workload can be tested by experimentally increasing brood size (185) or by attaching small weights to the provisioning bird (244). Such increased effort can reduce survival or subsequent fecundity (98). However, comparable experimental tests have not been widely performed on cooperative birds (137). Adult survivorship was not measured in a brood size manipulation experiment in bell miners (190). Increased brood size prompted pied kingfishers, *Ceryle rudis*, to recruit helpers that they would otherwise have rejected (208), and experimental supplementation of food to white-winged choughs increased the rate of provisioning of both adult and helper birds (7), so in both these cases the predictions for adult survivorship are obscure. Predictions are also muddied by the long lifespans of some cooperative breeders.

One way to demonstrate load-lightening is to measure survivorship in birds in which only one gender lightens their load. If enhanced survivorship is confined to that gender, there is better evidence of direct causation. Few studies have been presented in an appropriate way to test this prediction. Increased group size in groove-billed anis has no impact on male survival but slightly improves female survival, apparently because females suffer fewer costs of incubation (232). Female acorn woodpeckers suffer much higher mortality than males during reproduction (139), and they provide much more care (180). The prediction can be unequivocally rejected in some species. In both riflemen and European bee-eaters (*Merops apiaster*), males substantially reduce their load in the presence of helpers without any obvious effect on survival (154, 218). However, females survive better without reducing their load. No explanation has ever been offered for these latter results. I suggest that the results in females are a consequence of the effect of territory quality on both number of helpers and female survival. The male results remain unexplained. It is possible that dominants could perceive helpers as a threat to paternity and reduce their workload accordingly. Regardless of the cause, the absence of experimental tests of load-lightening and these counterintuitive results weakens our ability to apply this group of explanations of helping behavior.

Payment of Rent, or Mutualism

Load-lightening could benefit adults without allowing helpers deferred fitness benefits. Green et al (94) showed a novel benefit for dominant males with helpers in superb fairy-wrens, *Malurus cyaneus*, in which the majority of fertilizations are obtained by birds from outside the group that rears the young (59, 176). The male devotes large amounts of time to extra-group courtship, even leaving his own female during the period she is fertile (174). Helpers do not affect the probability of extra-group courtship by dominant males when a mate is fertile or incubating, but dominant males with helpers court to a greater extent during the nestling phase and gain more extra-group fertilizations as a consequence. This helps explain the paradox that dominants also lose more paternity to extra-group males when they have helpers (176). This hypothesis is also consistent with the observation that temporary experimental removal of helpers during the nestling phase triggers prolonged escalated aggression from the dominant male when the helper is returned after 24 hours, but return of a temporarily removed helper during the non-breeding season, fertile period, or incubation provokes no response or a much lower level of aggression (177).

These fairy-wrens provide the strongest support for the hypothesis that helping is a payment of rent (87) and in this case may be enforced by parental coercion. The role of coercion in animal societies is attracting new attention and has probably been underestimated (33, 85). In related observations, white-fronted bee-eaters are sometimes prevented from breeding independently by harassment from their parents, apparently to induce them to return and provide care to nestlings (79), and white-winged choughs will "kidnap" helpers from other groups, with the consequence that they provide care to young to which they are completely unrelated (111). In addition to assistance with reproduction, adults could obtain many benefits from the presence of supernumeraries, such as energetic benefits from huddling (63, 64) and improved attractiveness to mates (243, 247).

WHY SHOULD HELPERS PAY RENT?   Helpers could also benefit in a variety of ways from living in a group. All authors concede that an eventual benefit for subordinates in many societies is that they are well-placed to succeed the dominant when it dies, but there is considerable disagreement over whether this opportunity drives the evolution of help. In discussions of the evolution of philopatry, there has been debate over whether young remain on a territory because it provides intrinsic benefits or whether the absence of dispersal is shaped by extrinsic pressures such as a lack of territories or mates elsewhere (19, 69, 141, 224, 225). Empirical evidence provides excellent support for both effects (143).

Much attention has been paid to comparing the fitness of birds that are philopatric with the option of dispersing and seeking a breeding vacancy. Several studies suggest that philopatry is valuable for the young, even when any benefits arising from production of nondescendant kin are excluded from calculations (19, 235). It may be worth paying rent to enjoy these benefits.

Territorial inheritance may be less widely available to female helpers than to male helpers in many species where territorial philopatry predominates. The death of the breeding male can often lead to the eviction of the breeding female, probably because of an incest taboo that precludes sons from mating with their mother (8, 241). It is unclear whether this disadvantage for females will also be true in cooperative species in which females are larger and potentially dominant over males, such as laughing kookaburras, *Dacelo gigas*, and raptors. Societies wherein unrelated coalitions of males breed together have attracted less long-term analysis, yet incest avoidance presents no difficulty for these birds. Nakamura (183) presented convincing evidence for the importance of queuing in alpine accentors, *Prunella collaris*.

A further and universal benefit comes from the ability to exploit rare vacancies when they become available (246). Detailed radiotelemetric studies that have examined dispersal behavior suggest that vacancies are constantly investigated to an extent that has been dramatically underestimated in ordinary observational studies (142). Helping can again be viewed as rent payment, though benefits are not constrained by incest taboos or intersexual dominance and should be equally accessible to both male and female subordinates.

PARENTAL FACILITATION AND MUTUALISM   An alternative approach views philopatry and territorial inheritance as extended investment by parents in philopatric young, either by incorporating a proportion of young into the group as breeders [parental facilitation (20)] or by ensuring that the territory is occupied by their descendants when the parents die (15). This suggestion has not received the attention it deserves, as it remains a common assumption that dispersal is entirely under juvenile control. Long-term studies of cooperative breeders reveal considerable heterogeneity in individual success (137, 164, 212, 240). I suspect that dissection of reproductive success in many societies will reveal that less successful individuals leave descendants only via territorial inheritance, increasing the importance of juvenile philopatry to adults. One problem with this notion of parental facilitation is that if this is the sole factor driving the system, there is a limit to the benefit parents receive from accumulating helpers (5, 20), suggesting that dispersal will be facultative.

Access to Mates

MATING WITHIN THE GROUP   Another benefit that may make rent payment worthwhile is direct access to reproduction. In many cooperative societies, subordinates gain a direct benefit through fertilizations, shared nesting, or plural breeding. It has been possible to assess the apportionment of such sharing only with the advent of molecular techniques for assignment or exclusion of parentage (Table 3). Some complex societies have proved extremely difficult to dissect (88), but reproductive sharing appears to be very important both in species in which cooperation arises through natal philopatry, and in those where it occurs via establishment of coalitions of unrelated individuals. For reasons that are not understood, sharing is sometimes confined to one gender.

The extent of reproductive sharing in families has been interpreted in three main ways (72): (*a*) parents are dominant over their young and exclude them from copulations; (*b*) incest avoidance; and (*c*) reproductive skew models.

Dominance is easily understood, though reproduction is shared in some societies with quite rigid dominance hierarchies (201). Incest is selected against because inbreeding depression arises when deleterious recessives are expressed or when the expression of favorable overdominant allele combinations is reduced. These costs could be mitigated if inbreeding were frequent, purging deleterious recessives by exposure to selection. Incest is not routinely avoided in two mammalian cooperative breeders, and it confers little cost (129, 206), though there are many counterexamples of incest taboos (3, 195). Behavioral and some physiological evidence supports the existence of strong incest taboos in a large number of cooperatively breeding birds (e.g. Table 3; see also 8, 214). Only rails appear to copulate incestuously at a high rate (37, 168). Moorhen exhibit evidence of strong inbreeding depression (168), as do other free-living birds (195). Claims that incestuous matings occur commonly in other cooperatively breeding birds have not withstood paternity analysis (14), though the situation in Galapagos mockingbirds (*Neosimius* spp.)

Table 3 Molecular studies of parentage in cooperative breeders[a]

Species	Extra-group paternity	Within-group sharing	Behavioral evidence of incest taboo	Reference
Coalitions of unrelated males and females				
Dunnock *Prunella modularis*	0.7%	40% of broods between males; females nest separately	Not applicable	23
Alpine accentor *Prunella collaris*	0%	50% to 60% of broods between males; females nest separately	Not applicable	103, 106
Henderson's reed warbler *Acrocephalus vaughani taiti*	0%; incompatibility ascribed to egg-dumping	0% of brood between males; 100% sharing between females	Not applicable	12
Male and female helpers (relatedness high but not universal)				
Guira cuckoos *Guira guira*	Not tested	Sharing by both females and males, probably at very high levels	Unknown	199
Tasmanian native hen *Tribonyx mortierii*	Possibly, but attributed to band-scoring errors	None between females (small sample); probably 17% of families among males	Unknown	88
Moorhen *Gallinula chloropus*	0%, but brood parasitism by females	Sharing among males in the one case where two were reproductively active; 100% among females	Some, but incest prevalent	168, 169
Pukeko *Porphyrio porphyrio*	0%	Sharing among males in 92% of broods where males were unrelated; sharing among females	Incestuous matings common	37, 124, 126, 149
Acorn woodpecker *Melanerpes formicivorous*	0% to 2.2%	100% sharing between males (small sample); multiple egg-laying by females	Yes	52, 128, 140, 179, 182
White-fronted bee-eater *Merops bullockoides*	≥9% attributable to either extra-pair copulation or brood parasitism	Not tested	Yes	242
Gray-breasted jay *Aphelocoma ultramarina*	0%	At least 7% to 16% of broods between males; none between females	Unknown	9
Splendid fairy-wren *Malurus splendens*	>65% of young	Not tested	Incestuous social pairing very common	14

(Continued)

Table 3 (*Continued*)

Species	Extra-group paternity	Within-group sharing	Behavioral evidence of incest taboo	Reference
Bicolored wren *Campylorhynchus griseus*	2.3% of young	2.3% of young sired by subordinate male; only one female reproductive	Yes	105
Stripe-backed wren *Campylorhynchus nuchalis*	1.4% of young	8.7% of young sired by subordinate male; only one female reproductive	Yes	186, 187, 188, 202
Male philopatry with some mixing with non-relatives				
White-browed scrubwren *Sericornis frontalis*	24% for pairs; 6% for groups	31% of broods	Yes	162, 238
European bee-eater *Merops apiaster*	1%	0%	Unknown	127
Superb fairy-wren *Malurus cyaneus*	76%	5.6% of young	Unknown	60, 176
Red-cockaded woodpecker *Picoides borealis*	0%	0%	Unlikely due to sex-biased dispersal	99
Noisy miner *Manorina melanocephala*	5.8% of young	0%	Unknown	193
Unrelated coalitions of males				
Brown skua *Catharacta lonnbergi*	0%	15% of broods; different fathers may sire sequential broods	Not applicable	170
Galapagos hawk *Buteo galapagoensis*	0%	100% of broods; different fathers may sire sequential broods	Not applicable	82
Bushtit *Psaltriparus minimus*	0%	0% (small sample); one case where different males sire consecutive broods	Not applicable	22
Smith's longspur *Calcarius pictus*	0%	75% of broods	Not applicable	11
Females sharing a nest (relatedness unclear)				
Magpie goose *Anseranus semipalmata*	Not known	Yes, but some females may not contribute to large broods	Unknown	120

[a]Note: Only one published study has used techniques employing hypervariable single-locus probes that can reliably detect mother-son incest in nuclear families (149).

and green woodhoopoes, *Phoeniculus purpureus*, warrants molecular analysis (40, 160).

The molecular methods of paternity assignment that have been applied thus far are not as well suited to resolve this question as has generally been believed. Most seriously for published studies, multilocus fingerprinting methods cannot easily distinguish whether a mother has allowed her helper son to cuckold his genetic father, or whether a daughter is laying in her mother's nest, unless additional information such as egg shape and color is available (e.g. 168). This is because, in the absence of mutation, all the restriction fragments in the helper will have a maternal or paternal origin. By definition, all non-maternal bands in the helper will occur in its father, so the father can never be excluded when the son mates with his mother. While some authors have recognized this problem, they have excluded the possibility of incest for reasons of parsimony (e.g. 105), and therefore cannot be used to conclude that incest is rare. Single-locus methods can resolve this difficulty if polymorphism is sufficiently high, but they require testing at more loci than is usually realized in order to exclude close male relatives of the true sire reliably (54). Genetic markers with sufficient polymorphism have not been used in any population of cooperatively breeding birds, though the appropriate markers are now available for some species (55, 157). A combination of single- and multilocus methods was used to good effect in a study of pukeko, *Porphyrio porphyrio* (149), and contained hints of incest avoidance that could not have been inferred from behavioral data (37). I predict that the application of appropriate molecular techniques will demonstrate a pervasive role for incest avoidance.

The second set of arguments, originally proposed by Emlen (70) and explored more fully by Vehrencamp (230, 231), have been called reproductive skew models and are enjoying a surge of renewed interest (e.g. 131, 203). According to this view, dominants in group-living societies will receive an obvious benefit from monopolizing reproduction and will usually do so (despotic reproduction). However, when they gain from the presence of subordinates, it may pay them to sacrifice reproduction to encourage the subordinates to remain in the group. This reproductive bribing (204) leads to more egalitarian reproduction (230). From an ecological perspective, a critical component of these models will be the benefits received by the subordinate if it disperses; if dispersal to a breeding vacancy is easy then the extent of reproductive sharing should increase as the dominant increases the bribe (230). The most sophisticated attempt to test skew models in cooperatively breeding birds comes from molecular studies of pukeko living under different constraints to dispersal, and it supports these predictions (124).

Incest avoidance and skew models are not mutually contradictory, and may both operate in concert with age-related dominance in determining the distribution of paternity in families. The process of discriminating their relative importance in avian families has scarcely commenced (72). Reproductive skew models were originally focused predominantly on interactions among families, though the same problems arise among unrelated coalitions. Studies of such societies demonstrate an important feature currently lacking in skew models: there can be conflicts of interest between the dominant male and female where the supernumeraries are male (42, 43, 46). Females may benefit more from increased provisioning because they gain additional care without any sacrifice.

REPRODUCTION OUTSIDE THE GROUP   An alternative direct benefit occurs through access to extra-pair copulations. Living on a territory may allow helpers to attract copulations from females in a way that would not be available to a floater (91). Superb fairy-wren (*Malurus cyaneus*) helpers obtain little paternity in their own group, but older helpers can be very successful at siring extra-group young (176). Some subordinates are more successful at obtaining extra-pair young than the dominants they help. It seems unlikely that extra-group paternity is a fitness benefit for helpers in any well-studied birds other than fairy-wrens, as extra-group paternity is often very low (Table 3), though identifying the extra-pair fathers in species with an appreciable rate of extra-pair copulation will prove interesting (e.g. 238, 242).

FUTURE ACCESS TO MATES   Provisioning could be a way of inducing the female to divorce her present mate or to mate with the helper male in the future. Direct competition for mates occurs via helping in the moustached warbler, in which unrelated floater males assist the females in incubation, feeding of nestlings, and defense of chicks (83). Females sometimes switch mates to pair with the helper between the first and second nest of the season, but they never switch mates if they have not received assistance.

In pied kingfishers, unrelated "secondary" helpers are tolerated by adult males only under conditions in which the adults are having difficulty minimizing brood reduction (207). They provide less care than philopatric "primary" helpers, and they also engage in courtship feeding of the female and have an excellent chance of pairing with they female they court. Perhaps the most remarkable evidence comes from the riflemen I have already discussed, wherein regular helpers are likely to pair with females from the broods they provision. This is probably the sole reason for help, as there is no enhancement of productivity, helpers are unrelated to the young for which they provide care, and they care for only one of the two broods that are raised on the territory each year. In other species it may even be possible for the subordinate male to take over the breeding vacancy by ousting the resident male—e.g bushtit, *Psaltriparus minimus* (222).

I suspect that future access to mates also provides the best explanation for virtually all cases where male helpers feed at several nests, a phenomenon most pronounced among *Manorina* honeyeaters. While *Manorina* were once thought to be polyandrous (56), this interpretation has been rejected by more detailed observations and molecular studies that confirm that these birds are genetically monogamous (29, 192, 193). Naturally widowed female bell miners, *Manorina melanophrys*, pair preferentially with the unrelated, unmated male that provided them with the most aid in previous broods (30), and there is also evidence of courtship feeding (189). The behavior of casual helpers in riflemen and male subordinates in white-winged trumpeters is also consistent with this view.

Improvement of Local Conditions

Birds might gain an advantage from living in groups because increased numbers might lead to territorial expansion and hence to increased productivity. Helping could be beneficial because it increased productivity and hence allowed further territorial expansion, leading to inheritance of a superior territory, or fission and accelerated possibilities of attaining dominant status (e.g. 239, 240).

This explanation receives strong support from some descriptive studies (e.g. 13, 240) but suffers from all the difficulties that plague the hypotheses I have described previously. If reproductive output increases as territory size increases as helper number increases, what is the direction of causation? This hypothesis is also irrelevant unless provisioning enhances helper number, which, as I have already shown, is by no means obvious in many species. Only careful experimental analysis seems likely to further our understanding of this hypothesis.

RECIPROCITY OR HELP IN THE FUTURE   A related argument suggests that helping is important because it increases the likelihood that new helpers will be present when the old helper inherits the dominant position or disperses with other younger birds to take over a vacancy (159). Both territorial inheritance and dispersal in groups occur in several species (e.g. 137, 156, 201, 234). I know of no evidence for the most restricted version of this hypothesis, which suggests that young birds will only help a bird that has provisioned them. For reasons of circularity, it does not seem likely that the prospect of future help can be the initial selective force for helping behavior. It is also unclear how this model could be tested. Where helpers substantially augment productivity, they produce not only potential helpers or dispersal partners but also competitors for that role. Ligon (158) suggested that benefits might be restricted to cases where mortality of breeders is quite high, but appropriate comparative analyses have not been attempted.

PRISONER'S DILEMMA   An alternative way of looking at the benefits of grouping was suggested by Craig (36). Pukeko in trios have a much lower per capita
output than birds in pairs. However, if three birds group together for some mutual benefit such as expanding their territory, other birds may be forced to form
groups to defend their own space, forcing all birds into a suboptimal pay-off.
Some reproductive bribery may be necessary to stabilize group membership
and may lead to helping where reproduction is distributed asymmetrically (124).
Comparable benefits have been suggested for some other species. In Henderson's reed warbler, *Acrocephalus vaughani taiti*, about one third of birds breed
in trios, which often have two males but more rarely two females (12). Trio
members are not first order relatives. Food delivery does not differ between
pairs and trios. There is no evidence of reproductive sharing between males,
and the male who does not sire any young provisions at the same rate as the true
parents. Brooke & Hartley (12) hypothesized that trio membership increases
the ability to secure a nesting territory.

Formation of Alliances

Zahavi has been the main champion of the view that helping behavior serves
predominantly to enhance "privilege" within the group (24, 248, 249), accelerating attainment of a breeding position by facilitating the formation of coalitions
at the time of dispersal or on the death of a breeder. The establishment of coalitions and alliances in avian society has attracted almost no empirical attention in
avian biology, though it is entrenched in a comparable literature on mammalian
sociality (101, 109, 166).

    In contexts other than cooperative breeding, birds are known to form coop-
erative alliances. These can be temporary, as illustrated by territorial and satellite pied wagtails seeking to maximize exploitation of renewable prey (48). A
longer-term example concerns the joint display of pairs of lekking long-tailed
manakins, *Chiroxiphia linearis*, which enhances the access of the dominant
male to fertilizations, but provides no benefits to the subordinate male through
direct copulatory access, inclusive fitness, or via later reciprocity (165). Subordinates may gain an eventual advantage because of female loyalty to the lek
site once the dominant has died.

    Subtle alliances and agreements may explain many of the paradoxical and
little explored anecdotes that litter the literature on cooperative breeding but
provoke no discussion because they fall outside the framework that governs most
analysis. These include the disruption caused by helper removal experiments
in complex cooperative societies (59, 125, 177), the joint dispersal by members
of groups, and the failure of some helpers to disperse when confronted with a
breeding vacancy in a neighboring territory. For example, male brown skuas,
Catharacta lonnbergi, will remain in unrelated polyandrous trios rather than

move to pair with a neighboring female when she is widowed (117), even though there is no appreciable difference in the productivity of pairs and trios (116). Helping could demonstrate the suitability of an individual as a coalition partner. If two breeding female Galápagos mockingbirds share a territory, they normally use the same nest if one has previously helped the other but nest separately if there has been no prior help (41).

It may be advantageous to other group members to break up or test the strength of alliances (249). In several societies, helpers and/or dominants will intervene to prevent helpers delivering food to the nest (226, 245, 248) or seize the food and deliver the prey themselves (186). Violence and fatal aggression between young birds is occasionally reported (57). A particularly interesting case occurs in white-winged choughs, wherein young helpers attempt to deceive group members by carrying food to the nest and placing it in the gaping mouths of the begging nestlings, but then consume it themselves, suggesting that they are trying to minimize the costs of participation in help (6), though being seen to be helpful. They often follow such cheating by ostentatious grooming. Deception can be reduced by providing supplemental food. Similar behavior has been described anecdotally for other cooperative breeders (e.g. 191).

I suggest that the study of alliances in avian cooperative breeders may prove as profitable as has been the case for primates. While relatedness may facilitate the formation of alliances (e.g. 61), there are numerous alternative hypotheses for the formation of cooperative associations between individuals (58). The only current reason for assuming that interactions in the more complex avian societies (e.g. acorn woodpeckers and superb fairy-wrens) are simpler than those that occur in primate societies is anthropocentrism.

Acquisition of Skills

Early discussion of helping behavior suggested that birds waiting for a vacancy could benefit from helping behavior by practicing parenting skills (216, 221). However, strong evidence for this hypothesis has only recently been uncovered. Seychelles warblers with prior helping experience are superior breeders, when they acquire a breeding vacancy, to birds without helping experience (147), though it seems that enhanced parenting is not the primary factor favoring the evolution of help (144, 145). There is no detectable effect in descriptive studies of some other species (76, 134), and the opposite appears to be true in western bluebirds (*Sialia mexicana*), perhaps because helping is mainly expressed in failed breeders or low quality birds that are intrinsically poorer parents (53, 84). Once again, this negative correlation is unhelpful, as all that is necessary for a fitness benefit is an improvement of ability in the individual in question. However, I am unaware of any author who believes that enhancement of parenting skills is the main explanation of helping.

A related hypothesis suggests a link between slow acquisition of foraging skills, prolonged parental investment, and obligate coexistence of parents and young: the skills hypothesis sensu Brown (19). The skills hypothesis may be restricted in its application. White-throated magpie-jays acquire foraging skills more rapidly when they have an experienced group member from which they can copy (152), and take considerable time to develop adult foraging skills, but this slow development does not influence the timing of dispersal of males, the dispersive sex (151). Indeed, there is no evidence for the importance of this hypothesis where dispersal is strongly sex-biased. The strongest evidence comes from white-winged choughs, which require several years before adult foraging skills are attained (114), dramatically extending the duration of parental care (112) and encouraging a dependence on help for successful reproduction (6, 110, 113). It would be interesting to investigate patterns in other cooperative breeders with prolonged dependence on parents, such as the ground hornbill, *Bucorvus leadbeateri*, which may feed the young for more than two years (132).

The skills hypothesis has been criticized most strongly by Koenig et al (141) because of a potential confusion over difficulty in interpreting the sequence of evolution. Do offspring fail to disperse because of foraging difficulties, or do extrinsic constraints on dispersal relax selection for rapid acquisition of skills, leading to the observed slow development? An alternative suite of arguments proposed for the evolution of eusociality in social insects may help suggest a link with extended parental care and skill acquisition in cooperatively breeding birds. One possibility is that prolonged dependence by young increases the prospect that parents will die before investment is complete, so helpers are valuable to parents because they can rear young to independence if the parent dies (196). Similarly, investment by helpers is valuable as they can acquire gains via inclusive fitness if they are able to invest for only part of the period that a brood requires to reach maturity (86).

WHY IS HELP SEX-BIASED?

The tendency that help will be delivered exclusively or predominantly by only one sex has recently been described (141:143) as a "nagging question," particularly for advocates of the view that philopatry is a consequence of ecological constraints to dispersal. The most sophisticated analyses of this question focus on attempts to explain the idiosyncrasies of acorn woodpeckers, wherein helper males provide little provisioning (139), and Florida scrub jays, wherein males are the major provisioners (241), although with both the sexes are philopatric. Several hypotheses have been proposed to account for sex bias in help. Some of these attempt to explain the overall pattern among birds that males exhibit helping behavior more often than females (19), while others attempt to explain the idiosyncrasies of a single species. Variation in help could arise because it is an epiphenomenon (*a*) of dispersal dynamics; (*b*) of different variance in the lifetime reproductive success of the sexes; (*c*) of role division between the sexes; (*d*) of uncertainty over parentage; or because (*e*) females are less able to share reproduction because sharing leads to a large and suboptimal brood size, or (*f*) birds exhibit female heterogamety.

Sex-Biased Help Is an Epiphenomenon of Sex-Biased Dispersal

The simplest explanation for male-biased care is that it is an unselected epiphenomenon reflecting the tendency for dispersal to be female-biased in birds (28, 96). The most convincing hypothesis for this bias remains Greenwood's (96) suggestion that sex-biased dispersal is selected to minimize inbreeding. According to Greenwood, females disperse in birds because male reproductive strategies are often based around the defense of resources that promote the attraction of mates, while females benefit from the opportunity to assess several males by dispersing before settlement. This hypothesis applies to many species of birds and has been supported in one study of male-biased dispersal in cooperative breeders (150). I suspect it is a powerful explanation of many cases of helping by failed breeders, and incidental helping sensu Brown (19). However, it may be of dubious relevance to many cooperative breeders, in which both sexes are philopatric and contribute to resource defense (139), or in which the sexes each have separate interests in the defense of territories (35, 42, 49). There are also cases wherein helping is not correlated with the usual dispersal bias. For example, a high incidence of helping by females was uncovered in a population of peregrine falcons, *Falco peregrinus*, from which helping is otherwise unknown, and wherein dispersal is also strongly female-biased (172).

Helping in one sex could be restricted either because of dispersal initiated by the juvenile, or by the direct action of the adult. Adult breeders can decline to accept help either by expulsion of young or by preventing supernumeraries from joining the group. They can also prevent provisioning by refusing young or supernumeraries access to the nest. The first cause has been overlooked as a result of the conceptual focus on asking why young remain philopatric, but growing evidence suggests that parental aggression may be a primary cause of interspecific and intersexual differences in the manifestation of help (173), e.g. *Cormobates* versus *Climacteris* treecreepers (184). For example, in superb fairy-wrens, territorial groups at the start of the breeding season comprise an adult female, a dominant male, one to four helper males, and supernumerary females that may have been born on the territory or joined the group from elsewhere (173). Although these supernumerary females occasionally persist long enough to help with the first nest of the season, they are inevitably driven forcibly from the territory by the adult female.

In these fairy-wrens, there is excellent experimental evidence that young helper males do not disperse and breed independently because females are in short supply (194). However, adult enforcement of dispersal consigns as many as 40% of females to their doom each year, and is therefore a primary contributor to the shortfall (173). This anomalous behavior is unlikely to be an artefact of any phylogenetic tendency for females to be the predominant dispersers, as their closest relative exhibits high levels of female philopatry and help (213), as may most other members of the genus (210). Similar anomalies plague our understanding of even the best-studied species with male-only helping (e.g. 207, 233) or female-biased helping (e.g. 143).

A more specialized version of the dispersal epiphenomenon hypothesis suggests that the sex most likely to live adjacent to relatives has a reduced investment in territorial defense because of mutual interests with its neighbors, and can therefore afford to invest in care (32). This hypothesis has been rejected in one non-cooperative species (95) and one cooperative species (156).

Lessells (156) examined the joint dispersal of brothers in the European bee-eater, and showed that males with relatives nearby were more likely to provide help if their own nest failed. She also uncovered a cost of joint dispersal, because predation was clustered, so the nests of both male relatives were likely to be destroyed as a consequence of dispersing together.

Variance in Lifetime Reproductive Success

Koenig et al (139) introduced a hypothesis they attributed to Anne Barrett Clark, which suggests that the sex with the greater variance in success should be more likely to help, as more individuals are likely to die without breeding, and hence gain fitness only via investment in collateral kin. Koenig et al suggested that variance should be higher in females because of the mortality they suffer through dispersal, and therefore females should invest more heavily. I believe that this hypothesis has quite general implications and should be explored further both theoretically and empirically. Unfortunately, no molecular study has yet measured lifetime reproductive success of a cooperatively breeding bird (Table 3).

INTERSEXUAL CONFLICTS   In contrast to this argument, Woolfenden & Fitzpatrick (241) suggested that social dominance by males in Florida scrub jays precludes females from inheriting the territory on the death of the male breeder, probably because of incest taboos that inhibit a son from mating with its mother. This in turn increases the value of male help. Eviction of a breeder female on the death of its mate certainly occurs in several cooperative breeders (see above) and removes one very strong incentive to remain philopatric. However, I suggest a link between this difficulty and the Clark hypothesis. The incentive to help may be different, with females relying on augmentation through inclusive fitness (see Table 2 and associated discussion) while males act to maximize inheritance of the territory.

Variation in Help Reflects a Division of Labor Between the Sexes

In many birds, and some cooperative breeders, females undertake a greater range of parental activities (notably nest-building and incubation), while males may contribute more to territorial defense. Koenig et al (139) considered two ways that this could influence the measurement of help. These hypotheses apply only where both sexes help. On one hand, because care at the nest is more easily measured than territorial defense, a false impression could arise of predominant female care, though overall investment may be similar. This hypothesis can be explicitly rejected for both acorn woodpeckers, wherein females contribute vigorously to territory defense (139), and Florida scrub jays, wherein males provide the most territory defense and the most food (241).

Alternatively, acquisition of skills could be more important to the gender providing the wider range of care (139). It is certainly true that the best studied species with predominantly female care—the Seychelles warbler—also provides the best evidence for the skills hypothesis (147), but any generality for this hypothesis remains unexplored.

Parentage Uncertainty

Trivers (227) suggested that increased care by one sex could also be explained by uncertainty over parentage. Contrary to some earlier theoretical results, recent theory supports this view (198). In general, conflict over parentage is likely to be worse for males than females (227). Sensitivity to the cost of lost paternity is widespread and is indicated by refusal to accept secondary helpers in the pied kingfisher (207), nest destruction by male acorn woodpeckers denied access to the female during the fertile period (136), use of copulation frequency to adjust the level of paternal care in dunnocks and alpine accentors (45, 47), and adaptations for sperm competition in several cooperative breeders (10, 25, 175, 228). However, in some societies there is true egalitarian mating, and all males contribute to feeding regardless of their paternity (51, 126).

Charnov (27) noted the asymmetries in relatedness that arise when males share paternity. If multiple unrelated males contribute to a clutch, a female on average will be more related to her offspring than to her sibs. By contrast, if a male sires proportion p of the offspring of his mate, his average coefficient of relatedness to these young will be $p/2$. However, his coefficient of relatedness to his sibs will be $(1 + p^2)/4$, which is greater than $p/2$ whenever $p < 1$. According to Charnov, this very strong facilitation of male alloparental care is restrained by the existence of biparental care. If a helper produces x offspring, then in the simplest case he should have been able to produce x offspring independently. However, by helping, he also forgoes the x offspring produced by his mate.

However, in cooperatively breeding birds, conflict is overt in both sexes. Hatching failure appears to be unusually prevalent among cooperative breeders, perhaps as a consequence of conflict or the difficulties that arise when more than one bird incubates at the same time (2, 135). Reproductive conflict between females, such as egg-destruction, laying runt eggs as a counterploy, and even infanticide, has been reported, suggesting a cost that could restrain female helping in other societies (140, 161, 179, 229). Female help is often prevented by violence from other females (241) or by enforced dispersal (173). Indeed, Clutton-Brock (31) suggested that one stimulus for male help with incubation may be an attempt to minimize the probability that females will destroy each other's eggs.

Suboptimal Brood Size

Koenig et al (139) also proposed the interesting hypothesis that female sharing is fundamentally more difficult than male sharing because it is typically achieved by increasing the size of the brood, which might therefore be greater than is optimal. This effect has been recently modelled and polyandry confirmed as a more stable system of reproductive sharing than monogamy or polygyny (26). I suspect that this is the most promising avenue for further understanding of the greater frequency of systems where help is male-biased.

Female Heterogamety

Two studies have proposed analogies between sex-linked genes in diploid organisms, and inclusive fitness arguments derived to explain the frequent evolution of eusociality in haplodiploid organisms. Whitney (237) pointed out that in birds, where the female is heterogametic, sex-linked alleles on the large Z chromosome are more likely to be shared between brothers than between mother and offspring, facilitating cooperation between males. An alternative "protected invasion" hypothesis suggests that beneficial new sex-linked mutations that favor male-biased help enjoy relative protection from loss via genetic drift if they occur on the large, genetically more active Z chromosome (205). Both these hypotheses should be viewed as facilitating sex bias and not as explanations in themselves for the incidence of helping.

The value of these arguments is weakened by new data on the incidence and importance of female-biased helping (see above). Greenwood (96) has previously argued against the importance of an association between sex-biased dispersal in birds and mammals and the pattern of heterogamety.

CONCLUSIONS

Recent evidence from cooperatively breeding birds suggests that we are still some way from understanding the adaptive significance of helping behavior. First, experimental and well-controlled tests of the benefits of help for dominants suggests that male help may often be driven by considerations other than the production of nondescendant kin, while the significance of female helpers in the societies in which they occur may have been greatly underestimated. Second, the only differences between cooperative societies based on families and those based on unrelated individuals revealed in this review are the potential importance of incest avoidance within families and the advantages that parents obtain from facilitating inheritance of the territory by their offspring. Because of reliance on multilocus fingerprinting, evidence for incest taboos in cooperatively breeding birds remains based on field observations rather than on molecular reconstruction of genealogies. Natural history observations do not support an abrupt dichotomy among cooperative societies based on kinship. Third, numerous additional hypotheses have been suggested to explain helping behavior, but only reproductive sharing enjoys support from a wide range of species, perhaps because the importance of alliances between individuals has been ignored or because appropriate tests of other hypotheses have not been devised.

Even more difficulty surrounds attempts to explain why help is often provided exclusively or largely by only one sex. Most hypotheses are motivated by the observation that helping behavior is male-biased, and several are based around the common interests shared by kin. Hence they struggle to deal with the empirical evidence summarized in this review, which suggests that female help may often be explained by inclusive-fitness arguments, while male help depends on direct benefits. Among available hypotheses, I suspect that most sex bias will ultimately be explained as an epiphenomenon of sex-biased dispersal to avoid inbreeding, as due to the separate interests of males and females that arise because of different variance in lifetime reproductive success, and as due to the suboptimal brood size that can arise when females share a nest. These possibilities have not been adequately distinguished in any species, hindering interpretation of recent studies that indicate that sex allocation in cooperative breeders can be biased to an extent not seen in other birds (50, 108, 146, 148), possibly because one gender repays more of the cost of parental investment than the other (74, 155).

An important problem that I have avoided thus far in this review is whether understanding the benefits and sex bias of help can contribute to an explanation of the ecological and taxonomic distribution of helping behavior among birds. In his classical review of cooperative breeding in birds, Brown (19) dismissed the importance of phylogenetic influences. However, it is now clear that complex cooperative systems are often not the product of local ecological constraints but may be deeply seated and ancestral characteristics of many important avian clades (34, 67, 211). Group living may occur in more than 30% of all species in some large clades of birds such as hornbills and hoopoes (115, 132), and the passerine group Corvida (34). As for social insects (236), there is evidence that pair-dwelling often evolves from more complex ancestral systems, rather than vice versa, as is usually assumed (34). It is salutary that the best two experimental demonstrations of an ecological constraint to dispersal and independent breeding (143, 194) occur in *Acrocephalus* and *Malurus*, two genera that display a bizarre diversity of complex mating and social systems that have no obvious relationship to ecological constraints (83, 210, 215). Arguably the best understood cooperative breeder, the dunnock, also exhibits a social system that may reflect its ancestry among accentors living in high alpine habitats, rather than the habitats it has recently invaded (44). As emphasized some time ago for another hypercooperative group, the New World jays (15), dissecting the evolutionary pathways among these taxa may hasten a general understanding of helping behavior among birds.

ACKNOWLEDGMENTS

My ideas have been forged in innumerable conversations with the "cooperative breeders" at the Australian National University: Chris Boland, Rosie Cooney, Nina Cullen, Mike Double, Peter Dunn, Daniel Ebert, Janet Gardner, David Green, Michelle Hall, Elsie Krebs, Naomi Langmore, Ashley Leedman, Milton Lewis, Sarah Legge, Rob Magrath, Michael Magrath, Raoul Mulder, Anjeli Nathan, Penny Olsen, Helen Osmond, Anne Peters, David Westcott, and Linda Whittingham. In particular, I am indebted to my first student and longtime collaborator, Rob Heinsohn. The review would not have got off the ground without encouragement from Ross Crozier. My research on cooperative breeding has been consistently supported by the Australian Research Council.

Visit the *Annual Reviews home page* at
http://www.AnnualReviews.org

Literature Cited

1. Austad SN, Rabenold KN. 1985. Reproductive enhancement by helpers and an experimental inquiry into its mechanism in the bicolored wren. *Behav. Ecol. Sociobiol.* 17:19–27
2. Bednarz JC. 1987. Pair and group reproductive success, polyandry, and cooperative breeding in Harris' hawks. *Auk* 104:393–404
3. Bennett NC, Faulkes CG, Molteno AJ. 1996. Reproductive suppression in subordinate, non-breeding female Damaraland mole-rats: two components to a lifetime of socially induced infertility. *Proc.*

R. Soc. Lond. Ser. B 263:1599–603

4. Bennington CC, Thayne WV. 1994. Use and misuse of mixed model analysis of variance in ecological studies. *Ecology* 75:717–22

5. Blackwell PG. 1997. The *n*-player war of attrition and territorial groups. *J. Theor. Biol.* 189:175–81

6. Boland CRJ, Heinsohn R, Cockburn A. 1997. Deception by helpers in cooperatively breeding white-winged choughs and its experimental manipulation. *Behav. Ecol. Sociobiol.* 41:251–56

7. Boland CRJ, Heinsohn R, Cockburn A. 1997. Experimental manipulation of brood reduction and parental care in cooperatively breeding white-winged choughs. *J. Anim. Ecol.* 66:683–91

8. Bowen B, Koford R, Vehrencamp S. 1989. Dispersal in the communally breeding groove-billed ani (*Crotophaga sulcirostris*). *Condor* 91:52–64

9. Bowen BS, Koford RR, Brown JL. 1995. Genetic evidence for undetected alleles and unexpected parentage in the gray-breasted jay. *Condor* 97:503–11

10. Briskie JV. 1993. Anatomical adaptations to sperm competition in Smith's longspurs and other polygynandrous passerines. *Auk* 110:875–88

11. Briskie JV. 1993. Smith's longspur, *Calcarius pictus*. In *The Birds of North America*, ed. A Poole, P Stettenheim, F Gill. 34:1–16. Washington, DC: Am. Ornithol. Union

12. Brooke MD, Hartley IR. 1995. Nesting Henderson reed-warblers (*Acrocephalus vaughani taiti*) studied by DNA fingerprinting: unrelated coalitions in a stable habitat? *Auk* 112:77–86

13. Brooker M, Rowley I. 1995. The significance of territory size and quality in the mating strategy of the splendid fairy-wren. *J. Anim. Ecol.* 64:614–627

14. Brooker MG, Rowley I, Adams M, Baverstock PR. 1990. Promiscuity: an inbreeding avoidance mechanism in a socially monogamous species? *Behav. Ecol. Sociobiol.* 26:191–200

15. Brown JL. 1974. Alternative routes to sociality in jays—with a theory for the evolution of altruism and cooperative breeding. *Am. Zool.* 14:63–80

16. Brown JL. 1975. *The Evolution of Behavior*. New York: Norton

17. Brown JL. 1978. Avian communal breeding systems. *Annu. Rev. Ecol. Syst.* 9:123–55

18. Brown JL. 1983. Cooperation—a biologist's dilemma. *Adv. Stud. Behav.* 13:1–37

19. Brown JL. 1987. *Helping and Communal Breeding in Birds: Ecology and Evolution*. Princeton, NJ: Princeton Univ. Press

20. Brown JL, Brown ER. 1984. Parental facilitation: parent-offspring relations in communally breeding birds. *Behav. Ecol. Sociobiol.* 14:203–9

21. Brown JL, Brown ER, Brown SD, Dow DD. 1982. Helpers: effects of experimental removal on reproductive success. *Science* 215:421–22

22. Bruce JP, Quinn JS, Sloane SA, White BN. 1996. DNA fingerprinting reveals monogamy in the bushtit, a cooperatively breeding species. *Auk* 113:511–16

23. Burke T, Davies NB, Bruford MW, Hatchwell BJ. 1989. Parental care and mating behaviour of polyandrous dunnocks *Prunella modularis* related to paternity by DNA fingerprinting. *Nature* 338:249–51

24. Carlisle TR, Zahavi A. 1986. Helping at the nest, allofeeding and social status in immature Arabian babblers. *Behav. Ecol. Sociobiol.* 18:339–51

25. Castro I, Minot EO, Fordham RA, Birkhead TR. 1996. Polygynandry, face-to-face copulation and sperm competition in the hihi *Notiomystis cincta* (Aves: Meliphagidae). *Ibis* 138:765–71

26. Chao L. 1997. Evolution of polyandry in a communal breeding system. *Behav. Ecol.* 8:668–74

27. Charnov EL. 1981. Kin selection and helpers at the nest: effects of paternity and biparental care. *Anim. Behav.* 29:631–32

28. Clarke AL, Saether BE, Roskaft E. 1997. Sex biases in avian dispersal: a reappraisal. *Oikos* 79:429–38

29. Clarke MF. 1984. Cooperative breeding by the Australian Bell Miner *Manorina melanophrys* Latham: a test of kin selection theory. *Behav. Ecol. Sociobiol.* 14:137–46

30. Clarke MF. 1989. The pattern of helping in the Bell Miner (*Manorina melanophrys*). *Ethology* 80:292–306

31. Clutton-Brock TH. 1991. *The Evolution of Parental Care*. Princeton, NJ: Princeton Univ. Press

32. Clutton-Brock TH, Harvey PH. 1976. Evolutionary rules and primate societies. In *Growing Points in Ethology*, ed. PPG Bateson, RA Hinde, pp. 195–237. Cambridge, UK: Cambridge Univ. Press

33. Clutton-Brock TH, Parker GA. 1995. Punishment in animal societies. *Nature* 373:209–16

34. Cockburn A. 1996. Why do so many Australian birds cooperate: social evolution in the Corvida? In *Frontiers of Population*

Ecology, ed. RB Floyd, AW Sheppard, PJ De Barro, pp. 451–72. East Melbourne, Aust.: CSIRO

35. Cooney R, Cockburn A. 1995. Territorial defence is the major function of female song in the superb fairy-wren, *Malurus cyaneus. Anim. Behav.* 49:1635–47

36. Craig JL. 1984. Are communal pukeko caught in the Prisoner's Dilemma? *Behav. Ecol. Sociobiol.* 14:147–50

37. Craig JL, Jamieson IG. 1988. Incestuous mating in a communal bird: a family affair. *Am. Natur.* 131:58–70

38. Crick HQP. 1992. Load-lightening in cooperatively breeding birds and the cost of reproduction. *Ibis* 134:56–61

39. Cullen NJ, Heinsohn R, Cockburn A. 1996. Food supplementation induces provisioning of young in cooperatively breeding white-winged choughs. *J. Avian Biol.* 27:92–94

40. Curry RL. 1988. Influence of kinship on helping behavior in Galapagos mockingbirds. *Behav. Ecol. Sociobiol.* 22:141–52

41. Curry RL, Grant PR. 1990. Galapagos mockingbirds: territorial cooperative breeding in a climatically variable environment. In *Cooperative Breeding in Birds*, ed. PB Stacey, WD Koenig, pp. 291–329. Cambridge, UK: Cambridge Univ. Press

42. Davies NB. 1985. Cooperation and conflict among dunnocks, *Prunella modularis*, in a variable mating system. *Anim. Behav.* 33:628–48

43. Davies NB. 1989. Sexual conflict and the polygamy threshold. *Anim. Behav.* 38:226–234

44. Davies NB. 1992. *Dunnock Behaviour and Social Evolution*. Oxford, UK: Oxford Univ. Press

45. Davies NB, Hartley IR, Hatchwell BJ, Langmore NE. 1996. Female control of copulations to maximize male help: a comparison of polygynandrous alpine accentors, *Prunella collaris*, and dunnocks, *P. modularis. Anim. Behav.* 51:27–47

46. Davies NB, Hatchwell BJ. 1992. The value of male parental care and its influence on reproductive allocation by male and female dunnocks. *J. Anim. Ecol.* 61:259–72

47. Davies NB, Hatchwell BJ, Robson T, Burke T. 1992. Paternity and parental effort in dunnocks *Prunella modularis*: how good are chick-feeding rules? *Anim. Behav.* 43:729–45

48. Davies NB, Houston AI. 1983. Time allocation between territories and flocks and owner-satellite conflict in foraging pied

wagtails *Motacilla alba. J. Anim. Ecol.* 52:621–34

49. Davies NB, Lundberg A. 1984. Food distribution and a variable mating system in the dunnock, *Prunella modularis. J. Anim. Ecol.* 53:895–912

50. De Coux JP. 1997. Variation of secondary sex-ratio in birds and other tetrapodes. The case of *Colius striatus nigricollis* (Coliiformes). *Rev. Ecol. Terre et la Vie* 52:37–68

51. Delay LS, Faaborg J, Naranjo J, Paz SM, DeVries T, et al. 1996. Paternal care in the cooperatively polyandrous Galapagos hawk. *Condor* 98:300–11

52. Dickinson J, Haydock J, Koenig W, Stanback M, Pitelka F. 1995. Genetic monogamy in single-male groups of acorn woodpeckers, *Melanerpes formicivorus. Mol. Ecol.* 4:765–69

53. Dickinson JL, Koenig WD, Pitelka FA. 1996. Fitness consequences of helping behavior in the western bluebird. *Behav. Ecol.* 7:168–77

54. Double MC, Cockburn A, Barry SC, Smouse PE. 1997. Exclusion probabilities for single-locus paternity analysis when related males compete for matings. *Mol. Ecol.* 6:1155–66

55. Double MC, Dawson D, Burke T, Cockburn A. 1997. Finding the fathers in the least faithful bird: a microsatellite-based genotyping system analysis for the superb fairy-wren *Malurus cyaneus. Mol. Ecol.* 6:691–93

56. Dow DD. 1978. Breeding biology and development of the young of *Manorina melanocephala*, a communally breeding Honeyeater. *Emu* 78:207–22

57. Dow DD, Whitmore MJ. 1990. Noisy miners: variations on the theme of communality. In *Cooperative Breeding in Birds*, ed. PB Stacey, WD Koenig, pp. 561–92. Cambridge, UK: Cambridge Univ. Press

58. Dugatkin LA. 1997. *Cooperation Among Animals: an Evolutionary Perspective*. Oxford, UK: Oxford Univ. Press

59. Dunn PO, Cockburn A. 1996. Evolution of male parental care in a bird with almost complete cuckoldry. *Evolution* 50:2542–48

60. Dunn PO, Cockburn A, Mulder RA. 1995. Fairy-wren helpers often care for young to which they are unrelated. *Proc. R. Soc. London Ser. B* 259:339–43

61. Duplessis MA. 1993. Do group-territorial green woodhoopoes choose roosting partners on the basis of relatedness? *Anim. Behav.* 46:612–15

62. Duplessis MA. 1993. Helping behaviour

in cooperatively-breeding green wood-hoopoes: selected or unselected trait? *Behaviour* 127:49–65

63. Duplessis MA, Weathers WW, Koenig WD. 1994. Energetic benefits of communal roosting by acorn woodpeckers during the nonbreeding season. *Condor* 96:631–37

64. Duplessis MA, Williams JB. 1994. Communal cavity roosting in green wood-hoopoes: consequences for energy expenditure and the seasonal pattern of mortality. *Auk* 111:292–99

65. Eason PK, Sherman PT. 1995. Dominance status, mating strategies and copulation success in cooperatively polyandrous white-winged trumpeters, *Psophia leucoptera* (Aves: Psophiidae). *Anim. Behav.* 49:725–36

66. Eden SF. 1987. When do helpers help? Food availability and helping in the moorhen, *Gallinula chloropus*. *Behav. Ecol. Sociobiol.* 21:191–95

67. Edwards SV, Naeem S. 1993. The phylogenetic component of cooperative breeding in perching birds. *Am. Natur.* 141:754–89

68. Emlen ST. 1978. The evolution of cooperative breeding in birds. In *Behavioural Ecology: an Evolutionary Approach*, ed. JR Krebs, NB Davies, pp. 245–81. Oxford, UK: Blackwell

69. Emlen ST. 1982. The evolution of helping. I: an ecological constraints model. *Am. Natur.* 119:29–39

70. Emlen ST. 1982. The evolution of helping. II: the role of behavioural conflict. *Am. Natur.* 119:40–53

71. Emlen ST. 1995. An evolutionary theory of the family. *Proc. Natl. Acad. Sci. USA* 92:8092–99

72. Emlen ST. 1996. Reproductive sharing in different types of kin associations. *Am. Natur.* 148:756–63

73. Emlen ST. 1997. Predicting family dynamics in social vertebrates. In *Behavioural Ecology: an Evolutionary Approach*, ed. JR Krebs, NB Davies, pp. 228–53. Oxford, UK: Blackwell Sci. 4th ed.

74. Emlen ST, Emlen JM, Levin SA. 1986. Sex-ratio selection in species with helpers-at-the-nest. *Am. Natur.* 127:1–8

75. Emlen ST, Reeve HK, Sherman PW, Wrege PH, Ratnieks FLW, et al. 1991. Adaptive versus nonadaptive explanations of behavior: the case of alloparental helping. *Am. Natur.* 138:259–70

76. Emlen ST, Wrege P. 1989. A test of alternate hypotheses for helping in white-fronted bee-eaters of Kenya. *Behav. Ecol. Sociobiol.* 25:303–19

77. Emlen ST, Wrege P. 1991. Breeding biology of white-fronted bee-eaters at Nakuru: the influence of helpers on breeder fitness. *J. Anim. Ecol.* 60:309–26

78. Emlen ST, Wrege PH. 1988. The role of kinship in helping decisions among white-fronted bee-eaters. *Behav. Ecol. Sociobiol.* 23:305–15

79. Emlen ST, Wrege PH. 1992. Parent-offspring conflict and the recruitment of helpers among bee-eaters. *Nature* 356:331–33

80. Emlen ST, Wrege PH, Demong NJ. 1995. Making decisions in the family: an evolutionary perspective. *Am. Scient.* 83:148–57

81. Faaborg J, Bednarz JC. 1990. Galapagos and Harris' hawks: divergent causes of sociality in two raptors. In *Cooperative Breeding in Birds*, ed. PB Stacey, WD Koenig, pp. 359–83. Cambridge, UK: Cambridge Univ. Press

82. Faaborg J, Parker PG, Delay L, DeVries T, Bednarz JC, et al. 1995. Confirmation of cooperative polyandry in the Galapagos hawk (*Buteo galapagoensis*). *Behav. Ecol. Sociobiol.* 36:83–90

83. Fessl B, Kleindorfer S, Hoi H, Lorenz K. 1996. Extra male parental behaviour: evidence for an alternative mating strategy in the moustached warbler *Acrocephalus melanopogon*. *J. Avian Biol.* 27:88–91

84. Fitzpatrick JW, Woolfenden GE. 1989. Florida scrub jay. In *Lifetime Reproduction in Birds*, ed. I Newton, pp. 201–18. London: Academic

85. Frank SA. 1995. Mutual policing and repression of competition in the evolution of cooperative groups. *Nature* 377:520–22

86. Gadagkar R. 1990. Evolution of eusociality: the advantage of assured fitness returns. *Philos. Trans. R. Soc. London Ser. B* 329:17–25

87. Gaston AJ. 1978. The evolution of group territorial behaviour and cooperative breeding. *Am. Natur.* 112:1091–100

88. Gibbs HL, Goldizen AW, Bullough C, Goldizen AR. 1994. Parentage analysis of multi-male social groups of Tasmanian native hens (*Tribonyx mortierii*): genetic evidence for monogamy and polyandry. *Behav. Ecol. Sociobiol.* 35:363–71

89. Goodburn SF. 1991. Territory quality or bird quality? Factors determining breeding success in the magpie *Pica pica*. *Ibis* 13:85–90

90. Gowaty PA. 1996. Field studies of parental care in birds: new data focus

questions on variation among females. *Adv. Study Behav.* 26:477–531

91. Gowaty PA. 1996. Multiple mating by females selects for males that stay: another hypothesis for social monogamy in passerine birds. *Anim. Behav.* 51:482–84

92. Grafen A. 1982. How not to measure inclusive fitness. *Nature* 298:425–26

93. Grafen A. 1984. Natural selection, kin selection and group selection. In *Behavioural Ecology: an Evolutionary Approach*, ed. JR Krebs, NB Davies, pp. 62–84. Oxford, UK: Blackwell Sci. 2nd ed.

94. Green DJ, Cockburn A, Hall ML, Osmond HL, Dunn PO. 1995. Increased opportunities for cuckoldry may be why dominant male fairy-wrens tolerate helpers. *Proc. R. Soc. London Ser. B* 262:297–303

95. Greenwood PH, Harvey PH, Perrins CM. 1979. Kin selection and territoriality in birds? A test. *Anim. Behav.* 27:645–51

96. Greenwood PJ. 1980. Mating systems, philopatry and dispersal in birds and mammals. *Anim. Behav.* 28:1140–62

97. Griffiths R, Double MC, Orr K, Dawson RJG. 1998. A simple DNA test to sex most birds. *Mol. Ecol.* 7:In press

98. Gustafsson L, Pärt T. 1990. Acceleration of senescence in the collared flycatcher *Ficedula albicollis* by reproductive costs. *Nature* 347:279–81

99. Haig SM, Walters JR, Plissner JH. 1994. Genetic evidence for monogamy in the cooperatively breeding red-cockaded woodpecker. *Behav. Ecol. Sociobiol.* 34:295–303

100. Hailman JP, McGowan KJ, Woolfenden GE. 1994. Role of helpers in the sentinel behaviour of the Florida scrub jay (*Aphelocoma c. coerulescens*). *Ethology* 97:119–40

101. Harcourt AH. 1989. Social influences on competitive ability: alliances and consequences. In *Comparative Socioecology: the Behavioural Ecology of Humans and Other Mammals*, ed. V Standen, RA Foley, pp. 223–42. Oxford, UK: Blackwell

102. Hartley IR, Davies NB. 1994. Limits to cooperative polyandry in birds. *Proc. R. Soc. London Ser. B* 257:67–73

103. Hartley IR, Davies NB, Hatchwell BJ, Desrochers A, Nebel D, et al. 1995. The polygynandrous mating system of the alpine accentor, *Prunella collaris*. 2: multiple paternity and parental effort. *Anim. Behav.* 49:789–803

104. Hatchwell BJ, Russell AF. 1996. Provisioning rules in cooperatively breeding long-tailed tits *Aegithalos caudatus*: an experimental study. *Proc. R. Soc. London Ser. B* 263:83–88

105. Haydock J, Parker PG, Rabenold KN. 1996. Extra pair paternity uncommon in the cooperatively breeding bicolored wren. *Behav. Ecol. Sociobiol.* 38:1–16

106. Heer L. 1996. Cooperative breeding by alpine accentors *Prunella collaris*: polygynandry, territoriality and multiple paternity. *J. Ornithol.* 137:35–51

107. Heinsohn R, Cockburn A. 1994. Helping is costly to young birds in cooperatively breeding white-winged choughs. *Proc. R. Soc. London Ser. B* 256:299–303

108. Heinsohn R, Legge S, Barry S. 1997. Extreme bias in sex allocation in eclectus parrots. *Proc. R. Soc. London Ser. B* 264:1325–29

109. Heinsohn R, Packer C. 1995. Complex cooperative strategies in group-territorial African lions. *Science* 269:1260–62

110. Heinsohn RG. 1991. Evolution of obligate cooperative breeding in white-winged choughs: a statistical approach. *Acta XX Congr. Int. Ornithol.* 3:1309–16

111. Heinsohn RG. 1991. Kidnapping and reciprocity in cooperatively breeding white-winged choughs. *Anim. Behav.* 41:1097–1100

112. Heinsohn RG. 1991. Slow learning of foraging skills and extended parental care in cooperatively breeding white-winged choughs. *Am. Natur.* 137:864–81

113. Heinsohn RG. 1992. Cooperative enhancement of reproductive success in white-winged choughs. *Evol. Ecol.* 6:97–114

114. Heinsohn RG, Cockburn A, Cunningham RB. 1988. Foraging, delayed maturity and cooperative breeding in white-winged choughs (*Corcorax melanorhamphos*). *Ethology* 77:177–86

115. Heinsohn RG, Cockburn A, Mulder RA. 1990. Avian cooperative breeding: old hypotheses and new directions. *Trends Ecol. Evol.* 5:403–7

116. Hemmings AD. 1989. Communally breeding skuas: breeding success of pairs, trios and groups of *Catharacta lonnbergi* on the Chatham Islands, New Zealand. *J. Zool.* 218:393–405

117. Hemmings AD. 1994. Cooperative breeding in the skuas (*Stercorariidae*): history, distribution and incidence. *J. R. Soc. N. Z.* 24:245–60

118. Heppell SS, Walters JR, Crowder LB. 1994. Evaluating management alternatives for red-cockaded woodpeckers: a modeling approach. *J. Wildl. Manage.* 58:479–87

119. Högstedt G. 1980. Evolution of clutch size in birds: adaptive variation in relation to territory quality. *Science* 210:1148–50

120. Horn PL, Rafalski JA, Whitehead PJ. 1996. Molecular genetic (RAPD) analysis of breeding magpie geese. *Auk* 113:552–57

121. Innes KE, Johnston RE. 1996. Cooperative breeding in the white-throated magpie-jay: how do auxiliaries influence nesting success? *Anim. Behav.* 51:519–33

122. Jamieson IG. 1986. The functional approach to behaviour: is it useful? *Am. Natur.* 127:195–208

123. Jamieson IG. 1991. The unselected hypothesis for the evolution of helping behavior: too much or too little emphasis on natural selection? *Am. Natur.* 138:271–82

124. Jamieson IG. 1997. Testing reproductive skew models in a communally breeding bird, the pukeko, *Porphyrio porphyrio*. *Proc. R. Soc. London Ser. B* 264:335–40

125. Jamieson IG, Quinn JS. 1997. Problems with removal experiments designed to test the relationship between paternity and parental effort in a socially polyandrous bird. *Auk* 114:291–95

126. Jamieson IG, Quinn JS, Rose PA, White BN. 1994. Shared paternity is a result of an egalitarian mating system in a communally breeding bird, the pukeko. *Proc. R. Soc. London Ser. B* 257:271–77

127. Jones CS, Lessells CM, Krebs JR. 1991. Helpers-at-the-nest in European bee-eaters (*Merops apiaster*): a genetic analysis. In *DNA Fingerprinting: Approaches and Applications*, ed. T Burke, G Dolf, AJ Jeffreys, R Wolff, pp. 169–92. Basel: Birkhäuser Verlag

128. Joste N, Ligon JD, Stacey PB. 1985. Shared paternity in the acorn woodpecker (*Melanerpes formicivorus*). *Behav. Ecol. Sociobiol.* 17:39–41

129. Keane B, Creel SR, Waser PM. 1996. No evidence of inbreeding avoidance or inbreeding depression in a social carnivore. *Behav. Ecol.* 7:480–89

130. Keller L, Genoud M. 1997. Extraordinary lifespans in ants: a test of evolutionary theories of ageing. *Nature* 389:958–60

131. Keller L, Reeve HK. 1994. Partitioning of reproduction in animal societies. *Trends Ecol. Evol.* 9:98–102

132. Kemp A. 1995. *The Hornbills: Bucerotiformes*. Oxford, UK: Oxford Univ. Press

133. Ketterson ED, Nolan V. 1994. Male parental behavior in birds. *Annu. Rev. Ecol. Syst.* 25:601–28

134. Khan MZ, Walters JR. 1997. Is helping a beneficial learning experience for red-cockaded woodpecker (*Picoides bo-*realis) helpers. *Behav. Ecol. Sociobiol.* 41:69–73

135. Koenig WD. 1982. Ecological and social factors affecting hatchability of eggs. *Auk* 99:526–36

136. Koenig WD. 1990. Opportunity of parentage and nest destruction in polygynandrous acorn woodpeckers, *Melanerpes formicivorus*. *Behav. Ecol.* 1:55–61

137. Koenig WD, Mumme RL. 1987. *Population Biology of the Cooperatively Breeding Acorn Woodpecker*. Princeton, NJ: Princeton Univ. Press

138. Koenig WD, Mumme RL. 1990. Levels of analysis and the functional significance of helping behavior. In *Interpretation and Explanation in the Study of Animal Behavior. II: Explanation, Evolution and Adaptation*, ed. M Bekoff, D Jamieson, pp. 269–303. Boulder, CO: Westview

139. Koenig WD, Mumme RL, Pitelka FA. 1983. Female roles in cooperative breeding acorn woodpeckers. In *Social Behaviour of Female Vertebrates*, ed. SK Wasser, pp. 235–61. New York: Academic

140. Koenig WD, Mumme RL, Stanback MT, Pitelka FA. 1995. Patterns and consequences of egg destruction among joint nesting acorn woodpeckers. *Anim. Behav.* 50:607–21

141. Koenig WD, Pitelka FA, Carmen WJ, Mumme RL, Stanback MT. 1992. The evolution of delayed dispersal in cooperative breeders. *Q. Rev. Biol.* 67:111–50

142. Koenig WD, Van Vuren D, Hooge PN. 1996. Detectability, philopatry, and the distribution of dispersal distances in vertebrates. *Trends Ecol. Evol.* 11:514–17

143. Komdeur J. 1992. Importance of habitat saturation and territory quality for evolution of cooperative breeding in the Seychelles warbler. *Nature* 358:493–95

144. Komdeur J. 1994. The effect of kinship on helping in the cooperative breeding Seychelles warbler (*Acrocephalus sechellensis*). *Proc. R. Soc. London Ser. B* 256:47–52

145. Komdeur J. 1994. Experimental evidence for helping and hindering by previous offspring in the cooperative-breeding Seychelles warbler *Acrocephalus sechellensis*. *Behav. Ecol. Sociobiol.* 34:175–86

146. Komdeur J. 1996. Facultative sex ratio bias in the offspring of Seychelles warblers. *Proc. R. Soc. London Ser. B* 263:661–66

147. Komdeur J. 1996. Influence of helping and breeding experience on reproductive performance in the Seychelles warbler:

a translocation experiment. *Behav. Ecol.* 7:326–33

148. Komdeur J, Daan S, Tinbergen J, Mateman C. 1997. Extreme adaptive modification in sex ratio of the Seychelles warbler's eggs. *Nature* 385:522–25

149. Lambert DM, Millar CD, Jack K, Anderson S, Craig JL. 1994. Single- and multilocus DNA fingerprinting of communally breeding pukeko: do copulations or dominance ensure reproductive success? *Proc. Natl. Acad. Sci. USA* 91:9641–45

150. Langen TA. 1996. The mating system of the white-throated magpie-jay *Calocitta formosa* and Greenwood's hypothesis for sex-biased dispersal. *Ibis* 138:506–13

151. Langen TA. 1996. Skill acquisition and the timing of natal dispersal in the white-throated magpie-jay, *Calocitta formosa. Anim. Behav.* 51:575–88

152. Langen TA. 1996. Social learning of a novel foraging skill by white-throated magpie-jays (*Calocitta formosa,* Corvidae): a field experiment. *Ethology* 102: 157–66

153. Leonard ML, Horn AG, Eden SF. 1989. Does juvenile helping enhance breeder reproductive success? A removal experiment on moorhens. *Behav. Ecol. Sociobiol.* 25:357–62

154. Lessells CM. 1990. Helping at the nest in European bee-eaters: who helps and why? In *Population Biology of Passerine Birds: An Integrated Approach,* ed. J Blondel, A Gosler, J-D Lebreton, R McLeery, pp. 357–68. Berlin: Springer-Verlag

155. Lessells CM, Avery MI. 1987. Sex-ratio selection in species with helpers at the nest: some extensions of the repayment model. *Am. Natur.* 129:610–20

156. Lessells CM, Avery MI, Krebs JR. 1994. Nonrandom dispersal of kin: why do European bee-eater (*Merops apiaster*) brothers nest close together. *Behav. Ecol.* 5:105–13

157. Li SH, Huang YJ, Brown JL. 1997. Isolation of tetranucleotide microsatellites from the Mexican jay *Aphelocoma ultramarina. Mol. Ecol.* 6:499–501

158. Ligon JD. 1983. Cooperation and reciprocity in avian social systems. *Am. Natur.* 121:366–84

159. Ligon JD, Ligon SH. 1978. Communal breeding in green woodhoopoes as a case for reciprocity. *Nature* 276:496–98

160. Ligon JD, Ligon SH. 1990. Green woodhoopoes: life history traits and sociality. In *Cooperative Breeding in Birds,* ed. PB Stacey, WD Koenig, pp. 33–65. Cambridge, UK: Cambridge Univ. Press

161. Macedo RHF, Bianchi CA. 1997. When birds go bad: circumstantial evidence for infanticide in the communal South American guira cuckoo. *Ethol. Ecol. Evol.* 9:45–54

162. Magrath RD, Whittingham LA. 1997. Subordinate males are more likely to help if unrelated to the breeding female in cooperatively breeding white-browed scrubwrens. *Behav. Ecol. Sociobiol.* 41:185–92

163. Magrath RD, Yezerinac SM. 1997. Facultative helping does not influence reproductive success or survival in cooperatively breeding white-browed scrubwrens. *J. Anim. Ecol.* 66:658–70

164. Marzluff JM, Balda RP. 1990. Pinyon jays: making the best of a bad job by helping. In *Cooperative Breeding in Birds,* ed. PB Stacey, WD Koenig, pp. 199–237. Cambridge, UK: Cambridge Univ. Press

165. McDonald DB, Potts WK. 1994. Cooperative display and relatedness among males in a lek-mating bird. *Science* 266:1030–32

166. McDonald DW, Carr GM. 1989. Food security and the rewards of tolerance. In *Comparative Socioecology: the Behavioural Ecology of Humans and Other Mammals,* ed. V Standen, RA Foley, pp. 75–99. Oxford, UK: Blackwell

167. McGowan KJ, Woolfenden GE. 1989. A sentinel system in the Florida scrub jay. *Anim. Behav.* 37:1000–6

168. McRae SB. 1996. Family values: costs and benefits of communal nesting in the moorhen. *Anim. Behav.* 52:225–45

169. McRae SB, Burke T. 1996. Intraspecific brood parasitism in the moorhen: parentage and parasite-host relationships determined by DNA fingerprinting. *Behav. Ecol. Sociobiol.* 38:115–29

170. Millar CD, Anthony I, Lambert DM, Stapleton PM, Bergmann CC, et al. 1994. Patterns of reproductive success determined by DNA fingerprinting in a communally breeding oceanic bird. *Biol. J. Linn. Soc.* 52:31–48

171. Monadjem A, Owen-Smith N, Kemp AC. 1995. Aspects of the breeding biology of the arrowmarked babbler *Turdoides jardineii* in South Africa. *Ibis* 137:515–18

172. Monerret R-J. 1983. L'aid à l'élevage chez le faucon pèlerin. *Alauda* 51:241–20

173. Mulder RA. 1995. Natal and breeding dispersal in a co-operative, extra-group-mating bird. *J. Avian Biol.* 26:234–40

174. Mulder RA. 1997. Extra-group courtship displays and other reproductive tactics of superb fairy-wrens. *Aust. J. Zool.* 45:131–43

175. Mulder RA, Cockburn A. 1993. Sperm competition and the reproductive anatomy of male superb fairy-wrens. *Auk* 110:588–93

176. Mulder RA, Dunn PO, Cockburn A, Lazenby-Cohen KA, Howell MJ. 1994. Helpers liberate female fairy-wrens from constraints on extra-pair mate choice. *Proc. R. Soc. London Ser. B* 255:223–29

177. Mulder RA, Langmore NE. 1993. Dominant males punish helpers for temporary defection in superb fairy-wrens. *Anim. Behav.* 45:830–33

178. Mumme RL. 1992. Do helpers increase reproductive success—an experimental analysis in the Florida scrub jay. *Behav. Ecol. Sociobiol.* 31:319–28

179. Mumme RL, Koenig WD, Pitelka FA. 1983. Reproductive competition in the acorn woodpecker: sisters destroy each other's eggs. *Nature* 306:583–84

180. Mumme RL, Koenig WD, Pitelka FA. 1990. Individual contributions to cooperative nest care in the acorn woodpecker. *Condor* 92:360–68

181. Mumme RL, Koenig WD, Ratnieks FLW. 1989. Helping behaviour, reproductive value, and the future component of direct fitness. *Anim. Behav.* 38:331–43

182. Mumme RL, Koenig WD, Zink RM, Marten JA. 1985. Genetic variation and parentage in a California population of acorn woodpeckers. *Auk* 102:305–12

183. Nakamura M. 1998. Multiple mating and cooperative breeding in polygynandrous alpine accentors. II: Male mating tactics. *Anim. Behav.* 55:277–89

184. Noske RA. 1991. A demographic comparison of cooperative and non-cooperative treecreepers (Climacteridae). *Emu* 91:73–86

185. Pettifor RA, Perrins CM, McCleery RH. 1988. Individual optimization of clutch size in great tits. *Nature* 336:160–62

186. Piper WH. 1994. Courtship, copulation, nesting behavior and brood parasitism in the Venezuelan stripe-backed wren. *Condor* 96:654–71

187. Piper WH, Parker PG, Rabenold KN. 1995. Facultative dispersal by juvenile males in the cooperative stripe-backed wren. *Behav. Ecol.* 6:337–42

188. Piper WH, Slater G. 1993. Polyandry and incest avoidance in the cooperative stripe-backed wren of Venezuela. *Behaviour* 124:227–47

189. Poiani A. 1992. Feeding of the female breeder by male helpers in the bell miner *Manorina melanophrys*. *Emu* 92:233–37

190. Poiani A. 1993. Effects of clutch size manipulations on reproductive behaviour and nesting success in the cooperatively breeding bell miner (*Manorina melanophrys*). *Evol. Ecol.* 7:329–56

191. Poiani A. 1993. Social structure and the development of helping behaviour in the bell miner (*Manorina melanophrys*, Meliphagidae). *Ethology* 93:62–80

192. Pöldmaa T, Holder K. 1997. Behavioural correlates of monogamy in the noisy miner, *Manorina melanocephala*. *Anim. Behav.* 54:571–78

193. Pöldmaa T, Montgomerie R, Boag P. 1995. Mating system of the cooperatively breeding noisy miner *Manorina melanocephala*, as revealed by DNA profiling. *Behav. Ecol. Sociobiol.* 37:137–43

194. Pruett-Jones SG, Lewis MJ. 1990. Sex ratio and habitat limitation promote delayed dispersal in superb fairy-wrens. *Nature* 348:541–42

195. Pusey A, Wolf M. 1996. Inbreeding avoidance in animals. *Trends Ecol. Evol.* 11:201–6

196. Queller DC. 1994. Extended parental care and the origin of eusociality. *Proc. R. Soc. London Ser. B* 256:105–11

197. Queller DC. 1996. The measurement and meaning of inclusive fitness. *Anim. Behav.* 51:229–32

198. Queller DC. 1997. Why do females care more than males? *Proc. R. Soc. London Ser. B* 24:1555–57

199. Quinn JS, Macedo R, White BN. 1994. Genetic relatedness of communally breeding guira cuckoos. *Anim. Behav.* 47:515–29

200. Rabenold KN. 1984. Cooperative enhancement of reproductive success in tropical wren societies. *Ecology* 65:871–85

201. Rabenold KN. 1990. *Campylorhynchus* wrens: the ecology of delayed dispersal and cooperation in the Venezuelan savanna. In *Cooperative Breeding in Birds*, ed. PB Stacey, WD Koenig, pp. 159–96. Cambridge, UK: Cambridge Univ. Press

202. Rabenold PP, Rabenold KN, Piper WH, Haydock J, Zack SN. 1990. Shared paternity revealed by genetic analysis in cooperatively breeding tropical wrens. *Nature* 348:538–40

203. Reeve HK, Keller L. 1995. Partitioning of reproduction in mother-daughter versus sibling associations: a test of optimal skew theory. *Am. Natur.* 145:119–32

204. Reeve HK, Keller L. 1997. Reproductive bribing and policing: evolutionary mechanisms for the suppression of within-group selfishness. *Am. Natur.* 150:S42–58

205. Reeve HK, Shellman-Reeve JS. 1997. The general protected invasion theory: sex biases in parental and alloparental care. *Evol. Ecol.* 11:357–70

206. Reeve HK, Westneat DF, Noon WA, Sherman PW, Aquadro CF. 1990. DNA 'fingerprinting' reveals high levels of inbreeding in colonies of the eusocial naked mole-rat. *Proc. Natl. Acad. Sci. USA* 87: 2496–500

207. Reyer H-U. 1990. Pied kingfishers: ecological causes and reproductive consequences of cooperative breeding. In *Cooperative Breeding in Birds*, ed. PB Stacey, WD Koenig, pp. 529–57. Cambridge, UK: Cambridge Univ. Press

208. Reyer H-U, Westerterp K. 1985. Parental energy expenditure: a proximate cause of helper recruitment in the pied kingfisher (*Ceryle rudis*). *Behav. Ecol. Sociobiol.* 17:363–69

209. Richner H, Christie P, Oppliger A. 1994. Paternal investment affects prevalence of malaria. *Proc. Natl. Acad. Sci. USA* 92:1192–94

210. Rowley I, Russell E. 1997. *Fairy-wrens and Grasswrens: Maluridae.* Oxford, UK: Oxford Univ. Press

211. Russell EM. 1989. Cooperative breeding: a Gondwanan perspective. *Emu* 89:61–62

212. Russell EM, Rowley I. 1993. Demography of the cooperatively breeding splendid fairy-wren, *Malurus splendens* (Maluridae). *Aust. J. Zool.* 41:475–505

213. Russell EM, Rowley I. 1993. Philopatry or dispersal: competition for territory vacancies in the splendid fairy-wren, *Malurus splendens. Anim. Behav.* 45:519–39

214. Schoech SJ, Mumme RL, Wingfield JC. 1996. Delayed breeding in the cooperatively breeding Florida scrub-jay (*Aphelocoma coerulescens*): inhibition or the absence of stimulation? *Behav. Ecol. Sociobiol.* 39:77–90

215. Schulze-Hagen K, Swatschek I, Dyrcz A, Wink M. 1993. Multiple vaterschaften in bruten des Seggenrohrsängers *Acrocephalus paludicola*: erste ergebnisse des DNA-fingerprintings. *J. Ornithol.* 134: 145–54

216. Selander RK. 1964. Speciation in wrens of the genus *Campylorhynchus. Univ. Calif. Publ. Zool.* 74:1–224

217. Sherley GH. 1989. Benefits of courtship-feeding for rifleman (*Acanthisitta chloris*) parents. *Behaviour* 109:309–18

218. Sherley GH. 1990. Co-operative breeding in rifleman (*Acanthisitta chloris*): benefits to parents, offspring and helpers. *Behaviour* 112:1–22

219. Sherman PT. 1995. Breeding biology of white-winged trumpeters (*Psophia leucoptera*) in Peru. *Auk* 112:285–95

220. Sherman PT. 1995. Social organization of cooperatively polyandrous white-winged trumpeters (*Psophia leucoptera*). *Auk* 112:296–309

221. Skutch AF. 1961. Helpers among birds. *Condor* 63:198–226

222. Sloane SA. 1996. Incidence and origins of supernumeraries at bushtit (*Psaltriparus minimus*) nests. *Auk* 113:757–70

223. Stacey PB, Koenig WD, eds. 1990. *Cooperative Breeding in Birds: Long-Term Studies of Ecology and Behavior.* Cambridge, UK: Cambridge Univ. Press

224. Stacey PB, Ligon JD. 1987. Territory quality and dispersal options in the acorn woodpecker and a challenge to the habitat-saturation model of cooperative breeding. *Am. Natur.* 130:654–76

225. Stacey PB, Ligon JD. 1991. The benefits-of-philopatry hypothesis for the evolution of cooperative breeding: variation in territory quality and group size effects. *Am. Natur.* 137:831–46

226. Stallcup JA, Woolfenden GE. 1978. Family status and contributions to breeding by Florida scrub jays. *Anim. Behav.* 26:1144–56

227. Trivers RL. 1972. Parental investment and sexual selection. In *Sexual Selection and the Descent of Man*, ed. BG Campbell, pp. 136–79. Chicago: Aldine

228. Tuttle EM, Pruett-Jones S, Webster MS. 1996. Cloacal protuberances and extreme sperm production in Australian fairy-wrens. *Proc. R. Soc. London Ser. B* 263: 1359–64

229. Vehrencamp SL. 1978. The adaptive significance of communal nesting in groove-billed anis (*Crotophaga sulcirostris*). *Behav. Ecol. Sociobiol.* 4:1–33

230. Vehrencamp SL. 1983. A model for the evolution of despotic versus egalitarian societies. *Anim. Behav.* 31:667–82

231. Vehrencamp SL. 1983. Optimal degree of skew in cooperative societies. *Am. Zool.* 23:327–35

232. Vehrencamp SL, Koford RR, Bowen BS. 1988. The effect of breeding-unit size on fitness components in groove-billed anis. In *Reproductive Success: Studies on Individual Variation in Contrasting Breeding Systems*, ed. TH Clutton-Brock, pp. 291–304. Chicago: Univ. Chicago Press

233. Walters JR. 1990. Red-cockaded woodpeckers: a primitive cooperative breeder. In *Cooperative Breeding in Birds*, ed. PB Stacey, WD Koenig, pp. 69–101. Cambridge, UK: Cambridge Univ. Press

234. Walters JR. 1991. Application of ecological principles to the management of endangered species: the case of the red-cockaded woodpecker. *Annu. Rev. Ecol. Syst.* 22:505–23

235. Walters JR, Doerr PD, Carter JH. 1992. Delayed dispersal and reproduction as a life-history tactic in cooperative breeders: fitness calculations from red-cockaded woodpeckers. *Am. Natur.* 139:623–43

236. Wcislo WT, Danforth BN. 1997. Secondarily solitary: the evolutionary loss of social behaviour. *Trends Ecol. Evol.* 12:468–74

237. Whitney G. 1976. Genetic substrates for the initial evolution of human sociality. I: Sex chromosome mechanisms. *Am. Natur.* 110:867–75

238. Whittingham LA, Dunn PO, Magrath RD. 1997. Relatedness, polyandry and extragroup paternity in the cooperatively-breeding white-browed scrubwren (*Sericornis frontalis*). *Behav. Ecol. Sociobiol.* 40:261–70

239. Woolfenden GE. 1975. Florida scrub jay helpers at the nest. *Auk* 92:1–15

240. Woolfenden GE, Fitzpatrick JW. 1984. *The Florida Scrub Jay: Demography of a Cooperative Breeding Bird.* Princeton, NJ: Princeton Univ. Press

241. Woolfenden GE, Fitzpatrick JW. 1986. Sexual asymmetries in the life history of the Florida scrub jay. In *Ecological Aspects of Social Evolution:*
Birds and Mammals, ed. DI Rubenstein, RW Wrangham, pp. 87–107. Princeton, NJ: Princeton Univ. Press

242. Wrege PH, Emlen ST. 1987. Biochemical determination of parental uncertainty in white-fronted bee-eaters. *Behav. Ecol. Sociobiol.* 20:153–60

243. Wrege PH, Emlen ST. 1994. Family structure influences mate choice in white-fronted bee-eaters. *Behav. Ecol. Sociobiol.* 35:185–91

244. Wright J, Cuthill I. 1989. Manipulation of sex differences in parental care. *Behav. Ecol. Sociobiol.* 25:171–81

245. Yamagishi S, Urano E, Eguchi K. 1995. Group composition and contributions to breeding by rufous vangas *Schetba rufa* in Madagascar. *Ibis* 137:157–61

246. Zack S. 1990. Coupling delayed breeding with short-distance dispersal in cooperatively breeding birds. *Ethology* 86:265–86

247. Zack S, Rabenold KN. 1989. Assessment, age and proximity in dispersal contests among cooperative wrens: field experiments. *Anim. Behav.* 38:235–47

248. Zahavi A. 1990. Arabian babblers: the quest for status in a cooperative breeder. In *Cooperative Breeding in Birds*, ed. PB Stacey, WD Koenig, pp. 105–30. Cambridge, UK: Cambridge Univ. Press

249. Zahavi A. 1995. Altruism as a handicap—the limitations of kin selection and reciprocity. *J. Avian Biol.* 26:1–3

Annu. Rev. Ecol. Syst. 1998. 29:179–206

THE ECOLOGICAL EVOLUTION OF REEFS

Rachel Wood

Department of Earth Sciences, University of Cambridge, Downing Street,
Cambridge CB2 3EQ, United Kingdom

KEY WORDS: reef, ecology, evolution, predation, photosymbiosis

ABSTRACT

Many groups of extinct and extant organisms have aggregated to form reefs for over 3.5 billion yr (Ga). Most of these communities, however, grew under ecological and environmental controls profoundly different from those that govern modern coral reefs. Not only has the global distribution of reefs varied considerably through geological time—determined largely by sea level, and latitudinal temperature/saturation gradients—but more importantly the trophic demands of reef-building organisms have changed, as has the degree of biological disturbance faced by sessile biota in shallow marine environments.

Reefs differentiated into open surface and cryptic communities as soon as open frameworks developed in the Proterozoic, some 1.9 million yr ago (mya) and diverse and complex ecosystems were established by the early Cambrian (~520 mya). Calcified heterotrophs were conspicuous in reefs during the Paleozoic and early Mesozoic, but considerable rigidity was imparted to these often otherwise fragile communities both by indirect microbial processes that induced the formation of carbonate and by rapid early cementation. While photosymbiosis was probably acquired by scleractinian corals early in their history (~210 mya), this does not appear to have immediately conferred a superior reef-building ability. Large, modular corals and coralline algae showed notable powers of regeneration after partial mortality but poor ability to compete with macroalgae for limited substrate in the absence of intense herbivory; they did not rise to prominence in reef communities until the early to mid-Cenozoic. This may be related to the appearance at this time of major predator groups such as echinoids, limpets, and particularly fish that are capable of rapid algal denudation and excavation. The existence of this reciprocal relationship is corroborated by the observation that branching corals, which appear to flourish because and not in spite of breakage,

0066-4162/98/1120-0179$08.00

show a particularly dramatic increase in diversity coincident with the increase in predation pressure from the late Mesozoic onwards.

INTRODUCTION

Although estimates suggest that modern coral reefs occupy only 0.2% of Earth's ocean area (47), their influence is global and multifaceted. As substantial topographic structures, coral reefs protect coastlines from erosion and help create sheltered harbors. Reefs and their associated carbonate sediments are also important as storehouses of organic carbon and as regulators of atmospheric CO_2, which in turn may influence climate and sea-level fluctuations (61). Being highly porous structures, ancient subsurface reefs provide extensive reservoirs for oil and gas. From a biological standpoint, however, the greatest significance of reefs lies in the fact that they generate and maintain a substantial proportion of tropical marine biodiversity.

It is widely assumed that the complex and varied photosymbiotic association of scleractinian corals with dinoflagellate algae provides both driving energy and physical structure for the whole coral reef community. Photosymbiosis confers a variety of advantages to reef-building organisms, including rapid rates of growth and calcification, and the potential for host selection of optimal symbionts in the highly dynamic tropical marine environment (2, 11–13, 77). In addition, a substantial and increasing body of research continues to underline the importance of predation, in particular herbivory, in maintaining the coral reef ecosystem (see summaries in 32, 37). Yet the fossil record of reef-building shows that both the acquisition of photosymbiosis and the appearance of modern predator, herbivore, and bioeroding groups are relatively recent geological occurrences: Many ancient reefs clearly grew under profoundly different ecological controls than those that govern the functioning of modern coral reefs. Moreover, the global distribution of reefs has varied considerably through geological time, determined largely by sea-level, geochemical, and climatic fluctuations. Present day sea level is relatively low compared to that in much of the geological record such that the area of shallow water tropical seas is small, resulting not only in a reduced volume of shallow-water carbonates being formed, but also in an absence of analogs for the very extensive shallow seas (known as epeiric or epicontinental seas) that were common when sea levels were high. The extensive carbonate platform reefs and atolls of today are also the product of an unusually prolonged period of stable sea level, which, together with relict topography, have exerted a strong influence over modern reef form and style of sedimentation. Present-day climate is also relatively cool, with well-developed polar ice caps. During significant periods of geological time, ice caps were not present, and at times such as the mid-Cretaceous, northern hemisphere mean annual surface temperatures may have been 15 to 25° warmer than today (4, 62). In addition, 60% of modern carbonate production is accounted for by calcareous plankton that produce pelagic ooze deposited on the deep ocean floor, but before the evolution of such plankton during the mid-Mesozoic, shallow marine carbonate deposits represented up to 90% of global production (27). Consequently, for much of geological time, marine carbonate distribution, and possibly carbonate saturation levels, were very different from those found in modern seas.

It is clear then that the origins of the ecosystems that dominate Earth today cannot be fully understood without a historical perspective. Reefs are highly susceptible to environmental change, and a substantial proportion of modern coral reefs are currently under threat (5). In the past few years, we have learned how reefs respond to global change and catastrophe on a human timescale, but the origins of many characters we seek to explain are in the past. Only study of the geological record can reveal the dynamics of long-term evolutionary change.

This review details some of the recent advances in the reconstruction of ancient reef ecologies and explores the relationship between evolutionary innovation and environmental opportunity. The last few years have seen some radical developments of our understanding of how modern reef communities and reef corals function, but ancient reefs have too often been studied solely as geological phenomena with little regard to biological interactions. Yet, being the result of in situ growth, reefs frequently preserve exquisite details of past ecological interactions lost from most other fossil communities. Such ecological interactions are of profound importance and interest, as they are the ultimate determinants of reef growth. The theme developed here is to explore the evolution of reefs as biological phenomena: Reefs concern the biological occupation of space, but the demands on sessile reef organisms to achieve this have changed over the course of geological time.

What Is a Reef?

If only the characteristics of modern coral reefs are taken as diagnostic, then very few fossil examples would qualify as reefs. This is due to two interrelated reasons: Modern coral reefs appear to possess unique ecological features and environmental requirements, but these characteristics are notoriously difficult to detect in the geological record.

Attempts to understand ancient ecologies by direct reference to modern communities is hampered by several obstacles. First, processes of physical and biological destruction, the incomplete nature of the fossil record, and diagenesis all obscure the original reef community structure and form. As a result, detection of a wave-resistant framework, determination of the depth at which a reef community grew, and reconstruction of original topographic relief can all be problematic in ancient examples. Second, the impossibility of dating and correlating geological strata at the fine temporal and spatial scale appropriate to modern ecological analysis makes the direct transfer of information and techniques inappropriate. This problem is compounded by the general coarsening of temporal resolution with increased time back from the present. Third, there are clearly no modern analogs for many ancient ecologies, not only because many of the constituent organisms are extinct, but also because the environmental conditions under which they grew have changed radically. A full appreciation of ancient reef ecology clearly both calls for a nonuniformitarian approach and also necessitates the formulation of universally applicable criteria for the recognition of reefs.

Reefs are here considered to form as the direct or indirect result of organic activity, developing due to the aggregation of sessile epibenthic marine organisms, with the resultant higher rate of in-situ carbonate production than in surrounding sediments.

The Diversity of Modern Reefs

Some understanding of the processes common to all reef formation can be gained through an appreciation of the diversity of communities that form reefs in modern seas. In addition to corals, microbes, algae, and many invertebrates aggregate to form reefs in modern seas, in an apparently wide range of environmental settings. But as reef formation is contingent upon the successful occupation of space by sessile organisms, it is not surprising that all modern reef-building communities develop under specific environmental conditions that allow such growth. Possession of a sessile nature makes organisms vulnerable to disturbance—both to physical destruction and biological attack by predation—which is probably the major control on both the distribution and the morphology, of epibenthic organisms (19). The distribution of modern reefs shows that their development is dictated by avoidance of competition and predation, or by specialized adaptation to physical or biological attack.

Modern coral reefs are specialized communities that develop in tropical and subtropical environments characterized by warm temperatures, aragonite supersaturation (that is where concentrations of calcium and carbonate exceed the thermodynamic mineral solubility product), and high light intensities, where stable, elevated substrates are available for colonization, and typically where nutrient levels are low. Except for some reefs built by cyanobacteria and calcified algae, many modern reefs are constructed by organisms that possess little inherent stability under agitated conditions and, being nonphotosynthetic, are not dependent upon light. Consequently, they often develop in low-energy settings, either in relatively deep waters or in marginal, shallow water environments. These include sabellariid and serpulid polychaetes, vermetid gastropods, bryozoans, and cyanobacterial communities (60a, 78a). Deep-water habitats, such as the area supporting bryozoan-sponge reef growth off the shelf of south-eastern Australia, receive minimal or no light but offer abundant nutrients (7). These environments are relatively benign in that predation pressure and turbulence are often low. Shallow water noncoral reefs are usually constructed by organisms able to grow under conditions (such as nonmarine salinity) that exclude normal marine competitors or predators. Such reefs often grow in embayments protected from wave destruction.

What Is a Reef Community?

Current ecological thought considers that biological communities are not fixed entities with precise boundaries, but are chance associations of species with similar ecological requirements. This notion is supported by several models that adequately predict community composition only on the basis of immigration and extinction, spatial distribution of environments, and the size of the species pool (23, 39). Communities constantly change through local extinction and recruitment of component species, and new communities, developed in previously unoccupied habitats, are composed of species from the available population that have geographic access to the new area (16). It has been well documented that formerly co-occurring species are now found in associations entirely different from those they occupied in previous times, e.g. in Indo-Pacific reefs, corals responded to changing shelf area and water depth as sea level fluctuated during the Pleistocene by changing community membership (68). Associations within local reef communities have clearly changed as new species have evolved while order taxa persist. Clearly, community traits are not heritable: Natural selection cannot operate upon communities, only upon their constituent species.

Such observations undermine the notion of strong cohesion with communities and support the view that communities are chance associations. While it is now commonly accepted that many marine communities are not discrete entities, this has not yet been universally accepted for reefs (43, 63). The sheer abundance of specialized interactions and symbioses as well as diversity of ecological niches seem to indicate long-lived coexistence between organisms. However, it is now apparent (49a) that such seemingly specific interactions are, in fact, frequently modified as species membership of a reef community changes so that, like the constituent species, the way in which organisms interact is not fixed. This suggests an enormous ecological redundancy of species in reefs, that is, many species can occupy a broadly similar niche.

PHYSICOCHEMICAL CONTROLS ON
THE DISTRIBUTION OF REEFS

Much debate has concerned the relative importance of biologically and chemically controlled factors in predicting suitable sites for carbonate production and, in particular, reef growth (cf 53a, 62). It is now clear that elevated ambient seawater supersaturation with respect to calcium carbonate broadly corresponds to sites of carbonate sediment generation, and this is related to a number of factors, most importantly surface seawater temperature (12, 14, 15, 47, 62). Most notable is that the broad documented patterns of change in accumulation rate with latitude are very similar to those defined by saturation values, and those calculated from laboratory growth of carbonate minerals are also in very close agreement (62). These data support the contention that the latitudinal variation of seawater composition is therefore indistinguishable from biological factors related to the ecological demands of carbonate-producing biota in controlling the broad global distribution, rates, and composition of shallow water carbonate accumulation (62).

Rates of shallow-water carbonate deposition are strongly dependent upon the latitudinal temperature/saturation gradient, with carbonate deposition greatest between 15° and 20°, and decreasing both polewards and toward equatorial regions. The latitudinal distribution of carbonate deposits thus records ancient climatic gradients in the temperature and saturation state of seawater. For example, the latitudinal extent of Cretaceous limestone deposition occurred some 10° more poleward that at present, suggesting that the temperature between the equator and 38°N was some 5°C warmer than at present (62). However, higher atmospheric pCO_2 most likely existed during the Cretaceous (4), which may have controlled the dominant mineralogy of carbonate-secreting marine organisms, favoring those with calcitic rather than aragonitic skeletons (36). This might explain the dominance of rudist bivalves with outer calcitic layers, rather than wholly aragonitic corals at this time.

THE COMPLEXITY OF ANCIENT
REEF ECOSYSTEMS

In principle, any sessile organism with a sufficiently stable habit capable of producing carbonate production has the ability to form a reef, and indeed, the geological record shows that many microbial communities, algae, and skeletal metazoans (most now extinct) have formed reefs since the early Archean, some 3.5 Ga. An understanding of reef evolution—such as the response of biota to environmental change, and the way in which reefs control sedimentation across shelf margins—is wholly dependent upon accurate description of ancient reef ecology. Study of a wide variety of Paleozoic reefs is now revealing hitherto unexpected levels of ecological complexity. Moreover, although some Paleozoic reefs achieved rates of accretion similar to modern coral reefs (e.g. 3–4 mm yr^{-1} for the Permian Capitan reef; 30), the trophic structure and relative contributions of inorganic and organic carbonate were profoundly different. In particular, Paleozoic reefs appear to have grown in the absence of the photosymbiosis, and in some examples, much of the preservable biodiversity was housed within cryptic communities. Reef construction by relatively fragile organisms was made possible by the absence of biological destruction and by rapid inorganic lithification. Such detailed ecological analyses confound the long-held belief (29, 86) that photosymbiosis and large, skeletal metazoans are vital prerequisites for successful reef building. I illustrate these new-found complexities with two examples: the oldest metazoan reefs and reefs from the Upper Permian.

The Oldest Metazoan Reefs (~570–543 mya)

The oldest known reefs were constructed by microbialites—calcareous organosedimentary deposits formed by the interaction of benthic microbial communities and detrital or chemical sediments. From their first appearance 3.5 Ga and for the next 2.5 Gyr, these microbialites were expressed solely as stromatolites, which may be a reflection of their cyanobacterial origin. Prior to 2.3 Ga, oceans and the atmosphere were essentially anoxic, and atmospheric oxygen levels probably continued to be very low during the Paleoproterozoic (2–1.65 Ga). Reefs were differentiated into distinct open surface and cryptic communities as soon as crypts formed in the Proterozoic 1.9 Ga (38).

Coincident with the decline of stromatolites at the end of the Proterozoic was a rise in abundance and diversity of thrombolites and calcified cyanobacteria, and the appearance of metazoans. The early evolution of animals probably proceeded as a single, protracted evolutionary radiation lasting some 55 myr (34) before culminating in the Cambrian explosion, which records widespread skeletonization and the expansion of behavioral repertoires, although there is evidence to support the presence of a terminal Proterozoic extinction event (49). The first metazoan reefs are known from the latest Neoproterozoic, and are particularly diverse in the widespread carbonates of the Nama Group of southern Namibia. Indeed, some of the earliest skeletal metazoans known were sessile, gregarious, and probably heterotrophic organisms capable of forming limited topographic relief. These include weakly skeletonized cup- or goblet-like solitary, sessile organisms, such as the globally distributed *Cloudina* (possibly a cnidarian or worm tube; 33), as well as seven other as yet undescribed forms (AH Knoll, personal communication). The stratigraphic range of these sessile organisms overlaps with the most diverse Ediacaran soft-bodied fossil assemblages (34). Probable borings have also been recorded from this time (3), and a detritivorous fauna associated with reef cavities had developed by the earliest Cambrian (102).

Although stromatolites remained common in intertidal-supratidal environments, the first biotically diverse metazoan-algal reefs formed subtidally at the base of the Tommotian (530 Ma; 9) on the Siberian Platform. These reef communities comprised the highly gregarious calcified sponges known as archaeocyaths, which lived in cyanobacterial-thrombolitic communities, became globally distributed, and persisted until the virtual demise of the archaeocyaths at the end of the Toyonian, some 520 mya (9).

Contrary to earlier reports (77a), the earliest metazoan reefs at the base of the Tommotian, as exposed on the Aldan River, Siberia, were already ecologically complex (75). While possessing low diversity, they had erect framework elements of branching archaeocyaths, with a cryptic biota of archaeocyaths and calcified cyanobacteria. These reefs were associated with skeletal debris of a diverse associated fauna; microburrowing deposit-feeders continued to proliferate within the sheltered areas of the framework.

By the middle Tommotian, archaeocyath-cyanobacterial reefs became more diverse and ecologically complex (52) due to the appearance of other sessile, calcified organisms inferred to have been suspension- or filter-feeders. These organisms included radiocyaths, a variety of simple cup-shaped forms known as "coralomorphs," globally rare but locally abundant large skeletal tabulate corals and other cnidarians (53, 78), and stromatoporoid sponges (69). Possible calcarean sponges appeared in the early mid-Tommotian (52); probable sponge borings have been noted within coralomorph skeletons from the Canadian Rocky Mountains (70), and silt-sized microspar grains resembling "chips" from clionid-type sponges have been identified within Lower Cambrian reef cavities (50).

These reefs were taxonomically diverse and ecologically complex, and they were differentiated into distinct open surface and cryptic communities: This differentiation may have been promoted by intense competition for limited hard substrates. Indeed, crypts housed a substantial proportion of overall reef biodiversity: solitary archaeocyath sponges, calcified cyanobacteria, and a microburrowing (?)metazoan were the most ubiquitous and abundant elements; coralomorphs, spiculate sponges, and various skeletal problematica were also common (102). The rigidity of Lower Cambrian reefs was enhanced substantially by the growth of synsedimentary cements—but this may also have contributed to the limited longevity of the cryptic community by rapid occlusion of the reef framework. These elements produced one of the most diverse and ecologically complex reef ecosystems known from the Paleozoic (Figure 1).

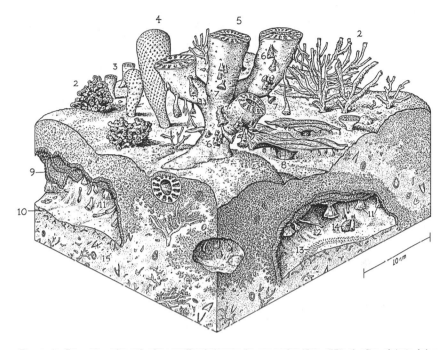

Figure 1   Reconstruction of a Lower Cambrian reef community (from 97). 1. *Renalcis* (calcified cyanobacterium); 2: branching archaeocyath sponges; 3: solitary cup-shaped archaeocyath sponges; 4: chancellorid (?sponge); 5: radiocyath (?sponge); 6: small, solitary archaeocyath sponges; 7: cryptic 'coralomorphs'; 8: *Okulitchicyathus* (archaeocyath sponge); 9: early fibrous cement forming within crypts; 10: microburrows (traces of a deposit-feeder) within geopetal sediment; 11: cryptic archaeocyaths and coralomorphs; 12: cryptic cribricyaths (problematic, attached skeletal tubes); 13: trilobite trackway; 14: cement botryoid; 15: sediment with skeletal debris.

Upper Permian Reefs (~260 Ma)

Upper Permian reefs are characterized by frondose bryozoan (e.g. *Gonipora*, *Polypora*) sponges and calcified sponges, as well as by the problematicum *Tubiphytes* and various algae including *Archaeolithoporella*. Volumetrically, late Permian reefs are dominated by bioclastic sediments and early cements, but they were capable of forming well-developed rimmed margins with zonation from fore-reef talus to reef slope, crest, reef flat, and back-reef lagoon. In many examples, including the Zeichstein reef of northern England, and the Capitan Reef of west Texas and New Mexico, the stabilization of sediments and aggregating communities of frondose bryozoans was aided by rapid cementation and/or encrusting microbialite formation that created a wave-resistant reef rock along the shelf edge and slope (89, 98, 99).

The Capitan reef forms one of the finest examples of an ancient rimmed carbonate shelf, marking a prominent topographic boundary between deep-water basinal deposits and shallow shelf sediments. Established ecological reconstructions have emphasized the role of various baffling branching or solitary organisms (sphinctozoan calcified sponges, bryozoans, and *Tubiphytes*) and massive putative algae (*Collenella*, *Parachaetetes*, and *Solenopora*) in the construction of the Capitan reef, together with the binding and encrusting contribution of *Archaeolithoporella* and extensive early marine cementation (29). However, the reef was, in fact, strongly differentiated into distinct open surface and cryptic communities. Unlike modern phototrophic coralgal reefs, most of the preservable epibenthos was housed within the cryptos, and zonation developed only in the shallow parts of the reef. Most sphinctozoan sponges did not grow upright to form a baffling framework but rather were pendent cryptobionts as were nodular bryozoans and rare solitary rugose corals and crinoids (98, 99). Indeed, many members of the cryptos appear to have been obligate cryptobionts. Much of the Middle Capitan reef framework was constructed by a scaffolding of large frondose bryozoans (Figure 2*a*). Bathymetrically shallow areas of both the Middle and Upper Capitan reef, however, were characterized by abundant platy sponges. In parts of the Upper Capitan, some forms reached up to 2 m in diameter and formed the ceilings of huge cavities that supported an extensive cryptos (Figure 2*b*).

In the absence of destructive forces (both biotic and physical) prevalent on modern reefs, the relatively fragile Capitan reef remained intact after death of the constructing organisms. Rigidity was imparted to this community by a post-mortem encrustation of *Tubiphytes* and *Archaeolithoporella*, together with microbialite. The resultant cavernous framework was partially infilled with sediment and preserved by synsedimentary intergrowth of aragonitic botryoid cements and *Archaeolithoporella* (Figure 2). Extensive cement precipitation was favored by factors including deep anoxia, which generated upwelling waters with elevated alkalinity (35). So although the accumulation rate of the Capitan may be comparable to that of modern coral reefs, both the trophic structure and relative contributions of inorganic and organic carbonate were profoundly different (99).

THE IMPORTANCE OF MICROBIAL COMMUNITIES

Although the recognition of microbialites in the geological record is problematic and controversial (for example, see 58), presumed microbialites are common as primary framebuilders and secondary encrusters within many Phanerozoic reefs; indeed, it is probable that they are more important than currently

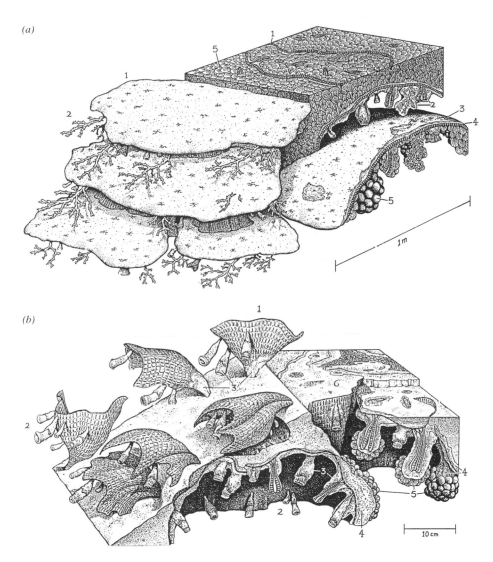

Figure 2 Reconstruction of an Upper Permian reef: the Capitan Reef, Texas and New Mexico (260 Ma) (from 97). (*a*) Platy sponge community. 1. *Gigantospongia discoforma* (platy sponge); 2: solitary and branching sphinctozoan sponges; 3: *Archaeolithoporella* (encrusting ?algae); 4: microbial micrite; 5: cement botryoids. (*b*) Frondose bryozoan-sponge community. 1. Frondose bryozoans (*Polypora sp.; Goniopora sp.*) 2: solitary sphinctozoan sponges; 3: *Archaeolithoporella* (encrusting ?algae); 4: microbial micrite; 5: cement botryoids; 6: sediment (grainstone-packstone).

recognized (71, 93, 99). Likewise, they were probably widespread in the Proterozoic but are difficult to identify unequivocally (48).

Most living cyanobacteria and other microbes are unable to produce directly a calcareous skeleton (64), so lithification within microbial communities is due either to the indirect post-mortem inorganic cementation processes (induced by decay or indirect metabolic processes) prevalent in hypersaline or brackish waters (with cements precipitated either on or within microbial sheaths), or to direct carbonate particle entrapment. For example, modern stromatolites appear to form only where two environmental criteria are satisfied (72):

1. High sedimentation rates (e.g. Exuma Cays, Bahamas) or low nutrient levels (e.g. Shark Bay) exclude the growth of potential macroalgal algae competitors for substrate space, and

2. Oceanographic conditions create a water chemistry favorable for carbonate precipitation, such as high levels of supersaturation of carbonate, rapid degassing (loss of CO_2) rates, or local elevations of sea water temperature.

The geological distribution of microbialites may therefore be controlled by physicochemical factors including the saturation state of seawater driven by changes in pCO_2 or Ca/Mg ratios, and global temperature distribution (48, 93). The decline in abundance of reefal microbialite after the Jurassic may be the result of low saturation states of sea water due to increased sequestration of carbonate by the newly evolved calcareous plankton (31, 93). However, although the relative importance of microbialites within modern coral reefs is not clear, the increasing numbers of examples recognized also suggest some role (e.g. 73, 74). Two examples are given below of how recognition of microbial fabrics has radically altered our understanding of ancient reef formation.

Upper Devonian Reefs (~360 Ma)

The Devonian, in particular the mid-late Devonian (Givetian-Frasnian), is considered to represent possibly the largest global expansion of reefs in the Phanerozoic (21). During this interval, the climate was equable and sea level high. Extensive reef tracts, in size exceeding those of the present day, are known from Canada and central Asia, and large reefs are also found throughout Europe, western North Africa, South China, southeast Asia, and most famously from the Canning Basin, Western Australia.

As so many Devonian reef tracts are subsurface and/or dolomitized, exact reef geometry, ecology, and zonation details are poorly known for many examples. However, mid-late Devonian reefs are generally assumed to have been built by large, heavily calcified metazoans—stromatoporoid sponges and tabulate corals—together with calcified cyanobacteria, and to a far lesser extent rugose corals. Receptaculitids and lithistid sponges are common in foreslope or deeper water reef communities or in low energy shallow settings. The faunal diversity of shallow water reefs is high, with metazoan reef-builders showing a tremendous variety of unusual morphologies (Figure 3) and growth rates as high as some modern scleractinian corals. Brachiopods were common, sometimes nestling within crypts, but especially attached to the undersurface of laminar/tabular stromatoporoids or tabulate (often alveolitid) corals (22, 85). Some were cementers (e.g. *Davidsonia* and *Rugodavidsonia*); others such as atrypids attached by means of spines (97).

Microbialite, as free-standing mounds, heads and columns, or encrusting components (Figure 3), were volumetrically important components of many Frasnian reefs (18, 60, 88, 97), as were the calcified cyanobacteria *Renalcis* and *Rothpletzella* [*Sphaerocodium*] (60, 93). Many stromatoporoids and other calcified metazoans grew attached to these lithified substrates, only rarely intergrowing themselves to form a reef framework (97). Early marine cements were also volumetrically important in many Frasnian reefs, in part reflecting the abundance of substantial and intact framework cavities formed beneath large skeletal metazoans; up to 50% cement by volume is recorded from the Golden Spike Reef, Canada, and synsedimentary radiaxial cements account for 20–50% by volume for the reefs of the Canning Basin (41, 45). The importance of early cementation both within cavities and of microbial communities in the Canning Basin is manifest by the numerous, huge reef-talus blocks (up to 100 m) incorporated into fore-reef strata and basin debris flows, and the extensive development of neptunian dykes and other fractures subparallel to the reef front. The presence of spur-and-groove structures (67) is testament to the growth to sea level and wave-resistant capacity of these reefs.

Waulsortian Mud Mounds

The Carboniferous was a time of great climatic and sea-level fluctuations with polar glacial events and depressed global temperatures (82). Continents were fused into one landmass that straddled the equator, thus restricting circulation and excluding the possibility of any equatorial currents (21). Deep-water mud mounds, known as Waulsortian mounds, form a group of reefs that flourished during the Early Carboniferous, although Waulsortian-like reefs are known from the Late Carboniferous of Ellesmere and Axel Heiberg Islands of the Canadian Arctic (26).

Waulsortian mounds are distinctive reefs characterized by a core facies containing many generations of micrite (lime mud), complex micrite-supported cavity systems infilled by marine cements (including the enigmatic stromatactis), and fenestrate bryozoans, that are generally considered to lack any framework. These mounds are proposed to have originated in deep waters below

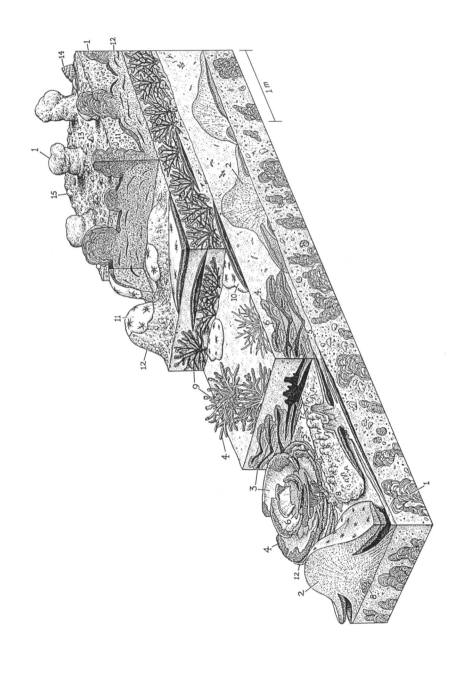

wave base (up to 280 m) (on the basis of the evidence of regional sedimentology and general absence of photic organisms), but they could grow into fairly
shallow waters in higher energy regimes (81). Waulsortian mounds bear steep
depositional slopes and were commonly surrounded by flanking beds rich in
disarticulated crinoids.

Although bryozoans and crinoids were probably capable of baffling and trapping locally fine-grained mud (especially from slightly turbid water) due to unidirectional cilia-generated currents (57), they were not ubiquitous components
of Waulsortian reefs and were often equally common in level bottom sediments.
Moreover, it is clear that sediment-baffling alone could not create slopes up to
50° and reefs up to 100 m high; such steep slopes suggest that unrecognized
reef framework must have been present. Sediments surrounding deep-water
mounds are typically thinner and contain significantly higher quantities of fine-
grained siliciclastic sediment than those of the mound itself; mounds in shallow
settings show a massive, muddy appearance at variance with the coarser-grained

←———————————————————————————————

Figure 3 Reconstruction of an Upper Devonian back-reef community: Canning Basin, western
Australia (360 Ma) (from 97). Back-reef sediments show distinctive shallowing-upwards cycles,
which are interpreted to reflect the lateral zonation of four communities:

1. The onset of carbonate sedimentation, probably induced by deepening, is marked by the colonization of large stromatoporoids on stabilized coarse clastic sediment. These include domal (*Actinostroma* sp.), and inferred whorl-forming foliaceous (*?Actinostroma sp.*), and platy columnar
growth forms. Many appear to have initiated upon crinoids, and their considerable elevation
above the substrate is reflected by the accumulation of the same geopetal infill within their
tiered growth. These stromatoporoids were heavily encrusted by *Renalcis* and microbialite,
particularly on sheltered undersurfaces.

2. The next zone characterized by thickets of the branching stromatoporoid (*Stachyodes* sp.), and
thin, laminar stromatoporoids that arched over the sediment. These show either encrusting
collars of *Renalcis*, or cryptic *Renalcis* attached to sheltered undersurfaces.

3. This was followed by the extensive growth of large mounds of microbialite, which were encrusted by the stromatoporoid (*?Clathrocoilona spissa*).

4. Columnar heads of stromatolites develop as the sediments became more shallow, energetic and
dominated by very coarse sands, together with patches of large oncolites (coated grains with
irregular and overlapping laminae, which were probably also microbially-mediated) and large
gastropods.

1: stromatolites; 2: domal stromatoporoid (*Actinostroma*); 3: inferred whorl-forming foliaceous
stromatoporoids (*?Actinostroma sp.*); 4: calcified cyanobacterium (*Renalcis*); 5: fibrous cement;
6: geopetal sediment infill; 7: platy stromatoporoid; 8: crinoids; 9: branching stromatoporoid
(*Stachyodes* sp.); 10: laminar stromatoporoid; 11: Encrusting stromatoporoid (*?Clathrocoilona
spissa*); 12: microbialite; 13: coarse clastic sediment; 14: gastropods; 15: oncolites.

surrounding sediments, indicative of mobile, winnowed sediments and high or turbulent water energies (65). Such observations indicate that the carbonate mud within the mounds was generated in situ under a stabilizing biological control.

There is now agreement that diverse and complex microbial processes, involving auto- and heterotrophic communities, were important in Waulsortian reef formation, especially in the shallow water facies (e.g. 54, 65), while the contribution of skeletal organisms was variable and probably relates to depositional setting and opportunistic colonization. A classic Waulsortian mound, known as Muleshoe, in the Sacramento Mountains, New Mexico, has been revealed to be constructed by bulbous, micrite masses with thrombolitic fabrics that are lined by early marine cements (46). In the upper parts of the mound, inferred to have grown in shallow waters, there may be a pronounced high-angle orientation of digitate micrite masses and intervening in-situ bryozoan fronds that matches the regional orientation of other current indicators such as crinoid segments.

Encrusting, often cryptic stromatolites and thrombolites, and occasional *Renalcis* colonies, offer direct evidence of microbial activity (46, 65). Peloidal and clotted micrites (thought to have been primarily calcite) are also common in many Tournasian and Viséan mounds, and they are thought to have formed within surficial microbial mats or biofilms that trapped bioclasts and stabilized the accreting reef surface, allowing the development of steep depositional slopes. Encrusting and boring organisms are often associated with the microbialites (e.g. 92); such features confirm the primary origin and early lithification of the micrites.

The initiation of deep-water mud-mound growth remains a mystery, but some have suggested that reduced sedimentation rates may favor the growth of microbial communities, as mounds seem to form preferentially during transgressions and high sea-level stands (10). Cold, nutrient-rich waters may also have favored inorganic cement precipitation and the growth of microbial communities and suspension-feeding metazoans; indeed, some early Carboniferous mud-mound development coincides with areas influenced by oceanic upwelling (101). The intermound and basin strata of Muleshoe, as well as other build-ups in the Lake Valley area and in other Lower Carboniferous mound complexes in Alberta and Montana, are dominated by dysaerobic and anaerobic strata alternating with thin oxygenated horizons. This inferred ocean stratification, which suggests a tendency to ocean anoxia during the Tournasian, may also be related to the ecology or diagenesis of mound growth (46).

THE ORIGINS OF THE MODERN CORAL REEF ECOSYSTEM

This section explores the evolutionary origin of two major ecological aspects of modern coral reefs: photosymbiosis and predation.

The Appearance of Photosymbiosis: Invasion of New Habitats?

Symbiosis can be better understood as an evolutionary innovation to acquire access to novel metabolic capabilities rather than by means of the more traditionally cited notion of mutual benefit (28). Microbial symbionts often possess a capability lacking in their host, and access to a new capability can enable the host to exploit a new food resource or invade a previously inhospitable environment.

Contrary to the widely accepted belief that corals harbor only symbionts of one type, recent work on many Caribbean and Indo-Pacific corals has shown that colonies growing at different depths can contain more than one type of zooxanthella (2, 76, 77). This suggests that symbionts exist as complex communities that can track gradients in environmental radiance within a colony. Although bleaching is a poorly understood response to environmental stress, it has also been suggested that it may be an adaptive strategy that allows corals to recombine with a different algal type (13). That recombination with different algal types can occur is supported by the observation that symbiont polymorphism can explain both the depth distribution of bleaching (which predominates at intermediate depths) and within-colony patterns, as a symbiont type associated with low irradiance is preferentially eliminated from the brightest parts of its distribution, presumably due to reaching some limit of physiological tolerance (77). This is clear evidence that the patterns of symbiont distribution are highly dynamic. Although host-symbiont partnerings may be specific at any one time, if recombination with a different type is possible, then specificity may alter as environmental conditions change (13).

We are only just beginning to appreciate the complexity of many symbiotic relationships. Future development of molecular taxonomic methods will certainly revolutionize our understanding of specificity, such as the extent of intracolony and intraspecific geographic variability within different taxa and the way this might be determined by local symbiont availability. Biogeographic constraints on symbiont diversity may have profound evolutionary implications. But the apparent fragility of photosymbiotic relationship also holds the key to its success: The dynamic nature of symbiotic combinations in corals may well have allowed them to persist through hundreds of millions of years of rapid, and sometimes extreme, environmental change (12).

Were Ancient Reef Communities Photosymbiotic?

The clear correlation between photosymbiosis and the success of living scleractinian corals in reef construction has led to the widespread supposition that the ability of metazoans to build reefs is contingent upon possession of photosymbionts. Many (e.g. 24, 25, 86) have argued, therefore, that extinct groups of reef-building metazoans also possessed this metabolic capability. In addition, the widespread loss and subsequent slow re-establishment of photosymbiotic relationships have been implicated in the long recovery time (up to 10 myr) of reef communities after mass extinction events (e.g. 29, 84). Such assumptions have not, however, been rigorously tested.

Inference of ancient photosymbiosis is highly problematic on morphological grounds alone (95). Such reasoning relies heavily upon modern analogies, but, not only are none available for several important extinct groups of reef-associated metazoans, such uniformitarian reasoning may be invalid. Collection of valid isotopic data is only possible where unaltered skeletal material is available, and such material is limited (84).

Contrary to received opinion, current evidence suggests that photosymbiotic metazoans have not always been present in reef communities. Indeed, with the probable exception of fusilinid foraminifera (Carboniferous-Permian) and alatoconchid bivalves (Permian), there are no clear data to support the presence of photosymbiotic reef-associated faunas before the Triassic (96, 97). A photosymbiotic isotopic signal has been detected in latest Triassic scleractinian corals (84): older material has not been analyzed. Based on morphological criteria, a few, nonaggregating rudists were probably photosymbiotic, but the evidence for Paleozoic tabulate corals and stromatoporoids is equivocal (20). That photosymbiosis was present in other extinct reef-building groups is, at best, unconfirmed: It is not necessary to conclude that, by virtue of their reef-building abilities, all major Phanerozoic reef-building groups possessed symbionts.

Photosymbiosis has not been a necessary prerequisite for reef-building in the past, nor did its appearance notably increase carbonate platform accumulation rates (8). Moreover, the loss of photosymbioses cannot alone account for periods in Earth's history when there was no widespread reef building, nor does its inferred presence correlate with phases of abundant reef distribution. As branching, phaceloid growth forms in Upper Triassic scleractinian corals show apparent zooxanthellate isotopic signatures (84), inferences derived from analysis of typical growth morphology in living zooxanthellate corals would appear to be a poor predictor of zooxanthellate status in ancient representatives. The need for more comparative growth rate and isotopic data within and between fossil assemblages is critical.

The evolutionary appearance and diversification of symbioses may be determined by the availability of an appropriate symbiont, together with the selection pressure to acquire the metabolic capability of that symbiont. It is highly likely that dinoflagellate symbionts—particularly the *Symbiodinium* group that is so successful at overcoming the defense systems of a host—did not evolve until the Triassic (59). Once freely available, however, such potential symbionts might infect many unrelated organisms that are in some way preadapted to acquire photosymbiosis. Before the Triassic, groups were presumably symbiotic with either chlorophytes, rhodophytes, or cyanophytes. Such symbioses would not necessarily have conferred adaptation to clear, well-lit, and low nutrient environments characteristic of the many modern reef-associated hosts symbiotic with dinoflagellates and diatoms (95).

The Rise of Predation and Bioerosion

Although there is no evidence that routine physical processes have changed significantly over the Phanerozoic, considerable data now suggest that biologically induced disturbance has increased dramatically in shallow marine seas, especially since the Mesozoic. The Mesozoic Marine Revolution (MMR; 91) involved the origin and diversification of many groups of bioturbators, predators, and bioeroders. The differential effects of the MMR are extremely difficult to isolate, as many organisms cause disturbance in more than one way: Some predators are also bioturbators; others are capable of significant bioerosion. The MMR thus involved coincident developments that might be predicted to have had many common evolutionary consequences (87).

Grazers and carnivores throughout the Paleozoic and early Mesozoic were relatively small individuals with limited foraging ranges (91, 91a). They were incapable of excavating calcareous substrates (83). By the early Mesozoic, sessile organisms had to contend with an increasing battery of novel feeding methods as well as sediment disruption due to deep bioturbating activity (87). Biological disturbance reached new heights of intensity from the Cretaceous to early Tertiary with the appearance of deep-grazing limpets, sea urchins with camerodont lanterns (Cretaceous), and especially the highly mobile reef fishes (Eocene); with them came the ability to excavate substantially large areas of hard substrata. A concurrent radiation of endoliths began in the Triassic; deep borings are known only from the mid-Mesozoic and Cenozoic (91). The first live-borers are described from the Eocene (51), and by the Oligo-Miocene, reef bioerosion had gained a modern cast (66) (Figure 4). How did post-Paleozoic reef communities respond to these new threats?

The importance of herbivory and predation in regulating modern reef community structure suggests that profound changes must have occurred with the appearance of these new consumer groups. Tropical marine hard substrata are usually sparsely vegetated, but a richer filamentous and macroalgal flora develops when herbivores are excluded and/or nutrient input increases (6, 37, 40, 55, 83). The presence of abundant coralline algae and corals in a reef community is therefore indicative of moderate or intense, and often specialist, herbivory. In waters < 20 m deep, corals and coralline algae may cover in excess of 80% of the substratum (17). In addition, grazers also contribute notably to carbonate sediment production and redistribution (79), to algal ridge formation (1),

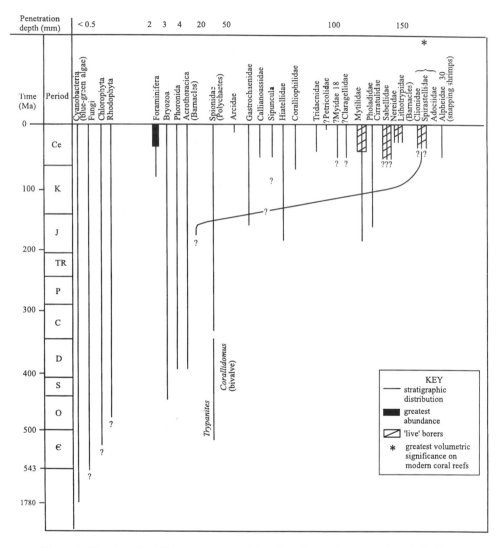

Figure 4 The rise of endolith groups arranged in order of increasing penetration depth (Data modified from 91a, from 97).

and to amelioration of the effects of competition for the maintenance of high diversity; they interact with physical controls to produce the characteristic zonation of modern coral reefs (94).

Many coral reef organisms show a vast array of supposed antipredation mechanisms, but details of their evolutionary origin and development are poorly known: The fossil record is silent on defenses concerning behavior and physiology. Sessile organisms have a relatively restricted range of antipredatory options at their disposal as they must be based upon passive constructional defences; they are also vulnerable to predators that do not rely upon prey manipulation for successful predation. Moreover, susceptibility to partial mortality and reliance upon herbivory to remove competitors or foulers usually entails loss of the prey's own tissues. This means that particular anatomies are required that allow resumption of normal growth as quickly as possible. One might predict, therefore, that the rise of excavatory herbivory would select for organisms with structural or chemical defenses and the ability to recover from partial mortality by rapid regeneration of damaged tissue.

Demonstration that habits have been acquired polyphyletically over a very short space of geological time is compelling evidence for the operation of an extrinsic selective force (80). Many of the characteristics of modern reef-building corals have antipredatory characteristics, including rapid regeneration from partial mortality (42). These traits have been present in the Scleractinia from the early origins of the group but proliferated as they subsequently proved useful for withstanding partial predation (95–97). Particularly dramatic is the spectacular rise of multiserial, branching forms in the late Cretaceous (Tithonian) coincident with the appearance of new groups of predatory excavators (Figure 5a). Such forms are easily broken as a result of both high wave activity and bioerosion, especially by boring sponges that infest the colony bases. However, branching corals are able to reanchor fragments and rapidly regenerate and grow, often fusing with other colonies, at rates up to 120 mm yr^{-1} (90). Branching corals appear to have turned adversity into considerable advantage, and appear to flourish because, and not in spite of, breakage. Taxa such as *Acropora*, *Porites*, and *Pocillopora* appeared during the Eocene, coinciding with the radiation of reef fish and with a major expansion and reorganization of the coral reef ecosystem.

Coralline algae are able to withstand the most intense herbivore onslaught by virtue of distinct morphological structures that have been shown experimentally to serve an antipredation function, including a heavily calcified thallus that is resistant to attack, intercellular conduits (fusion cells and secondary pits) for translocating photosynthates, and armored reproductive structures (conceptacles) (83). These anatomical features were present in the oldest abundant

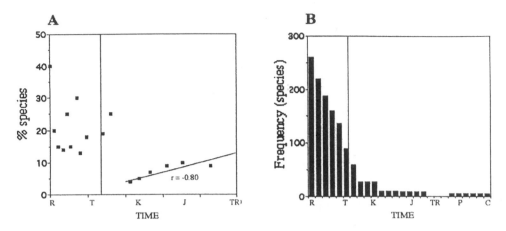

Figure 5   The rise of the modern coral reef biota. (*a*) Percent erect species of scleractinian corals. (replotted from 43a, data from sources in 19). (*b*) Species diversity of coralline algae (data from 83). C, Carboniferous; P, Permian; TR, Triassic; J, Jurassic; K, Cretaceous; T, Tertiary; R, Recent.

coralline algae, *Archaeolithophyllum*, known from the late Carboniferous, which formed a thin, leafy ("phylloid") crust (100).

The temporal distribution of reef-associated algae through the Phanerozoic broadly follows a trend of increasing ability to withstand disturbance. Poorly defended calcified microalgae declined at the end of the Paleozoic. Many Paleozoic reef-associated algae (including algal mats) were free-lying forms that were able to cover soft-substrates, whereas most post-Paleozoic reef floras encrusted hard substrates. Solenopores (red algae with raised conceptacles and an undifferentiated thallus) sharply declined in the late Mesozoic/early Cenozoic with the onset of excavatory grazing, and this was coincident with a reciprocal rise in the abundance and diversity of coralline algae (83). Delicately branched corallines reduced in importance in the tropics after the Eocene (44), when thick crusts became more abundant (83), and the first algal ridges are known from the Eocene-Miocene—both coincident with the evolution of excavatory herbivorous fish. Escalating herbivory is also conjectured to have resulted in the progressive disappearance of fleshy macroalgae from shallow marine environments through the Mesozoic (83).

The well-developed hypothallus of corallines initially allowed rapid lateral growth required for life on an unstable substratum. The hypothallus, together with the presence of fusion cells, also allowed the corallines to encrust, acquire branching morphologies, produce conceptacles, and regenerate from deep wounds—features that probably enabled corallines to radiate as herbivory intensified through the mid to late Mesozoic and Cenozoic. More than 100 myr elapsed between the first appearance of coralline algae and the subsequent radiation of the group (Figure 5*b*), strongly suggesting that these distinctive features of coralline algae are all exaptations (traits whose benefits are secondary or incidental to the primary function to which they are adapted). One exception might be multilayered epithallial growth, which might be an adaptation that arose in response to intense grazing by particular species of limpets in the Pleistocene (44).

The late Paleozoic decline of immobile epifauna also coincides with the rise of major bulldozing taxa, which had passed through the end Permian extinction unscathed (87). Although the hypothesis that the loss of unattached epifauna was caused by increased bioturbation remains untested, the fact that the modern deep seas appear to suffer the same degree of bioturbation as early Paleozoic shelves, and indeed harbor an immobile soft-substrate fauna of shallow marine Paleozoic caste, is supportive of it. Stalked crinoids, articulate brachiopods, hexactinellid sponges, and free-living immobile bryozoans are all concentrated, often in considerable abundance, in the deep sea. It has long been suggested (55a, 79a) that archaic Cambrian and Paleozoic faunas migrated to deep-sea environments from shallow shelves. But alternatively, the migration from near shore to deeper waters might equally reflect competitive displacement.

Of these surveyed trends in bryozoan morphology, encrusting morphologies, forms with continuous budding of new modules (zooidal and frontal), rapid growth, and good powers of regeneration have all markedly increased since the Mesozoic, especially from the late Cretaceous-Eocene. In contrast, the incidence of erect growth in relatively small and fragile bryozoans plummeted in the Mesozoic (see summaries in 56). These trends also coincide with the rise of biological disturbance.

DISCUSSION

This review has highlighted some recent advances in our understanding of the ecological evolution of reefs and has considered this evolution as a history of the changing way in which reef organisms have successfully occupied space. To understand the record of reef-building, it is clear that we must consider not only the influence of extrinsic factors in the control of in situ carbonate production, but also how the biological environment has changed. Such integrative studies are in their infancy, but data are now becoming available that are enabling researchers to document evolutionary processes at a level capable of generating testable hypotheses.

ACKNOWLEDGMENTS

The continued support of the Royal Society is most appreciated. The paleo-ecological reconstructions were drawn by John Sibbick and Figure 4 was drafted by Hilary Alberti. Earth Sciences Publication No. 5232.

Visit the *Annual Reviews home page* at
http://www.AnnualReviews.org

Literature Cited

1. Adey WH, Steneck RS. 1985. Highly productive eastern Caribbean reefs: synergistic effects of biological, chemical, physical and geological factors. In *The Ecology of Coral Reefs*, ed. ML Reaka, 3:163–87. NOAA's Undersea Res. Prog.
2. Baker AC. 1997. Biogeography and ecology of symbiont diversity in scleractinian corals. Soc. Compar. Integrative Biologists, Final Program & Abstracts, Jan. 3–7, 1998. *Am. Zool.* 37:73A (Abstr.)
3. Bengtson S, Zhao Y. 1992. Predational borings in late Precambrian mineralized skeletons. *Science* 257:367–69
4. Berner RA, Lasaga AC, Garrels RM. 1983. The carbonate-silicate geochemical cycle and its effects upon atmospheric carbon dioxide over the past 100 million years. *Am. Jour. Sci.* 282:641–83
5. Birkeland C. 1997. Introduction. In *Life and Death of Coral Reefs*, ed. C Birkeland, pp. 1–12. New York: Chapman & Hall. 536 pp.
6. Birkeland C, Nelson SG, Wilkins S, Gates P. 1985. Effects of grazing of herbivorous fishes on coral reef community metabolism. *Proc. Fifth Int. Coral Reef Cong., Tahiti:* 4:47–51
7. Boreen JN, James NP. 1993. Holocene sediment dynamics in a cool-water carbonate shelf: Otway, southeastern Australia. *J. Sed. Petrol.* 63:574–88
8. Bosscher H, Schlager W. 1993. Accumulation rates of carbonate platforms. *J. Geol.* 10:345–55
9. Bowring SA, Grotzinger JP, Isachsen CE, Knoll AH, Pelechaty SM, Kolosov P. 1993. Calibrating rates of early Cambrian evolution. *Science* 261:1293–98
10. Bridges PH, Gutteridge P, Pickard NAH. 1995. The environmental setting of Early Carboniferous mud-mounds. *Spec. Publ. Int. Assoc. Sedimentol.* 23:171–90
11. Brown BE. 1997. Adaptations of reef

corals to physical environmental stress. *Adv. Mar. Biol.* 31:221–99
12. Buddemeier RW. 1997. Making light work of adaptation. *Nature* 388:229–30
13. Buddemeier RW, Fautin DG. 1993. Coral bleaching as an adaptive mechanism. *Bio. Sci.* 43:320–26
14. Buddemeier RW, Fautin DG. 1996. Saturation state and the evolution and biogeography of symbiotic calcification. *Bull. Inst. Ocean.* 14(4):23–32
15. Buddemeier RW, Fautin DG. 1996. CO_2 and evolution among the Scleractinia. *Bull. Inst. Ocean.* 14(4):33–38
16. Buzas MA, Culver SJ. 1994. Species pool and dynamics of marine paleocommunities. *Science* 264:1439–41
17. Carpenter RC. 1986. Partitioning herbivory and its effects on coral reef algal communities. *Ecol. Mono.* 56:345–63
18. Clough JG, Blodgett RG. 1989. Silurian-Devonian algal reef mound complex of southwest Alaska. In *Reefs, Canada and Adjacent Areas*, ed. HHJ Geldsetzer, NP James, GE Tebbutt. *Mem. Can. Soc. Petrol. Geol.* 13:404–07
19. Coates AG, Jackson JBC. 1985. Morphological themes in the evolution of clonal and aclonal marine invertebrates. In *Population Biology and Evolution of Clonal Organisms*, ed. JBC Jackson, LW Buss, RE Cook, pp. 67–106. New Haven: Yale Univ. Press. 530 pp.
20. Coates AG, Jackson JBC. 1987. Clonal growth, algal symbiosis, and reef formation in corals. *Paleobiology* 13:363–78
21. Copper P. 1989. Enigmas in Phanerozoic reef development. *Mem. Assoc. Austr. Palaeontol.* 8:371–85
22. Copper P. 1996. *Davidsonia* and *Rugo-davidsonia* (new genus), cryptic Devonian atrypid brachiopods from Europe and south China. *J. Paleontol.* 70:588–602
23. Cornell HV, Karlson RH. 1996. Species

richness of reef-building corals determined by local and regional processes. *J. Anim. Ecol.* 65:233–41

24. Cowen R. 1983. Algal symbiosis and its recognition in the fossil record. In *Biotic Interactions in Recent and Fossil Benthic Communities*, ed. MJS Tevesz, PL McCall, pp. 432–78. New York: Plenum. 837 pp.

25. Cowen R. 1988. The role of algal symbiosis in reefs through time. *Palaios* 3:221–27

26. Davies GR, Nassichuk WW, Beauchamp B. 1989. Upper Carboniferous 'Waulsortian' reefs, Canadian Arctic Archipelago. In *Reefs, Canada and Adjacent Areas*, ed. HHJ Geldsetzer, NP James, GE Tebbutt. *Mem. Can. Soc. Petrol. Geol.* 13:658–66

27. Davies TA, Worsley TR. 1981. Paleoenvironmental implications of oceanic carbonate sedimentation rates. In *The Deep Sea Drilling Project: A Decade of Progress*, ed. JE Warme, RG Douglas, EL Winterer. *Spec. Publ. SEPM* 32:169–79

28. Douglas AE, Smith DC. 1989. Are endosymbioses mutualistic? *Trends Ecol. Evol.* 4:350–52

29. Fagerstom JA. 1987. *The Evolution of Reef Communities.* New York: Wiley. 600 pp.

30. Garber RA, Grover GA, Harris PM. 1989. Geology of the Capitan Shelf Margin—subsurface data from the northern Delaware Basin. In *Subsurface and Outcrop Examination of the Capitan Shelf Margin, Northern Delaware Basin*, ed. PM Harris, GA Grover. SEPM Core Workshop 13:3–269

31. Gebelein CD. 1976. The effects of physical, chemical, and biological evolution of the Earth. In *Stromatolites*, ed. MR Walter, pp. 499–515. Amsterdam: Elsevier

32. Glynn PW. 1988. Predation on coral reefs: some key processes, concepts and research directions. *Proc. Sixth Int. Coral Reef Symp.* 1:51–62

33. Grant SWF. 1990. Shell structure and distribution of *Cloudina*, a potential index fossil for the terminal Proterozoic. *Am. J. Sci.* 290A:261–94

34. Grotzinger JP, Bowring SA, Saylor BZ, Kaufman AJ. 1995. Biostratigraphic and geochronological constraints on early animal evolution. *Science* 270:598–604

35. Grotzinger JP, Knoll AH. 1995. Anomalous carbonate precipitates: Is the Precambrian the key to the Permian? *Palaios* 10:578–96

36. Harper EM, Palmer TJ, Alphey JR. 1997. Evolutionary response by bivalves to changing Phanerozoic sea-water chemistry. *Geol. Mag.* 134:403–7

37. Hay ME. 1997. The ecology and evolution of seaweed-herbivore interactions on coral reefs. *Coral Reefs, Suppl.* 16:67–76

38. Hoffmann HJ, Grotzinger JP. 1985. Shelf-facies microbiotas from the Odjick and Rocknest formations (Epworth Group: 1.89 Ga), northwest Canada. *Can. J. Earth Sci.* 22:1781–92

39. Hubbell SP. 1997. A unified theory of biogeography and relative species abundance and its application to tropical rain forests and coral reefs. *Coral Reefs, Suppl.* 16:9–21

40. Hughes TP. 1994. Catastrophes, phase shifts, and large-scale degradation of a Caribbean coral reef. *Science* 265:1547–51

41. Hurley NF, Longman KC. 1989. Diagenesis of the Devonian reefal carbonates of the Oscar Range, Canning Basin, western Australia. *J. Sediment. Petrol.* 59:127–46

42. Jackson JBC. 1983. Biological determinants of present and past sessile animal distributions. In *Biotic Interactions in Recent and Fossil Benthic Communities*, ed. M Tevesz, PW McCall, pp. 39–120. New York: Plenum. 837 pp.

43. Jackson JBC. 1992. Pleistocene perspectives on coral reef community structure. *Am. Zool.* 32.719–31

43a. Jackson JBC, McKinney FK. 1991. Ecological processes and progressive macroevolution of marine clonal benthos. In *Causes of Evolution*, ed. RM Ross, WD Allmon, pp. 173–209. *Chicago*: Univ. Chicago Press. 479 pp.

44. Johnson JH. 1961. *Limestone-Building Algae and Algal Limestones.* Boulder, Co: Johnson. 143 pp.

45. Kerans C, Hurley NF, Playford PE. 1986. Marine diagenesis in Devonian reef complexes of the Canning Basin, western Australia. In *Reef Diagenesis*, ed. JH Schroeder, BH Purser, pp. 357–80. Berlin: Springer-Verlag

46. Kirkby KC, Hunt D. 1996. Episodic growth of a Waulsortian buildup: the Lower Carboniferous Muleshoe mound, Sacramento Mountains, New Mexico, USA. In *Recent Advances in Carboniferous Geology*, ed. P Strogen, ID Sommerville, GL Jones, pp. 107:97–110. London: Geol. Soc. Spec. Publ.

47. Kleypas JA. 1997. Modeled estimates of global reef habitat and carbonate production since the last glacial maximum. *Paleoceanography* 12:533–45

48. Knoll AH, Fairchild IJ, Swett K. 1993. Calcified microbes in Neoproterozoic carbonates: implications for our understanding on the Proterozoic/Cambrian transition. *Palaios* 8:512–25

49. Knoll AH. 1996. Daughter of time. *Paleobiology* 22:1–7

49a. Knowlton N, Lang JC, Rooney MC, Clifford P. 1981. Evidence for delayed mortality in hurricane-damaged Jamaican staghorn corals. *Nature* 294:251–52

50. Kobluk DR. 1981. The record of cavity-dwelling (coelobiontic) organisms in the Paleozoic. *Can. J. Earth Sci.* 18:181–90

51. Krumm DK, Jones DS. 1993. A new coral-bivalve association (*Actinastrea-Lithophaga*) from the Eocene of Florida: *J. Paleontol.* 67:945–51

52. Kruse PD, Zhuravlev A Yu, James NP. 1995. Primordial metazoan-calcimicrobial reefs: Tommotian (early Cambrian) of the Siberian platform. *Palaios.* 10:291–321

53. Lafuste J, Debrenne F, Gandin A, Gravestock DI. 1991. The oldest tabulate coral and the associated Archaeocyatha, Lower Cambrian, Flinders Ranges, South Australia: *Geobios* 24:697–718

53a. Lees A. 1975. Possible influences of salinity and temperature on modern shelf carbonate sediments contrasted. *Mar. Geology* 13:1767–73.

54. Lees A, Miller J. 1985. Facies variation in Waulsortian buildups. Part 1: A model from Belgium. *Geol. J.* 20:133–58

55. Lewis SM. 1985. Herbivory on coral reefs: algal susceptibility to herbivorous fishes. *Oecologia* 65:370–75

55a. Macintyre IG. 1985. Submarine cements—the peloidal question. In *Carbonate Cements*, ed. N Schneidermann, PM Harris, pp. 109–16. *SEPM Special Publ. 36.* 379 pp.

55b. McKerrow WS. 1978. *The Ecology of Fossils.* London: Duckworth. 384 pp.

56. McKinney FK, Jackson JBC. 1988. *Bryozoan Evolution.* Boston: Unwin Hyman. 238 pp.

57. McKinney FK, McKinney MJ, Listokin MRA. 1987. Erect bryozoans are more than baffling: enhanced sedimentation rate by a living unilaminate branched bryozoan and possible implications for fenestrate bryozoan mudmounds. *Palaios* 2:41–47

58. Deleted in proof

59. Moldowan JM, Dahl J, Jacobson SR, Huizinga BJ, Fago FJ, et al. 1996. Chemostratigraphic reconstruction of biofacies: molecular evidence linking cyst-forming dinoflagellates with pre-Triassic ancestors. *Geology* 24:159–62

60. Mountjoy EW, Riding R. 1981. Foreslope stromatoporoid-renalcid bioherm with evidence of early cementation, Devonian Ancient Wall reef complex, Rocky Mountains. *Sedimentology* 28:299–319

60a. Multer HG, Milliman JD. 1967. Geologic aspects of sabellarian reefs, southeastern Florida. *Bull. Mar. Sci.* 17:257–67

61. Opdyke BN, Walker JCG. 1992. Return of the coral reef hypothesis: basin to shelf partitioning of $CaCO_3$ and its effect on atmospheric CO_2. *Geology* 20:733–36

62. Opdyke BN, Wilkinson BH. 1993. Carbonate mineral saturation state and cratonic limestone accumulation. *Am. J. Sci.* 293:217–34

63. Pandolfi JM. 1996. Limited membership in Pleistocene reef coral assemblages from the Huon Peninsula, Papua New Guinea: constancy during global change. *Paleobiology* 22:152–76

64. Pentacost A, Riding R. 1986. Calcification in cyanobacteria. In *Biomineralization in Lower Plants and Animals*, ed. BSA Leadbeater, R Riding, 73–90. Oxford: Clarendon

65. Pickard NAH. 1996. Evidence for microbial influence on the development of Lower Carboniferous buildups. In *Recent Advances in Lower Carboniferous Geology*, ed. P Strogen, ID Sommerville, GLI Jones, 107:65–82. London: Geol. Soc. Spec. Publ.

66. Pleydell SM, Jones B. 1988. Boring of various faunal elements in the Oligocene-Miocene Bluff Formation of Grand Cayman, British West Indies. *J. Paleontol.* 62:348–67

67. Playford PE. 1981. Devonian Reef complexes of the Canning Basin, western Australia: *Geol. Soc. Aust. Fifth Aust. Geol. Conv. Field Excursion Guidebook.* 64 pp.

68. Potts DC. 1984. Generation times and the Quaternary evolution of reef-building corals. *Paleobiology* 10:48–58

69. Pratt BR. 1990. Lower Cambrian reefs of the Mural Formation, southern Canadian Rocky Mountains. *13th Int. Sed. Congr.:* 436 (Abstr.)

70. Pratt BR. 1994. Lower Cambrian reefs of the Mural Formation, southern Rocky Mountains. *Terra Nova* 3, Abstract Supplement 6:5 (Abstr.)

71. Pratt BR. 1955. The origin, biota and evolution of deep-water mud-mounds. In *Carbonate Mud Mounds; Their Origin and Evolution*, ed. CLV Monty, DWJ

Bosence, PH Bridges, BR Pratt, 23:49–123. Spec. Publ. Int. Ass. Sedimentol. Oxford: Blackwell Science. 537 pp.

72. Reid RP, Macintyre IG, Browne KM, Steneck RS, Miller T. 1995. Modern marine stromatolites in the Exuma Cays, Bahamas: uncommonly common. *Facies* 33:1–18

73. Reitner J. 1993. Modern cryptic microbialite/metazoan facies from Lizard Island (Great Barrier Reef)—formation and concepts. *Facies* 29:3–40

74. Reitner J, Neuweiler F, Gautret P. 1995. Modern and fossil automicrites: implications for mud mound genesis. *Facies* 32:4–17

75. Riding R, Zhuravlev AYu. 1995. Structure and diversity of oldest sponge-microbe reefs: Lower Cambrian, Aldan River, Siberia. *Geology* 23:649–52

76. Rowan R, Knowlton N. 1995. Interspecific diversity and ecological zonation in coral-algal symbiosis. *Proc. Natl. Acad. Sci. USA* 92:2850–53

77. Rowan R, Knowlton N, Baker A, Jara J. 1997. Landscape ecology of algal symbionts creates variation in episodes of coral bleaching. *Nature* 388:265–69

77a. Rowland SM, Gangloff RA. 1988. Structure and paleoecology of Lower Cambrian reefs. *Palaios* 3:111–35

78. Savarese M, Mount JF, Sorauf JE, Bucklin L. 1993. Paleobiologic and paleoenvironmental context of coral-bearing Early Cambrian reefs: implications for Phanerozoic reef development. *Geology* 21:917–20

78a. Scheltema RS, Williams IP, Shaw MA, Loudon C. 1981. Gregarious settlement by the larvae of *Hydroides dianthus* (Polychaeta: Serpulidae). *Mar. Ecol. Prog. Ser.* 5:59–74

79. Scoffin TP, Stearn CW, Boucher D, Frydl P, Hawkins CM, et al. 1980. Calcium carbonate budget of a fringing reef on the west coast of Barbados. Pt. II. Erosion, sediments and internal structure. *Bull. Mar. Sci.* 32:457–508

79a. Sepkoski JJ, Sheehan PM. 1983. Effects of interactions on community evolution. In *Biotic Interactions in Recent and Fossil Benthic Communities*, ed. M. Tevesz, PW McCall, pp. 480–625. New York: Plenum. 837 pp.

80. Skelton PW. 1991. Morphogenetic versus environmental cues for adaptive radiations. In *Constructional Morphology and Evolution*, ed. N Schmidt-Kittler, K Voegel, pp. 375–88. Berlin: Springer

81. Somerville ID, Pickard NAH, Strogen P, Jones GL. 1992. Early to mid-Viséan shallow water platform buildups, north Co. Dublin, Ireland. *Geol. J.* 27:151–72

82. Stanley SM. 1988. Climatic cooling and mass extinctions of Paleozoic reef communities. *Palaios* 3:228–32

83. Steneck RS. 1983. Escalating herbivory and resulting adaptive trends in calcareous algal crusts. *Paleobiology* 9:44–61

84. Stanley GD, Swart PW. 1995. Evolution of the coral-zooxanthellae symbiosis during the Triassic: a geochemical approach. *Paleobiology* 21:179–99

85. Szulczewski M, Racki G. 1981. Early Frasnian bioherms in the Holy Cross Mts. *Acta Geol. Polon* 31:147–62

86. Talent J. 1988. Organic reef-building: episodes of extinction and symbiosis? *Seneck. Leth.* 69:315–68

87. Thayer CW. 1983. Sediment-mediated biological disturbance and the evolution of marine benthos. In *Biotic Interactions in Recent and Fossil Benthic Communities*, ed. M Tevesz, PW McCall, pp. 480–625. New York: Plenum. 837 pp.

88. Tsein HH. 1994. Contributions of reef-building organisms in reef carbonate construction. *Cour. Forsch.-Inst. Seneck.* 172:95–102

89. Tucker ME, Hollingworth NTJ. 1986. The Upper Permian reef complex (EZ) of north east England. In *Reef Diagenesis*, ed. JH Shroeder, BH Purser, pp. 270–90. Berlin: Springer Verlag

90. Tunnicliffe V. 1981. Breakage and propagation of the stony coral *Acropora cervicornis. Proc. Natl. Acad. Sci. USA* 78:2427–31

91. Vermeij GJ. 1977. The Mesozoic marine revolution: evidence from snails, predators and grazers. *Paleobiology* 3:245–58

91a. Vermeij GJ. 1987. *Evolution and Escalation*. Princeton, NJ: Princeton Univ. Press. 527 pp.

92. Webb GE. 1987. Late Mississippean thrombolite bioherms from the Pitkin Formation of northern Arkansas. *Bull. Geol. Soc. Am.* 99:686–98

93. Webb GE. 1996. Was Phanerozoic reef history controlled by the distribution of non-enzymatically secreted reef carbonates (microbial carbonate and biologically induced cement)? *Sedimentology* 43:947–71

94. Wellington GM. 1982. Depth zonation of corals in the Gulf of Panama: control and facilitation by resident reef fish. *Ecol. Monogr.* 52:223–41

95. Wood R. 1993. Nutrients, predation and the history of reefs. *Palaios* 8:526–43

96. Wood R. 1995. The changing biology of reef-building. *Palaios* 10:517–29

97. Wood RA. 1998. *Reef Evolution*. Oxford: Oxford Univ. Press. In press
98. Wood R, Dickson JAD, Kirkland-George B. 1994. Turning the Capitan reef upside down: a new appraisal of the ecology of the Permian Capitan Reef, Guadalupe Mountains, Texas and New Mexico. *Palaios* 9:422–27
99. Wood R, Dickson JAD, Kirkland-George B. 1996. New observations on the ecology of the Permian Capitan Reef, Guadalupe Mountains, Texas and New Mexico: *Palaeontology* 39:733–62
100. Wray JL. 1964. *Archaeolithophyllum*, an abundant calcareous alga in limestones of the Lansing Group (Pennsylvania), southeastern Kansas. *State Geol. Surv. Kans. Bull.* 170:1–13
101. Wright VP. 1991. Comment on "Probable influence of Early Carboniferous (Tournasian-early Viséan) geography on the development of Waulsortian-like mounds". *Geology* 19:413
102. Zhuravlev A Yu, Wood RA. 1995. Lower Cambrian reefal cryptic communities. *Palaeontology* 38:443–70

Annu. Rev. Ecol. Syst. 1998. 29:207–31

ROADS AND THEIR MAJOR ECOLOGICAL EFFECTS

Richard T. T. Forman and Lauren E. Alexander

Harvard University Graduate School of Design, Cambridge, Massachusetts 02138

KEY WORDS: animal movement, material flows, population effects, roadside vegetation, transportation ecology

ABSTRACT

A huge road network with vehicles ramifies across the land, representing a surprising frontier of ecology. Species-rich roadsides are conduits for few species. Roadkills are a premier mortality source, yet except for local spots, rates rarely limit population size. Road avoidance, especially due to traffic noise, has a greater ecological impact. The still-more-important barrier effect subdivides populations, with demographic and probably genetic consequences. Road networks crossing landscapes cause local hydrologic and erosion effects, whereas stream networks and distant valleys receive major peak-flow and sediment impacts. Chemical effects mainly occur near roads. Road networks interrupt horizontal ecological flows, alter landscape spatial pattern, and therefore inhibit important interior species. Thus, road density and network structure are informative landscape ecology assays. Australia has huge road-reserve networks of native vegetation, whereas the Dutch have tunnels and overpasses perforating road barriers to enhance ecological flows. Based on road-effect zones, an estimated 15–20% of the United States is ecologically impacted by roads.

INTRODUCTION

Roads appear as major conspicuous objects in aerial views and photographs, and their ecological effects spread through the landscape. Few environmental scientists, from population ecologists to stream or landscape ecologists, recognize the sleeping giant, road ecology. This major frontier and its applications to planning, conservation, management, design, and policy are great challenges for science and society.

0066-4162/98/1120-0207$08.00

This review often refers to The Netherlands and Australia as world leaders with different approaches in road ecology and to the United States for especially useful data. In The Netherlands, the density of main roads alone is 1.5 km/km^2, with traffic density of generally between 10,000 and 50,000 vehicles per commuter day (101). Australia has nearly 900,000 km of roads for 18 million people (66). In the United States, 6.2 million km of public roads are used by 200 million vehicles (85). Ten percent of the road length is in national forests, and one percent is interstate highways. The road density is 1.2 km/km^2, and Americans drive their cars for about 1 h/day. Road density is increasing slowly, while vehicle kilometers (miles) traveled (VMT) is growing rapidly.

The term road corridor refers to the road surface plus its maintained roadsides and any parallel vegetated strips, such as a median strip between lanes in a highway (Figure 1; see color version at end of volume). "Roadside natural strips" of mostly native vegetation receiving little maintenance and located adjacent to roadsides are common in Australia (where road corridors are called road reserves) (12, 39, 111). Road corridors cover approximately 1% of the United States, equal to the area of Austria or South Carolina (85). However, the area directly affected ecologically is much greater (42, 43).

Theory for road corridors highlights their functional roles as conduits, barriers (or filters), habitats, sources, and sinks (12, 39). Key variables affecting processes are corridor width, connectivity, and usage intensity. Network theory, in turn, focuses on connectivity, circuitry, and node functions (39, 71).

This review largely excludes road-construction-related activities, as well as affiliated road features such as rest stops, maintenance facilities, and entrance/exit areas. We also exclude the dispersed ecological effects of air pollution emissions, such as greenhouse gases, nitrogen oxides (NOX), and ozone, which are reviewed elsewhere (85, 135). Bennett's article (12) plus a series of books (1, 21, 33, 111) provide overviews of parts of road ecology.

Gaping holes in our knowledge of road ecology represent research opportunities with a short lag between theory and application. Current ecological knowledge clusters around five major topics: (*a*) roadsides and adjacent strips; (*b*) road and vehicle effects on populations; (*c*) water, sediment, chemicals, and streams; (*d*) the road network; and (*e*) transportation policy and planning.

Figure 1 Road corridor showing road surface, maintained open roadsides, and roadside natural strips. Strips of relatively natural vegetation are especially characteristic of road corridors (known as road reserves) in Australia. Wheatbelt of Western Australia. Photo courtesy of BMJ Hussey. See color version at end of volume.

ROADSIDE VEGETATION AND ANIMALS

Plants and Vegetation

"Roadside" or "verge" refers to the more-or-less intensively managed strip, usually dominated by herbaceous vegetation, adjacent to a road surface (Figure 1). Plants on this strip tend to grow rapidly with ample light and with moisture from road drainage. Indeed, management often includes regular mowing, which slows woody-plant invasion (1, 86). Ecological management may also maintain roadside native-plant communities in areas of intensive agriculture, reduce the invasion of exotic (non-native) species, attract or repel animals, enhance road drainage, and reduce soil erosion.

Roadsides contain few regionally rare species but have relatively high plant species richness (12, 139). Disturbance-tolerant species predominate, especially with intensive management, adjacent to highways, and exotic species typically are common (19, 121). Roadside mowing tends to both reduce plant species richness and favor exotic plants (27, 92, 107). Furthermore, cutting and removing hay twice a year may result in higher plant species richness than does mowing less frequently (29, 86). Native wildflower species are increasingly planted in dispersed locations along highways (1).

Numerous seeds are carried and deposited along roads by vehicles (70, 112). Plants may also spread along roads due to vehicle-caused air turbulence (107, 133) or favorable roadside conditions (1, 92, 107, 121, 133). For example, the short-distance spread of an exotic wetland species, purple loosestrife (*Lythrum salicaria*), along a New York highway was facilitated by roadside ditches, as well as culverts connecting opposite sides of the highway and the median strip of vegetation (133). Yet few documented cases are known of species that have successfully spread more than 1 km because of roads.

Mineral nutrient fertilization from roadside management, nearby agriculture, and atmospheric NOX also alter roadside vegetation. In Britain, for example, vegetation was changed for 100–200 m from a highway by nitrogen from traffic exhaust (7). Nutrient enrichment from nearby agriculture enhances the growth of aggressive weeds and can be a major stress on a roadside native-plant community (19, 92). Indeed, to conserve roadside native-plant communities in Dutch farmland, fertilization and importing topsoil are ending, and in some places nutrient accumulations and weed seed banks are reduced by soil removal (86; H van Bohemen, personal communication).

Woody species are planted in some roadsides to reduce erosion, control snow accumulation, support wildlife, reduce headlight glare, or enhance aesthetics (1, 105). Planted exotic species, however, may spread into nearby natural ecosystems (3, 12). For example, in half the places where non-native woody species were planted in roadsides adjacent to woods in Massachusetts (USA), a species had spread into the woods (42).

Roadside management sometimes creates habitat diversity to maintain native ecosystems or species (1, 86, 131). Mowing different sections along a road, or parallel strips in wide roadsides, at different times or intervals may be quite effective (87). Ponds, wetlands, ditches, berms, varied roadside widths, different sun and shade combinations, different slope angles and exposures, and shrub patches rather than rows offer variety for roadside species richness.

In landscapes where almost all native vegetation has been removed for cultivation or pasture, roadside natural strips (Figure 1) are especially valuable as reservoirs of biological diversity (19, 66). Strips of native prairie along roads and railroads, plus so-called beauty strips of woodland that block views near intensive logging, may function similarly as examples. However, roadside natural strips of woody vegetation are widespread in many Australian agricultural landscapes and are present in South Africa (11, 12, 27, 39, 66, 111). Overall, these giant green networks provide impressive habitat connectivity and disperse "bits of nature" widely across a landscape. Yet they miss the greater ecological benefits typically provided by large patches of natural vegetation (39, 41).

In conclusion, roadside vegetation is rich in plant species, although apparently not an important conduit for plants. The scattered literature suggests a promising research frontier.

Animals and Movement Patterns

Mowing, burning, livestock grazing, fertilizing, and planting woody plants greatly impact native animals in roadsides. Cutting and removing roadside vegetation twice a year in The Netherlands, compared with less frequent mowing, results in more species of small mammals, reptiles, amphibians, and insects (29, 86). However, mowing once every 3–5 y rather than annually results in more bird nests. Many vertebrate species persist better with mowing after, rather than before or during, the breeding period (86, 87). The mowing regime is especially important for insects such as meadow butterflies and moths, where different species go through stages of their annual cycle at different times (83). Roadsides, especially where mowed cuttings are removed, are suitable for ~80% of the Dutch butterfly fauna (86).

Planting several native and exotic shrub species along Indiana (USA) highways resulted in higher species richness, population density, and nest density for birds, compared with nearby grassy roadsides (105). Rabbit (*Sylvilagus*) density increased slightly. However, roadkill rates did not differ next to shrubby versus grassy roadsides.

In general, road surfaces, roadsides, and adjacent areas are little used as conduits for animal movement along a road (39), although comparisons with null models are rare. For example, radiotracking studies of wildlife across the landscape detect few movements along or parallel to roads (35, 39, 93). Some exceptions are noteworthy. Foraging animals encountering a road sometimes move short distances parallel to it (10, 106). At night, many large predator species move along roads that have little vehicular or people traffic (12, 39). Carrion feeders move along roads in search of roadkills, and vehicles sometimes transport amphibians and other animals (11, 12, 32). Small mammals have spread tens of kilometers along highway roadsides (47, 60). In addition, migrating birds might use roads as navigational cues.

Experimental, observational, and modeling approaches have been used to study beetle movement along roadsides in The Netherlands (125–127). On wide roadsides, fewer animals disappeared into adjacent habitats. Also, a dense grass strip by the road surface minimized beetle susceptibility to roadkill mortality (126, 127). Long dispersals of beetles were more frequent in wide (15–25 m) than in narrow (<12 m) roadsides. Nodes of open vegetation increased, and narrow bottlenecks decreased, the probability of long dispersals. The results suggest that with 20–30-m-wide roadsides containing a central suitable habitat, beetle species with poor dispersal ability and a good reproductive rate may move 1–2 km along roadsides in a decade (127).

Adjacent ecosystems also exert significant influences on animals in corridors (39). For example, roadside beetle diversity was higher near a similar patch of sandy habitat, and roadsides next to forest had the greatest number of forest beetle species (127). In an intensive-agriculture landscape (Iowa, USA), bird-nest predation in roadsides was highest opposite woods and lowest opposite pastures (K Freemark, unpublished data). Finally, some roadside animals also invade nearby natural vegetation (37, 47, 54, 60, 63, 127).

The median strip between lanes of a highway is little studied. A North Carolina (USA) study found no difference in small-mammal density between roadsides on the median and on the outer side of the highway (2). This result was the same whether comparing mowed roadside areas or unmowed roadside areas. Also, roadkill rates may be affected by the pattern of wooded and grassy areas along median strips (10).

In conclusion, some species move significant distances along roadsides and have major local impacts. Nevertheless, road corridors appear to be relatively unimportant as conduits for species movement, although movement rates should be better compared with those at a distance and in natural-vegetation corridors.

ROAD AND VEHICLE EFFECTS ON POPULATIONS

Roadkilled Animals

Sometime during the last three decades, roads with vehicles probably overtook hunting as the leading direct human cause of vertebrate mortality on land. In addition to the large numbers of vertebrates killed, insects are roadkilled in prodigious numbers, as windshield counts will attest.

Estimates of roadkills (faunal casualties) based on measurements in short sections of roads tell the annual story (12, 39, 123): 159,000 mammals and 653,000 birds in The Netherlands; seven million birds in Bulgaria; five million frogs and reptiles in Australia. An estimated one million vertebrates per day are killed on roads in the United States.

Long-term studies of roadkills near wetlands illustrate two important patterns. One study recorded >625 snakes and another >1700 frogs annually roadkilled per kilometer (8, 54). A growing literature suggests that roads by wetlands and ponds commonly have the highest roadkill rates, and that, even though amphibians may tend to avoid roads (34), the greatest transportation impact on amphibians is probably roadkills (8, 28, 34, 128).

Road width and vehicle traffic levels and speeds affect roadkill rates. Amphibians and reptiles tend to be particularly susceptible on two-lane roads with low to moderate traffic (28, 34, 57, 67). Large and mid-sized mammals are especially susceptible on two-lane, high-speed roads, and birds and small mammals on wider, high-speed highways (33, 90, 106).

Do roadkills significantly impact populations? Measurements of bird and mammal roadkills in England illustrate the main pattern (56, 57). The house sparrow (*Passer domesticus*) had by far the highest roadkill rate. Yet this species has a huge population, reproduces much faster than the roadkill rate, and can rapidly recolonize locations where a local population drops. The study concluded, based on the limited data sets available, that none of the >100 bird and mammal species recorded had a roadkill rate sufficient to affect population size at the national level.

Despite this overall pattern, roadkill rates are apparently significant for a few species listed as nationally endangered or threatened in various nations (~9–12 cases) (9, 39, 43; C Vos, personal communication). Two examples from southern Florida (USA) are illustrative. The Florida panther (*Felis concolor coryi*) had an annual roadkill mortality of approximately 10% of its population before 1991 (33, 54). Mitigation efforts reduced roadkill loss to 2%. The key deer (*Odocoileus virginianus clavium*) has an annual roadkill mortality of ~16% of its population. Local populations, of course, may suffer declines where the roadkill rate exceeds the rates of reproduction and immigration. At least a dozen local-population examples are known for vertebrates whose total populations are not endangered (33, 39, 43).

Vehicles often hit vertebrates attracted to spilled grain, roadside plants, insects, basking animals, small mammals, road salt, or dead animals (12, 32, 56, 87). Roadkills may be frequent where traffic lanes are separated by impermeable barriers or are between higher roadside banks (10, 106).

Landscape spatial patterns also help determine roadkill locations and rates. Animals linked to specific adjacent land uses include amphibians roadkilled near wetlands and turtles near open-water areas (8). Foraging deer are often roadkilled between fields in forested landscapes, between wooded areas in open landscapes, or by conservation areas in suburbs (10, 42, 106). The vicinity of a large natural-vegetation patch and the area between two such patches are likely roadkill locations for foraging or dispersing animals. Even more likely locations are where major wildlife-movement routes are interrupted, such as roads crossing drainage valleys in open landscapes or crossing railway routes in suburbs (42, 106).

In short, road vehicles are prolific killers of terrestrial vertebrates. Nevertheless, except for a small number of rare species, roadkills have minimal effect on population size.

Vehicle Disturbance and Road Avoidance

The ecological effect of road avoidance caused by traffic disturbance is probably much greater than that of roadkills seen splattered along the road. Traffic noise seems most important, although visual disturbance, pollutants, and predators moving along a road are alternative hypotheses as the cause of avoidance.

Studies of the ecological effects of highways on avian communities in The Netherlands point to an important pattern. In both woodlands and grasslands adjacent to roads, 60% of the bird species present had a lower density near a highway (102, 103). In the affected zone, the total bird density was approximately one third lower, and species richness was reduced as species progressively disappeared with proximity to the road. Effect-distances (the distance from a road at which a population density decrease was detected) were greatest for birds in grasslands, intermediate for birds in deciduous woods, and least for birds in coniferous woods.

Effect-distances were also sensitive to traffic density. Thus, with an average traffic speed of 120 km/h, the effect-distances for the most sensitive species (rather than for all species combined) were 305 m in woodland by roads with a traffic density of 10,000 vehicles per day (veh/day) and 810 m in woodland by 50,000 veh/day; 365 m in grassland by 10,000 veh/day and 930 m in grassland by 50,000 veh/day (101–103). Most grassland species showed population decreases by roads with 5000 veh/day or less (102). The effect-distances for both woodland and grassland birds increased steadily with average vehicle speed up to 120 km/h and also with traffic density from 3000 to 140,000 veh/day (100, 102, 103). These road effects were more severe in years when overall bird population sizes were low (101).

Songbirds appear to be sensitive to remarkably low noise levels, similar to those in a library reading room (100, 102, 103). The noise level at which population densities of all woodland birds began to decline averaged 42 decibels (dB), compared with an average of 48 dB for grassland species. The most sensitive woodland species (cuckoo) showed a decline in density at 35 dB, and the most sensitive grassland bird (black-tailed godwit, *Limosa limosa*) responded at 43 dB. Field studies and experiments will help clarify the significance of these important results for traffic noise and birds.

Many possible reasons exist for the effects of traffic noise. Likely hypotheses include hearing loss, increase in stress hormones, altered behaviors, interference with communication during breeding activities, differential sensitivity to different frequencies, and deleterious effects on food supply or other habitat attributes (6, 101, 103, 130). Indeed, vibrations associated with traffic may affect the emergence of earthworms from soil and the abundance of crows (*Corvus*) feeding on them (120). A different stress, roadside lighting, altered nocturnal frog behavior (18). Responses to roads with little traffic may resemble behavioral responses to acute disturbances (individual vehicles periodically passing), rather than the effects of chronic disturbance along busy roads.

Response to traffic noise is part of a broader pattern of road avoidance by animals. In the Dutch studies, visual disturbance and pollutants extended outward only a short distance compared with traffic noise (100, 103). However, visual disturbance and predators moving along roads may be more significant by low-traffic roads.

Various large mammals tend to have lower population densities within 100–200 m of roads (72, 93, 108). Other animals that seem to avoid roads include arthropods, small mammals, forest birds, and grassland birds (37, 47, 73, 123). Such road effect zones, extending outward tens or hundreds of meters from a road, generally exhibit lower breeding densities and reduced species richness compared with control sites (32, 101). Considering the density of roads plus the total area of avoidance zones, the ecological impact of road avoidance must well exceed the impact of either roadkills or habitat loss in road corridors.

Barrier Effects and Habitat Fragmentation

All roads serve as barriers or filters to some animal movement. Experiments show that carabid beetles and wolf spiders (*Lycosa*) are blocked by roads as narrow as 2.5 m wide (73), and wider roads are significant barriers to crossing for many mammals (11, 54, 90, 113). The probability of small mammals crossing lightly traveled roads 6–15 m wide may be <10% of that for movements within adjacent habitats (78, 119). Similarly, wetland species, including amphibians and turtles, commonly show a reduced tendency to cross roads (34, 67).

Road width and traffic density are major determinants of the barrier effect, whereas road surface (asphalt or concrete versus gravel or soil) is generally a minor factor (34, 39, 73, 90). Road salt appears to be a significant deterrent to amphibian crossing (28, 42). Also, lobes and coves in convoluted outer-roadside boundaries probably affect crossing locations and rates (39).

The barrier effect tends to create metapopulations, e.g. where roads divide a large continuous population into smaller, partially isolated local populations (subpopulations) (6, 54, 128). Small populations fluctuate more widely over time and have a higher probability of extinction than do large populations (1, 88, 115, 122, 123). Furthermore, the recolonization process is also blocked by road barriers, often accentuated by road widening or increases in traffic. This well-known demographic threat must affect numerous species near an extensive road network, yet is little studied relative to roads (6, 73, 98).

The genetics of a population is also altered by a barrier that persists over many generations (73, 115). For instance, road barriers altered the genetic structure of small local populations of the common frog (*Rana temporaria*) in Germany by lowering genetic heterozygosity and polymorphism (97, 98). Other than the barrier effect on this amphibian and roadkill effects on two southern Florida mammals (20, 54), little is known of the genetic effects of roads.

Making roads more permeable reduces the demographic threat but at the cost of more roadkills. In contrast, increasing the barrier effect of roads reduces roadkills but accentuates the problems of small populations. What is the solution to this quandary (122, 128)? The barrier effect on populations probably affects more species, and extends over a wider land area, than the effects of either roadkills or road avoidance. This barrier effect may emerge as the greatest ecological impact of roads with vehicles. Therefore, perforating roads to diminish barriers makes good ecological sense.

WATER, SEDIMENT, CHEMICALS, STREAMS, AND ROADS

Water Runoff

Altering flows can have major physical or chemical effects on aquatic ecosystems. The external forces of gravity and resistance cause streams to carve channels, transport materials and chemicals, and change the landscape (68). Thus, water runoff and sediment yield are the key physical processes whereby roads have an impact on streams and other aquatic systems, and the resulting effect-distances vary widely (Figure 2).

Roads on upper hillslopes concentrate water flows, which in turn form channels higher on slopes than in the absence of roads (80). This process leads to smaller, more elongated first-order drainage basins and a longer total length of the channel network. The effects of stream network length on erosion and sedimentation vary with both scale and drainage basin area (80).

Water rapidly runs off relatively impervious road surfaces, especially in storm and snowmelt events. However, in moist, hilly, and mountainous terrain, such

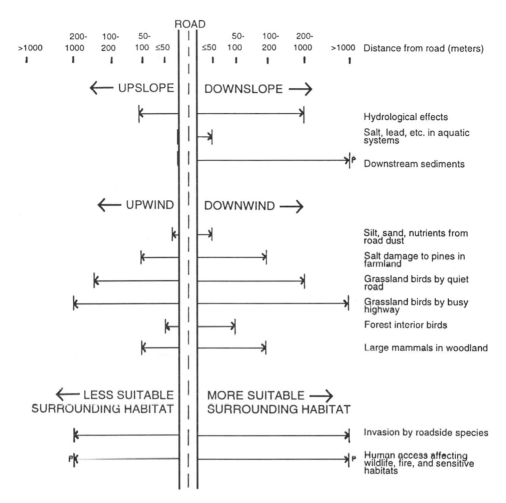

Figure 2   Road-effect zone defined by ecological effects extending different distances from a road. Most distances are based on specific illustrative studies (39); distance to left is arbitrarily half of that to right. (*P*) indicates an effect primarily at specific points. From Forman et al (43).

runoff is often insignificant compared with the conversion of slow-moving groundwater to fast-moving surface water at cutbanks by roads (52, 62, 132). Surface water is then carried by roadside ditches, some of which connect directly to streams while others drain to culverts with gullies incised below their outlets (132). Increased runoff associated with roads may increase the rates and extent of erosion, reduce percolation and aquifer recharge rates, alter channel morphology, and increase stream discharge rates (13, 14). Peak discharges or floods then restructure riparian areas by rearranging channels, logs, branches, boulders, fine-sediment deposits, and pools.

In forests, the combination of logging and roads increases peak discharges and downstream flooding (62, 132). Forest removal results in lower evapotranspiration and water-storage capabilities, but roads alone may increase peak discharge rates (62). Also, flood frequency apparently correlates with the percentage of road cover in a basin (52, 62, 110).

Roads may alter the subsurface flow as well as the surface flow on wetland soils (116). Compacted saturated or nearly saturated soils have limited permeability and low drainage capacity. Wetland road crossings often block drainage passages and groundwater flows, effectively raising the upslope water table and killing vegetation by root inundation, while lowering the downslope water table with accompanying damage to vegetation (116, 118).

Streams may be altered for considerable distances both upstream and downstream of bridges. Upstream, levees or channelization tend to result in reduced flooding of the riparian zone, grade degradation, hydraulic structural problems, and more channelization (17). Downstream, the grade change at a bridge results in local scouring that alters sedimentation and deposition processes (17, 49). Sediment and chemicals enter streams where a road crosses, and mathematical models predict sediment loading in and out of reaches affected by stream crossings (5). The fixed stream (or river) location at a bridge or culvert reduces both the amount and variability of stream migration across a floodplain. Therefore, stream ecosystems have altered flow rates, pool-riffle sequences, and scour, which typically reduce habitat-forming debris and aquatic organisms.

Sediment

The volume of sediment yield from a road depends on sediment supply and transport capacity (5). Sediment yield is determined by road geometry, slope, length, width, surface, and maintenance (5, 51), in addition to soil properties and vegetation cover (59). Road surfaces, cutbanks, fillslopes, bridge/culvert sites, and ditches are all sources of sediment associated with roads. The exposed soil surfaces, as well as the greater sediment-transport capacity of increased hydrologic flows, result in higher erosion rates and sediment yields (99).

Road dust as a little-studied sediment transfer may directly damage vegetation, provide nutrients for plant growth, or change the pH and vegetation (109). Effect-distances are usually <10–20 m but may extend to 200 m downwind (Figure 2). In arid land, soil erosion and drainage are common road problems (61).

Arctic roads also are often sources of dust. Other ecological issues include change in albedo, flooding, erosion and thermokarst, weed migration, waterfowl and shorebird habitat, and altered movement of large mammals (129).

Landsliding or mass wasting associated with roads may be a major sediment source (13, 117). Some of this sediment accumulates on lower slopes and is subject to subsequent erosion. The rest reaches floodplains or streams, where it alters riparian ecosystems, channel morphology, or aquatic habitat. Although gradual sediment transport and episodic landslides are natural processes affecting streams, elevated levels caused by roads tend to disrupt aquatic ecosystems. Indeed, logging roads commonly produce more erosion and sediment yield, particularly by mass wasting, than do the areas logged (45, 51, 104, 117).

Buffer strips between roads and streams tend to reduce sediments reaching aquatic ecosystems (77, 91). Buffers may be less effective for landslides than for arresting water and sediment from culverts and roadside ditches. Good road location (including avoiding streamsides and narrow floodplains for many ecological reasons), plus good ecological design of roadsides relative to slope, soil, and hydrology, may be a better strategy than depending on wide buffers to absorb sediment.

Water from road ditches tends to deposit finer sediment in streams, whereas landslides generally provide coarser material. Fine sediment increases turbidity (51), which disrupts stream ecosystems in part by inhibiting aquatic plants, macro-invertebrates, and fish (14, 16, 31, 99). Coarse deposits such as logs and boulders help create deep pools and habitat heterogeneity in streams. During low-flow periods, fine-sediment deposits tend to fill pools and smooth gravel beds, hence degrading habitats and spawning sites for key fish (13, 14, 31). During high discharge events, accumulated sediment tends to be flushed out and redeposited in larger water bodies.

In short, roads accelerate water flows and sediment transport, which raise flood levels and degrade aquatic ecosystems. Thus, local hydrologic and erosion effects along roads are dispersed across the land, whereas major impacts are concentrated in the stream network and distant valleys.

Chemical Transport

Most chemical transport from roads occurs in stormwater runoff through or over soil. Runoff pollutants alter soil chemistry, may be absorbed by plants, and affect stream ecosystems, where they are dispersed and diluted over considerable distances (16, 50, 66, 137, 138). Deicing salt and heavy metals are the two main categories of pollutants studied in road runoff.

The primary deicing agent, NaCl, corrodes vehicles and bridges, contaminates drinking water supplies, and is toxic to many species of plants, fish, and other aquatic organisms (4, 16, 84). Calcium magnesium acetate (CMA) is a more effective deicer, less corrosive, less mobile in soil, biodegradable, and less toxic to aquatic organisms (4, 84, 89). Also, $CaCl$ used to decrease dust may inhibit amphibian movement (28).

Airborne NaCl from road snowplowing may cause leaf injury to trees (e.g. *Pinus strobus*) up to 120 m from a road, especially downwind and downslope (58, 84). Trees seem to be more sensitive to chloride damage than are common roadside shrubs and grasses. Sodium accumulation in soils, mainly within 5 m of a road, alters soil structure, which affects plant growth (84). Road salt has facilitated the spread of three coastal exotic plants as much as 150 km in The Netherlands (1).

Deicing agents tend to increase the mobility of chemical elements in soil, such as heavy metals (by NaCl) and Na, Cl, Ca, and Mg (by CMA) (4). This process facilitates contamination of groundwater, aquifers, and streams. Because of dilution, the chemical effects of road runoff on surface water ecosystems may be primarily confined to small streams, particularly where they run adjacent to roads (36, 84).

Heavy metals are relatively immobile and heterogeneously distributed in roadsides, especially due to drainage ditch flows (15, 55, 80). Soils adjacent to the road surface typically contain the greatest mass (136). Elevated concentrations in grass tissue may occur within 5–8 m of a road, although high lead levels have been found in soil out to 25 m (30, 65, 82). Elevated lead concentrations were found in tissue of several small-mammal species in a narrow zone by roads, with higher lead levels by busy roads (48).

Highway roadsides of 5–15-m width next to traffic densities of 11,000–124,000 veh/day in The Netherlands had somewhat higher heavy-metal accumulations on the downwind side, but no correlation with traffic density was found (30). All average levels of Pb, Cd, Zn, Cr, Ni, and As in cut grass (hay) from these roadsides were below the Dutch maximum-acceptable-levels for livestock fodder and "clean compost." Only Zn in some roadsides studied exceeded the maximum for "very clean compost."

Many other chemicals enter roadsides. Herbicides often kill non-target plants, particularly from blanket applications in drifting air. For polycyclic aromatic hydrocarbons from petroleum (136), the preliminary conclusion for the Dutch highways was that levels in roadside hay "do not seem to give cause for alarm" (30). Fertilizer nutrients affect roadside vegetation (19, 86, 92), and nitrogen from vehicular NOX emission altered vegetation up to 100–200 m from a highway in Britain (7). Acidic road runoff may have impacts on stream ecosystems (36, 81). Of the hazardous materials transported on roads, e.g. >500,000 shipments moved each day in the United States, a small fraction is spilled, although occasional large spills cause severe local effects (85).

Typical water-quality responses to road runoff include altered levels of heavy metal, salinity, turbidity, and dissolved oxygen (16, 23, 81). However, these water-quality changes, even in a wetland, tend to be temporary and localized due to fluctuations in water quantity (23). Road runoff is a major source of heavy metals to stream systems, especially Pb, Zn, Cu, Cr, and Cd (16, 50, 64, 137). Fish mortality in streams has been related to high concentrations of Al, Mn, Cu, Fe, or Zn, with effects on populations reorded as far as 8 km downstream (81). Both high traffic volume and high metal concentration in runoff have correlated with mortality of fish and other aquatic organisms (59). Floodplain soil near bridges may have high heavy-metal concentrations (138). Although highway runoff generally has little adverse effect on vegetation or plant productivity, it may change the species composition of floodplain plant communities, favoring common species (138).

Overall, terrestrial vegetation seems to be more resistant than aquatic organisms to road impacts (59, 138). Drainage of road runoff through grassy channels greatly reduces toxic solid- and heavy-metal concentrations (59). Furthermore, dense vegetation increases soil infiltration and storage. Therefore, instead of expensive detention ponds and drainage structures to reduce runoff impacts, creative grassland designs by roads, perhaps with shrubs, may provide both sponge and biodiversity benefits. The wide range of studies cited above lead to the conclusion that chemical impacts tend to be localized near roads.

THE ROAD NETWORK

New Roads and Changing Landscape Pattern

Do roads lead to development, or does development lead to roads? This timeless debate in the transportation community has greater ramifications as environmental quality becomes more important in the transportation–land-use interaction (114). For example, new roads into forested landscapes often lead to economic development as well as deforestation and habitat fragmentation (22).

At the landscape scale, the major ecological impacts of a road network are the disruption of landscape processes and loss of biodiversity. Interrupting horizontal natural processes, such as groundwater flow, streamflow, fire spread, foraging, and dispersal, fundamentally alters the way the landscape works (40, 53). It truncates flows and movements, and reduces the critical variability in natural processes and disturbances. Biodiversity erodes as the road network impacts interior species, species with large home ranges, stream and wetland species, rare native species, and species dependent on disturbance and horizontal flows.

A new road system in the Rondonia rainforest of Brazil illustrates such effects. By 1984, road construction and asphalt paving had stimulated a major influx of people and forest clearing (25). A regular pattern of primary roads plus parallel secondary roads 4 km apart was imposed on the land. Commonly, small forest plots of ~5 ha were gradually converted to grass and joined with neighboring plots to form large pastures (26). Simulation models of this typical scenario were compared with models of a worst-case scenario and an "innovative farming" scenario with perennial crops and essentially no fire or cattle (25, 26). Species requiring a large area and having poor "gap-crossability" disappeared in all model scenarios after road construction. Species with moderate requirements for area and gap-crossability persisted only in the innovative farming scenario. Reestablishing the connectivity of nature with a network of wildlife corridors was proposed as a solution to maintain the first group of species, which are of conservation importance.

The closure and removal of some roads in the grid is an alternative ecological approach. Natural landscape processes and biodiversity are both inhibited by a rigid road grid. Closing and eliminating some linkages would permit the reestablishment of a few large patches of natural rainforest (39, 41, 43). Such a solution helps create a road network with a high variance in mesh size. Large natural-vegetation patches in areas remote from both roads and people are apparently required to sustain important species such as wolf, bear, and probably jaguar (*Canis lupus, Ursus, Felis onca*) (39, 76). Temporary road closures could, for example, enhance amphibian migration during the breeding phase (67). In contrast, road closure and removal could eliminate motorized vehicle use, thus reducing numerous disturbance effects on natural populations and ecosystems.

A general spatial-process model emphasizes that roads have the greatest ecological impact early in the process of land transformation (39, 41). They dissect the land, leading to habitat fragmentation, shrinkage, and attrition. Forest road networks may also create distinctive spatial patterns, such as converting convoluted to rectilinear shapes, decreasing core forest area, and creating more total edge habitat by roads than by logged areas (79, 96).

Forest roads as a subset of roads in general are characterized as being narrow, not covered with asphalt, lightly traveled, and remote (98). Among the wide range of ecological effects of roads (39, 95), forest roads have a distinctive set of major ecological effects: (*a*) habitat loss by road construction, (*b*) altered water routing and downstream peak flows, (*c*) soil erosion and sedimentation impacts on streams, (*d*) altered species patterns, and (*e*) human access and disturbance in remote areas (43, 45, 62, 132). Thus, an evaluation of logging regimes includes the ecological effects of both the road network and forest spatial patterns (45, 69). In conclusion, a road network disrupts horizontal natural processes, and by altering both landscape spatial pattern and the processes, it reduces biodiversity.

Road Density

Road density, e.g. measured as km/km^2, has been proposed as a useful, broad index of several ecological effects of roads in a landscape (39, 43, 44, 95). Effects are evident for faunal movement, population fragmentation, human access, hydrology, aquatic ecosystems, and fire patterns.

A road density of approx. 0.6 km/km^2 (1.0 mi/mi^2) appears to be the maximum for a naturally functioning landscape containing sustained populations of large predators, such as wolves and mountain lions (*Felis concolor*) (43, 76, 124). Moose (*Alces*), bear (*Ursus*) (brown, black, and grizzly), and certain other populations also decrease with increasing road density (11, 43, 72). These species are differentially sensitive to the roadkill, road-avoidance, and human-access dimensions of road density. Species that move along, rather than across, roads presumably are benefitted by higher road density (12, 39).

Human access and disturbance effects on remote areas tend to increase with higher road density (39, 72, 76). Similarly, human-caused fire ignitions and suppressions may increase, and average fire sizes decrease (111).

Aquatic ecosystems are also affected by road density. Hydrologic effects, such as altered groundwater conditions and impeded drainage upslope, are sensitive to road density (116, 118). Increased peak flows in streams may be evident at road densities of 2–3 km/km^2 (62). Detrimental effects on aquatic ecosystems, based on macro-invertebrate diversity, were evident where roads covered 5% or more of a watershed in California (75). In southeastern Ontario, the species richness of wetland plants, amphibians/reptiles, and birds each correlated negatively with road density within 1–2 km of a wetland (38).

Road density is an overall index that averages patterns over an area. Its effects probably are sensitive to road width or type, traffic density, network connectivity, and the frequency of spur roads into remote areas. Thus network structure, or an index of variance in mesh size, is also important in understanding the effect of road density (39, 76, 79, 96). Indeed, although road density is a useful overall index, the presence of a few large areas of low road density may be the best indicator of suitable habitat for large vertebrates and other major ecological values.

TRANSPORTATION POLICY AND PLANNING

Environmental Policy Dimensions

Ecological principles are increasingly important in environmental transportation policy, and Australia, The Netherlands, and the United States highlight contrasting approaches. Australian policy has focused on biodiversity, including wildflower protection. An enormous network of road reserves with natural-vegetation strips 10–200 m wide (Figure 1) stands out in many agricultural landscapes (66, 92, 111). Public pressure helped create this system, which "helps to prevent soil erosion" and "where wildflowers can grow and flourish in perpetuity" (111). Diverse experimental management approaches involve burning, weed control, planting native species, and nature restoration. Ecological scientists commonly work side by side with civil engineers in transportation departments at all levels of government.

In contrast, Dutch policy has focused on the open roadside vegetation, road-kills, animal movement patterns, and nature restoration (1, 21, 29, 86). This approach reflects the stated national objectives of (*a*) recreating "nature, including natural processes and biodiversity; and (*b*) enhancing the national ecological network," mainly composed of large natural-vegetation patches and major wildlife and water corridors (1, 46). An impressive series of mitigation overpasses, tunnels, and culverts provide for animal and water movement where interrupted by road barriers (24, 43, 44, 86). Environmental activities in transportation revolve around a group of environmental scientists in the national Ministry of Transport who work closely with engineers and policy-makers at both local and national levels.

In the United States, environmental transportation policy focuses on vehicular pollutants, as well as engineering solutions for soil erosion and sedimentation (85). A few states have built wildlife underpasses and overpasses to address local roadkill or wildlife movement concerns. A 1991 federal law—the Intermodal Surface Transportation Efficiency Act (ISTEA)—establishes policy for a transportation system that is "economically efficient and environmentally sound," considers the "external benefits of reduced air pollution, reduced traffic congestion and other aspects of the quality of life," considers transportation in a region-wide metropolitan area, and links ecological attributes with the aesthetics of a landscape (43). Thus, US transportation policy largely ignores biodiversity loss, habitat fragmentation, disruption of horizontal natural processes, natural stream and wetland hydrology, streamwater chemistry, and reduction of fish populations, a range of ecological issues highlighted in the transportation community in 1997 (85).

Of course, many nations use ecological principles in designing transportation systems (21, 33, 63, 95, 134), and environmental scientists, engineers, and policy-makers in Europe have united to "conserve biodiversity and reduce ...fauna casualties" at the international level (21; H van Bohemen, personal communication). The successful removal of lead from petroleum led to less lead in roadside ecosystems worldwide. Nevertheless, the huge Australian road-reserve system and the Dutch mitigation system for animal and water flows are especially ambitious and pioneering.

Spatial Planning and Mitigation

Most existing roads were built before the explosion in ecological knowledge, and many are poorly located ecologically (43). Yet the Dutch have developed a promising transportation planning process for the movement of both people and natural processes across the land (40, 44, 86). In essence, the ecological network, consisting of large natural-vegetation patches plus major corridors for water and wildlife movement, is mapped. The road network is then superimposed on the ecological network to identify bottlenecks. Finally, mitigation or compensation techniques are applied to eliminate designated percentages of bottlenecks in a time sequence. The earlier such spatial planning begins, the greater its effect (41).

Compensation is proposed where bottlenecks apparently cannot be overcome by mitigation. The principle of no-net-loss has been used internationally for wetlands with varying success (46, 94), whereas no-net-loss of natural processes and biodiversity by roads is a concept only beginning to be applied (1, 24, 46, 86). The loss, e.g. of biodiversity or groundwater flow, is compensated by increasing an equivalent ecological value nearby. Options include protection of an equivalent amount of high-quality habitat, reestablishment of another wildlife corridor, or creation of new habitat. Mitigation, on the other hand, attempts to minimize detrimental ecological impacts and is illustrated by the varied wildlife passages (tunnels, pipes, underpasses, overpasses) operating for animal movement (21, 86).

Diverse tunnel designs focus on small and mid-sized animals. Amphibian tunnels, generally 30–100 cm wide and located where roads block movement to breeding ponds or wetlands, are widespread in Europe and rare in the United States (33, 43, 67). "Ecopipes," or badger tunnels, are pipes ~40 cm in diameter mainly designed for movement of mid-sized mammals across Dutch roads, and located where water can rarely flow through (9, 44, 46, 86). In contrast, Dutch wildlife culverts are ~120 cm wide, with a central channel for water flow between two raised 40-cm-wide paths for animal movement. "Talus tunnels" are designed for a mid-sized mammal that lives and moves in rock talus slopes in Australia (74).

Wildlife underpasses, generally 8–30 m wide and at least 2.5 m high, have been built for large mammals in southern Florida and scattered locations elsewhere in the United States, Canada, and France (33, 43, 54, 63, 93, 113). Wildlife overpasses, also designed for large mammals, range in width up to 200 m and are scarce: approximately 6 in North America (New Jersey, Utah, Alberta, British Columbia) (33, 43; BF Leeson, personal communication) and 17 in Europe (Germany, France, The Netherlands, Switzerland) (44, 46, 63, 86, 134). The minimum widths for effectiveness may be 30–50 m in the center and 50–80 m on the ends (33, 46, 86). The two Swiss overpasses of 140-m and 200-m width remind us that ultimately the goal should be "landscape connectors" that permit all horizontal natural processes to cross roads (43, 44).

These mitigation structures are normally combined with fencing and vegetation to enhance animal crossing (86). Almost all such passages are successful in that the target species crosses at least occasionally, and most are used by many other species. Florida underpasses are used by the Florida panther (*Felis concolor coryi*), nearly the whole local terrestrial fauna, and groundwater as well (33, 54). Underpasses and overpasses are used by almost all large mammal species of a region. Yet, little information exists on crossing rates relative to population sizes, movement rates away from roads, predation rates, home range locations, and so forth. Nevertheless, mitigation passages are effective in perforating road barriers to maintain horizontal natural processes across the land.

The Road-Effect Zone

Roads and roadsides cover 0.9% of Britain and 1.0% of the United States, while road reserves (Figure 1) cover 2.5% of the State of Victoria, Australia (12, 85, 131). Yet how much of the land is ecologically impacted by roads with vehicles?

The road-effect zone is the area over which significant ecological effects extend outward from a road and typically is many times wider than the road surface plus roadsides (Figure 2) (39, 95, 101, 134). The zone is asymmetric with convoluted boundaries, reflecting the sequence of ecological variables, plus unequal effect-distances due to slope, wind, and habitat suitability on opposite sides of a road (40, 43). Knowing the average width of the road-effect zone permits us to estimate the proportion of the land ecologically affected by roads (43). For example, based on the traffic noise effect-distances of sensitive bird species described above, road-effect zones cover ~10–20% of The Netherlands (101).

Finally, a preliminary calculation for the United States was made based on nine water and species variables in Massachusetts (USA), plus evidence from the Dutch studies (42). An estimated 15–20% of the US land area is directly affected ecologically by roads. These estimates reemphasize the immensity and pervasiveness of ecological road impacts. Moreover, they challenge science and society to embark on a journey of discovery and solution.

ACKNOWLEDGMENTS

Virginia H Dale, Robert D Deblinger, Malcolm L Hunter, Jr, and Julia A Jones provided terrific reviews, which we deeply appreciate.

> **Visit the *Annual Reviews home page* at**
> **http://www.AnnualReviews.org**

Literature Cited

1. Aanen P, Alberts W, Bekker GJ, van Bohemen HD, Melman PJM, et al. 1991. *Nature Engineering and Civil Engineering Works*. Wageningen, Netherlands: PUDOC
2. Adams LW. 1984. Small mammal use of an interstate highway median strip. *J. Appl. Ecol.* 21:175–78
3. Amor RL, Stevens PL. 1976. Spread of weeds from a roadside into sclerophyll forests at Dartmouth, Australia. *Weed Res.* 16:111–18

4. Amrhein C, Strong JE, Mosher PA. 1992. Effect of deicing salts on metal and organic matter mobilization in roadside soils. *Environ. Sci. Technol.* 26:703–9

5. Anderson B, Simons DB. 1983. Soil erosion study of exposed highway construction slopes and roadways. *Transp. Res. Rec.* 948:40–47

6. Andrews A. 1990. Fragmentation of habitat by roads and utility corridors: a review. *Aust. J. Zool.* 26:130–41

7. Angnold PG. 1997. The impact of road upon adjacent heathland vegetation: effects on plant species composition. *J. Appl. Ecol.* 34:409–17

8. Ashley EP, Robinson JT. 1996. Road mortality of amphibians, reptiles and other wildlife on the Long Point causeway, Lake Erie, Ontario. *Can. Field Nat.* 110:403–12

9. Bekker H, Canters KJ. 1997. The continuing story of badgers and their tunnels. See Ref. 21, pp. 344–53

10. Bellis ED, Graves HB. 1971. Deer mortality on a Pennsylvania interstate highway. *J. Wildl. Manage.* 35:232–37

11. Bennett AF. 1988. Roadside vegetation: a habitat for mammals at Naringal, southwestern Victoria. *Victorian Natur.* 105:106–13

12. Bennett AF. 1991. Roads, roadsides and wildlife conservation: a review. See Ref. 111, pp. 99–117

13. Beschta R. 1978. Long-term patterns of sediment production following road construction and logging in the Oregon Coast Range. *Water Resour. Res.* 14:1011–16

14. Bilby RE, Sullivan K, Duncan SH. 1989. The generation and fate of road-surface sediment in forested watersheds in southwestern Washington. *For. Sci.* 35:453–68

15. Black FM, Braddock JN, Bradow R, Ingalls M. 1985. Highway motor vehicles as sources of atmospheric particles: projected trends 1977–2000. *Environ. Int.* 11:205–33

16. Brown KJ. 1994. River-bed sedimentation caused by off-road vehicles at river fords in the Victorian Highlands, Australia. *Water Resour. Bull.* 30:239–50

17. Brown SA. 1982. Prediction of channel bed grade changes at highway stream crossings. *Transp. Res. Rec.* 896:1–11

18. Buchanan BW. 1993. Effects of enhanced lighting on the behaviour of nocturnal frogs. *Anim. Behav.* 43:893–99

19. Cale P, Hobbs RJ. 1991. Condition of roadside vegetation in relation to nutrient status. See Ref. 111, pp. 353–62

20. Calvo RN, Silvy NJ. 1996. Key deer mortality, U.S. 1 in the Florida Keys. See Ref. 33, pp. 337–48

21. Canters K, ed. 1997. *Habitat Fragmentation & Infrastructure.* Minist. Transp., Public Works & Water Manage., Delft, Netherlands. 474 pp.

22. Chomitz KM, Gray DA. 1996. Roads, land use, and deforestation: a spatial model applied to Belize. *World Bank Econ. Rev.* 10:487–512

23. Cramer GH, Hopkins WC. 1982. Effects of a dredged highway construction on water quality in a Louisiana wetland. *Transp. Res. Rec.* 896:47–51

24. Cuperus R, Piepers AAG, Canters KJ. 1997. Elaboration of the compensation principle in the Netherlands: outlines of the draft manual *Ecological Compensation in Road Projects.* See Ref. 21, pp. 308–15

25. Dale VH, O'Neill RV, Southworth F, Pedlowski M. 1994. Modeling effects of land management in the Brazilian Amazonian settlement of Rondonia. *Conserv. Biol.* 8:196–206

26. Dale VH, Pearson SM, Offerman IIL, O'Neill RV. 1994. Relating patterns of land-use change to faunal biodiversity in the Central Amazon. *Conserv. Biol.* 8:1027–36

27. Dawson BL. 1991. South Africa road reserves valuable conservation reserves? See Ref. 111, pp. 119–30

28. deMaynadier PG, Hunter ML Jr. 1995. The relationship between forest management and amphibian ecology: a review of the North American literature. *Environ. Rev.* 3:230–61

29. Dienst Weg- en Waterbouwkunde. 1994. *Managing Roadside Flora in The Netherlands.* DWW Wijzer 60. Delft, Netherlands. 4 pp.

30. Dienst Weg- en Waterbouwkunde. 1994. *The Chemical Quality of Verge Grass in The Netherlands.* DWW Wijzer 62. Delft, Netherlands. 4 pp.

31. Eaglin GS, Hubert WA. 1993. Effects of logging and roads on substrate and trout in streams of the Medicine Bow National Forest, Wyoming. *North Am. J. Fish. Manage.* 13:844–46

32. Ellenberg H, Muller K, Stottele T. 1991. Strassen-Okologie. In *Okologie und Strasse*, pp.19–115. Broschurenreihe de Deutschen Strassenliga, Bonn, Ger.

33. Evink GL, Garret P, Zeigler D, Berry J, eds. 1996. *Trends in Addressing Transportation Related Wildlife Mortality.* No. FL-ER-58-96. Florida Dep. Transp. Tallahassee, FL. 395 pp.

34. Fahrig L, Pedlar JH, Pope SE, Taylor PD, Wegner JF. 1995. Effect of road traffic on amphibian density. *Biol. Conserv.* 73:177–82

35. Feldhamer GA, Gates JE, Harman DM, Loranger AJ, Dixon KR. 1986. Effects of interstate highway fencing on white-tailed deer activity. *J. Wildl. Manage.* 50:497–503

36. Fennessey TW. 1989. Guidelines for handling acid-producing materials on low-volume roads. *Transp. Res. Rec.* 1291: 186–89

37. Ferris CR. 1979. Effects of Interstate 95 on breeding birds in northern Maine. *J. Wildl. Manage.* 43:421–27

38. Findlay CS, Houlahan J. 1997. Anthropogenic correlates of species richness in southeastern Ontario wetlands. *Conserv. Biol.* 11:1000–9

39. Forman RTT. 1995. *Land Mosaics: The Ecology of Landscapes and Regions.* Cambridge, UK: Cambridge Univ. Press

40. Forman RTT. 1998. Horizontal processes, roads, suburbs, societal objectives, and landscape ecology. In *Landscape Ecological Analysis: Issues and Applications*, ed. JM Klopatek, RH Gardner. New York: Springer-Verlag. In press

41. Forman RTT, Collinge SK. 1997. Nature conserved in changing landscapes with and without spatial planning. *Landsc. Urban Plan.* 37:129–35

42. Forman RTT, Deblinger RD. 1998. The ecological road-effect zone for transportation planning, and a Massachusetts highway example. In *Proc. Int. Conf. Wildlife and Transportation, Rep. FL-ER-69-98*, ed. GL Evink, P Garrett, D Zeigler, J Berry. Florida Dep. Transp. Tallahassee, FL. In press

43. Forman RTT, Friedman DS, Fitzhenry D, Martin JD, Chen AS, Alexander LE. 1997. Ecological effects of roads: toward three summary indices and an overview for North America. See Ref. 21, pp. 40–54

44. Forman RTT, Hersperger AM. 1996. Road ecology and road density in different landscapes, with international planning and mitigation solutions. See Ref. 33, pp. 1–22

45. Forman RTT, Mellinger AD. 1998. Road networks and forest spatial patterns in ecological models of diverse logging regimes. In *Nature Conservation in Production Environments: Managing the Matrix*, ed. DA Saunders, J Craig, N Mitchell. Chipping Norton, Australia: Surrey Beatty. In press

46. Friedman DS. 1997. *Nature as Infrastructure: The National Ecological Network and Wildlife-crossing Structures in The Netherlands.* Rep. 138, DLO Winand Staring Centre. Wageningen, Netherlands. 49 pp.

47. Getz LL, Cole FR, Gates DL. 1978. Interstate roadsides as dispersal routes for *Microtus pennsylvanicus. J. Mammal.* 59:208–12

48. Getz LL, Verner L, Prather M. 1977. Lead concentrations in small mammals living near highways. *Environ. Pollut.* 13:151–57

49. Gilje SA. 1982. Stream channel grade changes and their effects on highway crossings. *Transp. Res. Rec.* 895:7–15

50. Gilson MP, Malivia JF, Chareneau RJ. 1994. Highway runoff studied. *Water Environ. Technol.* 6:37–38

51. Grayson RB, Haydon SR, Jayasuriya MDA, Enlayson BC. 1993. Water quality in mountain ash forests—separating the impacts of roads from those of logging operations. *J. Hydrol.* 150:459–80

52. Harr RD, Harper WC, Krygier JT, Hsieh FS. 1975. Changes in storm hydrographs after road building and clear-cutting in the Oregon Coast Range. *Water Resour. Res.* 11:436–44

53. Harris LD, Hoctor TS, Gergel SE. 1996. Landscape processes and their significance to biodiversity conservation. In *Population Dynamics in Ecological Space and Time*, ed. O Rhodes Jr, R Chesser, M Smith, pp. 319–47. Chicago: Univ. Chicago Press

54. Harris LD, Scheck J. 1991. From implications to applications: the dispersal corridor principle applied to the conservation of biological diversity. See Ref. 111, pp. 189–220

55. Hewitt CN, Rashed MB. 1991. The deposition of selected pollutants adjacent to a major rural highway. *Atmos. Environ.* 35A(5–6):979–83

56. Hodson NL. 1962. Some notes on the causes of bird road casualties. *Bird Study* 9:168–73

57. Hodson NL. 1966. A survey of road mortality in mammals (and including data for the grass snake and common frog). *J. Zool., Lond.* 148:576–79

58. Hofstra G, Hall R. 1971. Injury on roadside trees: leaf injury on pine and white cedar in relation to foliar levels of sodium chloride. *Can. J. Bot.* 49:613–22

59. Horner RR, Mar BW. 1983. Guide for assessing water quality impacts of highway operations and maintenance. *Transp. Res. Rec.* 948:31–39

60. Huey LM. 1941. Mammalian invasion via

the highway. *J. Mammal.* 22:383–85
61. Iverson RM, Hinckley BS, Webb RM. 1981. Physical effects of vehicular disturbances on arid landscapes. *Science* 212:915–17
62. Jones JA, Grant GE. 1996. Peak flow responses to clearcutting and roads in small and large basins, Western Cascades, Oregon. *Water Resour. Res.* 32:959–74
63. Keller V, Pfister HP. 1997. Wildlife passages as a means of mitigating effects of habitat fragmentation by roads and railway lines. See Ref. 21, pp. 70–80
64. Kerri KD, Racin JA, Howell RB. 1985. Forecasting pollutant loads from highway run off. *Transp. Res. Rec.* 1017:39–46
65. Lagerwerff JV, Specht AN. 1970. Contamination of roadside soil and vegetation with cadmium, nickel, lead, and zinc. *Environ. Sci. Technol.* 4:583–86
66. Lamont DA, Blyth JD. 1995. Roadside corridors and community networks. In *Nature Conservation 4: The Role of Networks*, ed. DA Saunders, JL Craig, EM Mattiske, pp. 425–35. Chipping Norton, Australia: Surrey Beatty
67. Langton TES, ed. 1989. *Amphibians and Roads*. ACO Polymer Products, Shefford, Bedfordshire, UK. 202 pp.
68. Leopold LB, Wolman MG, Miller JP. 1964. *Fluvial Processes in Geomorphology*. San Francisco: Freeman
69. Li H, Franklin JF, Swanson FJ, Spies TA. 1993. Developing alternative forest cutting patterns: a simulation approach. *Landsc. Ecol.* 8:63–75
70. Lonsdale WM, Lane AM. 1994. Tourist vehicles as vectors of weed seeds in Kakadu National Park, northern Australia. *Biol. Conserv.* 69:277–83
71. Lowe JC, Moryadas S. 1975. *The Geography of Movement*. Boston: Houghton Mifflin
72. Lyon LJ. 1983. Road density models describing habitat effectiveness for elk. *J. For.* 81:592–95
73. Mader HJ. 1984. Animal habitat isolation by roads and agricultural fields. *Biol. Conserv.* 29:81–96
74. Mansergh IM, Scotts DJ. 1989. Habitat-continuity and social organization of the mountain pygmy-possum restored by tunnel. *J. Wildl. Manage.* 53:701–7
75. McGurk B, Fong DR. 1995. Equivalent roaded area as a measure of cumulative effect of logging. *Environ. Manage.* 19:609–21
76. Mech LD. 1989. Wolf population survival in an area of high road density. *Am. Midl. Nat.* 121:387–89
77. Megahan WF, Ketcheson GL. 1996. Predicting downslope travel of granitic sediments from forest roads in Idaho. *Water Resour. Bull.* 32:371–82
78. Merriam G, Kozakiewiez M, Tsuchiya E, Hawley K. 1989. Barriers as boundaries for metapopulations and demes of *Peromyscus leucopus* in farm landscapes. *Landsc. Ecol.* 2:227–35
79. Miller JR, Joyce LA, Knight RL, King RM. 1996. Forest roads and landscape structure in the southern Rocky Mountains. *Landsc. Ecol.* 11:115–27
80. Montgomery D. 1994. Road surface drainage, channel initiation, and slope instability. *Water Resour. Res.* 30:192–93
81. Morgan E, Porak W, Arway J. 1983. Controlling acidic-toxic metal leachates from southern Appalachian construction slopes: mitigating stream damage. *Transp. Res. Rec.* 948:10–16
82. Motto HL, Daines RL, Chilko DM, Motto CK. 1970. Lead in soils and plants: its relationship to traffic volume and proximity to highways. *Environ. Sci. Technol.* 4:231–37
83. Munguira ML, Thomas JA. 1992. Use of road verges by butterfly and burnet populations, and the effect of roads on adult dispersal and mortality. *J. Appl. Ecol.* 29:316–29
84. National Research Council. 1991. *Highway Deicing: Comparing Salt and Calcium Magnesium Acetate*. Spec. Rep. 235, Transp. Res. Board, Washington, DC. 170 pp.
85. National Research Council. 1997. *Toward a Sustainable Future: Addressing the Long-term Effects of Motor Vehicle Transportation on Climate and Ecology*. Washington, DC: Natl. Acad. Press
86. *Natuur Over Wegen (Nature Across Motorways)*. 1995. Dienst Weg- en Waterbouwkunde, Delft, Netherlands. 103 pp.
87. Oetting RB, Cassel JF. 1971. Waterfowl nesting on interstate highway right-of-way in North Dakota. *J. Wildl. Manage.* 35:774–81
88. Opdam P, van Apeldoorn R, Schotman A, Kalkhoven J. 1993. Population responses to landscape fragmentation. In *Landscape Ecology of a Stressed Environment*, ed. CC Vos, P Opdam, pp. 147–71. London: Chapman & Hall
89. Ostendorf DW, Pollack SJ, DeCheke ME. 1993. Aerobic degradation of CMA in roadside soils: field simulations from soil microcosms. *J. Environ. Qual.* 22:299–304
90. Oxley DJ, Fenton MB, Carmody GR. 1974. The effects of roads on populations

of small mammals. *J. Appl. Ecol.* 11:51–59

91. Packer PE. 1967. Criteria for designing and locating logging roads to control sediment. *For. Sci.* 13:1–18

92. Panetta FD, Hopkins AJM. 1991. Weeds in corridors: invasion and management. See Ref. 111, pp. 341–51

93. Paquet PC, Callaghan C. 1996. Effects of linear developments on winter movements of gray wolves in the Bow River Valley of Banff National Park, Alberta. See Ref. 33, pp. 51–73

94. Race MS, Fonseca MS. 1995. Fixing compensatory mitigation: What will it take? *Ecol. Appl.* 6:94–101

95. Reck H, Kaule G. 1993. Strassen und Lebensraume: Ermittlung und Beurteilung strassenbedingter Auswirkungen auf Pflanzen, Tiere und ihre Lebensraume. *Forschung Strassenbau und Strassenverkehrstechnik*, Heft 654. Herausgegeben vom Bundesminister fur Verkehr, Bonn-Bad Godesberg, Ger. 230 pp.

96. Reed RA, Johnson-Barnard J, Baker WL. 1996. Contribution of roads to forest fragmentation in the Rocky Mountains. *Conserv. Biol.* 10:1098–106

97. Reh W. 1989. Investigations into the influences of roads on the genetic structure of populations of the common frog *Rana temporaria*. See Ref. 67, pp. 101–13

98. Reh W, Seitz A. 1990. The influence of land use on the genetic structure of populations of the common frog *Rana temporaria*. *Biol. Conserv.* 54:239–49

99. Reid LM, Dunne T. 1984. Sediment production from forest road surfaces. *Water Resour. Res.* 20:1753–61

100. Reijnen MJSM, Veenbaas G, Foppen RPB. 1995. *Predicting the Effects of Motorway Traffic on Breeding Bird Populations.* DLO Inst. For. Nat. Res., Ministry Transp. Public Works, Delft, The Netherlands. 92 pp.

101. Reijnen R. 1995. *Disturbance by car traffic as a threat to breeding birds in The Netherlands.* PhD thesis, DLO Inst. For. Nat. Res., Wageningen, Netherlands. 140 pp.

102. Reijnen R, Foppen R, Meeuwsen H. 1996. The effects of traffic on the density of breeding birds in Dutch agricultural grasslands. *Biol. Conserv.* 75:255–60

103. Reijnen R, Foppen R, ter Braak C, Thissen J. 1995. The effects of car traffic on breeding bird populations in woodland. III. Reduction of density in relation to the proximity of main roads. *J. Appl. Ecol.* 32:187–202

104. Rice RM, Lewis J. 1991. Estimating erosion risks associated with logging and forest roads in northwestern California. *Water Resour. Bull.* 27:809–18

105. Roach GL, Kirkpatrick RD. 1985. Wildlife use of roadside woody plantings in Indiana. *Transp. Res. Rec.* 1016:11–15

106. Romin LA, Bissonette JA. 1996. Temporal and spatial distribution of highway mortality of mule deer on newly constructed roads at Jordanelle Reservoir, Utah. *Gt. Basin Nat.* 56:1–11

107. Ross SM. 1986. Vegetation change on highway verges in south-east Scotland. *J. Biogeogr.* 13:109–13

108. Rost GR, Bailey JA. 1979. Distribution of mule deer and elk in relation to roads. *J. Wildl. Manage.* 43:634–41

109. Santelman MV, Gorham EV. 1988. The influence of airborne road dust on the chemistry of *Sphagnum* mosses. *J. Ecol.* 76:1219–31

110. Sauer VB, Thomas EO Jr, Stricker VA, Wilson KV. 1982. Magnitude and frequency of urban floods in the United States. *Transp. Res. Rec.* 896:30–33

111. Saunders DA, Hobbs RJ, eds. 1991. *Nature Conservation 2: The Role of Corridors.* Chipping Norton, Australia: Surrey Beatty

112. Schmidt W. 1989. Plant dispersal by motor cars. *Vegetatio* 80:147–52

113. Singer FJ, Langlitz WL, Samuelson EC. 1985. Design and construction of highway underpasses used by mountain goats. *Transp. Res. Rec.* 1016:6–10

114. Skinner RE Jr, Moore T, Cervero R, Landis J, Giuliano G, Leinberger CB, Epstein LR. 1996. The transportation–land use interaction. *TR News (Washington)* 187:6–17

115. Soule ME, ed. 1987. *Viable Populations for Conservation.* Cambridge, UK: Cambridge Univ. Press

116. Stoeckeler JH. 1965. Drainage along swamp forest roads: lessons from Northern Europe. *J. For.* 63:771–76

117. Swanson FJ, Dyrness CT. 1975. Impact of clear-cutting and road construction on soil erosion by landslides in the western Cascade Range, Oregon. *Geology* 3:393–96

118. Swanson GA, Winter TC, Adomaitis VA, LaBaugh JW. 1988. *Chemical characteristics of prairie lakes in south-central North Dakota—their potential for influencing use by fish and wildlife. Tech. Rep. 18,* US Fish Wildl. Serv., Washington, DC

119. Swihart RK, Slade NA. 1984. Road crossing in *Sigmodon hispidus* and *Microtus ochrogaster*. *J. Mammal.* 65:357–60

120. Tabor R. 1974. Earthworms, crows, vibrations and motorways. *New Sci.* 62: 482–83
121. Tyser RW, Worley CA. 1992. Alien flora in grasslands adjacent to road and trail corridors in Glacier National Park, Montana (U.S.A.). *Conserv. Biol.* 6:253–62
122. van Apeldoorn RC. 1997. Fragmented mammals: What does that mean? See Ref. 21, pp. 121–26
123. van der Zande AN, ter Keurs J, van der Weijden WJ. 1980. The impact of roads on the densities of four bird species in an open field habitat—evidence of a long distance effect. *Biol. Conserv.* 18:299–321
124. van Dyke FB, Brocke RH, Shaw HG, Ackerman BB, Hemker TP, Lindzey FG. 1986. Reactions of mountain lions to logging and human activity. *J. Wildl. Manage.* 50:95–102
125. Vermeulen HJW. 1993. The composition of the carabid fauna on poor sandy road-side-verges in relation to comparable open areas. *Biodiv. Conserv.* 2:331–50
126. Vermeulen HJW. 1994. Corridor function of a road verge for dispersal of stenotopic heathland ground beetles (*Carabidae*). *Biol. Conserv.* 69:339–49
127. Vermeulen HJW, Opdam PFM. 1995. Effectiveness of roadside verges as dispersal corridors for small ground-dwelling animals: a simulation study. *Landsc. Urban Plan.* 31:233–48
128. Vos CC. 1997. Effects of road density: a case study of the moor frog. See Ref. 21, pp. 93–97
129. Walker DA, Walker MD. 1991. History and pattern of disturbance in Alaskan arctic terrestrial ecosystems: a hierarchical approach to analysing landscape change. *J. Appl. Ecol.* 28:244–76
130. Wasser S, Bevis K, King G, Hanson E. 1997. Noninvasive physiological measures of disturbance in the northern spotted owl. *Conserv. Biol.* 11:1019–22
131. Way JM. 1977. Roadside verges and conservation in Britain: a review. *Biol. Conserv.* 12:65–74
132. Wemple BC, Jones JA, Grant GE. 1996. Channel network extension by logging roads in two basins, Western Cascades, Oregon. *Water Resour. Bull.* 32:1195–207
133. Wilcox DA. 1989. Migration and control of purple loosestrife (*Lythrum salicaria* L.) along highway corridors. *Environ. Manage.* 13:365–70
134. *Wildtiere, Strassenbau und Verkehr.* 1995. Schweiz. Ges. Wildtierbiol., Zurich. 53 pp.
135. Winner WE. 1994. Mechanistic analysis of plant responses to air pollution. *Ecol. Appl.* 4:651–61
136. Wust W, Kern U, Hermann R. 1994. Street wash-off behavior of heavy metals, polyaromatic hydrocarbons and nitrophenols. *Sci. Total Environ.* 147:457–63
137. Yousef YA, Wanielista MP, Harper HH. 1985. Removal of highway contaminants by roadside swales. *Transp. Res. Rec.* 1017:62–68
138. Yousef YA, Wanielista MP, Harper HH, Skene ET. 1983. Impact of bridging on floodplains. *Transp. Res. Rec.* 948:26–30
139. Zwaenepoel A. 1997. Floristic impoverishment by changing unimproved roads into metalled roads. See Ref. 21, pp. 127–37

Annu. Rev. Ecol. Syst. 1998. 29:233–61

SEX DETERMINATION, SEX RATIOS, AND GENETIC CONFLICT

John H. Werren

Biology Department, University of Rochester, Rochester, New York 14627;
e-mail: werr@uhura.cc.rochester.edu

Leo W. Beukeboom

Institute of Evolutionary and Ecological Sciences, University of Leiden, NL-2300 RA Leiden, The Netherlands; e-mail: beukeboom@rulsfb.LeidenUniv.nl

KEY WORDS: genomic conflict, meiotic drive, cytoplasmic male sterility, sex-ratio distorter, sex chromosome

ABSTRACT

Genetic mechanisms of sex determination are unexpectedly diverse and change rapidly during evolution. We review the role of genetic conflict as the driving force behind this diversity and turnover. Genetic conflict occurs when different components of a genetic system are subject to selection in opposite directions. Conflict may occur between genomes (including paternal-maternal and parental-zygotic conflicts) or within genomes (between cytoplasmic and nuclear genes or sex chromosomes and autosomes). The sex-determining system consists of parental sex-ratio genes, parental-effect sex determiners, and zygotic sex determiners, which are subject to different selection pressures because of differences in their modes of inheritance and expression. Genetic conflict theory is used to explain the evolution of several sex-determining mechanisms, including sex chromosome drive, cytoplasmic sex-ratio distortion, and cytoplasmic male sterility in plants. Although still limited, there is growing evidence that genetic conflict could be important in the evolution of sex-determining mechanisms.

PERSPECTIVES AND OVERVIEW

Sex-determining mechanisms in plants and animals are remarkably diverse. A brief synopsis illustrates the point. In hermaphroditic species, both male (microgamete) and female (macrogamete) function reside within the same

233

234 WERREN & BEUKEBOOM

individual, whereas dioecious (or gonochoristic) species have separate sexes. Within these broad categories there is further diversity in the phenotypic and genetic mechanisms of sex determination. In dioecious species, various mechanisms exist, including haplodiploidy (males derived from haploid eggs, females from diploid eggs), paternal genome loss (sex determined by loss of paternal chromosomes after fertilization), male heterogamety (males with heteromorphic XY sex chromosomes and females with homomorphic XX), female heterogamety (ZW females and ZZ males), polygenic sex determination, environmental sex determination, and a variety of other mechanisms (reviewed in 17, 175). Sex determination can even differ markedly within a species and between closely related species. For example, platyfish (*Xiphophorus maculatus*) can have either male heterogamety or female heterogamety (104). In addition, mechanisms that appear to be the same can differ markedly in the underlying genetics. For example, male heterogametic systems can be based on dominant male determiners on the Y (e.g. in mammals) or on a genic balance between factors on the X and autosomes (e.g. in *Drosophila*). Molecular studies have shown that genes involved in primary sex determination evolve rapidly (48, 111, 166, 169, 170, 176) and that sex-determining genes in one species may not be involved in sex determination in related species (67, 100).

In this diversity lies a quandary. Although one would assume that such a basic aspect of development as sex determination would be highly stable in evolution, the opposite is the case. This observation leads to two important evolutionary questions: "Why are sex-determining mechanisms so diverse, and how do sex-determining mechanisms change, i.e. how do transitions occur from one sex-determining mechanism to another?" Presumably, sex-determining systems change when some factor (or factors) destabilizes an existing sex-determining mechanism, leading to the evolution of a new mechanism. Therefore, the focus should be on factors that potentially destabilize sex-determining mechanisms and whether some features of sex determination make it inherently unstable over evolutionary time.

In this review, we consider the role of genetic conflict in the evolution of sex-determining systems. Genetic conflict occurs when different genetic elements within a genome are selected to "push" a phenotype in different directions. There are two basic forms of genetic conflict. Intragenomic conflict involves conflicting selective pressures between different genetic elements within an individual organism (e.g. between cytoplasmic genes and autosomal genes). Intergenomic conflict occurs between genetic elements in different individuals that interact over a particular phenotype.

Genetic conflict is an inherent feature of sex-determining systems. For example, cytoplasmically inherited genetic elements (e.g. mitochondria, cytoplasmic microorganisms, plastids) are typically inherited through the egg cytoplasm but not through sperm. As a result, these elements are selected to produce strongly female-biased sex ratios, which increases their transmission to future generations (42, 55). In contrast, autosomal genes (those residing on non-sex chromosomes) are generally selected to produce a balance in the sex ratio (57). As a result, cytoplasmic and autosomal genes are selected to push sex determination in different directions. There is considerable evidence that conflict between autosomal and cytoplasmic genes is widespread (86, 170). Genetic conflict over sex determination can also occur between sex chromosome and autosomal genes and between parental- and offspring-expressed genes. Coevolutionary interactions among these conflicting selective components may provide a "motor" for evolutionary change in sex determination.

We discuss various models for the evolution of sex determination, focusing on the potential role of genetic conflict. We argue that genetic conflict is the most likely general explanation for the diversity of sex-determining mechanisms. However, although the evidence for its role in sex determination is mounting, unequivocal examples of genetic conflict causing evolutionary transitions in sex determination have yet to be made. In light of this, possible directions for future research are discussed.

The reader is also referred to reviews on the diversity of sex-determining mechanisms (17, 175), sex-ratio evolution (3, 31, 171), the evolution of heteromorphic sex chromosomes (27, 142), and somatic and germline sex determination in fruitflies (35, 137, 156), vinegar worms (36, 80), mammals (64, 82, 100), and plants (66).

BRIEF HISTORICAL SKETCH

Genetic Conflict

The concept of genetic conflict is intimately associated with two closely related developments in evolutionary biology—the idea that selection operates on individual genetic elements rather than just on the individual organism (levels of selection) and the observation that some genetic elements can be selfish or parasitic (e.g. they gain a transmission advantage despite being detrimental to the organism in which they occur). Among the first publications on what is now known as intragenomic conflict were theoretical studies by Lewis (110), who considered the fate of cytoplasmic male sterility (CMS) genes in plants, and Howard (81), who investigated cytoplasmic factors causing all-female families in animals. Both showed that cytoplasmic factors producing female biases can spread through a population, even though they may potentially cause extremely female-biased sex ratios and population extinction. Thus, the idea of intragenomic conflict was associated with questions concerning sex determination from its very inception. However, the implications of these models to the then-current views of natural selection were not widely recognized.

The botanist Östergen (135) was among the first to recognize that selection may operate in different directions on different parts of the genome. In his studies on B chromosomes [supernumerary chromosomes that occur in a wide range of species (102)], he realized that these genetic elements were parasitic, gaining a transmission advantage relative to the rest of the host's genome. Although long opposed (127), the idea that B chromosomes are selfish elements is now widely accepted (7, 102, 153, 173). The discovery of meiotic drive chromosomes (chromosomes that are transmitted to greater than 50% of gametes) (150) also stimulated consideration of the gene as the level of selection. Evolution of such systems can be understood by invoking conflicting selective pressures between the driving genes and unlinked repressors (56, 73, 94, 115, 180).

Dawkins (47) played an instrumental role in promoting the concept that selection operates at the level of the gene. Cosmides & Tooby (42) introduced the term intragenomic conflict and published a comprehensive paper on the possible role of intragenomic conflict in evolutionary processes including cytoplasmic inheritance, the evolution of anisogamy, the transition of hermaphroditism to dioecy, and the evolution of sex and sex determination. Several studies addressed the role of genetic conflict in evolution (1, 11, 55, 79). The idea that DNA could be selfish or parasitic started to receive attention through simultaneous publications by Doolittle & Sapienza (50) and Orgel & Crick (133), and an accumulating number of discoveries of selfish non-Mendelian elements such as transposons, B chromosomes, and cytoplasmic sex-ratio distorters. Werren et al (173) formally defined selfish genetic elements and reviewed existing evidence.

The concept of genetic conflict is now widely accepted in evolutionary biology (e.g. 89, 118; reviewed in 84, 91, 143). Recent theoretical and empirical work has focused on genetic conflict between cytoplasmic and autosomal sex-ratio factors (43, 59, 76, 145, 148, 160, 170), conflict between sex-chromosome–drive factors and repressors of drive (71, 72, 178), the potential importance of genetic conflict in the evolution of sex (78, 89), and paternal-maternal genome conflict over allocation of resources to progeny (70). Although the evidence of its importance is mounting, the role of genetic conflict in evolution remains to be established for many phenomena.

Sex Determination

An important early development in the study of sex determination was the discovery of sex chromosomes (77) and development of the theory of heterogametic sex determination (120). Subsequent research focused on the basic mechanisms of sex determination in a wide range of organisms (reviewed in 175) and revealed considerable diversity. Detailed genetic studies of sex determination were limited to a few organisms, most notably *Drosophila melanogaster*, which has male heterogamety. In genic balance systems, sex depends on a balance between female-determining factors on the X chromosome and male-determining factors on the autosomes. This system was uncovered in early genetic experiments by Bridges (12), who varied the number of X chromosomes in *Drosophila* and suggested that sex in *Drosophila* is determined by the ratio between X chromosomes and autosomes. In dominant-Y systems (e.g. in some mammals), a dominant male determiner is present on the Y chromosome. Bull (17), in an important treatise of the evolution of sex-determining mechanisms, considered transitions between sex-determining systems and, among other forces, the possible role of genetic conflict. Evolution of sex chromosomes and heterogamety has also been widely considered (18, 20, 26, 27, 142).

Currently, the molecular regulation of sex determination is known in detail from only a few organisms, including the house mouse (*Mus*), the fruitfly (*D. melanogaster*), and a nematode (*Caenorhabditis elegans*) (reviews in 80, 156). These systems serve as a basis for comparisons with other systems. However, it is difficult to extrapolate on the evolutionary changes leading to the differences between these species, due to their phylogenetic distance.

There is an extensive theoretical and empirical literature on the evolution of sex ratios (31, 37, 63, 73, 171, 179), but most of these studies focus on how selection acts on the parent to manipulate sex ratio of offspring under various circumstances. There has been little consideration of the coevolutionary interactions between sex-ratio genes acting in the parent and sex-determination genes acting within the zygote (but see 19).

CONCEPTUAL FRAMEWORK

The Sex Determining System

As pointed out by Bull (17) sex-ratio selection is the underlying force shaping the evolution of sex-determining systems. Sex-ratio selection concerns the transmission success of genetic factors through male function (sperm or pollen) versus female function (eggs or ovules). When a particular genetic element has higher transmission through one sexual function than the other, selection will favor variants of that element that bias sex ratio (or sex determination) toward the transmitting sex.

To understand the evolution of sex determination, it is necessary to consider how selection acts on each of the components of the overall sex-determining system. This system consists not only of the genes acting within an individual to determine its sex but also of genes acting within the parents that influence either sex ratio or sex determination (Figure 1). Components of the sex-determining

The sex determining system

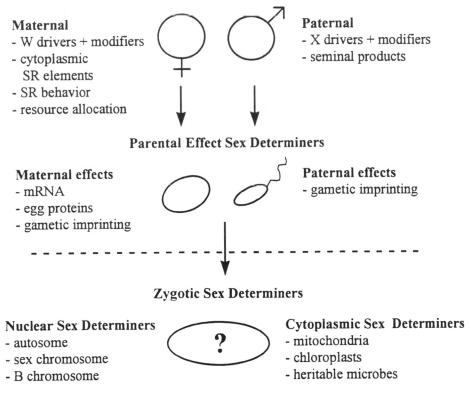

Parental Sex Ratio Genes

Maternal
- W drivers + modifiers
- cytoplasmic
  SR elements
- SR behavior
- resource allocation

Paternal
- X drivers + modifiers
- seminal products

Parental Effect Sex Determiners

Maternal effects
- mRNA
- egg proteins
- gametic imprinting

Paternal effects
- gametic imprinting

Zygotic Sex Determiners

Nuclear Sex Determiners
- autosome
- sex chromosome
- B chromosome

?

Cytoplasmic Sex Determiners
- mitochondria
- chloroplasts
- heritable microbes

Zygotic sex determiners and parental effect products
interact in the developing zygote

Figure 1 The sex-determining system, showing the different interacting components of the sex-determining system, including parental sex-ratio (SR) genes, maternal- and paternal-effect sex determiners, and zygotic sex determiners.

system can be further categorized based on their mode of inheritance. The mode of inheritance of a genetic element has a major influence on how sex-ratio selection acts on it.

Classically, genetic studies of sex determination have focused on genes that act within the developing zygote to influence its sex. However, the evolution of sex determination is influenced by selection acting on three broad categories of genes—(a) sex-ratio genes, which are genes that act within the parent to influence the sex ratio among its progeny, (b) sex-determination genes, which are genes that act within the developing zygote to influence its sex, and (c) parental-effect sex determiners, genes that are expressed in the parent (i.e. they depend on parental genotype) but that act in the developing zygote to influence sex. Examples of the latter category are maternal-effect sex-determining genes in D. melanogaster (34, 156), in the housefly (Musca domestica) (51, 93), in species demonstrating monogeny-production of all-male or all-female families (e.g. Chrysomya, 167), and in coccids that show paternal genome loss early in development (17, 130). These three categories of sex-determination genes are briefly discussed below, and examples are given in Table 1.

PARENTAL SEX-RATIO GENES   Parental influences over sex ratio occur in a broad range of species. One category of parental sex-ratio genes are those causing sex-chromosome meiotic drive. Sex-chromosome drive is a parental phenotype that alters the ratio of gametes bearing X and Y (or Z and W) genes but does not directly affect the zygotic sex-determining mechanism. X-chromosome drive has been documented in a wide range of species with male heterogamety, including fruitflies, mosquitoes, and lemmings (see below). Parental influences on sex ratio are common in haplodiploid insects, in which females manipulate the sex ratio among progeny by altering the probabilities that an egg is fertilized (63). Unfertilized eggs develop into males, and fertilized eggs develop into females. Genetic variation for fertilization proportion has been documented in some species (135) and is inferred in many others (63). Another mechanism of parental effects on sex-ratio selection is differential allocation of resources to male and female progeny. By allocating more resources to offspring of one sex, parental phenotypes could alter selection acting on zygotic sex determiners. In species with environmental sex determination, the parent can influence sex among progeny by selectivity in oviposition sites, as shown in terrapins (146) and western painted turtles (96). This, in turn, will affect how selection operates on environmental sex-determining genes expressed in the zygote. Some birds [e.g. the Seychelles warbler (108)] alter sex ratio among progeny based on available resources. This is due to either preferential segregation of Z or W chromosomes during meiosis (a parental sex-ratio effect) or to maternal modification of zygotic sex determination (see below).

Table 1 Categories of genetic elements involved in sex determination

Category	Expression	Action	Examples	Reference
Sex-ratio genes				
Maternal	Maternal	Maternal	Sex-ratio control in parasitic wasps	63
			Oviposition site selection (ESD systems)	96
			Sex ratio meiotic drive (ZW females)	17
			msr cytoplasmic factor in *Nasonia vitripennis*	154
Paternal	Paternal	Paternal	X-chromosome drive in many species	92, 1
			Suppressors of sex chromosome drive	122
Parental-effect genes				
Maternal effect	Maternal	Zygotic	Maternal-effect SD in coccids	128
			da in *Drosophila melanogaster*	156
			f factor in *Musca domestica*	44, 9
			Monogeny in *Chrysomia rufifacies*	167
			Cytoplasmic sex-ratio distorters	86
Paternal effect	Paternal	Zygotic	Paternal imprinting of *sd* genes (hypothetical)	8
			psr chromosome in *N. vitripennis*	131
Zygotic sex-determining genes	Zygotic	Zygotic	*D. melanogaster sd* cascade (X:A balance genes, *Sxl, tra, dsx*)	156
			Caenorhabditis elegans sd cascade (*sdc, her, tra, fem*)	80
			SrY in humans, housemice	100
Other (social) interactions				
Sex ratio	Individual 1	Individual 2	Worker sex-ratio manipulation in social insects	164
Social effect	Individual 1	Individual 2	Social influences in ESD species (*Heterodera* nematodes, *Mytilicola* copepods)	17

ESD, Environmental sex determination; SD, segregation distorter; *da*, daughterless; *f* factor, feminizing factor of unknown etiology; CMS, cytoplasmic male sterility; *psr*, paternal sex-ratio chromosome; *Sxl, Sex lethal* gene; *SrY*, dominant male determiner.

PARENTAL-EFFECT SEX DETERMINERS   Functionally, parental-effect sex-determining genes are similar to zygotic sex determiners because their products act within the developing zygote. However, they are subject to the same selection pressures as sex-ratio genes because they are expressed in the parent and depend on parental genotype. Both maternal- and paternal-effect sex determiners exist (Table 1).

Most maternal effects are due to maternal products (e.g. mRNA or proteins) placed in the developing egg. Maternal effects are typically important in early development because in most organisms the zygotic genotype is not expressed during early mitotic divisions, and early development is therefore dependent on products placed in the egg. Thus, gene products placed in the egg by the mother could have major effects on sex determination in the developing zygote. Molecular genetic studies of sex determination have revealed several interesting maternal effects. In *D. melanogaster*, daughterless (*da*) is a maternal-effect nuclear gene that produces a transcription factor involved in sex determination (34, 156). Similar maternal effects on zygotic sex determination have been detected in *M. domestica* (51, 93) and *Chrysomya rufifacies* (167). Nur (128) modeled maternal control of sex determination.

One example of a paternal-effect sex determiner appears to be the paternal sex-ratio chromosome (psr), a supernumerary (B) chromosome, that occurs in the parasitic wasp *Nasonia vitripennis* (131). Normally these wasps control sex among their progeny by either fertilizing eggs (diploid female progeny) or withholding fertilization (haploid male progeny). After fertilization of the egg by psr-bearing sperm, the paternal chromosomes (except psr) fail to condense properly in the first mitotic division and are eventually lost. Thus, the fertilized egg is haploid and develops into a male. Indirect evidence suggests that psr acts during spermatogenesis to modify the developing sperm, although its expression occurs in the fertilized egg (10). Despite few current examples of paternal-effect sex determiners, they may be more common than appreciated. One mechanism could be paternal imprinting of sex-determining genes, thus influencing their expression in the developing zygote (8, 125).

ZYGOTIC SEX DETERMINERS   Studies of sex determination classically consider genes acting in the zygote to determine its sex. Examples of zygotic sex determiners include *SrY* in mice and humans (64), *Sex lethal* in *D. melanogaster* (156), and the *xol* and *sdc* genes in *C. elegans* (80). In both *D. melanogaster* and *C. elegans*, the primary sex-determining signal is the X:A ratio. Multiple X numerator elements are present on the X chromosome, and a regulatory cascade involving several genes determines somatic sex (80). The evolution of X:A systems appears to be associated with the evolution of dosage compensation. An unresolved evolutionary question is how X:A sex determination evolved from an ancestral state presumably involving a major sex determiner on a nascent sex chromosome. In other words, why did the system evolve from a major-effect gene to multiple female-determining elements on the X chromosome and male determiners on the autosomes? Wilkins (177) proposed, based on the molecular genetic structure of these systems, that *C. elegans* and *D. melanogaster* sex determination evolved by a sequential addition of genetic switches, each reversing sex determination of the previous. He further proposed that the process was driven by frequency-dependent sex-ratio selection. The model is consistent with strong sex-ratio selection induced by genetic conflict or other mechanisms (see below). A dominant male determiner exists in mice and humans (*SrY*), although it is still unclear whether *SrY* is the primary signal or whether other signals induce the *SrY* testis-determining cascade (100).

Genetic Conflict Over Sex Determination

Genetic conflict will occur when the various components of the sex-determining system are selected to push zygotic sex determination or parental sex ratios in different directions. Given the divergent selective pressures acting on genes with different inheritance patterns (cytoplasmic, autosomal, and sex chromosomal) and different sites of expression (maternal, paternal, and zygotic), genetic conflict is an inherent feature of sex-determining systems. Here we list the general arenas of conflict over sex determination and sex ratios.

CYTO-NUCLEAR CONFLICT   Conflict between cytoplasmic and nuclear genes over sex determination and sex ratios is obvious and appears to be common and widespread. Many cytoplasmic sex-ratio distorters are microorganisms that are transmitted through the egg cytoplasm but not through sperm (reviewed in 86). In plants, cyto-nuclear conflict has been documented between maternally inherited organelles inducing CMS and autosomal suppressors of cytoplasmic male sterility (CMS) (reviewed in 39, 148). In the absence of suppression or other counterbalancing forces, cytoplasmic sex-ratio distorters can spread near or to fixation, potentially driving the population (and species) to extinction (81, 160). Cyto-nuclear conflict is discussed in more detail below.

SEX-CHROMOSOME DRIVE AND B-CHROMOSOME DRIVE CONFLICT   Sex-chromosome drive is just one manifestation of selection favoring meiotic drive loci, which also occur on autosomes (reviewed in 114). However, the sex-ratio distortion resulting from it can create intense sex-ratio selection. There is considerable evidence that X-chromosome drive selects for repressors on the Y chromosome and autosomes (see below). In species with recombination on the sex chromosomes, selection on linked genes can favor either enhancement of drive or suppression of drive, depending on how tightly linked the gene is and whether linkage disequilibria are maintained (180). However, the possibility that sex-ratio distortion induced by X-drive favors compensatory shifts in zygotic sex determination (or maternal-effect sex determiners) has not been extensively explored. Sex-chromosome drive can also potentially cause population extinction (73, 112, 113).

Many B chromosomes are parasitic genetic elements that have an increased transmission in gametes (transmission drive), by which the chromosomes are maintained within populations despite the fitness costs they impose on the host (127, 129). In many cases, transmission of Bs through males and females (or male and female function in hermaphrodites) is asymmetric. Under this circumstance, selection is expected to lead to the accumulation of sex ratio and sex-determining genes that bias sex toward the transmitting sex. However, detailed studies in a few coccid species with biased transmission of B chromosomes have failed to show an effect of B on sex determination (U Nur, personal communication). One striking example of a sex-ratio distorting B chromosome is the *psr* chromosome in *N. vitripennis* described previously (131).

PARENT-OFFSPRING CONFLICT   Trivers (163) originally formulated the idea that parents and offspring can have divergent genetic interests due to the fact that they are genetically related but not genetically identical. Studies of parent-offspring conflict usually concern conflict over the amount of resources allocated to offspring. However, Trivers & Hare (164) proposed that conflict should exist between a queen social insect and her worker progeny over sex ratios in social insects. Empirical studies provide strong support that such conflict exists (159).

Given the growing evidence for maternal-effect sex-determining genes, the possibility of conflict over sex determination needs to be considered more thoroughly. There are two situations in which such conflict is likely: (*a*) when fitness costs to a parent of a son and daughter differs, and (*b*) under partial inbreeding or local mate competition. When one sex is more costly to the parent to produce than the other, natural selection will favor the parent overproducing the less-costly sex (57). However, selection acting on the zygote will generally favor a more-balanced sex ratio. This is particularly true when the cost to the mother is in terms of future survival and reproduction. For example, in red deer (*Cervus elaphus*), producing a male is more reproductively costly to the mother than producing a daughter, and the mother often fails to reproduce in the year following a male birth (38). The dynamics of this interaction have not been explored theoretically. Depending on the mating system, paternal-effect sex determiners will have genetic interests more concordant with either zygotic or maternal genes.

Under partial inbreeding or local mate competition, maternal-effect genes will be selected to produce a more female-biased sex ratio. Zygotic-effect sex determiners will also be selected to produce a female bias, but the equilibrium ratio should be less biased because of asymmetries in genetic relatedness. The result will be conflicting selective pressures. A possible outcome would be the accumulation of maternal modifiers and zygotic modifiers pushing in opposite directions. Again, the interacting system has not been explored theoretically. Conflict also clearly occurs between parental sex-chromosome drivers and zygotic sex-determining genes. In principle, the sex-ratio distortion resulting from driving sex chromosomes should lead to compensatory shifts in sex determination to the underrepresented sex (113).

MATERNAL-PATERNAL CONFLICT   Interest has focused primarily on intragenomic conflict between maternally derived and paternally derived genes over resource allocation to developing zygotes and on intergenomic male-female conflict over female reproductive effort (70). Nevertheless, there are some interesting applications to sex determination evolution. Brown (13) and Bull (15, 17) have shown that maternal gene/paternal gene conflict can lead to the evolution of paternal genome loss and haplodiploid sex determination. Basically, there is a selective advantage to maternal genes that "eliminate" the paternal genome. This advantage (termed the automatic frequency response by Brown) results from a higher maternal genome transmission in the next generation in haploid males relative to diploid males (i.e. no reduction due to meiosis). The advantage accrues as long as haploid males have a fitness greater than one half that of diploid males.

In addition, intergenomic maternal-paternal conflict clearly occurs in species with haplodiploid and paternal genome–loss sex determination (71). In haplodiploids, males are under selection to increase the proportion of fertilized eggs (proportion of females) produced by their mates. However, it is unclear what opportunities are available to males for affecting female sex ratios. In paternal genome–loss systems [e.g. coccids (130)], paternal genes will be selected to escape or suppress paternal genome loss. Some supernumerary chromosomes have evolved escape mechanisms from paternal genome loss, such as in the mealy bug (127) and the flatworm *Polycelis nigra* (9).

Alternative Models for Sex-Determination Evolution

Genetic conflict is an inherent feature of sex-determining systems. However, a number of models have been proposed for the evolution of sex determination besides that of genetic conflict. We briefly review some models currently in the literature, focusing on factors that destabilize sex-determining systems and cause evolutionary transitions in the sex-determining mechanism.

TRANSIENT COVARIANCE OF FITNESS AND SEX (HITCHIKING)   Bull (17) has proposed that transient linkage disequilibrium between sex-determining alleles and genes under strong positive selection could destabilize sex determination by causing distorted population sex ratios. These distorted sex ratios would create counter-selection for sex-determining loci producing the opposite sex. Such an affect may explain the diversity of sex determination found in *M. domestica*, in which some sex-determination variants appear to be linked to pesticide-resistance alleles (106, 124, 147). In the platyfish, several body-color genes are tightly linked to sex-determining loci (104).

ACCUMULATION-ATTRITION   Graves (67) proposed an "addition-attrition" model to explain the evolution of mammalian sex determination. According to the model, mammalian sex determination evolves by a series of autosomal additions (translocations) to the Y chromosome followed by degeneration of these pseudo-autosomal regions. Only genes that evolve functions in male sex determination escape mutational degradation that results when crossing over is suppressed between X and Y chromosomes. A series of translocation events could result in turnover of sex-determining genes on the Y. The model is consistent with the view that sexually antagonistic genes can accumulate on the sex chromosomes (e.g. Y-linked genes that enhance male fitness and diminish female fitness) (141, 142) and the idea that male growth enhancers will accumulate on the Y (87, 88).

POPULATION STRUCTURE AND INBREEDING   Hamilton (73) pointed out that subdivided populations with local mating (and inbreeding) select for parents that have female-biased sex ratios. There is considerable empirical evidence that local mate competition does lead to female-biased sex ratios (reviewed in 3, 74). However, there has been little consideration of how inbreeding and local mate competition shape the zygotic sex-determining mechanism in species without parental sex-ratio control.

Two other population-structure effects relevant to sex-determination evolution are local resource competition (33) and local resource enhancement (152). Whenever fitness returns differ through males and females (or male and female function for hermaphrodites) as a function of amount of investment in that sex (e.g. because of differential dispersal), biased sex ratios will be selected (58, 59). However, most models of these effects implicitly assume parental sex-ratio control. The same selective force should also select for biases in the zygotic sex-determining genes, although less strongly than for parental sex ratio and parental-effect sex-determining genes. Such effects have not been investigated theoretically.

VARIABLE FITNESS OF MALES AND FEMALES   Facultative adjustments in sex ratio and sex determination are expected when male and female fitness are differentially affected by some environmental factors. For example, Trivers &

Willard (165) pointed out that when maternal condition varies, and this variation translates into a greater fitness effect on sons versus daughters, then selection will favor mothers in good condition overproducing sons and mothers in bad condition overproducing daughters. Variable fitness affects are also invoked to explain age-specific sex change in sequential hermaphrodites and host-size effects on sex in parasitic wasps (31).

Variable fitness effects almost certainly are important in the evolution of environmental sex determination (16, 32). Environmental sex determination is observed in some marine worms and molluscs, in parasitic nematodes such as mermithids, in some plants (136), in a few fish, and in some lizards, turtles, and crocodillians (reviewed in 17, 97). In invertebrates, crowding or poor nutrition is typically associated with increased male determination. Sex determination is temperature sensitive in a variety of reptiles, although the selective factors favoring such sex determination are still unclear.

Locked-In Sex Determination?

Some sex-determining systems may be more rigid than others, reducing or pre-cluding further evolution of the system. Heteromorphic sex chromosomes are believed to evolve primarily by mutational degeneration of chromosomes in the heterozygous state (the Y in XY males and W in ZW females) following sup-pression of recombination between homomorphic sex chromosomes (27, 142). Once heteromorphic sex chromosomes have evolved, further changes in sex determination may be constrained by sterility or inviability of XX males, XY females, and/or YY individuals of either sex (21). For instance, in mice and humans, male fertility factors are present on the Y chromosome, restricting the potential fitness of XX males (67). Phylogenetic patterns support the view that evolution of sex chromosome heteromorphisms increases conservation of sex-determining mechanisms (17, 133).

Pleiotropic effects of sex-determining genes can constrain sex determination evolution. For example, complicated interactions between sex determination and dosage compensation likely restrict the ability of heteromorphic XX/XY and ZW/ZZ sex-determining systems to change. Because dosage compensa-tion and primary sex determination are intimately entangled in the X:A balance system of *D. melanogaster*, mutants in the central sex-determining gene, *Sex lethal (Sxl)*, are typically lethal for one sex (hence the name) due to disruptions in dosage compensation (149). In humans, *SrY* and related sex-determining genes (*DAX1*, *SF1*) have pleiotropic effects on other developmental processes, such as skeletal, nervous, and adrenal development (105, 140).

Arguing against the notion that sex-determination mechanisms can become locked in is the mounting evidence that superficially similar sex-determination mechanisms can differ in underlying genetic structure (41, 111, 126). For example, murine rodent species differ in the number of *SrY* genes (111) and *SrY* can differ in potency even between different geographic strains of *Mus musculus*, resulting in the production of hermaphroditic and XY females in interstrain crosses (126). Furthermore, it is clear that even groups believed to be conserved by heteromorphic sex chromosomes (e.g. mammals) show variation in this feature. Some vertebrates previously believed to have genetic sex-determining systems actually have a mixture of genetic and environmental sex determination (44), and transitions between these mechanisms may be relatively easy (40, 45, 101).

GENETIC CONFLICT SYSTEMS

Sex-Chromosome Drive

Meiotic-drive chromosomes are inherited in a non-Mendelian fashion, typically ending up in 70–100% of gametes (150). The best known examples are Segregation Distorter in *Drosophila* (46, 162) and the *t*-locus in *Mus* (112). Meiotic-drive sex chromosomes are easily recognized because they have an immediate effect on the progeny sex ratio. They are known from several mammals and insect groups, including fruitflies, mosquitoes, and butterflies (reviewed in 92). Most examples are driving X chromosomes typically referred to as Sex-Ratio (SR) chromosomes. Driving Y chromosomes are rare, probably because of their stronger drive capacity leading to fast extinction in the absence of counter-selection (73).

Recent evidence (5, 94, 122) accords with predictions by Frank (60) and Hurst & Pomiankowsky (92) that driving sex chromosomes are much more common than was previously thought. Without countering selection, meiotic drive of sex chromosomes would quickly lead to extinction of carrier populations (73). Counter-selection can occur at the gene, individual, and group levels (see 94). At the individual level, driving sex chromosomes often reduce male fertility (115), the result of their mode of action that typically involves dysfunction of gametes carrying the nondriving sex chromosome homolog (138). If driver genes are associated with chromosomal inversions, females may have reduced fitness as well (see 95). Wilkinson et al (178) found that the frequency of Y drive increased as a correlated response in populations selected for increased stalk-eye size, which suggests that genes involved in this male character are Y linked.

Selection generally favors alleles on the autosomes and the nondriving sex chromosome that suppress the meiotic drive of the SR chromosome. Theoretical models have shown that a system of sex chromosome drive is most likely to evolve into a two-locus polymorphism with linkage disequilibrium (114, 180). The drive allele is expected to show coupling with enhancer alleles and repulsion with suppressor alleles, which might be further promoted by chromosome inversions (115). The evolution of autosomal suppressors to drive is not inevitable and depends on the specific fitness effects of driver chromosomes in males and females (180). Jaenike (94) has invoked frequency-dependent selection in the absence of linkage. Modifiers of SR chromosomes occur in a number of organisms (5, 23, 24, 114, 122, 139, 155). For example, Cazemajor et al (25) showed that in *Drosophila simulans* drive results from the action of several X-linked loci and the modification of drive from drive suppressors on each major autosome as well as on the Y chromosome. Similarly, in the plant *Silene alba*, restorer loci on the Y chromosome balance the sex-ratio bias caused by a postulated driving X (161). Hurst (85, 90) has argued that the Stellate locus in *D. melanogaster* is a relict driver gene on the X chromosome that has been silenced by modifier genes on the Y chromosome.

Driving sex chromosomes clearly illustrate intragenomic conflict. However, does sex-chromosome drive select for compensatory changes in the zygotic sex-determining mechanism? There is not strong evidence for this in nature. All known modifier genes appear to counteract the action of the driver within the parent. In contrast, Lyttle (113) constructed laboratory populations of driving Y chromosomes containing segregation distorter (SD) genes in *D. melanogaster*. In most populations, suppressors of drive evolved, but in one population, the sex-ratio distortion was counterbalanced by the accumulation of sex chromosome aneuploids (XXY females and XYY males). This example shows that a new sex-determining system (although the X:A ratio is maintained) may evolve in response to a driving sex chromosome. More such experimental studies are needed to explore the possible evolutionary outcomes of sex-chromosome drive. Whether sex-chromosome drive selects for changes in the zygotic sex-determining system will likely depend on the severity of sex-ratio distortion in the population and on the nature of standing genetic variation for the relevant traits.

Cytoplasmic Sex-Ratio Distorters in Animals

Cytoplasmically inherited sex-ratio distorters are widespread in animals (reviewed in 54, 83, 86). In most cases, cytoplasmic sex-ratio distortion is caused by maternally inherited microorganisms that distort sex ratio toward females. Cytoplasmic sex-ratio distorters include male-killers, primary sex-ratio distorters, feminizers, and parthenogenesis inducers. Examples of male-killing microbes include spiroplasms in *Drosophila willistoni* (69), gamma proteobacteria in *Nasonia* wasps (174), rickettsia, spiroplasms and flavobacteria in ladybird beetles (83, 172), and microsporidia in mosquitoes (2). Feminization of genetic males is caused by *Wolbachia* rickettsia in isopods (144) and microsporidia in amphipods (52). *Wolbachia*-induced parthenogenesis is found in an array of hymenoptera (158; reviewed in 157) and is implicated in other organisms. Primary sex-ratio distortion toward females is caused by the msr element in *Nasonia* (154); although the causative agent is unknown, it is possibly due to a mitochondrial variant.

Coevolutionary interactions between cytoplasmic sex-ratio distorters and nuclear genes can be complex. When transmission of the sex-ratio distorter is incomplete, selection for compensatory shifts in the parental sex ratio can lead to a positive feedback that results in monogeny—some females producing all-female progeny (cytoplasmic control) and some producing all-male progeny (nuclear control following compensation) (17, 170). This effect does not occur when transmission of the distorter is near 100% (170). A similar effect was shown for cytoplasmic sex determiners (53). Autosomal repressors of cytoplasmic distorters are generally favored, both in the parent and in the zygote (168) because of sex-ratio selection. Theoretical studies indicate there is no selection for compensatory sex-ratio alleles in response to male-killing microorganisms, at least in panmictic populations (170), although repressors to male-killers are expected to evolve.

Taylor (160) investigated the coevolution of nuclear zygotic sex determiners (compensatory genes), zygotic suppressors, and cytoplasmic feminizing elements. He found that compensatory nuclear male determiners will increase. However, in the presence of nuclear restorers, sex ratios will often evolve back to 1·1, with suppression of the cytoplasmic element. If this process is common in nature, interspecies crosses may reveal cytoplasmic sex-ratio distorters due to their release from suppressing genotypes. It has been proposed that hybrid lethality and sterility can result (60, 92). One interesting feature of cytoplasmic sex-ratio distorters is hitchhiking by associated mitochondria. If transmission of the distorter is incomplete or restorer genes are present, the mitochondrial variant associated with the cytoplasmic distorter can become fixed in the population. Similar arguments apply to cytoplasmic sterility in plants (see below). Features that can limit the spread of cytoplasmic distorters include reduced fitness of YY individuals (in male heterogametic systems) (160) and interdemic selection against local populations with male scarcity (17).

Although it is expected, there is not extensive empirical evidence for nucleo-cytoplasmic conflict over sex determination in animals. However, few systems have been investigated in detail. The best example occurs in the isopod *Armadillidium vulgare*, populations of which can harbor a feminizing *Wolbachia*, a second feminizing factor of unknown etiology (f), masculinizing autosomal genes, and suppressors of the feminizing factors (145; reviewed in 144). The f factor shows a complex inheritance pattern, with primarily cytoplasmic transmission but also some paternal transmission. An apparent association between *Wolbachia* and f led Legrand & Juchault (109) to propose that f was a bacterial phage carrying feminizing elements from the *Wolbachia* that occasionally incorporated into the isopod genome. It is unclear whether this is the case or whether *f* is actually a nuclear gene showing variable penetrance and expression. A dominant masculinizing gene has been characterized that can restore males in the presence of *f* but only weakly so in the presence of the feminizing *Wolbachia* (mostly resulting in functional intersexes). Populations differ considerably in frequencies of these elements, although the presence of feminizing factors is associated with the masculinizing autosomal gene.

A. *vulgare* normally has female heterogamety (ZZ males:ZW females). However, in populations harboring the feminizing factors, the female-determining chromosome (W) can be driven from the population because of sex-ratio selection. Juchault & Mocquard (103) proposed a cycle where presence of the *Wolbachia* with incomplete transmission causes loss of the W chromosome, leading to ZZ males and ZZ+WO females followed by integration of the *f* factor onto an autosome, which results in a neo-W (female determining) chromosome. This process would effectively prevent the evolution of degenerate (heteromorphic) sex chromosomes. What is less clear, is whether nucleocytoplasmic conflict could result in a shift of sex determination from female heterogamety to male heterogamety (i.e. due to the spread of an autosomal masculinizer and repressors of feminizing elements). The sequence of events is likely to strongly influence the outcome of this genetic conflict, although the full spectrum of possibilities has not been explored theoretically. Rigaud (144) pointed out that the physiological mechanism of sex determination (production of an androgenic gland) may make isopods particularly vulnerable to "hijacking" of sex determination by cytoplasmic elements.

Cytoplasmic Male Sterility in Plants

CMS is the failure of anther or pollen development caused by a cytoplasmically inherited factor. CMS is widespread (e.g. in maize, *Petunia*, rice, the common bean, and sunflower), occurring as a polymorphism in species with a mixture of hermaphroditic and male-sterile individuals (referred to as gynodioecy). Lewis (110) first pointed out that male sterility is much more readily selected for when caused by a cytoplasmic rather than a nuclear gene. CMS will be selectively favored as long as a male-sterile plant produces more effective ovules than does a hermaphroditic plant. This can occur, for example, when there is resource allocation to ovule production or (even slight) outbreeding advantage to ovules in male steriles. In contrast, a dominant nuclear male sterility gene is favored only when more than twice as many effective ovules are produced. The result is nucleo-cytoplasmic conflict, and there is now overwhelming evidence that such conflict occurs in plants (43, 49).

This conflict is manifested by complex interactions between CMS genes and nuclear repressors of CMS. Because many plant species showing CMS are of economic importance, extensive molecular genetic analyses of CMS have been conducted (reviewed in 148). In all cases examined, CMS genes occur within the mitochondria and are chimeras resulting from genetic rearrangements. Nuclear restorers of male fertility have been shown to function by elimination of CMS sequences (in *Phaseolus vulgaris*) and modification of CMS transcripts (in maize) or transcript abundance (in *Petunia*).

Genetic studies indicate a specificity between CMS genes and nuclear restorers in many systems (107). Most gynodioecious species harbor more than one CMS cytotype and multiple interacting nuclear restorers segregating within populations. For example, there are three different CMS types in *Plantago lanceolata*, each with a set of specific nuclear restorer loci (49). These range from dominant to recessive to epistatically interacting restorers. It is likely that the occurrence of restorers restrains the spread of CMS cytotypes in many species, although other processes such as deme level selection may also be involved (59, 119). Under some circumstances, CMS cytotypes can go to fixation within a species but be repressed by restorer alleles and therefore be cryptic. Such situations can subsequently be detected in interspecies crosses, in which the CMS cytotype escapes its nuclear suppression. Consistent with this scenario, CMS is a common source of hybrid sterility in plants (110).

There is an extensive theoretical literature on the coevolutionary dynamics of CMS and nuclear genes (e.g. 28–30, 59, 65, 116, 119, 151). Among the interesting questions is whether gynodioecy is a transitional stage to the evolution of dioecy, i.e. whether nucleo-cytoplasmic conflict promotes the evolution of dioecy. Consistent with this view, Maurice et al (116) documented a taxonomic association of gynodioecy and dioecy. One modeling approach involves investigating the fate of a nuclear female-sterile allele in a gynodioecious population (an extreme form of a compensatory gene). Results generally show that evolution of dioecy is restrictive but possible (28, 30, 116, 151). More models are needed to determine if dioecy can evolve by sequential shifts of sex allocation to male function in gynodioecious populations rather than by large-effect female sterile alleles. Consistent with the view that sex allocation shifts toward male function can be favored, Atlan (4) observed such sex allocation shifts in gynodioecious populations of *Thymus vulgaris*. Explicit genetic models (e.g. 30) are necessary for investigating these complex processes because phenotypic models do not capture the nonrandom association of alleles (gametic phase disequilibria) that can be crucial to the ultimate fate of different genotypes. Nevertheless the clearest and most compelling examples of genetic conflict causing turnover in sex-determining alleles occur within these plant systems.

Other Systems

Genetic conflict has been invoked as a driving force in the evolution of sex-determining systems in the cases described below. These systems show that the role of genetic conflict (a) is still hypothetical in most cases, (b) cannot be fully interpreted because of lack of information in some cases, and (c) is worth considering because it could help to explain the genetic structure of the sex determination system.

LEMMINGS   The evolution of aberrant sex-chromosome systems of lemmings has been extensively considered (61, 62, 68). The wood lemming (*Myopys schisticolor*) has three types of individuals: XX are normal females, XY are normal males, and X*Y are females. The variant X chromosome [X* (considered to suppress the male determining effect of the Y so that X*Y individuals are female)] shows drive in X*Y females, which results in a strongly biased sex ratio toward females in carrier populations. X*Y females have X*X* oocytes through nondisjunction (YY cells die) and produce nearly all daughters. A somewhat similar system has been described from various lemmings (*Dicrostonyx groenlandicus* and *D. torquatus*) (62). In these species, X*Y females also occur but have sons and daughters, presumably through production of both X* and Y eggs. The X* is considered to suppress the male sex determiner on the Y chromosome.

Several authors have modeled the evolutionary dynamics of these systems and considered how selection might lead to modifications of the reproductive biology (6, 19, 22, 117). Most work deals with how effects of inbreeding and reduced fertility under subdivided population structure may influence the spread of the driving X* chromosome and its potential suppressors (see 19 for a comprehensive treatment). To what extent can X*Y females select for changes in the sex-determining system? One of the most straightforward means of eliminating XY females would be evolution of a Y-linked suppressor gene of X*, but invasion of a suppressor Y appears restricted under inbreeding. This is consistent with the fact that there is little empirical evidence for the existence of resistant Ys in lemming populations. Models based on structured populations further show that selection for autosomal restorer genes is even weaker than for Y-linked suppression. The actual path that evolution has taken in the wood lemming seems to have involved evolution of an X* that feminized X*Y males followed by evolution of a modifier of the segregation ratio so that X*Y females produce exclusively X* oocytes, which overcame their reduced fertility (half of the Y oocytes die when they are fertilized by Y sperm). As an alternative scenario, McVean & Hurst (121) suggested that the current situation is a response to a driving Y chromosome, i.e. X*Y females counteract the spread of the driving Y by suppressing its male-determining gene and producing only

X* oocytes. There is, however, no empirical evidence for the existence of a driving Y. In conclusion, the evolution of aberrant sex-chromosome systems in lemmings may be interpreted from a genetic conflict perspective, but its exact role is unclear.

SCIARA COPROPHILA   In the fungal gnat *Sciara coprophila*, sex determination is associated with paternal genome loss (14, 123). All zygotes are initially XXX, and sex is determined by maternal factor causing somatic loss of X chromosomes. In addition, certain chromosomes (so-called limited, or L, chromosomes) are present in the germline but not in the somatic line. During spermatogenesis, all paternally derived chromosomes (i.e. both X and all autosomes) are eliminated except for the L chromosomes and the maternal X, which is doubled. Thus, males transmit only maternally derived chromosomes (i.e. two Xs and all autosomes). Females transmit all paternally and maternally derived chromosomes except for one paternal X that is eliminated during early development. Haig (72) suggested an evolutionary scenario based on genomic conflict to explain this unusual sex-determining mechanism. He envisaged the following steps: (*a*) origin of a driving X chromosome causing female-biased sex ratios, counteracted by (*b*) conversion of XX daughters into sons by elimination of one paternal X, and (*c*) origin of dispensable L chromosomes derived from X chromosomes that favor male-biased sex ratios, followed by (*d*) origin of an X' chromosome that suppresses the effect of L chromosomes. The conflicting parties are the driving X chromosome and L chromosomes that gain a transmission advantage by biasing the sex ratio toward females and the maternal autosomes and variant (doubling) X' that counteract their effects. A weak test of this scenario is the prediction that the L chromosomes are derived from X chromosomes.

COCCIDS   The evolution of unusual chromosome systems of scale insects (Cocoidea) (130) has been described in the context of genetic conflict (13, 15, 71, 75). Using similar reasoning as for *Sciara*, Haig (71) attempted to explain the origin of paternal genome loss from heterogamety through a number of evolutionary transitions. Several of these transitory stages are found in the scale insects, many species of which exhibit paternal genome loss [paternally derived chromosomes are not transmitted by males because they are eliminated from their germ lines at different developmental stages (see 14, 130 for details and references)]. Haig's model involves three steps: (*a*) meiotic drive by the X chromosome in XO males causing female-biased sex ratios; (*b*) linkage of the maternal set of autosomes in males to exploit X-drive; and (*c*) conversion of XX daughters into sons by autosomal genes expressed in mothers. One outcome could be mothers that determine the sex of their offspring by controlling the elimination of X chromosomes during embryogenesis, as observed in *Sciara*. Conflict between sex chromosome drive and autosomal suppressors is considered the driving force. Haig's model illustrates how genetic conflict may lead to novel sex-determining mechanisms, but although evolutionarily plausible, there is currently no supportive empirical evidence that the observed system is indeed the outcome of conflict between sex-determining genes.

MOLES   Using genetic conflict theory, McVean & Hurst (121) proposed three evolutionary pathways to explain the high frequency of intersexes in moles (*Talpa europaea* and *T. occidentalis*) (98, 99). Males are XY and have only testes, but females are XX and have ovotestes, i.e. functional ovaries and a variable amount of nonfunctional testicular tissue. In their first model, McVean & Hurst (121) consider the evolution of a Y-linked factor (in our terminology, a paternal-effect sex-determining gene) that masculinizes XX embryos and that is counteracted by a modifier on the autosomal or X-chromosome. In their second model, they considered intersex XX individuals as the outcome of a balance between a driving X chromosome with a masculinization effect in females and an autosomal modifier that restores functional femaleness. Their third alternative is a driving Y chromosome in males that is counteracted by an X-linked suppressor that causes partial sterility when present in the homozygous state, followed by invasion of an autosomal modifier that restores fertility in XX intersexes. We agree with McVean & Hurst (121) that there is currently little empirical evidence for any of these models.

CONCLUSIONS

Evidence for the role of genetic conflict in the evolution of sex-determining systems is growing but still circumstantial. Genetic conflict theory is consistent with much of the observed diversity, including sex-chromosome drive systems, cytoplasmic sex-ratio distorters in animals, and CMS in plants. Plausible scenarios have been developed for specific systems. However, more convincing evidence for the role of genetic conflict exists in only a few cases, notably the genetic diversity in sex determination of *A. vulgare* and CMS in plants. In many systems, the invoked role of genetic conflict is speculative and future empirical research is needed. There is also ample scope for further theoretical investigation.

Interesting issues concerning genetic conflict and the evolution of sex-determining systems include (*a*) how X:A balance systems evolve from major sex-determining gene systems and whether genetic conflict is involved, (*b*) whether sex-chromosome drive and cytoplasmic sex ratio distortion cause compensatory changes in zygotic sex-determination mechanisms, and (*c*) whether parental gene/zygotic gene conflict plays a role in sex-determination evolution. We believe that genetic conflict will eventually be shown to be an important force shaping sex-determining mechanisms, but this has yet to be demonstrated.

ACKNOWLEDGMENTS

Special thanks are extended to U Nur for many fascinating discussions on sex determination and sex-ratio evolution through the years. We also thank D Charlesworth and P Samitou-Laprade for helpful comments and discussion on the evolution of CMS, and D Bopp and A Dübendorfer on the sex-determining system of *Musca*. J Bull's seminal work on sex determination evolution played an important role in the evolution of our thinking on this topic. This review was initiated during a stay of JHW in the research group of N Michiels at the Max-Planck-Institute for Behavioural Physiology at Seewiesen (Germany), made possible by a Research Award from the Alexander von Humboldt-Stiftung. Support for JHW was also provided by the US NSF. The research of LWB has been made possible by a fellowship of the Royal Netherlands Academy of Arts and Sciences.

> Visit the *Annual Reviews* home page at
> http://www.AnnualReviews.org

Literature Cited

1. Alexander RD, Borgia G. 1978. Group se lection, altruism, and the levels of the organisation of life. *Annu. Rev. Ecol. Syst.* 9:449–74
2. Andreadis TG. 1985. Life cycle and epizootiology and horizontal transmission of *Amblyospora* (Microspora: Amblyosporidae) in a univoltine mosquito *Aedes stimulans*. *J. Invertebr. Pathol.* 46:31–46
3. Antolin MF. 1993. Genetics of biased sex ratios in subdivided populations: models, assumptions and evidence. In *Oxford Surveys in Evolutionary Biology*, ed. D Futuyma, J Antonovics, 9:239–81. Oxford, UK: Oxford Univ. Press
4. Atlan A. 1992. Sex allocation in an hermaphroditic plant: the case of gynodioecy in *Thymus vulgaris* L. *J. Evol. Biol.* 5:189–203
5. Atlan A, Merçot H, Lamdre C, Montchamp-Moreau C. 1997. The sex-ratio trait in *Drosophila simulans*: geographical distribution of distortion and resistance. *Evolution* 51:1886–95
6. Bengtsson BO. 1977. Evolution of the sex ratio in the wood lemming, *Myopus schis-*

ticolor. In *Measuring Selection in Natural Populations*, ed. FB Christiansen, TM Fenchel, pp. 333–43. Berlin: Springer-Verlag
7. Beukeboom LW. 1994. Bewildering Bs: an impression of the 1st B-chromosome conference. *Heredity* 73:328–36
8. Beukeboom LW. 1995. Sex determination in Hymenoptera: a need for genetic and molecular studies. *BioEssays* 17:813–17
9. Beukeboom LW, Seif M, Mettenmeyer T, Plowman AB, Michiels NK. 1996. Paternal inheritance of B chromosomes in a parthenogenetic hermaphrodite. *Heredity* 77:646–54
10. Beukeboom LW, Werren JH. 1993. Transmission and expression of the parasitic paternal sex ratio (PSR) chromosome. *Heredity* 70:437–43
11. Birky CWJ. 1983. Relaxed cellular controls on organelle heredity. *Science* 222:468–75
12. Bridges CB. 1921. Triploid intersexes in *Drosophila melanogaster*. *Science* 54:252–54
13. Brown SW. 1964. Automatic frequency response in the evolution of male hap-

loidy and other coccid chromosome systems. *Genetics* 49:797–817

14. Brown SW, Chandra HS. 1977. Chromosome imprinting and the differential regulation of homologous chromosomes. In *Cell Biology: A Comprehensive Treatise*, Vol. 1, ed. L Goldstein, DM Presscott, 109–89 New York: Academic Press

15. Bull JJ. 1979. An advantage for the evolution of male haploidy and systems with similar genetic transmission. *Heredity* 43:361–81

16. Bull JJ. 1981. Sex ratio evolution when fitness varies. *Heredity* 46:9–26

17. Bull JJ. 1983. *The Evolution of Sex Determing Mechanisms*. Menlo Park, CA: Benjamin/Cummings

18. Bull JJ. 1985. Sex determining mechanisms: an evolutionary perspective. *Experientia* 41:1285–96

19. Bull JJ, Bulmer MG. 1981. The evolution of XY females in mammals. *Heredity* 47:347–65

20. Bull JJ, Charnov EL. 1977. Changes in the heterogametic mechanism of sex determination. *Heredity* 39:1–14

21. Bull JJ, Charnov EL. 1985. On irreversible evolution. *Evolution* 39:1149–55

22. Carothers AD. 1980. Population dynamics and the evolution of sex-determination in lemmings. *Genet. Res., Camb.* 36:199–209

23. Carvalho AB, Klaczko LB. 1993. Autosomal suppressors of *sex-ratio* in *Drosophila mediopunctata*. *Heredity* 71:546–51

24. Carvalho AB, Klaczko LB. 1994. Y-linked suppressors of the *sex-ratio* trait in *Drosophila mediopunctata*. *Heredity* 73:573–79

25. Cazemajor M, Landré C, Montchamp-Moreau C. 1997. The sex-ratio trait in *Drosophila simulans*: Genetic analysis of distortion and suppression. *Genetics* 147:635–42

26. Charlesworth B. 1991. The evolution of sex chromosomes. *Science* 251:1030–33

27. Charlesworth B. 1996. The evolution of chromosomal sex determination and dosage compensation. *Curr. Biol.* 6:149–62

28. Charlesworth D. 1981. A further study of the problem of the maintenance of females in gynodioecious species. *Heredity* 46:27–39

29. Charlesworth D. 1989. Allocation to male and female functions in sexually polymorphic populations. *J. Theor. Biol.* 139:327–42

30. Charlesworth D, Ganders FR. 1979. The population genetics of gynodioecy with cytoplasmic male-sterility. *Heredity* 43:213–18

31. Charnov EL. 1982. *The Theory of Sex Allocation*. Princeton, NJ: Princeton Univ. Press

32. Charnov EL, Bull JJ. 1979. When is sex environmentally determined. *Nature* 266:828–30

33. Clarke AB. 1978. Sex ratio and local resource competition in a prosimian primate. *Science* 201:163–65

34. Cline TW. 1980. Maternal and zygotic sex-specific gene interactions in *Drosophila melanogaster*. *Genetics* 96:903–26

35. Cline TW. 1993. The *Drosophila* sex determination signal: How do flies count to two. *Trends Genet.* 9:385–90

36. Cline TW, Meyer BJ. 1996. Vive la difference: males vs females in flies vs worms. *Annu. Rev. Genet.* 30:637–702

37. Clutton-Brock TH. 1986. Sex ratio variation in birds. *Ibis* 128:317–29

38. Clutton-Brock TH, Albon SD, Guinness FE. 1981. Parental investment in male and female offspring in polygynous mammals. *Nature* 289:487–89

39. Conley CA, Hanson MR. 1995. How do alterations in plant mitochondrial genomes disrupt pollen development? *J. Bioener. Biomem.* 27:447–57

40. Conover DO, Van Voorhees DA, Ehtisham A. 1992. Sex ratio selection and the evolution of environmental sex determination in laboratory populations of *Menidia menidia*. *Evolution* 46:1722–30

41. Cook JM. 1993. Sex determination in the Hymenoptera: a review of models and evidence. *Heredity* 71:421–35

42. Cosmides ML, Tooby J. 1981. Cytoplasmic inheritance and intragenomic conflict. *J. Theor. Biol.* 89:83–129

43. Couvet D, Atlan A, Belhassen E, Gliddon C, Gouyon PH, Kjellberg F. 1990. Co-evolution between two symbionts: the case of cytoplasmic male sterility in higher plants. In *Oxford Surveys in Evolutionary Biology*. Vol. 7, ed. D Futuyma, J Antonovics, pp. 225–47 Oxford, UK: Oxford Univ. Press

44. Craig JK, Foote CJ, Wood CC. 1996. Evidence for temperature-dependent sex determination in sockeye salmon (*Oncorhynchus nerka*). *Can. J. Fish. Aq. Sci.* 53:141–47

45. Crews D. 1996. Temperature-dependent sex determination: the interplay of steroid hormones and temperature. *Zool. Sci.* 13:1–13

46. Crow JF, Dove WF. 1988. Anecdotal, historical and critical commentaries on genetics the ultraselfish gene. *Genetics* 118:389–91

47. Dawkins R. 1976. *The Selfish Gene*. Oxford, UK: Oxford Univ. Press

48. De Bono M, Hodgkin J. 1996. Evolution of sex determination in *Caenorhabditis*: unusually high divergence of *tra–1* and its functional consequences. *Genetics* 144:587–95

49. De Haan AA, Koelewijn HP, Hundscheid MPJ, Van Damme JJM. 1997. The dynamics of gynodioecy in *Plantago lanceolata* L. II. Mode of action and frequencies of restorer genes. *Genetics* 147:1317–28

50. Doolittle WF, Sapienza C. 1980. Selfish genes, the phenotype paradigm and genome evolution. *Nature* 284:601–3

51. Dübendorfer A, Hilfiker-Kleiner D, Nöthiger R. 1992. Sex determination mechanisms in dipteran insects: the case of *Musca domestica*. *Dev. Biol.* 3: 349–56

52. Dunn AM, Adams J, Smith JE. 1993. Transovarial transmission and sex ratio distortion by a microsporidian parasite in a shrimp. *J. Invertebrate Pathol.* 61:248–52

53. Dunn AM, Hatcher MJ, Terry RS, Tofts C. 1995. Evolutionary ecology of vertically transmitted parasites: transovarial transmission of a microsporidian sex-ratio distorter in *Gammarus duebeni*. *Parasitology* 111:S91–109

54. Ebbert MA. 1993. Endosymbiotic sex ratio distorters in insects and mites. In *Evolution and Diversity of Sex Ratio in Insects and Mites*, ed. DL Wrensch, MA Ebbert, pp. 150–91. New York: Chapman Hall

55. Eberhard WG. 1980. Evolutionary consequences of intracellular organelle competition. *Q. Rev. Biol.* 55:231–49

56. Edwards AWF. 1961. The population genetics of "sex-ratio" in *Drosophila pseudoobscura*. *Heredity* 16:291–304

57. Fisher RA. 1930. *The Genetical Theory of Natural Selection*. Oxford, UK: Oxford Univ. Press

58. Frank SA. 1987. Individual and population sex allocation patterns. *Theor. Pop. Biol.* 31:47–74

59. Frank SA. 1989. The evolutionary dynamics of cytoplasmic male-sterility. *Amer. Nat.* 133:345–76

60. Frank SA. 1991. Divergence of meiotic drive-suppression systems as an explanation for sex-biased hybrid sterility and inviability. *Evolution* 45:262–67

61. Fredga K, Gropp A, Winking H, Frank

F. 1977. A hypothesis explaining the exceptional sex ratio in the wood lemming (*Myopus schisticolor*). *Heredity* 85:101–4

62. Gileva EA. 1980. Chromosomal diversity and an aberrant genetic system of sex determination in the Arctic lemming, *dicrostonyx torquatus* Pallas (1779). *Genetica* 52/53:99–103

63. Godfray HCJ. 1994. *Parasitoids. Behavioral and Evolutionary Ecology*. Princeton, NJ: Princeton Univ. Press

64. Goodfellow PN, Lovell-Badge R. 1993. *SRY* and sex determination. *Annu. Rev. Genet.* 27:71–92

65. Gouyon PH, Vichot F, Van Damme JJM. 1991. Nuclear-cytoplasmic male sterility: single point equilibria vs limit cycles. *Amer. Nat.* 137:498–514

66. Grant S, Houben A, Vyskot B, Siroky J, Pan W-H, Macas J, et al. 1994. Genetics of sex determination in flowering plants. *Dev. Genet.* 15:214–30

67. Graves JAM. 1995. The evolution of mammalian sex chromosomes and the origin of sex determining genes. *Phil. Trans. R. Soc. London, B.* 350:305–11

68. Gropp A, Fredga K, Winking H, Frank F. 1976. Sex-chromosome aberrations in wood lemmings (*Myopus schisticolor*). *Cytogenet. Cell. Genet.* 17:343–58

69. Hackett KJ, Lynn DE, Williamson DL, Ginsberg AS, Whitcomb RF. 1985. Cultivation of the *Drosophila* spiroplasm. *Science* 232:1253–55

70. Haig D. 1992. Genomic imprinting and the theory of parent-offspring conflict. *Sem. Dev. Biol.* 3:153–60

71. Haig D. 1993. The evolution of unusual chromosomal systems in coccoids: extraordinary sex-ratios revisited. *J. Evol. Biol.* 6:69–77

72. Haig D. 1993. The evolution of unusual chromosomal systems in sciarid flies: intragenomic conflict and the sex ratio. *J. Evol. Biol.* 6:249–61

73. Hamilton WD. 1967. Extraordinary sex ratios. *Science* 156:477–88

74. Hardy ICW. 1994. Sex ratio and mating structure in the parasitoid Hymenoptera. *Oikos* 69:3–20

75. Hartl DL, Brown SW. 1970. The origin of male haploid genetic systems and their expected sex ratios. *Theor. Pop. Biol.* 1:165–90

76. Hatcher MJ, Dunn AM. 1995. Evolutionary consequences of cytoplasmically inherited feminizing factors. *Phil. Trans. R. Soc. London, B.* 348:445–56

77. Henking H. 1891. Untersuchungen über die ersten Entwicklungsvorgänge in den

Eiern der Insekten. *Zeitschr. Wiss. Zool.* 51:685–736

78. Hickey DA, Rose MR. 1988. The role of gene transfer in the evolution of eukaryotic sex. In *The Evolution of Sex,* ed. RE Michod, BR Levin, pp. 161–75 Sunderland, MA: Sinauer

79. Hickey DH. 1982. Selfish DNA: a sexually transmitted nuclear parasite. *Genetics* 101:519–31

80. Hodgkin J. 1990. Sex determination compared in *Drosophila* and *Caenorhabditis. Nature* 344:721–28

81. Howard HW. 1942. The genetics of *Armadillidium vulgare* Latr. II. Studies on the inheritance of monogeny and amphogeny. *J. Genet.* 44:143–59

82. Hunter RHF. 1995. *Sex Determination, Differentiation and Intersexuality in Placental Mammals.* Cambridge, UK: Cambridge Univ. Press

83. Hurst GDD, Hurst LD, Majerus MEN. 1997. Cytoplasmic sex-ratio distorters. In *Influential Passengers: Inherited Microorganisms and Arthropod Reproduction,* ed. SL O'Neill, AA Hoffmann, JH Werren, pp. 125–54 Oxford, UK: Oxford Univ. Press

84. Hurst LD. 1992. Intragenomic conflict as an evolutionary force. *Proc. R. Soc. London, B.* 248:135–40

85. Hurst LD. 1992. Is stellate a relict meiotic driver? *Genetics* 130:229–30

86. Hurst LD. 1993. The incidences, mechanisms and evolution of cytoplasmic sex-ratio distorters in animals. *Biol. Rev. Camb. Phil. Soc.* 68:121–94

87. Hurst LD. 1994. Embryonic growth and the evolution of the mammalian Y chromosome. I. The Y as an attractor for selfish growth factors. *Heredity* 73:223–32

88. Hurst LD. 1994. Embryonic growth and the evolution of the mammalian Y chromosome. II. Suppression of selfish Y-linked growth may explain escape from X-inactivation and for rapid evolution of *Sry. Heredity* 73:233–43

89. Hurst LD. 1995. Selfish genetic elements and their role in evolution: the evolution of sex and some of what that entails. *Phil. Trans. R. Soc. London, B.* 349:321–32

90. Hurst LD. 1996. Further evidence consistent with stellate's involvement in meiotic drive. *Genetics* 142:641–3

91. Hurst LD, Atlan A, Bengtsson BO. 1996. Genetic conflicts. *Q. Rev. Biol.* 71:317–64

92. Hurst LD, Pomiankowsky A. 1991. Maintaining Mendelism: Might prevention be better than cure? *BioEssays* 13:489–90

93. Inoue H, Hiroyoshi T. 1986. A maternal-effect sex-transformation mutant of the housefly, *Musca domestica. Genetics* 112:469–82

94. Jaenike J. 1996. Sex-ratio meiotic drive in the *Drosophila quinaria* group. *Amer. Nat.* 148:237–54

95. James AC, Jaenike J. 1990. "Sex ratio" meiotic drive in *Drosophila testacea. Genetics* 126:651–56

96. Janzen FJ. 1997. Vegetational cover predicts the sex-ratio of hatchling turtles in natural nests. *Ecology* 75:1593–9

97. Janzen FJ, Paukstis GL. 1991. Environmental sex determination in reptiles: ecology, evolution, and experimental design. *Q. Rev. Biol.* 66:149–79

98. Jiménez R, Burgos M, Caballero L, Diaz de la Guardia R. 1988. Sex reversal in a wild population of *Talpa occidentalis* (Insectivora, Mammalia). *Genet. Res. Camb.* 52:135–40

99. Jiménez R, Burgos M, Sánchez A, Sinclair AH, Alarcón J, Marin JJ, et al. 1993. Fertile females of the mole *Talpa occidentalis* are phenotypic intersexes with ovotestes. *Development* 118:1303–11

100. Jiménez R, Sanchez A, Burgos M, Díaz de la Guardia R. 1996. Puzzling out the genetics of mammalian sex determination. *Trends Genet.* 12:164–6

101. Johnston CM, Barnett M, Sharpe PT. 1995. The molecular biology of temperature-dependent sex determination. *Phil. Trans. R. Soc. London, B.* 350:297–303

102. Jones RN. 1985. Are B chromosomes 'selfish'? In *The Evolution of Genome Size,* ed. T Cavalier-Smith, pp. 397–425 Wiley

103. Juchault P, Mocquard JP. 1993. Transfer of a parasitic sex factor to the nuclear genome of the host: a hypothesis on the evolution of sex-determining mechanisms in the terrestrial isopod *Armadillidium vulgare* Latr. *J. Evol. Biol.* 6:511–28

104. Kallman KD. 1973. The sex determining mechanism of the Platyfish, *Xiphophorus maculatus,* In *Genetics and Mutagenesis of Fish,* ed. JH Schröder, pp. 19–28, Berlin/Heidelberg/New York: Springer-Verlag

105. Kent J, Wheatley SC, Andrews JE, Sinclair AH, Koopman P. 1996. A male-specific role for SOX9 in vertebrate sex determination. *Behav. Ecol. Sociobiol.* 122:2813–22

106. Kerr RW. 1970. Inheritance of DDT resistance in a laboratory colony of the housefly, *Musca domestica. Aust. J. Biol. Sci.* 23:377–400

107. Koelewijn HP, Van Damme JMM. 1995. Genetics of male sterility in gynodioe-

cious *Plantago coronopus* II. Nuclear genetic variation. *Genetics* 139:1759–75

108. Komdeur J, Daan S, Tinbergen J. 1997. Extreme adaptive modification in sex ratio of the Seychelles warbler's eggs. *Nature* 385:522–25

109. Legrand JJ, Juchault P. 1984. Nouvelles données sur le déterminisme génétique et épigénétique de la monogénie chez le crustacé isopode terrestre *Armadillidium vulgare* Latr. *Gén. Sél. Evol.* 16:57–84

110. Lewis D. 1941. Male-sterility in natural populations of hermaphrodite plants. *New Phytol.* 40:56–63

111. Lundrigan BL, Tucker PK. 1997. Evidence for multiple functional copies of the mole sex-determining locus, *Sry*, in African murine rodents. *J. Mol. Evol.* 45:60–65

112. Lyon MF. 1991. The genetic basis of transmission-ratio distortion and male sterility due to the *t*-complex. *Amer. Nat.* 137:349–58

113. Lyttle TW. 1981. Experimental population genetics of meiotic drive systems. III. Neutralisation of sex ratio distortion in *Drosophila* through sex chromosome aneuploidy. *Evolution* 98:317–34

114. Lyttle TW. 1991. Segregation distorters. *Annu. Rev. Genet.* 25:511–57

115. Lyttle TW, Wu C-I, Hawley RS. 1993. Molecular analysis of insect meiosis and sex ratio distortion. In *Molecular Approaches to Fundamental and Applied Entomology*, ed. J Oakeshott, MJ Whitten, pp. 357–406. New York/Berlin/Heidelberg: Springer Verlag

116. Maurice S, Belhassen E, Couvet D, Gouyon PH. 1994. Evolution of dioecy: can nuclear-cytoplasmic interactions select for maleness? *Heredity* 73:346–54

117. Maynard Smith J, Stenseth NC. 1978. On the evolutionary stability of the female-biased sex ratio in the wood lemming (*Myopus schisticolor*): the effect of inbreeding. *Heredity* 41:205–14

118. Maynard Smith J, Szathmáry E. 1995. *The Major Transitions in Evolution*. Oxford/New York: Freeman

119. McCauley DE, Taylor DR. 1997. Local population structure and sex ratio evolution in gynodioecious plants. *Amer. Nat.* 150:406–19

120. McClung CE. 1902. The accessory chromosome sex determinant? *Biol. Bull.* 3:43

121. McVean G, Hurst LD. 1996. Genetic conflicts and the paradox of sex determination: three paths to the evolution of female intersexuality in a mammal. *J. Theor. Biol.* 179:199–211

122. Merçot H, Atlan A, Jacques M, Montchamp-Moreau C. 1995. Sex-ratio distortion in *Drosophila simulans*: co-occurrence of drive and suppressors of drive. *J. Evol. Biol.* 8:283–300

123. Metz CW. 1938. Chromosome behavior, inheritance and sex determination in *Sciara*. *Amer. Nat.* 72:485–520

124. Milani R, Rubini PG, Franco MG. 1967. Sex determination in the house fly. *Genet. Agrar.* 21.385–411

125. Monk M. 1995. Epigenetic programming of differential gene expression in development and evolution. *Dev. Genet.* 17:188–97

126. Nagamine CM, Shiroishi T, Miyeshita N, Tsuchiya K, Ikeda H, Takao N, et al. 1994. Distribution of the Molossinus allele of *SrY*, the testit determining gene, in wild mice. *Mol. Biol. Evol.* 11:864–74

127. Nur U. 1966. Harmful supernumerary chromosomes in a mealy bug population. *Genetics* 54:1225–38

128. Nur U. 1974. The expected changes in the frequency of alleles affecting the sex ratio. *Theor. Pop. Biol.* 5:143–7

129. Nur U. 1977. Maintenance of a parasitic B chromosome in the grasshopper *Melanoplus femur-rubrum*. *Genetics* 87:499–512

130. Nur U. 1980. Evolution of unusual chromosome systems in scale insects (Coccoidea: Homoptera). In *Insect Cytogenetics*, ed. RL Blackman, GM Hewitt, M Ashburner, pp. 97–177 Oxford, UK: Blackwell Scientific

131. Nur U, Werren JH, Eickbush DG, Burke WD, Eickbush TH. 1988. A Selfish B chromosome that enhances its transmission by eliminating the paternal genome. *Science* 240:512–4

132. Ohno S. 1967. *Sex Chromosomes and Sex-Linked Genes*. Berlin: Springer-Verlag

133. Orgel LE, Crick FHC. 1980. Selfish DNA: the ultimate parasite. *Nature* 284:604–7

134. Orzack SH, Parker ED. 1990. Genetic variation for sex ratio traits within a natural population of a parasitic wasp. *Genetics* 124:373–84

135. Östergren G. 1945. Parasitic nature of extra fragment chromosomes. *Botaniska Notiser* 2:157–63

136. Pannel J. 1997. Variation in sex ratio and sex allocation in androdioecious *Mecurialis annua*. *J. Ecol.* 85:57–69

137. Pauli D, Mahowald AP. 1990. Germline sex determination in *Drosophila melanogaster*. *Trends Genet.* 6:259–64

138. Policansky D, Ellison J. 1970. Sex ratio in *Drosophila pseudoobscura*: spermiogenic failure. *Science* 169:888–89

139. Presgraves DC, Severance E, Wilkinson GS. 1997. Sex chromosome meiotic drive in stalk-eyed flies. *Genetics* 147:1169–80

140. Ramkisson Y, Goodfellow P. 1996. Early steps in mammalian sex determination. *Curr. Opinion Genet. Dev.* 6:316–21

141. Rice WR. 1992. Sexually antagonistic genes: experimental evidence. *Science* 256:1436–39

142. Rice WR. 1996. Evolution of the Y sex-chromosome in animals. *Bioscience* 46:331–43

143. Rice WR, Holland B. 1997. The enemies within: intergenomic conflict, interlocus contest evolution (ICE), and the intraspecific Red Queen. *Behav. Ecol. Sociobiol.* 41:1–10

144. Rigaud T. 1997. Inherited microorganisms and sex determination of arthropod hosts. In *Influential Passengers: Inherited Microorganisms and Arthropod Reproduction*, ed. SL O'Neill, AA Hoffmann, JH Werren, pp. 81–101 Oxford, UK: Oxford Univ. Press

145. Rigaud T, Juchault P. 1993. Conflict between feminizing sex-ratio distorters and an autosomal masculinizing gene in the terrestrial isopod *Armadillidium vulgare* Latr. *Genetics* 133:247–52

146. Roosenburg WM, Kelley KC. 1996. The effect of egg size and incubation temperature on growth in the turtle, *Malaclemys terrapin*. *J. Herpetology* 30:198–204

147. Rupeš V, Pinterová J. 1975. Genetic analyses of resistance to DDT, methyoxychlor and fenotrothrion in two strains of the housefly (*Musca domestica*). *Ent. Exp. Appl.* 18:480–91

148. Samitou-Laprade P, Cuguen J, Vernet P. 1994. Cytoplasmic male sterility in plants: molecular evidence and the nucleocytoplasmic conflict. *Trends Ecol. Evol.* 99:431–35

149. Sanchez L, Granadino B, Torres M. 1994. Sex determination in *Drosophila melanogaster*: X-linked genes involved in the initial step of Sex-lethal activation. *Dev. Genet.* 15:251–64

150. Sandler L, Novitski E. 1957. Meiotic drive as an evolutionary force. *Amer. Nat.* 41:105–10

151. Schultz ST. 1994. Nucleo-cytoplasmic male sterility and alternative routes to dioecy. *Evolution* 48:1993–45

152. Schwarz MP. 1988. Local resource enhancement and sex ratios in a primitively social bee. *Nature* 331:346–48

153. Shaw MW, Hewitt GM. 1990. B chromosomes, selfish DNA and theoretical models: Where next? In *Oxford Surveys in Evolutionary Biology*, Vol. 7, ed. D Futuyma, J Antonovics, pp. 197–223 Oxford, UK: Oxford Univ. Press

154. Skinner SW. 1982. Maternally inherited sex ratio in the parasitoid wasp *Nasonia vitripennis*. *Science* 215:1133–34

155. Stalker HD. 1961. The genetic systems modifying meiotic drive in *Drosophila paramelanica*. *Genetics* 46:177–202

156. Steinemann-Zwicky M, Amrein H, Nöthiger R. 1990. Genetic control of sex determination in *Drosophila*. *Adv. Genet.* 27:189–237

157. Stouthamer R. 1997. *Wolbachia*-induced parthenogenesis. In *Influential Passengers: Inherited Microorganisms and Arthropod Reproduction*, ed. SL O'Neill, AA Hoffmann, JH Werren, pp. 102–24 Oxford, UK: Oxford Univ. Press

158. Stouthamer R, Werren JH. 1993. Microbes associated with parthenogenesis in wasps of the genus *Trichogramma*. *J. Invertebr. Pathol.* 61:6–9

159. Sundström L. 1995. Sex allocation and colony maintenance in mongyne and polygyne colonies of *Formica truncorum* (Hymenoptera: Formicidae): the impact of kinship and mating structure. *Amer. Nat.* 146:182–201

160. Taylor DR. 1990. Evolutionary consequences of cytoplasmic sex ratio distorters. *Evol. Ecol.* 4:235–48

161. Taylor DR. 1996. The genetic basis of sex ratio in *Silene alba* (= *S. latifolia*). *Genetics* 136:641–51

162. Temin RG, Ganetzky B, Powers PA, Lyttle TW, Pimpinelli S, Dimitri P, et al. 1991. Segregation distortion in *Drosophila melanogaster*. Genetic and molecular analyses. *Amer. Nat.* 137:287–331

163. Trivers RL. 1974. Parent-offspring conflict. *Amer. Zool.* 14:249–64

164. Trivers RL, Hare H. 1976. Haplodiploidy and the evolution of the social insects. *Science* 191:249–63

165. Trivers RL, Willard DE. 1973. Natural selection of parental ability to vary the sex ratio of offspring. *Science* 179:90–92

166. Tucker PK, Lundrigan BL. 1993. Rapid evolution of the sex determining locus in Old World mice and rats. *Nature* 374:715–17

167. Ullerich F-H. 1984. Analysis of sex determination in the monogenic blowfly *Chrysomya rufifacies* by pole cell transplantation. *Mol. Gen. Genet.* 193:479–87

168. Uyenoyama MK, Feldman MW. 1978. The genetics of sex ratio distortion by

cytoplasmic infection under maternal and contagious transmission: an epidemiological study. *Theor. Pop. Biol.* 14:471–97

169. Walthour CS, Schaeffer SW. 1994. Molecular population genetics and sex determination genes: The transformer gene of *Drosophila melanogaster*. *Genetics* 136:1367–72

170. Werren JH. 1987. The coevolution of autosomal and cytoplasmic sex ratio factors. *J. Theor. Biol.* 124:317–34

171. Werren JH. 1987. Labile sex ratios in wasps and bees. *Bioscience* 37:498–506

172. Werren JH, Hurst GDD, Zhang W, Breeuwer JAJ, Stouthamer R, Majerus MEN. 1994. Rickettsial relative associated with male-killing in the ladybird beetle (*Adalia bipunctata*). *J. Bacteriol.* 176:388–94

173. Werren JH, Nur U, Wu C-I. 1988. Selfish genetic elements. *Trends Ecol. Evol.* 3:297–302

174. Werren JH, Skinner SK, Huger A. 1986. Male-killing bacteria in a parasitic wasp. *Science* 231:990–92

175. White MJD. 1973. *Animal Cytology and Evolution*. Cambridge, UK: Cambridge Univ. Press

176. Whitfield SL, Lovell-Badge R, Goodfellow PN. 1993. Rapid sequence evolution of the mammalian sex-determining gene *SRY*. *Nature* 364:713–15

177. Wilkins AS. 1995. Moving up the hierarchy: a hypothesis on the evolution of a genetic sex determination pathway. *BioEssays* 17:71–77

178. Wilkinson GS, Presgraves DC, Crymes L. 1998. Male eye span in stalk-eyed flies indicates genetic quality by meiotic drive suppression. *Nature* 391:276

179. Wrensch DL, Ebbert MA. 1993. *Evolution and Diversity of Sex Ratio*. New York/London: Chapman Hall

180. Wu C-I. 1983. The fate of autosomal modifiers of the sex-ratio trait in *Drosophila* and other sex-linked meiotic drive systems. *Theor. Pop. Biol.* 24:107–20

Annu. Rev. Ecol. Syst. 1998. 29:263–92

EARLY EVOLUTION OF LAND PLANTS: Phylogeny, Physiology, and Ecology of the Primary Terrestrial Radiation

Richard M. Bateman,[1] *Peter R. Crane,*[2] *William A. DiMichele,*[3]
Paul R. Kenrick,[4] *Nick P. Rowe,*[5] *Thomas Speck,*[6]
and William E. Stein[7]

[1]Royal Botanic Garden, 20A Inverleith Row, Edinburgh EH3 5LR, United Kingdom;
e-mail: r.bateman@rbge.org.uk; [2]Department of Geology, The Field Museum,
Chicago, Illinois 60605-2496; [3]Department of Paleobiology, National Museum of
Natural History, Smithsonian Institution, Washington, DC 20560; [4]Department of
Palaeontology, The Natural History Museum, London SW7 5BD, United Kingdom;
[5]Laboratoire de Paléobotanique, Institut des Sciences de l'Evolution (UMR 5554
CNRS), Université de Montpellier II, 34095 Montpellier cedex 05, France;
[6]Botanischer Garten der Albert-Ludwigs-Universität, D79104 Freiburg, Germany;
[7]Department of Biological Sciences, Binghamton University, Binghamton, New York
13902-6000

KEY WORDS: biomechanics, cladistics, evolutionary radiation, novelty radiation, paleobotany, systematics

ABSTRACT

The Siluro-Devonian primary radiation of land biotas is the terrestrial equivalent of the much-debated Cambrian "explosion" of marine faunas. Both show the hallmarks of novelty radiations (phenotypic diversity increases much more rapidly than species diversity across an ecologically undersaturated and thus low-competition landscape), and both ended with the formation of evolutionary and ecological frameworks analogous to those of modern ecosystems. Profound improvements in understanding early land plant evolution reflect recent liberations from several research constraints: Cladistic techniques plus DNA sequence data from extant relatives have prompted revolutionary reinterpretations of land plant phylogeny, and thus of systematics and character-state acquisition patterns. Biomechanical and physiological experimental techniques developed for extant

0066-4162/98/1120-0263$08.00

plants have been extrapolated to fossil species, with interpretations both aided and complicated by the recent knowledge that global landmass positions, currents, climates, and atmospheric compositions have been profoundly variable (and thus nonuniformitarian) through the Phanerozoic. Combining phylogenetic and paleoecological data offers potential insights into the identity and function of key innovations, though current evidence suggests the importance of accumulating within lineages a critical mass of phenotypic character. Challenges to further progress include the lack of sequence data and paucity of phenotypic features among the early land plant clades, and a fossil record still inadequate to date accurately certain crucial evolutionary and ecological events.

INTRODUCTION

Within paleobotany, there are few more popular review topics than the origin and initial radiation of vascular land plants in the Silurian (438–410 mya) and Devonian (410–355 mya) periods (20, 21, 23, 25, 34, 37, 39, 40, 51, 56, 57, 66, 67, 70, 95, 97, 132). Fortunately, each crop of reviews is separated by remarkable empirical and conceptual advances in a wide range of fields that amply justify frequent reappraisals.

Recent Advances

Any uniformitarian views of the Earth's environment that have survived the paradigm shift of plate tectonics and continental drift have since been undermined by evidence of dramatic changes in global climate and atmospheric composition through the Phanerozoic. Thus, paleoecologists must now deal with profound changes in the environmental theater as well as the evolutionary play. During the Siluro-Devonian there was a strong concentration of land masses in the Southern Hemisphere, with only North America, northern Europe, and parts of China straddling the equator (115). Consequently, both atmospheric and oceanic currents contrasted starkly with modern patterns. Atmospheric CO_2 levels were falling precipitously and O_2 levels rising rapidly; both phenomena, driven at least in part by the "greening"of the continents, had profound implications for the physiological competence of land plants (3, 13, 14, 44, 80, 85, 97).

Building on earlier intuitive advances (5), the systematics of early land plants has been revolutionized by the integration of morphological data from living and fossil species to generate cladistic phylogenies (50, 64–66). These not only define putative monophyletic (and thus natural) groups of species but also elucidate the sequence of acquisition of features and functions within specific lineages (9, 75). Molecular phylogenies of extant species have further clarified evolutionary relationships and tested the supposed primitive nature of some "living fossils" (9, 139).

Reexamination of exceptionally preserved biotas, notably the Rhynie Chert (64, 98, 99), has elucidated the novel life histories of several early land plants as well as revealed abundant interspecific interactions within their communities. Biomechanical and physiological models erected around extant plant species have been extrapolated onto fossils, often with surprising results (12, 87, 96, 97, 122).

Advances in understanding the genetic underpinnings of major phenotypic changes of extant plants (2, 22) offer deeper understanding of the nature of speciation in general and radiations in particular (54). Plants are viewed increasingly as evolutionary models in their own right, rather than religiously shoe-horned into pre-existing theories built around higher animals (11). The resulting insights permit fresh comparisons between radiations on land and those in marine environments (cf. 25, 55).

Radiations in general are being defined more precisely and categorized usefully according to pattern of diversification and inferred causal mechanisms (10, 45, 138). More specifically, attempts are being made to tease apart the supposedly explosive Siluro-Devonian terrestrial radiation into a better-defined nested set of evolutionary bursts (40, 66, 67).

Four Phases of Plant Evolution

Bateman (7) attempted to simultaneously categorize and interpret ca. 1800 million years of plant evolution as four successive phases.

1. The Biochemical Phase characterized the extensive history of life prior to the Ordovician (510–438 mya; 97, 113). During this period, fundamental biochemical pathways such as those facilitating respiration and photosynthesis were established in anatomically simple cyanobacteria and algae that primarily occupied aqueous environments. Also, more sophisticated life histories followed the advent of meiosis.

2. The Anatomical Phase spanned the Ordovician and Silurian. Erstwhile pioneering land plants struggled with the physical and physiological problems posed by a terrestrial existence, challenged more by the hostile environment than by competition with one another. Most of the tissue types that characterize modern land plants evolved, together with the alternation between independent sporophytic and gametophytic generations that defines the pteridophytic life history.

   Focusing more directly on the fossil evidence, Edwards & Selden (40) recognized phases corresponding largely to those of Bateman, but they effectively subdivided his anatomical phase into an Upper Ordovician-Silurian (quasi)bryophytic phase (Phase 2a) and a later Silurian rhyniophytic phase

(Phase 2b). These broadly correspond to the liverwort-dominated "eoembryophytic" and explosive "eotracheophytic" phases, respectively, of Gray (57; see also 67).

3. The Morphological Phase of Bateman ("eutracheophytic" phase sensu 57) reached an acme in the Devonian. Fully terrestrialized land plants experimented with various arrangements of tissue types, apparently engendering morphological and architectural escalation. This greatly increased the range and maximum complexity of gross morphological form as well as maximum body sizes; species of several clades adapted to exploit the third dimension far more effectively.

4. The Behavioral Phase increased exponentially through the Carboniferous and Permian, building on the previous evolutionary phases and establishing the ecosystem dynamics that continue to control modern vegetation. In particular, interactions between individual plants and their abiotic environment were increasingly supplemented with biotic interactions among individuals, which facilitated coevolution with mycorrhizal and pathogenic fungi, and with animal pollinators, dispersers, and herbivores.

This review concerns primarily the intermediate anatomical and morphological phases, which conveniently present the strongest evidence in the fossil record. Phase 1—the presumed earlier transition to land of prokaryotic and simple eukaryotic oxygen producers and fungi—is taken as an essential precursor to the embryophytic life history and subsequent vascularization (40, 66, 97). The well-integrated, mainly seed plant–dominated communities of Phase 4 (25) are deemed too modern in aspect to merit discussion here; their phylogeny is reviewed elsewhere in this volume (28).

In this chapter we examine the anatomical and morphological phases of terrestrialization by reviewing recent advances in land plant phylogeny, physiology–biomechanics, and ecology. We then attempt to draw together the patterns evident in these disparate sources of data to infer potential processes underlying this most profound of all terrestrial radiations.

LAND PLANT PHYLOGENY

First we summarize present evidence for land plant phylogeny, beginning with the presumed Mid-Ordovician colonization of the land and ending with the Late Devonian origin of the seed plants; examples of these taxa are described in greater detail in recent paleobotany texts (131, 136). This section focuses on the relationships of higher taxa (primarily classes and orders) and uses a novel informal nomenclature to describe some recently delimited monophyletic groups (66). Selected character-state transitions between potentially pivotal groups are elucidated in subsequent sections.

Phylogenetic studies based on comparative morphology (32, 56, 84) and molecular genetics (73, 78, 79, 81) provide compelling evidence for a close relationship between land plants and green algae, specifically the Charophyceae. Living Charophyceae are a small group of predominantly freshwater plants comprising simple unicellular and filamentous species (e.g. Klebsormidiales and Zygnematales), as well as highly differentiated forms (e.g. Charales) that include some of the most complex green algae (56). Charophyceae are a paraphyletic group, but the identity of the living land plant sister group remains unresolved. The most likely sister taxa to land plants are Coleochaetales (ca. 15 living species), Charales (ca. 400 living species), or a clade containing both.

The fossil record of charophycean algae is relatively poor, and the earliest evidence post-dates that of land plants (67). Fossils are limited mostly to decay-resistant or calcified parts of the life cycle, and only two groups (Charales, Zygnematales) are well represented. The earliest and most abundant charophycean algae are Charales (46), which first appear in the fossil record in the Late Silurian but probably had a considerably earlier origin, given that early fossils have well-developed and highly distinctive gametangia that resemble those of modern forms. The appearance of Charales in the fossil record may be linked to the evolution of calcification in the more derived members of this group. Zygnematales occur more rarely in the fossil record (58), and the group is first recognized in the Middle Devonian. The phylogenetically important Coleochaetales have not been recognized unequivocally in the fossil record. Some cuticular compressions in the Lower Devonian (notably *Parka*) resemble the delicate, filamentous thalli of living *Coleochaete orbicularis*, though other aspects of their morphology are inconsistent with this interpretation (cf. 61, 86).

Land Plants

Monophyly of land plants is strongly supported by comparative morphology (16, 32, 56, 66, 84) and nucleic acid sequences (62, 72, 73). Although relationships among the major basal living groups remain uncertain (4, 9, 32, 48, 56, 59, 66, 84), the hypothesis currently supported by the broadest range of data resolves "bryophytes" as paraphyletic, with liverworts basal in land plants and either mosses or hornworts as the living sister group to vascular plants. Liverworts themselves may be paraphyletic to other land plants, with marchantialeans basal and jungermannialeans more closely related to hornworts, mosses, or vascular plants (4, 15).

An alternative hypothesis, suggested by 18S rRNA sequences, places hornworts as basal and a liverwort-moss clade as sister group to vascular plants (59). Less parsimonious hypotheses recognize bryophyte monophyly and either a sister group relationship with vascular plants (66) or an origin from within basal vascular plants (32, 48, 62, 132). Phylogenetic evidence suggests that "bryophytes"[1] in general, and liverwort-like plants in particular, should have been important components of early terrestrial floras (67).

Evidence from fossil spores indicates that land plants originated in the mid-Ordovician and that the divergence of the four major living clades (liverworts, hornworts, mosses, vascular plants) may have occurred during the Late Ordovician and Silurian (52, 57); this hypothesis is consistent with phylogenetic data that resolve "bryophytes" as a basal grade within land plants. In contrast, the megafossil record documents a Late Silurian origin and Early Devonian diversification of vascular plants and a much later origin of "bryophytes." Kenrick & Crane (66, 67) argued that the spore record provides a more accurate picture of the time of origin and pattern of early diversification of land plants than the megafossil record because spores are more numerous and less influenced by taphonomic biases. They suggested that the late appearance of "bryophytic" megafossils probably reflects the combined effects of under-representation of this group in the fossil record and also the difficulties of recognizing early plants at the "bryophyte" grade that may have lacked the distinctive features of living groups.

However, the late appearance also offers comfort to a minority of phylogeneticists who view as credible scenarios of "bryophyte" origins via sporophytic reduction from isomorphic "pretracheophytes" (9), particularly if the Ordovician supposed bryophytic spores, cuticular sheets of cells, and tubes of nematophytes were in fact derived from free-living or lichenized fungi (128).

Liverworts

The most inclusive phylogenetic studies of liverworts resolve a jungermannialean (Metzgeriales, Jungermanniales, Calobryales) clade (4, 15, 83), a pattern consistent with traditional systematic treatments (114). Morphological studies indicate that Metzgeriales are paraphyletic to Jungermanniales (83), and this hypothesis has some support from 18S rRNA sequences (19). Recent molecular and morphological analyses also support monophyly of marchantialean liverworts (Sphaerocarpales, Marchantiales, Monocleales; 15). Within marchantialeans, the enigmatic Monocleales (two species) are nested within Marchantiales, and Sphaerocarpales are probably sister group to a Marchantiales–Monocleales clade.

Despite widespread support for an early origin of liverworts from phylogenetic studies, the group has a poor fossil record. Some of the earliest land plant spores possess features that are consistent with a sphaerocarpalean affinity

---

[1] This use of quotes is a convention in phylogenetics for identifying taxa as paraphyletic grades rather than monophyletic clades.

(57, 137), but there are insufficient characters to substantiate an unequivocal affiliation with liverworts. Many early spores could also belong to extinct taxa in the stem groups of land plants or major basal land plant clades. Most Paleozoic megafossils are related to Metzgeriales, including the earliest unequivocal liverwort (Upper Devonian). The precise relationships of earlier liverwort-like megafossils (38, 49) require further clarification. Jungermanniales first appear in the Mesozoic, and there are only a handful of well-substantiated Jurassic and Cretaceous records. Sphaerocarpales are first documented in the Triassic, and Marchantiales are clearly present in the Mid–Late Triassic.

Hornworts

Hornworts are a small, divergent group of land plants comprising ca. 400 living species. Monophyly of hornworts is well supported and uncontroversial (32, 83, 84), though generic limits and relationships among genera are poorly resolved. *Notothylas* may be sister group to an *Anthoceros–Dendroceros–Megaceros–Phaeoceros* clade (83), though a marginally less parsimonious alternative solution interprets the small, simple sporophytes of *Notothylas* as derived, and taxa with larger sporophytes such as *Dendroceros* and *Megaceros* as basal in the group. The fossil record of hornworts is poor and has not yet contributed important information to cladistic studies of this group.

Mosses

Monophyly of mosses has broad support in recent phylogenetic studies (15, 59, 60, 82). Within mosses, Sphagnales and Andreaeales are consistently resolved as basal groups. Molecular data support two major clades of peristomate mosses: (*a*) a nematodontous clade comprising Buxbaumiales, Tetraphidales, Polytrichales, and perhaps Andreaeales and (*b*) an arthrodontous clade containing Bryales (59, 60). New morphological data on sporophytes and gametogenesis in *Takakia* provide compelling evidence for an affinity with basal mosses (Andreaeales, Sphagnales), rather than with liverworts (Calobryales), as previously hypothesized (32, 48, 101); this relationship is also supported by similarities in 18S rRNA sequences (59).

Mosses have a poor Paleozoic and Mesozoic megafossil record. Arthrodontous groups such as Dicranales, Pottiales, Funariales, Leucodontales, and Hypnales have been documented in the Late Permian and Early Triassic (118). Sphagnales are also known from the Late Permian (*Protosphagnum*). Putative Polytrichales have been reported from the Carboniferous, but the first unequivocal record is from the Late Cretaceous (71). The earliest megafossils of possible moss affinity include *Sporogonites* (Lower Devonian) and *Muscites* (Lower Carboniferous), though the latter is more likely to be a lycopsid (NP Rowe, unpublished data).

Vascular Plants

Monophyly of vascular plants is supported by comparative morphology (66, 130) and by data from 18S rRNA (72) and 16S rDNA sequences (77). The inclusion of fossils results in the recognition of several additional clades that cannot be discriminated among living taxa alone. Phylogenetic analyses (65, 66) interpolate two Early Devonian Rhynie Chert plants, *Aglaophyton* and *Horneophyton*, as paraphyletic between "bryophytes" and basal vascular plants because they possess some features unique to vascular plants sensu lato (e.g. branched, nutritionally independent sporophyte) but also retain several plesiomorphic, bryophyte-like characteristics (e.g. terminal sporangia, columella in *Horneophyton*, and the absence of leaves, roots, and tracheids with well-defined thickenings).

The discovery of previously unrecognized diversity in extinct *Cooksonia* and similar early fossils (e.g. *Tortilicaulis, Uskiella, Caia*; 38) also suggests that simple early land plants—once grouped as rhyniophytes (5)—are not a monophyletic assemblage (66). *Rhynia* and a few related fossils (*Taeniocrada, Stockmansella*) form a small but distinctive clade in the vascular plant stem group (rhyniopsids; 65, 66). Some *Cooksonia* species may be among the paraphyletic precursors of vascular plants ("protracheophytes"), whereas others are true vascular plants apparently allied to the clubmoss lineage. The eutracheophyte clade (vascular plant crown group) contains all living and most fossil vascular plants.

Lycophytes

The Lycophytina constitutes a distinctive basal clade within eutracheophytes (16, 62, 66, 72, 107). Monophyly of lycophytes has broad support from comparative morphology (66, 130) and molecular studies (72, 76, 139). Within lycophytes, the extinct, leafless zosterophylls are resolved as a basal grade. Derived zosterophylls with marked bilateral symmetry of their axes form the clade Zosterophyllopsida (66). Within the sister group of leafy lycopsids, there is strong support for monophyly of the ligulate and heterosporous clades (66, 76, 139). Also, Isoetaceae constitutes clearly the most closely related living group to the far more diverse arborescent lycopsids of the Late Paleozoic.

Small herbaceous lycophytes (Zosterophyllopsida, Drepanophycaceae) are among the earliest recognizable land plant megafossils, and the group was a prominent component of Early Devonian floras. Several major clades evolved during the Devonian, including the three living groups: Lycopodiaceae, Selaginellaceae, and Isoetaceae. All living lycopsids are herbaceous or pseudo-herbaceous, but substantial trees evolved within Isoetales sensu lato during the Late Devonian (8). These arboreous species dominated tropical lowland coastal swamps during the Carboniferous.

Euphyllophytes

In eutracheophytes, monophyly of a euphyllophyte clade comprising living horsetails, ferns, and seed plants has broad support (28, 29, 62, 66, 72, 94, 107, 130), whereas the widely recognized "trimerophytes" are viewed as paraphyletic or polyphyletic. Some conflicting molecular data sets place elements of lycopsids within euphyllophytes (77) or resolve lycopsids as sister group to seed plants (76), but internal consistency of the data is low and neither hypothesis is supported by comparative morphology. The euphyllophyte stem group contains early fossils such as the "trimerophytes" *Psilophyton* and *Pertica* (66).

Within euphyllophytes there is strong support for monophyly of leptosporangiate ferns (93), lignophytes (paraphyletic "progymnosperms" plus monophyletic seed plants), and seed plants (72, 76, 109, 130), as well as for a horsetail clade comprising living *Equisetum* plus extinct *Calamites* and *Archaeocalamites* (129). Relationships among these groups and other smaller living taxa (Ophioglossales, Marattiales, Psilotales) and extinct taxa (Cladoxylales, Zygopteridales, Iridopteridales, Stauropteridales) remain highly ambiguous. For morphological data, one hypothesis views ferns sensu lato as monophyletic. The basal dichotomy in ferns is between a clade containing living eusporangiate ferns plus Filicales (fern crown-group) and a clade comprising Upper Devonian–Lower Carboniferous fernlike fossils of the Cladoxylales, Zygopteridales, and Stauropteridales (106). An alternative hypothesis shows ferns as paraphyletic to seed plants. Certain fernlike fossils, and possibly also living eusporangiate ferns, are depicted as more closely related to seed plants than to leptosporangiate ferns (107, 130).

PHYSIOLOGY AND BIOMECHANICS

Bryophytes

Mechanical constraints on early land plants were probably negligible, given their small body size, limited height, and presumed prostrate (thalloid) organization. Biophysical constraints were governed primarily by direct exposure to the atmosphere, prompting the acquisition of outer envelopes to restrict dehydration of the plant body (cuticle) and spores (sporopollenin) (87, 90, 95–97, 124). If the earliest terrestrial embryophytes morphologically resembled extant liverworts and hornworts, growth forms probably included relatively thin, dorsiventrally organized thalloid structures lacking specialized conducting or supportive tissues. Despite several reports of thalloid organisms in pre-Devonian sediments (36, 136), their status as embryophytes remains equivocal as a result of limited preservation.

Possibly the earliest upright cylindrical structures among embryophytes were erect gametophores embedded in thalloid structures. Because the cylindrical structures lacked specialized vascular tissues or hypodermal steromes, the diameter and height of such columns would have been severely restricted by potential conductance (105) and relied on turgor pressure to maintain an upright stance. The height of female sporophyte-bearing gametophores may have conferred greater dispersal potential than that possessed by forms sporulating directly from the thallus surface, allowing spores to reach uncolonized areas beyond the dense, extensive clonal mat.

Extant mosses reflect a transition from a thalloid to an axial growth form bearing leaflike appendages. Although some gametophyte axes show tissue differentiation into conducting cells (hydroids and leptoids), small axial diameters conferred considerable mechanical constraints. Nevertheless, such forms would have exceeded the height and vertical complexity of thalloid communities, in addition to developing photosynthetic leaflike appendages that were in some species connected to the water-conducting strand. In many extant moss species, dispersal of spores is effected by an elongate sporophyte seta that releases the spores slightly above the gametophyte layer. If such morphologies existed among "bryophytic" early land plants, the tiny columns constituting gametophyte axes and sporophytic setae represent mechanical innovations for trapping light and elevating the height of release of spores into the air (87, 90). The largely cylindrical design of such structures shows the earliest evidence of plant organs adapted to resisting bending forces in all lateral directions.

"Protracheophytes" and Rhyniopsids

Anatomically preserved early vascular plants, notably from the Rhynie Chert, provide sufficient anatomical information (cf. 35, 100) to construct accurate models of the biomechanical properties and growth forms of fossil plant stems (121–123). Prerequisites for such studies are reliable data on the distribution of contrasting tissues in axial transverse sections and well-preserved cell walls for comparison with those few living plant tissues already subjected to detailed biomechanical investigation (for methods see 120–123).

In the "protracheophytes" *Aglaophyton major* and *Horneophyton lignieri*, the flexural stiffness of upright axes and rhizomes was achieved by the maintenance of turgor pressure. The central conducting strands of such stems gave insignificant support, whereas the combined inner and outer cortex provided 98% of stem flexural stiffness in *Aglaophyton*. Quantitative estimates (12, 13) indicate that *Aglaophyton* could have reached a height of 19–33 cm before failing mechanically, compared with previous estimates of ca. 20 cm (42) and 50–60 cm (68). In *Horneophyton*, the parenchymatous cortex would have provided 98% of the flexural stiffness in stems possessing extrapolated maximum heights of 12–20 cm.

Both plants relied on the maintenance of turgor to retain an upright posture. Sculptured and banded cylindrical elements comprising the central conducting strand probably emulated true xylem tissue in facilitating water transport and thus maintaining the turgor pressure of the entire axis. However, these elements were positioned too centrally in the axis to be of direct mechanical significance, and the thickened and banded walls of the elements were more mechanically suited to withstanding collapse by internal negative pressures of the lumen than increasing the stiffness of the tissue.

Among rhyniopsids, biomechanical models of *Rhynia gwynne-vaughanii* indicate that the stele of derived, S-type tracheids (65) similarly offers little direct contribution to flexural stiffness. The stem could have reached a maximum height of 13–22 cm without mechanical failure, supported largely by the parenchymatous cortex [99% of the stem flexural stiffness (123)]. In none of the rhyniopsids tested does the stele contribute significantly to flexural stiffness of the stem. The phylogenetically heterogeneous fossils assigned to *Cooksonia* (e.g. *Cooksonia pertonii*) are also turgor systems, although those showing differentiation of an outer hypodermal sterome may be predominantly supported by this tissue. The banded tracheid elements that characterize the earliest tracheophytes, like those of "protracheophytes," would not have been suitable for mechanical support against bending forces, though they were better designed for resisting internal negative pressures and facilitating maintenance of turgor pressure.

In summary, the earliest land plant axes tested indicate that upright axes of protracheophytes and rhyniopsids were dependent on a maintained turgor pressure to remain upright and to prevent wilting. This must have represented an important constraint on stem height and on the ability to support both terminal and lateral appendages (i.e. end-loads and branches). The appearance of conducting tissues represented a marked physiological and mechanical innovation for maintaining turgor pressure. It made possible a self-supporting axial growth habit that could far exceed in height the light-trapping and spore dispersal capabilities observed among thalloid gametophores and smaller-bodied bryophytic gametophytes and sporophytes. Despite these innovations, early terrestrial plants with turgor-stabilized axes would have been confined to habitats with a continuous and sufficient water supply that provided relatively high humidity (100).

Eutracheophytes

Mechanical investigations of "zosterophylls" sensu lato and basal lycopsids reveal few further innovations for improving axial mechanical stability. The turgescent cortex represented the predominant tissue contributing to the flexural stiffness of the stem in the lycopsids *Asteroxylon mackiei* (>95%) and *Drepanophycus spinaeformis* (84–98%), and a similar figure is calculated for those primitive species of *Zosterophyllum* that lack hypodermal steromes (121, 123). In both lycopsid species, the lobed steles probably exerted little influence in direct mechanical support with likely values of 2–5%, and in *Drepanophycus*, maximum calculated values of up to only 16% toward flexural stiffness of the whole stem (87, 88, 105, 123).

During early terrestrialization, columnar growth forms reliant on maintenance of turgor pressure probably saturated habitats with unlimited water availability. The potential complexity of growth forms and communities would have been severely constrained to sparsely branched forms no higher than 1 m. Further morphological innovations would have been necessary to colonize water-limited habitats. We hypothesize that this next step in terrestrial colonization also involved lineages possessing turgor-stabilized upright stems. The high selective pressures needed to maintain high turgor pressures under even temporary water stress eventually drove the evolution of more complex and effective root systems for water uptake as well as an increase in cutinization of the epidermis and cell wall thickening of subepidermal tissues for reducing water loss via transpiration. The evolution of subepidermal layers in response to selective pressure for colonizing areas with temporary water stress was also a fortuitous preaptation for mechanical stability.

The appearance of a hypodermal sterome in zosterophylls probably marks the first mechanical innovation away from support based mostly on maintenance of turgor pressure. The production, and modulation during growth, of a ring of cortical fibers in an otherwise parenchymatous cortex would have permitted a wider range of mechanically viable structures and generated diverse small-bodied growth habits analogous to those of extant herbaceous lycopsids (110).

In the primitive "trimerophyte" *Psilophyton dawsonii*, the hypodermal sterome of collenchymatous-sclerenchymatous elements contributes significantly to the flexural stiffness of the entire stem (values of 96–99% result from calculations inputting either collenchyma or sclerenchyma as the outer tissue of the biomechanical model) (123). The predicted maximum height for a stem of *P. dawsonii* of basal stem diameter 6 mm is 75–200 cm, with the central steles contributing little to the flexural stiffness of the stems (<5%). Stems of the derived (sawdonialcan) zosterophyll *Gosslingia breconensis* (basal diameter 4 mm) yield critical buckling lengths of 51–140 cm, depending on the tissue type employed in the model to represent the newly acquired hypodermal sterome (over 96% when modeled as collenchyma and over 99% when modeled as sclerenchyma) (123). Similar values characterize derived *Zosterophyllum* species possessing hypodermal steromes (123).

Among the Devonian plants analyzed, perhaps the earliest empirical evidence of a significant contribution to flexural stiffness from xylem tissue is observed in the protolepidodendralean lycopsid *Leclercqia complexa*. Modeling the outer sterome as collenchymatous tissue yielded a critical buckling height of 85–142 cm (basal diameter 7 mm) (123). However, the contribution toward flexural stiffness of the xylem was still only 42%, compared with 56% from the collenchymatous hypodermal sterome.

Thus, prior to the appearance of extensive woody cylinders and periderm, mechanical stability among early land plants shifted from organizations relying on turgor systems to those employing hypodermal steromes. As well as affording the possibility of greater height, cylinders of collenchymatous or sclerenchymatous tissues were almost certainly important for supporting increasingly complex branch systems. Larger evapotranspiration surfaces would have required yet more efficient water conductance than was supplied by the primary steles in small-bodied early land plants. This may partially explain the relatively larger and mechanically more significant steles observed in *Leclercqia* and similar taxa.

Lignophytes

The appearance of secondary growth in the Middle Devonian, following the initial phase of land plant evolution, influenced water conductance, canopy formation, and mechanical support, and also prompted diversification in growth forms. The appearance of secondary xylem in eutracheophytes was probably linked to the water supply of megaphylls and selection for enlarging canopy surfaces, mechanically supported by hypodermal steromes. Biomechanical analyses indicate that secondary xylem in some early seed plants (and probably many early lignophytes) did not provide significant mechanical support; for example, the outer sparganum cortex in *Calamopitys* contributed over 85% to flexural stiffness of the stem and was essential for supporting the large megaphyllous leaves (111). Secondary xylem was confined within the primary body of the stem in many aneurophyte "progymnosperms" and basal seed plants, offering little mechanical strength but probably enhancing water conduction. Significant mechanical contributions from secondary xylem, such as that observed among archaeopteridalean "progymnosperms," were possible only following additional developmental innovations; the most notable was periderm formation, which permitted the wood cylinder to exceed the limits of the primary body of the stem (119).

PALEOECOLOGY

The evolution of ecological patterns during the Devonian parallels the appearance of morphological and phylogenetic structure and diversity. The Devonian record suggests a steady increase in ecological complexity at all spatial scales, from an alien simplicity at the beginning to nearly modern organization by the onset of the Carboniferous (25). Early Devonian ecosystems were composed of structurally simple plants with dynamically simple interactions (39). The differences in dynamics at local and landscape scales were slight and difficult to differentiate. By the end of the Devonian, landscapes were varied and local assemblages of plants were structurally complex, with much greater diversity of body plans, life histories, and survival strategies (112).

Early and Middle Devonian

Studies of Early Devonian landscapes (34, 39, 63) indicate that supposed communities consisted of patches of opportunistic clonal plants. Given that typical coeval plants were characterized by rhizoids or rudimentary true roots, were supported by turgor pressure, and showed homosporous life histories, most vascular plants were probably constrained to wetter parts of the landscape. Within these humid habitats there may have been more niche partitioning than generally supposed. Recent work on paleosols (63) implies that vascular plants had gained the capacity to colonize some habitats with seasonal moisture availability, pointing to the evolution of physiological drought tolerance. Furthermore, the recognition of probable roots in weakly developed paleosols of streamside environments (44) indicates that some groups of Early Devonian plants may have been more complex morphologically than previously believed, having the ability to tap into deeper sources of groundwater.

The Rhynie Chert flora, now one of the best-understood floras of the entire Paleozoic, offers a remarkable window into an Early Devonian ecosystem. Although sporophyte architecture was simple, many sporophyte ecological strategies clearly coexisted, such as the ability of *Rhynia gwynne-vaughanii* to spread rapidly over a substrate via deciduous lateral branches (41). Sporophyte diversification was matched by a wide array of gametophyte morphologies that record many variations on aids to syngamy (64, 98), and fungi played a "modern" spectrum of roles in the ecosystem (133, 135). Unfortunately, the supposedly archetypal Rhynie Chert flora appears to be an unusual assemblage specialized for life in a low-pH, periodically flooded habitat (9); the flora may have included species secondarily reduced for aquatic life habits. The Chert thus allows only a small and potentially relictual perspective on the ecology of this crucial time. Recent paleosol studies revealed evidence for prototype "forests" as early as the Middle Devonian from waterlogged soils of New York State (31) and even in well-drained habitats from Antarctica (103). Lowland wetland macrofossil assemblages also demonstrate increasing plant-animal interactions; the evolution of terrestrial arthropods was proceeding rapidly and may have included herbivory (74, 117).

Late Devonian

The evolution of community and landscape complexity escalated dramatically during the Late Devonian. Empirical studies of the relationship between megafloras and environments of deposition (17, 112) have revealed the initial phases of landscape partitioning by the major plant clades. Floras typical of swamps that were dominated by the fernlike plant *Rhacophyton*, and periswamp areas that included lycopsids, were distinct from floras of interfluves and drier parts of flood plains, dominated by the arboreous "progymnosperm" *Archaeopteris*. Seed-bearing "pteridosperms" appear to have originated in wetter parts of the landscape but then spread as opportunists into areas of disturbance and physical stress, including relatively arid habitats (108).

Pioneering studies using paleosols to resolve vegetational patterns across Late Devonian landscapes (17, 102) revealed a range of conditions on vegetated floodplains, from fully saturated to well drained and apparently dry. More recent investigations of in situ tree stumps and root casts in Late Devonian paleosols (31) indicated growth of *Archaeopteris* trees in seasonally wet but well-drained habitats. They further suggested that limited evidence of tree growth in the drier parts of floodplains may reflect both biological and physical processes that remove critical paleosol evidence rather than the absence of open forests. The development of rooting, and the evolution of forests and complex landscapes, apparently played a major role in determining global climatic and geochemical balances (3, 13, 44, 97).

Clearly, ecological and evolutionary changes were strongly linked and included both positive and negative feedback systems. The simple systems that characterized the Early and Middle Devonian apparently offered low resistance to invasion by species possessing major evolutionary innovations (25). Chaloner & Sheerin (21) documented the origin of nearly all major tissue and organ types during the Devonian, yet studies of Devono-Carboniferous plant biomechanics (87, 90, 119, 121, 122) indicate that the early plants were far from the biomechanical optima permitted by their tissues and organs (69). Studies of the developmental controls on Devonian plants (127) similarly suggest to some of us that initial increases in the complexity of development created many opportunities for morphological diversification with relatively few constraints. Maximum plant stature increased in many clades, along with the average diversity of organ types borne by any single species and the frequency of compound reproductive structures (70). Yet DiMichele et al (26) argued that the rate of increase in diversity of body plans slowed during the Late Devonian, as those architectures that characterize modern taxonomic classes and orders became clearly recognizable.

The Late Devonian encapsulated all the major body plans of vascular plants, each characterizing one of the major "modern" clades: seed plants, ferns, sphenopsids, and several groups of lycopsids (the lack of high-level innovation subsequent to the Early Carboniferous, with the arguable exception of the rise of the angiosperms, presumably reflects intensifying morphological and ecological constraints). Available (albeit limited) evidence indicates that each major lineage of plants had established a distinct ecological centroid by the Early Carboniferous (24, 26) and that together they had occupied a wide range of terrestrial habitats (31, 103). The result was the development of strong incumbency effects, or "home-field advantage" (see 53, 92, 104), whereby ecological resource occupation suppresses the likelihood of survival of new variants (138).

CHARACTERS AND PHYLOGENY RECONSTRUCTION

Strengths of the Phylogenetic Framework

Cladogram topology determines the relationships of higher taxa (and hence their delimitation into monophyletic groups) and the relative (but not absolute) timings of the speciation events that correspond to the lineage divergences. But more importantly, a rooted cladogram also provides an explicit evolutionary hypothesis that describes not only sister-group relationships but also character-state transitions. The relative position of character-state transitions determines branch lengths—effectively, the amount of evolution between the speciation events encompassed by the cladogram. This allows measurement of the phylogenetic distance among analyzed taxa as disparity (the number of characters separating taxa through their most recent shared divergence point) rather than as raw similarity (10, 18, 47).

The enhanced ability to understand character evolution is at least as valuable as recognizing clades, particularly where the analysis includes morphological data (9–11). Advantages include the replacement of statistical correlations by phylogenetic correlations among characters, so that the active origination of a character state can be distinguished from mere passive inheritance from a shared ancestor. Also, the co-occurrence of transitions in two or more characters on the same branches can be interpreted in terms of the underlying evolutionary mechanism. Given a cladistic branch of several character-state transitions, two extremes of interpretation are possible. The saltation model (9–11) argues for a null hypothesis that the co-transitional characters are developmentally linked (pleiotropic) and reflect a single speciation event. In contrast, the adaptive model (27, 75) assumes that the accumulation of developmentally independent character states in response to selection pressures is gradual, and probably involves phylogenetically intermediate species absent from the sampled terminal taxa (either deliberately excluded or not yet known to science).

More recently, morphological phylogenies have been supplemented (or, in many cases, supplanted) by molecular phylogenies based on nucleic acid base sequences. These have proved especially valuable for comparing taxa that are highly morphologically divergent, plesiomorphically simple, or secondarily simplified by reduction (and hence have insufficient clearly homologous structures) and for elucidating cases of parallel evolution (9). Often, nonmolecular characters are "mapped" across molecular phylogenies rather than included in the parsimony analysis (54); this approach is preferable for ecological and continuously variable phenotypic characters but is a suboptimal way of analyzing discrete phenotypic characters (10).

Constraints on Molecular Phylogenies

The insights gained by applying phylogenetic techniques to the Siluro–Devonian radiation are remarkable given the many severe handicaps. First, extinct Paleozoic plants do not yield DNA, and large swathes of pioneering land plants have left no close extant relatives—half of the Paleozoic plant groups traditionally regarded as taxonomic classes (albeit most paraphyletic) are "extinct." Thus, molecular attempts to understand the evolutionary origins of bryophytes would be greatly assisted by sequencing extinct rhyniophytes (9). Similarly, studies of fern origins need DNA from extinct cladoxylaleans, stauropteridaleans, and zygopteridaleans; studies of equisetaleans need DNA from extinct iridopteridaleans; and studies of gymnosperms need DNA from extinct "trimerophytes" and "progymnosperms."

Second, the untestability of such groups extends to experimental approaches that (a) separate ecophenotypic from genetically controlled variation, (b) directly observe ontogeny, and (c) test for pleiotropic and epigenetic behavior in key developmental genes.

Third, the Siluro-Devonian radiation occurred either quickly ($\leq$100 million years; see 66) or very quickly (35–50 million years; see 11) relative to the much longer period separating the radiation from present-day floras testable by sequencing. Such deep, rapid radiations are difficult to capture using clocklike molecules; those changing fast enough to capture the relationships of the classes and orders emerging during the radiation are now oversaturated with mutations, and slower molecules not oversaturated changed too slowly to capture key events (10).

Within vascular plants, molecular and morphological assessments of phylogeny at the level of orders and below give similar results (93), but at deeper levels—for example, the divergence of major groups of ferns, horsetails, and seed plants—phylogenetic resolution is poor. These difficulties highlight the weakness of analyses based solely on living species (cf. 1, 76). Rather, future progress hinges on solving the relationships of several fossil groups of uncertain status (e.g., "trimerophytes," Cladoxylales, Zygopteridales) with respect to the living ferns, horsetails, and seed plants (66, 107). Also, combined analyses of molecular sequences from multiple loci, and large-scale structural characteristics of the genome (e.g. introns, inversions; 94), may prove more informative than oversaturated base mutations when assessing deep phylogenetic patterns in land plants.

Thus, molecular data have been less helpful that might be supposed in unravelling the Siluro-Devonian radiation, though they have usefully revealed the fallacy of viewing certain pteridophytes as "living fossils" unchanged since the radiation. The most notable examples are the phenotypically simple Psilotaceae and Ophioglossaceae, which are actually secondarily reduced "pseudoplesiomorphs" rather than truly primitive relicts (9, 93). Sequencing has also revealed that among truly primitive lineages, such as the homosporous lycopsid *Huperzia*, most of the extant species may nonetheless be of recent origin and trivially distinct (139); longevity of clades does not necessarily equate with longevity of their constituent species.

Constraints on Morphological Phylogenies

Drawbacks to molecular analyses place unusually strong emphasis on morphological studies, particularly those that successfully integrate extant "living fossils" and decidedly dead fossils. If the plant fossil record is taken at face value, a punctuational pattern is evident—long periods of stasis are separated by much shorter periods of rapid change (43). Although this model has found favor with few neo-Darwinians, it is nonetheless congruent with the neo-Darwinian tenet that the dominant mode of selection is stabilizing selection, which precludes morphological change. The periods of change can be viewed either as (*a*) the result of strong directional selection or as (*b*) drift or saltation in the absence of directional selection. Whatever the underlying cause, the key point is that there is no "morphological clock," and as morphology is the direct manifestation of evolution, it is best placed to resolve rapid, deep radiations (10).

Unfortunately, two factors seriously limit our ability to resolve radiations morphologically. The first is the patchy fossil record. This can be exaggerated—the best of the conceptual whole plants, painstakingly reconstructed during over a century of paleobotanical research, are remarkably well understood (for example, see the near-complete evidence of morphology and reproductive morphology in presumed primitive taxa such as *Aglaophyton* in References 66, 67, 99). However, for less readily preserved groups such as the "bryophytes," and preservationally challenging periods such as the all-important Late Silurian–Earliest Devonian, the plants and the resulting data are fragmentary. Consequently, discussions of the Siluro-Devonian frequently focus on the overall assemblage of phenotypic characters available to land plants during particular time slices (6, 21, 40, 70). Although this approach can encompass a much greater proportion of the fossil record, only suites of characters packaged in a single plant (and thus the expression of a single genome) offer meaningful evolutionary interpretations. The pivotal role played by fossils in many plant phylogenies (29, 65–67, 91, 109) is often due to unique combinations of characters in a single species (9, 126). Doyle (30) famously compared and contrasted species trees with gene trees, but we could equally well contrast species trees with "organ trees" based on very limited data. Given the propensity of plants for mosaic evolution, such trees have poor probabilities of accuracy (12).

The second limiting factor is the relative paucity, simplicity, and high variability of features observed in the early land plants. This constraint reduces the number of potential synapomorphic characters. Also, the combination of morphological–anatomical simplicity and the impossibility of direct experimentation on extinct species renders primary homology more difficult to test before phylogeny reconstruction, and the small number of characters renders secondary homology difficult to test after phylogeny reconstruction by the congruence test of parsimony. These problems weaken key phylogenetic assertions. For example, the now widely accepted separation of the lycophytes from the remaining eutracheophytes (5, 6) leaves a group—the euphyllophytes of Kenrick & Crane (66)—diagnosed primarily by homologues to megaphyllous leaves, yet these are questionably present in the basal members of the clade (the paraphyletic "trimerophytes." Similarly, the synapomorphies of the ferns appear disconcertingly "retrospective" (125) when sought in basal fernlike taxa such as Iridopteridales, Cladoxylopsida, Stenokoleales, Stauropteridales, and Zygopteridaceae sensu lato, wherein characters such as branching patterns and stelar anatomy are ill defined and highly homoplastic (107).

Not surprisingly, homoplasy levels appear high in cladistic studies of Devonian plants; they may in part reflect genuine evolutionary processes rather than erroneous prior assertions of homology, given the presumed relatively poor developmental canalization (9, 26, 127). Stein (127) recently advocated a "strong-inference" approach, modeled on cladistics, for identifying homology at the level of developmental processes. Expanding on the "telome theory" of Zimmermann (140), units of developmental dynamic defining a relationship between external and/or internal environmental cues and developmental outcomes are termed evolutionary developmental gates (EDGs). Each hypothesized EDG employs a "logical conditional" (by analogy with logic gates in programming theory) and is assembled into networks specifying a causal relationship between developmental processes and resultant morphological structures. EDG networks can be tested by comparison with evidence of known developmental processes, computer modeling, and evolutionary-phylogenetic comparisons between hypothesized ancestor-descendant pairs (or, potentially, among sister groups as three-item statements).

To illustrate this approach, Stein proposed four developmental modules underlying the morphology of a primitive Devonian shoot system (telome): (a) establishment of self-recognition and auxin activity at the shoot apex; (b) normal cell division, modeled as an iterative and recursive process; (c) developmental switches and cascades leading to normal tissue of epidermis, cortex, and vascular tissues; and (d) establishment of new shoot apices by reassignment of self-recognition of the apex, during either bifurcation or de novo apex formation. Although more detail is required to enable important evolutionary comparisons with more derived taxa, conceptualizing homology as a developmental dynamic (as opposed to static, end-result morphology) may significantly improve our understanding of both phylogenetic relationships and underlying developmental/evolutionary causal agents during the primary land plant radiation (127). The resulting phylogenetic characters would be truly transformational and their degrees of dependency truly tested (8, 9).

INTERPRETING THE SILURO-DEVONIAN RADIATION(S)

Defining and Categorizing Radiations

Evolutionary radiations pose three primary challenges: (a) satisfactorily defining a radiation; (b) distinguishing among the "unholy trinity" of clade origination, radiation, and migration; and (c) identifying the underlying cause(s) of the radiation. There are many definitions of radiations, despite the fact that most authors use no explicit definition. Bateman (10) argued that a radiation could be defined most effectively using one or both of two properties: species diversity and phenotypic character diversity (the latter being strongly positively correlated with higher taxonomic diversity). Using either criterion, the best measure is the net surfeit of rate of gain over rate of loss in a specified clade during a specified time interval. And using either criterion, the fossil record clearly suggests that the first major radiation of vascular land plants peaked during the Early Devonian (66, 70).

Timing originations is more problematic; the major Devonian clades emerge, apparently fully formed at ca. 400 mya, from a putative period of preservationally discouraging worldwide marine regression in the Late Silurian (66). Extrapolation from their cladogram suggests that all of the major clades between liverworts and euphyllophytes evolved during that period, offering a maximum window of 35 my for a multiclass level radiation from a single putative terrestrialized ancestor (11). However, Kenrick & Crane (66, Figure 7.15; 67, Figure 4) controversially depicted the liverworts as evolving at the beginning of the preservational hiatus, but the remaining embryophytes diverging close to the end. Middle Devonian origins for the primitive, arguably fernlike clade of cladoxylopsids and relatives, and of the "progymnosperms," were followed by Late Devonian origins of the "pteridosperms," sphenopsids, and bona fide derived ferns, together spanning at least a further 35 my. Other workers also support a more prolonged radiation (40, 57; WE Stein, personal communication, 1998).

Proportionally, the increase in species number during the Early Devonian radiation is great, but the total numbers involved are small relative to the radical increase in phenotypic diversity. Rapid, highly divergent increases in complexity accompanied by relatively low speciation rates (and thus many vacant niches) constitute a novelty radiation sensu Erwin (45; see also 24, 26). This contrasts strongly with a niche-filling adaptive radiation, which also occurs within a single clade but involves slower and less profound phenotypic diversification and much greater species diversification. Given that this contrast relies on decoupling of phylogenetic disparity and species diversity (the latter correlating more closely with niche differentiation), it is theoretically possible that an adaptive radiation could be mistaken for a novelty radiation—but only in the unlikely event of massive extinction of phenotypically intermediate species that left no fossil record (11).

Environmental Conquests and Ecological Constraints

In retrospect (and in comparison with the Cambrian radiation of metazoan animals in the marine realm; 33), a novelty radiation should have been expected for the initial diversification of vascular plants; competition (and thus selection) has little role to play in habitats profoundly undersaturated in species. The very high rate of generation of profound phenotypic mutants evident in the modern flora is almost wholly filtered out by competition in saturated ecosystems (11). However, in the early Devonian ecosystems, saltational mutants would have been produced even more frequently, given their weaker developmental canalization and, in the case of "bryophytes" and "protracheophytes" with long-lived haploid phases, their lack of buffering by second alleles of specific genes. Also, mere economic establishment (preferably but not essentially accompanied by reproductive success; 24) would have been accomplished relatively easily in the undersaturated habitats.

Theory would then require subsequent radiations within the land plants at lower taxonomic levels to increasingly approach the adaptive mode, as developmental canalization strengthened and ecospace became increasingly crowded, encouraging competition and thus discouraging the establishment of radically novel phenotypes; ecological constraints generate a negative feedback loop (11, 24, 26, 45, 138). At this threshold of phenotypic and ecological complexity, other evolutionary processes come into play. Decreases in phenotypic complexity, such as those generating many "living fossils," can drive evolution by resetting the phenotypic clock, allowing the lineage to exploit a different (relatively uncompetitive) niche (9). In a fully occupied landscape, incumbent advantage can exclude theoretically fitter organisms, denied a foothold through the happenstance of prior occupation by other species (24, 53, 92, 104).

Once primary vacancy of niches in a habitat has been eliminated (saturation), opportunities for further evolution within that habitat focus on subdivision of existing niches or the occupation of secondary vacancies created by extrinsic environmental perturbations (ecological radiations). Preaptation (specifically, exaptation) becomes increasingly evolutionarily credible, as the repertoire of phenotypic characters increases and extrinsically driven niche vacancies allow potential functional switching (if we transfer focus from the physical landscape to the theoretical adaptive landscape of gene frequency variation, the latter temporarily becomes a "seascape," changing too rapidly to be tracked by changes in gene frequencies through populations; 10). Alternatively, new habitats must be invaded, a challenge generally requiring additional key innovations.

Overall, this model predicts a nested and fractal pattern of radiations, each generating more species and fewer higher taxa than the last as the average degree of phenotypic divergence between ancestor and descendant decreases. It also implies that attempts to explain the Early Devonian increases in diversity using adaptive landscapes (69, 89) may be misplaced; competition among plants was restricted by their tenuous hold on the abiotic landscape, which acted as a passive environmental filter for any viable novel phenotypes.

When compared with the above model, observed patterns of diversity suggest that Early and Middle Devonian ecology offered weak constraints to evolution relative to the Late Devonian, given that much of the land surface was either uncolonized or minimally occupied by vascular plants. Potential for positive feedbacks was also strong in areas such as nutrient cycling, creation of new resource spaces by morphological innovation, and the development of mutualistic interactions with the concomitantly expanding faunas and mycotas. The constriction of such opportunities was probably scale-dependent, beginning within certain resource pools while greater flexibility persisted in others. Landscape-scale opportunities for evolutionary innovation may have remained permissive, even when more local opportunies were becoming increasingly constrained.

Elusive Key Innovations

Erwin's (45) caution against uncritical acceptance of assertions of key innovations could have been written specifically for the Siluro-Devonian terrestrial radiation; most authors (e.g. 6, 21, 40, 57, 69, 70, 90, 95, 117, 124, 132) have either explicitly or implicitly identified one or more key innovations as crucial to the success of a particular taxonomic group. A bona fide key innovation should be a synapomorphy that was acquired immediately prior to an equally bona fide evolutionary radiation and can be shown to have been a far greater stimulus to that radiation than any other synapomorphies acquired on the same phylogenetic branch.

Following the preferred phylogeny of Kenrick & Crane (66), embryophyte synapomorphies include multicellular sporophytes and the desiccation resistance conferred by both cuticles and sporopollenin-walled spores. The functional values of these characters for terrestrialization are clear, yet there is no evidence that any or all of these characters prompted an immediate radiation. This conclusion also applies to the stomates that are present in most stomatophytes (tracheophytes plus "bryophytes" excluding liverworts), and the pronounced axial gametophyte, terminal gametangia, and well-developed sporangiophore of the mosses plus tracheophytes clade. The basal members of the Polysporangiomorpha clade are "protracheophyte"-grade genera such as *Horneophyton* and *Aglaophyton*. They exhibit branched, independent sporophytes isomorphic with gametophytes that possess sunken archegonia. Similarly, the rhyniopsids, basal to the tracheophytes sensu lato, provide the first evidence of vascular tissue more sophisticated than bryophytic leptomes and of diversification in sporangial morphology and function, together with stronger sporophytic dominance in the life history. The unique combinations of characters in these extinct taxa provide us with key information about the sequence of character acquisition within the land plant clade, but again they did not clearly engender a profound increase in the diversity of either phenotypic characters or species.

Even the subsequent dichotomy into the Lycophytina and Euphyllophytina, and within the Lycophytina into the Lycopsida and more derived zosterophylls of the Sawdoniales, did not immediately add greatly to the overall diversity of characters. The relative indistinctness of these groups is well illustrated by the paraphyletic or even polyphyletic nature of genera such as *Cooksonia* ("protracheophyte" grade) and *Zosterophyllum* ("zosterophyll" grade sensu lato) and by the ambiguous placements of the basal members of the "trimerophyte" grade within the euphyllophytes (66).

Rather, overall character diversity and complexity increased more rapidly when the more derived members of these first-formed vascular plant lineages became sufficiently distinct to show parallel evolution. Examples include the transition from unipolar (rhizomatous) to bipolar (upright) growth and the associated development of vascularized roots; the acquisition of leaves sensu lato by lineages as phylogenetically disparate as mosses, lycopsids, and various euphyllophyte groups; increasingly contrasting maturation patterns and cross-sectional complexity in vascular strands, and their consequences for biomechanical properties; and the much greater diversity and complexity of meristems that allowed the emergence of distinct orders of branching and a wider variety of (often disposable) organs.

Later in the Devonian, these characters were supplemented with the development of at least the early stages of secondary growth in perhaps five lineages (66) and with at least the early stages of heterospory in perhaps ten lineages (9, 24). Secondary thickening conferred the ability to exploit the vertical dimension, which in turn allowed a switch from a patchwork of monotypic "lawns" to more diverse nonclonal communities. Heterospory was a key precursor for more effective resourcing of propagules, facilitating more K-selective strategies. Both are tempting as key innovations, but neither has been tested effectively in this role.

Summarizing the biomechanical data, developmental processes leading to improvement of mechanical properties were driven primarily by selection for sustaining water supply, following the colonization of new biotopes or after reaching a critical height or sustainable level of branch complexity. Examples of such innovations include physiologically inactive water-conducting elements, the hypodermal sterome, and secondary xylem. These features served as important preaptations for improving mechanical properties, permitting greater sustainable height and more complex branched architectures. Among early land plants, biomechanical analyses indicate that structures evolved primarily for improving water relations were repeatedly co-opted for mechanical support. This resulted in relatively simple but multifunctional tissues, which significantly improved the potential for successfully colonizing new habitats and occupying an increasing variety of niches.

The appearance of secondary growth (arborescence sensu 8) did not only have significance for increasing stability of upright stems and the possibility of producing truly large-bodied (arboreous) growth forms; it also allowed a far wider spectrum of growth architectures, the basis for niche-filling habits that ranged from fully self-supporting plants to lianas. Secondary growth also conferred the ability to adjust conductance and mechanical properties of the axial system in accordance with local environmental conditions. For this reason, many of the early land plants probably faced severe constraints in exploiting new areas and niches compared with later phases of the primary land plant radiation.

CONCLUSIONS

No one character can be accused of having engendered the Siluro-Devonian radiation. Even if attempts are made to tease apart the radiation into a nested sequence of smaller-scale radiations, key innovations are still not readily identified. It seems more likely that a critical mass of phenotypic characters accumulated in several clades, eventually offering sufficient flexibility to define and divide many niches. This process eventually generated the threshold number of niches necessary to form communities that exhibited broadly modern ecological dynamics (if not modern species diversity; 25, 26).

There can be little doubt that, as this threshold approached, the physiological adaptations that enabled the putative (if very poorly preserved) original terrestrialization event into ever-wet soils continued to evolve, aiding the invasion of habitats of increasing degrees of abiotic hostility. This process may well have been aided by profound changes in the environment, viewed increasingly as fundamentally nonuniformitarian in character. In particular, the precipitous decrease in atmospheric CO_2 and concomitant (if slightly delayed) increase in O_2 (toward its Late Carboniferous maximum; 13, 85) should have greatly increased the effective balance between photosynthesis and respiration. Thus, the didactic distinction made by Bateman (7) between the physiological, anatomical, and morphological phases of plant evolution is revealed as simplistic.

Moreover, Bateman's (7) attribution of the ecologically driven "behavioral phase" of plant evolution to the post-Devonian is being progressively undermined as it becomes increasingly clear that paleobiologists have underestimated the role of interkingdom coevolution in early terrestrial ecosystems. For example, many enigmatic fossil taxa such as the Nematophytales (putative liverworts of the all-important Late Silurian preservation gap) and *Prototaxites* are increasingly perceived as fungal, and *Spongiophyton* (from the Early Devonian of Gaspé; 128) has a fungal architecture strongly comparable with modern lecanoralean lichens. Also, the recognition of both mycorrhizal (135) and saprophytic (133) fungi in the Rhynie Chert strongly supports arguments that fungi played an important mediating role allowing plants to accommodate to the rigors of terrestrial life (97, 116). When its diverse carnivorous and phytophagous arthropods (74) and representatives of the embryophyte sister group, the aquatic charophytes (134), are also considered, the Chert graphically illustrates that understanding the origin and early diversification of the land flora requires consideration of relationships among kingdoms, as well as relationships among classes and orders within Plantae.

Despite recent successes, additional reconstructions of fossil species, and genuinely worldwide floristic treatments of Siluro-Devonian plant communities and habitats, are badly needed. Nonetheless, even the available data are considerably better than those underpinning the much-vaunted studies of the marine Cambrian "explosion" of animal life, and at the level of exceptionally preserved Lagerstätten, the terrestrial Rhynie Chert undoubtedly ranks alongside the marine Burgess Shale (11, 18, 33, 55). Devonian ecosystems evidently constitute an excellent working laboratory for studying the relationships between the form and function of organisms during profound evolutionary radiations—especially primary radiations that occur across an ecologically undersaturated landscape and in a strongly nonuniformitarian environment.

Literature Cited

1. Albert VA, Backlund A, Bremer K, Chase MW, Manhart JR, et al. 1994. Functional constraints and *rbc*L evidence for land plant phylogeny. *Ann. Mo. Bot. Gard.* 81:534–67
2. Albert VA, Gustafsson MHG, Di Laurenzio L. 1998. Ontogenetic systematics, molecular developmental genetics, and the angiosperm petal. In *Molecular Systematics of Plants 2*, ed. D Soltis, P Soltis, JJ Doyle. London: Chapman & Hall. In press
3. Algeo TJ, Berner RA, Maynard JB, Scheckler SE. 1995. Late Devonian oceanic anoxic events and biotic crises: "rooted" in the evolution of vascular plants? *GSA Today* 5:45, 64–66
4. Arrington JM, Mishler BD, Lewis LA, Vilgalys RJ, Manos PS. 1997. A molecular phylogeny of the bryophytes and their relationships to the tracheophytes based on chloroplast 16S and 23S ribosomal-coding genes. *Am. J. Bot.* 84:12 (Abstr.)
5. Banks HP. 1975. Reclassification of Psilophyta. *Taxon* 24:401–13
6. Banks HP. 1981. Time of appearance of some plant biocharacters during Siluro-Devonian time. *Can. J. Bot.* 59:1292–96
7. Bateman RM. 1991. Palaeoecology. In *Plant Fossils in Geological Investigation: The Palaeozoic*, ed. CJ Cleal, pp. 34–116. Chichester UK: Horwood
8. Bateman RM. 1994. Evolutionary–developmental change in the growth architecture of fossil rhizomorphic lycopsids: scenarios constructed on cladistic foundations. *Biol. Rev.* 69:527–97
9. Bateman RM. 1996. Nonfloral homoplasy and evolutionary scenarios in living and fossil land plants. In *Homoplasy: The Recurrence of Similarity in Evolution*, ed. MJ Sanderson, L Hufford, pp. 91–130. London: Academic
10. Bateman RM. 1998. Integrating molecular and morphological evidence for evolutionary radiations. In *Advances in Plant Molecular Systematics*, ed. PM Hollingsworth, RM Bateman, RJ Gornall. London: Chapman & Hall. In press
11. Bateman RM, DiMichele WA. 1994. Saltational evolution of form in vascular plants: a neoGoldschmidtian synthesis. In *Shape and Form in Plants and Fungi*, ed. DS Ingram, A Hudson, pp. 63–102. Linnean Society Symposium Series 16. London: Academic
12. Bateman RM, Simpson NJ. 1998. Comparing phylogenetic signals from reproductive and vegetative organs. In *Advances in Plant Reproductive Biology*, ed. S Owens, P Rudall. London: Royal Botanic Gardens Kew. In press
13. Berner RA. 1993. Palaeozoic atmosphere CO_2: importance of solar radiation and plant evolution. *Science* 261:68–70
14. Berner RA, Canfield DE. 1989. A new model for atmospheric oxygen over Phanerozoic time. *Am. J. Sci.* 289:333–61
15. Bopp M, Capesius I. 1996. New aspects of bryophyte taxonomy provided by a molecular approach. *Bot. Acta* 109:1–5
16. Bremer K, Humphries CJ, Mishler BD, Churchill SP. 1987. On cladistic relationships in green plants. *Taxon* 36:339–49
17. Bridge JS, van Veen PM, Matten LC. 1980. Aspects of sedimentology, palynology and palaeobotany of the Upper Devonian of southern Kerry Head, Co. Kerry, Ireland. *Geol. J.* 15:143–70
18. Briggs DEG, Fortey RA, Wills MA. 1992. Morphological disparity in the Cambrian. *Science* 256:1670–73
19. Capesius I, Bopp M. 1997. New classification of liverworts based on molecular and morphological data. *Plant Syst. Evol.* 207:87–97
20. Chaloner WG. 1988. Early land plants: the saga of a great conquest. In *Proc. 14th Int. Bot. Congr.*, ed. W Greuter, B Zimmer, pp. 301–6. Königstein, Germany: Koeltz
21. Chaloner WG, Sheerin A. 1979. Devonian macrofloras. *Spec. Pap. Palaeont.* 23:145–61
22. Coen E, Meyerowitz EM. 1991. The war of the whorls: genetic interactions controlling flower development. *Nature* 353:31–37
23. Collinson ME, Scott AC. 1987. Factors controlling the organisation and evolution of ancient plant communities. In *Organisation of Communities Past and Present*, ed. GHR Gee, PS Giller, pp. 399–420. Oxford: Blackwell
24. DiMichele WA, Bateman RM. 1996. Plant paleoecology and evolutionary

inference: two examples from the Paleozoic. *Rev. Palaeobot. Palynol.* 90:223–47

25. DiMichele WA, Hook RW, et al. 1992. Paleozoic terrestrial ecosystems. In *Terrestrial Ecosystems Through Time*, ed. AK Behrensmeyer, JD Damuth, WA DiMichele, et al, pp. 205–325. Chicago: Univ. Chicago Press

26. DiMichele WA, Stein WE Jr, Bateman RM. 1998. Evolution of primordial patterns of resource partitioning among vascular land plant classes during the Late Paleozoic. In *Anatomy of Major Radiations*, ed. W Allmon. New York: Columbia Univ. Press. In press

27. Donoghue MJ. 1989. Phylogeny and the analysis of evolutionary sequences, with examples from seed plants. *Evolution* 43:1137–56

28. Doyle JA. 1998. Evolutionary radiation of the earliest seed plants. *Annu. Rev. Ecol. Syst.* 29:

29. Doyle JA, Donoghue MJ. 1986. Seed plant phylogeny and the origin of angiosperms: an experimental cladistic approach. *Bot. Rev.* 52:321–431

30. Doyle JJ. 1992. Gene trees and species trees: molecular systematics as one-character taxonomy. *Syst. Bot.* 17:144–63

31. Driese SG, Mora CI, Ellick JM. 1997. Morphology and taphonomy of root and stump casts of the earliest trees (Middle to Late Devonian), Pennsylvania and New York, U.S.A. *Palaios* 12:524–37

32. Duncan TM, Renzaglia KS, Garbary DJ. 1997. Ultrastructure and phylogeny of the spermatozoids of *Chara vulgaris* (Charophyceae). *Plant Syst. Evol.* 204:125–40

33. Eble GJ. 1998. The role of development in evolutionary radiations. In *Biodiversity Dynamics: Turnover of Populations, Taxa, and Communities*, ed. ML McKinney. New York: Columbia Univ. Press. In press

34. Edwards D. 1980. Early land floras. In *The Terrestrial Environment and the Origin of Land Vertebrates*, ed. AL Panchen, pp. 55–85. New York: Academic

35. Edwards D. 1993. Cells and tissues in the vegetative sporophytes of early land plants. *New Phytol.* 125:225–47

36. Edwards D. 1996. New insights into early land ecosystems: a glimpse of a Lilliputian world. *Rev. Palaeobot. Palynol.* 90:159–74

37. Edwards D, Davies KL. 1990. Interpretations of early land plant radiations: 'facile adaptationist guesswork' or reasoned speculation? In *Major Evolutionary Radiations*, ed. PD Taylor, GP Larwood, pp. 351–76. Oxford: Oxford Univ. Press

38. Edwards D, Duckett JG, Richardson JB. 1995. Hepatic characters in the earliest land plants. *Nature* 374:635–36

39. Edwards D, Fanning U. 1985. Evolution and environment in the Late Silurian–Early Devonian: the rise of the pteridophytes. *Philos. Trans. R. Soc. London Ser. B* 309:147–65

40. Edwards D, Selden P. 1992. The development of early terrestrial ecosystems. *Bot. J. Scotl.* 46:337–66

41. Edwards DS. 1980. Evidence for the sporophyte status of the Lower Devonian plant *Rhynia gwynne-vaughnii* Kidston and Lang. *Rev. Palaeobot. Palynol.* 29:177–88

42. Edwards DS. 1986. *Aglaophyton major*, a non-vascular land-plant from the Devonian Rhynie Chert. *Bot. J. Linn. Soc.* 93:173–204

43. Eldredge N, Gould SJ. 1972. Punctuated equilibria: an alternative to phyletic gradualism. In *Models in Paleobiology*, ed. TJM Schopf, pp. 82–115. San Francisco: Freeman

44. Ellick J, Driese SG, Mora CI. 1998. Very large plant and root traces from the Early to Middle Devonian: implications for early terrestrial ecosystems and atmospheric p(CO_2). *Geology* 26:143–46

45. Erwin D. 1992. A preliminary classification of evolutionary radiations. *Hist. Biol.* 6:133–47

46. Feist M, Feist R. 1997. Oldest record of a bisexual plant. *Nature* 385:401

47. Foote M. 1994. Morphological disparity in Ordovician–Silurian crinoids and the early saturation of morphological space. *Paleobiology* 20:320–44

48. Garbary DJ, Renzaglia KS, Duckett JG. 1993. The phylogeny of land plants: a cladistic analysis based on male gametogenesis. *Plant Syst. Evol.* 188:237–69

49. Geng B-Y. 1992. Studies on Early Devonian flora of Sichuan. *Acta Phytotax. Sinica* 30:197–211

50. Gensel PG. 1992. Phylogenetic relationships of the zosterophylls and lycopsids: evidence from morphology, paleoecology, and cladistic methods of inference. *Ann. Mo. Bot. Gard.* 79:450–73

51. Gensel PG, Andrews HN. 1984. *Plant Life in the Devonian*. New York: Praeger

52. Gensel PG, Johnson NG, Strother PK. 1991. Early land plant debris: Hooker's "waifs and strays"? *Palaios* 5:520–47

53. Gilinsky NL, Bambach RK. 1987. Asymmetrical patterns of origination and

extinction in higher taxa. *Paleobiology* 13:427–45
54. Givnish TJ, Sytsma KJ, ed. 1997. *Molecular Evolution and Adaptive Radiation.* Cambridge: Cambridge Univ. Press
55. Gould SJ. 1989. *Wonderful Life: The Burgess Shale and the Nature of History.* New York: Norton
56. Graham LE. 1993. *Origin of Land Plants.* New York: Wiley
57. Gray J. 1993. Major Paleozoic land plant evolutionary bio-events. *Palaeogeogr. Palaeoclimatol. Palaeoecol.* 104:153–69
58. Grenfell HR. 1995. Probable fossil zygnematacean algal spore genera. *Rev. Palaeobot. Palynol.* 84:201–20
59. Hedderson TA, Chapman RL, Rootes WL. 1996. Phylogenetic relationships of bryophytes inferred from nuclear encoded rRNA gene sequences. *Plant Syst. Evol.* 200:213–24
60. Hedderson TA, Cox CJ, Goffinett B, Chapman RL. 1997. Phylogenetic relationships among the main moss lineages inferred from 18S rRNA gene sequences. *Am. J. Bot.* 84:17 (Abstr.)
61. Hemsley AR. 1994. The origin of the land plant sporophyte: an interpolational scenario. *Biol. Rev.* 69:263–74
62. Hiesel R, von Haeseler A, Brennicke A. 1994. Plant mitochondrial nucleic acid sequences as a tool for phylogenetic analysis. *Proc. Natl. Acad. Sci. USA* 91:634–38
63. Hotton CL, Hueber FM, Griffing DH, Bridge JS. 1998. Early terrestrial plant paleoenvironments: an example from the Emsian of Gaspé, Canada. In *Early Land Plants and Their Environments*, ed. PG Gensel, D Edwards. New York: Columbia Univ. Press. In press
64. Kenrick P. 1994. Alternation of generations in land plants: new phylogenetic and palaeobotanical evidence. *Biol. Rev.* 69:293–330
65. Kenrick P, Crane PR. 1991. Water-conducting cells in early fossil land plants: implications for the early evolution of tracheophytes. *Bot. Gaz.* 152:335–56
66. Kenrick P, Crane PR. 1997a. *The Origin and Early Diversification of Land Plants: A Cladistic Study.* Smithsonian Series in Comparative Evolutionary Biology. Washington, DC: Smithsonian Inst. Press
67. Kenrick P, Crane PR. 1997b. The origin and early evolution of plants on land. *Nature* 389:33–39
68. Kidston R, Lang WH. 1921. On Old Red Sandstone plants showing structure, from the Rhynie chert bed, Aberdeenshire. Part

IV. Restorations of the vascular cryptogams, and discussion of their bearing on the general morphology of the Pteridophyta and the origin of the organisation of land-plants. *Trans. R. Soc. Edinburgh* 52:831–54
69. Knoll AH, Niklas KJ. 1987. Adaptation, plant evolution, and the fossil record. *Rev. Palaeobot. Palynol.* 50:127–49
70. Knoll AH, Niklas KJ, Gensel PG, Tiffney BH. 1984. Character diversification and patterns of evolution in early vascular plants. *Paleobiology* 10:34–47
71. Konopka AS, Herendeen PS, Smith Merrill GL, Crane PR. 1997. Sporophytes and gametophytes of Polytrichaceae from the Campanian (Late Cretaceous) of Georgia, U.S.A. *Int. J. Plant Sci.* 158:489–99
72. Kranz HD, Huss VAR. 1996. Molecular evolution of pteridophytes and their relationships to seed plants: evidence from complete 18S rRNA gene sequences. *Plant Syst. Evol.* 202:1–11
73. Kranz HD, Miks D, Siegler M-L, Capesius I, Sensen W, Huss VAR. 1995. The origin of land plants: phylogenetic relationships among charophytes, bryophytes, and vascular plants inferred from complete small-subunit ribosomal RNA gene sequences. *J. Mol. Evol.* 41: 74–84
74. Labandeira CC, Phillips TL. 1996. Insect fluid-feeding on Upper Pennsylvanian tree ferns (Palaeodictyoptera, Marattiales) and the early history of the piercing-and-sucking functional feeding group. *Ann. Entomol. Soc. Am.* 89:157–83
75. Lauder GV. 1996. The argument from design. In *Adaptation*, ed. MR Ruse, GV Lauder, pp. 55–91. New York: Academic
76. Manhart JR. 1994. Phylogenetic analysis of green plant *rbc*L sequences. *Molec. Phylogeny Evol.* 3:114–27
77. Manhart JR. 1995. Chloroplast 16S rDNA sequences and phylogenetic relationships of fern allies and ferns. *Am. Fern J.* 85: 182–92
78. Manhart JR, Palmer JD. 1990. The gain of two chloroplast tRNA introns marks the green algal ancestors of land plants. *Nature* 345:268–70
79. McCourt RM, Karol KG, Guerlesquin M, Feist M. 1996. Phylogeny of extant genera in the family Characeae (Charales, Charophyceae) based on *rbc*L sequences and morphology. *Am. J. Bot.* 83:125–31
80. McElwain JC, Chaloner WG. 1995. Stomatal index and density of fossil plants track atmospheric CO_2 in the Palaeozoic. *Ann. Bot.* 76:389–95

81. Melkonian M, Surek B. 1995. Phylogeny of the Chlorophyta: congruence between ultrastructural and molecular evidence. *Bull. Soc. Zool. France* 120:191–208

82. Mishler BD, Churchill SP. 1984. A cladistic approach to the phylogeny of the "bryophytes." *Brittonia* 36:406–24

83. Mishler BD, Churchill SP. 1985. Transition to a land flora: phylogenetic relationships of the green algae and bryophytes. *Cladistics* 1:305 28

84. Mishler BD, Lewis LA, Buchheim MA, Renzaglia KS, Garbary DJ, et al. 1994. Phylogenetic relationships of the "green algae" and "bryophytes." *Ann. Mo. Bot. Gard.* 81:451–83

85. Mora CI, Driese SG, Colarusso LA. 1996. Middle to Late Paleozoic atmospheric CO_2 levels from soil carbonate and organic matter. *Science* 271:1105–7

86. Niklas KJ. 1976. Morphological and ontogenetic reconstruction of *Parka decipiens* Fleming and *Pachytheca* Hooker from the Lower Old Red Sandstone, Scotland. *Trans. R. Soc. Edinburgh* B69:483–99

87. Niklas KJ. 1992. *Plant Biomechanics: An Engineering Approach to Plant Form and Function.* Chicago: Chicago Univ. Press

88. Niklas KJ. 1994a. *Plant Allometry: The Scaling of Form and Process.* Chicago: Chicago Univ. Press

89. Niklas KJ. 1994b. Morphological evolution through complex domains of fitness. *Proc. Natl. Acad. Sci. USA* 91:6772–79

90. Niklas KJ. 1997. *The Evolutionary Biology of Plants.* Chicago: Chicago Univ. Press

91. Nixon KC, Crepet WL, Stevenson D, Friis E-M. 1994. A reevaluation of seed plant phylogeny. *Ann. Mo. Bot. Gard.* 81:484–533

92. Pimm SL. 1991. *The Balance of Nature?* Chicago: Chicago Univ. Press

93. Pryer KM, Smith AR, Skog JE. 1995. Phylogenetic relationships of extant ferns based on evidence from morphology and *rbc*L sequences. *Am. Fern J.* 85:205–82

94. Raubeson LA, Jansen RK. 1992. Chloroplast DNA evidence on the ancient evolutionary split in vascular land plants. *Science* 255:1697–99

95. Raven JA. 1984. Physical correlates of the morphology of early vascular plants. *Bot. J. Linn. Soc.* 88:105–26

96. Raven JA. 1994. Physiological analyses of aspects of the functioning of vascular tissue in early land plants. *Bot. J. Scotl.* 47:49–64

97. Raven JA. 1995. The early evolution of land plants: aquatic ancestors and atmospheric interactions. *Bot. J. Scotl.* 47:151–75

98. Remy W, Gensel PG, Hass H. 1993. The gametophyte generation of some early Devonian land plants. *Int. J. Plant Sci.* 154:35–58

99. Remy W, Hass H. 1996. New information on gametophytes and sporophytes of *Aglaophyton major* and inferences about possible environmental adaptations. *Rev. Palaeobot. Palynol.* 90:175–93

100. Remy W, Remy D, Hass H. 1997. Organisation, Wuchsform und Lebensstrategien früher Landpflanzen des Unterdevons. *Bot. Jahrb. Syst.* 119:509–62

101. Renzaglia KS, McFarland KD, Smith DK. 1997. Anatomy and ultrastructure of the sporophyte of *Takakia ceratophylla* (Bryophyta). *Am. J. Bot.* 84:1337–50

102. Retallack GJ. 1985. Fossil soils as grounds for interpreting the advent of large plants and animals on land. *Philos. Trans. R. Soc. London Ser. B* 309:105–42

103. Retallack GJ. 1997. Early forest soils and their role in Devonian global change. *Science* 276:583–85

104. Rosenzweig M, McCord R. 1991. Incumbent replacements: evidence for long-term evolutionary progress. *Paleobiology* 17:202–13

105. Roth A, Mosbrugger V, Neugebauer J. 1994. Efficiency and evolution of water transport systems in higher plants—a modelling approach. *Philos. Trans. R. Soc. London Ser. B* 345:137–62

106. Rothwell GW. 1994. Phylogenetic relationships among ferns and gymnosperms: an overview. *J. Plant Res.* 107:411–16

107. Rothwell GW. 1996. Phylogenetic relationships of ferns: a paleobotanical perspective. In *Pteridology in Perspective,* ed. JM Camus, M Gibby, RJ Johns, pp. 395–404. London: Royal Botanic Gardens Kew

108. Rothwell GW, Scheckler SE. 1988. Biology of ancestral gymnosperms. In *Origin and Evolution of Gymnosperms,* ed. CB Beck, pp. 85–134. New York: Columbia Univ. Press

109. Rothwell GW, Serbet R. 1994. Lignophyte phylogeny and the evolution of spermatophytes: a numerical cladistic analysis. *Syst. Bot.* 19:443–82

110. Rowe NP, Speck T. 1997. Biomechanics of *Lycopodiella cernua* and *Huperzia squarrosa*: implications for inferring growth habits of fossil small-bodied lycopsids. *Med. Nederlands Inst. Toegep. Geowetensch. TNO* 58:293–302

111. Rowe NP, Speck T, Galtier J. 1993. Biomechanical analysis of a Palaeozoic

gymnosperm stem. *Proc. R. Soc. London (B)* 252:19–28

112. Scheckler SE. 1986. Geology, floristics and paleoecology of Late Devonian coal swamps from Appalachian Laurentia (USA). *Ann. Soc. Geol. Belg.* 109:209–22

113. Schopf JW. 1994. Disparate rates, differing fates: tempo and mode of evolution changed from the Precambrian to the Phanerozoic. *Proc. Natl. Acad. Sci. USA* 91:6735–42

114. Schuster RM. 1984. Evolution, phylogeny and classification of the Hepaticae. In *New Manual of Bryology*, ed. RM Schuster, pp. 892–1070. Japan: Hattori Botanical Laboratory

115. Scotese CR, McKerrow WS. 1990. Revised world maps and introduction. In *Palaeozoic Palaeogeography and Biogeography*, ed. WS McKerrow, CR Scotese, pp. 1–12. Geol. Soc. Lond. Mem. 12

116. Selosse M-A, Le Tacon F. 1998. The land flora: a phototroph-fungus partnership? *Trends Ecol. Evol.* 13:15–20

117. Shear W. 1991. The early development of terrestrial ecosystems. *Nature* 351:283–89

118. Smoot EL, Taylor TN. 1986. Structurally preserved fossil plants from Antarctica: II. A Permian moss from the Transantarctic Mountains. *Am. J. Bot.* 73:1683–91

119. Speck T, Rowe NP. 1994. Biomechanical analysis of *Pitus dayi*: early seed plant vegetative morphology and its implications on growth habit. *J. Plant Res.* 107:443–60

120. Speck T, Rowe NP. 1998. Biomechanical analysis. In *Fossil Plants and Spores: Modern Techniques*, ed. TP Jones, NP Rowe. London: Geol. Soc. Lond. Spec. Vol. In press

121. Speck T, Vogellehner D. 1988. Biophysical examinations of the bending stability of various stele types and the upright axes of early 'vascular' land plants. *Botanica Acta* 101:262–68

122. Speck T, Vogellehner D. 1992. Biomechanics and maximum height of some Devonian land plants. In *Palaeovegetational Development in Europe*, ed. J Kovar-Eder, pp. 413–22. Vienna: Museum of Natural History

123. Speck T, Vogellehner D. 1994. Devonische Landpflanzen mit und ohne hypodermales Sterom—eine biomechanische Analyse mit Überlegungen zur Frühevolution des Leit- und Festigungssystems. *Palaeontographica* B233:157–227

124. Spicer RA. 1989. Physiological characteristics of land plants in relation to environment through time. *Trans. R. Soc. Edinburgh* B80:321–29

125. Stein WE. 1983. *Iridopteris eriensis* from the Middle Devonian of North America, with sytematics of apparently related taxa. *Bot. Gaz.* 143:401–16

126. Stein WE. 1987. Phylogenetic analysis and fossil plants. *Rev. Palaeobot. Palynol.* 50:31–61

127. Stein WE. 1998. Developmental logic: establishing a relationship between developmental process and phylogenetic pattern in primitive vascular plants. Submitted

128. Stein WE, Harmon GD, Hueber FM. 1993. Lichens in the Lower Devonian of North America. *Geol. Soc. Amer. Abstr.* 25:82

129. Stein WE, Wight DC, Beck CB. 1984. Possible alternatives for the origin of Sphenopsida. *Syst. Bot.* 9:102–18

130. Stevenson DW, Loconte H. 1996. Ordinal and familial relationships of pteridophyte genera. In *Pteridology in Perspective*, ed. JM Camus, M Gibby, RJ Johns, pp. 435–67. London: Royal Botanic Gardens Kew

131. Stewart WN, Rothwell GW. 1993. *Paleobotany and the Evolution of Plants*. Cambridge: Cambridge Univ. Press. 2nd ed.

132. Taylor TN. 1988. The origin of land plants: some answers, more questions. *Taxon* 37:805–33

133. Taylor TN. 1990. Fungal associations in the terrestrial palaeoecosystem. *Trends Ecol. Evol.* 5:21–25

134. Tayor TN, Remy W, Hass H. 1992. Fungi from the Lower Devonian Rhynie Chert: Chytridiomycetes. *Am. J. Bot.* 79:1233–41

135. Taylor TN, Remy W, Hass H, Kerp H. 1995. Fossil arbuscular mycorrhizae from the early Devonian. *Mycologia* 87:560–73

136. Taylor TN, Taylor EL. 1993. *The Biology and Evolution of Fossil Plants*. Englewood Cliffs, NJ: Prentice-Hall

137. Taylor WA. 1996. Ultrastructure of Lower Paleozoic dyads from southern Ohio. *Rev. Palaeobot. Palynol.* 92:269–80

138. Valentine JW. 1980. Determinants of diversity in higher taxonomic categories. *Paleobiology* 6:444–50

139. Wikström N, Kenrick P. 1997. Phylogeny of Lycopodiaceae (Lycopsida) and the relationships of *Phylloglossum drummondii* Kunze based on *rbc*L sequences. *Int. J. Plant Sci.* 158:862–71

140. Zimmermann W. 1965. *Die Telomtheorie*. Stuttgart: Fischer

Annu. Rev. Ecol. Syst. 1998. 29:293–318

POSSIBLE LARGEST-SCALE TRENDS IN ORGANISMAL EVOLUTION: Eight "Live Hypotheses"

Daniel W. McShea

Department of Zoology, Duke University, Box 90325, Durham, North Carolina 27708-0325; e-mail: dmcshea@acpub.duke.edu

KEY WORDS: entropy, evolutionary trends, evolutionary progress, trend mechanisms, versatility

ABSTRACT

Historically, a great many features of organisms have been said to show a trend over the history of life, and many rationales for such trends have been proposed. Here I review eight candidates, eight "live hypotheses" that are inspiring research on largest-scale trends today: entropy, energy intensiveness, evolutionary versatility, developmental depth, structural depth, adaptedness, size, and complexity. For each, the review covers the principal arguments that have been advanced for why a trend is expected, as well as some of the empirical approaches that have been adopted. Also discussed are three conceptual matters arising in connection with trend studies: 1. Alternative bases for classifying trends: pattern versus dynamics; 2. alternative modes in which largest-scale trends have been studied: "exploratory" versus "skeptical"; and 3. evolutionary progress.

INTRODUCTION

The history of life is full of trends. Horses increased in size in the Cenozoic; shell sutures in ammonoids became more convoluted in the late Paleozoic and Mesozoic; in gastropods, the number of independent parameters controlling shell shape increased in the early Paleozoic; and clones of eukaryotic cells became integrated to form the first multicellular organisms sometime during the Proterozoic. In this 3.6-billion-year riot of small- and moderate-scale change—in which billions of species have been careening about in a structure and function space of huge and changing dimensionality—many have wondered whether any

293

large-scale order is expected. Is there some feature of organisms that we can expect to have changed directionally, on average, over the entire history of life as a whole, at the largest temporal and taxonomic scale? To put it another way, do we have any reason in theory to believe that organisms in later time periods will be different in some consistent way—in some aspect of their individual development, morphology, physiology, behavior, and so on—from those in earlier times?

Quite a few candidates have been proposed, both for features showing such a trend and for possible causes. Lamarck (48) thought complexity increased as a result of the activity of invisible fluids within organisms. Huxley (37) raised the possibility that selection favors increases in ability to modify the environment. More recently, Vermeij (117; see below) has proposed that organisms become more escalated, or energy intensive. And Maynard Smith & Szathmáry (53; see below) have argued that the salient trend has been in what might be called structural depth, with increases occurring in a small number of key transitions, each involving changes in the way that organisms transmit information. Other features said to increase include entropy, evolutionary versatility, developmental entrenchment, intelligence, independence from the environment, specialization, and ability to sense the environment. (For others, see 5, 6, 23, 37, 79, 94.)

Many of these hypotheses have nearly dropped out of evolutionary discourse, some because the feature identified has so far proved empirically intractable and the hypothesis has therefore failed to generate much research (e.g. perhaps Huxley's notion that ability to modify the environment increases). Others are not now taken seriously because their theoretical bases have been undermined (e.g. Lamarck) or because the language in which they were presented is incommensurate with modern discourse (perhaps Lamarck again).

Nevertheless, some are still, or have become, what the psychologist William James (42) would have called "live hypotheses." They are the candidates for large-scale trends that are intelligible today and that now capture the imagination of evolutionists and inspire research. In the first part of the discussion below, I list eight such candidates—entropy, energy intensiveness, evolutionary versatility, developmental depth, structural depth, adaptedness, size, and complexity—and for each discuss some of the rationales for increase that have been proposed. Some, such as Cope's rule, are well known and need little exposition. Others, such as the thermodynamic view, are not well understood, or have been largely overlooked, by the evolutionary community and therefore need more. As will be seen, the features are interrelated. For example, four on the list—evolutionary versatility, developmental depth, structural depth, and complexity—are aspects of complexity (65).

The focus is on causes, on theoretical arguments for why a given feature might be expected to increase. A major goal is to identify the testable predictions that each hypothesis makes. In some cases, a sample of the relevant evidence is also presented, mainly to show how the concepts involved have been operationalized or applied in the study of real trends. Summarizing and evaluating the available evidence is a larger project than can be attempted here.

The point of the list is only to review possibilities, to see where we stand theoretically in our understanding of the forces at work in evolution at the largest scale. Thus, inclusion on this list carries no endorsement of any kind (nor does omission imply rejection). Indeed, in the present state of our understanding, the possibility that no forces have operated at the largest scale, that the history of life has been dominated by chance events (32), and that no largest-scale trend has occurred in any feature of organisms, probably ought to count as a live hypothesis. However, the difficulties involved in testing these possibilities are not trivial (13, 57, 78, 80, 81), and they too merit longer discussions than are possible here.

A certain amount of arbitrariness is unavoidable in choosing the live hypotheses. A reasonable case could be made for including certain other candidates for trends, such as specialization or ecophenotypy. [Bambach's (8) recent suggestion that "physiological resilience" increases might soon belong on the list, but the idea has not yet been fully explicated.] In any event, the list is intended to be preliminary.

Three Issues

I also discuss certain methodological and philosophical issues arising in connection with trend studies, and for each I offer or defend a conceptual distinction that may help to resolve them. First, discussions of the causes of trends have been overconcerned with two sorts of pattern: rising maxima and stable minima. For size, an increase in the maximum for some group would be an increase over time in the size of the largest species in the group (12). A stable minimum might be the persistence of the group's smallest species. What sometimes confounds discussions of causes is that an increase in the maximum has seemed to support the existence of a directed, pervasive force (perhaps selection), tending to drive a trend, while a stable minimum has suggested that no driving force exists. However, a close look at the dynamics of large-scale trends reveals that both patterns are the expected result of a number of very different mechanisms, and that neither is conclusive evidence for or against driving forces. More generally, I argue that distinguishing trends based on their dynamics, rather than on the patterns they produce, would reduce confusion and encourage theorists to address certain fundamental issues.

Second, I propose that studies of largest-scale trends can be classified into two types, or modes of inquiry, based on differences in the way they frame questions and use data. One is the "exploratory" or hypothesis-generation mode and the other is the "skeptical" or hypothesis-testing mode. Both modes are essential to progress, but the skeptical mode has been underemphasized, I argue.

Finally, trends at the largest scale have always been associated with notions of evolutionary progress (86). I conclude by arguing that the term progress might be appropriate in this context, if it is understood in a certain sense, but that using it in this sense could place us in an awkward position: That is, we might be forced to acknowledge that a trend in a feature we do not value constitutes a kind of progress.

Scope

As indicated, I am concerned only with trends at the largest scale, those thought to characterize life or the evolutionary process as a whole. Topics not covered include trends unique to particular groups (e.g. in animals, 59; in plants, 72), trends in single lineages (13, 57, 80), and the analysis of evolutionary rates (29). Also omitted are methods for establishing that trends have occurred, such as recent cladistic methods for inferring ancestral character states (20a).

Finally, the discussion is limited to features at the organism level, in other words, the size, complexity, fitness, and so on, of individual organisms. (Occasionally, it is convenient to refer to features of species, but in these cases species are understood to be characterized by the typical or mean values of their member individuals.) Excluded are trends in global diversity (9a, 92), in space (40, 91), in ecosystem structure (102), and in ecospace occupation (7).

Of course, taxonomic levels may be independent, and as will be seen, this potential independence creates two sorts of difficulties. On the one hand, a hypothesis may not specify the level at which a trend is expected to be manifest, in which case testing is difficult, because we do not know whether to expect a trend in individuals (e.g. possibly entropy; see below). On the other hand, theory may unambiguously predict a trend at the level of the individual, but observed trends may be interpretable as the result of higher-level sorting (e.g. energy intensiveness; see below). Neither problem is insurmountable in principle (e.g. 50, 120).

LIVE HYPOTHESES

For each candidate, I give some of the principal arguments that have been advanced for why a large-scale increase is expected. The ordering of the list is not entirely arbitrary; I give priority to the hypotheses that are likely to be less familiar to biologists or conceptually more difficult.

Entropy

In recent years, a number of biologists have been developing a new outlook based on thermodynamics (16, 126). [Thermodynamic views have surfaced in biology on other occasions (e.g. 90).] Advocates of this outlook have struggled to render it in terms accessible to mainstream biologists but have not been successful on the whole. One problem has been, I believe, that the high level of generality at which thermodynamics offers explanations is rarely encountered in biology. More concretely, thermodynamic principles do not address the particulars of change in specific groups of organisms or in specific ecological circumstances, but rather they concern the principles of change in general, in all life (indeed, in all "dissipative structures;" see below), everywhere it occurs.

Here, I attempt to distill from the writings of the outlook's major advocates the predictions that are relevant to large-scale trends. In doing so, I distinguish two schools of thermodynamic thought, corresponding to the two classical formulations of the second law of thermodynamics in physics, the information school (16, 17, 20, 87, 128) and the energy school (99, 101, 125–127). A synthesis of the two schools may be possible (e.g. 87), but here I offer separate treatments. I should note that on account of theoretical disagreements within each school, it is likely that some who hold these views would not concur with my distillation. Finally, for both schools, my presentation is necessarily somewhat circuitous: I begin with the relevant thermodynamic principles and then explain their significance for evolutionary trends.

THE INFORMATION SCHOOL   The relevant thermodynamic principle for the information school is the second law itself, which for present purposes can be stated as follows: Change in a closed system that is not at equilibrium will occur in such a way as to increase the total amount of disorder, or entropy, of the system. To see what this means, imagine a gas that is confined (at a low density) inside a rectangular box, and further confined to one half of the box by a removable partition. If we remove the partition, the gas rapidly diffuses into the empty half of the box, filling it. The disorder, or entropy, of the gas has increased. This increase in entropy can be interpreted as informational in that the uncertainty, or lack-of-information, associated with the location of every gas molecule increased in the expansion. Before the partition was removed, we could at least say of any given gas molecule that we knew which half of the box it was in; afterward, we know less about its location. Consistent with the formulation favored by the information school, the diffusion could be said to satisfy the demand of the second law that informational entropy increase, and therefore, in a sense, could also be said to be a consequence of the second law.

The information school takes this example as illustrative of the dynamics of evolving taxa, noting that taxa can also split or branch as well as diffuse. That is, evolution can be understood as a kind of entropically driven production (branching) and diffusion of taxa in multidimensional descriptive spaces. Consider a genetic space in which each dimension corresponds to a genetic locus and coordinates in each dimension correspond to the various alleles available at that locus at a given time. In such a space, a species could be plotted as a single point, perhaps at its mean or modal genotype, and the evolution of a species would correspond to movement of the point in genotype space. One claim of the information school is that, under the influence of mutation alone, and therefore in the absence of selection, a group of related taxa, or clade, is expected to grow and to diffuse throughout its genotype space. In doing so, its component species are expected to diverge from each other, thereby increasing the clade's entropy (in the informational sense) over time and satisfying the demands of the second law. More simply (although less precisely), the second law predicts that divergence will occur, even without selection.

A parallel argument could be made at the scale of organisms. If the axes define a morphospace, with coordinates corresponding to the dimensions (or other characteristics) of organs or body parts, then an individual organism becomes a cluster of points, and the evolution of form becomes the movement of the cluster in morphospace. Here the second law predicts that morphology should tend to become more differentiated, more complex, even in the absence of selection.

In clade diffusion, it might seem that a clade should eventually fill the space, but in fact the space does not fill for two reasons. First, constraints are present, such as natural selection, which disfavor many allelic combinations. Second, the space itself is expanding over time, as mutation and gene duplication introduce new alleles and new loci, producing an on-going combinatorial explosion of new possibilities. Another prediction of the information school is that the size of the space will tend to increase faster than diffusion within it. In the gas analogy, it is as though the box were expanding faster than the gas inside. The entropy of the gas increases, but to an observer in the frame of reference of the expanding box, the gas appears to collapse down into one corner. For diffusion of form, the effect is to create the appearance of increased "organization" or structuring of form (Figure 1).

These predictions seem testable. Indeed, the prediction that species diversity and complexity of form will increase might seem trivial, but for complexity at least, no trend has been documented in any rigorous way. (For explicitly thermodynamic empirical treatments at a small scale, see 54, 55; for a broader review of empirical work, see 65.) But even if complexity does increase, the trend mechanism has not been established and is crucial in this context (see below). The second prediction, that the disparity between realized and potential genome space or morphospace should increase over time, is decidedly not trivial, but no tests have been done to my knowledge.

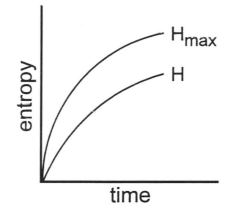

Figure 1   *H* represents the actual informational entropy of an organism, which tends to increase over time as morphology "diffuses;" see text. H_{max} represents its potential entropy, which is a function of the size of the morphological space. Both increase, but *H* lags behind. Also, the difference between *H* and H_{max} increases, which accounts for the appearance of increasing "organization," or specificity of structure, in organisms.

THE ENERGY SCHOOL   The predictions of the energy school follow from the phenomenological principle that, in open, far-from-equilibrium systems, structures emerge whose effect is to dissipate free-energy gradients. The paradigmatic example for the energy school is the Bénard convection cell (99, 113). Imagine a thin layer of viscous fluid spread evenly over a flat dish or plate. The plate is heated uniformly from below, creating a temperature gradient across the fluid. If the gradient is maintained below a critical threshold, heat flows by disordered collisions among fluid molecules (i.e. by conduction) from source below to sink above. If the gradient is increased above the threshold, however, the fluid flow spontaneously becomes structured at a large scale. Viewed from above, the surface of the fluid is no longer a smooth sheet but rather a honeycomb of closely-packed, nearly hexagonal, cells. The cells are convection structures, in which the centers are regions of mass fluid flow upward and the edges mark regions of downward flow. Importantly, with the emergence of these convection cells, the rate of flow of heat from the lower surface to the air above increases enormously. The cells are said to be "dissipative structures," or structures that exist and are maintained because of the contribution they make to the dissipation of the increased free-energy gradient.

Similar structures include natural vortices, such as tornados and hurricanes. Dissipative structures also arise in certain far-from-equilibrium chemical systems, typically as self-catalyzing cycles or networks. Indeed, the origin of life can be seen as the spontaneous formation of autocatalytic chemical cycles, which were thermodynamically favored on account of their contribution to dissipating the free energy gradient between the earth (heated by the sun) and colder surrounding space.

Both physical and chemical dissipative structures share a number of properties. First, growth and multiplication are both favored (in the Bénard apparatus described above, only initially), in that they augment the flow through the system, increasing the rate at which free energy is dissipated. Second, variations or perturbations occurring in these structures that tend to enhance the flow are favored over those that do not, a purely physical principle that produces and accounts for a kind of "natural selection" in dissipative systems that contain a number of such structures. Thus, in this view, organisms are distinguished from nonliving dissipative structures, not in their ability to spontaneously acquire definite structure, to grow, to reproduce, or even to evolve by natural selection. Rather, they are distinguished by possessing DNA, whose principal role is to confer stability and persistence, not only in ontogeny but on phylogenetic timescales as well.

Many dissipative structures, including organisms, undergo a continuous development throughout their existence. In particular, the energy school has drawn attention to four phenomenological rules for dissipative structures, mostly drawn from the principles of ecological succession developed by Lotka, Odum, Margelef, and others. Under certain initial conditions and constraints, dissipative structures show: 1. After an initial increase, a decrease in the rate of energy flow (per unit mass); 2. an increase in internal complexity, or degree of differentiation, although at a decreasing rate; 3. a general decrease in internal rates of change; and 4. increasing vulnerability to external perturbations, i.e. senescence.

These principles can be understood as predictions for trends, all eminently testable, at least in principle, and indeed some have been shown to occur in organismal development and in ecosystems. But for evolutionary—as opposed to ontogenetic—change in organisms, testing is problematic at certain scales where identification of the relevant dissipative structure is difficult. For example, it is not clear that the second rule predicts increasing complexity of individual morphology; whether or not it does depends on whether an evolving lineage constitutes a dissipative structure (whose physical parts are the organs and other elements we usually think of as constituting morphology). Further consideration of such issues is needed to make the predictions testable in practice.

Energy Intensiveness

Vermeij (117, 118) has argued that growth and reproduction in organisms is limited by their ability to find, consume, and defend resources, and that the principal obstacles to success are competitors, predators, and dangerous prey. Such a selective regime, he maintains, is expected to favor evolutionary changes that augment or improve such organismal features as offensive weapons, defensive armor, locomotor performance (including ability to attack and to escape), toxicity, crypsis, ability to gather and process information about the environment, growth rate, and metabolic rate. The common requirement for all of these changes is an increased flow of energy into and through an organism, or an increase in its energy intensiveness (what Vermeij calls "escalation").

Vermeij points out that evolutionary modification is typically constrained by functional trade-offs, so that, for example, a change in some structure that improves one function is likely to interfere with function in another. Consequently, he argues, increases in energy intensiveness are most likely to occur when resources are plentiful and such constraints are relaxed. He points to two periods in the Phanerozoic in particular, the early Paleozoic and the late Mesozoic, when nutrient and energy fluxes through the biosphere were substantially elevated by increases in submarine volcanism, and he argues that these (and possibly other such episodes as well) drove increases in the production of evolutionary innovations and in diversity during those times (119).

In his 1987 book, Vermeij discusses the various routes to increase in energy intensiveness that have been discovered in a variety of taxa, but focuses on molluscs, especially on improvements in drilling capability in certain gastropods and in armor in both gastropods and bivalves. Empirical studies have followed his lead: for example, Kelley & Hansen (45) found that naticid gastropods became more selective in their choice of prey, arguably as a route to increasing the net return on their energy investment in drilling. Miller (68–70) found increases in ornamentation in muricine gastropods that are consistent with escalation, although alternative mechanisms could not be excluded (e.g. 69). One major empirical problem has been determining to what extent apparent escalation is the result of individual-level selection rather than differential speciation and extinction, as well as determining to what extent the two levels are coupled (120).

Evolutionary Versatility

Evolutionary versatility is a function of number of degrees of independence in development, or number of independent dimensions along which variation can occur in evolution (77). Vermeij (114–116) raised the possibility that versatility should tend to increase in evolution, because it increases the range of adaptive possibilities, which permits greater homeostasis, mechanical efficiency, and ability to exploit resources. His argument also suggests that versatility should be most favored when energy resources are plentiful, adaptive constraints are reduced, and therefore selection for energy intensiveness is strongest.

Riska (84) proposed that new organismal traits tend to be highly correlated developmentally, or with other traits when they arise. [This might be expected if new traits commonly begin as duplications of parts and of their associated developmental pathways (49).] If so, selection might favor decreases in developmental integration, thereby allowing new combinations of trait values and new adaptive possibilities.

Wagner & Altenberg (123; see also 122) have argued that genotype-phenotype maps in organisms tend to be modular, or organized in such a way that groups of genes interact strongly among themselves in the development of single characters, but pleiotropic interactions among characters are fewer and weaker. Modularity enables characters to vary independently in evolution, and it is this independence that makes phenotypes evolvable. In evolution, modularization can occur by integration, if genotype-phenotype maps tend to be unintegrated initially, or by disintegration, what they call "parcellation," if maps tend to be highly integrated from the start. They argue that evolution in the Metazoa, at least, seems to proceed by parcellation, with the result that selection for evolvability should tend to increase numbers of modules, in other words, to increase numbers of independent developmental units.

Finally, Riedl (83) has made an argument (see below) that could be construed as leading to the opposite prediction. He contended that covariation in development is advantageous because it allows a rapid response to selection, which would seem to imply that versatility is expected to decrease. Interestingly, Vermeij (116) devised a route whereby both versatility and integration may increase alternately in stepwise fashion. He suggested that new dimensions of variation are added in the origin of new adaptive designs, new higher taxa, and that these enlarged suites of variability become more integrated in later evolution within higher taxa; subsequent innovations add more dimensions, which then in turn become integrated, and so on.

Various approaches to measuring degree of integration (e.g. 19, 73, 121, 132) and its opposite, developmental independence (see 65), have been devised. Also, there has been considerable interest (especially recently) in examining patterns of change in a phylogenetic context (133, and papers following it in the same journal issue). Trends have also been examined using fossils, both in single-species lineages (46, 67) and in certain higher taxa (114–116).

Developmental Depth

Development is to some degree hierarchically structured, in that early in development an organism consists of a relatively small number of structures, which interact to give rise to more structures, which in turn give rise to more, in a widening cascade (3, 4, 83, 87, 130; although see 36, 76, 131 for exceptions). Thus, early developmental steps have more consequences than later steps, on average, and deleterious natural variation occurring in early steps will be more strongly opposed by selection. In Wimsatt's terms (130), early steps and the structures arising from them are "generatively entrenched."

Arguably, developmental cascades should become longer, and entrenchment should deepen as organisms evolve. The reason is that, at any given time, the existing steps and structures in a cascade are likely to be at least somewhat integrated (that is, connected with and dependent on each other), so that their removal would tend to be disruptive in a way that addition of a new step or structure would not (88, 89). Thus, deletions are more likely to be deleterious than additions, on average, and developmental steps and structures should tend to accumulate. If so, then to the extent that additions occur developmentally downstream of existing steps, existing steps will tend to become entrenched, and the overall lability of development should tend to decrease.

Other arguments point to the same conclusion. As discussed, Riedl (83) has argued that covariation in development is advantageous, because it allows a rapid response to selection. He further suggests that covariation may arise by the addition of hierarchical controls in development, and that successive levels of control become superimposed on each other, which increases hierarchical depth. Also, Kauffman's (44) n-k model predicts that long-jumps to higher adaptive peaks in fitness landscapes become increasingly improbable over time, a prediction consistent with increasing developmental depth. And in Salthe's (87) version of the thermodynamic argument, developmental systems become more extended, or in his terms, more specified, in evolving lineages.

Finally, Valentine et al (106; see also 22, 105) propose that, at least within the Metazoa, developmental evolution occurred in two phases. First, the developmental control systems required for basic metazoan architectures (bodyplans) arose, probably in the earliest Cambrian or just before. Thereafter, diversification within bodyplans occurred as a result of elaboration and reorganization of these control systems. In present terms, this view seems to predict a large increase in entrenchment associated with the origin of the Metazoa. And although the subsequent elaboration of developmental pathways might result in occasional further increases in entrenchment, there would seem to be no expectation of an ongoing trend after the Cambrian.

Empirical treatments have focused on changes within higher taxa in disparity, or the degree to which organisms differ from each other morphologically (33, reviewed in 26). Various metrics have been developed (e.g. 15, 25, 26, 43, 129). The developmental arguments above allow that disparity should increase as diversity increases and morphologies diverge, but they also predict that the rate of increase should fall over time, as developmental programs become more constrained (27, 62; cf. 28). This pattern has been documented in certain groups (e.g. 24, 124; but see 15), but in others, the opposite pattern occurs (e.g. 25). In any case, an alternative hypothesis predicts the same trend: increases in the intensity of natural selection, perhaps the result of an increase in ecological packing that accompanies diversification, will tend to reduce rates of morphological divergence (82, 103; but see 27, 34, 62). A major challenge for research in this area is to devise ways to test the developmental hypothesis more directly (e.g. 95, 124).

Structural Depth

Maynard Smith & Szathmáry (53; see also 100) have argued that the major trend in the history of life is an increase in complexity, which is manifest in a series of eight "major transitions": 1. Early replicating molecules to populations of molecules in compartments; 2. unlinked replicators to replicator linkage in chromosomes; 3. the origin of the genetic code and of translation; 4. prokaryote to eukaryote; 5. asexual to sexual reproduction; 6. single-celled existence to multicellularity; 7. solitary existence to coloniality; and 8. the emergence of human social organization based on language.

A common theme in most of these transitions is the emergence of a higher level of nesting of parts within wholes (see also 74, 98), and therefore an increase in what might be called hierarchical complexity (64, 65; see below). In their own terms, what we see is the emergence of new "levels of organization" or new levels of selection (52). "Entities that were capable of independent replication before the transition can replicate only as part of a larger whole after it" (53, p. 6). They further argue that each of these transitions was accompanied by an increase in differentiation and division of labor within levels, as well as a change in the way that information is transmitted.

Maynard Smith & Szathmáry maintain that each of these transitions should be understood as the result of a series of chance events, in particular, a sequence of preadaptations, each of which was undoubtedly favored by short-term selection but for reasons having nothing to do with the transition itself or its longer-term consequences. Consistent with this description, one might further argue, although they do not, that if all or most of the transitions ultimately produced a significant radiation of new forms, as at least some apparently did, the entire series of transitions could be the result of selection for ever higher levels of inclusiveness, albeit selection at a high taxonomic level and on a very long timescale.

Finally, increasing hierarchical complexity need not involve the addition of new levels. It could instead take the form of increasing individuation of existing ones. An example might be the increase in autonomy, or individuation, at the colony level in certain colonial invertebrates (e.g. 11; reviewed in 65).

Adaptedness

One of the most problematic candidates for a trend at the largest scale is adaptedness, or in certain senses, fitness. The difficulty is that it is unclear whether or not the operation of natural selection predicts a large-scale trend, even if we are only interested in selection at the level of the individual, and even if a very general understanding of adaptedness is adopted, such as "ability to survive and reproduce in a given environment" (14). Here is the sort of problem that arises:

On the one hand, adaptedness of individuals in a species should increase in absolute terms if environments deteriorate directionally, perhaps as a result of relentless improvement by other species (107). On the other hand, if environments are complexly varied in time, then arguably selection (on a short time scale) should be able to produce adaptation only to local or transitory environments. To the extent that adaptation is local, or "context sensitive" (23), in this way, it should not be cumulative (35, 96).

However, the problem of context sensitivity should be reduced on sufficiently long time scales, in that species surviving numerous changes in the environment could be understood to have been selected for survival in an average of all of the environments experienced (111). (Of course, as time scales lengthen, selection presumably weakens.) Also, the problem of context sensitivity is removed if adaptedness can be shown to have context-independent aspects (23). To identify such an aspect we would have to establish a feature or combination of features of organisms that corresponds to adaptedness in all of its manifestations; Van Valen's (108, 110) suggestion that the amount of energy that an organism has available for expansion constitutes a kind of "universal currency" for fitness may be useful here, although it has not yet been operationalized.

In principle, trends in adaptedness can be studied empirically even if theoretical issues have not yet been worked out. One approach is to sidestep the problem of identifying universal features and use the fact of success as a proxy for adaptedness. One tactic within this approach is to use decreasing probability of extinction as an indicator of increasing adaptedness (79, 109). Another tactic is to use replacement (10, 112): The fact that one organismal design replaced another in the fossil record, at least raises the possibility of the successor's adaptive superiority (e.g. 41; but see 50). A case can be made even stronger if the replacement can be shown to have occurred multiple times in different environments (e.g. 85); if abundances of the replaced taxon and of its successor are found to be negatively correlated in environments where they overlap, thereby suggesting competition (112); if the superiority of the successor can be demonstrated in direct-competition experiments (e.g. 56); or if the pattern of replacement can be shown to fit a model based on competitive interaction (e.g. 93). The overall tactic would seem to be appropriate for documenting improvement in a single transition (even a long-term one); for a series of transitions, the problem of establishing transitivity arises.

A possible difficulty with this tactic is that replacement is often documented at a high taxonomic level. This leaves open the possibility that the increase in adaptedness was the result of a clade-level property, which may or may not have arisen from properties at the level of the individual organism. For example, in principle, one group might supplant another on account of having a higher speciation rate, even if the loser is superior by individual-level design criteria.

Size

Cope's rule is an empirical generalization that the size of individuals tends to increase in most evolutionary groups. Newell (71) discusses a number of possible rationales, such as the advantages of large size in intra- and interspecific competition (18; but see 21). Bonner (12) argues that increases in size may allow, and also be required by, increases in internal division of labor (complexity). The subject has been reviewed by LaBarbera (47) and McKinney (58), and more recently by Jablonski (38). (For recent empirical treatments see 1, 2, 39.)

Complexity

If complexity is defined narrowly (65) as some function of number of different types of structure or number of independent steps in a process, then four types can be identified for organisms (64); three have already been discussed under different names (in parentheses): 1. Hierarchical developmental complexity, or the number of steps in developmental pathways (developmental depth); 2. hierarchical morphological complexity, or the number of levels of nesting of parts within wholes (structural depth); 3. nonhierarchical developmental complexity, or the number of independent steps at a given developmental stage (versatility); and 4. nonhierarchical morphological complexity, or the number of different part types at a given hierarchical level. Elsewhere, I have reviewed theoretical and empirical aspects of this fourth type (61, 65).

Common Themes and Connections

None of the above list of candidates for largest-scale trends invokes contingent properties of life on earth as the principal cause of increase, and none depends (at least primarily) on long-term *directional* changes in the abiotic environment, such as changes in solar output, meteorite flux, or distribution of environments. Curiously, all are ordinarily framed as *increasing* trends, although, in principle, every increasing trend in some variable could equally well be described as a decreasing trend in its inverse variable. This may reflect nothing more than the general optimism of scientists, but it may also reflect the common association between largest-scale trends and progress (86).

Connections among the candidates are too numerous to list exhaustively. A sample will suffice here: an increase in size, or in structural depth, may also constitute an increase in rate of entropy production; an increase in versatility may also be an increase in informational entropy; increases in energy intensiveness may be correlated with increases in adaptedness. More connections are made explicitly in the theoretical literature cited, and doubtless even more can be imagined. So many are the connections that the possibility of a grand unification should not be ignored (and may already be implicit in some theoretical treatments).

TYPES OF TRENDS

This section suggests a classification of trend causes according to their underlying mechanism or dynamics. The point is to try to clarify the relationship between the hypotheses above and observed, or potentially observable, patterns of change in the mean, maximum, and minimum for the relevant organismal features.

Passive and Driven

Figure 2A shows the diversification of a group. The vertical axis is time and the horizontal axis is what might be called the "state variable" (57): It represents size, complexity, fitness, or any organismal feature in which a trend is thought to occur. The one-dimensional space defined by the state variable and within which diversification occurs is called the "state space." In the figure, the group originates as a single species at some value of the state variable. The assumption is that a species can be characterized by a single value of the state variable at any given time, perhaps the average value for the species. As time passes, new species arise, some become extinct, but diversity increases on average. Also over time, state variable values change, both in the origin of new species and anagenetically between speciation events. (The figure is the output of a computer model; for details, see 63.)

For purposes of discussion, let us suppose that the state variable is size. Figure 2A can be understood as a kind of null model, a picture of the expected evolution of size in a diversifying group in the absence of external forces acting in any systematic way on size in the lineages. Importantly, the suggestion is not that changes occurring in lineages are uncaused; rather, it is that, as in diffusion, the causes of change are many and various, perhaps different in each lineage, and most crucially, not dependent in any systematic way on the lineage's location in state space. In this null model, the group simply diversifies, with increases in size occurring as often as decreases; mean size therefore has no tendency to change.

In Figure 2B, however, a trend in the mean occurs as a result of diffusion away from a boundary, a lower limit on size (vertical line) for the group. For the evolution of all life, the lower boundary might correspond to the size of the

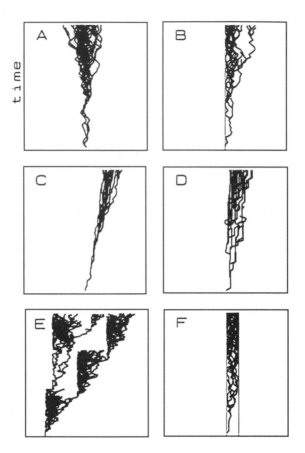

Figure 2 Output of a computer model for simulating the diversification of a group. The vertical axis is time and the horizontal axis is the state variable. In each figure, a group begins as a single species. In every time step, each lineage may increase (move right) or decrease (move left) in state space, speciate, and/or become extinct, each according to fixed probabilities. Boundaries (vertical lines in *B* and *F*) are "cushioning," meaning changes that would cause lineages to cross them are nullified. Biases are introduced (*C* and *D*) by setting the probability of moving right higher than that for moving left. (See 63 for further details.) (*A*) No trend—no boundary, no bias. (*B*) Passive trend—lower boundary, no bias. (*C*) Driven trend—no boundary, strong bias. (*D*) Weakly driven—no boundary, weak bias. (*E*) Driven, at the large scale (in that an increasing bias is present in the origin of groups, although change within groups is passive)—no boundary, strong bias. (*F*) No trend—upper and lower boundary, no bias.

smallest possible organism. The resulting trend mechanism might be called "passive," to indicate that the trend occurs in the absence of any driving forces operating over most of the state space (63). In Figure 2C, no boundary is present, but instead there is a bias, a pervasive tendency for size increases to occur more often than decreases in most lineages; such a trend mechanism might be called "driven." (See also 23, 31, 35, 51, 57, 97.) Figure 2D shows a special kind of driven trend that mimics a passive trend in certain respects to be discussed shortly.

Both the passive and driven mechanisms are consistent with a variety of different lower-level causes. A bias might result from selection favoring size increase in all or most lineages, or even—in principle—from internal, developmental tendencies of some kind that are present in most lineages. Likewise, a boundary might result from selection against small species or from developmental constraints on size (97). Also, boundaries may take a number of different forms, occurring as abrupt or gradual changes in speciation or extinction rate, or even as *local* biases in the direction of change. Finally, Figure 2 shows only classic or ideal cases; various intermediate and ambiguous cases can be imagined (66).

The passive-driven scheme has a number of virtues, notably the fact that several tests are available to distinguish empirically between the two categories (63). Applying such tests to data that show a large-scale trend in some feature of organisms, it will sometimes be possible to eliminate one of the two categories and thereby narrow the range of possible causes considerably. In order for such tests to be decisive, however, we must be able to classify the trend mechanism invoked by a proposed cause as either passive or driven.

Classifying the Live Hypotheses

Classification is straightforward in some cases. For example, Vermeij clearly imagines a driven mechanism for energy intensiveness (117; although a broadly similar passive version of his argument could be devised). The energetic version of the thermodynamic view certainly predicts a driven mechanism. The informational view would seem to be driven also: For morphology, complexity measures are generally variances or variance analogues (61), which are expected to increase in all lineages, at least in the absence of selection. (Of course, variances may increase imperceptibly slowly, so that simple forms may persist, and the expected driven pattern for informational entropy would be more like Figure 2D than Figure 2C; see below.) Cope's rule has been acknowledged to have both passive and driven versions, with different predictions for each (38).

For the remaining hypotheses, theory makes no commitment to either a passive or a driven mechanism, and both versions can be imagined. For example, Maynard Smith & Szathmáry's mechanism would be driven if they predicted that transitions to higher levels of organization occurred more frequently than to lower, as in Figure 2*E* perhaps. But it would be passive if reductions were about equally frequent and if, as seems plausible, decreases were limited at some low level by a boundary, a lower limit on hierarchical depth.

What the passive-driven distinction does in these and other ambiguous cases is raise questions for theoreticians about the sort of dynamical mechanism envisioned. Does the hypothesis predict merely that increases in the relevant state variable will occur (Figure 2, all cases), or does it predict that they occur preferentially, more frequently than decreases (Figure 2*C*, 2*D*, and at a larger scale, 2*E*)? Are lower limits to change expected (Figure 2*B*), and perhaps upper limits as well (Figure 2*F*)? Also, the scheme shows us why a theorist's acknowledgment that *decreases* will occur occasionally, and are permitted by the proposed hypothesis, is relatively uninformative. Occasional decreases are expected in all systems (Figure 2), except the most extreme driven ones (not shown), which for the most part are implausible anyway.

Behavior of Maxima

The scheme represented in Figure 2 can also help to train our intuition about the behavior of maxima and minima. In the null model (Figure 2*A*), the maximum increases. Indeed, the maximum is expected to increase for all but one of the mechanisms illustrated in Figure 2. Thus, a hypothesis which predicts only that the maximum in a diversifying group will increase makes a very weak claim in dynamical terms. The fact that maximum size has increased since the origin of life (12), or that the largest nonclonal animal ever exists today, requires no special explanation (31, 35), either in terms of selection or boundaries, at least not at the scale of evolution as a whole. (Of course, as discussed, every change in every individual lineage will have its own unique explanation.) Likewise for complexity: Even if it were granted that maximum complexity has increased, and that some extant species, such as humans, was the most complex in some sense (cf 65), this pattern could well be the simple consequence of diffusion in a complexity space (104), in which case no special higher-level explanation would be required.

This is not to say that maxima are irrelevant, or that stronger claims about them cannot be made. For example, since diversity has increased on the whole (although with significant interruptions), a long-term leveling off of a maximum would certainly require special explanation, perhaps one that predicts existence of an upper bound of some sort (Figure 2*F*). Also, Maynard Smith & Szathmáry's suggestion that the maximum in hierarchical complexity increased episodically, in occasional revolutions, could be a strong claim if it implies the existence of two distinct causal regimes, one operating during revolutions and another between.

Behavior of Minima

The null model (Figure 2A) also shows that, in the absence of boundaries, the minimum is expected to decrease. However, no theoretical treatment I know predicts a long-term decrease in the minimum, for any feature of organisms, over the history of life. Rather, the assumption in most hypotheses (often implicit) is that the minimum has not changed. (Some driven mechanisms are exceptions; see above.) Notice first that a static minimum can come about in two ways. One possibility is that the earliest forms had very low values of the state variable and that these forms have persisted more or less unchanged. For example, for complexity, cyanobacteria might have been very simple at the time of their origin in the Archean and might not have changed much in complexity since then. The other possibility is that there has been considerable turnover at the lowest values in state space. For complexity, this might mean that cyanobacteria have in fact become more complex, but that other more complex species have evolved simpler forms, and thus the lowest levels of the complexity space have remained occupied.

More importantly in this context, notice that while a stable minimum might seem to suggest the existence of a boundary and therefore a passive mechanism, this is not the case: A stable minimum can be produced by either a passive (Figure 2C) or a weakly driven system (Figure 2D). Again, the virtue of this scheme is that it forces the theorist to consider not just the pattern of behavior of the minimum but the sort of lower-level dynamics that is supposed to account for that pattern.

COMPLEMENTARY MODES OF INVESTIGATION

The study of trends in all of life can be (and has been) undertaken in two modes, what might be called the "exploratory" and the "skeptical." In the exploratory mode, we ask what features of organisms *might* have increased over the history of life? Some candidates may follow from theory, as above. Some may emerge from observation, whether from data collected in a deliberate and organized way or from intuitions—perhaps gestalts that emerge from a survey in the imagination of the history of life. For example, modern organisms might simply *seem* more complex than ancient ones. (A certain amount of vagueness is sometimes tolerable in this mode; we may not be able to specify *yet* precisely what we mean by "complexity.") In this mode, the goal is the production of hypotheses, and in pursuit of promising ones, we draw on all our resources.

In the exploratory mode, a common tactic is to investigate the promise of a candidate variable by trying to build a case for it. Arguments in favor are proposed, and possible defenses against counter-arguments are devised. Evidence in favor is marshaled, and apparently contradictory evidence is explained. In this mode, the goal is to investigate whether, and ultimately to show why, the candidate for a trend ought to be considered a viable one, to show why further investigation is worthwhile. Hypotheses produced in this way are what Popper (75) called bold "conjectures."

In the skeptical mode, we ask whether a certain hypothesized trend actually occurred. The hypothesis is formulated in a testable way, with all ambiguity eliminated from the relevant terms, such as fitness or complexity: The terms are operationalized. Cases appropriate for testing, perhaps groups of organisms, are chosen without regard to the likelihood that an increase will be found (except in testing for trends thought to be produced by unique or episodic changes), and to the maximum extent possible, testing is a blind application of operational measures to those cases. Ideally, in the skeptical mode, we approach testing in a neutral frame of mind, with no particular preference for one outcome over another.

Part of the skeptic's job is to be relentless. If the result of a single test is negative, the hypothesis is not necessarily refuted. For one thing, a larger sample may be needed, perhaps application of the tests in more groups of organisms. For another, the possibility needs to be considered that a somewhat different formulation of the hypothesis, or a different approach to operationalizing its terms, might yield a different result. For example, if an increase in versatility is operationalized as a reduction in the linear correlation between developmental variables, and if a test for a trend in the tooth row of mammals is negative (e.g. 67), then first, other structures and other taxa need to be examined, and second, other senses of versatility—perhaps those based on various nonlinear correlations—need to be examined. This program of intense hypothesis-testing corresponds to what Popper (75) called "refutation."

The exploratory and skeptical modes, hypothesis production and hypothesis testing, conjecture and refutation, are obviously complementary, even if—in addition to their methodological differences—the spirit in which their practitioners approach a problem often seems to be at odds. And in practice, most thoughtful research operates to some degree in both modes at once. In the study of trends, considerable skepticism can be found even in the purely theoretical treatments listed above. And the skeptical approach is exploratory, in a sense, in its methodical consideration of, and elimination of, alternatives.

Still, it is clear that the study of trends at the largest scale has been dominated by the exploratory mode. The thermodynamic treatments of both schools; the hypotheses of increasing energy intensiveness, degree of entrenchment, versatility, and adaptedness; and the notion that revolutions have occurred in the way that information is managed and transmitted all seem to have been proposed mainly in the spirit of exploration. And for the most part, such skeptical testing as has been done has been limited to single formulations of hypotheses, in a small number of organismal structures and groups. This imbalance might seem to be an artifact of the way the subject has been presented here: The emphasis has been on theory, and empirical treatments have been discussed only so far as they show how the various candidate variables might be operationalized. However, this organization was itself partly a result of the imbalance in the field. Very little hypothesis testing has been done.

EVOLUTIONARY PROGRESS

Ayala (5, 6) has argued that progress contains both a descriptive component and an axiological or evaluative component. Thus, a statement that the evolution of some feature of organisms has been progressive is implicitly both a claim that directional change has occurred in that feature, a description, and also that the change was good or valuable by some standard, an evaluation. Accepting this decomposition of the term progress, I would add that the value component has two senses: First, a feature of organisms can be valued by human beings. [It was the cultural embeddedness of our values on the matter of change that led Gould (30) to reject the notion of progress.] Second, it can be valued by the evolutionary process itself. Obviously, the process does not value anything consciously, but it might value a feature in the sense that it tends to generate or preserve it. (This second sense is somewhat weaker because it is metaphorical.) Undoubtedly, it is valuation in this sense that is implicit in most contemporary discussions of evolutionary progress (41, 85).

However, I do not frame the issue that way here; that is, I disallow the second sense of value. The reason is that doing so could force us to accept some odd claims, to allow that evolutionary changes of a sort that we do not value at all nevertheless constitute progress. The risk might seem slight, because historically the principal candidates for evolutionary progress have been features that we often do value, in some senses, at least in ourselves: intelligence, adaptability, ability to modify and control the environment, efficiency, and so on (5, 6, 37, 94). But no largest-scale trend has yet been demonstrated in any of these, and in the meantime we should consider the possibility that the features in which largest-scale trends are eventually found—if any are found—will not turn out to be so agreeable. For example, if developmental entrenchment increases, it should produce a long-term reduction in evolvability. Or a long-term tendency for higher levels of organization (such as the colony or the society) to emerge and to become more individuated might result in a continual reduction in autonomy for entities at lower levels (such as individuals; see 87). Trends that sound even more unfortunate, from our perspective, can doubtless be imagined. And it would, I think, sound very strange to call any of these progress.

Finally, a comment on complexity: In discussions of largest-scale trends, it has been common to substitute the word complexity for progress, as though they were synonyms (discussed further in 60, 65). The substitution may be appropriate if, in the exploratory mode, a temporary label is needed to describe a not-yet-completely-specifiable "something" that is thought to progress in the second sense above; complexity, at least in its colloquial sense, is sufficiently vague to make it suitable for that purposes. However, in making this substitution, we should be aware, first, that progress (in the second sense above), and therefore complexity, may not be a good thing, by human standards. And further, there is a risk of some confusion developing between a deliberately vague usage of complexity and the very precise, technical usages of the term that have been devised in physics (9), as well as in biology (65, 104). I expect that as these technical approaches mature, and as complexity in its various technical senses becomes better understood, using the term as a placeholder will become less and less appropriate.

ACKNOWLEDGMENTS

I thank D Brooks, M Foote, D Jablonski, S Salthe, R Swenson, and G Vermeij for discussions, and J Collier, D Erwin, J Mercer, S Salthe, L Van Valen, and G Vermeij for reading and commenting on all or part of the manuscript.

Visit the *Annual Reviews home page* at
http://www.AnnualReviews.org

Literature Cited

1. Alroy J. 1997. Cope's rule revisited: the evolution of body mass in North American mammals. *Geol. Soc. Am. Abstr. with Programs* 29:A108
2. Arnold AJ, Kelly DC, Parker WC. 1995. Causality and Cope's Rule: evidence from the planktonic foraminifera. *J. Paleontol.* 69:203–10
3. Arthur W. 1984. *Mechanisms of Morphological Evolution.* New York: Wiley. 275 pp.
4. Arthur W. 1988. *A Theory of the Evolution of Development.* New York: Wiley. 94 pp.
5. Ayala FJ. 1974. The concept of biological progress. In *Studies in the Philosophy of Biology*, ed. FJ Ayala, T Dobzhansky, 19:339–55. New York: Macmillan. 390 pp.
6. Ayala FJ. 1988. Can "progress" be defined as a biological concept? In *Evolutionary Progress*, ed. M Nitecki, pp. 75–96.

Chicago: Univ. Chicago Press. 354 pp.
7. Bambach RK. 1993. Seafood through time: changes in biomass, energetics, and productivity in the marine ecosystem. *Paleobiology* 19:372–97
8. Bambach RK. 1997. Fundamental physiological control on patterns of diversification in the marine biosphere. *Geol. Soc. Am. Abstr. Progr.* 29:A31
9. Bennett CH. 1990. How to define complexity in physics, and why. In *Complexity, Entropy, and the Physics of Information*, ed. WH Zurek, pp. 137–48. Redwood City, CA: Addison-Wesley. 530 pp.
9a. Benton MJ. 1993. *The Fossil Record 2.* London: Chapman & Hall. 845 pp.
10. Benton MJ. 1996. On the nonprevalence of competitive replacement in the evolution of tetrapods. In *Evolutionary Paleobiology*, ed. D Jablonski, DH Erwin, JH Lipps, 8:185–210. Chicago: Univ. Chicago Press. 484 pp.

11. Boardman RS, Cheetham AH. 1973. Degrees of colony dominance in stenolaemate and gymnolaemate bryozoa. In *Animal Colonies: Development and Function through Time*, ed. RS Boardman, AH Cheetham, WA Oliver, pp. 121–220. Stroudsburg, PA: Dowden, Hutchinson, & Ross. 603 pp.

12. Bonner JT. 1988. *The Evolution of Complexity*. Princeton, NJ: Princeton Univ. Press. 260 pp.

13. Bookstein FL. 1988. Random walk and biometrics of morphological characters. *Evol. Biol.* 23:369–98

14. Brandon RN. 1996. *Concepts and Methods in Evolutionary Biology*. Cambridge, UK: Cambridge Univ. Press. 221 pp.

15. Briggs DEG, Fortey RA, Wills MA. 1992. Morphological disparity in the Cambrian. *Science* 256:1670–73

16. Brooks DR, Collier J, Maurer BA, Smith JDH, Wiley EO. 1989. Entropy and information in evolving biological systems. *Biol. Philos.* 4:407–32

17. Brooks DR, Wiley EO. 1988. *Evolution as Entropy*. Chicago: Univ. Chicago Press. 415 pp. 2nd ed.

18. Brown JH, Maurer BA. 1986. Body size, ecological dominance and Cope's rule. *Nature* 324:248–50

19. Cheverud JM. 1995. Morphological integration in the saddle-back tamarin (*Saguinus fuscicollis*) cranium. *Am. Naturalist* 145:63–89

20. Collier J. 1986. Entropy in evolution. *Biol. Philos.* 1:5–24

20a. Cunningham CW, Omland KE, Oakley TH. 1998. Reconstructing ancestral character states: a critical reappraisal. *Trends Ecol. Evol.* In press

21. Damuth J. 1993. Cope's rule, the island rule and the scaling of mammalian population density. *Nature* 365:748–50

22. Erwin DE, Valentine J, Jablonski D. 1997. The origin of animal body plans. *Am. Sci.* 85:126–37

23. Fisher DC. 1986. Progress in organismal design. In *Patterns and Processes in the History of Life*, ed. DM Raup, D Jablonski pp. 99–117. Berlin: Springer. 447 pp.

24. Foote M. 1992. Paleozoic record of morphological diversity in blastozoan echinoderms. *Proc. Nat. Acad. Sci. USA* 89:7325–29

25. Foote M. 1993. Discordance and concordance between morphological and taxonomic diversity. *Paleobiology* 19:185–204

26. Foote M. 1997. The evolution of morphological diversity. *Annu. Rev. Ecol. Syst.* 28:129–52

27. Foote M, Gould SJ. 1992. Cambrian and Recent morphological disparity. *Science* 258:1816

28. Fortey RA, Briggs DEG, Wills MA. 1996. The Cambrian evolutionary 'explosion': decoupling cladogenesis from morphological disparity. *Biol. J. Linn. Soc.* 57:13–33

29. Gingerich PD. 1993. Quantification and comparison of evolutionary rates. *Am. J. Sci.* 293A:453–78

30. Gould SJ. 1988. On replacing the idea of progress with an operational notion of directionality. In *Evolutionary Progress*, ed. M Nitecki, pp. 319–38. Chicago: Univ. Chicago Press. 354 pp.

31. Gould SJ. 1988. Trends as changes in variance: a new slant on progress and directionality in evolution. *J. Paleontol.* 62:319–29

32. Gould SJ. 1989. *Wonderful Life*. New York: Norton. 347 pp.

33. Gould SJ. 1991. The disparity of the Burgess Shale arthropod fauna and the limits of cladistic analysis: Why we must strive to quantify morphospace. *Paleobiology* 17:411–23

34. Gould SJ. 1993. How to analyze the Burgess Shale—a reply to Ridley. *Paleobiology* 19:522–23

35. Gould SJ. 1997. *Full House: The Spread of excellence from Plato to Darwin*. New York: Harmony. 244 pp.

36. Hall BK. 1996. Baupläne, phylotypic stages, and constraint. *Evol. Biol.* 29:215–61

37. Huxley JS. 1942. *Evolution: The Modern Synthesis*. New York: Harper. 645 pp.

38. Jablonski D. 1996. Body size and macroevolution. In *Evolutionary Paleobiology*, ed. D Jablonski, DH Erwin, JH Lipps, 10:256–89. Chicago: Univ. Chicago Press. 484 pp.

39. Jablonski D. 1997. Body-size evolution in Cretaceous molluscs and the status of Cope's rule. *Nature* 385:250–52

40. Jablonski D, Bottjer DJ. 1990. Onshore-offshore trends in marine invertebrate evolution. In *Causes of Evolution*, ed. RM Ross, WD Allmon, pp. 21–75. Chicago: Univ. Chicago Press. 479 pp.

41. Jackson JBC, McKinney FK. 1990. Ecological processes and progressive macroevolution of marine clonal benthos. In *Causes of Evolution*, ed. RM Ross, WD Allmon, pp. 173–209. Chicago: Univ. Chicago Press. 479 pp.

42. James W. 1956. *The Will to Believe and Other Essays*. New York: Dover. 332 pp.

43. Jernvall J, Hunter JP, Fortelius M. 1996. Molar tooth diversity, disparity, and

ecology in Cenozoic ungulate radiations. *Science* 274:1489–95

44. Kauffman SA. 1993. *The Origins of Order.* New York: Oxford Univ. Press. 709 pp.

45. Kelley PH, Hansen TA. 1996. Naticid gastropod prey selectivity through time and the hypothesis of escalation. *Palaios* 11:437–45

46. Kurtén B. 1988. *On Evolution and Fossil Mammals.* New York: Columbia Univ. Press. 301 pp.

47. LaBarbera M. 1986. The evolution and ecology of body size. In *Patterns and Processes in the History of Life*, ed. DM Raup, D Jablonski, pp. 69–98. Berlin: Springer

48. Lamarck JBPAM. [1809] 1984. *Zoological Philosophy.* Chicago: Univ. Chicago Press. 453 pp.

49. Lauder GV. 1981. Form and function: structural analysis in evolutionary morphology. *Paleobiology* 4:430–42

50. Lidgard S, McKinney FK, Taylor PD. 1993. Competition, clade replacement, and a history of cyclostome and cheilostome bryozoan diversity. *Paleobiology* 19:352–71

51. Maynard Smith J. 1970. Time in the evolutionary process. *Studium Generale* 23:266–72

52. Maynard Smith J. 1988. Evolutionary progress and levels of selection. In *Evolutionary Progress*, ed. M Nitecki, pp. 219–30. Chicago: Univ. Chicago Press. 354 pp.

53. Maynard Smith J, Szathmáry E. 1995. *The Major Transitions in Evolution.* Oxford: Freeman. 346 pp.

54. Maze J, Bohm LR. 1997. Studies into abstract properties of individuals. I. Emergence in grass inflorescences. *Int. J. Plant Sci.* 158:685–92

55. Maze J, Scagel RK. 1982. Morphogenesis of the spikelets and inflorescence of *Andropogon gerardii* Vitt. (Gramineae) and the relationship between form, information theory, and thermodynamics. *Can. J. Bot.* 60:806–17

56. McKinney FK. 1995. Taphonomic effects and preserved overgrowth relationship among encrusting marine organisms. *Palaios* 10:279–82

57. McKinney ML. 1990. Classifying and analyzing evolutionary trends. See Ref. 59, 2:28–58

58. McKinney ML. 1990. Trends in body-size evolution. See Ref. 59, 4:75–118

59. McNamara KJ. 1990. *Evolutionary Trends.* Tucson: Univ. Ariz Press. 368 pp.

60. McShea DW. 1991. Complexity and evolution: what everybody knows. *Biol. Philos.* 6:303–24

61. McShea DW. 1993. Evolutionary change in the morphological complexity of the mammalian vertebral column. *Evolution* 47:730–40

62. McShea DW. 1993. Arguments, tests and the Burgess Shale—a commentary on the debate. *Paleobiology* 19:399–402

63. McShea DW. 1994. Mechanisms of large-scale trends. *Evolution* 48:1747–63

64. McShea DW. 1996. Complexity and homoplasy. In *Homoplasy: The Recurrence of Similarity in Evolution*, ed. MJ Sanderson, L Hufford, pp. 207–25. San Diego, CA: Academic. 339 pp.

65. McShea DW. 1996. Metazoan complexity and evolution: Is there a trend? *Evolution* 50:477–92

66. McShea DW. 1998. Dynamics of diversification in state space. In *Biodiversity Dynamics*, ed. ML McKinney. New York: Columbia Univ. Press. In press

67. McShea DW, Hallgrimsson B, Gingerich PD. 1995. Testing for evolutionary trends in non-hierarchical developmental complexity. *Geol. Soc. Am. Abstr. Progr.* 27:53–54

68. Miller DJ. 1994. Large scale evolutionary patterns in muricine gastropods: passive or driven? *Geol. Soc. Am. Abstr. Progr.* 26:A52

69. Miller DJ. 1996. Differences in the incidence of armor between Indo-West Pacific and western Atlantic/Caribbean muricine gastropods: escalation or passive evolution? *Paleontol. Soc. Spec. Publ. No. 8*, p. 273

70. Miller DJ. 1996. Temporal and spatial heterogeneities in the morphological evolution of muricine gastropods. *Geol. Soc. Am. Abstr. Progr.* 28:A431

71. Newell ND. 1949. Phyletic size increase, an important trend illustrated by fossil invertebrates. *Evolution* 3:103–24

72. Niklas KJ. 1997. *The Evolutionary Biology of Plants.* Chicago: Univ. Chicago Press. 449 pp.

73. Olson E, Miller R. 1958. *Morphological Integration.* Chicago: Univ. Chicago Press. 317 pp.

74. Pettersson M. 1996. *Complexity and Evolution.* Cambridge, UK: Cambridge Univ. Press. 143 pp.

75. Popper KR. 1989. *Conjectures and Refutations.* London: Routledge. 431 pp. 5th ed.

76. Raff RA. 1996. *The Shape of Life.* Chicago: Univ. Chicago Press. 520 pp.

77. Raup DM. 1966. Geometric analysis of

shell coiling: general problems. *J. Paleontol.* 40:1178–90

78. Raup DM. 1977. Stochastic models in evolutionary palaeontology. In *Patterns of Evolution*, ed. A Hallam, 3:59–78. Amsterdam: Elsevier. 591 pp.

79. Raup DM. 1988. Testing the fossil record for evolutionary progress. In *Evolutionary Progress*, ed. M Nitecki, pp. 293–317. Chicago: Univ. Chicago Press. 354 pp.

80. Raup DM, Crick RE. 1981. Evolution of single characters in the Jurassic ammonite *Kosmoceras. Paleobiology* 7:200–15

81. Raup DM, Gould SJ. 1974. Stochastic simulation and evolution of morphology—towards a nomothetic paleontology. *Syst. Zool.* 23:305–22

82. Ridley M. 1993. Analysis of the Burgess Shale. *Paleobiology* 19:519–21

83. Riedl R. 1977. A systems-analytical approach to macro-evolutionary phenomena. *Q. Rev. Biol.* 52:351–70

84. Riska B. 1986. Some models for development, growth, and morphometric correlation. *Evolution* 40:1303–11

85. Rosenzweig ML, McCord RD. 1991. Incumbent replacement: evidence for long-term evolutionary progress. *Paleobiology* 17:202–13

86. Ruse M. 1996. *Monad to Man: The Concept of Progress in Evolutionary Biology* Cambridge, MA: Harvard Univ. Press. 628 pp.

87. Salthe SN. 1993. *Development and Evolution* Cambridge, MA: Bradford/MIT Press. 357 pp.

88. Saunders PT, Ho MW. 1976. On the increase in complexity in evolution. *J. Theoret. Biol.* 63:375–84

89. Saunders PT, Ho MW. 1981. On the increase in complexity in evolution II: the relativity of complexity and the principle of minimum increase. *J. Theoret. Biol.* 90:515–30

90. Schrödinger E. 1956. *What Is Life?, and Other Scientific Essays.* Garden City, NY: Doubleday. 263 pp.

91. Sepkoski JJ Jr. 1991. A model of onshore-offshore change in faunal diversity. *Paleobiology* 17:58–77

92. Sepkoski JJ Jr. 1993. Ten years in the library: new data confirm palcontological patterns. *Paleobiology* 19:43–51

93. Sepkoski JJ Jr. 1996. Competition in macroevolution: the double wedge revisited. In *Evolutionary Paleobiology*, ed. D Jablonski, DH Erwin, JH Lipps, 9:211–55. Chicago: Univ. Chicago Press. 484 pp.

94. Simpson GG. 1967. *The Meaning of Evolution.* New Haven, CT: Yale Univ. Press. 368 pp.

95. Smith LH. 1996. Developmental integration in trilobites. *Geol. Soc. Am. Abstr. with Programs* 28:A53

96. Sober E. 1984. *The Nature of Selection.* Cambridge, MA: Bradford/MIT Press. 383 pp.

97. Stanley SM. 1973. An explanation for Cope's rule. *Evolution* 27:1–26

98. Stebbins GL. 1969. *The Basis of Progressive Evolution.* Chapel Hill: Univ. N. Carolina Press. 150 pp.

99. Swenson R, Turvey MT. 1991. Thermodynamic reason for perception-action cycles. *Ecol. Psychol.* 3:317–48

100. Szathmáry E, Maynard Smith J. 1995. The major evolutionary transitions. *Nature* 374:227–32

101. Ulanowicz RE. 1997. *Ecology: The Ascendent Perspective.* New York: Columbia Univ. Press. 201 pp.

102. Valentine JW. 1969. Patterns of taxonomic and ecological structure of the shelf benthos during Phanerozoic time. *Palaeontology* 12:684–709

103. Valentine JW. 1995. Why no new phyla after the Cambrian? Genome and ecospace revisited. *Palaios* 10:190–94

104. Valentine JW, Collins AG, Meyer CP. 1993. Morphological complexity increase in metazoans. *Paleobiology* 20:131–42

105. Valentine JW, Erwin DH. 1987. Interpreting great developmental experiments: the fossil record. In *Development as an Evolutionary Process*, ed. RA Raff, EC Raff, pp. 71–107. New York: AR Liss. 329 pp.

106. Valentine JW, Erwin DH, Jablonski D. 1996. Developmental evolution of metazoan bodyplans: the fossil evidence. *Devel. Biol.* 173:373–81

107. Van Valen LM. 1973. A new evolutionary law. *Evol. Theory* 1:1–30

108. Van Valen LM. 1978. Evolution as a zero-sum game for energy. *Evol. Theory* 4:289–300

109. Van Valen LM. 1984. A resetting of Phanerozoic community evolution. *Nature* 307:50–52

110. Van Valen LM. 1989. Three paradigms of evolution. *Evol. Theory* 1:1–17

111. Van Valen LM. 1989. General adaptation, escalation, and phagy. *Condor* 91:1014–18

112. Van Valen LM, Sloan RE. 1966. The extinction of the multituberculates. *Systemat. Zool.* 15:261–78

113. Velarde MG, Normand C. 1980. Convection. *Sci. Am.* July:93–108

114. Vermeij GJ. 1971. Gastropod evolution

and morphological diversity in relation to shell geometry. *J. Zoology* 163:15–23

115. Vermeij GJ. 1973. Biological versatility and earth history. *Proc. Nat. Acad. Sci. USA* 70:1936–38

116. Vermeij GJ. 1974. Adaptation, versatility, and evolution. *Systemat. Zool.* 22:466–77

117. Vermeij GJ. 1987. *Evolution and Escalation.* Princeton, NJ: Princeton Univ. Press. 527 pp.

118. Vermeij GJ. 1994. The evolutionary interaction among species: selection, escalation, and coevolution. *Annu. Rev. Ecol. Syst.* 25:219–36

119. Vermeij GJ. 1995. Economics, volcanoes, and Phanerozoic revolutions. *Paleobiology* 21:125–52

120. Vermeij GJ. 1996. Adaptation and clades: resistance and response. In *Adaptation*, ed. MR Rose, GV Lauder, 11:363–80. San Diego, CA: Academic. 511 pp.

121. Wagner GP. 1990. A comparative study of morphological integration in *Apis melifera* (Insecta, Hymenoptera) *Z. Zool. Syst. Evolutionsforsch.* 28:48–61

122. Wagner GP. 1996. Homologues, natural kinds and evolution of modularity. *Am. Zool.* 36:36–43

123. Wagner GP, Altenberg L. 1996. Complex adaptations and evolution of evolvability. *Evolution* 50:967–76

124. Wagner PJ. 1995. Testing evolutionary constraint hypotheses with early Paleozoic gastropods. *Paleobiology* 21:248–72

125. Weber BH, Depew DJ, Smith JD, eds. 1988. *Entropy, Information, and Evolution.* Cambridge, MA: Bradford/MIT Press. 376 pp.

126. Weber BH, Depew DJ, Dyke C, Salthe SN, Schneider ED, et al. 1989. Evolution in thermodynamic perspective: an ecological approach. *Biol. Philos.* 4:373–405

127. Wicken JS. 1987. *Evolution, Thermodynamics, and Information.* New York: Oxford Univ. Press. 243 pp.

128. Wiley EO. 1988. Evolution, progress, and entropy. In *Evolutionary Progress*, ed. M Nitecki, pp. 275–91. Chicago: Univ. Chicago Press. 354 pp.

129. Wills MA, Briggs DEG, Fortey RA. 1994. Disparity as an evolutionary index: a comparison of Cambrian and Recent arthropods. *Paleobiology* 20:93–130

130. Wimsatt WC. 1986. Developmental constraints, generative entrenchment, and the innate-acquired distinction. In *Integrating Scientific Disciplines*, ed. W Bechtel, pp. 185–208. Dordrecht: Martinus-Nijhoff

131. Wray GA, Raff RA. 1990. Novel origins of lineage founder cells in the direct developing sea urchin *Heliocidaris erythrogramma. Dev. Biol.* 141:41–54

132. Zelditch ML. 1988. Ontogenetic variation patterns of phenotypic integration in the laboratory rat. *Evolution* 42:28–41

133. Zelditch ML. 1996. Introduction to the symposium: historical patterns of developmental integration. *Am. Zool.* 36:1–3

Annu. Rev. Ecol. Syst. 1998. 29:319–43

FUNGAL ENDOPHYTES: A Continuum of Interactions with Host Plants

K. Saikkonen,[1] S. H. Faeth,[2] M. Helander,[1] and T. J. Sullivan[2]

[1]Department of Biology, University of Turku, FIN-20014 Turku, Finland and
[2]Department of Biology, Arizona State University, Tempe, Arizona 85287-1501;
e-mail: karisaik@utu.fi

KEY WORDS: mutualism, antagonism, grasses, woody plants

ABSTRACT

Endophytic fungi living asymptomatically within plant tissues have been found in virtually all plant species. Endophytes are considered plant mutualists: They receive nutrition and protection from the host plant while the host plant may benefit from enhanced competitive abilities and increased resistance to herbivores, pathogens, and various abiotic stresses. Limited evidence also indicates that endophytes may influence population dynamics, plant community diversity, and ecosystem function. However, most of the empirical evidence for this mutualism and its ecological consequences has been based on a few agronomic grass endophytes. More recent studies suggest that endophyte-host plant interactions are variable and range from antagonistic to mutualistic. A more comprehensive view of the ecology and evolution of endophytes and host plants is needed. This article discusses how life history traits—such as fungal reproduction and pattern of infections and genotypic variation and ecological factors—influence the direction and strength of the endophyte-host plant interaction.

INTRODUCTION

Fungal endophytes live internally, either intercellularly or intracellularly, and asymptomatically (i.e. without causing overt signs of tissue damage) (14, 139) within plant tissues (113; sensu 151, 152, 116, but see 147). Endophytes usually occur in above-ground plant tissues, but also occasionally in roots, and are distinguished from mycorrhizae by lacking external hyphae or mantels. The meaning of the term endophyte has undergone various transformations in the

319

last decade, and there is still considerable disagreement as what constitutes an endophyte (e.g. 50, 113, 126, 147, 152). We refer here broadly to fungi that live for all, or at least a significant part, of their life cycle internally and asymptomically in plant parts. We thus include a wide range of fungi, from fungal plant pathogens and saprophytes that have extended latency periods before disease or external signs of infection appear (139, 140) to specialized fungi in grasses that are considered obligate mutualists. Our rationale, as well as that of others (e.g. 140), is that the distinction between classical plant fungal pathogens and mutualists is not clear (e.g. 50, 139) and interactions between fungi and host plant are often variable among and within populations and communities. We contend that the blurring of boundaries between fungal pathogen and mutualists is supported by evolutionary and ecological theory and by accumulating empirical studies.

Although mycologists have long known that plants harbor fungal endophytes (49, 88, 149) and that endophytes in certain grass species in the subfamily Pooideae are associated with toxicity to grazing livestock (e.g. 11, 62, 149), the causal link between fungal endophytes and toxicity to herbivores was not firmly established until the 1970s (10, 125). Since then, ecological, evolutionary, mycological, and agronomic studies involving fungal endophytes in pooid grasses have proliferated (e.g. see reviews by 19, 27, 28, 33–35, 92, 135, 136) (Table 1). It is now well-established that clavicipitaceous fungal endophytes in cultivated turf and pasture, as well as some naturally occurring grasses, can have wide-ranging and often dramatic biological effects on growth and reproduction of host grasses, pathogens and herbivores of grasses, and natural enemies of herbivores (e.g. 19, 22, 26, 33, 34, 41, 47, 84, 122, 148).

Endophytic fungi have been found in all woody plants that have been examined for endophytes (14, 87, 113, 126). Studies of fungal endophytes in trees, shrubs, and ferns show that individual species and even individual plants typically harbor scores of fungal species (29, 54, 59, 66, 75, 76, 98, 106, 112, 114–117, 127, 140, 145, 150, 155). Despite the great diversity and abundances of endophytes in woody plants, these endophytes and interactions with their host plants have received less attention than those of grass endophytes (see below). Furthermore, there has been little attempt to integrate ecological and evolutionary aspects of the endophytes of grasses and woody plants (but see 26, 27, 71, 126).

Separate treatment of grass and tree endophytes may have its origins in the agricultural importance of grass endophytes, which became the domain mainly of agronomists and applied mycologists. Furthermore, real or perceived differences in taxonomy, biology, and life history between grass and woody plant endophytes may have also served to widen the research gap. We argue, however,

Table 1   Outcome of interactions among endophyte-infected grasses and herbivores[a]

Plant and plant part	Herbivore	Direction of effect (number of responses in bioassays)			Reference
		Positive	Neutral	Negative	
Cenchrus echinatus					
Leaves	*Spodoptera frugiperda*	—	—	1	39
Cyperus virens					
Leaves	*S. frugiperda*	—	—	4	40
Cyperus pseudovegetus					
Leaves	*S. frugiperda*	—	2	3	40
Danthonia spicata					
Leaves	*S. frugiperda*	—	—	1	30
Festuca arizonia					
Leaves	Other herbivores tested	1	2	—	97, 128
Festuca arundinacea					
Leave[b]	Other herbivores tested	1	13	26	17, 18, 23, 39, 42, 48, 52, 73, 81, 82, 85, 86, 90, 107, 138
	S. frugiperda	—	8	6	17, 23, 39, 48, 73
	Rhopalosiphu padi	—	—	2	81
	Schizaphis graminum	—	—	9	18, 40, 52, 81, 138
Seeds		—	—	1	30
Roots		—	7	8	32, 48, 84, 110, 119, 148
Festuca gigantea					
Leaves	*R. padi*	—	—	1	138
	S. graminum	—	—	1	138
Festuca glauca					
Leaves	Other herbivores tested	—	1	—	18
	S. frugiperda	—	1	—	17
	R. padi	—	2	—	18, 138
	S. graminum	—	1	1	18, 138
Festuca longifolia					
Leaves	*R. padi*	—	1	—	138
	S. graminum	—	—	4	18, 138
Festuca obtusa					
Leaves	*S. frugiperda*	—	2	3	30
	S. graminum	—	—	1	138
Festuca ovina					
Leaves	Other herbivores tested	—	1	—	18
	R. padi	—	1	—	18
	S. graminum	—	—	1	18
Festuca paradoxa					
Leaves	*S. frugiperda*	—	—	1	30
Festuca rubra					
Leaves	Other herbivores tested	—	1	—	18
	S. frugiperda	—	—	5	30, 17
	R. padi	—	3	—	18, 90, 138
	S. graminum	—	1	1	18, 138

(Continued)

Table 1   *(Continued)*

Plant and plant part	Herbivore	Direction of effect (number of responses in bioassays)			Reference
		Positive	Neutral	Negative	
Festuca versuta					
Leaves	S. frugiperda	—	—	1	30
	S. graminum	—	—	1	138
Glyceria striata					
Leaves	S. frugiperda	—	—	1	30
Hordeum spp.	Other herbivores tested	—	—	1	44
Lolium perenne					
Leaves[c]	Other herbivores tested	—	1	7	1, 2, 13, 18, 43, 67, 85, 86, 104, 120, 121
	S. frugiperda	—	3	10	17, 39, 72
	R. padi	—	4	—	18, 90, 138
	S. graminum	—	—	3	18, 138
Lolium perenne					
Seeds	Other herbivores tested	—	—	1	30
Paniculum agrostides					
Leaves	S. frugiperda	—	2	2	30
Paspalum dilatatum					
Leaves	S. frugiperda	—	1	4	39
Puspalum notatum					
Leaves	S. frugiperda	—	3	2	30
Poa ampla					
Leaves	R. padi	—	1	—	138
Poa autumnalus					
Leaves	R. padi	—	—	1	138
	S. graminum	—	—	1	138
Stipa leucotricha					
Leaves	S. frugiperda	—	5	1	39
Tridens flavus					
Leaves	S. frugiperda	—	3	2	30

[a]Response variables in bioassays were relative growth rate, time to pupation, pupal weight, time to eclosion, development time, food consumption, preference, survival, or reproduction. Of 22 other grass species, 45/80 (56%) showed negative effects on herbivores.

[b]*F. arundinacea* leaves: 43/65 (66%) showed negative effects on herbivores.

[c]*L. perenne* leaves: 20/28 (71%) showed negative effects on herbivores.

that characteristics frequently associated with either grass or woody plant endophytes often apply to both groups. Based on this overlap in life histories and biologies, we expect a continuum of ecological consequences for both grass and woody plant endophytes at the individual, population, and community levels. We first summarize the taxonomic, biological, and evolutionary differences and similarities of grass and woody plant endophytes. We then (*a*) discuss their ecological consequences at the individual, population, and community levels and (*b*) predict direction and strength of interactions of endophytes with host plants based on life history traits and on interactions with other species and abiotic environments.

Taxonomy and Specificity

Systemic grass endophytes are restricted to clavicipitaceous members of the tribe Balansiae (Ascomycotina) and infect at least 80 genera and 300 species (91). We focus here on *Neotyphodium* and its sexual or teleomorphic stage *Epichloë*, which are found almost exclusively in the grasses of the Pooidae. A few other genera of systemic endophytes occur in pooid (e.g. 3, 4) or non-pooid grasses (e.g. 92), but most studies of endophyte-host grass interactions have involved *Neotyphodium* and *Epichloë*. Species of *Neotyphodium* and *Epichloë* tend to be host specific (130) and often form genetically distinct races among populations of the same host species (130). Recent molecular evidence suggests that *Epichloë typhina*, considered ancestral to asexual clavicipitaceous endophytes, is a complex of diverse biological species, each with a high degree of host specificity (130, 132). Taxonomically, endophytes from woody plants are usually members of the Ascomycotina but may also include members of the Basidiomycotina, Deuteromycotina, and Oomycetes (112, 139). Endophytes in woody plants are more diverse than grass endophytes in terms of genera and species, as expected given the much broader taxonomic range of host plants (113, 116). At least at the generic level, woody plant endophytes have wider host ranges than the grass endophytes and, thus, appear to be more generalized than grass endophytes (113, 116). However, Petrini et al (116) argue that many woody plant endophytes are likely to be highly specific at the host-species level and among host populations, much like systemic endophytes of grasses. Furthermore, little is known about nonsystemic grass endophytes, which may be as diverse and generalized as those found in woody plants (133, 134).

Systemic Versus Localized Infections

Grass endophytes, such as *Neotyphodium*, are generally thought to form systemic infections throughout the host, although mycelial biomass may be unevenly distributed within plants (133, 134, 146). However, few systemic endophyte-grass associations have been studied intensively for within-plant distribution. In *Neotyphodium*-infected Arizona fescue plants, for example, infection of individual tillers is variable, ranging from 10% to 100% (133, 134). Even in the intensively studied, *Neotyphodium*-infected cultivars of tall fescue and perennial ryegrass, uninfected seeds are often produced by infected plants, which suggests less than complete systemic infections within plants (146).

Infections by endophytes of woody plants are usually highly localized within leaves, petioles, bark, or stems (26, 27, 61, 116, 151, 153, 155). However, under certain conditions, such as when leaves age (e.g. 65) or senesce (59, 60, 157), localized infections can become more widespread, although the term endophyte may no longer apply because the infections become external. Nevertheless,

Petrini et al (116) caution that far too little is known about endophytes of woody plants to assume that they all nonsystemic. Likewise, little is known about nonsystemic endophytic infections in grasses because virtually all studies have focused on systemic endophytes. However, a recent study indicates that in addition to systemic infections, endophytes that form localized infections in grasses may be as diverse and abundant as those found in woody plants (133, 134).

Reproduction

Asexual or anamorphic forms of grass endophytes are transmitted vertically and maternally by hyphae growing into seeds, whereas sexual stages like *Epichloë* can be transmitted either clonally and vertically (by growing into seeds) or sexually and horizontally (via spores) (26, 130). Horizontal transmission of asexual endophytes like *Neotyphodium*, however, cannot be ruled out because artificial infection of grass seedlings can be accomplished in the laboratory (89). We are unaware of any studies of horizontal transmission of asexual forms in natural grass populations, but it would not be surprising if invertebrate herbivores act as vectors of asexual hyphae and thus effect horizontal transmission. If so, then it may be premature to consider systemic, asexual endophytes as "trapped" pathogens under strict control by the host grass (131, 130).

Many nonsystemic endophytes in woody plants are transmitted horizontally via asexual spores (116). However, the frequency of asexual and sexual reproduction of endophytic fungi has not been extensively studied in woody plants. Because endophytes of woody plants have been found in seeds (116) and acorns (155), endophytes of woody plants may also be transmitted vertically and maternally via seeds. This possibility is supported by at least anecdotal evidence in non-grass hosts (15, 16). Horizontally transmitted endophytes in grasses are also probably very diverse and common (e.g. 133, 134), although, like woody plants, little is known about whether sexual or asexual spore production is the most common mode of reproduction.

Evolutionary Origins and Ecological Consequences

Endophytes of both grasses and woody plants are thought to have evolved from parasitic or pathogenic fungi (25–28, 80). Woody plant endophytes are closely related to pathogenic fungi, and presumably evolved from them via an extension of latency periods and a reduction of virulence (116, 149). The *Neotyphodium* grass endophytes are also thought to have evolved from fungal choke grass pathogens in the genus *Epichloë* (137). Evolution does not appear, however, to follow a strict coevolutionary pathway between host plant and endophyte. Although some strains of *E. typhina* have likely cospeciated with their grass hosts (140), a coevolutionary hypothesis does not completely explain the distribution of asexual *Neotyphodium* endophytes. *Neotyphodium* lineages within a host species may have independent evolutionary origins (3, 131). Additionally, DNA sequence evidence shows that some *Neotyphodium* lineages in tall fescue (145a, 150a) and perennial ryegrass (132a) are actually hybrids of *Neotyphodium* and *Epichloë*. From an evolutionary perspective, the relationship between *Neotyphodium* and its hosts may be somewhat dynamic in that novel genetic combinations of endophytes and hosts may be formed regularly. *Neotyphodium* species are usually considered plant mutualists, and hybridization may increase genetic diversity and thus heterogeneity in defenses against herbivores and competing microorganisms (130).

Mutualism has been the prevailing conceptual framework under which the evolution and ecology of endophytes have been viewed and interpreted. Discovery of severe biological effects of endophytes in grasses on livestock, such as toxicosis and hoof gangrene (12, 33, 34), and on invertebrate pest species (24, 42–44) led to the concept of grass endophytes as plant mutualists, primarily by deterring herbivores as "acquired defenses" (30). Defense against vertebrate and invertebrate herbivores, and also against plant pathogens, purportedly results from production of multiple alkaloid compounds by endophytes, at least in agronomic grasses (135, 136, 138). In turn, host plants provide endophytic fungi with a protective refuge, nutrients, and—in the case of vertically transmitted endophytes—dissemination to the next generation of hosts. Other studies of agronomic grasses show that endophytes provide other fitness-enhancing properties for their hosts, such as increasing plant competitive abilities, mainly by increasing efficiency of water use (8, 53). Furthermore, because alkaloids from endophytes are often concentrated in seed, vertically transmitted grass endophytes also may deter seed predators and increase seed dispersal (85a, 158, 166). Fungal endophytes in grasses were thus considered prototypical mutualists. "Endophyte" quickly became synonymous with "mutualist" (e.g. 9, 116) at least in grasses (140, 147), and the primary driving selective force behind the mutualist was (e.g. 30, 33), and still is (e.g. 37, 92), considered defense against herbivores.

Similarly, Carroll (26) proposed that endophytes of woody plants provide a defensive role for the host plant because they produce a wide array of mycotoxins and enzymes that can inhibit growth of microbes and invertebrate herbivores (116, 140). Because endophytes of woody plants are diverse and have shorter life cycles than their perennial host plants, defense via endophytes is considered a mechanism by which long-lived woody plants could keep pace evolutionarily with shorter generational and, hence, presumably more rapidly evolving invertebrate herbivores (26).

If endophytes, as ubiquitous and abundant residents of plants, increase plant fitness mainly by deterring herbivores, then individual effects could be expected to translate into discernible effects at the population, community, and ecosystem levels. We first review evidence for the assumption that endophytes provide a defensive mutualism against herbivores. We then discuss potential effects of endophytes at the population, community, and ecosystem levels and the factors that may moderate these effects. Finally, we predict when endophytes should mediate the effects of herbivores based on life history features, such as mode of transmission and patterns of infection.

VARIABILITY IN INTERACTIONS AMONG ENDOPHYTES, PLANTS, AND HERBIVORES

We found more than 200 references dealing with ecological or evolutionary aspects of endophytic fungi-plant-herbivore associations. Nearly 75% of these focus on grass-endophyte interactions and 8% represent reviews, definitions of endophytes, and comments; only 17% report studies of endophytic fungi in plants other than pooid grasses, even though endophytic fungi are found in almost all plant species (112). Most studies have focused on alkaloid production by grass endophytes and how these confer resistance to herbivores (Table 1). Only three papers involve competitive interactions among grasses (41, 77, 99). Furthermore, most grass endophyte-herbivore studies have focused on a few agricultural cultivars of two agronomically important grass species, *Festuca arundinacea* (tall fescue) and *Lolium perenne* (perennial ryegrass) (Table 1).

The majority of studies indicate that endophytes associated with tall fescue (66%) and perennial ryegrass (71%) negatively affect herbivores (Table 1). Studies with other grass species show more variable effects: 56% of studies demonstrate negative effects, whereas others indicate no effect or even positive effects on herbivores (Table 1). Endophytes of woody plants also have variable effects on herbivores, ranging from negative to positive, although the majority of studies tend to show negative effects (see e.g. 58–61, 65, 122, 129). However, where endophytes of woody plants do adversely affect herbivores, the herbivores are usually either leafminers or gallformers whose life stages are closely linked to the chemical and phenological traits of an individual leaf or stem, and whose dispersal in feeding stages is restricted within and among plants (58–60, 155–157).

Clearly, the mutualistic nature of plant-endophyte association via herbivore resistance is not universal, even for systemic endophytes of the tall fescue and perennial ryegrass systems. We should expect that endophyte-plant-herbivore interactions are more variable in natural grass systems for the following reasons.

(*a*) Tall fescue and perennial ryegrass are introduced Eurasian grasses. Cultivars of these grass species, and probably their endophytes, have limited genotypic diversity because of selective breeding and cultivation relative to grasses and endophytes in natural populations. (*b*) The number of alkaloid types and concentrations produced by these two introduced grasses may be atypically high relative to natural populations of grasses, again because of limited genotypic diversity and artificial selection but also because of anomalously high nutrient conditions of agricultural and turfgrass systems (see below). (*c*) Most studies of endophyte-grass-herbivore interactions have used either introduced livestock or non-native, generalist invertebrate pests such as fall armyworm (*Spodoptera frugiperda*), green bug (*Schizaphis graminum*), or oat bird cherry aphid (*Rhopalosiphum padi*) (Table 1). Conversely, in natural plant communities, most herbivorous invertebrate species are specialists, restricted to feeding on one or a few host plant species (46, 64, 141). These specialist herbivores may be less sensitive to allelochemicals such as alkaloids. Therefore, endophytes and their mycotoxins probably confer less herbivore resistance than that commonly found in agricultural turf grass–pest interactions. (*d*) Herbivory may not be the primary selective pressure on plant traits or associations. Herbivory may have weak, nonexistent, or even positive effects on plant fitness (100, 102, 123). Perennial grasses, in particular, may be adapted to tolerate high and consistent levels of herbivory without appreciable declines in fitness because of the location of meristematic tissue near the base of the grass, where it is protected from most grazers (46).

Systemic endophytes, such as *Neotyphodium*, alter other properties that increase plant fitness, such as drought resistance, competitive abilities and seed dispersal, and germination success (19, 33–35). Endophyte mediation of seed dispersal and survival may have more direct fitness consequences for both endophyte and host than does reduction of herbivory of adult plants. Endophytes may also increase host fitness by altering attack by natural enemies of herbivores (e.g. 22) or by increasing fire resistance (134) or resistance to competing microorganisms. These latter effects of endophytes have not been well studied in natural grass populations. For example, Arizona fescue, like perennial woody plants, seasonally accumulates a diverse array of horizontally transmitted endophytic fungal species. Plants that harbor *Neotyphodium* infections appear to resist colonization by other fungi (133, 134). Thus, asexual endophytes may deter potential plant pathogens, perhaps by inducing chemical responses of their hosts or by directly inhibiting colonization via interference competition (134, 154). These interactions among endophyte, host, herbivores, and other interacting species in natural populations provide a rich opportunity for future studies.

In sum, concepts of plant-endophyte interactions may be biased by focusing on mutualistic aspects of atypical endophyte-grass cultivars and their generalist herbivores. Endophyte-plant interactions in natural populations and communities are probably more variable in direction and strength of interaction and less likely to be based on herbivore resistance.

We next examine possible effects of endophytes at the population and community level.

POPULATION AND COMMUNITY EFFECTS OF ENDOPHYTES

Systemic Grass Endophytes

If endophytes increase fitness of plants—by increasing resistance to herbivores, seed predators, or abiotic stresses or by increasing competitive abilities—then frequencies of infected plants should increase over time in plant populations (e.g. 92). Evidence from agricultural grasses generally supports this prediction (36, 92). Seminatural and natural grass populations (30, 36), however, seem more variable. Although relatively few native grass populations and their endophytes and herbivores have been examined, in those that have been, endophyte frequencies and interactions with their host plants and herbivores are more variable (see 20, 36, 91 and references therein). For example, frequency of grasses of one *Holcus* species infected by *Epichloë* (the sexual or teleomorphic form of *Neotyphodium*) decreased in insecticide-treated plots in English pastures where presumably insect herbivory declined, and thus infected grasses should not be at a selective advantage relative to uninfected plants. However, frequencies of *Epichloë* in another *Holcus* species were not different in insecticide-treated and untreated plots (38). High infectivity by *Epichloë* and *Neotyphodium* in fescues in native habitats in Europe has been reported (91; A Leuchtmann, personal communication), but typically these fescues are found in mixed culture with other uninfected grasses. This suggests that unlike agronomic fescues, infected fescues in natural habitats do not necessarily become competitive dominants because of benefits associated with the endophyte. In general, the range and variability of endophyte infections in native grasses is higher (from 5% to 100%; 36, 133, 134), and the frequency of infected plants is sometimes lower than that expected from agricultural grasses (94, 95, 108, 124), which suggests either variability in beneficial effects of endophyte infection or imperfect transmission of seed-borne endophytes (124) (see below).

Our studies of *Neotyphodium* in native populations of Arizona fescue also indicate high but variable frequencies of infected plants among populations, ranging from 76% to 97% among six populations (133, 134). However, the frequency of *Neotyphodium* is not related to degree of grazing by vertebrates. Plots within populations where native and introduced vertebrates intensively grazed did not have greater frequencies of infected plants than those plots where long-term grazing was prevented, which would be expected if *Neotyphodium* infections increased fitness via herbivore resistance. Presence of *Neotyphodium* also does not increase resistance to several native and non-native invertebrate herbivores (97, 128, 134). Instead, *Melanoplus femurrubrum* (97) and *Xanthipus corallipes*, two native grasshopper species, show increased growth rates and assimilation when feeding on *Neotyphodium*-infected plants relative to uninfected ones (97, 128, 134).

A possible explanation for the lack of endophyte-conferred resistance in Arizona fescue may be limited alkaloid production. Preliminary evidence indicates that infected Arizona fescue plants from natural populations contain only peramines, one of the four alkaloidal types known to be produced in *Neotyphodium*-infected tall fescue (136). Furthermore, peramine levels are highly variable among infected plants even in the same population, ranging from trace amounts to 1.9 ppm (MR Siegel, SH Faeth, unpublished data), and these levels increase twofold in plants grown in the greenhouse with supplemented soil nutrients. This variability in alkaloid variety and concentration appears to be representative of most natural grass-endophyte populations. For example, the toxicity of *Neotyphodium* infections in sleepygrass, *Stipa robusta*, which can cause horse toxicosis (11), is widely variable among populations at high elevations in a southwestern US forest-grassland (83, 101), presumably because of variation in fungal strain or environmental factors that alter alkaloid production (118). Similarly, Siegel & Bush (136) (Table 1) report that of seven *Festuca* grass species (excluding tall fescue) infected with *Neotyphodium* or *Epichloë*, one species produces no alkaloids, six produce one type, and none produces either three or four of the four possible types of alkaloids. Similarly, of 11 non-*Festuca* species (excluding perennial ryegrass) infected with *Neotyphodium* or *Epichloë*, two species produce zero, six species produce one, three produce two, and none produces four of the four possible types of alkaloids. On the other hand, most cultivars of tall fescue and perennial ryegrass that harbor *Neotyphodium* or *Epichloë* endophytes produce high levels of two or three alkaloids (136) and therefore are unlikely to be representative of natural grass populations. Low genetic diversity of these introduced grasses and their endophytes because of selective breeding and artificial selection has likely resulted in an unusually high diversity and concentration of alkaloids and, consequently, in a high degree of herbivore resistance. In addition, tall fescue and perennial ryegrass are usually grown under artificially high soil nutrient levels of agricultural or turfgrass systems, which may inflate concentrations of nitrogen-rich alkaloids (31). Thus, the assumption that high frequencies of seed-borne endophytes in natural grass populations are evidence for endophyte-grass mutualism via primarily herbivore deterrence (92) may be incorrect.

Community and ecosystem effects of grass endophytes are largely unknown. Clay (37) reports that plant diversity is reduced and biomass is increased in experimental plots with *Neotyphodium*-infected tall fescue compared with plots with uninfected tall fescue. He suggests that endophytes may alter plant and herbivore diversity and ecosystem function (e.g. productivity), by antiherbivore effects, competitive dominance, or allelopathy that inhibits seed germination. The tall fescue system, however, may not be typical of endophyte-grass interactions (see above), and thus extrapolation to natural communities may be premature. Endophytes in agronomic grasses can also influence small mammal, insect, and soil microbial abundances and diversity, which in turn potentially alter plant community structure (see 37 and references therein).

In natural grass populations of Arizona fescue, the presence of *Neotyphodium* is negatively associated with the presence of most other endophytic fungal species, both among and within (individual leaves) plants (134). Thus, seed-borne, systemic endophytes potentially alter the composition of the microbial community.

Nonsystemic Endophytes of Woody Plants and Grasses

Less is known about effects of nonsystemic endophytes in woody plants and grasses at the population or community level. Accumulating evidence suggests that endophyte–host tree interactions are variable and range from antagonistic to mutualistic in terms of herbivory (e.g. 58–61, 65, 122, 129, 153–155), sometimes within the life history stages of the fungus (e.g. 65). Thus, effects of endophytes on plant fitness, populations, and communities, and on interacting populations or communities of herbivores, may also be variable, though perhaps predictable (see below). The assumption that herbivory has detrimental effects on endophytes because herbivores reduce plant fitness may not hold generally for endophytes of woody plants (or horizontally transmitted endophytes of grasses). Insect herbivory may increase colonization of horizontally transmitted endophytes of woody plants by breaching the leaf surface and allowing spore and hyphal penetration (59–61). In addition, insect herbivores are well-known vectors of fungal pathogens (e.g. conifer-bark beetles) (74), and spores or hyphae of fungal endophytes may be dispersed via feeding activities or passage through the gut (106a; K Craven, D Wilson, SH Faeth, unpublished data). If herbivores mediate dispersal and colonization of endophytes in woody plants, then these endophytic fungi may be selected to tolerate, or even promote, herbivory by reducing resistance (58–61). In cases in which herbivores facilitate spore or hyphal dispersal, nonsystemic endophyte interactions with their woody host plants should fall near the antagonistic end of the interaction continuum (Figure 1).

The high diversity and abundance of nonsystemic fungal endophytes in woody plants and grasses increases potential for interactions among endophyte

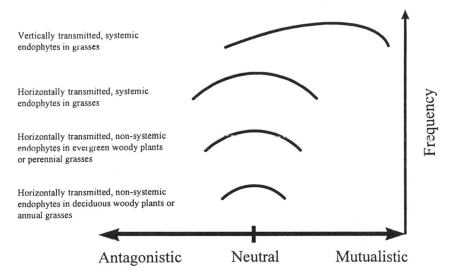

Figure 1   Probability of endophyte-plant interactions occurring along the continuum from antagonistic to mutualistic interactions, based on mode of transmission, pattern of infections, and life span
of the host plant. (*top line*) Vertically transmitted, systemic endophytes in grasses; (*line second
from top*) horizontally transmitted, systemic endophytes in grasses; (*line second from bottom*) horizontally transmitted, nonsystemic endophytes in evergreen woody plants or perennial grasses;
and (*bottom line*) horizontally transmitted, nonsystemic endophytes in deciduous woody plants or
annual grasses.

species, especially as diversity and abundances accumulate seasonally (see
below). Although direct interactions may be limited because of localized infections of horizontally transmitted fungi, limited evidence suggests that endophytic species in woody plants and grasses may compete with each other
(*a*) by production of mycotoxins or induction plant defenses in leaves that inhibit
colonization by other endophytes (61, 134) or (*b*) by increasing premature leaf
abscission (154). These changes could also affect herbivores (e.g. 60, 61, 157),
so indirect interactions between woody plant endophytes and herbivores remain
an interesting, yet largely untested, area for future research.

EVOLUTIONARY AND ECOLOGICAL FACTORS
INFLUENCING ENDOPHYTE INTERACTIONS

Endophyte-plant symbioses represent a broad continuum of interactions, from
strong antagonisms to obligate mutualisms. Mutualisms are generally thought
to have evolved from antagonistic interactions, mainly parasitic (7, 27, 93, 142,

143), and the same has been assumed for fungal endophytes of grasses and woody plants (9, 10, 19, 33, 35, 130). However, most mutualisms, even specialized and obligate ones, have antagonistic components (33, 45, 78, 109, 111, 142, 143, 150a). The continuum of antagonistic-mutualistic interactions for any two interacting species depends on phylogenetic and life history constraints, geography, interactions with other species in the community, and abiotic factors (143, 144, 150a). Similarly, the complex microbial mutualisms with host plants vary along a continuum from pathogenic to mutualistic, even within the lifespan of the microorganism and host plant (e.g. 51, 63, 65, 68, 69). Despite the complexity and variability of endophytic fungal-host plant interactions, evolutionary traits—such as mode of transmission and patterns of infection—and ecological factors—such as condition of host, competition with other microorganisms, spatial structure of populations, and prevailing abiotic factors—permit predictions of where endophyte-plant associations are likely to fall along the continuum.

Mode of Transmission

Modes of reproduction and transmission to other hosts are now well recognized as important factors related to virulence and aggressiveness in disease-causing microorganisms and parasites (e.g. 21, 55, 79, 96, 103). Similarly, mode of transmission may affect degree of antagonism and mutualism of grass endophytes (130, 150a). Schardl et al (131), for example, show that *Epichloë*, the sexual form of *Neotyphodium*, is more likely to be antagonistic to the host when transmitted horizontally via spores and more mutualistic when reproducing vertically by growing into seeds. Grass endophytes that occur as non-sporulating, systemic infections and are transmitted vertically from maternal plant to offspring via seeds (19, 34) probably fall nearer the mutualistic end of the interaction continuum (Figure 1) because reproduction, and thus fitness, of systemic fungi and plants are closely linked (55–57).

Most endophytes of woody plants, and probably many grasses, however, are usually transmitted horizontally by spores from plant to plant (134). Because reproduction of host plant and fungus is typically decoupled by horizontal transmission, mutualistic interactions are less favored by natural selection (21, 57, 71). These endophytes are thus more likely to vary in the continuum of interactions with host plants (Figure 1). This variation is confirmed by empirical studies: Interactions range from negative to positive (e.g. 58–61, 65, 122, 129, 153–155). However, far too little is known about the great diversity of nonsystemic grass endophytes (e.g. 134) and woody plant endophytes. If some woody plant endophytes are transmitted vertically via seeds (see above), then these endophytes should interact more mutualistically with their hosts by increasing some aspect of host fitness at the seed, seedling, or adult stage.

Spatial Patterns of Infection

Infection patterns of nonsystemic endophytes in woody plants and grasses should be more variable than systemic grass endophytes in dispersion and density because colonization of tissues depends on availability and viability of spores, which, in turn, are influenced by surrounding vegetation, ground topography, plant density and architecture, weather conditions, and microclimate within or near the plant (76, 127, 129, 134, 154). Furthermore, infections are usually highly localized within stems and leaves. Herbivores may be unable to avoid these localized infections while selecting or feeding on host plants, especially if among- and within-host preferences by herbivores operate at different spatial or temporal scales than the localized infections (e.g. 58, 60, 156, 157). Carroll (26) proposes that the great diversity and high spatio-temporal variation in distribution of horizontally transmitted endophytes may create sufficient heterogeneity within long-lived woody plants to keep pace evolutionarily with short-cycled insect herbivores. However, because many of these endophytes may be transmitted via asexual spores, genetic variability of endophytes within and among trees may be correspondingly low relative to insect herbivores that reproduce sexually. Future studies should consider distribution patterns and genotypic diversity of endophytes in grasses and woody plants to ascertain the importance of endophytes to herbivore performance.

Seasonal Accumulation of Endophytes

Colonization by sporulating tree and grass endophytes results in a seasonal accumulation of local, and perhaps independent, endophytic fungal colonies (59, 75, 133, 134). This accumulation depends on the number of spore sources, such as infected leaves within the same plant, leaf litter, or nearby plants. Seasonal and yearly climatic factors such as humidity and rainfall may determine spread and germination success of the spores (59, 75, 134). Horizontally transmitted endophytes reinfect the foliage of deciduous trees and annual grasses yearly and accumulate seasonally, whereas the foliage of evergreen woody plants and perennial grasses may remain infected year round and infections usually accumulate in older leaf or needle year classes (60, 140). Thus, for evergreen woody plants or perennial grasses, there may be greater variability in the costs and benefits of harboring endophytes than for deciduous woody plants or annual grasses (59, 75, 76) (Figure 1).

The high diversity and abundance of nonsystemic endophytes but overall low mycelial biomass suggest that costs of harboring these endophytes are probably relatively low or diffuse in terms of plant fitness relative to systemic grass endophytes. Nevertheless, small metabolic costs, or indirect costs due to premature leaf abscission and thus loss of photosynthetic tissue (60, 154, 157), may accumulate over the life span of the host. Furthermore, because many woody plant endophytes appear to provide little, if any, protection against herbivores (59, 60, 129), most of these interactions will probably fall near the antagonistic or neutral regions of the interaction continuum (Figure 1).

In contrast, among-plant infection of vertically transmitted, systemic grass endophytes like *Neotyphodium* cannot increase seasonally or annually by colonization of adult plants (unless horizontal transmission occurs, see above). Instead, infection frequency among plants changes with time only by differential survival and reproduction of infected and noninfected plants and seeds. Infections may be lost from seeds produced by infected plants by failure of hyphae to grow into tillers that produce seed heads, by failure of hyphae to grow into seeds from infected tillers, or by loss of viability of hyphae in infected seeds (124). Furthermore, mycelial biomass per unit volume of host tissue is much greater for systemic grass endophytes than for most woody plant endophytes. Thus, if asexual endophytes are neutral or even slightly antagonistic to their host grasses, infected plants should decline rapidly within a few generations (124). Although there are few long-term studies of endophytes in natural grass populations, most of these studies show consistently high but variable frequencies of infected plants (36, 95, 128, 133, 134). Maintenance of high but variable frequencies of infected plants suggests that interactions of asexual, vertically transmitted endophytes generally fall near the mutualistic end of the interaction continuum (Figure 1).

Spatial and Genotypic Changes in Endophyte-Plant Interactions

It is increasingly clear that endophyte-plant associations are more spatially variable in natural populations than is suggested by studies of tall fescue and perennial ryegrass. For asexual endophytes of grasses, infected plants are widespread (e.g. 36), but most natural grass populations include varying fractions of noninfected plants, and occasionally populations contain no infected plants at all (95, 124, 133, 134, but see 92). If asexual endophytes provide benefits to the host, strong selection for infected plants should result in increasing frequencies of infected plants within a few generations, (124) and frequencies of infected plants should be 100% (92). Limited evidence from agricultural grasses supports increase in frequency of infected grasses, but rarely to 100% (94, 137). Leuchtmann & Clay (92) proposed that less than 100% frequency of asexual, systemic endophytes in grass populations indicates variation in the mutualistic interaction. There are, however, several possible explanations for the maintenance of noninfected and infected plants in natural populations, of which at least one does not directly involve mutualistic interactions.

First, uninfected and infected hosts may coexist as interconnected populations or metapopulations (70). For plant pathogens, metapopulation models predict coexistence even when subpopulations of infected hosts become extinct (e.g. 6). These models may thus also explain persistence and coexistence of noninfected grasses within populations, even if noninfected plants are at a selective disadvantage relative to infected plants. Spatial arrangement of maternal plants and patterns of seed dispersal may be critical in determining frequency and distribution of infected and uninfected plants. For horizontally transmitted endophytes, density and dispersion of the plant population may be particularly important (e.g. 5, 6) if colonization of hosts is frequency or density dependent. Disturbances in communities, such as fire or drought, may also serve to maintain patchy distributions of infected and noninfected hosts (92, 133, 134).

Second, Ravel et al (124) provided an explanation for disequilibrium of endophyte-infected and uninfected grass hosts based on imperfect transmission of asexual endophytes. Their mathematical model showed that uninfected hosts could be maintained in a population assuming that loss of infection in seeds from infected plants, due to either hyphae inviability or failure to propagate into seeds, was greater than 10%. This level of loss of infection is apparently common in agronomic grasses (146), but little is known about reduced infection in seeds in natural populations due to imperfect transmission. Tests of this hypothesis should provide for interesting future research.

Third, the costs and benefits of endophyte infections may vary spatially and temporally in natural populations, and thus selection and frequency of infected and uninfected hosts should vary accordingly. Even asexual endophytes in genetically uniform, agronomic grasses can become parasitic under varying conditions of water and nutrient availability, plant competition, and degree of herbivory (e.g. 31, 36). In natural populations, uninfected plants may persist where patchily distributed abiotic and biotic factors select against plants that harbor asexual endophytes. This should be especially true if there are fitness costs to harboring the endophyte as a defense against herbivores (e.g. 27), as predicted from studies of maintenance of plant chemical defenses against herbivores (e.g. 105). Maintenance of uninfected and infected hosts may be additionally promoted by variation in endophyte and host genotypes. For example, Saikkonen et al (128) showed that host genotype of Arizona fescue, in addition to the presence of the endophyte, determines performance under various nutrient and water regimes.

These three explanations are not mutually exclusive (e.g. 124). For example, metapopulation processes, varying selective pressures, and imperfect transmission may combine to maintain infected and uninfected hosts in natural populations. Further, most natural populations may be mosaics of unique endophyte-host plant genotypic combinations that are adapted to local biotic and abiotic environments. Consideration of metapopulation processes, spatially and temporally varying selective pressures, and endophyte-host genotypic combinations will become increasingly important in unraveling interactions between endo-phytes, host plants, and other interacting species.

PERSPECTIVE

Endophyte-plant interactions may vary from antagonistic to mutualistic (Figure 1). Systemic, vertically transmitted endophytes in grasses have a higher probability of mutualistic interactions with their host plants than nonsystemic, horizontally transmitted endophytes in woody plants and grasses (132). How-ever, because life history characters, such as mode of transmission and patterns of infection, are not unique to either group, there may be considerable overlap in the direction and strength of interaction with the host plant. Attention has cen-tered on endophyte protection of host plants against invertebrate and vertebrate herbivores via physiologically active alkaloids. However, both inter- and intraspecific variation in amount and composition of chemicals in infected plants in natural populations is high, and herbivore sensitivity to different alkaloid pat-terns may vary among species and feeding guilds. Alternatively, endophytes provide defenses against herbivores and pathogens by altering plant physiology, morphology, and allelochemistry. Endophytes may also be an important source of within-plant variation that makes host plant quality more unpredictable for herbivores and pathogens. However, many endophyte-host plant interactions likely have neutral effects on herbivores and some may even promote herbivory.

Endophyte-plant associations are undoubtedly affected by genotypic inter-actions between host and endophyte and local abiotic environments. Systemic grass-endophytes are tightly linked to host plant genotype because they are vertically transmitted from host plant to its offspring, but little is known about genotypic influences of relationships between woody plants and their endo-phytes. To understand evolutionary relationships in plant-endophyte interac-tions and how the associations are modified by environmental changes, the effects of environment, genetic background, and genotype/environment inter-actions should be considered. These considerations should include the follow-ing: (a) the extent of genetic diversity of endophytes in grasses and woody plants, (b) how different endophyte-plant genotype combinations affect plant fitness, and (c) how environmental factors alter the strength and direction of endophyte-plant interactions.

Endophytes may profoundly affect population dynamics and community structure of plants and their associated species (see e.g. 92). Endophytes, how-ever, may be important in natural systems because of their effects on host plant demography via alterations of seed and seedling survival, and competitive abil-ities of adult plants in varying environments, rather than through mediating interactions with herbivores.

ACKNOWLEDGMENTS

We thank K Craven, N Hill, J Rango, I Saloniemi, F Schulthess, M Siegel, and D Wilson for comments or access to unpublished data. This research was supported by National Science Foundation grants DEB 9406934 and 9727020 to SHF, by an anonymous donor to TJS, and by Finnish Academy grants 33351 and 33324 and the Maytag Postdoctoral Fellowship (ASU) to KS and MH.

Visit the *Annual Reviews home page* at
http://www.AnnualReviews.org

Literature Cited

1. Ahmad S, Govindarajan JM, Funk CR, Johnson-Cicalese JM. 1985. Fatality of house crickets on perennial ryegrasses infected with a fungal endophyte. *Entomol. Exp. Appl.* 39:183–90
2. Ahmad S, Johnson-Cicalese JM, Dickson WK, Funk CR. 1986. Endophyte-enhanced resistance in perennial ryegrass to the bluegrass bill-bug *Sphenophorus parvalus*. *Entomol. Exp. Appl.* 41:3–10
3. An Z-Q, Liu J-S, Siegel MR, Bunge G, Schardl CL. 1992. Diversity and origins of endophytic fungal symbionts of North American grass *Festuca arizonica*. *Theor. Appl. Genet.* 85:366–71
4. An Z-Q, Siegel MR, Hollin W, Tsai H-F, Schmidt D, Schardl CL. 1993. Relationship among non-*Acremonium* sp. fungal endophytes in five grass species. *Appl. Environ. Microbiol.* 59:1540–48
4a. Andrews JH, Hirano SS, ed. 1991. *Microbial Ecology of Leaves*. New York: Springer-Verlag
5. Antonovics J. 1994. The interplay of numerical and gene-frequency dynamics in host-pathogen systems. See Ref. 124a, pp. 129–45
6. Antonovics J, Thrall P, Jarosz A, Stratton D. 1994. Ecological genetics of metapopulations: the *Silene-Ustilago* plant-pathogen system. See Ref. 124a, pp. 146–70
7. Atstatt P. 1988. Are vascular plants 'inside-out' lichens? *Ecology* 69:17–23
8. Bacon CW. 1993. Abiotic stress tolerances (moisture, nutrients) and photosynthesis in endophyte-infected tall fescue. *Agric. Ecosyst. Environ.* 44:123–41
9. Bacon CW, Hill NS. 1996. Symptomless grass endophytes: products of co-evolutionary symbioses and their role in the ecological adaptations of infected grasses. See Ref. 126, pp. 155–78
10. Bacon CW, Porter JK, Robbins JD, Luttrell ES. 1977. *Epichloë typhina* from toxic tall fescue grasses. *Appl. Environ. Microbiol.* 34:576–81
11. Bailey V. 1903. Sleepy grass and its effects on horses. *Science* 17:392–93
12. Ball DM, Pedersen JF, Lacefield GD. 1993. The tall-fescue endophyte. *Am. Sci.* 81:370–79
13. Barker GM, Pottinger RP, Addison PJ, Prestidge RA. 1984. Effect of *Lolium* endophyte fungus infections on behaviour of adult Argentine stem weevil. *NZ J. Agric. Res.* 27:271–77
14. Bills GF. 1996. Isolation and analysis of endophytic fungal communities from woody plants. See Ref. 126, pp. 31–65
15. Boursnell JG. 1950. The symbiotic seed-borne fungus in the Cistaceae. I. Distribution and function of the fungus in the seedling and in the tissues of the mature plant. *Ann. Bot.* 14:217–42
16. Bose SR. 1947. Hereditary (seed-borne) symbiosis in *Casuarina equisetifolia*. *Nature* 159:512–14
17. Breen JP. 1993. Enhanced resistance to fall armyworm (Lepidoptera: Noctuidae) in *Acremonium* endophyte-infected turfgrasses. *J. Econ. Entomol.* 86:621–29
18. Breen JP. 1993. Enhanced resistance to three species of aphids (Homoptera: Aphididae) in *Acremonium* endophyte-infected turfgrasses. *J. Econ. Entomol.* 86:1279–86
19. Breen JP. 1994. *Acremonium* endophyte interactions with enhanced plant resistance to insects. *Annu. Rev. Entomol.* 39:401–23
20. Bucheli E, Leuchtmann A. 1996. Evidence for genetic differentiation between choke-inducing and asymp-

tomatic strains of the *Epichloë* grass endophyte from *Brachypodium sylvaticum*. *Evolution* 50:1879–87

21. Bull JJ, Molineux IJ, Rice WR. 1991. Selection of benevolence in a host-parasite system. *Evolution* 45:875–82

22. Bultman TL, Borowicz KL, Schneble RM, Coudron TA, Bush LP. 1997. Effect of a fungal endophyte on the growth and survival of two *Euplectrus* parasitoids. *Oikos* 78:170–76

23. Bultman TL, Ganey DT. 1995. Induced resistance to fall armyworm (Lepidoptera: Noctuidae) mediated by a fungal endophyte. *Environ. Entomol.* 24:1196–200

24. Bultman TL, Murphy JC. 1998. Do fungal endophytes mediate wound-induced resistance? In *The Evolution of Endophytism*, ed. JF White Jr. New York: Dekker. In press

25. Carroll GC. 1986. The biology of endophytism in plants with particular reference to woody perennials. See Ref. 61a, pp. 205–22

26. Carroll GC. 1988. Fungal endophytes in stems and leaves: from latent pathogen to mutualistic symbiont. *Ecology* 69:2–9

27. Carroll GC. 1991. Beyond pest deterrence. Alternative strategies and hidden costs of endophytic mutualisms in vascular plants. See Ref. 4a, pp. 358–75

28. Carroll GC. 1992. Fungal mutualism. In *The Fungal Community. Its Organization and Role in the Ecosystem*, ed. GC Carroll, DT Wicklow, pp. 327–54. New York: Dekker

29. Carroll GC, Carroll FE. 1978. Studies on the incidence of coniferous needle endophytes in the Pacific Northwest. *Can. J. Bot.* 56:3034–43

30. Cheplick GP, Clay K. 1988. Acquired chemical defenses of grasses: the role of fungal endophytes. *Oikos* 52:309–18

30a. Carroll GC, Tudzynski P, eds. 1997. *The Mycota.* V. *Plant Relationships, Part B.* Berlin: Springer-Verlag

31. Cheplick GP, Clay K, Marks S. 1989. Interactions between infection by endophytic fungi and nutrient limitation in the grasses *Lolium perenne* and *Festuca arundinacea*. *New Phytol.* 111:89–97

32. Chu-Chou M, Guo B, An Z-Q, Hendrix JW, Ferriss RS, et al. 1992. Suppression of mycorrhizal fungi in fescue by the *Acremonium coenophialum* endophyte. *Soil Biol. Biochem.* 24:633–37

33. Clay K. 1988. Fungal endophytes of grasses: a defensive mutualism between plants and fungi. *Ecology* 69:10–16

34. Clay K. 1990. Fungal endophytes of grasses. *Annu. Rev. Ecol. Syst.* 21:275–79

35. Clay K. 1992. Fungal endophytes of plants: biological and chemical diversity. *Nat. Toxins* 1:147–49

36. Clay K. 1996. Fungal endophytes, herbivores, and the structure of grassland communities. In *Multitrophic Interactions in Terrestrial Systems*, ed. AC Gange, pp. 151–69. Oxford, UK: Blackwell Sci.

37. Clay K. 1997. Consequences of endophyte-infected grasses on plant biodiversity. In *Neotyphodium/Grass Interactions*, ed. CW Bacon, NS Hill, pp. 109–24. New York: Plenum

38. Clay K, Brown VK. 1997. Infection of *Holcus lanatus* and *H. mollis* by *Epichloë* in experimental grasslands. *Oikos* 79:363–70

39. Clay K, Hardy TN, Hammond AM Jr. 1985. Fungal endophytes of grasses and their effects on an insect herbivore. *Oecologia* 66:1–5

40. Clay K, Hardy TN, Hammond AM Jr. 1985b. Fungal endophytes of *Cyperus* and their effect on an insect herbivore. *Am. J. Bot.* 72:1284–89

41. Clay K, Marks S, Cheplick GP. 1993. Effects of insect herbivory and fungal endophyte infection on competitive interactions among grasses. *Ecology* 74:1767–77

42. Clement SL, Lester DG, Wilson AD, Johnson RC, Bouton JH. 1996. Expression of Russian wheat aphid (Homoptera: Aphididae) resistance in genotypes of tall fescue harboring different isolates of *Acremonium* endophyte. *J. Econ. Entomol.* 89:766–70

43. Clement SL, Lester DG, Wilson AD, Pike KS. 1992. Behavior and performance of *Diuraphis noxia* (Homoptera: Aphididae) on fungal endophyte-infected and uninfected perennial ryegrass. *J. Econ. Entomol.* 85:583–88

44. Clement SL, Wilson AD, Lester DG, Davitt CM. 1997. Fungal endophytes of wild barley and their effects on *Diuraphis noxia* population development. *Entomol. Exp. Appl.* 82:275–81

45. Connor RC. 1995. The benefits of mutualism: a conceptual framework. *Biol. Rev.* 70:427–57

46. Crawley MJ. 1983. *Herbivory, the Dynamics of Animal-Plant Interactions.* Berkeley, CA: Univ. Calif. Press. 437 pp.

47. Dahlman DL, Eichenseer H, Siegel MR. 1991. Chemical perspectives of endophyte-grass interactions and their

implications to insect herbivory. In *Microbial Mediation of Plant-Herbivore Interactions*, ed. P Barbosa, VA Krischik, CL Jones, pp. 227–52. New York: Wiley

48. Davison AW, Potter DA. 1995. Responses of plant-feeding, predatory, and soil-inhabiting invertebrates to *Acremonium* endophyte and nitrogen fertilization in tall fescue turf. *J. Econ. Entomol.* 88:367–79

49. De Bary A. 1866. *Morphologie und Physiologie der Pilze, Flechten und Myxomyceten*. Leipzig, Germany: Engelmann

50. Dorworth CE, Callan BE. 1996. Manipulation of endophytic fungi to promote their utility as vegetation biocontrol agents. See Ref. 126, pp. 209–16

51. Douglas AE, Smith DC. 1983. The cost of symbionts to their host in green hydra. In *Endocytobiology II*, ed. HEA Schenk, W Schwemmler, pp. 631–48. Walter de Gruyer

52. Eichenseer H, Dahlman DL. 1992. Antibiotic and deterrent qualities of endophyte-infected tall fescue to two aphid species (Homoptera: Aphididae). *Environ. Entomol.* 21:1046–51

53. Elmi AA, West CP. 1995. Endophytic infection effects on stomatal conductance, osmotic adjustment and drought recovery of tall fescue. *New Phytol.* 131:61–67

54. Espinosa-Garcia FJ, Langenheim JH. 1990. The endophytic fungal community in leaves of a costal redwood population—diversity and spatial patterns. *New Phytol.* 116:89–97

55. Ewald PW. 1983. Host-parasite relations, vectors, and the evolution of disease severity. *Annu. Rev. Ecol. Syst.* 14:465–85

56. Ewald PW. 1987. Transmission modes and evolution of the parasitism-mutualism continuum. *Ann. NY Acad. Sci.* 503:295–306

57. Ewald PW. 1994. *Evolution of Infectious Disease*. Oxford, UK: Oxford. Univ. Press. 298 pp.

58. Faeth SH, Hammon KE. 1996. Fungal endophytes and phytochemistry of oak foliage: determinants of oviposition preference of leafminers? *Oecologia* 108:728–36

59. Faeth SH, Hammon KE. 1997. Fungal endophytes in oak trees. Long-term patterns of abundance and associations with leafminers. *Ecology* 78:810–19

60. Faeth SH, Hammon KE. 1997. Fungal endophytes in oak trees: experimental

analyses of interactions with leafminers. *Ecology* 78:820–27

61. Faeth SH, Wilson D. 1996. Induced responses in trees: mediators of interactions between macro and microherbivores. In *Multitrophic Interactions in Terrestrial Systems*, ed. AC Gange, pp. 201–15. Oxford, UK: Blackwell Sci.

61a. Fokkema NJ, van den Heuvel J, eds. 1986. *Microbiology of the Phyllosphere*. Cambridge, MA: Cambridge Univ. Press

62. Freeman EM. 1904. The seed fungus of *Lolium temulentum* L., the darnel. *Philos. Trans. R. Soc. London Ser. B.* 196:1–27

63. Freeman S, Rodriguez RJ. 1993. Genetic conversion of a fungal plant pathogen to a nonpathogenic, endophytic mutualist. *Science* 260:75–78

64. Futuyma DJ, Moreno G. 1988. The evolution of ecological specialization. *Annu. Rev. Ecol. Syst.* 19:207–33

65. Gange AC. 1996. Positive effects of endophyte infections on sycamore aphids. *Oikos* 75:500–10

66. Gaylord ES, Preszler RW, Boecklen WJ. 1996. Interactions between host plants, endophytic fungi, and a phytophagous insect in an oak (*Quercus grisea × Quercus gambelii*) hydrid zone. *Oecologia* 105:336–42

67. Gaynor DL, Hunt WF. 1983. The relationship between nitrogen supply, endophytic fungus, and Argentine stem weevil resistance in ryegrass. *Proc. NZ Grassland Assoc.* 44:257–63

68. Gehring CA, Cobb NS, Whitham TG. 1997. Three-way interactions among ectomycorrhizal mutualists, scale insects, and resistant and susceptible pinyon pines. *Am. Nat.* 149:824–41

69. Gehring CA, Whitham TG. 1994. Interactions between above ground herbivores and the mycorrhizal mutualists of plants. *Trends Ecol. Evol.* 9:251–55

70. Gilpin M, Hanski I. 1991. *Metapopulation Dynamics: Empirical and Theoretical Investigations*. New York: Academic

71. Hammon KE, Faeth SH. 1992. Ecology of plant-herbivore communities: a fungal component. *Nat. Toxins* 1:197–208

72. Hardy TN, Clay K, Hammond AM Jr. 1985. Fall armyworm (Lepidoptera: Noctuidae): a laboratory bioassay and larval preference study for the fungal endophyte of perennial ryegrass. *J. Econ. Entomol.* 78:571–75

73. Hardy TN, Clay K, Hammond AM Jr. 1986. Leaf age and related factors affecting endophyte-mediated resistance

to fall armyworm (Lepidoptera: Noctu-diae) in tall fescue. *Environ. Entomol.* 15:1083–89

74. Hatcher 1995. Three-way interactions between plant pathogenic fungi, herbivorous insects and their host plants. *Biol. Rev.* 70:639–94
75. Helander ML, Neuvonen S, Sieber T, Petrini O. 1993. Simulated acid rain affects birch leaf endophyte populations. *Microb. Ecol.* 26:227–34
76. Helander ML, Sieber TN, Petrini O, Neuvonen S. 1994. Endophytic fungi in Scots pine needles: spatial variation and consequences of simulated acid rain. *Can. J. Bot.* 72:1108–13
77. Hill NS, Belesky DP, Stringer WC. 1991. Competitiveness of tall fescue as influenced by *Acremonium coenophialum. Crop Sci.* 31:185–90
78. Howe HF, Westley LC. 1988. *Ecological Relationships of Plants and Animals.* Oxford, UK: Oxford Univ. Press. 273 pp.
79. Jarosz AM, Davelos AL. 1995. Effects of disease in wild plant populations and the evolution of pathogen aggressiveness. *New Phytol.* 129:371–87
80. Johnson JA, Whitney NJ. 1994. Cytotoxicity and insecticidal activity of endophytic fungi from black spruce (*Picea mariana*) needles. *Can. J. Microbiol.* 40:24–27
81. Johnson MC, Dahlman DL, Siegel MR, Bush LP, Latch GCM, et al. 1985. Insect feeding deterrents in endophyte-infected tall fescue. *Appl. Environ. Microbiol.* 49:568–71
82. Johnson-Cicalese JM, White RH. 1990. Effects of *Acremonium* endophytes on four species of billbug (Coleoptera: Curculionidae) found on New Jersey turfgrasses. *J. Am. Soc. Hortic. Sci.* 115:602–4
83. Kaiser WJ, Breuhl GW, Davitt CM, Klein RE. 1996. *Acremonium* isolates from *Stipa robusta. Mycologia* 88:539–47
84. Kimmons CA, Gwinn KD, Bernard EC. 1990. Nematode reproduction on endophyte-infected and endophyte-free tall fescue. *Plant Dis.* 74:757–61
85. Kindler SD, Breen JP, Springer TL. 1991. Reproduction and damage by Russian wheat aphid (Homoptera: Aphididae) as influenced by fungal endophytes and cool-season turfgrasses. *J. Econ. Entomol.* 84:685–92
85a. Knoch TR, Faeth SH, Arnott DL. 1993. Endophytic fungi alter foraging and dispersal by desert seed-harvesting ants. *Oecologia* 95:470–75

86. Koga H, Hirai Y, Kanda K, Tsukiboshi T, Uematsu T. 1997. Successive transmission of resistance to bluegrass webworm to perennial ryegrass and tall fescue plants by artificial inoculation with *Acremonium* endophytes. *Jpn. Agric. Res. Q.* 31:109–15
87. Kowalski T, Kehr RD. 1996. Fungal endophytes of living branch bases in several European tree species. See Ref. 126, pp. 67–86
88. Large EC. 1940. *The Advance of the Fungi,* Vol. 7. New York: Dover. 488 pp.
89. Latch GCM, Christensen MJ. 1985. Artificial infection of grasses with endophytes. *Ann. Appl. Biol.* 107:17–24
90. Latch GCM, Christensen MJ, Gaynor DL. 1985. Aphid detection of endophyte infection in tall fescue. *NZ J. Agric. Res.* 28:129–32
91. Leuchtmann A. 1992. Systematics, distribution, and host specificity of grass endophytes. *Nat. Toxins* 1:150–62
92. Leuchtmann A, Clay K. 1997. The population biology of grass endophytes. See Ref. 30a, pp. 185–204
93. Lewis DH. 1985. Symbiosis and mutualism: crisp concepts and soggy semantics. In *The Biology of Mutualisms,* ed. DH Boucher, pp. 29–39. London: Helm
94. Lewis GC, Clements RO. 1986. A survey of ryegrass endophyte (*Acremonium loliae*) in the U.K. and its apparent ineffectuality on a seedling pest. *J. Agric. Sci.* 107:633–38
95. Lewis GC, Ravel C, Naffaa W, Astier C, Charmet G. 1997. Occurrence of *Acremonium*-endophytes in wild populations of *Lolium* sp. in European contries and a relationship between level of infection and climate in France. *Ann. Appl. Biol.* 130:27–38
96. Lipsitch M, Siller S, Nowak MA. 1996. The evolution of virulence in pathogens with vertical and horizontal transmission. *Evolution* 50:1729–41
97. Lopez JE, Faeth SH, Miller M. 1995. The effect of endophytic fungi on herbivory by redlegged grasshoppers (Orthoptera: Acrididae) on Arizona fescue. *Environ. Entomol.* 24:1576–80
98. Luginbühl M, Müller E. 1980. Endophytische Pilze in den oberirdischen Organen von 4 gemeinsam an gleichen Standorten wachsenden Pflanzen (*Buxus, Hedera, Ilex, Ruscus*). *Sydowia* 33:185–86
99. Marks S, Clay K, Cheplick CP. 1991. Effects of fungal endophytes on interspecific and intraspecific competition in the

grasses *Festuca arundinacea* and *Lolium perenne*. *J. Appl. Ecol.* 28:194–204

100. Marquis RJ. 1992. The selective impact of herbivores. In *Plant Resistance to Herbivores and Pathogens. Ecology, Evolution, and Genetics*, ed. RS Fritz, EL Simms, pp. 301–25. Chicago: Univ. Chicago Press

101. Marsh CD, Clawson AB. 1929. Sleepy grass (*Stipa vaseyi*) as a stock-poisoning plant. *USDA Tech. Rep.* 114. Washington DC: USGPO. 19 pp.

102. Maschinki J, Whitham TG. 1989. The continuum of plant responses to herbivory. The influence of plant association, nutrient availability and timing. *Am. Nat.* 134:1–19

103. Massad E. 1987. Transmission rates and the evolution of pathogenicity. *Evolution* 41:1127–30

104. Mathias JK, Ratcliffe RH, Hellman JL. 1990. Association of an endophytic fungus in perennial ryegrass and resistance to the hairy chinch bug (Hemiptera: Lygaeidae). *J. Econ. Entomol.* 83:1640–46

105. Mauricio R. 1998. Cost of resistance to natural enemies in field populations of the annual plant *Arabidopsis thaliana*. *Am. Nat.* 151:20–28

106. McCutchcon TL, Carroll GC, Schwab S. 1993. Genotypic diversity of a fungal endophyte from Douglas fir. *Mycologia* 85:180–86

106a. Monk KA, Samuels GJ. 1990. Mycophagy in grasshoppers (Orthoptera: Acrididae) in Indo-Malayan rain forests. *Biotropica* 22:16–21

107. Murphy JA, Sun S, Betts LL. 1993. Endophyte-enhanced resistance to billbug (Coleoptera: Curculionidae), sod webworm (Lepidoptera: Pyralidae), and white grub (Coleoptera: Scarabeidae) in tall fescue. *Environ. Entomol.* 22:699–703

108. Deleted in proof

109. Parker MA. 1995. Plant fitness variation caused by different mutualist genotypes. *Ecology* 76:1525–35

110. Pedersen JF, Rodriguez-Kabana R, Shelby RA. 1988. Ryegrass cultivars and endophyte in tall fescue affect nematode in grass and succeeding soybean. *Agron. J.* 80:811–14.

111. Pellmyr O, Thompson JN, Brown JM, Harrison RG. 1996. Evolution of pollination in the yucca moth lineage. *Am. Nat.* 148:827–47

112. Petrini, O. 1986. Taxonomy of endophytic fungi of aerial plant tissues. See Ref. 61a, pp. 175–87

113. Petrini O. 1991. Fungal endophytes of tree leaves. See Ref. 4a, pp. 179–97

114. Petrini O, Carroll G. 1981. Endophytic fungi in foliage of some Cupressaceae in Oregon. *Can. J. Bot.* 59:629–36

115. Petrini O, Müller E. 1979. Pilzliche Endophyten, am Beispiel von *Juniperus communis* L. *Sydowia* 32:224–51

116. Petrini O, Sieber TH, Toti L, Viret O. 1992. Ecology, metabolite production, and substrate utilization in endophytic fungi. *Nat. Toxins* 1:185–96

117. Petrini O, Stone J, Carroll FE. 1982. Endophytic fungi in evergreen shrubs in western Oregon: a preliminary study. *Can. J. Bot.* 60:789–96

118. Petroski RJ, Powell RG, Clay K. 1992. Alkaloids of *Stipa robusta* (Sleepygrass) infected with an *Acremonium* endophyte. *Nat. Toxins* 1:84–88

119. Potter DA, Patterson CG, Redmond CT. 1992. Influence of turfgrass species and tall fescue endophyte on feeding ecology of Japanese beetle and southern masked chafer grubs (Coleoptera: Scarabaeidae). *J. Econ. Entomol.* 85:900–9

120. Prestidge RA, Gallagher RT. 1988. Endophyte fungus confers resistance to rye grass: Argentine stem weevil larval studies. *Ecol. Entomol.* 13:429–35

121. Prestidge RA, Pottinger RP, Barker GM. 1982. An association of *Lolium* endophyte with ryegrass resistance to Argentine stem weevil. *Proc. 35th NZ Weed Pest Control Conf.* 35:119–22

122. Preszler RW, Gaylord ES, Boecklen WJ. 1996. Reduced parasitism of a leaf-mining moth on trees with high infection frequencies of an endophytic fungus. *Oecologia* 108:159–66

123. Rausher MD. 1992. Natural selection and the evolution of plant-insect interactions. In *Insect Chemical Ecology: an Evolutionary Approach*, ed. RD Roitberg, MB Isman, pp. 20–88. New York: Chapman & Hall

124. Ravel C, Michalakis Y, Charmet G. 1997. The effect of imperfect transmission on the frequency of mutualistic seed-borne endophytes in natural populations of grasses. *Oikos* 80:18–24

124a. Real RA, ed. 1994. *Ecological Genetics.* Princeton, NJ: Princeton Univ. Press

125. Reddick BB, Collins MH. 1988. An improved method for detection of *Acremonium coenophialum* in tall fescue plants. *Phytopathology* 78:418–20

126. Redlin SC, Carris LM. 1996. *Endophytic Fungi in Grasses and Woody Plants. Systematics, Ecology and Evolution.* St. Paul: APS. 216 pp.

127. Rollinger JL, Langenheim JH. 1993. Geographic survey of fungal endophyte community composition in leaves of coastal redwood. *Mycologia* 85:149–56

128. Saikkonen K, Helander M, Faeth SH, Schulthess FM. *Neotyphodium* endophytes in native grass populations: against herbivore-driven mutualism. Submitted

129. Saikkonen K, Helander M, Ranta H, Neuvonen S, Virtanen T, et al. 1996. Endophyte-mediated interactions between woody plants and insect herbivores? *Entomol. Exp. Appl.* 80:269–71

130. Schardl CL, Clay K. 1997. Evolution of mutualistic endophytes from plant pathogens. See Ref. 30a, pp. 221–38

131. Schardl CL, Liu J-S, White JF, Finkel RA, An Z, Siegel MR. 1991. Molecular phylogenetic relationships of nonpathogenic grass mycosymbionts and clavicipitaceous plant pathogens. *Plant Syst. Evol.* 178:27–41

132. Schardl CL, Leuchtmann A, Chung K-R, Penny D, Siegel MR. 1997. Coevolution by common descent of fungal symbionts (*Epichloë* spp.) and grass hosts. *Mol. Biol. Evol.* 14:133–43

132a. Schardl CL, Leuchtmann A, Tsai H-F, Collett MA, Watt DM, Scott DB. 1994. Origin of a fungal symbiont of perennial ryegrass by interspecific hybridization of a mutualist with the ryegrass choke pathogen, *Epichloë typhina. Genetics* 136:1307–17

133. Schulthess FM. 1997. *Endophytes of Festuca arizonica: Distribution and effects on the host.* MS thesis. Ariz. State Univ., Tempe. 33 pp.

134. Schulthess FM, Faeth SH. 1998. Distribution, abundances and associations of the endophytic fungal community of Arizona fescue (*Festuca arizonica* Vasey). *Mycologia* 90:569–78

135. Siegel MR, Bush LP. 1996. Defensive chemicals in grass-fungal endophyte associations. *Recent Adv. Phytochem.* 30:81–118

136. Siegel MR, Bush LP. 1997. Toxin production in grass/endophyte association. See Ref. 30a, pp. 185–208

137. Siegel MR, Johnson MC, Varney DR, Nesmith WC, Buckner RC, et al. 1984. A fungal endophyte in tall fescue: incidence and dissemination. *Phytopathology* 74:932–37

138. Siegel MR, Latch GCM, Bush LP, Fannin FF, Rowan DD, et al. 1990. Fungal endophyte-infected grasses: alkaloid accumulation and aphid response. *J. Chem. Ecol.* 16:3301–15

139. Sinclair JB, Cerkauskas RF. 1996. Latent infection vs. endophytic colonization by fungi. See Ref. 126, pp. 3–29

140. Stone JK, Petrini O. 1997. Endophytes of forest trees: a model for fungus-plant interactions. See Ref. 30a, pp. 129–42

141. Strong DR, Lawton JH, Southwood R. 1984. *Insects on Plants. Community Patterns and Mechanisms.* Oxford, UK: Blackwell Sci. 313 pp.

142. Thompson JN. 1982. *Interaction and Coevolution.* New York: Wiley. 179 pp.

143. Thompson JN. 1994. *The Coevolutionary Process.* Chicago: Univ. Chicago Press. 376 pp.

144. Thompson JN, Pellmyr O. 1992. Multiple occurrences of mutualism in the yucca moth lineage. *Proc. Natl. Acad. Sci.* 89:2927–29

145. Todd D. 1988. The effects of host genotype, growth rate, and needle age on the distribution of a mutualistic, endophytic fungus in Douglas-fir plantations. *Can. J. For. Res.* 18:601–5

145a. Tsai H-F, Liu J-S, Staben C, Christensen MJ, Latch GCM, et al. 1994. Evolutionary diversification of fungal endophytes of tall fescue grass by hybridization with *Epichloë* species. *Proc. Natl. Acad. Sci. USA* 91:2542–46

146. Welty RE, Craig, AM, Azevedo, MD. 1994. Variability of ergovaline in seeds and straw and endophyte infection in seeds among endophyte-infected genotypes of tall fescue. *Plant Dis.* 78:845–49

147. Wennström A. 1994. Endophyte—the misuse of an old term. *Oikos* 71:535–36

148. West CP, Izekor E, Oosterhuis DM, Robbins RT. 1988. The effect of *Acremonium coenophialum* on the growth and nematode infestation of tall fescue. *Plant Soil* 112:3–6

149. White JF Jr, Morgan-Jones G, Morrow AC. 1993. Taxonomy, life cycle, reproduction and detection of *Acremonium* endophytes. *Agric. Ecosyst. Environ.* 44:13–37

150. Widler B, Müller, E. 1984. Untersuchungen über endophytische Pilze von *Arctostaphylos uva-ursi* (L.) Sprengel (Ericaceae). *Bot. Helv.* 94:307–17

150a. Wilkinson HH, Schardl CL. 1997. The evolution of mutualism in grass-endophyte associations. In *Neotyphodium/Grass Interactions*, ed. CW Bacon, NS Hill, pp. 13–25. New York: Plenum

151. Wilson D. 1993. Fungal endophytes: out of sight but should not be out of mind. *Oikos* 68:379–84

152. Wilson D. 1995. Endophyte—the evolution of a term, and clarification of its use and definition. *Oikos* 73:274–76

153. Wilson D. 1995. Fungal endophytes which invade insect galls: insect pathogens, benign saprophytes, or fungal inquilines? *Oecologia* 103:255–60

154. Wilson D. 1998. Fungal endophyte-insect interactions: endophyte-driven premature abscission of insect-damaged leaves. *Ecology*. In press

155. Wilson D, Carroll GC. 1994. Infection studies of *Discula quercina*, and endophyte of *Quercus garryana. Mycologia* 86:635–47

156. Wilson D, Carroll GC. 1997. Avoidance of high-endophyte space by gall-forming insects. *Ecology* 78:2153–63

157. Wilson D, Faeth S. 1998. Do fungal endophytes result in selection for leafminer ovipositional preference? *Ecology*. In press

158. Wolock-Madej C, Clay K. 1991. Avian seed preference and weight loss experiment: the role of fungal-infected fescue seeds. *Oecologia* 88:296–302

Annu. Rev. Ecol. Syst. 1998. 29:345–73

FLORAL SYMMETRY AND ITS ROLE IN PLANT-POLLINATOR SYSTEMS: Terminology, Distribution, and Hypotheses

Paul R. Neal

Department of Biology, University of Miami, Coral Gables, Florida 33124;
e-mail: pneal@fig.cox.miami.edu

Amots Dafni

Institute of Evolution, University of Haifa, Haifa 31905, Israel;
e-mail: adafni@research.haifa.ac.il

Martin Giurfa

Institute for Neurobiology, Free University of Berlin, Königin-Luise-Str.
28/30, 14195 Berlin, Germany; e-mail: giurfa@neuro.biologie.fu-berlin.de

KEY WORDS: actinomorphy, bilateral, flower, radial, zygomorphy

ABSTRACT

Floral symmetry has figured prominently in the study of both pollination biology and animal behavior. However, a confusion of terminology and the diffuse nature of the literature has limited our understanding of the role that this basic characteristic of flower form has played in plant-pollinator interactions. Here, we first contribute a classification scheme for floral symmetry that we hope will resolve some of the confusion resulting from the inconsistent application of terms. Next, we present a short review of the distribution of floral forms in angiosperm families. Finally, we provide a list of hypotheses and, when available, supporting evidence for the causes of the evolution of floral symmetry.

345

INTRODUCTION

Floral symmetry played an important role in Sprengel's (110) pioneering attempt to relate form to function in the pollination of flowering plants. He recognized two types of floral symmetry, regular and irregular. Regular flowers were those in which the pistil(s), stamens, and segments of the perianth radiated out uniformly from the central axis. Furthermore, all segments of each organ type were equal in size and form. Irregular flowers were those in which any parts of the perianth or sexual organs did not meet these criteria. Sprengel suggested that regularity should be the rule unless circumstances resulted in an advantage to irregularity. He went on to propose several hypotheses concerning circumstances that would result in such an advantage.

Several forms of irregular flowers [i.e. nonradially symmetric or nonactinomorphic in modern terminology (see below and Table 1)] have been recognized. The modern phylogenetic approach, although philosophically far from Sprengel's teleological and creationistic outlook (121), suggests that most of the species with irregular flowers were derived from species with regular flowers (i.e. radially symmetric or actinomorphic) (16, 112, 114). We are still looking for the circumstances that give the advantage to irregularity.

Study of pollination ecology has played an important role in investigations of floral symmetry. In particular, honeybees have been the subject of many studies related to symmetry preferences (69) and perception (39, 59). Honeybees can be trained to respond to various visual stimuli, allowing investigation of the perceptual and processing mechanisms associated with the responses to these stimuli.

As a result of a long history and interest by biologists in a variety of fields, the literature regarding floral symmetry is widely scattered in publications about botany, ecology, animal physiology, and behavior. Unfortunately, the diffuse nature of the literature and diversity of interests of the investigators have prevented an integrated understanding of floral symmetry. The problem is further exacerbated by inconsistent use of terminology regarding symmetry. We point out and reconcile some of the problems relating to the description and classification of floral symmetry. We summarize the hypotheses that have been used to explain evolutionary changes in symmetry and suggest some directions for new research in floral symmetry.

PLANES OF SYMMETRY AND FLORAL PHYLOGENY

Floral symmetry is the repeated pattern in structural units as assessed in relation to the principal axis of the flower (i.e. the line or vector emanating from the center of the receptacle) (129). Accordingly, we address symmetry of the flower

Table 1 Classification of floral symmetry types[a]

Planes of symmetry	Symmetry type[b]	Symmetrical images (principal axis of flower relative to subtending leaf)	Synonyms[c]	Examples
Usually more than two (polysymmetric)	Actinomorphy or radial symmetry	Rotational	Regular, pleomorphy, multisymmetry, stereomorphy	*Primula* (Primulaceae), *Narcissus* (Amaryllidaceae), *Pyrola* (Ericaceae)
Two (disymmetric)	Disymmetry	Reflectional in two perpendicular orientations	Bisymmetry, bilateral	*Dicentra* (Fumariaceae)
One (monosymmetric)	Zygomorphy	Reflectional	Bilabiate, bilateral, irregular, ligulate, medial zygomorphy	
	Medial zygomorphy or bilateral symmetry Transverse zygomorphy	Right-left reflectional Upper-lower reflectional (see text on morphogenesis vs functional position)	Bilabiate, equilateral, medial zygomorphy	*Salvia* (Lamiaceae), *Orchis* (Orchidaceae), *Scrophularia* (Scrophulariaceae) *Fumaria* and *Corydalis* (Fumariaceae)
	Diagonal zygomorphy	Right-left reflectional slightly off vertical	Oblique zygomorphy	*Aesculus* (Hippocastanaceae), found in Malpighiaceae, Sapindaceae, Trigoniaceae, Vochysiaceae
None (asymmetric)	Ancestral asymmetry or haplomorphy[d]	Translational; floral organs arranged in spirals rather than whorls	Actinomorphy, radial, regular	*Magnolia* (Magnoliaceae), *Nymphaea* (Nymphaeaceae)
	Derived asymmetry neo-asymmetry	None	Irregular, asymmetry Irregular, asymmetry	*Centranthus* (Valerianaceae), found in Cannaceae, Fabaceae, Marantaceae, Zingiberaceae
	enantiomorphy	Right-left reflectional perianth but with right- and left-styled morphs (stamens taking position opposite the style)	Enantiostyly, inequilateral	*Cassia* (Caesalpinaceae), *Cyanella* (Tecophilaeaceae), *Monochoria* (Pontederiaceae), *Solanum* (Solanaceae), *Barberetta* and *Wachendorffia* (Haemodoraceae)
	mono-enantiomorphy	Both style-stamen morphs on one plant		
	di-enantiomorphy	Style-stamen morphs on separate plants		

[a]Modified from References 32, 73, 102, 123, and 129.
[b]Symmetry based on overall pattern of flower, not solely on perianth; see text for discussion.
[c]Terms listed may be full or partial synonyms; many have been ambiguously applied.
[d]Although organs radiate from the principal axis, the spiral arrangement of organs results in no repetition of pattern for the flower as a whole.

only en face, with a two-dimensional perspective; the third dimension of depth (as in the tubular portion of the perianth of some flowers) is not considered. Repetition of pattern of floral structural units can be obtained through three symmetry operations: rotation, reflection, and translation (129). Symmetry arising by a translational operation, which is the repetition along a straight line (e.g. successive whorls of floral organs) along the principal axis of the flower (i.e. depth), has received little or no attention in regard to pollination biology and is not considered here. Symmetry arising by a rotational operation occurs when a pattern is repeated as a plane turns about the principal axis and results in two or more repetitions of the pattern over 360°. Symmetry arising by a reflectional operation occurs when a plane of symmetry through the principal axis produces a pattern of two mirror images.

In general, the primitive state of floral organs (e.g. petals or stamens) is (a) a spiral (helical) arrangement of organ members, (b) an indefinite number of each floral organ (e.g. many petals or stamens), and (c) similar morphology of all members of each organ type (30, 31, 114, 117). The derived state is (a) a whorled arrangement of floral organs, (b) a definite number of each floral organ (often multiples of three or five), and (c) dissimilar morphology among the members of an organ type (e.g. banner, keel, and wings of papilonaceous flowers). As a result of a spiral arrangement of floral organs (but often also of the overlapping arrangement of an indefinite number of members) many primitive flowers (e.g. many Magnoliaceae and Nymphaceae) are asymmetrical (i.e. there is no repetition of pattern for the flower as a whole; see Table 1). The whorled arrangement of floral organ members and the reduction in organ members creates rotationally arising symmetry (e.g. many Ranunculaceae, Liliaceae) with polysymmetry (i.e. two to many planes of symmetry). Monosymmetry results when the organs on the two sides of one plane develop differentially through reflectional operation (e.g. Scrophulariaceae, Lamiaceae). The term disymmetry is reserved for the case of two planes of reflectionally derived symmetry (e.g. *Dicentra*: Fumariaceae). Asymmetry has also been secondarily derived (e.g. *Centranthus*: Valerianaceae) (32, 49).

PROBLEMS IN THE TERMINOLOGY OF FLORAL SYMMETRY

Synonyms

Some confusion has been caused by the use of synonyms for symmetry types and, in some cases, the application of the same name to more than one form of floral symmetry (Table 1). Perhaps the biggest source of confusion has resulted from Leppik's (72–75) use of the terms actinomorphy and zygomorphy to describe overall floral form (i.e. three-dimensional shape) but radial and bilateral to describe floral symmetry. Most researchers have used the term radial symmetry as a synonym of actinomorphy, and the term bilateral symmetry as a synonym of zygomorphy. However, according to Leppik's classification scheme, not all radially symmetric flowers are actinomorphic, and bilateral flowers may or may not be zygomorphic. Leppik (72–75) also used similar terms to describe structures other than flowers (for example, paleomorphy for primitive nonangiosperm fossil forms and amorphy for flowers arranged in clusters (e.g. catkins in *Salix* and capitula in Asteraceae). In Table 1, we present a classification of floral symmetry aimed to remove much of the present ambiguity. In accordance with common usage in the literature, we equate radial symmetry with actinomorphy. Similarly, we equate bilateral symmetry with the most common type of zygomorphy (i.e. medial zygomorphy). We favor the term medial zygomorphy over dorsiventral zygomorphy because the latter implies a vertical en face orientation of the flower; the term is thus inappropriate for species that are oriented horizontally (e.g. most Apiaceae, Asteraceae, Dipsacaceae).

Orientation of Symmetry Planes

A second problem of terminology is establishing a point of reference for orienting symmetry planes in flowers with different types of symmetry. Ideally, the method of orientation should meet three criteria. First, the same method should be used for describing the symmetry planes for all flower types. Second, the description should not depend on the orientation of the flower in space, yet, third, it should be possible to relate the orientation of the planes of symmetry to the orientation of the flower. These criteria would allow the developmental or morphological aspects of symmetry (i.e. orientation in relation to growth of the plant) to be evaluated separately from the ecological or functional aspects of symmetry (i.e. orientation in the pollination process). In practice, none of these three criteria is met.

In an attempt to give orientation to floral diagrams, some authors (81, 129) define the median (or central) plane of the flower as the plane that passes through the principal axis (i.e. the line or vector emanating from the center of the receptacle) and the subtending leaf. Actinomorphic and disymmetric forms are characterized by most authors in relation to the principal axis of the flower. Symmetry is thus independent of flower orientation but could be related to it by indicating whether the flower is erect, pendulous, or obliquely or upwardly inclined. However, orientations of planes of symmetry are not discussed in relation to the median plane. This omission is not generally a problem in actinomorphic flowers because the orientation of the planes may have neither developmental nor ecological significance because of the large number of planes.

In most zygomorphic and most enantiomorphic forms, developmental and ecological planes of symmetry are coincident. This coincidence occurs because the én face surface of most zygomorphic flowers is oriented vertically or obliquely (inclining or declining). The plane of symmetry is, therefore, often designated as vertical (resulting in right and left mirror images). However, there are some exceptions. For example, in most, if not all, transversely zygomorphic species, the symmetry is vertically oriented at anthesis, even though the developmental plane of symmetry is transverse to the median axis. The vertical orientation of the plane of symmetry—the ecologically important plane—is obtained in these species by rotation of the pedicel shortly before the flower opens. Thus, in transverse zygomorphy the plane of symmetry in the mature flower is vertical as in medially zygomorphic flowers, not horizontal as might be expected from the name.

Whole Versus Parts of Flowers

A third problem of terminology arises when not all organs of the flower exhibit the same pattern of symmetry. When symmetry differs among organs, designation of symmetry is usually based on the form of the corolla (112). Differences may be trivial from the ecological perspective or, at least, in terms of pollination. For example, in many species of *Verbascum* the corolla is zygomorphic (although often weakly so), but the calyx is actinomorphic. It seems unlikely that the form of calyx would affect the pollination process in these species (114). Some flowers that appear to be actinomorphic may not actually be so in the strict sense. For example, each whorl of floral organs (petals, stamens, ovary locules) may be radially arranged but contains different numbers of member components (e.g. five petals and three stigmas in the Polemoniaceae). This inequality might have ecological consequences if pollinators align themselves according to the petals and contact with the stigmas is asymmetrical on the body of the pollinator. In other cases, the inequality obviously has important effects on the pollination process. For example, in many species of *Hibiscus* the corolla is radially symmetrical, but the style and anthers are upwardly curved, which results in the sternotribic deposition of pollen (i.e. on the ventral surface of the vector). Most authorities classify such flowers as radially symmetrical (i.e. actinomorphic) based on the form of the corolla. A similar situation arises in most enantiomorphic species (see Table 1), which are generally classified as bilaterally symmetrical (i.e. medially zygomorphic) based on the form of their corollas despite the asymmetric nature of the pistil and stamens.

Individual organs may be asymmetric but be arranged symmetrically. For example, in a condition that has been called pseudo-actinomorphy (116), individual petals of many Apocynaceae are asymmetrical but the flower as a whole appears to be actinomorphic. Similarly, some radially symmetrical flowers consist of units that function like bilaterally symmetrical flowers (e.g. *Iris*, *Moraea*). Focus on the symmetry of the corolla ignores other aspects of floral symmetry with important ecological consequences. Terminology must clearly indicate the organs to which the description of symmetry is applied. Published reports characterize symmetry by the corolla only, by the perianth (corolla and calyx taken together), or by the flower as a whole. Terminology is applied inconsistently even within some publications. We suggest that symmetry designations be applied to the pattern of the flower as a whole unless otherwise indicated. This usage would be consistent with Sprengel's usage in which he considered all forms of irregularity within a flower. We distinguish two types of regular forms based on the repetition of rotational patterns over all floral organs of the flower (i.e. planes of symmetry through sepals, petals, stamens, and carpels). Actinomorphic forms exhibit repeating patterns as a plane rotates the principal axis of the flower, while haplomorphic forms do not.

FREQUENCY OF FLORAL SYMMETRY TYPES

To assess the frequency of floral symmetry types, we consulted several taxonomic and morphological references (55, 56, 68, 98, 104, 105). Where symmetry designations differed within a family, the differences could usually be resolved by taking into account usage of terminology and taxonomic classification. We excluded from the survey those families with minute flowers that lacked a perianth because most references do not give a symmetry designation for these families, and most of the excluded families are thought to be wind pollinated. A well-developed perianth, or at least one that has not been extremely reduced, is generally thought to be an adaptation to anthophily (33, 100, 110). We consider symmetry at the level of the entire flower, not just the corolla or perianth, so we consider a flower with an actinomorphic corolla but medially zygomorphic pistil and stamens (e.g. *Hibiscus schizopetalus*: Malvaceae, *Adansonia digitata*: Bombacaceae, *Gloriosa superba*: Liliaceae) to be medially zygomorphic.

The survey resulted in symmetry designations for a total of 241 families (212 dicot, 29 monocot). Actinomorphy and medial zygomorphy were the most common symmetry types (Table 2). Actinomorphy was found in 83% of dicot and 72.4% of monocot families, while medial zygomorphy was found in 33% of dicot and 44.8% of monocot families. The more highly derived forms of symmetry (i.e. disymmetry, transverse and diagonal zygomorphy, and derived asymmetry) were found in only 7.1% of dicot and 13.8% of monocot families. Similarly, ancestral asymmetry is uncommon—4.7% of dicots and

Table 2   Frequency of symmetry types across angiosperm families[a]

Symmetry type	Number of families		
	All families	Dicots	Monocots
Actinomorphy	197	176	21
Disymmetry	1	1	0
Zygomorphy			
Medial	83	70	13
Transverse	1	1	0
Diagonal	5	5	0
Asymmetry			
ancestral	11	10	1
neo-asymmetry	4	3	1
mono-enantiomorphy	7	5	2
di-enantiomorphy	1	0	1
Total[b]	309	271	39

[a]Compiled from References 3, 55, 56, 68, 98, 104, and 105.
[b]212 dicots and 29 monocots were examined for symmetry type. Totals exceed these values because many families possess more than one symmetry type.

3.4% of monocots. Of families with actinomorphy, 54.7% of dicots and 44% of monocots were exclusively actinomorphic, while only 5.7% of dicots and 3.4% of monocots were exclusively medially zygomorphic (Table 3).

MOLECULAR GENETICS AND THE DEVELOPMENT OF FLORAL SYMMETRY

The development of zygomorphy occurs during different phases of floral ontogeny in various species (12, 117). This supports the suggestion of multiple, independent origins of monosymmetry (114). Species with late development of zygomorphy are usually found in taxonomic groups that are otherwise predominantly actinomorphic (32). In some flowers with late-developing zygomorphy, gravity appears to facilitate, directly or indirectly, the ontogeny of bilateral symmetry (32, 119, 129). Coen (12) suggested that a gravimetrically controlled system may have become coupled to internal cues then eventually evolved into the more typical, genetically controlled system with an early development of zygomorphy.

Most research on the ontogeny of non-actinomorphic floral symmetry has focused on *Antirrhinum majus* (Scrophulariaceae) and is based on the ABC model of organ identity. The ABC model postulates three overlapping regions of gene function, with each region affecting two adjacent whorls of floral organs

Table 3 Consistency of symmetry types across families[a]

Symmetry type	Number of families		
	All families	Dicots	Monocots
Ancestrally asymmetric			
Exclusively	10	9	1
Primarily	1	1	0
Actinomorphic			
Exclusively	129	116	13
Primarily	46	40	6
Equally actinomorphic and medially zygomorphic	14	12	2
Medially zygomorphic			
Exclusively	13	12	1
Primarily	28	22	6
Total	241	212	29

[a]Compiled from References 3, 55, 56, 68, 98, 104, and 105.

(e.g. petals and stamens, or stamens and pistil) (13, 133). The polar coordinate model for zygomorphy builds on the ABC model to hypothesize that another gene(s) varies in function in the upper and lower halves of the flower, generally with a gradient of increasing functional effect through the vertical axis of the flower (7, 12, 79). There is, thus, a unique polar coordinate specified for each floral organ and the result is reflectional symmetry. This model explains several phenomena of floral morphology and/or function observed in *A. majus* and other species. For example, the vestigial uppermost stamen in *A. majus* and other species in the Scrophulariaceae may result because it lies further up along the vertical axis than do the other stamens.

Position of flowers in the inflorescence also appears to play an important role in floral symmetry (12). The inflorescence of *A. majus* has indeterminant growth; however, there is a recessive allele that results in mutants with terminal flowers (62). These flowers have radial symmetry and all organs resemble those in the lower half of the typical zygomorphic flower, so they are peloric mutants. There are, however, other mutants with peloric flowers in axillary positions (12). The asymmetric environment of the axillary floral meristem of the inflorescence may be necessary for action of genes controlling zygomorphic floral symmetry (14, 79). That zygomorphy is a derived condition is suggested by the fact that there are many mutations producing actinomorphic flowers in normally zygomorphic species, but few mutations in the reverse direction (21, 136). Certainly, the evolution of zygomorphy and the evolution of the inflorescence may be intimately related (11).

SYMMETRY, ADAPTIVE SUITES, AND THE SYNDROME CONCEPT

The pollination syndrome concept holds that suites of floral characteristics such as corolla morphology and color, and reward quantity and quality cluster in phenotypic space and are associated with broad taxonomic groups of pollinators (33, 112). Pollination syndromes are generally designated by pollinator type. Monosymmetrical floral symmetry, particularly medial zygomorphy, has been associated with melittophily (pollination by bees) and ornithophily (pollination by birds) (33, 100, 103, 112).

However, neither melittophily nor ornithophily is a single phenomenon. For example, different suites of floral characteristics have been assigned to different groups of bees (17, 33, 120). Not all of these suites include zygomorphic floral symmetry. Moreover, many pollinators are often seen to visit many types of flowers, and flowers are often visited by many types of pollinators (108). Stebbins (113) suggested that the characteristics of flowers might be molded by the most common and most effective pollinators. More recently, Herrera (54) suggested that ecological factors may constrain both the occurrence of and the response to selection by pollinators on floral characteristics. Thus, the syndrome concept has received much criticism (54, 127).

It may be premature, however, to discard all aspects of the pollination syndrome. For example, monosymmetrical floral symmetry, particularly medial zygomorphy, has originated multiple times from actinomorphy (113). Furthermore, monosymmetry is often associated with particular states of other characteristics (e.g. herbaceous habit, increases in the number of ovules, reduction in the number of stamens, and sympetaly) (33, 100, 103, 112). Thus, floral characteristics may in fact cluster in phenotypic space. If so, we should not rule out the possibility that there are adaptive benefits to suites of floral characteristics that are independent of phylogeny. For example, Chittka (10, 127) found that the 154 plant species found in a nature reserve near Berlin can be grouped into distinguishable color clusters or color categories. Moreover, each of these categories seems to have a distinctive level of nectar reward (41). However, clustering of floral character states using a range of characters that are commonly employed to typify syndromes has not been studied. Whether suites cluster in phenotype space needs to be tested using modern phylogenetic approaches (1, 48).

HYPOTHESES OF FLORAL SYMMETRY

The hypotheses presented below vary in their approach to evolutionary trajectories, with some implying coevolution in the narrow sense [i.e. a series of reciprocal evolutionary changes with each change in one species caused by a change in the other (36)] and others suggesting that one member of the pollination system (i.e. flower or pollinator) adapts to the other. The plant may be seen to manipulate pollinators, or the plants may adapt to pollinator behavior or morphology. However, recent phylogenetically based studies of insect mouth parts (63) and photoreceptor types (8) suggest that plants have adapted to insect pollinators. Large changes in insect morphology as adaptations or compensations to floral morphology might be prohibited because they would interfere with flight efficiency, whereas flowers are not constrained in this way.

Most hypotheses relating to floral symmetry address only the evolution of bilateral (i.e. medial zygomorphy) from radial (i.e. actinomorphic) forms. The citations given here are not necessarily the first expression of the hypothesis but were chosen for their clear presentation, frequent citation, or supporting data. Some hypotheses were intended by their authors to explain the evolution of medial zygomorphy under specific conditions, while others were intended to be all-inclusive. There is much overlap among hypotheses; some are mutually exclusive, but others are not. The same or similar evidence has been used in support of opposing hypotheses. Some hypotheses are based on proximate factors (e.g. behavior patterns of pollen vectors) that may ultimately affect reproductive success (e.g. increased efficiency) Other hypotheses begin with the ultimate advantages (e.g. increased outcrossing is advantageous) and work toward proximate factors (e.g. pollen placement on the vector). Despite this, hypotheses can be placed into four operational groups (with significant overlap) based on the stage of the pollination process at which selection acts on floral symmetry: (a) environmental conditions, (b) perception by the pollinators, (c) information processing (i.e. learning abilities and innate preferences) by the pollinators, and (d) activity patterns (i.e. behavior and movement) of the pollinators.

Environmental Conditions

PROTECTION FROM RAIN HYPOTHESIS   As originally proposed by Sprengel (110), the upper lip of horizontally positioned flowers functions as a "nectar cover," protecting the nectar from rain. The lower lip, which does not serve this function, differs in shape. Sprengel suggested that the upper lip would be similar in form to the corolla of a pendulous flower, while the lower lip would resemble that of an erect flower. It has also been proposed that protection of the pollen from rain may be the driving force in the evolution of medial zygomorphic symmetry (118). We know of no systematic research on the relationships among symmetry, covers over nectar and/or pollen, dilution of nectar by rain, and effects of rain (water) on pollen viability and presentation.

Perception by the Pollinators

OPTICAL INFORMATION HYPOTHESIS   This hypothesis suggests that, according to information theory, bilateral symmetry should "give much greater possibilities for the transmission of visually mediated information than radial symmetry" (23, p. 241). This is based on the fact that fewer signals are required to transmit the information pertaining to an actinomorphic shape than to a zygomorphic one. For instance, in the former, information about only one petal would be information about the radial image, while for the latter at least half the pattern would be necessary. Furthermore, "the *difference* [italics in the original] between the two floral symmetries in ability to transmit information becomes even greater when they are in real or apparent motion" because motion increases the information in proportion to the original information content (23, p. 250). Davenport & Kohanzadeh (23) suspected that with more investigation, some correlation will be found between image complexity and pollinator specificity. However, such an analysis was discredited (24) because the nature of the processing of visual information in the pollinators was not taken into account.

Davenport & Lee (25) also proposed that increased complexity of the floral image in zygomorphic flowers has resulted in a greater opportunity for floral diversity (i.e. the more elements in an image, the more that image can vary). They argued that a greater diversity would allow recognition to be more specific: Zygomorphic flowers would increase discrimination possibilities of pollinators, resulting in greater foraging success. This in turn would favor increased information complexity of zygomorphy. However, this approach ignores that pollinators, including honeybees (111, 122, 130) as well as generalize features among patterns (127). Nevertheless, Davenport & Lee (25) were the first to use two-dimensional fast Fourier transform and convolution procedures of images to quantify flower patterns. This approach allows precise quantification of parameters that pollinators may use in evaluating a pattern [e.g. the total energy change across the different orientations of a pattern (39)] and provides an accurate tool for classifying flowers in different pattern dimensions.

UNEQUAL IMAGE PROJECTION HYPOTHESIS   Insects may memorize and recognize the shape of a flower by matching the actual image perceived by the eye with a memorized template (15, 38, 44–46, 131). According to this idea, choice is determined by degree of overlap between memorized image and observed flower shape (38, 46, 131). It has even been proposed that retinotopic matching is the visual strategy by which the memorized and the actual images are compared (26).

The algorithm used by insects in calculation of such overlap is unknown and seems to vary with the training schedule used (107). However, matching in the lower part of the visual field is critical for the recognition of a shape upon which an insect has been trained (9, 38, 131). For colored patterns, however, this applies only to long-wavelength colors; in ultraviolet light, the upper part of the visual field seems to have a more important role (83).

Flies remember the position of stimuli in the visual field and are able to distinguish two identical patterns displaced by 9° in space (26). Such displacement experiments reveal that position information is a prerequisite for recognition of a learned pattern. Thus, stimuli learned at one height relative to the én face axis of the flower would not be distinguished at a new height. Clearly, the restriction of the approach direction resulting from vertically presented bilateral symmetry contributes to position invariance and, thus, to efficient recognition of flower patterns. Visual fixation and scanning behavior would also greatly contribute to recognition. Indeed, before landing on a vertically presented pattern, honeybees fix it visually by adjusting all six degrees of locomotor freedom (roll, pitch, and yaw as well as forward-backward, sideways, and upward-downward movements) relative to the landing point (132) and scan it on suspended flight (70).

Thus, we postulate that manipulation of the landing of hymenopteran pollinators (and possibly other groups) will be enhanced in flowers with vertical advertising surfaces if the more complicated color pattern (e.g. nectar guides) as well as the dissected parts of the flower corolla (e.g. labellum or lobes) are concentrated at the lower part of the flower. Different patterns in the upper and lower halves of the flower will result in medial zygomorphic symmetry patterns. The position of the flower relative to an approaching visitor is critical in this hypothesis. It seems significant that, in contrast to radially symmetrical flowers, zygomorphic flowers are primarily vertical in én face orientation and tend to have less variable orientation both within and between plants (84, 97, 99, 112). Such a spatial orientation results in a restriction of the approach flight of the bee and thus facilitates the matching strategy.

It would be interesting to compare the visual complexity of the upper and lower halves of zygomorphic flowers pollinated by bees and by birds, which presumably do not have this visual field bias. However, ornithophilous flowers with hovering and nonhovering pollinators should be considered separately because flowers pollinated by hovering birds may have a reduced lower margin for other reasons (see the dangerous lower margin hypothesis below).

FLOWER DISTINCTIVENESS HYPOTHESIS   This hypothesis is based on reasoning similar to the optical information hypothesis. Zygomorphy provides more variation by which pollinators can distinguish and establish fidelity to plant species (94). An increase in fidelity would be especially important for plants in more diverse floras and for rare species.

Zygomorphy is positively correlated with plant diversity (number of zoophilous species/m^2) across 25 communities of the Wasatch Mountains of Utah and

Idaho (94). Harper (47) suggested that zygomorphy is more common among rare species (but see comments under the pollen position hypothesis below).

MARGINAL FLOWER–ATTRACTION AND MARGINAL FLOWER–LANDING PLATFORM HYPOTHESES   The marginal flowers of compact inflorescences of several families (e.g. Dipsacaceae, Apiaceae, Verbenaceae, Caprifoliaceae, Asteraceae) are bilaterally symmetrical, while the rest of the flowers are radially symmetrical. Although marginal flowers may have full or partial sexual function (e.g. male in Apiaceae, female in Asteraceae), commonly they are sterile and nectarless (74).

Two hypotheses have been proposed in which selection would favor the evolution of zygomorphic marginal flowers, but few, if any, critical studies have attempted to test or differentiate these hypotheses. In the marginal flower–attraction hypothesis, conspicuousness of the inflorescence as a whole would be heightened (43, 74, 134). Leppik (74) suggested that marginal flowers produce an overall form of the inflorescence that mimics the shape of solitary flowers. Good (43) suggested that marginal flowers enhance the "target-like effect" or result in a "more solid centre surrounded by a periphery of petaline rays" (43, p. 277).

The marginal flower–landing platform hypothesis postulates that in species with flat-topped, rounded cyme or corymb inflorescences (e.g. Asteraceae or some species of *Gentiana*), the zygomorphic, horizontally spreading petals or rays provide the platform for pollinators' landing (112, 113). In this case, the expanded marginal petals serve in the same capacity as the petals of a horizontal bowl-shaped flower.

Information Processing by the Pollinators

INNATE SYMMETRY PREFERENCE HYPOTHESIS   Although pollinators may not have as strong a fidelity to suites of characteristics as previously believed [see criticisms of the syndrome concept (54)], different taxonomic groups of pollinators might preferentially visit flowers of one symmetry type. For example, Leppik (71) found that beetles, honeybees, moths, and butterflies visited primarily actinomorphic forms, while bumblebees visited primarily zygomorphic forms. However, a close examination shows that Leppik's results were from casual (although systematic) observations and cannot be reliably subjected to statistical tests. Free (35) found that when given a choice between radially and bilaterally symmetrical models, honeybees trained on radial models preferred the radial models. Free also stated that training on oblong (i.e. bilateral) models "did not reverse the usual preference for radially symmetrical models" (35, p. 272). However, this conclusion cannot be unambiguously drawn from the data he presented. Moreover, since the previous experience of the honeybees in Free's experiments was not controlled, their choices may have reflected the information they learned in their encounters with flowers in the field and not an innate preference.

Nevertheless, it has been repeatedly suggested that pollinators "spontaneously" prefer symmetrical flowers and/or models (35, 69, 86, 88). The critical problem in testing these so-called spontaneous preferences is control of the previous experience of the animals. Most researchers tested these preferences without attention to this aspect. Insects, particularly honeybees, were usually trained to collect sucrose solution at a site that was not associated with any particular stimulus and then, at the same site, were presented with various stimuli that were, in principle, novel to them. The choices made may have reflected their previous experience with flower stimuli in the field rather than true spontaneous preferences. Only tests with naive pollinators [e.g. bumblebees (78), honeybees (41), butterflies (109), hawkmoths (61)] can provide evidence on innate preferences for particular stimuli.

Bees trained to discriminate bilaterally symmetric from nonsymmetric patterns learn the task and transfer it appropriately to novel stimuli, thus demonstrating a capacity to detect and generalize symmetry versus asymmetry (39), showing that bees use symmetry as an independent feature in pattern perception. Horridge (59) confirmed that bees can be trained to distinguish the axis of bilateral symmetry of a set of different patterns. Bees trained to select symmetrical patterns performed better than bees trained to select asymmetrical patterns (39). This result may reflect an innate predisposition to respond to stimuli that are biologically relevant (41, 82) and has important consequences in the field of pollination: If potential pollinators particularly beneficial to the plant (e.g. having high levels of constancy, pollen transfer efficiency, outcross pollen deposition) have an innate preference for zygomorphic forms, there should be strong selection for this morphology (but see 54).

Møller (86) suggested that impairing symmetry of a bilaterally symmetric flower (e.g. by cutting parts of flower petals) reduces the visitation rate of pollinators such as bumblebees and, thus, that bumblebees perform assortative pollination on the basis of symmetry as a flower feature. However, care must be exercised in interpreting experiments that damage a flower because the visitors may be responding to confounding factors such as changes in flower size, chemical signature, or optical properties that result from damaging the flower. Much more needs to be learned about the innate preferences of all pollinator groups as well as their relative qualities as pollinators.

FLUCTUATING ASYMMETRY HYPOTHESIS   As originally applied to animal mate choice (86), fluctuating asymmetry (i.e. small random deviations from symmetry, especially from bilateral symmetry) is thought to be the result of genetic or environmental stress (89, 96). Whether caused by genetic or environmental factors, individuals with higher levels of fluctuating asymmetry have been shown, in many cases, to be less fit and, therefore, tend to be discriminated against as potential mates (85, 95). With regard to pollination, the hypothesis suggests that the dependence of plants on pollinators allows the discriminatory properties of the pollen vectors to be interjected into the pollination process (87, 88). The suggested benefits of lower levels of fluctuating asymmetry to the plant are (a) higher visitation rates, which result in greater pollen removal and deposition and (b) receipt of pollen of superior quality.

Three modes of action are suggested (87). First, asymmetry, as in animals, may be an indication of genetic or environmental stress, and symmetry may be correlated with the amount of the reward (e.g. pollen or nectar) available (86). Second, in plant species with pollination involving sexual deception (e.g. *Ophrys* spp.), an innate preference for bilateral symmetry in mates by the pollinators will select for greater symmetry in the flowers. Third, pre-existing bias for bilateral symmetry may result from a selective advantage for pollinators to recognize predators, parasites, and/or competitors. The first of these modes applies to selection for symmetry in general, while the second and third apply to bilateral symmetry in particular.

Møller & Eriksson (88) found, after controlling for petal size, that bees preferentially visited flowers that were more symmetrical than their nearest neighbors in 7 of 10 species of plants studied (6 with radially symmetrical flowers, 4 with bilaterally symmetrical flowers). They also found significant assortative pollination with respect to fluctuating asymmetry in three species of plants in which it was tested. There was a tendency for the standing crop of nectar and rate of nectar production to decrease with fluctuating asymmetry, but there was no significant relationship with sugar content of the nectar. Although in this type of study it is difficult or impossible to control for correlated character states (e.g. intensity of floral odor correlated with the amount of nectar), the data do suggest that fluctuating asymmetry may play an important role in floral symmetry.

In animals, sexual characters usually exhibit greater levels of fluctuating asymmetry than nonsexual characters (89, 96). To test this in plants, Møller & Eriksson (87) compared levels of asymmetry in flowers and leaves of 19 species from Spain, Sweden, and Denmark. They found no significant difference in fluctuating asymmetry between floral and vegetative characters, suggesting that selection does not act differentially with regard to flowers and vegetative structures.

Comparing values given by Møller & Eriksson (87) for relative asymmetry in radial and bilateral flowers, we found that bilateral species have significantly lower levels of petal asymmetry (one-tailed t-test: $t = 2.34$, d.f. $= 17$, $p = 0.016$). The relative leaf asymmetry for the bilateral group also appears to be lower than for the radial group (one-tailed t-test: $t = 1.73$, d.f. $= 17$, $p = 0.051$). These results, limited as they are, suggest that the level of fluctuating asymmetry may be lower, in general, for species with bilaterally symmetrical flowers than for species with radially symmetrical flowers. The data are somewhat preliminary, but they do suggest that further investigation is warranted.

COMPLEXITY NEURONAL/BEHAVIORAL SOPHISTICATION HYPOTHESIS   According-ing to this hypothesis, complex flowers (bilateral symmetry being one type of floral complexity) require more sophisticated neuronal processing or behavioral versatility on the part of pollinators to attain floral rewards (51, 64, 67, 71, 74, 77). The notion of floral (usually morphological) complexity is a topic that can be interpreted in many ways. In general, a more complex flower is one likely to have bilateral symmetry and a narrow and/or long floral tube. There is a specific entrance indicated by (nectar) guides leading to an inaccessible reward. The entrance is likely to be closed, with a forced entry required. The complexity of a flower thus results from a combination of character states that may differentially affect pollinator taxa. Complexity is usually assessed in terms of the probability of or time required for successfully obtaining or learning to obtain the floral reward. Using these criteria, it is relatively straightforward to judge the relative complexity of different flowers for one pollinator. However, these criteria may not be good indicators when rates of energy needs differ among pollinators. Finally, the interaction of various floral characters with different pollinator taxa make it difficult to make unambiguous assessments of absolute levels (i.e. using a common unit of measurement) of floral complexity.

The use of the term "intelligence" to describe the sophistication level of pollinators (33, 51) is problematic because it is not unambiguously definable, quantifiable, or comparable across species. As such, this vague term should be discarded with regard to the abilities of pollinators to gain access to floral rewards.

The complexity–neuronal/behavioral sophistication hypothesis can be subdivided into three components: manipulation skills, learning ability, and sensory perception. Interaction among these components is certainly possible. For example, manipulation skills may be improved with learned experience, but the skill required for some flowers may be beyond the level of some pollinators regardless of learning ability.

Regarding manipulation skills, Heinrich (51) noted that zygomorphic flowers (e.g. many species in the Fabaceae and Scrophulariaceae) may require forced entry to gain access to the nectar reward. He concluded that only pollinators "such as higher or social bees can get entry, and many of the behaviorally less versatile foragers are excluded" (51, p. 172). Furthermore, Heinrich (51) noted that not all individuals of a species may acquire the ability to extract rewards from some complex flowers (e.g. bumblebees on *Aconitum napellus*).

Although this hypothesis applies generally to all taxonomic groups (71, 74), most research has focused on the learning ability of the bees (e.g. 27, 64, 67). Within the superfamily Apoidae, bees are often classified into two groups based on their behavioral sophistication or learning ability. Several different sets of contrasting terms have been used to distinguish between the groups (e.g. sophisticated, advanced, higher, specialist, social, and literate contrasted with less advanced, primitive, lower, generalist, unskilled, solitary, and illiterate). In addition, the terms generalist and specialist are sometimes used to describe diet breadth rather than neuronal properties of the bees (67). More sophisticated bees are said to be found on more complex flowers. The argument sometimes becomes circular because the bees are often classified by the flowers they visit, rather than by experimental tests of learning ability. Within the "higher" Apoidae (i.e. the Family Apidae and possibly Anthophoridae), Dukas & Real (27) claimed that social *Bombus* (Family Apidae) have better learning capacities than solitary *Xylocopa* (Family Anthophoridae) concerning flower reward, but we are not aware of any experimental evidence showing that "higher" Apoidae have better learning ability and/or better memory capabilities than "lower" Apoidae. Their study also raises the problem of the control of the rewarded trials for studies of learning abilities. If learning rates are to be characterized, a complete record of the rewarded and nonrewarded trials must be kept. The latter are critical for learning because they constitute extinction trials in which a bee learns that a given signal is not associated with reward. Usually, they are ignored, as in the study of Dukas & Real (27), but revisiting a just-depleted flower is as important as getting a reward from it.

Few data exist regarding other pollinator groups and their ability to obtain rewards from flowers of different "complexity." Lewis (77) determined discovery time (i.e. time from landing on a flower to finding nectar) in successive visits for the cabbage butterfly, *Pieris rapae*, on a variety of flowers varying in "complexity" as assessed by the human eye. Lewis concluded that it is difficult to determine "precisely how flower features influence learning time. The results do, however, suggest that flower morphology does influence learning time, with human judgements of morphological complexity having some predictive value" (77, p. 232). Despite this, it is clear that én face knowledge of floral morphology will not permit *specific* prediction about the relative difficulty that different pollinators will have in obtaining access to various flowers (65, 77). More data of this type need to be collected across a wide range of pollinators and flower types, quantifying complexity rigorously.

The complexity–neuronal/behavioral sophistication hypothesis also suggests that bilaterally symmetrical flowers require higher sensory perception in pollinators than radially symmetrical flowers and "other less complicated flow-ers," which would be expected to be associated with pollinators having less sensory development. Much of the evidence for this part of the hypothesis has been indirect or speculative.

A mechanism based on matching with neuronal filters has been proposed for the perception of symmetry (58, 69). Neuronal detectors that are specialized in detection of radial, circular (i.e. patterns of concentric rings), or spiral motion have been shown in humans (90), and it was proposed that bees (and other pol-linators) use similar detectors to categorize radial and circular patterns on the basis of symmetry, without attention to other local cues. In the same way, sim-ilar neuronal detectors for bilaterally symmetric patterns have been proposed to exist in bees and other pollinators (39). Alternatively, it has been postulated that symmetry might be detected by the interactive combination of a radial filter (58, 69) and an average-orientation filter (59) such as those found in dragonflies (92) and in honeybees (137). After being passed through a radial filter, many bilaterally symmetrical patterns are left with some preferred orientation, which would reveal the axis of symmetry. There is no evidence for deciding between the two possibilities. Regarding radial versus bilateral symmetry detection, there is no evidence to suggest that one class of detector should be considered perceptually more sophisticated than another. Moreover, that classes of detec-tors operate with different types of symmetry is not necessarily correlated with differences in learning ability.

COMPLEXITY-CONSTANCY HYPOTHESIS   According to this hypothesis, complex (e.g. zygomorphic) flowers promote floral constancy as a result of more efficient foraging by pollinators (64, 77) and, hence, greater outcrossing (22). Constancy in complex flowers (as compared to simple flowers in which access is easy or easily learned) can be promoted by two mechanisms. First, morphological complexity (e.g. long floral tubes) may require morphological adaptations of pollinators that limit them to flowers with similar morphology (77). However, as pointed out above, there is no evidence that zygomorphic flowers and bilateral symmetry are better learned than actinomorphic flowers and radial symmetry. This aspect of the hypothesis seems to be the inverse of the reward wastage hy-pothesis (see below) in that pollinators are limited to, rather than excluded by, a particular morphology. Moreover, the observation that many, if not most, polli-nators visit a wide range of flowers (108, 127) tends to discount this hypothesis. Waser (126) has termed this type of floral specialization "fixed preference" rather than "floral constancy."

The second mechanism hypothesized to promote constancy is that the cost of learning to extract the reward from complex new flowers may outweigh the benefits of obtaining additional sources of reward (22, 64, 77). There may be an energetic cost whereby time spent learning to handle the complex new form reduces the rate of reward intake compared to flowers with which the pollinator has experience (52, 66). The argument has also been made that cost of constancy and fully learning to handle a complex flower ultimately results in a higher return (22) as well as reduces competition from less constant pollinators or individuals (64). However, Laverty (66) found for bumblebees that the cost in time of switching was small and not likely to account for constancy. Another possible cost of learning to handle additional flowers might be interference with the efficiency on, or even elimination of, the ability to handle flowers already learned (76, 125, 126). There is some support for interference in several pollinator taxa (76, 77, 91), but interpretation of these results is difficult because many of the early studies lack appropriate controls for the motivation and experience of the animals tested. However, in a series of studies on bumblebees Laverty and his group have found no evidence of a complete elimination of an already learned skill, even as long as 24 h after learning to handle a new flower (37, 66, 67, 135).

Activity Patterns of the Pollinators

NATURAL POSITION HYPOTHESIS   Sprengel (110) observed that the natural position of insects in flight is upright. Similarly, although an inverted position is possible when walking or standing, it is usually avoided because it requires more effort. As a result of this "natural position," all horizontal or downward inclining flowers would be approached from only one direction. Sprengel suggested that, therefore, the anthers and stigma would be placed in the position most "suitable" for the natural position of the pollinators (i.e. irregular anthers and stigma(s) would result in better contact with the pollinator). Under the influence of gravity, changes in the orientation of a flower sometimes result in irregularity of stamens and style (e.g. *Epilobium angustifolium*) (129).

INFLORESCENCE TYPE–FLOWER ORIENTATION HYPOTHESIS   According to this hypothesis, actinomorphic flowers arranged in vertical racemes or spikes, especially those with tightly clustered flowers, would provide poor landing platforms (110, 112, 113). Therefore, in species visited by pollinators that land on the flower (as opposed to those that only hover, e.g. hummingbirds and hawkmoths), the lower lip of the flower is expanded to provide a platform. Sprengel (110) also suggested that, because pollinators will approach and contact the flower from one direction, the lower lip may be enlarged for attraction, in addition to its function as a landing platform.

POLLEN POSITION HYPOTHESIS   This hypothesis suggests that in bilaterally symmetrical flowers, the visitor is restricted to certain directions in its approach to and its movement on and/or within the flower. This is in contrast to radially symmetric flowers in which the visitor may approach the flower from any direction (33, 47, 74). The restricted approach results in increased precision of pollen placement on, and stigma contact with, the pollinator's body (2, 51, 64, 80). The increased precision thus results in a higher proportion of pollen reaching the stigma. Bowers (4) proposed a similar process for evolution of enantiomorphy.

In an unusual application of this hypothesis, Harper (47) suggested that "floral zygomorphy confers a reproductive advantage to rare plants" as a result of the enhanced pollination efficiency (47, p. 135). He found over representation of bilaterally symmetry in the rare flowers of five floras (one each in California and Colorado, and three in Utah), but only in one of these was the association significant. Therefore, his statement that there is a "universal over-representation of bilaterally symmetrical flowers among the rare taxa of all floras" needs further validation, especially in light of modern concepts concerning the definition of "rarity" in plants (101, p. 134).

Leppik (74) hypothesized that bilateral symmetry (i.e. medial zygomorphy) would make pollination be more efficient and also speculated (without giving a reason) that cross-pollination would thereby be more effective. Increased outcrossing has also been proposed as the driving force behind enantiomorphy (28, 29, 93), However, Fenster (34) suggested that enantiomorphy might actually increase selfing through geitonogamous visits to both morphs on the same plant. Fenster (34) found a small increase in the outcrossing rate of artificially non-enantiomorphous compared to enantiomorphous plants, and concluded that enantiomorphy is not a mechanism to promote outcrossing. Robertson (106, p. 344) hypothesized that "flower features that promote approach from all directions will lead to higher (pollen) carryover." Thus, in direct contrast to various versions of the pollen position hypothesis, actinomorphy should be the symmetry type that promotes outcrossing.

REWARD WASTAGE HYPOTHESIS   This hypothesis also relies on the restriction of approach by flower visitors (51, 74, 97, 115). However, in this case, morphology prevents (or at least limits) inefficient pollinating species or thieves (sensu 60) from obtaining and, thus, wasting the reward (usually nectar), which, in zygomorphic flowers, is often hidden and requires complicated behaviors and/or specialized morphology for access. Furthermore, in many zygomorphic flowers, the reward can be legitimately obtained only by a mechanical deformation of the flower requiring a minimum weight or strength of the visitor. Access is thus limited to "specialized" pollinators capable of more complicated behavior patterns and/or larger pollinators capable of mechanically forcing the flower (84). Such specialized and larger pollinators are thought to be more likely to transfer pollen between conspecific flowers because they tend to fly faster, farther, and more efficiently under adverse conditions (51, 115).

The pollen position and reward wastage hypotheses are difficult to separate. Both are thought to promote more efficient pollination (i.e. greater reproductive return from investment in pollen and reward). Although these arguments sound logical, experimental evidence is scarce and most of the aspects need validation, especially in comparison to actinomorphic flowers. Does bilateral symmetry result in more efficient pollination? For example, what proportion of pollen is actually deposited on stigmas in actinomorphic versus medial zygomorphic flowers? Does the proportion differ for generalist and specialist pollinators?

PRECISE STEERING—INDIVIDUAL FLOWER HYPOTHESIS   This hypothesis might be considered a variant of the pollen position hypothesis. For bird-pollinated species in western Australia, Holm (57) suggested that zygomorphic symmetry allows "precise steering" of the bird as it forages on single flowers. In contrast, in radially symmetric brush-type flowers, found in inflorescences where the birds do not discriminate individual flowers, the anthers brush the bird more or less at random (57). However, Holm also described a typical brush-type inflorescence with zygomorphic flowers. He noted that several species of *Banksia* have stiff, curved stigmas (and a reduced but zygomorphic perianth) that serve as perches for birds foraging on the inflorescence.

DANGEROUS LOWER MARGIN HYPOTHESIS   Zygomorphy is often associated with ornithophily (32, 33, 99, 115). In some bird-pollinated species, zygomorphy is thought to result from the elimination of the lower margin of the corolla (i.e. lower petals or lip of flowers with a horizontal or oblique én face orientation) (33). Faegri & van der Pijl (33) called the lower margin "dangerous" from the perspective of the plant and gave two reasons for its reduction: removal of a landing place for insects and elimination of an "obstacle" for the avian pollinators that are too large to alight on the flower itself. The benefits of these two effects to the plant were not stated explicitly. However, the benefits of excluding insects might be functionally identical to those proposed in the reward wastage hypothesis. Furthermore, in species visited primarily or exclusively by nonhovering birds, flowers (e.g. *Protea*, *Aloe*) often have an associated perch and are oriented toward this perch, while in species visited by hovering birds (hummingbirds or possibly sunbirds) the flowers (e.g. *Pedilanthus*, *Quassia*) are held away from potential perches (33). The absence of the lower margin but the presence of an associated perch provides easy visitor access in the nonhovering bird-pollinated species, while the absence of the perch requires hovering in the hovering bird-pollinated species. In both groups of plants, the direction of approach to the flower would be channeled, giving this hypothesis some aspects of the natural position, pollen position, and/or precise steering hypotheses. Finally, in some flowers enforcement of hovering may be the critical function leading to the loss of the dangerous lower margin. To meet the energetically higher costs of hovering, birds would have to visit more flowers (53). Hovering would also accelerate the rate at which flowers could be visited (50). Both of these effects would tend to increase outcrossing rates.

FEEDING ANTHER–POLLEN COLLECTION HYPOTHESIS   Faegri & van der Pijl (33) mentioned in passing that zygomorphy may be induced in families dominated by radially symmetrical flowers. This is thought to result from the tendency for feeding anthers to cluster in one part of the flower. Faegri & van der Pijl (33) implied that the manipulations required to remove pollen from feeding anthers (e.g. squeezing or vibrating anthers) may be the ultimate cause of the zygomorphy, but gave no specific selective forces. Dulberger (28) suggested that enantiomorphy may reduce the possibility of damage to the stigma during the vigorous manipulation required during buzz-pollination. Furthermore, separation of stigma and anthers in enantiomorphy may reduce the possibility of self-pollination resulting from the cloud of dry pollen released as the vector [usually a bee (6)] vibrates the anthers. Fenster (34) compared enantiomorphy to heterostyly and suggested that both may function to facilitate pollen collection and dispersal by reducing interference between male and female organs (128).

DISCUSSION AND CONCLUSIONS

What have we learned about floral symmetry since Sprengel? Several relatively uncommon types of symmetry have been described. The predominant type of irregular floral form, namely medial zygomorphy, has independently arisen many times (113). Multiple origins and the variety of hypotheses for these origins suggest that the evolution of non-actinomorphic floral symmetry is not a singular process. Apparently, reversion to actinomorphy is relatively uncommon. It would be instructive to compare pollination parameters (e.g. efficiency of pollination, constancy of pollinators) of pairs of closely related species differing in symmetry. Pairs should be chosen such that symmetry types could be represented in both the ancestral and the derived conditions.

A major trend in the evolution of floral symmetry has been the derivation of actinomorphic floral symmetry from ancestral forms with asymmetric flowers (namely haplomorphy). Given the rarity of ancestral haplomorphy, there must be advantages to actinomorphy over this primitive type of asymmetry. Floral organs radiate out from the central principal axis in both ancestrally asymmetric and actinomorphic flowers. Did selection for symmetry of pattern play a role in the transition from a spiral arrangement to a whorled arrangement of floral organs? We were unable to find hypotheses regarding the role that pollination might have played in this transition. Although of lesser importance in terms of the numbers of species, we were also unable to find hypotheses regarding the evolution of transverse and diagonal zygomorphy, disymmetry, and derived asymmetry.

With so many advantages to bilateral symmetry, the question then arises, why are not all flowers medially zygomorphic? There have been no good comparative studies showing the magnitude of the proposed advantages for most hypotheses. It may be difficult or impossible to quantify the importance of symmetry as proposed by some of the hypotheses. The hypotheses emphasize the advantages of zygomorphy, but little is said about the disadvantages. One disadvantage of zygomorphic flowers requiring specialized pollinators may be a risk of no pollination in years when the populations of pollinators are depressed (TM Laverty, personal communication). Laverty, therefore, suggested (personal communication) that the hypothesis of a greater tendency for zygomorphy in perennial species should be tested.

Is symmetry an absolute phenomenon? From most directions or orientations, flowers are perceived as asymmetrical no matter what type of symmetry they possess (19). It is only when pollinators are oriented along the principal axis of the flower that the "true" symmetry becomes apparent. What aspects of floral symmetry are important to the pollinators? In general, there is an accordance between the symmetry of external contour (of the flower outlines) and the "internal" one (of the "nectar guides") (18). When, if at all, does the symmetry of a flower become important to a pollinator (5)? For example, how close must a pollinator be to a flower before it perceives symmetry? To answer these questions, other flower parameters must also be taken into account—for example, size of the corolla and the visual angle that such a flower subtends at the pollinator eye (42). In the case of the honeybee, for instance, stimulus detection is organized in a sequential way, each stage mediated by a different visual subsystem. First, a target is detected by its achromatic contrast against the background, through the long-wave receptor system (40, 42). Then, chromatic information itself is perceived by the color vision system from a visual angle of approximately 15°, and finally, only very close to the target is the global form, and therefore the symmetry, perceived (124).

Similarly, is symmetry important only when the pollinator views the flower én face? If not, how far from the en face view is the pattern of symmetry recognized? Little or nothing is known about these aspects of flower and visitor interactions for most pollinator groups. Most of our knowledge about the role that symmetry plays in the behavior of pollinators (e.g. perception, recognition, innate preferences, learning) comes from the study of one species, the honeybee. While the honeybee is important in some pollination systems, its sensory capabilities and behavior may differ from those of some butterfly, fly, hummingbird, and even solitary bee species. Much more needs to be learned about the sensory capabilities and behavior patterns of other pollinator groups regarding all aspects of floral characters including symmetry.

Possibly the biggest gap in our understanding of the evolution of floral symmetry is the connection between the results of laboratory and field studies. In an attempt to control extraneous or unintended cues for the pollinators, laboratory studies have used somewhat simplified systems. These studies often use artificial patterns that have little resemblance to natural flowers. Similarly, the focus has been on flowers or patterns in isolation. Little is known about the role of floral symmetry in the morphological hierarchy of architectural complexity (20), that is, how pollinators react to individual flowers versus inflorescences versus patches of plants. For example, the distance at which floral symmetry becomes important to a particular pollinator may depend on whether flowers are in compact versus diffuse inflorescences, or are solitary. The trainability and ease of manipulation of the honeybee has, as mentioned above, limited most studies laboratory studies to this species. Field studies, on the other hand, often suffer from a lack of control. It is difficult to determine which floral characteristics pollinators are using, and it is impossible to know whether the behavior of pollinators is the result of innate preferences or learning. Finally, laboratory and field studies have tended to focus on different aspects of floral symmetry. Many laboratory studies are concerned with the behavior of the pollinators, while many field studies center on the details of floral morphology and its relationship to pollinator morphology and movement patterns.

ACKNOWLEDGMENTS

During the preparation of this review, PRN was supported by a fellowship from the Council of Higher Education of Israel and by grants from the Saloma Foundation; PRN and AD were generously supported by the Henk and Dorothy Schussheim Fund for Ecological Research on Mt. Carmel, Israel; MG was supported by the Alexander von Humboldt-Stiftung and the International Foundation for Science. We thank S Armbrutster, P Bernhardt, RI Bertin, L Chittka, R Cruden, P Endress, D Firmage, PG Kevan, TM Laverty, G LeBuhn, S Potts, C Schlichting, and S Vogel for discussions and/or for comments and criticisms on early drafts of the manuscript.

> **Visit the *Annual Reviews home page* at**
> **http://www.AnnualReviews.org**

Literature Cited

1. Armbruster WS. 1992. Phylogeny and the evolution of plant-animal interactions. *BioScience* 42:12–20
2. Armbruster WS, Edwards ME, Debevec EM. 1994. Floral character displacement generates assemblage structure of Western Australian triggerplants (*Stylidium*). *Ecology* 75:315–29
3. Bailey LH. 1951. *Manual of Cultiva-*
   ted Plants. New York: Macmillan. 1116 pp.
4. Bowers KAW. 1975. The pollination ecology of *Solanum rostratum* (Solanaceae). *Am. J. Bot.* 62:633–38
5. Bruter CP. 1994. On floral symmetries. *Acta Biotheor.* 42:181–86
6. Buchmann SL. 1983. Buzz pollination in angiosperms. In *Handbook of Exper-*

imental Biology, ed. CE Jones, RJ Little, pp. 73–117. New York: Van Nostrand & Reinhold

7. Carpenter R, Coen ES. 1990. Floral homeotic mutations produced by transposon-mutagenesis in *Antirrhinum majus*. *Genes Dev.* 4:1483–93

8. Chittka L. 1996. Does bee color vision predate evolution of flower color? *Naturwissenschaften* 83:136–38

9. Chittka L, Hoffmann M, Menzel R. 1988. Discrimination of UV-green patterns in honey bees. In *Sense Organs*, ed. N Elsner, FG Barth, pp. 218. Stuttgart: Thieme

10. Chittka L, Shmida A, Troje N, Menzel R. 1994. Ultraviolet as a component of flower reflections, and the colour perception of Hymenoptera. *Vision Res.* 34: 1489–508

11. Coen ES. 1996. Floral symmetry. *EMBO J.* 15:6777–88

12. Coen ES. 1991. The role of homeotic genes in flower development and evolution. *Annu. Rev. Plant Physiol. Plant Mol. Biol.* 42:241–80

13. Coen ES, Meyerowitz EM. 1991. The war of the whorls: genetic interactions controlling flower development. *Nature* 353: 31–37

14. Coen ES, Nugent JM, Lou DA, Bradley D, Cubas P, et al. 1995. Evolution of floral symmetry. *Proc. R. Soc. London Ser. B Biol. Sci.* 350:35–38

15. Collett TS, Cartwright BA. 1983. Eidetic images in insects: their role in navigation. *Trends Neurosci.* 6:101–5

16. Crepet WL, Friis EM, Nixon KC. 1991. Fossil evidence for the evolution of biotic pollination. *Philos. Trans. R. Soc. London B Biol. Sci.* 333:187–95

17. Dafni A, Kevan PG. 1995. Hypothesis on adaptive features of the compound eye of bees: flower-specific specializations. *Evol. Ecol.* 9:236–41

18. Dafni A, Kevan PG. 1996. Floral symmetry and nectar guides: ontogenetic constraints from floral development, colour pattern rules, and functional significance. *Bot. J. Linn. Soc.* 120:371–77

19. Dafni A, Kevan PG. 1997. Flower size and shape and their implications in pollination. *Israel J. Plant Sci.* 45:201–11

20. Dafni A, Neal PR. 1997. Size and shape in floral advertisement: measurement, concepts, and implications. *Acta Hortic.* 437: 121–40

21. Darwin CR. 1868. *The Variation of Animals and Plants Under Domestication*. London: Murray

22. Darwin CR. 1876. *The Effects of Cross and Self Fertilization in the Vegetable Kingdom*. London: Murray

23. Davenport D, Kohanzadeh Y. 1982. Orchids, bilateral symmetry and insect perception. *J. Theor. Biol.* 94:241–52

24. Davenport D, Kohanzadeh Y. 1982. Orchids, bilateral symmetry and insect perception: erratum. *J. Theor. Biol.* 97:541–42

25. Davenport D, Lee H. 1985. Image analysis of the Orchidaceae. *J. Theor. Biol.* 114:199–222

26. Dill M, Wolf R, Heisenberg M. 1993. Visual pattern recognition in *Drosophila* involves retinotopic matching. *Nature* 365: 751–53

27. Dukas R, Real LA. 1991. Learning foraging tasks by bees: a comparison between solitary and social species. *Anim. Behav.* 42:169–76

28. Dulberger R. 1981. The floral biology of *Cassia didymobotrya* and *C. auriculata* (Caesalpiniaceae). *Am. J. Bot.* 68:1350–60

29. Dulberger R, Ornduff R. 1980. Floral morphology and reproductive biology in four species of *Cyanella* (Tecophilaeaceae). *New Phytol.* 86:45–56

30. Endress PK. 1986. Reproductive structures and phylogenetic significance of extant primitive angiosperms. *Plant Syst. Evol.* 152:1–28

31. Endress PK. 1987. Floral phyllotaxis and floral evolution. *Bot. Jahrb. Syst. Pflanzengesch. Pflanzengeogr.* 108:417–38

32. Endress PK. 1994. *Diversity and Evolutionary Biology of Tropical Flowers*. Cambridge, UK: Cambridge Univ. Press. 511 pp.

33. Faegri K, van der Pijl L. 1979. *The Principles of Pollination Ecology*. Oxford: Pergamon. 244 pp.

34. Fenster CB. 1995. Mirror image flowers and their effect on outcrossing rate in *Chamaecrista fasciculata* (Leguminosae). *Am. J. Bot.* 82:46–50

35. Free JB. 1970. Effect of flower shapes and nectar guides on the behaviour of foraging honeybees. *Behavior* 37:269–85

36. Futuyma DJ. 1986. *Evolutionary Biology*. Sunderland, MA: Sinauer

37. Gegear RJ, Laverty TM. 1995. Effect of flower complexity on relearning flower-handling skills in bumble bees. *Can. J. Zool.* 73:2052–58

38. Giurfa M, Backhaus W, Menzel R. 1995. Color and angular orientation in the discrimination of bilateral symmetric patterns in the honeybee. *Naturwissenschaften* 82:198–201

39. Giurfa M, Eichmann B, Menzel R. 1996. Symmetry perception in an insect. *Nature* 382:458–61
40. Giurfa M, Núñez JA. 1992. Foraging by honeybees on *Carduus acanthoides*: pattern and efficiency. *Ecol. Entomol.* 17:326–30
41. Giurfa M, Núñez JA, Chittka L, Menzel R. 1995. Colour preferences of flower-naive honeybees. *J. Comp. Physiol. A* 177:247–59
42. Giurfa M, Vorobyev M, Kevan PG, Menzel R. 1996. Detection of coloured stimuli by honeybees: minimum visual angles and receptor specific contrasts. *J. Comp. Physiol. A* 178:699–710
43. Good R. 1974. *Features of Evolution in the Flowering Plants.* New York: Dover. 405 pp.
44. Gould JL. 1985. How bees remember flower shapes. *Science* 227:1492–94
45. Gould JL. 1986. Pattern learning by honey bees. *Anim. Behav.* 34:990–97
46. Gould JL. 1988. Resolution of pattern learning by honey bees. *J. Insect Behav.* 1:225–33
47. Harper KT. 1979. Some reproductive and life history characteristics of rare plants and implications of management. *Great Basin Natur. Mem.* 3:129–37
48. Harvey PH, Pagel MD. 1991. *The Comparative Method in Evolutionary Biology.* Oxford, UK: Oxford Univ. Press. 239 pp.
49. Hedström I, Thulin M. 1986. Pollination by a hugging mechanism in *Vigna vexillata* (Leguminosae-Papilionoideae). *Plant Syst. Evol.* 154:275–83
50. Heinrich B. 1975. Energetics of pollination. *Annu. Rev. Ecol. Syst.* 6:139–70
51. Heinrich B. 1979. *Bumblebee Economics.* Cambridge, MA: Harvard Univ. Press. 245 pp.
52. Heinrich B. 1984. Learning in invertebrates. In *The Biology of Learning*, ed. P Marler, HS Terrace, pp. 135–47. Berlin: Springer
53. Heinrich B, Raven PH. 1972. Energetics and pollination ecology. *Science* 176:597–602
54. Herrera CM. 1996. Floral traits and plant adaptation to insect pollinators: a devil's advocate approach. See Ref. 77a, pp. 65–87
55. Heywood VH, ed. 1993. *Flowering Plants of the World.* London: Batsford. 335 pp.
56. Hickey M, King C. 1988. *100 Families of Flowering Plants.* Cambridge, UK: Cambridge Univ. Press. 619 pp.
57. Holm E. 1988. *On Pollination and Pollinators in Western Australia.* Gedved, Denmark: Eigil Holms. 143 pp.
58. Horridge A, Zhang A. 1995. Pattern vision in honeybees: flower-like patterns with no predominant orientation. *J. Insect Physiol.* 41:681–88
59. Horridge GA. 1996. The honeybee (*Apis mellifera*) detects bilateral symmetry and discriminates its axis. *J. Insect Physiol.* 42:755–64
60. Inouye DW. 1980. The terminology of floral larceny. *Ecology* 61:1251–53
61. Kelber A. 1996. Colour learning in the hawkmoth *Macroglossum stellatarum*. *J. Exp. Biol.* 199:1127–31
62. Kuckuck H, Schick R. 1930. Die Erbfaktoren bei *Antirrhinum majus* und ihre Beeichnung. *Z. Induckt. Abstamm. Vererbungsl.* 56:51–83
63. Labandeira CC, Sepkoski JJ. 1993. Insect diversity in the fossil record. *Science* 261:310–315
64. Laverty TM. 1980. The flower-visiting behaviour of bumble bees: floral complexity and learning. *Can. J. Zool.* 58:1324–35
65. Laverty TM. 1994. Bumble bee learning and flower morphology. *Anim. Behav.* 47:531–45
66. Laverty TM. 1994. Costs to foraging bumble bees of switching plant species. *Can. J. Zool.* 72:1293–301
67. Laverty TM, Plowright RC. 1988. Flower handling by bumblebees: a comparison of specialists and generalists. *Anim. Behav.* 36:733–40
68. Lawrence GHM. 1951. *Taxonomy of Flowering Plants.* New York. Macmillan. 823 pp.
69. Lehrer M, Horridge GA, Zhang SW, Gadagkar R. 1995. Shape vision in bees: innate preference for flower-like patterns. *Philos. Trans. R. Soc. London B Biol. Sci.* 347:123–37
70. Lehrer M, Wehner R, Srinivasan M. 1985. Visual scanning behavior in honeybees. *J. Comp. Physiol. A* 157:405–15
71. Leppik EE. 1953. The ability of insects to distinguish number. *Am. Nat.* 87:229–36
72. Leppik EE. 1957. Evolutionary relationship between entomophilous plants and anthophilous insects. *Evolution* 11:466–81
73. Leppik EE. 1957. A new system for classification of flower types. *Taxon* 6:64–67
74. Leppik EE. 1972. Origin and evolution of bilateral symmetry in flowers. *Evol. Biol.* 5:49–85
75. Leppik EE. 1977. Floral evolution in relation to pollination ecology. *Intern. Bio.-Sci. Monogr.* 3:1–164

76. Lewis AC. 1986. Memory constraints and flower choice in *Pieriis rapae*. *Science* 232:863–65

77. Lewis AC. 1993. Learning and the evolution of resources: pollinators and flower morphology. In *Insect Learning: Ecology and Evolutionary Perspectives*, ed. DR Papaj, AC Lewis, pp. 219–42. New York: Chapman & Hall

77a. Lloyd DG, Barrett SCH, eds. 1996. *Floral Biology: Studies on Floral Evolution in Animal-Pollinated Plants*. New York: Chapman & Hall

78. Lunau K, Wacht S, Chittka L. 1996. Colour choices of naive bumble bees and their implications for colour perception. *J. Comp. Physiol. A* 178:477–89

79. Luo D, Carpenter R, Vincent C, Copsey L, Coen E. 1996. Origin of floral asymmetry in *Antirrhinum. Nature* 383:794–99

80. Macior LW. 1974. Behavioral aspects of coadaptations between flowers and insect pollinators. *Ann. Mo. Bot. Gard.* 61:760–69

81. McLean RC, Ivimey-Cook WR. 1956. *Textbook of Theoretical Botany*. Vol. 2. London: Longmans, Green. 2201 pp.

82. Menzel R. 1967. Untersuchungen zum Erlernen von Spektralfarben durch die Honigbiene, *Apis mellifica. Z. Vergl. Physiol.* 56:22–62

83. Menzel R, Lieke EE. 1983. Antagonistic color effects in spatial vision of honey bees. *J. Comp. Physiol. A* 151:441–48

84. Menzel R, Shmida A. 1993. The ecology of flower colors and the natural colour vision of insect pollinators: the Israeli flora as a study case. *Biol. Rev.* 68:81–120

85. Møller AP. 1990. Fluctuating asymmetry in male sexual ornaments may reliably reveal male quality. *Anim. Behav.* 40:1185–87

86. Møller AP. 1995. Bumblebee preference for symmetrical flowers. *Proc. Natl. Acad. Sci. USA* 92:2288–92

87. Møller AP, Eriksson M. 1994. Patterns of fluctuating asymmetry in flowers: implications for sexual selection in plants. *J. Evol. Biol.* 7:97–113

88. Møller AP, Eriksson M. 1995. Pollinator preference for symmetrical flowers and sexual selection in plants. *Oikos* 73:15–22

89. Møller AP, Pomiankowski A. 1993. Fluctuating asymmetry and sexual selection. *Genetica* 89:267–79

90. Morrone MC, Burr DC, Vaina LM. 1995. Two stages of visual processing for radial and circular motion. *Nature* 376:507–9

91. Nilsson LA, Jonsson L, Rolison L, Randrianjohany E. 1987. Angraecoid orchids and hawkmoths in central Madagascar: specialized pollination systems and generalist foragers. *Biotropica* 19:310–18

92. O'Carroll D. 1993. Feature detecting neurons in the dragonfly. *Nature* 362:541–43

93. Ornduff R, Dulberger R. 1978. Floral enantiomorphy and the reproductive system of *Wachendorfia paniculata* (Haemodoraceae). *New Phytol.* 80:427–34

94. Ostler WK, Harper KT. 1978. Floral ecology in relation to plant species diversity in the Wasatch Mountains of Utah and Idaho. *Ecology* 59:848–61

95. Palmer AR, Strobeck C. 1986. Fluctuating asymmetry: measurement, analysis, pattern. *Annu. Rev. Ecol. Syst.* 17:391–421

96. Parsons PA. 1990. Fluctuating asymmetry: an epigenetic measure of stress. *Biol. Rev.* 65:131–55

97. Polhill RM, Raven PH, Stirton CH. 1981. Evolution and systematics in the Leguminosae. In *Advances in Legume Systematics*, ed. RM Polhill, PH Raven, pp. 1–26

98. Porter CL. 1967. *Taxonomy of Flowering Plants*. San Francisco: Freeman. 472 pp.

99. Proctor M, Yeo P. 1973. *The Pollination of Flowers*. New York: Taplinger. 418 pp.

100. Proctor M, Yeo P, Lack A. 1996. *The Natural History of Pollination*. London: Collins. 479 pp.

101. Rabinowitz D, Rapp JK, Cairns S, Meyer M. 1981. Seven forms of rarity. In *The Biological Aspects of Rare Plant Conservation*, ed. H Synge, pp. 205–17. New York: Wiley

102. Radford AE, Dickison WC, Massey JR, Bell CR. 1974. *Vascular Plant Systematics*. New York: Harper & Row. 891 pp.

103. Rebelo AG, Siegfried WR, Crowe AA. 1984. Avian pollinators and the pollination syndromes of selected mountain fynbos plants. *S. Afr. J. Bot.* 2:285–396

104. Rendle AB. 1971. *The Classification of Flowering Plants*. Vol. I: *Gymnosperms and Monocotyledons*. Cambridge, UK: Cambridge Univ. Press. 412 pp.

105. Rendle AB. 1971. *The Classification of Flowering Plants*. Vol. II: *Dicotyledons*. Cambridge, UK: Cambridge Univ. Press

106. Robertson AW. 1992. The relationship between floral display size, pollen carryover and geitonogamy in *Myosotis colensoi* (Kirk) MacBride (Boraginaceae). *Biol. J. Linn. Soc.* 46:333–49

107. Ronacher B, Duft U. 1966. An image-matching mechanism describes a generalization task in honeybees. *J. Comp. Physiol. A* 178:803–12

108. Roubik DW. 1989. *Ecology and Natural History of Tropical Bees*. Cambridge, UK: Cambridge Univ. Press. 514 pp.

109. Scherer C, Kolb G. 1987. Behavioral experiments on the visual processing of color stimuli in *Pieris brassicae* L. (Lepidoptera). *J. Comp. Physiol. A* 160:645–56

110. Sprengel CK. 1793 (1996). Discovery of the secret of Nature in the structure and fertilization of flowers. See Ref. 77a, pp. 3–43

111. Srinivasan MV. 1994. Pattern recognition in the honeybee: recent progress. *J. Insect Physiol.* 40:183–94

112. Stebbins GL. 1951. Natural selection and differentiation of angiosperm families. *Evolution* 5:299–324

113. Stebbins GL. 1970. Adaptive radiation of reproductive characteristics in angiosperms. I. Pollination mechanisms. *Annu. Rev. Ecol. Syst.* 1:307–26

114. Stebbins GL. 1974. *Flowering Plants—Evolution Above the Species Level*. London: Edward Arnold. 399 pp.

115. Stiles FG. 1978. Ecological and evolutionary implications of bird pollination. *Am. Zool.* 18:715–27

116. Troll W. 1928. *Organisation und Gestalt im Bereich der Blüte*. Berlin: Springer

117. Tucker SC. 1984. Origin of symmetry in flowers. In *Contemporary Problems in Plant Anatomy*, ed. RA White, WC Dickison, pp. 351–95. Orlando, FL: Academic

118. van der Pijl L. 1972. Functional considerations and observations on the flowers of some Labiatae. *Blumea* 20:93–103

119. Vöchting H. 1886. Über Zygomorphie und der Ursachen. *Pringsheim's Jahrb. Wiss. Bot.* 17:297–346

120. Vogel S. 1954. *Blütenbiologische Typen als Elemente der Sippengliederung, dargestellt anhand der Flora Südafrikas*. Jena: Fischer

121. Vogel S. 1996. Christian Konrad Sprengel's theory of the flower: the cradle of floral ecology. See Ref. 77a, pp. 44–62

122. von Frisch K. 1965. *Tanzsprache und Orientierung der Bienen*. Berlin: Springer

123. von Wettstein RR. 1911 (1935). *Handbuch der Systematischen Botanik*. Leipzig & Vienna: Deuticke. 914 pp.

124. Vorobyev M, Gumbert A, Kunze J, Giurfa M, Menzel R. 1997. Flowers through insect eyes. *Israel J. Plant Sci.* 45:93–101

125. Waser NM. 1983. The adaptive nature of floral traits: ideas and evidence. In *Pollination Biology*, ed. L Real, pp. 241–85. Orlando, FL: Academic

126. Waser NM. 1986. Flower constancy: definition, cause, and measurement. *Am. Nat.* 127:593–603

127. Waser NM, Chittka L, Price MV, Williams NM, Ollerton J. 1996. Generalization in pollination systems, and why it matters. *Ecology* 77:1043–60

128. Webb CJ, Lloyd DG. 1986. The avoidance of interference between the presentation of pollen and stigmas in angiosperms. II. Herkogamy. *NZ J. Bot.* 24:163–78

129. Weberling F. 1989. *Morphology of Flowers and Inflorescences*. Cambridge, UK: Cambridge Univ. Press

130. Wehner R. 1971. The generalization of directional visual stimuli in the honey bee, *Apis mellifera*. *J. Insect Physiol.* 17:1579–91

131. Wehner R. 1972. Dorsoventral asymmetry in the visual field of the bee, *Apis mellifera*. *J. Comp. Physiol. A* 77:256–77

132. Wehner R, Flatt I. 1977. Visual fixation in free flying bees. *N. Naturforsch.* 32c:469–71

133. Weigel D, Meyerowitz EM. 1994. The ABCs of floral homeotic genes. *Cell* 78:203–9

134. Willemstein SC. 1987. *An Evolutionary Basis for Pollination Ecology*. Vol. 10. Leiden, Netherlands: Brill. 425 pp.

135. Woodward GL, Laverty TM. 1992. Recall of flower handling skills by bumble bees: a test of Darwin's interference hypothesis. *Anim. Behav.* 44:1045–51

136. Wordsell WC. 1915–1916. *The Principles of Plant-Teratology*. Cambridge, UK: Cambridge Univ. Press

137. Yang EC, Maddess T. 1997. Orientation-sensitive neurons in the brain of the honey bee. *J. Insect Physiol.* 43:329–36

Annu. Rev. Ecol. Syst. 1998. 29:375–403

VERTEBRATE HERBIVORES IN MARINE AND TERRESTRIAL ENVIRONMENTS: A Nutritional Ecology Perspective

J. H. Choat

Department of Marine Biology, James Cook University, Townsville, Australia 4811

K. D. Clements

School of Biological Sciences, University of Auckland, Private Bag 92019, Auckland, New Zealand

KEY WORDS: herbivory, vertebrates, nutritional ecology, digestive physiology

Abstract

The study of digestive physiology provides a framework for analyzing food re-sources, feeding patterns, and evolutionary trends in vertebrate herbivores. Most of the research in this field, nutritional ecology, has been focused on terrestrial herbivores, especially mammals. By integrating physiological, demographic, and evolutionary approaches, the study of terrestrial herbivores has generated several important hypotheses, notably on factors determining body mass. Marine verte-brate herbivores are abundant and locally diverse, but with the exception of reptiles and mammals, we lack information on digestive physiology and processing of plant foods, the key element in terrestrial studies. This review provides a founda-tion for a nutritional ecology of marine vertebrate herbivores, especially teleost fish, by summarizing the available information on their digestive physiology and identifying research priorities in the field.

INTRODUCTION

Food captured by carnivores is of relatively uniform quality and is readily digested and assimilated (159). This is not the case for herbivores. Although

0066-4162/98/1120-0375$08.00

plants may be harvested with little expenditure of energy, the processing and assimilation of plant materials dominated by structural carbohydrates presents a major challenge to the digestive system (28). The capacity to digest plant structural carbohydrates, which are usually impervious to endogenous enzymes, is a defining characteristic of herbivores, in terrestrial systems at least (158). The types of structural materials and the ratio of structural to storage compounds vary in plants and have a profound influence on the form and activity of the herbivore's digestive system (166).

Structural carbohydrates constitute most of the global plant biomass. A diet dominated by structural carbohydrates is associated with diagnostic herbivore features (28). These include a grinding dentition or other mechanisms for trituration, a complex alimentary architecture, and the symbiotic microbes necessary for fermentative digestion of such carbohydrates. These features provide an important starting point in the study of herbivores. However, the emphasis on processing rather than procurement of food requires an understanding of the digestive physiology of herbivores. By analyzing the components of plant materials that are assimilated, digestive physiology provides a key to identifying herbivore food resources and a basis for estimating the amount and type of such foods in different environments (47).

The study of the relationships among feeding structures, diet, and digestive physiology in terrestrial vertebrate herbivores has been termed nutritional ecology (43, 55). Nutritional ecology has helped researchers identify the food resources of herbivores, provide definitions of herbivory (159), and explore issues relating to herbivore ecology and evolution. For example, nutritional ecology has been important in our understanding of constraints affecting body size in herbivores and, by extension, the activity patterns, demography, and distribution of herbivores (42, 43, 55). Investigations involving studies at the physiological, demographic, and ecosystem levels have been the key to a robust and innovative research field.

Studies of marine herbivores have proceeded along different lines from those in terrestrial environments. To date there have been three major themes: (a) system-level analysis of plant productivity and yield to grazers (68, 69), (b) functional grouping of algal assemblages and their responses to herbivory (155, 156), and (c) the relationships among feeding patterns, algal structure, and secondary metabolites (74, 75, 116). Most of this work has focused on teleost fish, the dominant vertebrate herbivores of marine environments. Most studies of marine herbivorous fish have been conducted on coral reefs. However, this work has made little reference to the herbivores themselves. For example, the means by which most teleost fish process and assimilate plant food materials are unclear. With the exception of marine reptiles (13, 16, 168) and marine mammals (88, 92), we know little about the nutritional ecology of herbivorous marine vertebrates.

Our lack of knowledge about nutritional ecology in herbivorous marine vertebrates has caused two problems. First, a major strength of nutritional studies—a thematic approach that integrates individual-, demographic-, and ecosystem-level studies—is not available for marine herbivores. We lack a framework for developing hypotheses to explain the evolution of herbivore structural features, size distributions, and relative abundance patterns. Second, because we have little idea of the real nutritional targets of these fish, it is difficult to identify the food resources of many species. Uncertainty over dietary resources is especially problematic in marine environments because many dietary species are small and occur in complex mixtures (136, 150, 155, 156).

This review helps establish a nutritional ecology for marine herbivores, especially small ectothermic species. This nutritional ecology is needed for three reasons: (a) to more completely identify the food resources used by marine herbivores; (b) to better describe the process of digestion in marine herbivores, including the identification of novel digestive mechanisms; and (c) to obtain a better basis for comparisons between terrestrial and marine ecosystems. Studies of marine and terrestrial ecosystems have identified important differences in animal species diversity and size ranges (100), and in plant size, composition, and turnover rates (45, 71, 76). Nutritional ecology will provide a common framework for investigating the evolution and ecology of herbivores in each system.

This review compares various vertebrate herbivores. Terrestrial and marine systems support diverse and abundant vertebrate herbivores, dominated by mammals in terrestrial ecosystems, primarily on grasslands and savannas (102, 166), and teleost fish (145) on coral reefs. A recent review of the vertebrate digestive system (159) provided a wealth of background material for this comparison. Although numerous species of herbivorous freshwater fish exist (113), inclusion of all of these species, many of uncertain trophic status, is beyond the scope of this review.

The building blocks of a nutritional ecology for any group of animals are (a) details of food composition, (b) alimentary tract anatomy, and (c) mechanisms involved in digestion. This review focuses on the food materials and the digestive physiology of marine vertebrate herbivores, especially the abundant fish faunas of low latitudes. Such an approach has been advocated in previous reviews of marine herbivory (57, 81), but despite an earlier seminal study (108) there has been relatively little follow-up. We still lack knowledge about the mechanisms by which many marine herbivores process and digest food. This review seeks to examine these mechanisms.

MARINE AND TERRESTRIAL VERTEBRATE HERBIVORES

Species Diversity and Size Structure

Vertebrate herbivores are not distributed uniformly among taxa. Of the approximately 4100 species of terrestrial mammals, 1012 can be classified as herbivores (115). Nelson (113) identified a total of 23,637 species of teleosts, of which 13,765 are marine. These include approximately 650 species of nominal herbivores. Herbivores are not well represented among reptiles; an estimated 94 species are distributed among the chelonids (approximately 54 of 260 species) and the lizards (approximately 50 of 2800 species) (15, 126). In terms of species proportions, herbivorous teleost fish, reptiles, and birds (94) have minority representation (2–5%) compared with mammals (25%). Within teleost fish and reptiles, herbivores are found within restricted sets of families. Mammalian herbivores have a much broader taxonomic distribution (159).

Most herbivorous fishes are found on coral reefs (29, 81). On very local scales they achieve high densities (81, 155) and high species diversities (139, 140). Consistent estimates are on the order of 12–20 species per 500 m^2 (32, 101). Herbivorous reef fish also have extended geographical distributions (130) and relatively low global diversity. There are more species of herbivorous rodents (115) than herbivorous reef fish.

Size distributions of herbivorous mammals differ from those of fish (see Figure 1a). Mammalian faunas are dominated by large herbivores, and some species achieve sizes in the megaherbivore range (>1000 kg) (42, 115). This includes two extant sirenians that are aquatic herbivores. Among fish, the larger taxa are carnivores, and the herbivorous groups occur in the mid-size range (Figure 1b). These differences are more pronounced if development is taken into account, as in many teleosts, herbivory commences at settlement (29), and some species grow through two to three orders of magnitude in mass during the life cycle. Given the characteristic survivorship curves of coral reef fishes, populations are dominated by large numbers of small individuals (145). This is in contrast to mammals, in which high-density populations in grassland habitats are dominated by large adults (102, 103).

Figure 1   (a) Distribution of body mass by species in herbivorous mammals and herbivorous teleost fish. Mass is log scaled. Mammal data from Reference 115. (b) Distribution of body mass by species in herbivores relative to nonherbivores in coral reef fish from Indo-Pacific and Caribbean ocean systems and in eutherian mammals. Mass is log scaled. Indo-Pacific data from Reference 129, Caribbean data from Reference 18, and mammal data from Reference 24.

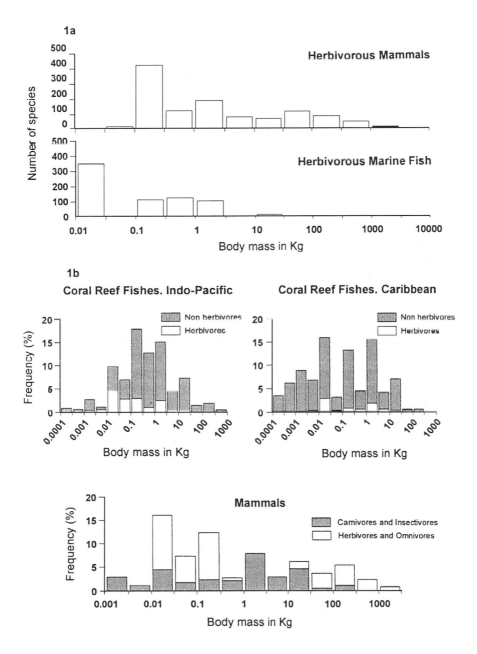

Tropical reef fish show similarities to mammals compared with other vertebrate groups. The relatively small proportion of herbivorous fish species on tropical reefs accounts for about 25–40% of the biomass (121). This is less than equivalent estimates of up to 90% biomass for mammalian grassland faunas (115) but substantially greater than other vertebrate groups. Both piscine and mammalian herbivores play a dominant role in determining the structure of plant assemblages and material cycling in their respective habitats.

Compared with mammals, most reptiles and fish are small. The restricted number of piscine and reptilian herbivores may result from constraints on the capacity of small ectotherms to digest plant starches and polysaccharides (81, 126). For example, White (167) claims that most vertebrate herbivores must pass through a carnivorous stage. However, some very small reptiles and fish can exist on a plant diet (5, 15, 29) by selection of specific dietary items. Although small reptiles may achieve elevated body temperatures through local habitat selection (15), fish with smaller size, laterally compressed bodies, and an external medium that rapidly disperses heat do not have this option. An ectothermic metabolism gives a 30-fold reduction in metabolic rates compared with that of endotherms (119), a basis for maintaining a small body size and high specific growth rates in teleost herbivores. Size-associated constraints may be a factor in temperate and cold water species in which carnivorous episodes occur during ontogeny (132). The mechanisms by which small mammals exploit dietary fiber is a rewarding and unifying research area (4, 43, 55). Equivalent studies in teleosts are required.

Habitat Comparisons

The marked differences in plant size, structure, and turnover rates between marine and terrestrial plants have been summarized elsewhere (71, 76). Both reef algal assemblages and grasslands yield substantial amounts of their daily production to herbivores—coral reefs yield 50–100% (23, 66, 70, 141), and tropical grasslands yield 60% (102). For terrestrial plants, the approximately 50% allocated to below-ground biomass is not generally available to herbivores. In contrast, even those aquatic species in which rhizomes penetrate the substratum will be harvested by excavating herbivores on coral reefs (9).

Estimates of primary production and detrital storage and flux in ecosystems predict the following patterns for coral reefs. A high proportion of the primary production will be removed by herbivores, which exhibit higher rates of herbivory than equivalent terrestrial animals (42). The carbon flux into the detrital pool will be high, with a low resident detrital mass and a high turnover rate (25). These observations are confirmed by numerous examples of high feeding rates by the abundant assemblages of small herbivorous fish on coral reefs and the high turnover of the digested material (122). However, our understanding of the contribution of detrital material, with its associated microbial and meiofaunal components, to coral reef systems remains equivocal (2, 49). Although high activities of microbes associated with algal and mucus-generated detritus have been demonstrated in some reef environments (49), it is not clear what reef organisms are involved in the high detrital flux and turnover.

For most terrestrial systems, plant turnover rate is lower and the accumulated detrital mass higher (25). The higher plant turnover rate in aquatic systems reflects both the structure of the vegetation and the high densities of small ectothermic herbivores that occur in biologically rich sites. The turnover rate in such aquatic systems is also attributable to the rapid passage of gut contents through most fish, which in many cases is associated with the ingestion of large amounts of inorganic material and indigestible organic material (8, 82, 114). Much of this material appears to be reincorporated rapidly into the feeding cycle of nominal herbivores after microbe colonization, resulting in high turnover rates of detrital carbon (25). No equivalent process has been found in terrestrial systems.

Cellulose is the major component of terrestrial plant cell walls, occurring as a complex of microfibrils embedded in a matrix of hemicellulose polymers and insoluble proteins (56). These polysaccharides are closely associated with lignin, which results in a structure of low porosity that requires the specialized yet widespread fermentation-based digestive system to achieve hydrolysis. In contrast, the skeletal and matrix components of algal cell walls are highly diverse. Unlike in vascular plants, the matrix components of marine algae are abundant compared with the skeletal elements (87). Skeletal α-cellulose makes up only 1–8% of thallus dry weight in phaeophytes and rhodophytes compared with the 30% frequently observed in vascular plants (87). The different rates of consumption and turnover of plant materials in terrestrial and marine environments (25, 45, 71) reflect these basic differences in plant structure.

Both terrestrial and marine environments support high numbers of vertebrate herbivores. These faunas and their associated plant foods are distinct. Terrestrial plants require a major investment in structural materials, which prevent desiccation, provide support, and transport nutrients. The herbivores are large endothermic species that must process substantial amounts of the structural polysaccharide cellulose. Marine plants appear to be more readily processed, and the large numbers of small ectothermic herbivores are associated with high turnover rates of plant materials.

DIET QUALITY: IMPLICATIONS FOR THE FEEDING OF HERBIVORES

Positive and Negative Choices

What determines the food choices of herbivores? Optimal foraging theory predicts that herbivores will choose particular food types in order to maximize energy intake. Though sometimes problematic (20), this approach has focused attention on the nutritional qualities of plants and the means by which herbivores select and process materials in order to satisfy their metabolic requirements. Plants also contain a variety of secondary compounds that may deter particular herbivores from feeding on those plants. The herbivore gut makes nutrients available through the action of symbiotic microbes (83), which digest plant fiber and detoxify secondary compounds. An understanding of herbivore feeding must take into account selection based on the nutritional properties of plants (positive choices) and the deterrent effects of secondary compounds (negative choices).

Marine and terrestrial studies have focused on different aspects of the relationship between plant composition and food choice. A major target for the study of herbivory in marine systems has been the occurrence and ecological significance of secondary compounds in algae (71, 72, 76). Responses of herbivores to plant foods differing in the amounts of different secondary compounds have been evaluated through feeding experiments. The results have provided a detailed taxonomy of algal secondary compounds (74, 75) and clarified the ecological consequences of secondary compounds as deterrents to herbivore feeding (72, 74, 116).

Steneck and coworkers (155, 156) have extended the analysis of marine plant structure and composition to develop a functional classification of algal assemblages. In this classification, herbivory is identified as a disturbance that has important consequences for the structural and demographic properties of the plants concerned. Steneck's approach has been a basis for correlating properties of plant assemblages, such as size and productivity, with a defined disturbance regime. Both approaches have concentrated on the properties of plants and the possible deterrent effects of various compounds but have provided relatively little information on the nutritional requirements of the herbivores.

Plant structural properties and secondary metabolites also deter feeding in terrestrial herbivores (79, 104); however, greater attention has been paid to the important nutrients in plant material and the digestive processes required for terrestrial herbivores to extract these nutrients. For example, the ratio of structural elements to storage elements and cell contents is seen as an important factor in herbivore food selection, with choice being dictated by the particular digestive system. The significance of structural elements, especially their role in modifying feeding patterns, is important in the understanding of herbivore nutrition.

Cellulose is a ubiquitous component of plants, and animals require specific modifications of the alimentary tract to achieve its digestion. For such animals cellulose is an important nutrient. For other herbivores an adequate energy intake cannot be achieved by processing of large amounts of cellulose. Is cellulose appropriately viewed as a plant defense? The answer depends on the questions posed and the particular organisms investigated. Regardless of the perspective taken, a comprehensive view requires details of plant structure as well as knowledge of digestive processing by herbivores.

Marine plants do not contain large amounts of cellulose. In some cases, especially in tropical algae, structure is provided by the inclusion of externally derived compounds such as $CaCO_3$. The incorporation of $CaCO_3$ into the matrix of coral reef algae strongly influences the feeding behavior of fish feeding on soft algal thalli but not of those that have calcium-rich diets (149). Is structural $CaCO_3$ most appropriately viewed as a defense against herbivores or as a structural innovation that provides the plant with morphological integrity? An understanding of tropical marine algae requires elements of both of these explanations. An important and neglected aspect of this question concerns how the digestive systems of fish meet the nutritional challenge of ingesting and processing large amounts of $CaCO_3$ in order to fulfill their energy requirements.

To date, our understanding of digestion in herbivorous marine fish has been based on alimentary tract morphology (82). Studies of mammalian herbivores have shown this approach to be problematic. Hofmann (80) used anatomical information to analyze the size-related distribution of feeding patterns in mammalian herbivores. However, both Gordon & Illius (59) and Robbins et al (134) demonstrated that perceived differences among herbivores were attributable to allometric scaling of critical digestive functions rather than anatomy per se.

In summary, terrestrial studies have focused on the nutritional qualities of plants and have explained diet quality and food selection as an interaction between plant nutritional status and the properties of the herbivore's digestive system. In marine systems, the focus has been on the structure and composition of the plants themselves, emphasizing their potentially deterrent properties in the context of herbivore feeding. With the exception of reptilian and mammalian studies, most of our information on digestion by marine vertebrate herbivores has been developed from anatomical studies.

Plant Assemblages and Feeding Patterns

The relatively large size and structural differentiation of terrestrial plants enables herbivores to make precise choices with respect to the parts of plants that are eaten. Herbivores can select parts of plants that are low in fiber content and high in nutrients or process large amounts of material that have a high fiber content (43, 80). The structure of marine plant assemblages, most of which are complex mixtures of small plants, does not allow a high degree of selection for individual plants or their parts (155). Although the algal assemblages of high latitudes contain large, structurally differentiated plants (156), most herbivores are found in low latitudes (81, 105) where individual plants are small. At these latitudes the major pathway to secondary production by the fish is presumed to be through nonselective grazing of algal turf assemblages (29, 67, 121, 125).

Although production by reef algae is high (69), the food quality for herbivores has been considered poor (29, 66, 121). Hatcher (68) summarized this situation in the following terms: "The food is plain, hard to find and at the end of the day nobody leaves with much." The situation is paradoxical. How are such high numbers of small herbivores with high specific metabolic and growth rates maintained on these resources? The problem is confounded by the cryptic and diffuse nature of some of the primary resources, modification of gut contents by mechanical trituration, and the lack of information on the composition and digestibility of dietary algae.

The principal resources used by reef herbivores are usually identified as algal turfs or the epilithic algal community (67, 121). However, analyses of gut contents (137), feeding observations, and morphological studies (2, 6, 8, 9, 114, 127, 170) indicate that processing of sediments and detritus is of equal or greater importance. Feeding behavior and food processing by the resident herbivore fauna, which involves high throughput rates (122), must contribute substantially to the recycling of detrital material.

Bowen and colleagues (19, 20) have clarified the contributions of detrital and algal production to fish nutrition. Their studies of the relative importance of protein and energy as nutritional constraints identified different patterns of constraint for fish feeding on invertebrates, algae, macrophytes, and detritus. Invertebrates clearly provide the highest food quality both in terms of protein and assimilable energy, but they are found in relatively low numbers. Protein is the important dietary constraint for detritivores. In contrast, energy is more important for fish on a diet of algae. An important conclusion is that protein and energy levels must be viewed relative to one another. In many cases, fish use a mixed feeding strategy of increased ingestion rates for energy and increased selection rates for protein. The effectiveness of these responses is shown by the fact that some of the highest growth rates reported for freshwater fish occur in herbivorous and detritivorous species (20).

Further analyses of the relative contributions of protein and energy to fish food resources are needed. The high defecation rates of reef fish suggest that partial digestion and subsequent colonization of feces by microbes will enrich the detrital pool. The rather different results with respect to the rate enrichment of algal-derived detritus that were obtained by Tenore et al (163) and Duggins & Eckman (50) indicate that algal source and subsequent patterns of colonization of detritus by microbes are important in this context. Bowen et al (20) have shown that wide ranges of protein content are seen in detritus, a reflection on its variable genesis. This observation contrasts with Polunin's statement (121) that detrital quality is so poor that only decomposer organisms can make extensive use of it. The capacity of fish to assess differences in diet quality in diffuse resources with small particle size is critical for the development of models of selective feeding.

Diet Quality and Food Selection

Are herbivorous fish able to qualitatively sample a diffuse, widely distributed resource? Finger (53) has described possible mechanisms for assessing dietary quality in diffuse resources. Innervation of the palatal organ and branchial surfaces by branches of the vagus nerve in *Carassius auratus* enables chemical properties of ingested substratum and detrital material to be assessed rapidly. A sensorimotor coupling of the nerve fibers allows analysis of ingested material by taste buds. Similar sensory organs in reef fish would allow continuous monitoring of the particulate material ingested and triturated.

Some fish species show distinct temporal patterning in daily feeding activities (34, 122, 123). The suggestion that daily feeding rates of herbivorous fish are structured around times of maximum photosynthesis (the diel feeding hypothesis) has two implications for nutritional ecology. The first concerns the ability of fish to detect the time of maximum photosynthesis. Fish can detect levels of photosynthate (photosynthetic product) accumulation in the algae, or they can detect the oxygen production resulting from photosynthesis, or both. The former is most likely, given that most herbivorous fish occupy shallow reef areas characterized by high levels of water movement. The second implication of the diel feeding hypothesis is that some nutritional benefit is to be gained by feeding on actively photosynthesizing algae. If so, then diel feeding periodicity could have come about by selection for feeding pattern alone (i.e. in the absence of the ability to detect photosynthates).

The main initial photosynthates produced by rhodophytes are digeneaside (2-D-glycerate-α-D-mannopyranoside) by the Ceramiales, and floridoside (2-O-D-glycerol-α-D-galactopyranoside) by the remainder of the Rhodophyta (86). The main chlorophyte and phaeophyte photosynthates are sucrose and mannitol, respectively (118). Of these four photosynthates, only sucrose is certain to be of direct nutritional benefit to herbivorous fish (i.e. in the absence of microbial fermentation). Sucrase activity is widespread among herbivorous fish (142), but mannitol is poorly utilized by vertebrates (147), and the susceptibility of digeneaside and floridoside to the endogenous enzymes of fish is unknown.

Initial photosynthates may be converted rapidly into useful nutrients such as amino acids and storage polysaccharides (90). Furthermore, the chlorophyte *Enteromorpha intestinalis* contains approximately an order of magnitude more starch than sucrose (52), indicating that starch is more important as an energy source. Given this observation, and the fact that the initial photosynthates of rhodophytes and phaeophytes are of questionable nutritional value, it is possible that fish use soluble sugars and amino acids as indicators of over-all algal nutrient status. Little experimental evidence supports the three main assumptions of the diel feeding hypothesis: (a) significant daily fluctuations in nutrient levels in dietary algae correlate with the photophase of the algae and the feeding activity of fish, (b) these daily fluctuations can be sensed by the fish, and (c) these daily fluctuations have some nutritional significance. Zoufal & Taborsky (171) documented a correlation between feeding activity and die-tary energy levels in two populations of *Parablennius sanguinolentus*, each of which fed on different algae. However, no significant relationship was found throughout the day between energy content and either starch or protein levels in the algae. Furthermore, the maximum daily variation in energy measured was only 7% for turf and 13% for *Ulva lactuca*.

At a more general level, the relationships between the time of day and the rate of algal photosynthesis and accumulation of photosynthates are extremely complex. Rates of photosynthesis in macroalgae can be lowest in the middle of the day (e.g. 64) or highest in the middle of the day (e.g. 77). Floridoside content reaches a peak around noon in *Porphyra perforata* (106), whereas a decrease in rate of photosynthesis at noon is accompanied by a decrease in soluble protein concentration in *Corallina elongata* (60). Research on phaeo-phytes and rhodophytes showed that the rate of photosynthesis decreases around noon and early afternoon because of photoinhibition, a mechanism that protects the photosynthetic apparatus from excessive light levels (61, 63). Photoinhibi-tion is influenced by algal species, tidal state, turbidity, algal depth, and water movement (61, 62). These findings indicate that caution should be taken when making generalizations about daily fluctuations in algal nutrient levels.

DIGESTION AND ASSIMILATION

Digestive Mechanisms Defined

Our current understanding of digestion in marine herbivorous fish is framed in the context of morphological and physiological specializations that allow access to nutrients inside algal cells. These digestive mechanisms or strategies were proposed by Lobel (91) and developed further by Horn (81, 82). Horn (82) characterized four basic types of digestive mechanism:

1. Type I: acid lysis in a thin-walled stomach, where low pH in the stomach is thought to weaken algal cell walls and allow digestive enzymes to come into contact with algal nutrients

2. Type II: trituration in a gizzard-like stomach, where algal cells are ingested with inorganic material and mechanically disrupted to release nutrients

3. Type III: trituration in pharyngeal jaws, where specialized pharyngeal jaws shred or grind algal material before it reaches the intestine

4. Type IV: microbial fermentation in the hindgut, where microbial symbionts in the hindgut assist in the breakdown of algal cells

These four mechanisms are not mutually exclusive, given that combinations of I and IV (Kyphosidae) and III and IV (Odacidae) are known. Digestion in marine herbivorous fish is customarily described solely in terms of these mechanisms (82). Thus, studies on digestion in marine herbivorous fish have been restricted largely to measurements of assimilation efficiency and gut pH and to descriptions of alimentary tract anatomy. Interpretations of digestive physiology and nutrition on the basis of anatomical differences (e.g. 80) are problematic for terrestrial herbivores (59, 134). Furthermore, digestion is usually defined as the breakdown and assimilation of nutrients. Given this definition, what are our current limitations in terms of explaining (a) how fish gain access to nutrients inside algal cells (cell lysis) and (b) the breakdown and assimilation of algal nutrients?

The ability of marine herbivorous fish to use nutrients found inside algal cells is thought to involve low gastric pH, a gizzard-like stomach, a pharyngeal mill, microbial symbionts in the posterior intestine, or a combination of these mechanisms. Interestingly, evidence that these mechanisms are capable of lysing algal cells is quite limited. For example, Lobel (91) is usually cited as demonstrating that low pH is effective in releasing cell contents; however, Lobel's results showed lysis in response to low pH in only three of seven species of algae, and two of those lysed were cyanobacteria. Lobel's cautious and sensible conclusion that pH has a direct effect on some algae but not others has been largely ignored.

Similarly, studies on the effects of trituration in a gizzard-like stomach and pharyngeal mill are limited. Payne (117) showed that bacterial and cyanobacterial cells were lysed by abrasion with sand grains in the pyloric stomach of the detritivorous mullet *Mugil cephalus*. Nelson & Wilkins (114) and Bellwood (8) showed that trituration in the stomach of *Ctenochaetus striatus* and in the pharyngeal mill of *Chlorurus* spp., respectively, reduced the size of ingested carbonaceous particles. None of these studies demonstrated cell lysis in chlorophyte, rhodophyte, or phaeophyte algae. To date, no study has examined cell lysis by microbial processes in the hindgut of marine herbivorous fish. Our intention is not to cast doubt upon the efficacy and importance of these mechanisms but rather to highlight the assumptions on which they are presently based. There is a clear need for detailed examination of the effect of pH, mechanical disruption, and microbial processes on different species of algae.

Digestive Enzymes

We have suggested that the evidence for lysis of algal cells by marine herbivorous fish is limited. What then is the status of knowledge on the process of digestion and assimilation? Digestion usually involves the enzymatic hydrolysis of carbohydrates, proteins, and lipids. In 1989, Horn (81) indicated our lack of knowledge about digestive enzymes in marine herbivorous fish. The situation has changed little since then. Marine herbivorous fish are thought to produce the standard complement of vertebrate carbohydrases, proteases, and lipases (81), but there are few demonstrations of this theory. An example of the research needed is Sabapathy & Teo's series of papers on the digestive enzymes of the rabbitfish *Siganus canaliculatus* (142–144); they examined the types of digestive enzymes present in an herbivorous fish. An interesting and important point is the degree to which the storage polysaccharides of chlorophytes, rhodophytes, and phaeophytes are susceptible to the endogenous digestive enzymes of algivorous fish. The starch of chlorophytes and the floridean glycogen of rhodophytes are likely to be hydrolyzed, at least partially, by endogenous amylase. However, little can be said until more is known about the structure of these carbohydrates in different species of algae and the specificity of amylases from different species of fish.

Evidence for endogenous utilization of the phaeophyte storage compounds laminarin, $\alpha(1\text{-}3)\text{-}\beta$-glucan, and mannitol is equivocal. Laminarin was hydrolyzed by extracts of the intestine, pyloric ceca, esophagus, and stomach of *S. canaliculatus* (144), and $(1\text{-}3)\text{-}\beta$-glucanase activity was detected in some carnivorous marine fish (120). However, some $(1\text{-}3)\text{-}\beta$-glucan hydrolysis in fish gut extracts may be the result of exogenous processes (160). Sugar alcohols such as mannitol are utilized poorly by higher vertebrates and serve as substrates for fermentation in the posterior intestine (110, 147). The fate of these compounds in the gut of fish remains to be determined.

The Role of Microbial Symbionts in Digestion

The lack of biochemical information on the relationship between the digestive enzymes of marine herbivorous fish and the carbohydrates of dietary algae stands in contrast to the terrestrial situation. The nutritional ecology of terrestrial vertebrate herbivores is underpinned by the concept of dietary fiber, which establishes a framework for the description of endogenous vs exogenous digestive mechanisms. The principal components of fiber in terrestrial herbivores are resistant starch and structural polysaccharides such as cellulose (166). Bacteria and fungi adhere to cellulose-rich plant cell wall fragments and cereal grains in the rumen (56, 146). In contrast, research on marine herbivorous fish suggests that many conspicuous elements of the microbiota are free swimming (35, 54, 109) and are thus unlikely to be involved in the degradation of particulate or cell wall material.

The diversity of algal storage and structural polysaccharides and the lack of information on digestive enzymes means we know little about substrates resistant to endogenous digestion by marine herbivorous fish, and thus we do not know which substrates are available for fermentation in the posterior intestine of these animals. However, Seeto et al (151) demonstrated that the hindgut microbiota of *Odax cyanomelas* and *Crinodus lophodon*, two species of Australian temperate water fish, differed in their ability to metabolize (i.e. ferment) the monosaccharides glucose, galactose, and fructose, and the sugar alcohol mannitol.

Although Seeto et al did not determine the dietary source of the monosaccharide substrates, that is, in terms of structural or storage polysaccharides, their results have two important implications. First, the finding that the metabolism of gut symbionts was geared to the substrates present in the diet of the host fish suggests that attention must be paid to the biochemical composition of different algal species. Second, if fish resemble other vertebrates in their inability to effectively assimilate sugar alcohols, the fermentation of mannitol to acetate in *O. cyanomelas* is likely to salvage energy for the host fish. It may be no accident that phaeophytes, which contain large amounts of mannitol, tend to be eaten by fish species (e.g. *Kyphosus sydneyanus*, *K. vaigiensis*, *O. cyanomelas*, *O. pullus*, *Naso lituratus*, and *N. unicornis*) that contain elevated levels of fermentation products in the posterior intestine (37–39, 133).

Our understanding of the importance of hindgut fermentation in marine herbivorous fish is hampered by the absence of data on the contribution of fermentation to host energy balance (36). Fish have the capacity to assimilate (165) and metabolize (39) acetate, the predominant short-chain fatty acid found in the fish hindgut (36). However, no data are available on the rate of fermentation in the hindgut, and information on the basal metabolic rates and daily energy requirements of marine herbivorous fish is extremely limited, although Benavides et al (10) estimated oxygen consumption and daily energy requirements of *Aplodactylus punctatus*. These parameters have been estimated for many terrestrial vertebrate herbivores (15, 83). Although hindgut fermentation likely contributes to the energy balance of many marine herbivorous fish (36), positive statements to this effect must await future research.

Assimilation of Nutrients

One area of digestion that has received considerable attention is digestibility or assimilation efficiency (e.g. 11, 21, 162). The importance of fish aquaculture has led to the development of a variety of methods to estimate the amount of nutrition derived by fish from a variety of foods (48). The most widely used methods for estimating digestibility are total collection methods, indigestible marker methods, and incorporation of radioactively labeled carbon. Galetto & Bellwood (58) compared and reviewed the first two methods and urged that caution be taken when comparing the results of studies that employed different techniques.

The samples used for digestibility estimates can be obtained from experiments on captive fish or wild-caught fish. In the former, captive fish are fed a known diet, which is analyzed in comparison to feces collected from aquaria. In the latter, material taken from the posterior intestine is used as a proxy for feces and is compared with the diet in terms of nutrient concentration. The fluid nature of the environment and in many cases the complex nature of the diet in marine herbivorous fish pose difficulties for both sources of material. Leaching of nutrients from food and feces in aquaria can bias estimates of digestibility unless appropriate techniques are used to present food and to isolate fecal material rapidly after defecation (58). Other problems with aquarium experiments involve recreating natural diets adequately and disruption or loss of gut flora because of the stress associated with capture and maintenance in captivity. Temporary cessation of feeding is known to lead to loss of epulo symbionts in surgeonfish (54), and functional changes in gut flora in response to captivity are well known in other animals, such as termites (M Slaytor, personal communication). Problems associated with the use of samples from the posterior intestine of wild-caught fish as a proxy for feces are (a) the assumption that digestion and assimilation are complete in this part of the gut and (b) the presence of autochthonous microbial biomass in the posterior intestine. Both of these problems will lead to an underestimate of digestibility when comparing nutrient levels in food and feces.

The effect of algal secondary metabolites on digestibility in a variety of marine herbivorous fish has recently been studied. In some of these studies, secondary metabolites reduced digestibility of dietary algae (e.g. 17, 84); however, in others they had no effect (161, 162). Boettcher & Targett (17) showed that the effects of phlorotannins on digestibility in the stichaeid fish *Xiphister mucosus* depend on their molecular size. Gut environment may influence the degree to which these compounds interfere with digestion in fish (17, 161). Targett & Targett (162) demonstrated the ability of fish to compensate for differences in digestibility by varying gut transit times and food intake rates. The ability to maximize the efficiency of nutrient extraction from different diets by varying intake rate, gut passage rate, and digestibility is also observed in some terrestrial herbivores, such as tortoises (14) and equids (51). Targett & Targett (162) concluded that the rate of energy gain cannot be determined simply by multiplying the energy content of the diet by its digestibility.

Food Intake Rate and Gut Passage Rate

Data on food intake rate by marine herbivorous fish are usually presented in the context of feeding rate expressed as bites per unit time, and the amount of material actually ingested has been estimated in only a few studies (e.g. 7, 22). Studies that have addressed the issue of intake rate of specific nutrients are even fewer (e.g. 22).

Gut passage rate (or gut retention time) has been estimated in a variety of marine herbivorous fish (81), but the methodology involved is generally crude and involves only the particulate fraction (36). Marnane & Bellwood (97) developed a particulate digestive marker that reliably replicated the passage rate of algal filaments in pomacentrids, although Bjorndal (15) indicated that the reliability of a similar method used to estimate passage time in herbivorous reptiles may depend on the complexity of gut structure. To date, no study has addressed the important issue of fluid retention times in marine herbivorous fish. Retention of fluid and fine particles is a critical feature of hindgut fermentation in mammals (83, 159). The production of reliable estimates for particle and fluid passage rates in the gut of marine herbivorous fish is an important area for future work.

A NUTRITIONAL ECOLOGY FOR PISCINE HERBIVORES?

For most piscine herbivores the small size of food items taken, the use of mechanical trituration (9, 81, 82), and the ingestion of large amounts of organic material (8, 114) make traditional methods of stomach content analysis inadequate. In such cases, identification of dietary substrata requires analysis at the biochemical level. For this purpose we have investigated short-chain fatty acids (SCFAs) produced through fermentation and known to be influenced by the proportions of dietary substrates. The data are derived from a variety of nominally herbivorous species (Figure 2), most of which have been shown to harbor intestinal microbes (36). Details of the species and their SCFAs are provided elsewhere (36–39, 151).

The predominant SCFA in all species was acetate, concentrated in the posterior gut section. The SCFA isovalerate was also present in several species. A highly systematic feature of the SCFA distributions was a negative correlation between the total SCFAs in the posterior gut and the proportion of isovalerate (Figure 2). This distribution also correlates with diet. Species with a high amount of acetate include the four kyphosid species sampled, *N. unicornis*, *N. lituratus*, a pomacanthid, and two species of the genus *Odax*. With the exception of the pomacanthid, these species are browsers on larger algae, and the kyphosids, *N. unicornis*, *N. lituratus*, and the odacids feed predominantly on fucoids and laminarians (33, 37, 38, 151). The central section of the plot consists of a complex of algal-grazing herbivores including species of *Acanthurus*, *Naso*, and members of the family Siganidae. Four species of the genus *Naso* (*N. annulatus*, *N. brevirostris*, *N. hexacanthus*, and *N. vlamingii*), which feed

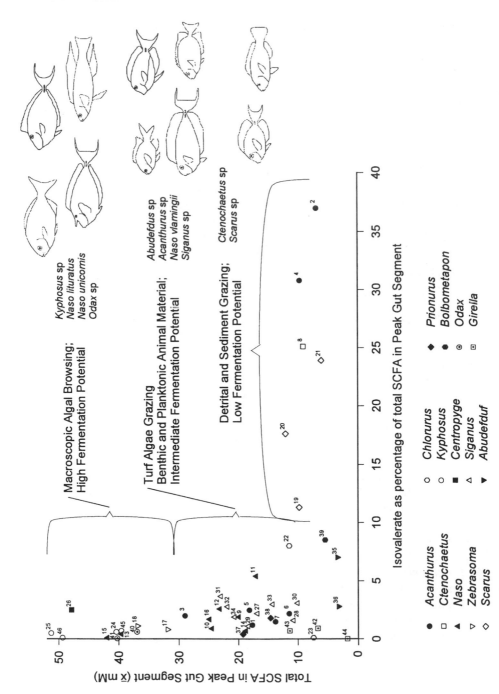

predominantly on planktonic and nektonic invertebrates as adults (85, 129), also occur here.

A third group of species was characterized by an unusually high proportion (>10%) of isovalerate in the hindgut. This group comprised fish of the genera *Scarus, Ctenochaetus*, and *Acanthurus*. These fish (with the exception of *Acanthurus mata*, which feeds on pelagic crustaceans) feed over reef and substrata and ingest substantial amounts of detritus and sediment. A smaller group of species with both low SCFA levels and a low proportion of isovalerate included excavating scarids (*Chlorurus, Bolbometapon*), girellids, and two pomacentrids of the genus *Abudefduf*.

The analysis of fermentation products indicates a variety of feeding and digestive mechanisms in nominally herbivorous fish. Factors contributing to this variety in other vertebrates include alimentary tract anatomy, gut retention time, dietary substrata, and endosymbiont populations (12, 93, 99, 157, 158). A variety of substrates are fermented in the gut of these fish. This group of fish includes phylogenetically similar groups such as nasiid acanthurids, which exhibit a range of dietary types (85, 169).

According to this analysis, one group of fish may derive an important portion of their nutritional energy from microbial biomass. With the exception of *A. mata*, this group includes acanthurids and scarids, which have gut contents characterized by high sediment and detrital loads and possess a mechanism for the trituration of ingested material. Isovalerate is known only as a metabolite of leucine catabolism in all organisms including bacteria (1, 98) and should be related to the amount of leucine available for fermentation. Species with high levels of isovalerate in the hindgut eat animal material or mixtures of detritus sediment and algal fragments. Reef sediments and detritus contain high levels of bacteria (2, 65, 153). None of the species with a high concentration

←——

Figure 2   Total short-chain fatty acids (SCFAs) in hindgut segment plotted against isovalerate as a percentage of total SCFAs in hindgut segment. The data cover 14 genera and 46 species of nominally herbivorous fish. Species key: 1 *Acanthurus lineatus*; 2 *Acanthurus mata*; 3 *Acanthurus nigricans*; 4 *Acanthurus nigricauda*; 5 *Acanthurus nigrofuscus*; 6 *Acanthurus olivaceus*; 7 *Acanthurus triostegus*; 8 *Ctenochaetus striatus*; 9 *Naso annulatus*; 10 *Naso brachycentron*; 11 *Naso brevirostris*; 12 *Naso hexacanthus*; 13 *Naso lituratus*; 14 *Naso tuberosus*; 15 *Naso unicornis*; 16 *Naso vlamingii*; 17 *Zebrasoma scopas*; 18 *Zebrasoma veliferum*; 19 *Scarus niger*; 20 *Scarus rivulatus*; 21 *Scarus schlegeli*; 22 *Chlorurus sordidus*; 23 *Chlorurus gibbus*; 24 *Kyphosus cinerascens*; 25 *Kyphosus vaigiensis*; 26 *Centropyge bicolor*; 27 *Siganus argenteus*; 28 *Siganus corallinus*; 29 *Siganus doliatus*; 30 *Siganus lineatus*; 31 *Siganus puellus*; 32 *Siganus punctissimus*; 33 *Siganus punctatus*; 34 *Siganus vulpinus*; 35 *Abudefduf sordidus*; 36 *Abudefduf septemfasciatus*; 37 *Prionurus microlepidotus*; 38; *Prionurus maculatus*; 39 *Bolbometapon muricatum*; 40 *Odax pullus*; 41 *Odax cyanomelas*; 42 *Girella cyanea*; 43 *Girella elevata*; 44 *Girella tricuspidata*; 45 *Kyphosus bigibbus*; 46 *Kyphosus sydneyanus*.

Table 1 Percent dry weight of leucine in dietary categories. Numbers in parentheses refer to number of species examined. Percent dry weight leucine values given as mean ± standard error

Category		% Dry weight leucine	Reference
Phaeophytes	(10)	0.48 ± 0.08	(111)
	(4)	0.79 ± 0.21	(89)
	(3)	0.88 ± 0.18	(107); (96)
Chlorophytes	(4)	1.32 ± 0.35	(111)
	(6)	1.44 ± 0.31	(89)
Rhodophytes	(10)	1.34 ± 0.21	(111)
	(2)	1.40 ± 0.07	(107); (96)
Marine diatoms	(6)	2.17 ± 0.29	(21)
	(4)	3.07 ± 0.23	(27); (46)
Planktonic crustacea	(3)	3.77 ± 0.38	(154)
	(5)	3.42 ± 0.45	(131)
Bacteria	(1)	5.61	(112)
(mixed sample)		6.01	(152)

of isovalerate in the gut is a strict algal feeder, although scarids are usually considered to be so (6, 123, 124). Table 1 shows published values of percent dry weight leucine content for selected dietary categories, resulting in the following ranking: phaeophytes < rhodophytes and chlorophytes < diatoms < bacteria. Our interpretation is that species with high levels of isovalerate in the gut feed on microbial biomass. The high levels in planktonic crustacea provide an explanation for the nutritional status of *A. mata*. Other species may ingest microbes but lack the mechanisms to digest them.

A greater diversity of feeding processes exists than has been revealed by ecological approaches, and on coral reefs such processes may have a more complex resource base than was previously thought. Abundance estimates of scraping scarids and detrital-feeding acanthurids (101, 139, 140) show that these fish are the most abundant elements of the fauna. Their small size suggests a relatively high mass-specific metabolic rate and a high turnover of the detrital material on which they graze. More explicitly herbivorous species may ingest algae such as phaeophytes (137, 150), which will be rapidly returned to the local reef substratum as a consequence of high gut turnover rates (81, 122, 137). This triturated and mucus-coated material will enhance bacterial colonization. Although these species may have relatively low assimilation rates (26, 81), they have initially high specific growth rates when small (30, 31). This observation confirms the general pattern of a negative correlation between assimilation efficiencies and growth efficiencies in aquatic grazing food chains (135).

The most abundant components of the nominally herbivorous fauna of coral reefs may not be functional herbivores in the sense that they target macroscopic plant material. Many fish, such as kyphosids and some nasiid acanthurids, are clearly herbivores grazing on complex macroscopic plants and exhibiting long food-processing times and microbial fermentation. Cold-water species such as odacids are explicitly herbivorous in these terms (36, 40). Most species found in nominally herbivorous fish faunas appear to target and process reworked biological material produced by the high-density grazing fish faunas.

CONCLUSIONS

In reviewing the comparative nutritional ecology of vertebrate herbivores, we focused on the composition of dietary algae and their digestion by different species of marine fish. This effort contrasts with attempts to define broad functional groups of species based on few characteristics (155, 156). The benefit of a functional group approach is that it provides a framework for ecological generalizations by putting site- and species-specific information into a broader context. In this sense, species might be regarded as noise in the system (73), noise that a functional group approach can filter out to give a clear view of the higher-level organization. Are we using the right set of filters?

In the case of marine herbivores, we argue that we are not. To reiterate, much of the work on marine plant-herbivore interactions has focused on the plant component. Activities of herbivores have been visualized in terms of morphological and behavioral features, an approach that has proved problematic in terrestrial (59, 80) and freshwater studies (95).

Functional groups differ from guilds, which are based on assumptions of common resource use (138, 156). The identification of resource use within a functional group such as herbivorous fish is the key issue. On coral reefs herbivores graze and browse a complex mixture of algae, microbes, meiofauna, and detritus (78, 150). Although fish may harvest a variety of algal species, we do not know which species are digested and which are passed through the alimentary tract and thus contribute to detrital food webs. In the absence of such knowledge we are unable to quantify the resources used by key groups.

A recurring question in marine plant-herbivore studies is the nature of constraints faced by herbivorous fish in higher latitudes. Physiological constraints associated with small size and an ectothermic metabolism have been suggested (57, 76, 81). It is ironic that the emerging picture of piscine herbivory suggests that cold-water species (Odacidae) and warm temperate species (Kyphosidae) may be the most explicitly herbivorous fish investigated to date. The capacity of these fish to digest and assimilate phaeophyte algae, which contain abundant secondary metabolites and storage products that may be resistant to endogenous digestion, suggests that they will be particularly important in the investigation of nutritional ecology.

Work on algal secondary metabolites has been extraordinarily productive in highlighting the biochemical disparity, sensu Raff (128), of algae (71, 74, 75, 116). However, the relationships among secondary metabolites, diet quality, and herbivore digestion are complex. The results of several recent studies caution against making generalizations concerning the role and function of secondary metabolites in algae: (a) Secondary metabolite levels are partially determined by nutrient levels (3, 170); (b) the effects of secondary metabolites depend on the digestive physiology of the herbivores concerned (162, 170); (c) algae differ in their ability to allocate secondary metabolites among different tissues (44); (d) secondary metabolites may have functions other than herbivore defense (148); and (e) the deterrent effects may be influenced by herbivore experience (164).

We agree with Hay (74) that the role of secondary metabolites should be viewed in the context of nutritional ecology and digestive physiology. Algal diets must be evaluated in terms of their susceptibility to endogenous and exogenous digestive processes, so that diet choice reflects both positive choices (levels of assimilable nutrients) and negative choices (levels of deterrent compounds).

For terrestrial vertebrates, the definition of herbivory and the identity of herbivores poses few problems. Although a variety of plant materials are consumed, the most important component of the definition is the ability to digest structural polysaccharides, the major component of plant fiber. For marine vertebrates there is limited consensus as to what constitutes a herbivore. From the viewpoint of functional ecology, herbivores simply remove plant material. A nutritional component is harder to define. We do not yet have the information needed to establish a nutritional component of this definition, a prerequisite for identifying resources and carrying capacities. If the definition includes consumption and assimilation of material from living plants, the notion of what constitutes a herbivore on coral reefs will need to be reconsidered.

The uncertainty concerning resource targets has implications for our understanding of the trophic structure of coral reefs. Estimates of secondary productivity usually assume relatively simple trophic links between primary producers (turfing algae) and nominal herbivores. However, if the nutritional targets include animal, bacterial, and detrital matter, then the trophic pathways on coral reefs may require evaluation. Future production estimates may require a more explicit acknowledgment of microbial contributions. More important, the study of herbivory in marine organisms requires an approach that recognizes its contrasts to terrestrial systems. In particular, the distinctive structure of marine plants, the ingestion of microbial biomass, and the prevalence of fermentative processes with a variety of dietary substrata raise the possibility of unique nutritional pathways in marine fish.

ACKNOWLEDGMENTS

DR Bellwood, GJ Cooper, DJ Crossman, CE Deibel, PJ Harris, ID Hume, AWD Larkum, WL Montgomery, TA Rees, AM Robertson, GR Russ, M Slaytor, CE Stevens, RMG Wells, and L Zemke-White discussed the ideas and provided constructive criticism. The Australian Museum made available critical resources through the Lizard Island Research Station. Support was provided by the Australian Research Council.

Visit the *Annual Reviews home page* at
http://www.AnnualReviews.org

Literature Cited

1. Allison MJ. 1978. Production of branched-chain volatile fatty acids by certain anaerobic bacteria. *Appl. Environ. Microb.* 35:872–77
2. Alongi DM. 1989. Detritus in coral reef ecosystems: fluxes and fates. In *Sixth Int. Coral Reef Symp.*, ed. JH Choat, 1:29–36. Townsville, Australia
3. Arnold TM, Tanner CE, Hatch WI. 1995. Phenotypic variation in polyphenolic content of the tropical brown alga *Lobophora variegata* as a function of nitrogen variability. *Mar. Ecol. Prog. Ser.* 123:177–83
4. Batzli GO, Broussard AD, Oliver RJ. 1994. The integrated processing response in herbivorous small mammals. See Ref. 28, pp. 324–36
5. Bellwood DR. 1988. Ontogenetic changes in the diet of early post-settlement *Scarus* species Pisces: Scaridae. *J. Fish Biol.* 33:213–19
6. Bellwood DR. 1995. Carbonate transport and within-reef patterns of bioerosion and sediment release by parrotfishes (family Scaridae) on the Great Barrier Reef. *Mar. Ecol. Prog. Ser.* 117:127–36
7. Bellwood DR. 1995. Direct estimate of bioerosion by parrotfish species, *Chlorurus gibbus* and *C. sordidus*, on the Great Barrier Reef, Australia. *Mar. Biol.* 121:419–29
8. Bellwood DR. 1996. Production and re-working of sediment by parrotfishes (family Scaridae) on the Great Barrier Reef, Australia. *Mar. Biol.* 125:795–800
9. Bellwood DR, Choat JH. 1990. A functional analysis of grazing in parrotfishes (family Scaridae): the ecological implications. *Environ. Biol. Fish* 28:189–214
10. Benavides AG, Cancino JM, Ojeda FP. 1994. Ontogenetic change in the diet of *Aplodactylus punctatus* (Pisces: Aplodactylidae): an ecophysiological explanation. *Mar. Biol.* 118:1–5
11. Benavides AG, Cancino JM, Ojeda FP. 1994. Ontogenetic changes in gut dimensions and macroalgal digestibility in the marine herbivorous fish, *Aplodactylus punctatus. Funct. Ecol.* 8:46–51
12. Bergman EN. 1990. Energy contributions of volatile fatty acids from the gastrointestinal tract in various species. *Physiol. Rev.* 70:567–90
13. Bjorndal KA. 1979. Cellulose digestion and volatile fatty acid production in the green turtle *Chelonia mydas. Comp. Biochem. Physiol.* 63A:127–33
14. Bjorndal KA. 1989. Flexibility of digestive responses in two generalist herbivores, the tortoises *Geochelone carbonaria* and *Geochelone denticulata. Oecologia* 78:317–21
15. Bjorndal KA. 1997. Fermentation in reptiles and amphibians. See Ref. 94, 1:199–230

16. Bjorndal KA, Suganuma H, Bolten AB. 1991. Digestive fermentation in green turtles *Chelonia mydas*, feeding on algae. *Bull. Mar. Sci.* 48:166–71

17. Boettcher AA, Targett NM. 1993. Role of polyphenolic molecular size in reduction of assimilation efficiency in *Xiphister mucosus*. *Ecology* 74:891–903

18. Bohlke JE, Chaplin CG. 1993. *Fishes of the Bahamas and Adjacent Tropical Waters*. Austin: Univ. Texas Press. 2nd ed.

19. Bowen SH. 1979. A nutritional constraint in detritivory by fishes: the stunted population of *Sarotherodon mossambicus* in Lake Sibaya, South Africa. *Ecol. Monogr.* 49:17–31

20. Bowen SH, Lutz EV, Ahlgren MO. 1995. Dietary protein and energy as determinants of food quality: trophic strategies compared. *Ecology* 76:899–907

21. Brown MR. 1991. The amino-acid and sugar composition of 16 species of microalgae used in mariculture. *J. Exp. Mar. Biol. Ecol.* 145:79–99

22. Bruggemann JH, Begeman J, Bosma EM, Verburg P, Breeman AM. 1994. Foraging by the stoplight parrotfish *Sparisoma viride*. II. Intake and assimilation of food, protein and energy. *Mar. Ecol. Prog. Ser.* 106:57–71

23. Carpenter RC. 1986. Partitioning herbivory and its effects on coral reef communities. *Ecol. Monogr.* 56:345–63

24. Caughley G, Krebs CJ. 1983. Are big mammals simply little animals writ large. *Oecologia* 59:7–17

25. Cebrian J, Duarte CM. 1995. Plant growth-rate dependence of detrital carbon storage in ecosystems. *Science* 268:1606–8

26. Chartock MA. 1983. The role of *Acanthurus guttatus* (Bloch & Schneider 1801) in cycling algal production to sediment. *Biotropica* 15:117–21

27. Chuecas L, Riley JP. 1969. The component combined amino acids of some marine diatoms. *J. Mar. Biol. Assoc. UK* 49:117–20

28. Chivers DJ, Langer P. 1994. *The Digestive System in Mammals: Food, Form and Function*. Cambridge, UK: Cambridge Univ. Press

29. Choat JH. 1991. The biology of herbivorous fishes on coral reefs. See Ref. 145, pp. 120–55

30. Choat JH, Axe LM. 1996. Growth and longevity in acanthurid fishes; an analysis of otolith increments. *Mar. Ecol. Prog. Ser.* 134:15–26

31. Choat JH, Axe LM, Lou DC. 1996. Growth and longevity in fishes of the family Scaridae. *Mar. Ecol. Prog. Ser.* 145:33–41

32. Choat JH, Bellwood DR. 1985. Interactions amongst herbivorous fishes on a coral reef: influence of spatial variation. *Mar. Biol.* 89:221–34

33. Choat JH, Clements KD. 1992. Diet in odacid and aplodactylid fishes in Australia and New Zealand. *Aust. J. Mar. Fresh. Res.* 43:1451–59

34. Choat JH, Clements KD. 1993. Daily feeding rates in herbivorous labroid fishes. *Mar. Biol.* 117:205–11

35. Clements KD. 1991. Endosymbiotic communities of two herbivorous labroid fishes, *Odax cyanomelas* and *O. pullus*. *Mar. Biol.* 109:223–29

36. Clements KD. 1997. Fermentation and gastrointestinal microorganisms in fishes. See Ref. 94, 1:156–98

37. Clements KD, Choat JH. 1995. Fermentation in tropical marine herbivorous fishes. *Physiol. Zool.* 68:355–78

38. Clements KD, Choat JH. 1997. Comparison of herbivory in the closely-related marine fish genera *Girella* and *Kyphosus*. *Mar. Biol.* 127:579–86

39. Clements KD, Gleeson VP, Slaytor M. 1994. Short-chain fatty acid metabolism in temperate marine herbivorous fish. *J. Comp. Physiol. B* 164:372–77

40. Clements KD, Rees D. 1998. Preservation of inherent contractility in isolated gut segments from herbivorous and carnivorous marine fish. *J. Comp. Physiol. B* 168:61–72

41. Deleted in proof

42. Clutton-Bock TH, Harvey PH. 1983. The functional significance of variation in body size among mammals. In *Advances in the Study of Mammalian Behaviour*, ed. J Eisenberg, pp. 632–63. Shippenberg, PA: Am. Soc. Mammologist Spec. Publ. No. 7

43. Cork SJ. 1994. Digestive constraints on dietary scope in small and moderately small mammals: How much do we really understand? See Ref. 28, pp. 337–69

44. Cronin G, Hay ME. 1996. Within-plant variation in seaweed palatability and chemical defenses: optimal defense theory versus the growth differentiation hypothesis. *Oecologia* 105:361–68

45. Cyr H, Pace ML. 1993. Magnitude and patterns of herbivory in aquatic and terrestrial systems. *Nature* 361:148–50

46. Darley WM. 1977. Biochemical composition. In *The Biology of Diatoms*, ed. D Werner, 13:198–223. Oxford, UK: Blackwell Sci.

47. Demment MW, Van Soest PJ. 1985. A nutritional explanation for body size patterns of ruminant and non-ruminant herbivores. *Am. Nat.* 125:641–72
48. De Silva SS, Anderson TA. 1995. *Fish Nutrition in Aquaculture.* London: Chapman & Hall
49. Ducklow HW. 1990. The biomass, production and fate of bacteria on coral reefs. In *Coral Reefs*, ed. Z Dubinsky, 25:265–90. New York: Elsevier
50. Duggins DO, Eckman JE. 1997. Is kelp detritus a good food for suspension feeders? Effects of kelp species, age and secondary metabolites. *Mar. Biol.* 128:489–95
51. Duncan P, Foose TJ, Gordon IJ, Gakahu CG, Lloyd M. 1990. Comparative nutrient extraction from forages by grazing bovids and equids: a test of the nutritional model of equid/bovid competition and coexistence. *Oecologia* 84:411–18
52. Edwards DM, Reed RH, Cliudek JA, Foster R, Stewart WDP. 1987. Organic solute accumulation in osmotically-stressed *Enteromorpha instestinalis. Mar. Biol.* 95:583–92
53. Finger TE. 1988. Sensorimotor mapping and oropharyngeal reflexes in goldfish, *Crassius aurautus. Brain Behav. Evol.* 31:17–24
54. Fishelson L, Montgomery WL, Myrberg AA. 1985. A unique symbiosis in the gut of tropical herbivorous surgeonfish (Acanthuridae: Teleostei) from the Red Sea. *Science* 229:49–51
55. Foley WJ, Cork SJ. 1992. Use of fibrous diets by small herbivores: How far can the rules be bent? *Trends Ecol. Evol.* 7:159–62
56. Fosberg CW, Cheng KJ, White BA. 1997. Polysaccharide degradation in the rumen and large intestine. See Ref. 94, 1:319–79
57. Gaines SD, Lubchenco J. 1982. A unified approach to marine plant-herbivore interactions. II. Biogeography. *Annu. Rev. Ecol. Syst.* 13:111–38
58. Galetto MJ, Bellwood DR. 1994. Digestion of algae by *Stegastes nigricans* and *Amphiprion akindynos* (Pisces: Pomacentridae), with an evaluation of methods used in digestibility studies. *J. Fish Biol.* 44:415–28
59. Gordon IJ, Illius AW. 1994. The functional significance of the browser-grazer dichotomy in African ruminants. *Oecologia* 98:167–75
60. Haeder DP, Lebert M, Flores Moya A, Jiminez C, Mercado J. 1997. Effects of solar radiation on the photosynthetic activity of the red alga *Corallina elongata* Ellis et Soland. *J. Photochem. Photobiol. B.* 37:196–202
61. Hanelt D. 1992. Photoinhibition of photosynthesis in marine macrophytes of the South China Sea. *Mar. Ecol. Prog. Ser.* 82:199–206
62. Hanelt D. 1996. Photoinhibition of photosynthesis in marine algae. *Sci. Mar.* 60(Suppl. 1):243–48
63. Hanelt D, Huppertz K, Nultsch W. 1993. Daily course of photosynthesis and photoinhibition in marine macroalgae investigated in the laboratory and field. *Mar. Ecol. Prog. Ser.* 97:31–37
64. Hanelt D, Li J, Nultsch W. 1994. Tidal dependence of photoinhibition of photosynthesis in marine macrophytes of the South China Sea. *Bot. Acta* 107:66–72
65. Hansen JA, Skilleter GA. 1994. Effects of the Gastropod *Rhinoclavis aspera* (Linnaeus, 1758) on microbial biomass and productivity in coral-reef-flat sediments. *Aust. J. Mar. Fresh Res.* 45:569–84
66. Hatcher BG. 1981. The interaction between grazing organisms and the epilithic algal community of a coral reef: a quantitative assessment. In *Proc. 4th Int. Coral Reef Symp., Manila,* Vol. 2, ed. E Gomez, Univ. Philippines
67. Hatcher BG. 1983. Grazing in coral reef ecosystems. In *Perspectives on Coral Reefs*, ed. DJ Barnes, pp. 164–79. Townsville, Australia: Aust. Inst. Mar. Sci.
68. Hatcher BG. 1988. Coral reef primary productivity: a beggar's banquet. *Trends Ecol. Evol.* 3:106–11
69. Hatcher BG. 1990. Coral reef primary productivity: a hierarchy of pattern and process. *Trends Ecol. Evol.* 5:149–55
70. Hatcher BG, Larkum AW. 1983. An experimental analysis of factors controlling the standing crop of the epilithic algal community on a coral reef. *J. Exp. Mar. Biol. Ecol.* 69:61–84
71. Hay ME. 1991. Marine-terrestrial contrasts in the ecology of plant chemical defenses against herbivores. *Trends Ecol. Evol.* 6:362–65
72. Hay ME. 1991. Fish-seaweed interactions on coral reefs: effects of herbivorous fishes and the adaptation of their prey. See Ref. 145, pp. 96–119
73. Hay ME. 1994. Species as "noise" in community ecology: Do seaweeds block our view of the kelp forest? *Trends Ecol. Evol.* 9:414–16
74. Hay ME. 1996. Marine chemical ecology: What's known and what's next? *J. Exp. Mar. Biol. Ecol.* 200:103–34

75. Hay ME. 1997. The ecology and evolution of seaweed-herbivore interactions on coral reefs. *Coral Reefs* 16(Suppl.):S67–76

76. Hay ME, Steinberg PD. 1992. The chemical ecology of plant-herbivore interactions in marine versus terrestrial environments. In *Herbivores: Their Interactions with Secondary Plant Metabolites*, Vol. II: *Evolutionary and Ecological Processes*, ed. GA Rosenthal, MR Berenbaum, pp. 371–411. San Diego, CA: Academic

77. Henley WJ, Levavasseur G, Franklin LA, Lindley ST, Ramus J, Osmond CB. 1991. Diurnal responses of photosynthesis and fluorescence in *Ulva rotundata* acclimated to sun and shade in outdoor culture. *Mar. Ecol. Prog. Ser.* 75:19–28

78. Hixon MA, Brostoff WN. 1996. Succession and herbivory: effects of differential fish grazing on Hawaiian coral-reef algae. *Ecol. Monogr.* 66:67–90

79. Hochuli DF. 1996. The ecology of plant/insect interactions: implications for digestive strategy for feeding by phytophagous insects. *Oikos* 75:133–41

80. Hofmann RR. 1989. Evolutionary steps of ecophysiological adaptation and diversification of ruminants: a comparative view of their digestive system. *Oecologia* 78:443–57

81. Horn MH. 1989. Biology of marine herbivorous fishes. *Oceanogr. Mar. Biol. Annu. Rev.* 27:167–272

82. Horn MH. 1992. Herbivorous fishes: feeding and digestive mechanisms. In *Plant-Animal Interactions in the Marine Benthos*, ed. DM John, SJ Hawkins, JH Price, 46:339–62. Systematics Assoc. Oxford: Clarendon

83. Hume ID. 1997. Fermentation in the hindgut of mammals. See Ref. 94, 1:84–115

84. Irelan CD, Horn MH. 1991. Effects of macrophyte secondary chemicals on food choice and digestive efficiency of *Cebidichthys violaceus* (Girard), a herbivorous fish of temperate marine waters. *J. Exp. Mar. Biol. Ecol.* 153:179–94

85. Jones RS. 1968. Ecological relationships in Hawaiian and Johnson Island Acanthuridae (surgeonfishes). *Micronesica* 4:309–61

86. Kirst GO. 1980. Low MW carbohydrates and ions in Rhodophyceae: quantitative measurement of floridoside and digeneaside. *Phytochemistry* 19:1107–10

87. Kloareg B, Quatrano RS. 1988. Structure of the cell walls of marine algae and ecophysiological functions of the matrix polysaccharides. *Oceanogr. Mar.*

Biol. Annu. Rev. 26:259–313

88. Lanyon JM, Marsh H. 1995. Digesta passage time in the dugong *Dugong dugong*. *Aust. J. Zool.* 43:119–27

89. Lewis EJ, Gonzalves EA. 1960. Amino acid contents of some marine algae. *New Phytol.* 59:109–15

90. Lobban CS, Harrison PJ. 1994. *Seaweed Ecology and Physiology*. Cambridge, UK: Cambridge Univ. Press

91. Lobel PS. 1981. Trophic biology of herbivorous reef fish: alimentary pH and digestive capabilities. *J. Fish Biol.* 19:365–97

92. Lomolino MV, Ewel KC. 1984. Digestive efficiencies of the West Indian manatee (*Trichechus manatus*). *Fla. Sci.* 47:176–79

93. Macfarlane GT, Macfarlane S. 1993. Factors affecting fermentation reactions in the large bowel. *Proc. Nutr. Soc.* 52:367–73

94. Mackie RI, White BA, ed. 1997. *Gastrointestinal Microbiology*. New York: Chapman & Hall

95. MacNeil CJ, Dick TA, Elwood RW. 1997. The trophic ecology of freshwater *Gammarus* spp. (Crustacea: Amphipoda): problems and perspectives concerning the functional feeding group concept. *Biol. Rev.* 72:349–64

96. Mai K, Mercer JP, Donlon J. 1994. Comparative studies on the nutrition of two species of abalone, *Haliotis tuberculata* Linnaeus and *Haliotis discus hannai* Ino. II. Amino acid composition of abalone and six species of macroalgae with an assessment of their nutritonal value. *Aquaculture* 128:115–30

97. Marnane M, Bellwood DR. 1997. Marker technique for investigating gut throughput rates in coral reef fishes. *Mar. Biol.* 129:15–22

98. Massey LK, Sokatch JR, Conrad RS. 1976. Branched-chain amino acid catabolism in bacteria. *Bacteriol. Rev.* 40:42–54

99. Mathers JC, Annison EF. 1993. Stoichiometry of polysaccharide fermentation in the large intestine. In *Dietary Fibre and Beyond—Australian Perspectives*, ed. S Samman, G Annison, 1:123–35. Sydney: Nutr. Soc. Aust. Occas. Publ.

100. May RM. 1994. Biological diversity: differences between land and sea. *Philos. Trans. R. Soc. B* 343:105–11

101. McClanahan TR. 1994. Kenyan coral reef lagoon fish: effects of fishing, substrate complexity, and sea urchins. *Coral Reefs* 13:231–41

102. McNaughton SJ. 1985. Ecology of a grazing system: the Serengeti. *Ecol. Monogr.* 55:259–94

103. McNaughton SJ, Georgiadis NJ. 1986. Ecology of African grazing and browsing mammals. *Annu. Rev. Ecol. Syst.* 17:39–65

104. McNaughton SJ, Tarrants JL. 1983. Grass leaf silification: natural selection for an inducible defense against herbivores. *Proc. Natl. Acad. Sci. USA* 80:790–91

105. Meekan MG, Choat JH. 1997. Latitudinal variation in the abundance of herbivorous fishes: a comparison of temperate and tropical reefs. *Mar. Biol.* 128:373–83

106. Meng J, Srivistava LM. 1993. Variations in the floridoside content and floridoside phosphate synthase activity in *Porphyra perforata* (Rhodophyta). *J. Phycol.* 29:82–84

107. Mercer JP, Mai KS, Donlon J. 1993. Comparative studies on the nutrition of two species of abalone, *Haliotis tuberculata* Linnaeus and *Haliotis discus hannai* Ino. I. Effects of algal diets on growth and biochemical composition. *Invertebr. Reprod. Dev.* 23:75–88

108. Montgomery WL, Gerking SD. 1980. Marine macroalgae as food for fishes: an evaluation of potential food quality. *Environ. Biol. Fish.* 5:143–53

109. Montgomery WL, Pollak PE. 1988. *Epulopiscium fishelsoni* N.G., n. sp., a protist of uncertain taxonomic affinities from the gut of a herbivorous reef fish. *J. Protozool. Res.* 35:565–69

110. Moroshita Y. 1994. The effect of dietary mannitol on the caecal microflora and short-chain fatty acids in rats. *Lett. Appl. Microbiol.* 133:31–36

111. Munda IM, Gubensek F. 1976. The amino acid composition of some common marine algae from Iceland. *Bot. Mar.* 19:85–92

112. Neidhardt FC. 1987. Chemical composition of *Escherichia coli*. In *Escherichia coli and Salmonella typhimurium. Cellular and Molecular Biology*, ed. FC Neidhardt, 1:3–6. Washington, DC: Am Soc. Microbiol.

113. Nelson JS. 1994. *Fishes of the World*. New York: Wiley. 3rd ed.

114. Nelson SG, Wilkins SD. 1988. Sediment processing by the surgeonfish *Ctenochaetus striatus* at Moorea, French Polynesia. *J. Fish Biol.* 32:817–24

115. Nowak RM. 1991. *Walker's Mammals of the World*. Baltimore, MD: Johns Hopkins Univ. Press. 5th ed.

116. Paul VJ, Hay ME. 1986. Seaweed susceptibility to herbivory: chemical, and morphological correlates. *Mar. Ecol. Prog. Ser.* 33:255–64

117. Payne AI. 1978. Gut pH and digestive strategies in estuarine grey mullet (Mugilidae) and tilapia (Cichlidae). *J. Fish Biol.* 13:627–29

118. Percival E, McDowell RH. 1967. *Chemistry and Enzymology of Marine Algal Polysaccharides*. New York: Academic

119. Peters RH. 1983. *The Ecological Implications of Body Size*. Cambridge, UK: Cambridge Univ. Press

120. Piavaux A. 1977. Distribution and localisation of the digestive laminarinases in animals. *Biochem. Syst. Ecol.* 5:231–39

121. Polunin NVC. 1996. Trophodynamics of reef fisheries productivity. In *Reef Fisheries*, ed. CM Roberts, NVC Polunin, pp. 113–35. New York: Chapman & Hall

122. Polunin NVC, Harmelin-Vivien M, Galzin R. 1995. Contrasts in algal food processing among five herbivorous coral reef fishes. *J. Fish Biol.* 47:455–65

123. Polunin NVC, Klumpp DW. 1989. Ecological correlates of foraging periodicity in herbivorous reef fishes of the Coral Sea. *J. Exp. Mar. Biol. Ecol.* 126:1–20

124. Polunin NVC, Klumpp DW. 1992. Algal food supply and grazer demand in a very productive coral-reef zone. *J. Exp. Mar. Biol. Ecol.* 164:1–15

125. Polunin NVC, Klumpp DW. 1992. A trophodynamic model of fish production on a windward reef tract. In *Plant Animal Interactions in the Marine Benthos*, ed. DM John, SJ Hawkins, JH Price, 46:213–33. Syst. Assoc. Oxford: Clarendon

126. Pough FH. 1973. Lizard energetics and diet. *Ecology* 54:837–44

127. Purcell SW, Bellwood DR. 1993. A functional analysis of food procurement in two surgeonfish species, *Acanthurus nigrofuscus* and *Ctenochaetus striatus* (Acanthuridae). *Environ. Biol. Fish.* 37:139–59

128. Raff RA. 1996. *The Shape of Life*. Chicago: Univ. Chicago Press

129. Randall JE, Allen GR, Steene RC. 1990. *Fishes of the Great Barrier Reef and Coral Sea*. Barthhurst, Aust.: Crawford House

130. Rapoport EH. 1994. Remarks on marine and continental biogeography: an areographical perspective. *Philos. Trans. R. Soc. B* 343:71–78

131. Raymont JEG, Morris RJ, Ferguson CF, Raymont JKB. 1975. Variation in the amino-acid composition of lipid-free residues of marine animals from the northeast Atlantic. *J. Exp. Mar. Biol. Ecol.* 17:261–67

132. Rimmer DW. 1986. Changes in the diet and the development of microbial digestion in juvenile buffalo bream *Kyphosus cornellii. Mar. Biol.* 93:443–48
133. Rimmer DW, Wiebe WJ. 1987. Fermentative microbial digestion in herbivorous fishes. *J. Fish Biol.* 31:229–36
134. Robbins CT, Spalinger DE, Van Hoven W. 1995. Adaptation of ruminants to browse and grass diets: Are anatomical-based browser-grazer interpretations valid? *Oecologia* 103:208–13
135. Robertson AI, Hatcher BG. 1994. Trophic relations, food webs and energy flow. In *Marine Biology*, ed. LS Hammond, RN Synnot, pp. 131–51. Melbourne, Aust.: Longman Cheshire
136. Robertson DR. 1982. Fish feces as fish food on a Pacific coral reef. *Mar. Ecol. Prog. Ser.* 7:253–65
137. Robertson DR, Gaines SD. 1986. Interference competition structures habitat use in a local assemblage of coral reef surgeonfishes. *Ecology* 67:1372–83
138. Root RB. 1973. Organisation of a plant-arthropod association in simple and diverse habitats: the fauna of collards (*Brassica oleracea*). *Ecol. Monogr.* 43:95–125
139. Russ GR. 1984. Distribution and abundance of herbivorous grazing fishes in the central Great Barrier Reef. I. Levels of variability across the entire continental shelf. *Mar. Ecol. Prog. Ser.* 20:23–34
140. Russ GR. 1984. Distribution and abundance of herbivorous grazing fishes in the central Great Barrier Reef. II. Patterns of zonation of mid-shelf and outershelf reefs. *Mar. Ecol. Prog. Ser.* 20:35–44
141. Russ GR. 1987. Is the rate of removal of algae by grazers reduced inside territories of tropical damsel fishes? *J. Exp. Mar. Biol. Ecol.* 110:1–17
142. Sabapathy U, Teo LH. 1993. A quantitative study of some digestive enzymes in the rabbitfish, *Siganus canaliculatus* and the sea bass, *Lates calcarifer. J. Fish Biol.* 42:595–602
143. Sabapathy U, Teo LH. 1994. Some kinetic properties of amylase from the intestine of the rabbitfish, *Siganus canaliculatus* (Park). *Comp. Biochem. Physiol. B* 109:139–44
144. Sabapathy U, Teo LH. 1995. Some properties of the intestinal proteases of the rabbitfish, *Siganus canaliculatus* (Park). *Fish Physiol. Biochem.* 14:215–21
145. Sale PF, ed. 1991. *The Ecology of Fishes on Coral Reefs.* San Diego, CA: Academic
146. Salyers A, Reeves A, D'Ella J. 1996.

147. Saunders DR, Wiggins HS. 1981. Conservation of mannitol, lactulose and raffinose by the human colon. *Am. J. Physiol.* 241:G397–402
148. Schmitt TM, Hay ME, Lindquist N. 1995. Constraints on chemically mediated co-evolution: multiple functions for seaweed secondary metabolites. *Ecology* 76:107–23
149. Schupp PJ, Paul VJ. 1994. Calcium carbonate and secondary metabolites in tropical seaweeds: variable effects on herbivorous fishes. *Ecology* 75:1172–85
150. Scott FJ, Russ GR. 1987. Effects of grazing on species composition of the epilithic algal community on coral reefs of the central Great Barrier Reef. *Mar. Ecol. Prog. Ser.* 39:293–304
151. Seeto GS, Veivers, PC, Clements KD, Slaytor M. 1996. Carbohydrate utilisation by microbial symbionts in the marine herbivorous fishes *Odax cyanomelas* and *Crinodus lophodon. J. Comp. Physiol. B* 165:571–79
152. Simon M, Azam F. 1989. Protein content and protein synthesis rates of planktonic marine bacteria. *Mar. Ecol. Prog. Ser.* 51:201–13
153. Sorokin YI. 1981. Periphytonic and benthic microflora on the reef: biomass and metabolic rates. In *Proc. 4th Int. Coral Reef Symp., Manila*, Vol. 2, ed. E Gomez, Univ. Philippines
154. Srinivasagam RT, Raymont JEG, Moodie CF, Raymont JKB. 1971. Biochemical studies on marine zooplankton. X. The amino acid composition of *Euphausia superba, Meganyctiphanes norvegica* and *Neomysis integer. J. Mar. Biol. Assoc. UK* 51:917–25
155. Steneck RS. 1988. Herbivory on coral reefs; a synthesis. In *Proc. 6th Int. Coral Reef Symp., Townsville*, ed. JH Choat, 1:37–49
156. Steneck RS, Dethier MN. 1994. A functional group approach to the structure of algal dominated communities. *Oikos* 69:476–98
157. Stevens CE. 1978. Physiological implications of microbial digestion in the large intestine of mammals: relation to dietary factors. *Am. J. Clin. Nutr.* 31:S161–68
158. Stevens CE. 1989. Evolution of vertebrate herbivores. *Acta Vet. Scand. Suppl.* 86:9–19
159. Stevens CE, Hume ID. 1995. *Comparative Physiology of the Vertebrate Diges-*

Solving the problem of how to eat something as big as yourself: diverse bacterial strategies for degrading polysaccharides. *J. Ind. Microbiol.* 17:470–76 *tive System.* Cambridge, UK: Cambridge Univ. Press. 2nd ed.

160. Stone BA, Clarke AE. 1992. *Chemistry and Biology of (1–3)-b-glucans.* Melbourne, Aust.: La Trobe Univ. Press

161. Targett NM, Boettcher AE, Targett TE, Vrolijk NH. 1995. Tropical marine herbivore assimilation of phenolic-rich plants. *Oecologia* 103:170–79

162. Targett TE, Targett NM. 1990. Energetics of food selection by the herbivorous parrotfish *Sparisoma radians*: roles of assimilation efficiency, gut evacuation rate, and algal secondary metabolites. *Mar. Ecol. Prog. Ser.* 66:13–21

163. Tenore KR, Hanson RB, McClain J, MacCubbin AE, Hodson RE. 1984. Changes in the composition and nutritional value to a benthic deposit feeder of decomposing detritus pools. *Bull. Mar. Sci.* 35:299–311

164. Thacker RW, Nagle DG, Paul VJ. 1997. Effects of repeated exposure to marine cyanobacterial secondary metabolites on feeding by juvenile rabbitfish and parrotfish. *Mar. Ecol. Prog. Ser.* 147:21–29

165. Titus E, Ahearn GA. 1992. Vertebrate gastrointestinal fermentation: transport mechanisms for volatile fatty acids. *Am. J. Physiol.* 262:R547–53

166. Van Soest PJ. 1994. *Nutritional Ecology of the Ruminant.* Ithaca, NY: Cornell Univ. Press. 2nd ed.

167. White TRC. 1993. *The Inadequate Environment: Nitrogen and the Abundance of Animals.* Berlin: Springer-Verlag

168. Wikelski M, Gall B, Trillmich F. 1993. Ontogenetic changes in food intake and digestion rate of the herbivorous marine iguana (*Amblyrhynchus cristatus* Bell). *Oecologia* 94:373–79

169. Winterbottom R, McLennan DA. 1993. Cladogram versatility: evolution and biogeography of acanthuroid fishes. *Evolution* 47:1557–71

170. Yates JL, Peckol P. 1993. Effects of nutrient availability and herbivory on polyphenolics in the seaweed *Fucus vesiculosus.* *Ecology* 74:1757–66

171. Zoufal R, Taborsky M. 1991. Fish foraging periodicity correlates with daily changes of diet quality. *Mar. Biol.* 108: 193–96

Annu. Rev. Ecol. Syst. 1998. 29:405–34

CARBON AND CARBONATE METABOLISM IN COASTAL AQUATIC ECOSYSTEMS

J.-P. Gattuso[1,*], M. Frankignoulle[2] and R. Wollast[3]

[1]Observatoire Océanologique Européen, Avenue Saint-Martin, MC-98000 Monaco, Principality of Monaco

[2]Mécanique des Fluides Géophysiques, Unité d'Océanographie Chimique (B5), Université de Liège, B-4000 Sart Tilman, Belgium; e-mail: michel.frankignoulle@ulg.ac.be

[3]Laboratoire d'Océanographie Chimique, Université Libre de Bruxelles, Campus Plaine, CP 208, Boulevard du Triomphe, B-1050 Brussels, Belgium; e-mail: rwollast@ulb.ac.be

*Present address and address for correspondence: Observatoire Océanologique, ESA 7076 CNRS-UPMC, B P 28, F-06234 Villefranche-sur-mer Cedex, France; e-mail: gattuso@obs-vlfr.fr

KEY WORDS: carbon cycle, calcification, primary production, community metabolism, coastal ecosystems

ABSTRACT

The coastal zone is where land, ocean, and atmosphere interact. It exhibits a wide diversity of geomorphological types and ecosystems, each one displaying great variability in terms of physical and biogeochemical forcings. Despite its relatively modest surface area, the coastal zone plays a considerable role in the biogeochemical cycles because it receives massive inputs of terrestrial organic matter and nutrients, is among the most geochemically and biologically active areas of the biosphere, and exchanges large amounts of matter and energy with the open ocean. Coastal ecosystems have therefore attracted much attention recently and are the focus of several current national and international research programs (e.g. LOICZ, ELOISE). The primary production, respiration, calcification, carbon burial and exchange with adjacent systems, including the atmosphere, are reviewed for the major coastal ecosystems (estuaries, macrophyte communities, mangroves, coral reefs, and the remaining continental shelf). All ecosystems

405

examined, except estuaries, are net autotrophic. The contribution of the coastal zone to the global carbon cycle both during pristine times and at present is difficult to assess due to the limited metabolic data available as well as because of major uncertainties concerning the magnitude of processes such as respiration, exchanges at the open ocean boundary, and air-sea fluxes of biogases.

INTRODUCTION

The world coastline, which extends over about 350,000 km, displays a wide diversity of geomorphological types and ecosystems. The coastal ocean—where land, ocean and atmosphere interact—is shallow (<200 m), covering approximately 7% (26×10^6 km^2) of the surface of the global ocean. Despite its relatively modest surface area, the coastal zone plays a considerable role in the biogeochemical cycles because it (a) receives massive inputs of terrestrial organic matter and nutrient through run-off and groundwater discharge; (b) exchanges large amounts of matter and energy with the open ocean; and (c) constitutes one of the most geochemically and biologically active areas of the biosphere. For example, it accounts for 14–30% of the oceanic primary production, 80% of organic matter burial, 90% of sedimentary mineralization, 75–90% of the oceanic sink of suspended river load, and ca. 50% of the deposition of calcium carbonate (87, 109). Additionally, it represents 90% of the world fish catch (107). Its overall economic value has been recently estimated as 43% of the value of the world's ecosystem services and natural capital (29). The coastal ocean is also the area of greatest human impact on the marine environment since approximately 37% of the human population currently live within 100 km of the coastline (27). The anthropogenic pressure on it is increasing steadily.

Despite its potential importance, the coastal ocean has been relatively neglected until recently, probably because of its intrinsic complexity. It is the focus of several national and international on-going research programs. The Land-Ocean Interactions in the Coastal Zone (LOICZ) program was established as part of the IGBP Global Change Programme in 1993, and the European Union has launched a coastal core project (European Land-Ocean Interaction Studies, ELOISE; Ref. 22).

Several reviews on the biogeochemistry of the coastal ocean have recently been published (e.g. 87, 132, 146, 157).[1] The aim of the present paper is to review the available information using an ecosystem approach, with special emphasis on primary production, respiration, calcification, carbon burial and

---

[1] A recent and exhaustive book on coastal ecosystem processes (5a) has been published too late for discussion in this chapter. Readers are strongly advised to refer to this authoritative book for additional information.

exchange with adjacent systems, including the atmosphere. Whereas the metabolism of the open ocean is by far dominated by phytoplankton primary production, the coastal ocean exhibits a great diversity of primary producers, often inhabiting the same area, which makes it difficult to subdivide this region into subdomains. We first provide some definitions of metabolic terms and, after a brief review of land inputs in the coastal zone, discuss separately, and somewhat arbitrarily, estuaries, macrophyte communities, mangroves, coral reefs, and the remaining continental shelves. Finally, the contribution of costal ecosystems to the marine carbon cycle is reviewed.

DEFINITIONS

The contribution of any biological system (e.g. organism, community, or ecosystem) to the global carbon cycle relies on (a) the balance between organic carbon production and consumption, and (b) the balance between calcium carbonate precipitation and dissolution. A simple model allows prediction of the potential air-sea CO_2 flux driven by these processes (52). A system is net autotrophic (in terms of organic carbon) when production is higher than consumption and is, conversely, net heterotrophic when consumption exceeds production. Note that autotrophy does not necessarily imply an air-to-sea CO_2 flux because the direction of this flux is driven by the sign of the CO_2 pressure gradient across the air-sea interface. For example, upwellings are net autotrophic but are a source of CO_2 to the atmosphere due to the high pCO_2 of upwelled water (higher than 360 μatm, the present average atmospheric pCO_2).

It is difficult to apply these production concepts to data compiled from the literature. First, there is some confusion about which type of production is reported. Net primary production (P_n) is the balance between gross primary production (P_g) and respiration of the autotrophic components of the system (R_a). Excess production (E) or net ecosystem production (NEP) is the difference between P_g and ecosystem respiration (R), which includes both the autotrophic and heterotrophic components. Therefore E ($= NEP$) is of interest for assessing the contribution of an ecosystem to net global processes. Another source of confusion is that it is not clear which type of production is measured by the ^{14}C technique (110). Last, metabolic data obtained on isolated photosynthetic organisms (P_n) are sometimes used instead of, or grouped with, ecosystem metabolic rate (NEP), which leads to overestimating NEP as the respiration rate of the heterotrophic components of the ecosystem is not taken into account.

Units of moles per m^2 and per year are used throughout this chapter. Some metabolic data expressed on a daily basis have been multiplied by 365 in order to get yearly rates; this, of course, neglects seasonal changes in the processes. The following abbreviations are used: $N =$ sample size; $p =$ probability;

$Mmol = 10^6$ moles; $Gmol = 10^9$ moles; $Tmol = 10^{12}$ moles; $Pmol = 10^{15}$ moles. Average data are reported as mean ± standard error of the mean (SE).

LAND INPUTS

River input constitutes the main flux of material from the continents to the ocean and can considerably influence the carbon metabolism of the coastal zone. Furthermore, both the riverine fluxes of nutrients and organic carbon have been significantly affected by human activities and have probably modified the autotrophic vs heterotrophic conditions in estuaries and locally on continental shelves.

Pristine and anthropogenic fluxes of dissolved and particulate carbon (C), nitrogen (N), and phosphorus (P) have been thoroughly investigated by Meybeck (92, 93). The global riverine flux of carbon is approximately 77 Tmol y^{-1}, among which are found 32 Tmol y^{-1} of dissolved inorganic carbon (DIC, 41%), 14 Tmol y^{-1} of particulate inorganic carbon (PIC, 18%), 17 Tmol y^{-1} of dissolved organic carbon (DOC, 22%), and 14 Tmol y^{-1} of particulate organic carbon (POC, 19%). The additional river fluxes of C due to human activity are estimated to be around 8 Tmol y^{-1} (DOC and POC; 93).

The natural flux of dissolved inorganic nitrogen is mainly due to nitrate (225 Gmol N y^{-1}), with a small contribution of ammonia (40 Gmol N y^{-1}). There is a significant contribution of dissolved and particulate organic nitrogen, estimated to be 700 and 2400 Gmol y^{-1}, respectively (93). Natural river fluxes of phosphorus are around 12 Gmol y^{-1} for the inorganic fraction, with a similar, but less well known, flux of dissolved organic phosphorus (93). The particulate flux of phosphorus is also poorly known, but is considerably higher than the dissolved flux—probably of the order of 600 Gmol y^{-1} (93). Meybeck's figures must be considered as minimal estimates because of the lack of data from rivers in developing nations, including the high standing islands in Oceania, which may have greater inputs than suggested by Meybeck (JD Milliman, personal communication).

River fluxes of N and P have been strongly affected by anthropogenic activities, leading to eutrophication in heavily disturbed areas. According to Meybeck (93), the anthropogenic riverine flux of total dissolved nitrogen and phosphorus are about 500 and 30 Gmol y^{-1}, respectively. Wollast (156) estimated the river flux of particulate nitrogen of anthropogenic origin at 500 Gmol y^{-1}. The origins of the increased nutrient fluxes are numerous and, for nitrogen, include washout of fertilizer and intensive livestock operations. Discharge of untreated or partially treated industrial and domestic waste water is responsible for the high fluxes of nutrients and organic carbon observed in heavily populated areas. The magnitude of fluxes of dissolved carbon, nitrogen, and phosphorus of anthropogenic and natural origins are similar (93). The increased riverine inputs of these elements due to anthropogenic activities have profoundly affected estuaries and the adjacent coastal zones (see next section). In the case of large rivers, the dissolved components are directly transferred to the coastal zone, whereas some of the particulate fraction accumulates in the delta. When river water passes across estuarine systems, large changes in the fluxes and speciation of the constituents typically occur.

ESTUARIES

Estuaries are difficult to define (see 69), and there is no consensus on their global surface area. We use a tentative estimate of 1.4×10^6 km^2, derived from areas where the salinity is lower than 34 (159). From a biogeochemical point of view, an estuary can be defined as a semi-enclosed zone where river water mixes with sea water. It should be noted, however, that in the case of large rivers, the flow is sufficiently great that water mixing mostly occurs on the continental shelf rather than in an embayment, and that the material carried by rivers is directly delivered to the shelf. The ecology and geochemistry of estuarine ecosystems, including salt marshes, have been the subject of recent reviews (39, 58, 68, 154).

Estuaries are pathways for the transfer of dissolved and particulate material from the continent to the marine system through rivers. They exhibit a wide range of diversity in terms of geomorphology, geochemistry of the drainage basin, river flow, and tidal influence. These affect physical attributes such as vertical stratification, longitudinal gradients, and residence time of fresh water. Estuaries are extremely dynamic systems usually characterized by strong physico-chemical gradients, enhanced biological activity, and intense sedimentation and resuspension (69). Profound changes are observed in the speciation of organic and inorganic compounds in response to these factors, particularly in macrotidal estuaries, where the tidal regime leads to an increased residence time of fresh water in the estuarine mixing zone and to the generation of a turbidity maximum.

One of the major changes due to human activity is the increased respiration of detrital organic carbon that usually occurs in the upper part of the estuaries, often in the turbidity maximum, and can lead to anoxic conditions that affect the behavior of other elements in the water (154). The turbidity maximum in macrotidal estuaries or deltas of large rivers is also an area of intense shoaling. Due to the flocculation of colloidal material transported by rivers when the salinity increases and to the presence of large quantities of particulate organic carbon, the sediments deposited are organically rich muds characterized by intense anaerobic processes. A large fraction of the particulate load transported by rivers can accumulate in these areas and never reach the continental shelf.

The balance between autotrophy and heterotrophy has been modified by human activity, but in a way still difficult to identify because of antagonistic responses (19, 58, 67, 94, 125, 132). The increased nutrient load leads to eutrophication, enhances net ecosystem production, and shifts the system toward increased autotrophy (e.g. 66, 85). On the other hand, respiration of the organic carbon leads to increased heterotrophy. Additionally, light may become limiting for primary production in the upper part of estuaries (e.g. 63); respiration is then the dominant metabolic process, and an oxygen-depleted zone may occur, stimulating various anaerobic processes. In such areas, the partial pressure of CO_2 in surface water may reach several thousand μatm, inducing very large fluxes of CO_2 to the atmosphere (45).

As a result of light limitation and nutrient availability (28), the maximum primary production in many estuaries occurs at intermediate salinities (125). *NEP* values in the literature display a wide range of variation (+2 to −23 mol C m^{-2} y^{-1}), which may be partly due to different methods used (58). The average value indicates that estuaries are net heterotrophic with a *NEP* significantly lower than 0 (−6 ± 2 mol C m^{-2} y^{-1}; $N = 21$, t-test, $p = 0.001$) and an average P_g/R ratio of 0.8 ± 0.1 (Table 1). Heterotrophy is more pronounced in the salinity range 0 to 30, at least in macrotidal (Figure 1A), where large CO_2 supersaturation is always observed (66; M Frankignoulle, unpublished observations) as a result of organic matter respiration. In the Scheldt estuary (North Sea), approximately 60% of the respiratory CO_2 is released to the atmosphere (45), 26% is transferred to the sediment (94, 101), and only 14% remains in the water column. Global P_g of estuaries is estimated to be approximately 31 Tmol C y^{-1}. The *NEP* is −8 Tmol C y^{-1}, a value that is in good agreement with a previous estimate (132) of −7 Tmol C y^{-1}. The benthic contribution to

Table 1 Surface area and metabolic data for the coastal zone (this paper) and open ocean (157)[a]

System	Surface area (10^6 km^2)	P_g (mol C m^{-2} y^{-1})	P_g (Tmol C y^{-1})	NEP (Tmol C y^{-1})	P_g/R
Coastal ecosystems					
Estuaries	1.4	22	31	−8	0.8 ± 0.05
Macrophyte-dominated	2.0	87	174	37	1.1 ± 0.1
Coral reefs	0.6	144	86	6	1.3 ± 0.2
Salt-marshes	0.4	185	74	7	1.2 ± 0.1
Mangroves	0.2	232	46	18	1.4 ± 0.4
Remaining shelf	21.4	18	377	171	1.4 ± 0.3
Sum	26		789	231	
Open ocean and slope	334	10	3396		

[a]The sources of the surface area are given in the text. The areal gross primary production and *NEP* are the average values of data collected from the literature (see legend of Figure 1). P_g and *NEP* of macrophyte-dominated ecosystems were adjusted empirically by a factor × 0.5 to take into account the bias in the data base (see text). The P_g/R ratio was estimated using a geometric regression technique (see Figure 1).

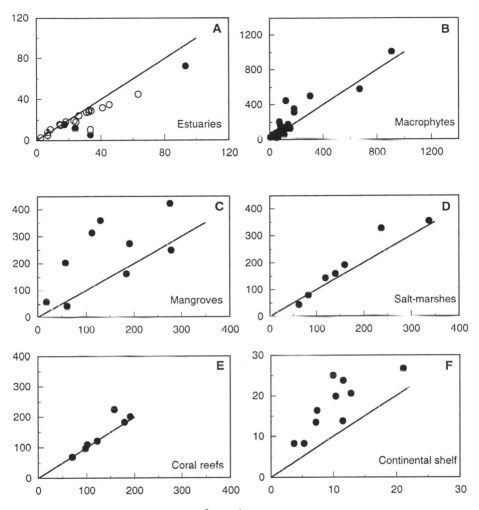

Figure 1 P_g vs R (both in mol C m^{-2} year^{-1}) in selected coastal ecosystems. *A*. Estuaries with macrotidal estuaries shown in dark symbols ($Y = 1.0 + 0.76X$; $r = 0.92$; $N = 21$). *B*. Macrophyte-dominated ecosystems ($Y = 19.7 + 1.13X$; $r = 0.85$; $N = 35$). *C*. Mangroves ($Y = 32.2 + 1.37X$; $r = 0.42$; $N = 9$). *D*. Salt-marshes ($Y = -11.7 + 1.22X$; $r = 0.94$; $N = 8$). *E*. Coral reefs ($Y = -24.6 + 1.28X$; $r = 0.85$; $N = 7$). *F*. Continental shelf ($Y = 3.9 + 1.36X$; $r = 0.648$; $N = 10$). The line shown is the 1:1 relationship; the regression equations were obtained using a geometric regression technique (114); r, correlation coefficient (all significantly different from 0 except for mangroves); N, sample size. The full data sets used, including the list of references, are available at the Annual Reviews web site (http://www.annurev.org/sup/material.htm) as well as at ftp://ccrv.obs-vlfr.fr/pub/gattuso/ares.xls.

total respiration is in the range of 25–50% (58). The ratio of benthic to planktonic respiration depends on the depth of the water, but planktonic respiration is higher than benthic respiration even in a shallow (ca. 3 m) estuary such as Tomales Bay, California (43).

Coastal eutrophication resulting from river inputs most often affects a relatively limited area in the immediate vicinity of the river mouth. The anoxic conditions in the water and/or in the sediments, associated with long residence times of the fresh water in estuaries, are extremely favorable for denitrification. A large part of the nitrate load is lost during the estuarine journey and never reaches the coastal zone. The importance of both the burial of nutrients in estuarine sediments and the denitrification process depends on the tidal prism and the depth of the estuary, two factors that affect the residence time of river water in the system (102).

Finally, carbon dioxide is not the only biogas produced in estuaries. The elevated nutrient loading enhances nitrous oxide (N_2O) production via denitrification of nitrate in the oxygen-depleted zones and nitrification of ammonia in more aerated waters (e.g. 136, 154). The photoproduction of carbon monoxide (CO) in surface water is probably stimulated by terrestrially derived dissolved organic matter. Anoxic sediments in the region of the turbidity maximum enhance hydrogen sulfide (H_2S) and methane (CH_4) production, with subsequent emission to the atmosphere. Eutrophic conditions are also very favorable for the production of gases of importance in climate regulation such as dimethyl sulfide (DMS) and carbonyl sulfide (COS). Despite their potential importance in biogas emission, very little is known about the coupling of estuaries to the atmosphere. On a global scale, estuaries may act as a significant source of these gases, and the magnitude of this source deserves further investigation.

MACROPHYTE-BASED ECOSYSTEMS

Macrophytes (seagrasses and macroalgae) do not constitute ecosystems by themselves and can be found in any shallow coastal or estuarine ecosystem. They cover approximately 2×10^6 km^2 worldwide (149), while the surface area available for micro- and macrophytobenthos has been estimated to be 6.8×10^6 km^2 (25). The areal biomass of macrophytes is about 400 times higher than that of phytoplankton, and their turnover time is much larger (ca. 1 year vs a few days). These attributes make them play a potentially significant role in the global carbon cycle (128). The macrophyte contribution to metabolism is highly variable among ecosystems, depending on their relative surface cover. For example, macrophytes account for less than 1% of net primary production in turbid and nutrient-rich estuaries, and more than 50% in non-turbid ones (58). Metabolism generally exhibits a strong seasonality in macrophyte-dominated ecosystems. The balance of organic carbon varies widely depending on factors such as the dominant species, interspecific competition, climatic conditions (temperature and light), nutrient availability, herbivore pressure, and anthropogenic disturbance (1, 41, 42a, 42b, 58, 73, 88, 89, 103). Data are available on net primary production of several species of macrophytes (see 25), but there is comparatively little information on P_g and R at the community or ecosystem level. Epiphytes, despite their comparatively low biomass, can significantly contribute (up to 20%) to macrophyte production (e.g. 25, 58).

Studies on tropical seagrass beds have suggested that their carbon cycle is balanced: low export is balanced by allochtonous inputs of organic carbon, and most biomass is either stored or remineralized within the system (41, 42a). Some seagrass communities are nutrient limited (1), whereas others are not (42b), demonstrating the variability of nutrients status depending on species and sediment types. Our compilation of data (Figure 1B) suggests that macrophyte-dominated ecosystems are net autotrophic with a NEP of 37 ± 13 mol C m^{-2} y^{-1} ($N = 35$), a value that is significantly different from 0 ($p = 0.008$). An earlier estimate was 42 mol C m^{-2} year^{-1} (128). The average P_g/R ratio is 1.1 ± 0.1 (Table 1).

The fate of P_n depends on the macrophyte ecosystem considered. P_n can be grazed by herbivores, exported outside the system, buried within the sediment, or enter the detrital pathway. Duarte & Cebrián (37) have compiled data from the literature on these pathways for several marine primary producers, including macroalgae, and seagrasses. Their major conclusions are that (a) decomposition within the system is an important process for each macrophyte system (>40% of P_n); (b) herbivore pressure is significant for macroalgae only (>30%); (c) export is significant (24–43%); and (d) storage within the sediment is negligible for macroalgal communities, but not for seagrass (>15%). These trends are quite variable from species to species; for example, 80% of the production of four Mediterranean seagrasses are consumed by detritivores (23).

Buried material within the sediment is estimated to be four times more abundant in higher plant than in algal communities. Marine angiosperm communities, which account for 4% of the oceanic net primary production, could store up to 30% of the total oceanic buried carbon (37). Moreover, seagrasses contain more carbon than N and P compared to pelagic communities: Their C:N:P ratios range from 204:4:1 to 3550:61:1 (126), with an average of 474:24:1 (36).

Posidonia oceanica is characterized by a large difference between above-ground (leaves) and below-ground (*matte* = roots and rhizomes) parts: the turnover of leaves is about 1 y compared to *matte* turnover on the order of a century (121). The latter behaves as a sink for biogenic material (120, 121), which has been estimated at 26% of the produced carbon (90). Large differences have, however, been observed from site to site, suggesting that accretion rate is controlled by local factors (91). Light also controls the transfer of C from shoots to roots in the eelgrass *Zostera marina*, underlining the importance of light in light-limited areas (162).

The global gross primary production and *NEP* of macrophyte-dominated ecosystems estimated from our compilation of the literature are 348 and 74 Tmol C y^{-1}. These figures are likely overestimates because (*a*) more data are available for the more productive tropical than for the less productive temperate ecosystems, and (*b*) more data were obtained in very shallow areas than in deeper, light-limited areas. An empirical adjusting factor of 0.5 can be tentatively and arbitrarily used to account for these biases; the resulting estimate for P_g and *NEP* are 174 and 37 Tmol C y^{-1}, respectively (Table). Previous estimates of *NEP* range from at least 83 (128, 140) to 254 Tmol C y^{-1} (25, 33). Macrophyte-dominated ecosystems appear to be net C sinks but Smith's (128) conclusion of 1981 that their quantitative significance in the global carbon budget was poorly known still stands today.

There is major concern about the survival of seagrasses worldwide due to anthropogenic disturbances. While large, presumably natural, changes in seagrass distribution have occurred (113), human activity severely disturbs seagrass communities in several ways. The eutrophication of coastal areas results in a higher pelagic activity, with subsequent light limitation to benthic communities that induces a decrease in primary production (14), or even seagrass mortality (103). In Chesapeake Bay, it has been suggested that the long-term survival of *Zostera marina* depends on water turbidity rather than on changes in the nutrient concentration or salinity (98). Brown tides, induced by coastal eutrophication, are major causes of seagrass decline (104, 141). However, primary production depends more on temperature than on light availability in some seagrasses species such as *Thalassia testudinum* and *Cymodocea nodosa* (80, 88). A strong correlation has been observed between the standing stock of *Zostera marina* in a Netherlands estuary and the concentration of dissolved silicon that may decrease due to coastal eutrophication (59). Large tidal ranges combined with stresses are responsible for the decline of *Z. marina* in Long Island Sound (73). Worldwide, the mortality of seagrasses is higher than growth rate (89).

MANGROVES

Mangroves are intertidal forests growing above mean sea level, distributed on sheltered shores of the tropics and subtropics (31°N to 39°S); they cover 0.18×10^6 km^2 (138). Mangrove ecosystem function has been the subject of several reviews (3, 4, 56, 86, 115, 144). The major primary producers are mangrove trees, but seedlings, macroalgae, periphyton, and phytoplankton also contribute. The respective contribution of these producers to total mangrove primary production varies with their relative surface cover, delivery of nutrients, and turbidity.

Net assimilation of leaves and above-ground P_n of mangrove trees have been estimated using indirect (allometric) methods, but P_n of whole trees is poorly known because there are no reliable estimates of the respiration rate of the stem and roots (including the above-ground portion known as prop roots) due to the unknown contribution to gas exchange by non-photosynthetic components (26).

Community metabolism of mangrove trees displays considerable variation at both the local and regional scales, primarily as a response to environmental forcings (tide, climate, and seawater composition); forest type is of secondary importance (16). The net primary production of mangrove trees derived from indirect measurements ranges from 12 to 142 mol C m^{-2} y^{-1} (mean $= 58 \pm 7$; $N = 22$). The only measurement of root production is an indirect estimate of about 9 mol C m^{-2} y^{-1}, i.e. ca. 10% of the above-ground P_n of that site (77). Twilley et al (144) compiled data on wood production and provided a global estimate of 13.3 Tmol C y^{-1}.

Most leaf production enters the detrital pathway as litter fall (119). Leaves and, to a lesser extent, twigs, branches, and bark are shed as litter throughout the year. Reproductive parts are shed seasonally. Litterfall, which is negatively correlated with latitude, ranges from 5 to 70 mol C m^{-2} y^{-1} (mean $= 32$; Ref. 124).

Submerged primary production is often limited by high turbidity and changes in salinity. Water column metabolism is largely heterotrophic (e.g. 57, 106, 116). Despite its generally low quantitative importance, phytoplankton production may play an important role in sustaining secondary production because of the poor nutritional quality of mangrove detrital material (117). Macrophytes are generally absent from mangrove ecosystems, but seagrass beds can thrive in areas adjacent to mangrove stands and significantly contribute to total primary production of lagoons (31). Prop root periphyton can be relatively important when shading is moderate ($P_n = 12–34$ mol C m^{-2} y^{-1}). Benthic microalgal production is generally very low or undetectable (e.g. 5) because of: (a) light limitation resulting from shading by the mangrove canopy (76), (b) inhibition by sedimentary organic compounds such as tannins (see 3), and (c) nutrient limitation (76). P_g of benthic microalgae ranges from 0 to 26 mol C m^{-2} y^{-1} (mean $= 6 \pm 2$; $N = 28$).

Sediment respiration is different in submerged and emergent conditions, and data obtained with the widely used O_2 technique are doubtful because anaerobic processes are of major importance in mangrove sediments (3). There is, however, little doubt that the sediment is largely net heterotrophic (average $P_g/R = 0.6$; $P_n = -3 \pm 1.5$ mol C m^{-2} y^{-2}; $N = 31$).

The major source of C for benthic heterotrophs is litter fall, followed by deposited phytoplankton and benthic micro- and macrophytes. The retention and processing of litter within the mangrove system is much greater than initially thought due to consumption and hiding by crabs (116). Mangrove leaf litter supports a very high benthic bacterial productivity (2, 15, 100, 139). It is now recognized, despite methodological uncertainties (2), that carbon flow through microbial pathways probably accounts for a large proportion of total C flow in mangrove ecosystems. Densities of micro-, meio-, and macrofauna are generally very low and are not correlated with bacterial production (3). The so-called 'carbon sink hypothesis' (3, 4) satisfactorily addresses this discrepancy by suggesting that only a small proportion of the large bacterial biomass is consumed in the sediment and that the remaining bacterial carbon is recycled very efficiently though natural mortality and carbon turnover within the sedimentary microbial food web. There are very few estimates of total community metabolism of mangrove systems (Figure $1C$). Most systems investigated are net autotrophs as shown by a NEP significantly different from 0 (89 ± 28 mol C m^{-2} y^{-1}; $N = 12$; $p = 0.008$), despite large variation of the P_g/R ratio (1.4 ± 0.4; $N = 9$; Table 1).

The net organic matter produced can be accumulated or exported to adjacent systems. The content of organic carbon in mangrove sediment varies widely depending on the type of forest and the geomorphology of the site (e.g. 0.21–18 wt%; Ref. 96a). The rate of sedimentation is 0.3–2.4 mm y^{-1}, and the average rate of C accumulation is 23 mol C m^{-2} y^{-1} (144). Burial leads to the accumulation of peat deposits containing up to 17 mmol C g sed.$^{-1}$ (145). The total carbon sequestered in mangrove peat is about 1.7 Tmol C y^{-1} (144). The quality and quantity of material exported from mangroves depend on forest type (riverine, fringe, or basin) and productivity, as well as on physical constraints (strength and frequency of tidal inundation, river flow, wind speed and direction) and biological forcings (e.g. consumption of litterfall by macrodetritivores). In open habitats subjected to tidal flushing (riverine mangroves), a large proportion of leaf litter is exported as debris to the adjacent systems (bays) where it is decomposed (ca. 30% in Pacific mangroves; Ref. 118). Inland habitats (basin forests) are comparatively less subject to tidal flushing, export is very low (e.g. <0.3%; Ref. 81), and decomposition primarily occurs within the mangrove. As a result, the material exported comprises little particulate matter but a greater proportion of dissolved organic compounds (e.g. 143). Consumption and hiding of detritus by macrodetritivores can greatly diminish (by up to 30%) the amount of litter available for export.

Physical forcing, and its effect on export, has been relatively well studied. Outwelling is favored by tidal flow, rates of which are higher during ebb tide than flood tide (152), but lateral trapping in forested tidal rivers (152) and high-salinity plugs (150) can limit export. Export of DOC can change seasonally with tidal inundation and rainfall. DOC is clearly the dominant form of total exported C in two basin forests in Florida (ca. 70%; 3 and 4 mol C m^{-2} y^{-1}; Ref. 143). To our knowledge, the contribution of land-derived DOC and in situ production to total DOC export has not been estimated, but a small net import of DOC ($<1\%$ of P_n) was measured in a tidally dominated (i.e. without terrestrial runoff or groundwater input) creek at Hinchinbrook Island, Australia (18). It is therefore possible that a large proportion of DOC export may actually originate with freshwater inputs (116), although there is also evidence of export in a non-estuarine mangrove forest (99). Export of DOC via groundwater seepage has received very little attention, although it represents 20% of the TOC exported from a Florida forest (143). Mangroves generally act as exporters of organic C, although some forests are net importers due to a limited inundation regime (82). Global export has been estimated at 4.2 Tmol C y^{-1} (144). Exported C can have a significant influence on nearshore benthic processes (6), especially where hydrodynamic features inhibit the mixing of estuarine and offshore waters. Its influence appears to be limited offshore (5–15 km) due to its dispersion and its refractory nature (116).

No data are available for air-water CO_2 fluxes in mangrove areas, and there are only limited data on seawater pCO_2, which seems to remain higher than atmospheric pCO_2 during most of a diurnal cycle (at ca. 1114 μatm) in two mangrove areas in India (53). Efflux of methane from mangrove sediment appears to be very small (<0.1 mmol m^{-2} y^{-1}; Ref. 95), but significant fluxes have been measured when pore water salinity is <1 (9). Mangrove forests are net autotrophic, with a global NEP of 18 Tmol C y^{-1} (Table 1). Marshes, another angiosperm-based ecosystem, bear some similarity to mangrove forests, but are not fully discussed here. The surface area of these temperate ecosystems is twice that of mangroves (0.4 vs 0.2 $\times$ 10^6 km^2), but marshes make a smaller contribution to the global carbon cycle ($NEP = 7$ vs 18 Tmol C y^{-1}) because they are less net autotrophic (P_g/R ratio: 1.2 vs 1.4; Figure 1D and Table 1). NEP data based on burial rates can be greatly overestimated in some marsh communities because import and storage of allochthonous carbon is not always accounted for (96b).

Mangroves are carbon sinks but are being increasingly cleared by humans for activities such as wood production, farming, mining, peat extraction, and other forms of land exploitation (56). It is estimated that 50% of mangrove ecosystems have been transformed or destroyed by human activities (160). The loss of mangrove forest not only diminishes fixation of atmospheric CO_2 and C burial, but also results in the oxidation and release to the atmosphere of the organic C stored in sediments. Approximately 39.3 Mmol C are released per ha of mangrove swamp cleared and excavated, and 31.3 Mmol C are released per 1000 t of dry peat combusted (32).

CORAL REEFS

Coral reefs are carbonate structures located at or near sea level, dominated by scleractinian corals and algae, that display high rates of organic carbon metabolism and calcification. They are mostly distributed in the tropics, although they can also reach higher latitudes (32.5°N; 31.5°S), and they cover approximately 0.6×10^6 km^2 (71, 127). Information on various aspects of reef ecology and metabolism can be found in recent reviews (12, 38, 137).

Most community metabolism data were obtained on reef flat communities, which are technically suited for measurement because they are relatively shallow, protected from the swells and subject, in many cases, to a unidirectional flow, which allows the use of flow respirometry techniques. Consequently, despite the numerous community metabolism data available, the database is biased, and there is comparatively little information available for reef slopes, lagoons, and complete reef systems. The contribution of reef slopes to reef metabolism cannot be ignored in principle, since slopes are generally the most actively growing part of the system. However, they represent a small contribution to the surface area of reef systems (15% in the central Kaneohe Bay sector; Ref. 135), and it is likely that a relatively large proportion of their communities are light-limited.

Reef metabolism is dominated by benthic processes. Phytoplankton community production is very minor, often dominated by picoplankton, and ranges from 0.3 to 22 mol C m^{-2} y^{-1} in atoll lagoons (reviewed in 24). The highest values were in the few sites with elevated nutrient concentrations.

Coral/algal reef flats display a wide range of P_g (79–584 mol C m^{-2} y^{-1}), R (76–538 mol C m^{-2} y^{-1}), and G (5–126 mol CaCO$_3$ m^{-2} y^{-1}). This variability is owing to the absolute and relative surface area covered by the major communities (e.g. corals, macrophytes, and sediments) as well as to seasonal and environmental conditions (see 84). Modal rates of metabolic performances have nevertheless been proposed (e.g. 70, 129). Such standards can be applied only to reefs having similar structure and zonation, and have little predictive value (111). The average P_g/R ratio, estimated using a geometric regression technique, is $1.07 \pm 0.1 (N = 43)$, perhaps indicating a slight net autotrophy.

Algal-dominated reef communities generally display higher rates of organic C metabolism ($P_g = 30-1369$ and $R = 6-910$ mol C m^{-2} y^{-1}) and lower rates of net calcification ($G = -0.4$ to 40 mol CaCO$_3$ m^{-2} y^{-1}) than coral/algal reef flats. Sediments represent the largest physiographic zone of many reef ecosystems (e.g. 95% of the surface area of the SW Caledonian reef complex; J Clavier, personal communication) but have received comparatively less attention than coral/algal reef flats, probably because carbon and carbonate fluxes in reef sediments are of lower magnitude ($P_g = 8-82$ and $R = 1-73$ mol C m^{-2} y^{-1}; $G = -1$ to 12 mol CaCO$_3$ m^{-2} y^{-1}). Sedimentary areas contribute 3–30% of the community excess production of a Pacific barrier reef flat (18).

There are only nine data sets for complete reef systems, some of which do not provide both organic and inorganic carbon metabolism. P_g and R are well correlated (Figure $1E$). The estimated P_g/R ratio (1.28 ± 0.2; Table 1) suggests a net autotrophy, although it is not statistically significant. It is, moreover, dominated by a single data set (75), which casts some doubt on that conclusion. Additionally, P_g and R are measured separately, and each has significant error terms, which are cumulative when calculating the P_g/R ratio or NEP. The approach of ecosystem stoichiometry (e.g. 130) enables estimation of NEP directly by upscaling changes in the concentration of dissolved inorganic phosphorus to NEP using the ecosystem C:N:P ratio. The overall average NEP obtained with both approaches is slightly higher than the previous estimate of Crossland et al (10 ± 7 vs 3 mol C m^{-2} y^{-1}; Ref. 30) but is not significantly different from 0 ($p = 0.21$, $N = 9$). It can therefore be concluded that the organic C metabolism of complete reef systems is essentially balanced. The average net calcification rate is 10 ± 3 mol CaCO$_3$ m^{-2} y^{-1} ($N = 7$).

Several sources of nutrients sustain reef primary production, but the contribution of each is largely site-dependent and generally poorly known. NEP of reef ecosystems is not significantly different from that of tropical oligotrophic oceans (30). It seems unlikely that reef productivity is sustained by the same nutrient source as the surrounding ocean (advective inputs from below) due to physical limitation, although active upwelling along the slope (e.g. 151) and internal tidal bores (83) can sometimes provide a significant nutrient supply. Smith (129) offered two alternative explanations. First, the C:N:P ratio of reef benthic plants (550:30:1; Ref. 7) is much higher than the typical Redfield ratio (106:16:1). Therefore, the production of organic carbon is much more efficient in reef systems per unit of nitrogen and phosphorus. Second, oceanic water impinging reefs is typically depleted in nitrogen relative to phosphorus. The well-established capacity of several reef organisms and physiographic zones to fix nitrogen enables reef communities to overcome nitrogen limitation. The convection resulting from upward geothermal heat flow drives circulation of nutrient-rich deep oceanic water within the reef matrix (endo-upwelling; Ref. 122). The magnitude of this nutrient source remains unknown, but it is probably relatively small (79) and not required to sustain excess primary production on reefs (142). Last, phytoplankton and planktonic microbial communities advected from the ocean are a significant source of nutrients; their retention rates are virtually identical to the net excess primary production of a Pacific reef (2.7 vs 3 mol C m^{-2} y^{-1}; Ref. 8).

The net organic and inorganic carbon production can be accumulated as biomass or reef structure, buried in sediments, or exported to adjacent ecosystems. The importance of these alternative fates is poorly known, but it is somewhat different for organic and inorganic carbon. The rate of $CaCO_3$ accumulation is limited by sea level on reef flats (ca. 1 mm y^{-1}; see 135) but not in lagoons. A budgetary approach (135) suggests that 65% of total carbonate production of an hypothetical atoll is stored in the lagoon, 25% is exported, and about 10% is accumulated in the fore reef and reef flat areas. The rate of $CaCO_3$ export increases sharply as lagoon area decreases. In contrast, there is little accumulation of organic carbon in reef sediments (ca. 15% of E), which contains typically less than 0.7 weight % C (reviewed in 137), and in biomass (ca. 2% of E; 55). Approximately 10% of E is available for human harvest, a potential not realized at present, but most of the very little net excess organic carbon produced appears to be exported as particulate or dissolved organic carbon or in the form of migrating organisms (50–75% of E; 30, 135).

Air-sea CO_2 fluxes in reef ecosystems and communities can be investigated using either direct measurements or a budgeting approach based on community metabolism data (52). The few direct measurements indicate that the sites studied were sources of CO_2 to the atmosphere at the time of measurement (46, 51). The same conclusion was reached using indirect estimates (50, 52, 131, 148). Data on organic C burial in reef sediments support such conclusion on longer time scales (21). A few recent reports suggest, however, that some fringing reefs are sinks for atmospheric CO_2 (62, 74, 75, 161), and localized results have been extended to a global scale (64). Although the interpretation and generalization of some of these data were inappropriate (21, 48), most studies suggesting that reefs may be sinks of CO_2 were carried out on fringing reefs, which are more likely subject to anthropogenic stresses than are other reef systems. It is well established that such disturbances shift reef communities from a coral-dominated to an algal-dominated state (e.g. 35, 61). The resulting changes in community metabolism (increased primary production and/or decreased calcification) can turn the systems from a CO_2 source into a CO_2 sink (49).

The global production of reef carbonate (6 Tmol $CaCO_3$ y^{-1}) represents, respectively, 26% and 11% of the coastal and total marine $CaCO_3$ precipitation estimated by Milliman (97). Crossland et al (30) have estimated the global significance of reef metabolism to oceanic processes. Gross primary production is 86 Tmol C y^{-1}, a value higher than the previous estimate of 58 Tmol C y^{-1} (30) due to the recent addition of metabolic data collected in algal-dominated fringing reefs. The *NEP* is probably better constrained than P_g as there are more data because of the inclusion of some direct determinations. Our estimate of *NEP* is slightly higher than the previous estimates of Crossland et al (6 vs 1.7 Tmol C y^{-1}; Ref. 30) for reasons outlined above. As pointed out by Ware et al (148) and Smith (131), coral reefs have a minor role in the present global carbon cycle, and their release of CO_2 is 0.4–1.4% of the current rate of anthropogenic CO_2 production. According to the so-called coral reef hypothesis, reefs may have played a significant role in the change of atmospheric pCO_2 that occurred during the last glacial-interglacial cycle (10, 20, 105).

CONTINENTAL SHELF

The continental shelf comprises the area between the continents and the open ocean. It is limited by the ocean margin, which corresponds to the abrupt bathymetric change that occurs between the shelf and the slope, at an average depth of 130 m. The world coastline is about 350,000 km long; the shelf has a mean width of 70 km. The total surface area of the coastal zone represents 26×10^6 km^2 or 7% of the total surface area of the ocean (360×10^6 km^2). Here we consider the main aspects of the carbon cycle occurring on the continental shelf, with the exception of the ecosystems discussed in the previous sections. Several recent reviews (87, 97, 132, 146, 157) provide detailed information on the carbon and carbonate cycling of continental shelves. There are distinct differences in physical, chemical, and biological properties of the neritic and oceanic provinces, leading to marked gradients that generate fluxes at the ocean margins. Because of the diversity of processes at the margins and the large variability of coastal systems, the exchanges of energy and of dissolved or particulate matter between the shelf and the open ocean remain poorly understood (87). The circulation and mixing of water are especially complicated by the steep bathymetric change introduced by the continental slope and rise (62). As a consequence, the exchange of organic matter and nutrients between the coastal zone and the open ocean is poorly known. Attempts are being made, in the framework of LOICZ, to establish mass balances of dissolved nutrients using a limited amount of field data for a variety of shelf environments (54). The difference between the input and output fluxes of phosphorus is scaled to carbon and is assumed to represent the *NEP* of the system considered. One of the critical data required in these calculations is the mixing rate of water masses at the open ocean boundary. It is estimated from the mass balance of salt, and usually assumes the system is at steady state, which may be a crude approximation on an annual basis. The development and improvement of models may soon provide a better evaluation of the role of the coastal zone on a global basis. We cite in this section the studies for which the fluxes of carbon linked to various elemental processes have been estimated for sufficiently long periods of time.

Unfortunately, in only a few studies have P_g and the total (benthic and pelagic) respiration of coastal waters been investigated simultaneously. The average areal P_g is 18 ± 2 mol C m^{-2} y^{-1}, and the corresponding global gross primary production is 377 Tmol C y^{-1}.

The relative importance of recycled production, resulting from the regeneration of nutrients by the bacterial degradation of dead biomass, and of new production sustained by nutrients from an external source, is different on the continental shelf and in the open ocean. New production represents between 5% and 15% of P_g in the oligotrophic central gyres of the open ocean (40), and its contribution is close to 50% on the shelves if the remineralization of the nutrients in the sediments is taken into account (72, 157). In the open ocean, new production is essentially due to upwelling and vertical mixing of deep, nutrient-rich water with surface water. The source and fluxes of nutrients required to sustain the high productivity of the coastal zone is still controversial. The origin of nutrients is much more complex than in the open ocean and involves fluxes at the margins of deep ocean water, in addition to riverine and atmospheric inputs. Furthermore, nutrients recycled in the sediments can be rapidly transferred to the overlying waters by diffusion.

Most attention has been devoted to the fluxes of nitrogen, which is often the limiting nutrient. The occurrence of two species, ammonium and nitrate, also enables one to distinguish between new and recycled production. The main source of nitrogen is the deep ocean reservoir, which is transferred to the shelf by upwelling and vertical mixing resulting from the shelf break (60, 147, 156). This flux represents about half the nitrogen required to sustain new production in the North Atlantic, the river input accounting for the other half (47). Atmospheric deposition (8%) and nitrogen fixation (1%) are negligible. On a global basis, Wollast (157) estimated that the contribution of the open ocean represents 80% of the nitrogen flux required to sustain new production of the continental shelf, an evaluation close to an earlier estimate (147). Riverine input of nitrogen, although heavily enhanced by anthropogenic activities, accounts for less than 15%, and atmospheric deposition and nitrogen fixation constitute about 5% of the total required N.

The behavior and fate of organic matter produced in the water column are also very different in the coastal zone and open ocean. First, the number of trophic levels decreases markedly with increasing primary production. As many as six trophic levels can be identified in oligotrophic waters; there are as few as three in upwelling areas (78). In addition, coastal phytoplankton is typically dominated by large cells, whereas micro- and picophytoplankton dominate in the open ocean (78). Fecal pellets produced by organisms grazing small phytoplankton in the open ocean are small and are not exported from the photic zone efficiently as a result of low settling velocities (108). Due to the large size of fecal pellets (157) and shallow depth of the coastal zone, a large fraction of primary production and detrital matter imported by the rivers may be deposited and stimulate biological activity in the sediments. Figure 1E compares total respiration (pelagic plus benthic) to gross primary production in various shelf areas. Approximately 30% of the production is respired in the water column, and an equivalent amount is mineralized in the sediments.

Total respiration is therefore 10 ± 2 mol C m^{-2} y^{-1} or 214 Tmol C y^{-1} for the global coastal zone, which would make the continental shelf net autotrophic with a P_g/R ratio of 1.4 ± 0.3 (Table 1) and a *NEP* of 171 Tmol C y^{-1} that must be exported (157). These observations are in good agreement with the high values of new production found for the coastal area (40, 72). Although coastal sediments accumulate about 90% of detrital organic carbon on a global basis, this represents only 3–4% of shelf production. The remaining 36% of total production must therefore be exported to the open ocean, re-exporting simultaneously particulate organic nutrients, which compensate for the transfer of dissolved nutrients from the ocean across the shelf (157). An alternative fate may exist for nitrogen if denitrification, which occurs mainly in shelf sediments, is significant. This hypothesis has been proposed for the North-Atlantic (102). Note that such denitrification must be balanced by an equivalent flux of nitrogen fixation in the open ocean to maintain a steady-state condition in the marine system. It must be emphasized that temperate shelf ecosystems can be net heterotrophic in winter and net autotrophic in summer, when high rates of photosynthesis occur (146).

Eutrophication resulting from the discharge of estuarine nutrient-rich water to the coastal sea can induce a wide range of ecological and societal consequences. For example, a correlation between primary production and the supply of inorganic nitrogen from the Mississippi River has been observed in the Gulf of Mexico (85). An increasingly large part of the Gulf becomes hypoxic or anoxic in summer, with considerable potential effect on catches in this leading US fishery area. In the southern Bight of the North Sea, which is under the influence of several macro-tidal and polluted estuaries (Rhine, Scheldt, and Thames), pCO_2 varies from 100 to 800 μatm depending on river flow, water temperature, and light availability (44).

SIGNIFICANCE OF COASTAL ECOSYSTEMS IN THE GLOBAL OCEANIC CARBON CYCLE

Metabolic data from coastal ecosystems are summarized in Table 1. The P_g/R ratios vary considerably but are, in most cases, not statistically different from 1. All coastal ecosystems are net autotrophs ($P_g/R > 1$; *NEP* > 0) except estuaries, which are net heterotrophs exhibiting a negative net ecosystem production (-8 Tmol C y^{-1}). These data can be integrated to provide an independent estimate of coastal metabolism for comparison with estimates obtained by other approaches, a method that has several limitations. The sites for which metabolic data are available are scarce for some ecosystems (e.g. $N = 7$ for coral reefs) and may not adequately represent the range of metabolic parameters. Average areal productions are not weighted averages, so large error can result when they are scaled up to derive global production estimates (see the section on macrophyte-dominated ecosystems). Additionally, most *NEP* data are estimated as the difference between P_g and R, each of which has an associated error. If, for example, these errors are 25% and are independent, P_g and R must differ by more than 35% for the *NEP* to be statistically different from 0 (130). Nevertheless, our estimate of P_g (789 Tmol C y^{-1}; i.e. 23% of the global marine gross primary production (see Table 1), is of the same order of magnitude as the previous estimate of 500 Tmol C y^{-1} (132, 157). The latter estimate (e.g. 157) did not specifically take into account systems such as macrophyte-dominated ecosystems and mangroves, which might partly explain the difference.

Although there is a consensus on the magnitude of P_g, there are differences in the estimates of the global coastal *NEP*. The ecosystem approach provides an estimate of 231 Tmol C y^{-1}, a value in good agreement with that provided by Wollast (200 Tmol C y^{-1}; 157) but much higher than the -7 Tmol C y^{-1} proposed by Smith & Hollibaugh (132) or the 12 Tmol C y^{-1} given by Rabouille (112). Smith & Hollibaugh (132) used a linear relationship between *NEP* and P_g based on 22 nearshore and estuarine sites, as well as the average P_g of estuaries and the remaining continental shelves, to predict the *NEP* of both systems. They concluded that estuaries are net heterotrophic (*NEP* $= -7$ Tmol C y^{-1}), that the remaining coastal ocean has a balanced organic carbon metabolism (*NEP* ≈ 0), and that the coastal ocean is thus net heterotrophic (*NEP* $= -7$ Tmol C y^{-1}). They therefore estimated R indirectly at 507 Tmol C y^{-1}. Most values of *NEP* used in the linear regression were obtained as the difference between separate estimates of P_g and R, a procedure that induces great uncertainty in *NEP*. Also, most data used to derive the predictive equation were from nearshore sites, the outer shelf areas being poorly represented. On the other hand, Wollast's estimate of R (and *NEP*) is based on the observed average remineralization rate of P_g (60%; 157) of 500 Tmol C y^{-1}, so R and *NEP* are respectively estimated to be 300 and 200 Tmol C y^{-1}. This approach is also limited by the average remineralization rate being calculated from a small data set ($N = 10$) exclusively based on temperate and boreal shelves of the northern hemisphere. The continental shelf proper (excluding specific ecosystems) is the major contributor to *NEP* of the coastal zone (75%) followed by macrophyte-dominated ecosystems (16%), mangroves (7%), marshes (3%), and coral reefs (2.6%). Respiratory processes are poorly known, not only on the continental shelves but also in the open ocean (34).

The coastal ocean contributes more than 40% of marine calcium carbonate production (23 vs 53 Tmol $CaCO_3$ y^{-1}; 97). The highest deposition occurs in coral reef habitats (9 Tmol y^{-1}, according to Ref. 97, and 6 Tmol y^{-1} according to our own estimate), followed by banks and embayments (4 Tmol y^{-1}), carbonate shelves (6 Tmol y^{-1}), and non-carbonate shelves (4 Tmol y^{-1}). However,

Milliman (97) has suggested that a significant fraction (4 Tmol y^{-1}) of the calcium carbonate produced on the shelf is exported and deposited on the continental slope and rise. This was confirmed by Sabine & Mackenzie (123), who observed abundant carbonate skeleton debris, characteristic of shallow water organisms, in traps deployed along the Hawaiian slope.

The delivery of carbon to the coastal ocean has been enhanced by human activities and is presently ca. 85 Tmol C y^{-1} (93). In the marine system, the riverine DIC is believed to be partitioned equally between deposition of carbonate minerals and CO_2 evasion to the atmosphere. The organic carbon delivered is either oxidized to CO_2, accumulated in coastal sediment, or exported to the deep open ocean. The importance of these various fates is poorly known and is one of the major source of uncertainty in the global carbon cycle. The extent of export from the ocean margin has recently been examined using radiocarbon (^{14}C) data and a mass balance approach by Bauer & Druffel (9a). Their results suggest that inputs of DOC and POC from ocean margins to the deep open ocean may be more than an order of magnitude greater than inputs of recently produced organic carbon derived from the surface ocean.

It is increasingly evident that the higher fertility of the coastal ocean compared to the open ocean and slope (20 vs 8 mol C m^{-2} y^{-1}; 11, 146, 157) is mainly due to the large fluxes of nutrients transferred from the deep ocean to the shelf by upwelling or vertical mixing. The recycled production is relatively low, and thus the new production related to the large nutrient input must be balanced by the export of an equivalent amount of these elements. Walsh (147) and Wollast (156) have suggested that a significant fraction of the primary production is exported to the open ocean and that the nutrients are re-exported as particulate organic matter. However, along the northwest Atlantic coast, only 5% of the primary production is exported from the shelf to the adjacent slope (13). Nixon et al (102) concluded that only phosphorus, presumably in dissolved form, is largely exported to the open ocean, but that most of the nitrogen, mainly imported from the open ocean, is lost by denitrification on the shelf. This implies a considerable rate of denitrification in the sediments, contrasting with a high lability and release of the phosphorus constituents from the bottom to the water column.

Until recently, models of the global carbon cycle did not incorporate the coastal ocean, but directly linked ocean and continents. The status of the coastal ocean in global models is still a matter of debate because the magnitude of the transfer of carbon between the coastal zone and the open ocean is poorly constrained. There is no doubt that, in pristine times, the total riverine input of organic carbon in the coastal zone was greater than organic carbon preserved in shelf sediments. The coastal ocean was net heterotrophic and a source of CO_2 to the atmosphere (7 Tmol C y^{-1}), assuming little or no transfer of C at the margins (132, 158; Ver et al, submitted). If these transfers are assumed to be significant (112; C. Rabouille, personal communication), it was slightly autotrophic and a net sink of atmospheric CO_2 (20 Tmol C y^{-1}).

Anthropogenic disturbance has led to an increased delivery of inorganic nutrients, organic carbon, and suspended matter into the coastal ocean. The excess nutrients may locally enhance planktonic primary production and carbon sequestration. Even though the total load of riverine N and P has more than doubled with respect to pristine conditions (93), it has not significantly affected the productivity of the coastal zone on a global basis. This perturbation is, however, responsible for the eutrophication in zones adjacent to polluted estuaries, especially in semi-enclosed areas. The increased amount of organic carbon delivered to the coastal zone can be stored in the sediment and/or oxidized to CO_2. In the latter case, remineralization releases nutrients and promotes primary production. The present rate of sedimentation in the coastal zone is probably twice that of preindustrial times because of increased continental erosion resulting from deforestation and changes in agricultural practices. This should increase the rate of C burial in coastal sediments (155). The balance between increased primary production and increased respiration may shift the coastal zone toward a more heterotrophic or a more autotrophic state relative to initial conditions (158). According to Smith & Hollibaugh (132), the present coastal zone remains heterotrophic (but a sink for fossil fuel CO_2). The changes in air-sea CO_2 flux relative to pristine conditions depend on the response of both ocean carbonate chemistry and atmospheric CO_2 to anthropogenic perturbations. Increased carbonate precipitation and increased heterotrophy (or decreased autotrophy) result in a source of CO_2 smaller than the rise of atmospheric pCO_2 due to anthropogenic activities. The CO_2 sink potential of the coastal ocean is therefore diminished. A model that assumes a low rate of C transfer at the margin estimates that CO_2 flux has changed both in direction (from net evasion to net invasion) and magnitude (by 6.7 Tmol C y^{-1}) relative to the year 1700 (Ver et al, submitted). Finally, Rabouille (112) has suggested that the various human-induced modifications of the coastal carbon cycle have resulted in decreased autotrophy of the coastal ocean (20 to 11 Tmol C y^{-1}).

CONCLUSIONS

Kempe (67) asserted that whether coastal seas are net sinks or sources of CO_2 for the atmosphere cannot be determined. There are currently few carbon budgets available for coastal ecosystems. An important research initiative was recently launched by the LOICZ program to develop modeling guidelines (54), compile 150–200 carbon (and nitrogen) budgets for coastal ecosystems in key regions, and extend these budgets to a global scale using a functional coastal zone classification system (109). Approximately 30 budgets are available (e.g. 133), and it is anticipated that the compilation will be completed in the near future (SV Smith, personal communication).

It is difficult to evaluate the autotrophic/heterotrophic character of the coastal zone on the basis of the balance between inputs and outputs because of the very limited knowledge of circulation and water exchange between the shelf and the open ocean. The net flux of material at this boundary is poorly constrained: it is the difference between two huge numbers, both of which are affected by large uncertainties. An additional difficulty lies with the extremely non-stationary conditions of the coastal zone. Hydrodynamically, river discharge exhibits strong seasonal and annual variations, and the shelf is periodically affected by storms that resuspend freshly deposited sediments, and favor export to the slope area and open ocean.

The available data suggest that riverine and atmospheric inputs of dissolved and particulate carbon represent a negligible fraction of the high primary production of the coastal zone on a global scale. New production on the shelf represents at least 50% of primary production and thus only 50% or less is respired and recycled (40, 72, 146, 155–157). Some of the production that is not recycled accumulates in the sediments, but most of the detrital organic matter—dissolved or particulate—must be exported to the slope and open ocean. Changes in riverine fluxes of organic matter and nutrients or suspended matter due to human activities are also small with respect to natural fluxes, and have probably affected the global carbon cycle only slightly.

ACKNOWLEDGMENTS

Thanks are due to the following colleagues for providing data, reprints and/or papers in press or submitted: L. Charpy, C. X. Duarte, A.-M. Leclerc, S. Kraines, F. T. Mackenzie, H. M. E. Nacorda, A. Ormond, and C. Rabouille. Numerous other colleagues kindly provided material that could not be included in the present review. S. Dallot assisted with statistical analysis. This paper has greatly benefitted from the assistance of M. Loijens and from insightful comments of D. M. Alongi, R. W. Buddemeier, J. Dolan, J. J. Middelburg, J. D. Milliman, M. Pichon, C. Rabouille, J. M. Shick, and S. V. Smith on early drafts. JPG thanks C. Emery and the staff of the library of the Oceanographic Museum (Monaco) for their invaluable help during preparation of this review. The research of the authors on coastal metabolism is supported by the *Programme National Récifs Coralliens* (PNRCO, France), FNRS (Belgium) with which MF is a research associate, the European Union programs Environment and Climate (BIOGEST contract ENV4-CT96-0213) and MAST (OMEX contracts MAS2-CT93-0069 and MAS3-CT97-0076), the Belgian State Prime Minister's Office (OSTC) in the framework of the Global

Change and Sustainable Development Programme (contract CG/DD/11C) and the Council of Europe. This is a contribution of PNRCO and ELOISE contribution 020.

Visit the *Annual Reviews home page* at
http://www.AnnualReviews.org

Literature Cited

1. Agawin NSR, Duarte CM, Fortes MD. 1996. Nutrient limitation of Philippine seagrasses (Cape Bolinao, NW Philippines): *in situ* experimental evidence. *Mar. Ecol. Prog. Ser.* 138:233–43
2. Alongi DM. 1988. Bacterial productivity and microbial biomass in tropical mangrove sediments. *Microb. Ecol.* 15:59–79
3. Alongi DM. 1989. The role of soft-bottoms benthic communities in tropical mangrove and coral reef ecosystems. *Rev. Aquat. Sci.* 1:243–80
4. Alongi DM. 1990. The ecology of tropical soft-bottom benthic ecosystems. *Oceanogr. Mar. Biol. Ann. Rev.* 28:381–496
5. Alongi DM. 1994. Zonation and seasonality of benthic primary production and community respiration in tropical mangrove forests. *Oecologia* 98:320–7
5a. Alongi DM. 1998. *Coastal Ecosystem Processes.* Boca Raton: CRC Press
6. Alongi DM, Boto KG, Tirendi F. 1989. Effect of exported mangrove litter on bacterial productivity and dissolved organic carbon fluxes in adjacent tropical nearshore sediments. *Mar. Ecol. Prog. Ser.* 56:133–44
7. Atkinson MJ, Smith SV. 1983. C:N:P ratios of benthic marine plants. *Limnol. Oceanogr.* 28:568–74
8. Ayukai T. 1995. Retention of phytoplankton and planktonic microbes on coral reefs within the Great Barrier Reef, Australia. *Coral Reefs* 14:141–47
9. Bartlett DS, Bartlett KB, Hartman JM, Harriss RC, Sebacher DI, et al. 1989. Methane emissions from the Florida Everglades: patterns of variability in a regional wetland ecosystem. *Glob. Biogeochem. Cycles* 3:363–73
9a. Bauer JE, Druffel ERM. 1998. Ocean margins as a significant source of organic matter to the deep open ocean. *Nature* 392:482–85
10. Berger WH. 1982. Increase of carbon dioxide in the atmosphere during

deglaciation: the coral reef hypothesis. *Naturwissenschaften* 69:87–88
11. Berger WH. 1989. Appendix. Global maps of ocean productivity. In *Productivity of the Ocean: Present and Past*, ed. WH Berger, VS Smetacek, G Wefer, pp. 429–55. Chichester, UK: Wiley & Sons
12. Birkeland C, ed. 1997. *Life and Death of Coral Reefs.* New York: Chapman & Hall. 536 pp.
13. Biscaye PE, Flagg CN, Falkowski PG. 1994. The Shelf Edge Exchange Processes experiment, SEEP-II: an introduction to hypotheses, results and conclusions. *Deep Sea Res.* 41:231–52
14. Bondsorff E, Blomqvist EM, Mattila J, Norkko A. 1997. Coastal eutrophication: causes, consequences and perspectives in the Archipelago areas of the Northern Baltic Sea. *Estuar. Coast. Shelf Sci.* 44(Suppl. A):63–72
15. Boto KG, Alongi DM, Nott ALJ. 1989. Dissolved organic carbon-bacteria interactions at sediment-water interface in a tropical mangrove system. *Mar. Ecol. Prog. Ser.* 51:243–51
16. Boto KG, Bunt JS, Wellington JT. 1984. Variations in mangrove forest productivity in northern Australia and Papua New Guinea. *Estuar. Coast. Shelf Sci.* 19:321–29
17. Deleted in proof
18. Boucher G, Clavier J, Hily C, Gattuso J-P. 1998. Contribution of soft-bottoms to the community metabolism (primary production and calcification) of a barrier reef flat (Moorea, French Polynesia). *J. Exp. Mar. Biol. Ecol.* 225:269–83
19. Boynton WR, Murray L, Hagy JD, Stokes C, Kemp WM. 1996. A comparative analysis of eutrophication patterns in a temperate coastal lagoon. *Estuaries* 19:408–21
20. Broecker WS, Lao Y, Klas M, Clark E, Bonani G, et al. 1993. A search for an early holocene $CaCO_3$ preservation event. *Paleoceanography* 8:333–9

21. Buddemeier RW. 1996. Coral reefs and carbon dioxide. *Science* 271:1298–89
22. Cadée N, Dronkers J, Heip C, Martin J-M, Nolan C, eds. 1994. ELOISE (European Land-Ocean Interaction Studies) Science Plan, Luxembourg: Off. Official Publ. Eur. Communities. 52 pp.
23. Cebrián J, Duarte CM, Marbà N, Enríquez S. 1997. Magnitude and fate of the production of four co-occurring western Mediterranean seagrass species. *Mar. Ecol. Prog. Ser.* 155:29–44
24. Charpy L. 1996. Phytoplankton biomass and production in two Tuamotu atoll lagoons (French Polynesia). *Mar. Ecol. Prog. Ser.* 145:133–42
25. Charpy-Roubaud C, Sournia A. 1990. The comparative estimation of phytoplanktonic, microphytobenthic and macrophytobenthic primary production in the oceans. *Mar. Microb. Food Webs* 4:31–57
26. Clough BF. 1992. Primary productivity and growth of mangrove forests. In *Tropical Mangrove Ecosystems*, ed. AI Robertson, DM Alongi, pp. 225–49. Washington, DC: Am. Geophys. Union
27. Cohen JE, Small C, Mellinger A, Gallup J, Sachs J. 1997. Estimates of coastal populations. *Science* 278:1211–12
28. Conley DJ, Smith WM, Cornwell JC, Fisher TR. 1995. Transformation of particle-bound phosphorus at the land-sea interface. *Estuar. Coast. Shelf Sci.* 40:161–76
29. Costanza R, d'Arge R, de Groot R, Farber S, Grasso M, et al. 1997. The value of the world's ecosystem services and natural capital. *Nature* 387:253–59
30. Crossland CJ, Hatcher BG, Smith SV. 1991. Role of coral reefs in global ocean production. *Coral Reefs* 10:55–64
31. Day JW, Day RH, Barreiro MT, Ley-Lou F, Madden CJ. 1982. Primary production in the Laguna de Terminos, a tropical estuary in the southern Gulf of Mexico. *Oceanol. Acta* 5 (Suppl.):269–76
32. De la Cruz AA. 1986. Tropical wetlands as a carbon source. *Aquat. Bot.* 25:109–15
33. De Vooys CGN. 1979. Primary production in aquatic environments. In *The Global Carbon Cycle*, ed. B Bolin, ET Degens, S Kempe, P Ketner, pp. 259–92. Chichester, UK: Wiley & Sons
34. del Giorgio PA, Cole JJ, Cimbleris A. 1997. Respiration rates in bacteria exceed phytoplankton production in unproductive aquatic systems. *Nature* 385:148–51
35. Done TJ. 1992. Phase shifts in coral reef communities and their ecological significance. *Hydrobiologia* 247:121–32
36. Duarte CM. 1990. Seagrass nutrients content. *Mar. Ecol. Prog. Ser.* 67:201–7
37. Duarte CM, Cebrián J. 1996. The fate of marine autotrophic production. *Limnol. Oceanogr.* 41:1758–66
38. Dubinsky Z, ed. 1990. *Coral Reefs*. New York: Elsevier. 550 pp.
39. Dyer KR, Orth DR, eds. 1994. *Changes in Fluxes in Estuaries: Implications from Science to Management*. Fredensborg: Olsen & Olsen. 485 pp.
40. Eppley RW. 1989. New production: history, methods, problems. In *Productivity of the Ocean: Present and Past*, ed. WH Berger, VS Smetacek, G Wefer, pp. 85–97. Chichester, UK: Wiley & Sons
41. Erftemeijer PLA, Middelburg JJ. 1993. Sediment-nutrient interactions in tropical seagrass beds: a comparison between a carbonate and a terrigenous sedimentary environment in South Sulawesi (Indonesia). *Oecologia* 99:45–59
42a. Erftemeijer PLA, Middelburg JJ. 1995. Mass balance constraints on nutrient cycling in tropical seagrass beds. *Aquat. Bot.* 50:21–36
42. Erftemeijer PLA, Stapel J, Smekens MJE, Drossaert WME. 1994. The limited effect of in situ phosphorus and nitrogen additions to seagrass beds on carbonate and terrigenous sediments in South Sulawesi, Indonesia. *J. Exp. Mar. Biol. Ecol.* 182:123–40
43. Fourqurean JW, Webb KL, Hollibaugh JT, Smith SV. 1997. Contributions of the plankton community to ecosystem respiration, Tomales Bay, California. *Estuar. Coast. Shelf Sci.* 44:493–505
44. Frankignoulle M, Bourge I, Canon C, Dauby P. 1996. Distribution of surface seawater partial CO_2 pressure in the English Channel and in the Southern Bight of the North Sea. *Cont. Shelf Res.* 16:381–95
45. Frankignoulle M, Bourge I, Wollast R. 1996. Atmospheric CO_2 fluxes in a highly polluted estuary (the Scheldt). *Limnol. Oceanogr.* 41:365–69
46. Frankignoulle M, Gattuso J-P, Biondo R, Bourge I, Copin-Montégut G, et al. 1996. Carbon fluxes in coral reefs. 2. Eulerian study of inorganic carbon dynamics and measurement of air-sea CO_2 exchanges. *Mar. Ecol. Prog. Ser.* 145:123–32
47. Galloway JN, Howarth RW, Michaels AF, Nixon SW, Prospero JM, et al. 1996. Nitrogen and phosphorus budgets of the North Atlantic Ocean. *Biogeochemistry* 35:3–25

48. Gattuso J-P, Frankignoulle M, Smith SV, Ware JR, Wollast R. 1996. Coral reefs and carbon dioxide. *Science* 271:1298

49. Gattuso J-P, Payri CE, Pichon M, Delesalle B, Frankignoulle M. 1997. Primary production, calcification, and air-sea CO_2 fluxes of a macroalgal-dominated coral reef community (Moorea, French Polynesia). *J. Phycol.* 33:729–38

50. Gattuso J-P, Pichon M, Delesalle B, Canon C, Frankignoulle M. 1996. Carbon fluxes in coral reefs. 1. Lagrangian measurement of community metabolism and resulting air-sea CO_2 disequilibrium. *Mar. Ecol. Prog. Ser.* 145:109–21

51. Gattuso J-P, Pichon M, Delesalle B, Frankignoulle M. 1993. Community metabolism and air-sea CO_2 fluxes in a coral reef ecosystem (Moorea, French Polynesia). *Mar. Ecol. Prog. Ser.* 96:259–67

52. Gattuso J-P, Pichon M, Frankignoulle M. 1995. Biological control of air-sea CO_2 fluxes: effect of photosynthetic and calcifying marine organisms and ecosystems. *Mar. Ecol. Prog. Ser.* 129:307–12

53. Ghosh S, Jana TK, Singh BN, Choudhury A. 1987. Comparative study of carbon dioxide system in virgin and reclaimed mangrove waters of Sundarbans during freshet. *Mahasagar* 20:155–61

54. Gordon DC Jr, Boudreau PR, Mann KH, Ong J-E, Silvert WL, et al. 1996. LOICZ biogeochemical modelling guidelines. *LOICZ Rep. & Stud.* 5:1–96

55. Hatcher BG. 1997. Organic production and decomposition. In *Life and Death of Coral Reefs*, ed. C Birkeland, pp. 140–74. New York: Chapman & Hall

56. Hatcher BG, Johannes RE, Robertson AI. 1989. Review of research relevant to the conservation of shallow tropical marine ecosystems. *Oceanogr. Mar. Biol. Ann. Rev.* 27:337–414

57. Healey MJ, Moll RA, Diallo CO. 1988. Abundance and distribution of bacterioplankton in the Gambia River, West Africa. *Microb. Ecol.* 16:291–310

58. Heip CHR, Goosen NK, Herman PMJ, Kromkamp J, Middleburg JJ, et al. 1995. Production and consumption of biological particles in temperate tidal estuaries. *Oceanogr. Mar. Biol. Ann. Rev.* 33:1–149

59. Herman PMJ, Heminnga MA, Nienhuis PH, Verschuure JM, Wessel EGJ. 1996. Wax and wane of eelgrass *Zostera marina* and water column silicon levels. *Mar. Ecol. Prog. Ser.* 144:303–7

60. Howarth RW, Billen G, Swaney D, Townsend A, Jaworski N, et al. 1996. Regional nitrogen budgets and riverine N&P fluxes for the drainages to the North Atlantic Ocean: natural and human influences. *Biogeochemistry* 35:75–139

61. Hughes TP. 1994. Catastrophes, phase shifts, and large-scale degradation of a Caribbean coral reef. *Science* 265:1547–51

62. Huthnance J. 1995. Circulation, exchange and water masses at the ocean margin: the role of physical processes at the shelf edge. *Prog. Oceanogr.* 35:353–431

63. Irigoien X, Castel JC. 1997. Light limitation and distribution of chlorophyll pigments in a highly turbid estuary: the Gironde (SW France). *Estuar. Coast. Shelf Sci.* 44:507–17

64. Kayanne H, Suzuki A, Saito H. 1995. Diurnal changes in the partial pressure of carbon dioxide in coral reef water. *Science* 269:214–16

65. Deleted in proof

66. Kempe S. 1982. Valdivia Cruise, October 1981: carbonate equilibria in the estuaries of Elbe, Weser, Ems and in the Southern German Bight. *Mitt. Goel.-Paläont. Inst. Univ. Hamburg* 52:719–42

67. Kempe S. 1995. Coastal seas: a net source or sink of atmospheric carbon dioxide? *LOICZ Rep. & Stud.* 1:1–27

68. Ketchum BH, ed. 1983. *Estuaries and Enclosed Seas*. Amsterdam: Elsevier. 500 pp.

69. Ketchum BH. 1983. Estuarine characteristics. In *Estuaries and Enclosed Seas*, ed. BH Ketchum, pp. 1–14. Amsterdam: Elsevier

70. Kinsey DW. 1985. Metabolism, calcification and carbon production. I. System level studies. *Proc. 5th Int. Coral Reef Congr.* 4:505–26

71. Kleypas J. 1997. Modeled estimates of global reef habitat and carbonate production since the last glacial maximum. *Paleoceanography* 12:533–45

72. Knauer GA. 1993. Productivity and new production of the oceanic system. *Interactions of C, N, P and S Biogeochemical Cycles and Global Change*, ed. R Wollast, FT Mackenzie, L Chou, pp. 211–31. Berlin: Springer-Verlag

73. Koch EW, Beer S. 1996. Tides, light and the distribution of *Zostera marina* in Long Island Sound, USA. *Aquat. Bot.* 53:97–107

74. Kraines S, Suzuki Y, Omori T, Shitashima K, Kanahara S, et al. 1997. Carbonate dynamics of the coral reef system at Bora Bay, Miyako Island. *Mar. Ecol. Prog. Ser.* 156:1–61

75. Kraines S, Suzuki Y, Yamada K, Komiyama H. 1996. Separating biologi-

cal and physical changes in dissolved oxygen concentration in a coral reef. *Limnol. Oceanogr.* 41:1790–9

76. Kristensen E, Andersen FØ, Kofoed LH. 1988. Preliminary assessment of benthic community metabolism in a south-east Asian mangrove swamp. *Mar. Ecol. Prog. Ser.* 48:137–45

77. Kristensen E, Holmer M, Banta GT, Jensen MH, Hansen K. 1995. Carbon, nitrogen and sulfur cycling in sediments of the Ao Nam Bor mangrove forest, Phuket, Thailand: a review. *Phuket Mar. Biol. Cent. Res. Bull.* 60:37–64

78. Lalli CM, Parsons TR. 1993. *Biological Oceanography: An Introduction.* Oxford: Pergamon Press. 301 pp.

79. Leclerc A-M, Broc D, Jean-Baptiste P, Texier D. Density-gradient induced water circulations in atoll coral reefs: a numerical study *Limnol. Oceanogr.* In press

80. Lee KS, Dunton KH. 1996. Production and carbon reserve dynamics of the seagrass *Thalassia testudinum* in Corpus Christi Bay, Texas, USA. *Mar. Ecol. Prog. Ser.* 143:201–10

81. Lee SY. 1989. Litter production and turnover of the mangrove *Kandelia candel* (L.) Druce in a Hong Kong tidal shrimp pond. *Estuar. Coast. Shelf Sci.* 29:75–87

82. Lee SY. 1995. Mangrove outwelling: a review. *Hydrobiologia* 295.203–12

83. Leichter JJ, Wing SR, Miller SL, Denny MW. 1996. Pulsed delivery of subthermocline water to Conch Reef (Florida Keys) by internal tidal bores. *Limnol. Oceanogr.* 41:1490–501

84. Lewis JB, Gladfelter EH, Kinsey DW. 1985. Metabolism, calcification and carbon production. III. Seminar discussion. *Proc. 5th Int. Coral Reef Congr.* 4:540–2

85. Lohrenz SE, Fahnenstiel GL, Redalje DG, Lang GA, Chen X, et al. 1997. Variations in primary production of northern Gulf of Mexico continental shelf waters linked to nutrients inputs from the Mississippi River. *Mar. Ecol. Prog. Ser.* 155:45–54

86. Lugo AE, Snedaker SC. 1974. The ecology of mangroves. *Annu. Rev. Ecol. Syst.* 5:39–64

87. Mantoura RFC, Martin J-M, Wollast R, eds. 1991. *Ocean Margin Processes in Global Change.* 469 pp. Chichester, UK: Wiley & Sons

88. Marbà N, Cebrián J, Enríquez S, Duarte CM. 1996. Growth patterns of Western Mediterranean seagrasses: species-specific responses to seasonal forcing. *Mar. Ecol. Prog. Ser.* 133:203–15

89. Marbà N, Duarte CM, Cebrián J, Gallegos ME, Olesen B, et al. 1996. Growth and population dynamics of *Posidonia oceanica* on the Spanish Mediterranean coast: elucidating seagrass decline. *Mar. Ecol. Prog. Ser.* 137:203–13

90. Mateo MA, Romero J. 1997. Detritus dynamics in the seagrass *Posidonia oceanica*: elements for an ecosystem carbon and nutrient budget. *Mar. Ecol. Prog. Ser.* 151:43–53

91. Mateo MA, Romero J, Pérez M, Littler DS. 1997. Dynamics of millenary organic deposits resulting from the growth of the Mediterranean seagrass *Posidonia oceanica. Estuar. Coast. Shelf Sci.* 44:103–10

92. Meybeck M. 1982. Carbon, nitrogen and phosphorus transport by world rivers. *Am. J. Sci.* 282:401–50

93. Meybeck M. 1993. C, N, P and S in rivers: from sources to global inputs. In *Interactions of C, N, P and S Biogeochemical Cycles and Global Change*, ed. R Wollast, FT Mackenzie, L Chou, pp. 163–93. Berlin: Springer-Verlag

94. Middelburg JJ, Klaver G, Nieuwenhuize J, Vlug T. 1995. Carbon and nitrogen cycling in intertidal sediments near Doel, Scheldt estuary. *Hydrobiologia* 311:57–69

95. Middelburg JJ, Nieuwenhuize J, Markusse R, Ohowa B. 1995. Some preliminary results on the biogeochemistry of mangrove sediments from Gazi Bay. In *Monsoons and Coastal Ecosystems in Kenya*, ed. CHR Heip MA, Hemminga, MJM de Bie, pp. 51–66. Leiden: Natl. Mus. Nat. Hist.

96a. Middelburg JJ, Nieuwenhuize J, Slim FJ, Ohowa B. 1996. Sediment biogeochemistry in an East African mangrove forest (Gazi Bay, Kenya). *Biogeochemistry* 34:133–55

96b. Middelburg JJ, Nieuwenhuize J, Lubberts RK, van de Plassche O, 1997. Organic carbon isotope systematics of coastal marshes. *Estuar. Coast. Shelf Sci.* 45:681–87

97. Milliman JD. 1993. Production and accumulation of calcium carbonate in the ocean: budget of a nonsteady state. *Glob. Biogeoch. Cycles* 7:927–57

98. Moore KA, Neckles HA, Orth RJ. 1996. *Zostera marina* (eelgrass) growth and survival along a gradient of nutrients and turbidity in the lower Chesapeake Bay. *Mar. Ecol. Prog. Ser.* 142:247–59

99. Moran MA, Wicks RJ, Hodson RE. 1991. Export of dissolved organic matter from a mangrove swamp ecosystem:

evidence from natural fluorescence, dissolved lignin phenols, and bacterial secondary production. *Mar. Ecol. Prog. Ser.* 76:175–84

100. Moriarty DJW. 1986. Measurement of bacterial growth rates in aquatic systems from rates of nucleic acid synthesis. *Adv. Microb. Ecol.* 8:245–92

101. Nedwell DB, Trimmer M. 1996. Nitrogen fluxes through the upper estuary of the Great Ouse, England: the role of the bottom sediments. *Mar. Ecol. Prog. Ser.* 142:273–86

102. Nixon SW, Ammerman JW, Atkinson LP, Berounsky VM, Billen G, et al. 1996. The fate of nitrogen and phosphorus at the land-sea margin of the North Atlantic Ocean. *Biogeochemistry* 35:141–80

103. Olesen B. 1996. Regulation of light attenuation and eelgrass *Zostera marina* depth distribution in a Danish embayment. *Mar. Ecol. Prog. Ser.* 134:187–94

104. Onuf CP. 1996. Seagrass responses to long-term light reduction by brown tide in upper Laguna Madre, Texas: distribution and biomass patterns. *Mar. Ecol. Prog. Ser.* 138:219–31

105. Opdyke BN, Walker JCG. 1992. Return of the coral reef hypothesis: shelf partitioning of $CaCO_3$ and its effect on atmospheric CO_2. *Geology* 20:733–36

106. Pant A, Dhargalkar VK, Bhosale NB, Untawale AG. 1980. Contribution of phytoplankton photosynthetic to a mangrove ecosystem. *Mahasagar* 13:225–34s

107. Pauly D, Christensen V. 1995. Primary production required to sustain global fisheries. *Nature* 374:255–57

108. Peinert R, von Bodungen B, Smetacek VS. 1989. Food web structure and loss rate. In *Productivity of the Ocean: Present and Past*, ed. WH Berger, VS Smetacek, G Wefer, pp. 35–48. Chichester, UK: Wiley & Sons

109. Pernetta JC, Milliman JD, eds. 1995. Land-ocean interactions in the coastal zone. Implementation plan. *IGBP Rep.* 33:1–215

110. Peterson BJ. 1980. Aquatic primary productivity and the ^{14}C-CO_2 method: a history of the productivity problem. *Annu. Rev. Ecol. Syst.* 11:359–85

111. Pichon M. 1997. Coral reef metabolism in the Indo-Pacific: the broader picture. *Proc. 8th Int. Coral Reef Symp.* 1:977–80

112. Rabouille C. 1997. *Human perturbation on carbon and nitrogen cycles in the global coastal ocean*. Presented at LOICZ Open Sci. Meet., 3rd, The Netherlands

113. Rasmussen E. 1977. The wasting disease of eelgrass (*Zostera marina*) and its effects on the environmental factors and fauna. In *Seagrass Ecosystems: A Scientific Perspective*, ed. CP McRoy, C Helffrich, pp. 1–51. New York: Dekker

114. Ricker WE. 1973. Linear regressions in fishery research. *J. Fish. Res. Bd. Canada* 30:409–34

115. Robertson AI, Alongi DM, ed. 1992. *Tropical Mangrove Ecosystems*. Washington, DC: Am. Geophys. Union. 329 pp.

116. Robertson AI, Alongi DM, Boto KG. 1992. Food chain and carbon fluxes. In *Tropical Mangrove Ecosystems*, ed. AI Robertson, DM Alongi, pp. 293–326. Washington, DC: Am. Geophys. Union

117. Robertson AI, Blaber SJM. 1992. Plankton, epibenthos and fish communities. In *Tropical Mangrove Ecosystems*, ed. AI Robertson, DM Alongi, pp. 173–224. Washington, DC: Am. Geophys. Union

118. Robertson AI, Daniel PA, Dixon P. 1991. Mangrove forest structure and productivity in the Fly River estuary, Papua New Guinea. *Mar. Biol.* 111:147–55

119. Robertson AI, Duke NC. 1987. Insect herbivory on mangrove leaves in North Queensland. *Aust. J. Ecol.* 12:1–7

120. Romero J, Pérez M, Mateo MA, Sala E. 1994. The belowground organs of the Mediterranean seagrass *Posidonia oceanica* as a biogeochemical sink. *Aquat. Bot.* 47:13–19

121. Romero J, Pergent G, Pergent-Martini C, Mateo MA. 1992. The detritic compartment in a *Posidonia oceanica* meadow: litter features, decomposition rates and mineral stocks. *PZNI: Mar. Ecol.* 13:69–83

122. Rougerie F, Wauthy B. 1993. The endo-upwelling concept: from geothermal convection to reef construction. *Coral Reefs* 12:19–30

123. Sabine CL, Mackenzie FT. 1995. Bank-derived carbonate sediment transport and dissolution in the Hawaiian Archipelago. *Aquat. Geochem.* 1:189–230

124. Saenger P, Snedaker SC. 1993. Pantropical trends in mangrove above-ground biomass and annual litterfall. *Oecologia* 96:293–99

125. Schlesinger WH. 1997. *Biogeochemistry. An Analysis of Global Change*. London: Academic. 588 pp.

126. Short FT. 1990. Primary elemental constituents. *Seagrass Research Methods*, ed. RC Phillips, CP McRoy, pp. 105–9. Paris: Unesco

127. Smith SV. 1978. Coral-reef area and the contributions of reefs to processes and

resources of the world's oceans. *Nature* 273:225–26

128. Smith SV. 1981. Marine macrophytes as a global carbon sink. *Science* 211:838–40

129. Smith SV. 1988. Mass balance in coral reef-dominated areas. In *Coastal-Offshore Ecosystem Interactions*, ed. B-O Jansson, pp. 209–26. Berlin: Springer-Verlag

130. Smith SV. 1991. Stoichiometry of C:N:P fluxes in shallow-water marine ecosystems. In *Analyses of Ecosystems: Patterns, Mechanisms and Theory*, ed. J Cole, G Lovett, S Findlay, pp. 259–86. New York: Springer-Verlag

131. Smith SV. 1995. Reflections on the measurements and significance of carbon metabolism on coral reefs. *Kans. Geol. Surv. Open-File Rep. Ser.* 95–96a:1–18. Lawrence: Kans. Geol. Surv.

132. Smith SV, Hollibaugh JT. 1993. Coastal metabolism and the oceanic organic carbon balance. *Rev. Geophys.* 31:75–89

133. Smith SV, Hollibaugh JT. 1997. Annual cycle and interannual variability of ecosystem metabolism in a temperate climate embayment. *Ecol. Monogr.* 67:509–33

134. Deleted in proof

135. Smith SV, Jokiel PL, Key GS. 1978. Biochemical budgets in coral reef systems. *Atoll Res. Bull.* 220:1–11

136. Soetaert K, Herman PMJ. 1995. Carbon flows in the Westerschelde Estuary (The Netherlands) evaluated by means of an ecosystem model (MOSES). *Hydrobiologia* 311:247–66

137. Sorokin YI. 1993. *Coral Reef Ecology*. Berlin: Springer-Verlag. 465 pp.

138. Spalding MD, Blasco F, Field CD, eds. 1997. *World Mangrove Atlas*. 178 pp. Okinawa, Japan: Inter. Soc. Mangrove Ecosystems

139. Stanley SO, Boto KG, Alongi DM, Gillian FT. 1987. Composition and bacterial utilization of free amino acids in tropical mangrove sediments. *Mar. Chem.* 22:13–30

140. Stevenson JC. 1988. Comparative ecology of submersed grass beds in freshwater, estuarine, and marine environments. *Limnol. Oceanogr.* 33:867–93

141. Street GT, Montagna PA, Parker PL. 1997. Incorporation of brown tide into an estuarine food web. *Mar. Ecol. Prog. Ser.* 152:67–78

142. Tribble GW, Atkinson MJ, Sansone FJ, Smith SV. 1994. Reef metabolism and endoupwelling in perspective. *Coral Reefs* 13:199–201

143. Twilley RR. 1985. The exchange of organic carbon in basin mangrove forests in a southwest Florida estuary. *Estuar. Coast. Shelf Sci.* 20:543–57

144. Twilley RR, Chen RH, Hargis T. 1992. Carbon sinks in mangroves and their implications to carbon budget of tropical coastal ecosystems. *Water Air Soil Poll.* 64:265–88

145. Twilley RR, Lugo AE, Patterson-Zucca C. 1986. Litter production and turnover in basin mangrove forests in southwest Florida. *Ecology* 67:670–83

146. Walsh JJ. 1988. *On the Nature of Continental Shelves*. San Diego, CA: Academic. 520 pp.

147. Walsh JJ. 1991. Importance of continental margins in the marine biogeochemical cycling of carbon and nitrogen. *Nature* 350:53–55

148. Ware JR, Smith SV, Reaka-Kudla ML. 1992. Coral reefs: sources or sinks of atmospheric CO_2? *Coral Reefs* 11:127–30

149. Whittaker RH, Likens GE. 1973. Carbon and the biota. *Brookhaven Symp. Biol.* 24:281–302

150. Wolanski E. 1986. An evaporation-driven salinity maximum zone in Australian tropical estuaries. *Estuar. Coast. Shelf Sci.* 22:415–24

151. Wolanski E, Delesalle B. 1995. Upwelling by internal waves, Tahiti, French Polynesia. *Cont. Shelf Res.* 15:357–68

152. Wolanski E, Mazda Y, Ridd P. 1992. Mangrove hydrodynamics. In *Tropical Mangrove Ecosystems*, ed. AI Robertson, DM Alongi, pp. 43–62. Washington, DC: Am. Geophys. Union

153. Deleted in proof

154. Wollast R. 1983. Interactions in estuaries and coastal waters. In *The Major Biogeochemical Cycles and Their Interactions*, ed. B Bolin, RB Cook, pp. 385–407. Chichester, UK: Wiley-Interscience

155. Wollast R. 1991. The coastal organic carbon cycle: fluxes, sources, and sinks. In *Ocean Margin Processes in Global Change*, ed. RFC Mantoura, J-M Martin, R Wollast, pp. 365–81. Chichester, UK: Wiley & Sons

156. Wollast R. 1993. Interactions of carbon and nitrogen cycles in the coastal zone. *Interactions of C, N, P and S Biogeochemical Cycles and Global changes*, ed. R Wollast, FT Mackenzie, L Chou, pp. 195–210. Berlin: Springer-Verlag

157. Wollast R. 1998. Evaluation and comparison of the global carbon cycle in the coastal zone and in the open ocean. In

The Sea, ed. KH Brink, AR Robinson, pp. 213–52. New York: Wiley & Sons

158. Wollast R, Mackenzie FT. 1989. Global biogeochemical cycles and climate. In *Climate and Geo-Sciences*, ed. A Berger, S Schneider, J-C Duplessy, pp. 453–73. Dordrecht: Kluwer

159. Woodwell GM, Rich PH, Hall CAS. 1973. Carbon in estuaries. In *Carbon and Biosphere*, pp. 223–40. Springfield, VA: U.S. Atomic Energy Commission

160. World Resour. Inst. 1996. *World Resources 1996–1997*. New York: Oxford Univ. Press. 384 pp.

161. Yamamuro M, Kayanne H, Minagawa M. 1995. Carbon and nitrogen stable isotopes of primary producers in coral reef ecosystems. *Limnol. Oceanogr.* 40:617–21

162. Zimmerman RC, Alberte RS. 1996. Effect of light/dark transition on carbon translocation in eelgrass *Zostera marina* seedlings. *Mar. Ecol. Prog. Ser.* 136:305–9

Annu. Rev. Ecol. Syst. 1998. 29:435–66

THE SCIENTIFIC BASIS
OF FORESTRY

David A. Perry

Department of Forest Science, Oregon State University, Corvallis, Oregon 97331,
and Ha o Ka 'Aina, Kapa'au, North Kohala, Hawai'i

KEY WORDS: forest management, sustainability, intensive forestry

> A sufficiently wise and flexible silvicultural art can be developed on the ground only by practitioners who understand the forest as a biological entity.

> F. S. Baker (10)

> ...the existing level of knowledge about forests is inadequate to develop sound forest management policies.

> National Research Council (114)

ABSTRACT

Over the past two decades forestry in the United States has diverged into two approaches with quite different objectives and scientific priorities. The management focus of most industrial lands is on increasing productivity of wood fiber via plantations and various cultural tools, especially genetic selection, fertilization, and control of noncrop vegetation. Federal forest management has shifted from a similar focus to greater emphasis on protecting diversity and water. Issues of long-term sustainability are important regardless of ownership. Science has played and continues to play a fundamental role in all aspects. Selection for fast-growing genotypes has increased yields on the order of 10% to 20% depending on species. Fertilization often increases growth significantly but responses are variable and difficult to predict. Significant questions remain concerning the sustainability of intensive forestry, particularly when practiced over wide areas. Soils are heavily impacted by some harvesting practices, and the degree to which damage can be repaired by fertilizers is an important scientific issue. Intensive forestry often results in increased pest problems. In at least one case (fusiform rust in southern pines), a pest has been contained by selecting resistant cultivars, a situation that may or may not be evolutionarily stable. Species diversity is clearly reduced under

435

0066-4162/98/1120-0435$08.00

intensive management, raising questions about the functional role of species with no commercial value. Many of the questions facing forestry science—particularly those dealing with the relation between complexity and function—are precisely the ones confronting basic ecology. Over the past decade scientists have labored to develop ecosystem-based management approaches that maintain system complexity and function, and scientists have increasingly played nontraditional roles at the interface between biology, sociology, and policy.

INTRODUCTION

Forestry has been defined as "the scientific management of forests for the continuous production of goods and services" (10), though, as with agriculture, forestry that reflects the specifics of place as well as the generalities of science necessarily involves art. Biologic, physical, social, management, and engineering sciences all play an important role in forestry. Over the past 20 years the scope of biologic and environmental sciences contributing to forest management has expanded beyond ecophysiology, genetics, and vegetation management to encompass soil processes, ecosystem structure and dynamics, hydrology, wildlife biology, fisheries, restoration ecology, conservation biology, and landscape ecology. The social sciences, once relegated to a backseat (except for economics), are now seen as critically important, at least on public lands in the United States. Research into innovative engineering techniques and the development of a broad array of forest products are essential parts of the contemporary package, and management science is playing an increasingly important role in helping to integrate science, economics, and politics. A new dialogue among science, philosophy, and religion is exploring the esthetic/spiritual dimensions of forests—and nature as a whole—that have been common to humans for millenia.

In this paper I deal mostly with biology, soils, and hydrology, but with the understanding that sociology, esthetics, ethics, spirituality, economics, and history intertwine with virtually all aspects of the biologic and physical sciences to produce the complex systems that foresters work with and within. Ignoring that basic truth during much of the twentieth century has resulted in social and political turmoil throughout the world over the way forests are managed—eventually triggering a fundamental reevaluation on the part of foresters and forest scientists of what forestry is all about (88). The need for integrating a broad array of scientific disciplines to guide forestry, indeed all natural resource management, has never been more acute.

This paper is basically a story of the changing role of science in forestry and an assessment of the current state of knowledge and priority needs. I discuss what I believe to be the most important challenges facing forestry science; however, the breadth of the topic does not allow all aspects to be given their rightful due. My choice of emphases reflects my own experience and biases; someone else may have made different choices. I focus mostly on North America—though both the practice and science of forestry are largely global affairs—because, until recently, the basis strategy of forestry has largely been the same everywhere during the latter half of the twentieth century.

A SHORT HISTORY

Forestry as a science originated in Europe during the latter eighteenth and early nineteenth centuries, largely in response to the poor condition of Europe's remaining forests and an impending wood famine (126). From early on there were competing philosophies about the proper approach. Aldo Leopold, who once supervised a US National Forest, described two types of forestry: Type A sees "... land as a commodity and trees as cellulose to be grown much like cabbages"; Type B "... treats land as a community of interacting and interdependent parts, all of which must be cared for" (quoted in 12). Type B, which I refer to as ecosystem-based forestry, has always had a strong philosophical representation within the forestry profession; however, Type A has predominated throughout the world during the twentieth century. Like modern agriculture, its focus has been on the properties of individual ideotypes rather than communities and ecosystems (119). Type A is commonly referred to as intensive forestry (where "intensive" refers to cultural inputs).

Intensive forestry had its beginnings in Germany during the mid-1800s. German forest scientists, motivated by the ideas of the English economist, Adam Smith, formulated an economic approach called soil rent theory, which held that interest should be earned on land, timber capital, and silvicultural expenses (in opposition, "forest rent" theory held that interest charges against these assets were inappropriate). Plochmann (126) describes the result. "The soil rent method furnished foresters with an ideal planning tool for calculating the species with the highest monetary return and the financial rotation with the highest internal rate of interest on a given site. It fit perfectly with classical liberal economic theory, which set the maximization of profit as the general objective of economic activities and therefore the general objective for forestry as well." In central Europe, native hardwood and mixed hardwood/conifer forests were cleared and planted to monocultures that grew fast and produced high value—mostly Norway spruce. Because interest was charged on land and other capital assets, the rotation length (time period between regeneration of a stand and final harvest) yielding the highest rate of return was much shorter than the normal life span of forests.

Despite some misgivings on the part of both foresters and the public (12), intensive forestry took root in North America following World War II, and by the 1950s was firmly established on both industrial and public lands. Over time it became the dominant theme in most or all of the nation's forestry schools. The reasons behind the ascendancy of intensive forestry were clear. A primary responsibility of foresters was to supply the fiber needs of a rapidly expanding economy, and this would best be done through plantation monocultures of fast-growing tree species. The full growth potential of land could be ultimately achieved by planting healthy, genetically superior seedlings, controlling non-crop vegetation, enhancing site quality through fertilization (and other techniques where necessary and affordable), and clear-cutting when either rate of tree growth or economic returns calculated for a series of rotations were maximized (e.g. 40). The responsibility for meeting the demands of a rapidly growing market without overharvesting or otherwise degrading future productive capacity fell to scientists, and intensive forestry came to be described by many of its proponents, and more recently some critics, as scientific forestry. In a narrow sense that characterization is accurate, as the tools of science are used, albeit in the service of a particular economic model. However, like crop-centered agriculture, intensive forestry was not derived in the context of testable hypotheses about alternative approaches, nor was there much open debate about alternatives (until lately); thus in a more fundamental sense it is not scientific. Not surprisingly, forestry shared one of the central distinguishing features of industrialism: uncritical acceptance of the untested hypothesis that maximizing economic efficiency in the short term was the path to maximizing social and economic benefits into the future.

The last 35 years have seen much change in forestry and a significant divergence in approaches. During the 1960s, Germany, for social, environmental, and economic reasons, began converting state lands back to the native hardwood/conifer mixed forests and growing high-value trees on long rotations (126). In the United States, beginning in the late 1960s and continuing to the present, public opposition to clear-cutting and herbicides joined mounting scientific evidence that, in contradiction to federal laws, native diversity was not being maintained on federal lands; together these led the US Forest Service (USFS) and the Bureau of Land Management (BLM), after significant prodding by the courts to adopt ecosystem management as official policy. Although some companies have adopted aspects of ecosystem management, intensive forestry remains by far the most commonly used approach on industrial lands. (They do not necessarily incorporate all aspects; fertilization and herbicides in particular may or may not be used, depending on perceived economic benefits.) Thus, over the last decade in the United States what was once a single approach has diverged into two paths that, while still sharing some aspects, differ significantly in objectives, approach, and scientific priorities.

STATE OF KNOWLEDGE

The biological sciences have played three distinct roles in forestry: improving growth through intensive cultural practices; researching the environmental impacts and sustainability of intensive forestry; and, most recently, developing a science of ecosystem-based management, which includes a broad array of basic and applied research dealing either directly or indirectly with alternative management approaches. Each of these will continue to be important in forestry, though their relative importance is shifting from crop-centeredness to a greater balance between traditional forestry research and the science of ecosystem-based management. Questions of sustainability remain of central importance regardless of management approach.

Improving Growth

The basic approaches to increasing stand growth are (*a*) rapid establishment of a new stand following harvest (nursery and planting practices); (*b*) maximizing the flow of site resources to the crop (controlling competing vegetation); (*c*) improving soil resources (fertilization, bedding, and drainage on some sites); (*d*) selecting and breeding fast-growing genotypes; and (*e*) minimizing losses due to insects, diseases, fire, and wind (stand protection). All approaches have been actively researched over the past 30 years, except that comparatively little research has been done on stand protection, and that almost exclusively in reaction to specific problems. The theoretical potential for productivity gains through intensive culture is significant. Farnum et al (40) calculated growth rates of unmanaged stands of Douglas-fir and loblolly pine (two of the most important commercial species in the United States) at, respectively, only 23% and 12% of what could be achieved. In this section I focus on genetics, vegetation control, and fertilization, which are the centerpieces of intensive forestry. I deal with vulnerability of managed stands to biotic and abiotic disturbances later.

GENETICS   Most commercially important tree species contain large within-population genetic variation, providing a rich source for selection. For example, approximately 92% of the genetic diversity in loblolly pine, the dominant timber species in the southern United States, occurs within stands (177). A further boon to selection programs is that many individual genotypes of commercially important species have wide ecological amplitude. Loblolly pine and Douglas-fir families tend to maintain a constant ranking across a wide range of environments (101, 157); when significant GXE interactions do occur in those species it is usually due to a few families.

Much tree breeding in the United States and Canada has been through co-operatives involving industry, universities, and in some cases public agencies (e.g. 164). Although there is growing interest in cloning (155), most genetically improved conifers (by far the most important commercial species in North

America) come from seed orchards. The basic strategy is to select trees in the field with desired characteristics and propagate them in both open-pollinated production orchards and breeding orchards where select trees are interbred through controlled pollination (40). At the end of the first breeding cycle, seed from the production orchard is outplanted, while seed from the breeding orchard is used (after progeny testing) to establish second-generation production and breeding orchards, and so on. The use of elite breeding populations (progressive selection and interbreeding of the best performing genotypes), a strategy adopted from agronomic and animal programs, is increasingly common in conifer breeding (177). Various technological advances have reduced breeding cycles considerably over the past two decades—in loblolly pine from 22 years in 1974 to 7.5 years in 1994 (177). In 1992, 90% of all plantings in the southern United States, amounting to 1.5 million acres, were with genetically improved seedlings (164). The first genetically improved trees are now being harvested in the southern United States, with estimated volume gains of 12% for loblolly pine (32% in harvest value) and 18% for slash pine (73, 102). Weyerhauser Company estimates a 10% gain in juvenile height growth for select Douglas-fir in Oregon and Washington (157).

The importance of maintaining genetic diversity is much discussed among forest geneticists (e.g. 38, 102, 110, 135). Erosion of genetic diversity through random drift is an especially significant concern in forest trees, which carry high levels of lethal recessive alleles and are particularly vulnerable to inbreeding depression (93, 111). Generally one of two approaches is used to maintain diversity within breeding populations (39): the hierarchical open-ended system (HOPE) and the multiple population breeding system (MPBS). The HOPE strategy, developed originally by agronomic breeders, periodically introduces genetic material from populations early in the selection cycle to populations at later stages of selection. MPBS, developed by forest geneticists (110), sets up a number of different breeding populations and exchanges genes in controlled crosses among these—establishing what is in essence a controlled metapopulation structure. Allozyme studies in loblolly pine show that, although both strategies maintain relatively high levels of allozyme diversity within elite breeding populations, neither maintains the diversity found in natural stands (176). As would be expected, rare alleles are the most heavily impacted. Moreover, because seedlings are culled in the various steps between controlled breeding populations and the eventual planting stock, the diversity of resulting plantations will be lower yet (177). Studies of other commercial species show similar patterns: Breeding programs to date have had little or no influence on overall allozyme diversity; however, rare alleles may be sharply reduced (37, 156). Major questions remain concerning effects of selection on genetic diversity at the landscape level, or what Friedman & Foster (48) refer to as beta genetic diversity, i.e. the allozyme diversity of an elite breeding population may differ little from that of wild stands (alpha diversity), but planted over large areas the elite population could replace a fine-scale genetic mosaic, resulting in genetically simplified landscapes (48).

Any reduction of genetic diversity raises the presently unanswerable question of how much is sufficient to maintain resistance to pests and the capacity to adapt to changing environments (102, 132, 177). The adaptive value of rare alleles is unknown at present (38). Some forest geneticists argue they contribute little to overall fitness; others, however, suggest that loss of rare alleles could compromise long-term adaptive flexibility (177). Inability to relate allozyme measurements to phenotypic traits with potential adaptive value seriously limits the usefulness of allozyme measures as predictors of ecological response (135).

VEGETATION MANAGEMENT   Grounded in the common view of plant community ecologists that competition is the most important interaction among plants, the focus of vegetation management in forestry has been on increasing crop yields through the use of chemicals and tillage. Considerable research (though not all) shows that controlling noncrop vegetation increases growth of crop trees during the early years of stand development, and in some cases establishing plantations can be difficult or even impossible without vegetation control (54, 171). A major thrust has been to develop competition indices relating levels of noncrop vegetation to potential growth response. However, the very short lifetime of most experiments limits the reliability of indices, which have been found to vary over time with shifting community structure (170).

The issues surrounding vegetation management are among the more complex in forestry, involving ecological effects of herbicides, long-term and indirect effects of altering community structure, and the functional roles of different plant species. As pointed out by some prominent researchers in the field, little attention has been paid to long-term and ecosystem-level effects of vegetation management (130, 170). The basic scientific issue is one of the relationship between community structure and ecosystem function, something ecologists in general are just beginning to grapple with.

In North American forestry, where conifers are the major commercial species, noncrop vegetation is virtually always broadleaved trees, forbs, shrubs, and grasses. A variety of studies have either conclusively demonstrated or strongly suggested that these plants perform numerous important ecological functions, including providing unique food (e.g. nuts, nectar), enhancing nutrient availability (9, 49), replenishing nitrogen capital through biological fixation (14, 67), stabilizing soil nutrients and biology following disturbances (5, 16), and increasing resistance of conifers to herbivorous insects, pathogens, and fire (e.g. 52, 123,

136, 146, 161, 181). The complexity of interactions among plant species is only beginning to be understood but goes far beyond simple competition. For example, it is now well established that different plant species within at least some communities—including broadleaved trees and conifers—participate in a network of shared resources mediated by mycorrhizal fungi (131, 147).

The major challenge for the science of vegetation management is to find balance between competition and other important ecological functions, and this will require a much improved understanding of interactions among plant species and between plants, animals, and microbes and how these interactions are affected by relative density, environmental factors, and time. The few studies that have manipulated relative density of crop and noncrop species show complex, shifting competitive relationships; as Shainsky & Rose (142) put it, "competition is not linear and unidirectional." Moreover, protective and stabilizing functions may come into play only during critical periods, such as during wildfire or recovery from disturbance, which means the importance of these functions could easily be missed in short-term studies (125).

Scientists working with vegetation management in forestry increasingly take an ecosystem view (the theme of the 1998 International Conference on Forest Vegetation Management is "Forest Vegetation Management and Ecosystem Sustainability"). Experiments that manipulate relative density or spatial patterns of noncrop plants or that separate shrub from herb competition are becoming more common (e.g. 103, 141). There is movement toward an "integrated vegetation management," which, like integrated pest management, seeks ways to reduce pesticide use or replace it altogether (170).

FERTILIZATION   Hundreds and perhaps thousands of fertilizer trials have been installed over the past 30 years in forests of the United States and Canada; these vary widely in experimental rigor, with later installations generally better designed and more intepretable than earlier ones. A substantial number in the United States are university-industry-USFS cooperatives. Fertilization experiments are frequently coupled with other cultural treatments such as thinning and vegetation control. Most research has focused on nitrogen and, in the south, phosphorus (15), though experiments have been installed using NPK (112) and, in at least one case, slow-release multinutrient tabs (168).

N fertilization frequently produces significant, occasionally spectacular, but often highly variable growth responses. Volume gains from N average 16% to 26% in coastal Oregon, Washington, and British Columbia (104). In a regional study employing N and P in a factorial design, loblolly pine volume growth in response to N and P averaged 25% greater than controls, with NP yielding 2–3 times greater response than did the addition of N and P alone (112). A recent research trend has been to employ frequent, small additions of N, a technique shown by Swedish researchers to better mimic natural processes and produce significantly greater growth response than infrequent, large additions (80). Using that approach in a North Carolina study resulted in exceptional growth responses in young loblolly pine, doubling leaf area over a four-year period and increasing stem volume 180% compared with controls (3). Part of the growth response was due to reallocation of CHO from fine roots to above-ground tissues, a dynamic seen in other studies and probably quite general. Despite a droughty site, irrigation had a relatively small effect in the North Carolina study, suggesting that other cases in which water is thought to be limiting may actually be N limited.

Predicting fertilizer response has been a problem area and the subject of considerable research. Although N has enhanced tree growth in 70% of trials in Oregon, Washington, and British Columbia, the magnitude of response varies widely (104). Similarly, researchers have had limited success in predicting growth response to N and P in the southeastern coastal plain (66). Standard measures of soil nutrient availability are seldom useful (66, 144), while foliar levels may or may not be. Evidence from both the south and the west indicates that limited response to N and P is often due to deficiencies of other nutrients, with potassium, sulfur, boron, magnesium, and iron identified in one or more studies (e.g. 112, 144, 168). Recently, DRIS norms have been employed with some success to classify the Douglas-fir stands with regard to their response to N fertilization (144). DRISS, originally developed by agronomists, uses nutrient ratios in foliage or soil to predict fertilizer response.

TREATMENT COMBINATIONS   Cultural practices are usually applied as sets, i.e. a single site may be drained, fertilized, herbicided, and planted with genetically improved trees. Some research has shown non-additive responses to treatment combinations. For example, trees may not respond to fertilization unless stands are thinned or competing vegetation is controlled (153). Even though GXE is minimal in loblolly pine, McKeand et al (101) argue that planting certain responsive families will significantly increase gains from cultural treatments. The degree to which intensive cultural factors interacts with ecologic factors such as long-term soil fertility and resistance to pests and pathogens is the subject of the next section.

Sustainability

Concern with sustainability in forestry dates back to the late seventeenth century; in fact, Wiersum (174) argues forestry was the first science in the western world to explicitly acknowledge the need to safeguard finite natural resources for future generations. However, achieving sustainability within an economic system that devalues the future is a particular challenge, especially for a long-term endeavor like forestry (13). Negative impacts of practices on future yields of wood can, in theory, be dealt with by incorporating them within economic calculations, though the further in the future such costs are realized the less influence they have on present net value. Impacts on nonmarketable goods (e.g. habitat, water) and risks with high uncertainty must be dealt with in other ways.

What To Sustain?

For much of the history of forestry, sustainability meant maintaining a steady supply of wood (174). In the United States considerable discussion occurred prior to and in the years following World War II about the role of forests in wildlife habitat, watershed protection, and even regional climate, but this had relatively little effect on practices (e.g. 92). The Multiple Use Sustained Yield Act of 1960 specified five things to be sustained on public lands: timber, fish and wildlife, outdoor recreation, range and fodder, and watersheds (174). More recently, sustaining biodiversity has been center stage, though that term is itself subject to various interpretations. Angermeier & Karr (9) argued that policies for achieving sustainability should focus on biological integrity, which they define as "a system's wholeness, including presence of all appropriate elements, and occurrence of all processes at appropriate rates." Franklin (44) defines sustainability as "... maintenance of the *potential* for our land and water ecosystems to produce the same quantity and quality of goods and services in perpetuity."

The following discussion focuses on four critical components of ecological sustainability: soils, water, species diversity (habitat), and resistance to disturbances (118). A fifth, evolutionary potential, was discussed in an earlier section.

Soils

German foresters knew as early as the mid-1800s that export of nutrients associated with removing forest litter reduced tree growth. Outside of Europe, however, little attention was given to the effects of forestry practices on soils and nutrient capital until the mid-1960s, when data from intensively managed *Pinus radiata* in Australia and New Zealand showed yields had declined significantly between the first and second rotations, a phenomenon eventually traced to loss of soil organic matter (SOM) and nutrients during site preparation for planting. As the same practices—hot slash burns or use of heavy equipment to push logging residues (and often topsoils) into piles—were widely used in North America, considerable research was stimulated that continues today (e.g. 11, 53, 68, 120). A group of leading forest soil scientists expressed the current view of many soil scientists and ecologists: "... intensive site preparation, increased utilization, and shortened rotations that accompany domestication carry high potential risks to the site's capability to sustain growth" (129).

Soil fertility (broadly, the capacity of soils to support plant growth) is an emergent property arising from dynamic interactions among nutrients, organic matter, soil structure (a function of both physical and biological processes), soil organisms, and plants (121). Most forestry-related research to date has focused on nutrient balances, comparing losses associated with harvest and site preparation with inputs to replace those losses. Nutrient losses associated with clear-cutting have been well documented and include three primary pathways: (a) direct removal in harvested biomass; (b) losses associated with site preparation techniques such as burning, which impacts volatile elements (especially carbon and nitrogen), or windrowing (also called raking or pile-and-burn), which displaces residual organic matter from the majority of a site onto small portions; and (c) export in erosion and leaching. In most cases, losses in harvested biomass and site preparation far exceed leaching and erosion (32, 97); an important exception is where early successional vegetation is prevented from recovering, especially in forest types characterized by low C/N organic matter (19).

Not surprisingly, the more biomass removed over a given period of time and the greater its nutrient content, the greater the probability that rates of nutrient export will exceed rates of input through natural processes. Short rotations combined with practices that remove a high proportion of site biomass result in exceedingly large nutrient drains. By far the largest losses of nutrients and either present or future soil carbon are associated with harvesting tree crowns and site preparation (81). Various studies show that removing crowns, which have relatively high nutrient concentration, increases nutrient loss by 50% to 400% over harvesting boles only (21, 87, 97, 149).

Throughout North America, sites are often prepared for planting by either windrowing or burning residues in place ("broadcast burning"). Windrowing as traditionally practiced compacts soils and removes large amounts of SOM and nutrients; Morris et al (108) estimated nutrient losses during windrowing as equal to six bolewood harvests. Hot broadcast burns, once the norm in western North America, volatilize large amounts of carbon and nitrogen (81).

Reductions in SOM and soil pore space (the latter due to compaction) are a significant concern (129). Soil organic matter (SOM)—including humus, forest floor, and both fine and coarse woody debris—stores nutrients (especially nitrogen), provides cation exchange and water-holding capacity, and serves as substrate for the belowground food chain; hence SOM is a keystone resource for numerous soil functions. SOM falls into two broad groups that differ both functionally and with respect to potential management impacts. Fine litter (foliage, small branches, epiphytes) in various stages of decay forms the majority of surface organic layers and (eventually) humus within the soil profile, and it functions in nutrient cycling, cation exchange capacity, and water retention.

Coarse debris (logs and large branches) is abundant in natural forests and functions as sites for N fixation, water reservoirs (decayed logs are sponge-like in their ability to retain water), and habitat for numerous fungi, invertebrates, and vertebrates such as salamanders, which are at the top of the soil food chain (62, 71). Immediate impacts of forest management on soil C are mainly associated with extreme site preparation techniques such as hot fires, windrowing, and tilling (70, 81), whereas practices that remove sources of future soil C (trees and litter layers) have longer-term effects that remain to be determined. Presumably losses in those components of soil C that derive from fine litter will be replenished so long as plant communities regrow vigorously, a condition tied in a self-referential loop to the status of soil nutrients, soil C, and soil biology after harvest. Components of soil C derived from large debris—especially tree boles—are more problematic. Until recently, forestry never considered leaving tree boles on site; in fact—quite the opposite—doing so was seen as waste. As I discuss later, retaining trees on site as future sources of large dead wood is a major component of new silvicultural approaches. However, the question of how many trees to leave turns on gaining a better understanding of the functions of large dead wood within the ecosystem.

Though soil biology is undoubtedly the key to understanding soil function (124, 138), the enormous diversity belowground has made research into functional aspects slow and expensive. Consequently, the science of the belowground is in its infancy. Existing studies related to forestry have focused either on total microbial biomass, particularly as nutrient sinks following harvest (27, 169), or on specific functional guilds such as mycorrhizal fungi, nitrogen-fixing and other rhizosphere bacteria, invertebrates, or some combination of these (e.g. 5, 7, 69, 71, 84, 105, 121, 122). Soil bacteria act as important nutrient sinks following clear-cutting, a phenomenon tied to the availability of labile C and therefore likely to be impacted by intensive site preparation (169). Management effects on specific guilds range from minor to severe, depending on the management practice employed and the guild of organisms studied. Significant change in soil biology as a result of clear-cutting and site preparation is always found, including increased bacterial relative to fungal biomass, greater rates of nitrification, reduced concentrations of microbially produced iron chelators, shifts in mycorrhizal types, and sharp declines in invertebrate numbers. However, interpretation is complicated by poor understanding of functional and dynamic aspects of the belowground ecosystem.

Translating impacts on soils to longer-term soil fertility, tree growth, and other ecosystem processes is not always straightforward. Nutrient inputs in precipitation are reasonably well known, but inputs from weathering—the major natural source of all essential elements except N, C, and O—are difficult to measure and poorly understood (31). Moreover, recent research shows conifers (and probably other non-nodulated trees) have some capacity to renew soil fertility. Bacteria associated with conifer rhizospheres and mycorrhizal fungi can fix appreciable amounts of N (17, 94). Both ectomycorrhizal fungi and rhizosphere bacteria significantly accelerate rock weathering and are likely to speed recovery of fertility in soils with unweathered rock in the rooting zone (18, 83). [Soils in the northern United States and Canada contain abundant unweathered rock, but that is not the case with some old soils in the southern United States (107).]

Better understanding the capacity of trees and their symbionts to renew soil fertility, and how that varies with site, species, and environmental conditions, promises to add a significant new dimension to our picture of ecosystem nutrient budgets. A key question that has received little research is whether site impacts can disrupt these biologically mediated renewal processes. The few studies that do exist suggest the renewal process can be influenced by management. N fixation associated with Douglas-fir and its mycorrhizae is stimulated by proximity to certain hardwood species (8), suggesting that in systems where that relationship occurs, excessive vegetation control will reduce associative N fixation. N fixation in ponderosa pine rhizospheres is depressed in both clear-cuts and patch-cuts (<0.2 ha) relative to undisturbed forest (79).

Predicting long-term impacts on soil fertility has been further complicated by the fact that pines, in particular, often grow better during their early years on sites that have been intensively prepared than on sites that have not (23). Although the ability of trees to renew fertility may be a partial explanation, most researchers agree better growth on intensively prepared sites is a transitory phenomenon due to various factors, especially reduced competition from herbs and grasses (23, 106, 129). As plantations have reached crown closure, relatively poor growth on intensively prepared sites has become widely evident (20, 33, 43, 128). In one of the few studies in which trees were approaching rotation age, stem volume per hectare of 31-year-old loblolly pine was 23% lower on plots that had been windrowed prior to planting than on plots that had been broadcast burned (43).

The multirotation predictions required for assessments of sustainability, coupled with the complexity and nonlinearity of forest resilience, require experiments and models based in ecosystem-level processes (89, 152). Several models exist that link nutrients (mainly nitrogen) to tree growth (1, 86, 112, 116). Although the importance of such models is clear, they are also potential traps because their long-term predictions cannot be validated in the short term, and there is a history of managers using models as substitutes for observation and critical thought. Yarie (178) compared two models and found they predicted different outcomes for the same management practices. He concluded "... neither (should) be used as a ... decision tool without the help of sufficient expertise in ecosystem ecology to correctly interpret the results."

Several long-term experiments dealing with management impacts on soils and tree growth have been installed within the past few years, each nearly unique in approach and objectives. Now 14 years old, a study installed by the North Carolina State University Nutrition Co-operative manipulated levels of harvest, site preparation, and vegetation control factorially (112). A widely replicated experiment sponsored by the US Forest Service includes different levels of compaction, biomass removal, and vegetation control in a nested design within clearcuts (129). Several studies in the United States and Canada have experimentally manipulated forest structure (via levels of harvest and size of harvest units) and are considering an array of soil chemical and biological variables (6, 46, 76).

Information on potential negative impacts of intensive site preparation has reached field foresters through symposia and other sources. Hot broadcast burns have become less common in the west, and managers in some areas have either abandoned windrowing or adopted windrowing techniques with lower impact. Intensive site preparation is still common, however, at least in part because of faith in the remedial effects of fertilizers. A central challenge for research is to determine the degree to which that faith is justified.

Water

Forests are keystone modulators of the water cycle; hence forestry and water are inseparable. Reflecting that fact, watershed protection was a primary rationale for the creation of the National Forest System, and hydrologic research by the USFS dates back at least 50 years. By the mid-1960s approximately 150 gaged, forested watersheds existed in the United States, many of which involved managed and unmanaged pairs (78). Other studies have either followed single basins as they are harvested over time, or, particularly in the case of large basins, compared data among basins operationally harvested to different levels (e.g. 82).

Hydrology exemplifies many other issues in forest science—large scale and high background variability (temporally and spatially) make generalizations that are both broad and accurate extremely difficult, expensive, and in some cases impossible. Only dramatic impacts on stream hydrology are statistically detectable in short-term studies. Moreover, effects of clear-cutting on stream hydrology tend to be highly variable for various reasons, including bedrock geology, topography, climate (e.g. rain or snow dominated, importance of fog-drip from canopies), and harvest practices. Clearest effects are from experimentally paired, small watersheds, and the ability to detect effects diminishes as one scales up to larger basins (133). Decades-long records are often necessary to discern trends, especially in larger basins.

Small watershed studies generally show clear-cutting increases water yield, though in areas with significant fog-drip (water raked from clouds or fog by tree crowns), yields may decline following harvest (133). Peak flows, which are of more concern environmentally and economically, are often increased by clear-cutting, though that depends on various factors, particularly the extent and rate of logging within a basin, how logging is done, and the road network. Intensive site preparation, vegetation control, extensive roading (especially when poorly designed), and disruption of riparian forests have the greatest likelihood of increasing the magnitude and duration of peak flows (133).

Studies in western Oregon show that clear-cutting and roads act synergistically to alter hydrology. Clear-cutting reduces evapotranspiration (ET) and increases snow accumulation and melt, resulting in increased deep soil water storage that persists until leaf area of deep-rooted trees and shrubs has fully recovered (on the order of decades) (64). In poorly drained areas, water tables rise in clear-cuts (23), and in some cases bog formation can be triggered (DA Perry, personal observation). Roads, on the other hand, alter hillslope flows by converting subsurface to surface flow; hence they provide pathways for deep soil water to reach the surface and flow rapidly to streams (63). On the HJ Andrews Experimental Forest (HJA) (Oregon Cascades), peak flows did not differ between a watershed that was 100% clear-cut without roads and one that was 25% clear-cut with roads, though seasonality of altered flows did differ (82). In both, average peak discharge increased by >50% (compared with pretreatment) for the first 5 years after treatment; 25 years after treatment, discharge remained 25% to 40% higher than pretreatment peak flows.

Large basins are a particular challenge because experimental controls do not exist and also because even dramatic events in small watersheds may be statistically undetectable at larger scales without many years of record (133). In the west, records are just now becoming sufficiently long to allow trends to be separated from noise in larger basins. Jones & Grant (82) examined 50-to-55-year records from three pairs of 60- to 600-km^2 basins in the western Cascades of Oregon (paired based on proximity and different rates of logging). Responses in all cases were statistically significant (though with low r^2): a 5% difference in cumulative area clear-cut translated to a difference in yearly peak flows ranging from 10% to 55%, depending on basin-pair. None of the basins were heavily clear-cut (maximum 25% of total basin area), yet Jones & Grant estimate that peak discharges in these basins have increased by 50% to 250% compared with prelogging. The extensive road network required where clear-cuts are in a dispersed patchwork, as is common on federal forests, means total cutover area significantly underestimates potential impacts on hydrology and on spatially mediated processes in general.

Effects of forest management on sedimentation have been easier to demonstrate than effects on water flows because background variability is much less; very little soil is eroded from undisturbed forests. As with water, the best documented studies have been in experimental watersheds, although clear evidence exists from other sources, such as comparisons of logged and unlogged watersheds and historical observations. Sediments associated with forestry come from four primary sources (159): surface erosion from roads, surface erosion from clear-cuts, mass transport during slash burns, and landslides associated with either roads or clear-cuts. Studies on the HJ Andrews Experimental Forest (HJA) found that landslides, especially from poorly designed roads during major storms, pulsed large amounts of sediment in brief episodes, while surface erosion from roads and clear-cuts was more chronic. Eleven years after treatment, suspended sediment from a roaded watershed, 25% clear-cut and burned, averaged 57 times higher (per year) than the unroaded, unlogged control watershed, whereas nine years after treatment, sediment from an unroaded watershed, 100% clear-cut and broadcast-burned, averaged 23 times higher than its control (159). Total increases in surface erosion following clear-cutting are most severe on steep slopes (generally >60%); however, proportional increases may be more severe on shallow slopes (159). Broadcast burning greatly exacerbates losses in both cases. Absolute losses vary widely depending on soil and rock type.

As with many issues in forestry, short-term studies may not be adequate to determine effects on sedimentation. Zeimer et al (179, 180) modeled the long-term cumulative effects of harvesting on sedimentation for coastal northern California to central Oregon (an area with high landslide potential). Their simulations predicted that sediment produced during the first century following harvest would be initially stored in small tributaries, to be washed into larger streams during the second century. That prediction was supported by data from Casper Creek in northern California, which is still adjusting to logging-related sedimentation from the late nineteenth and early twentieth centuries.

Diversity and Habitat

Intensive forestry significantly alters the spatial and temporal structure of forests and, applied widely, of forested landscapes. Aside from the fact that living trees are killed, it has little ecological similarity to natural disturbances, which leave a plethora of structural legacies and usually have a frequency distribution characterized by many small and few large events, hence leave a more variable patch mosaic than does intensive forestry (109, 143, 154). Stand reconstructions show that forest types once believed to be extensively even-aged due to infrequent, large natural disturbances are actually mosaics of age classes resulting from relatively frequent, small disturbances that established cohorts of young trees within a matrix of older trees (42, 47, 160). Because of their biological legacies and spatial patchiness, natural disturbances are likely to initiate different early successional patterns than does intensive forestry, which aims for rapid site capture by even-aged crop trees.

Fire exclusion, which along with roading is traditionally the first step in bringing a forested landscape under management, has significantly altered forest types with a history of frequent, low-intensity fires, especially ponderosa pine in the west and longleaf pine in the south, leading in many cases to increased problems with insect pests and pathogens, and paradoxically to a greater risk of catastrophic fires. Roads may affect diversity in a number of ways, principally as barriers to movement and by providing access for predators (including humans), noxious weeds, and pathogens (123, 148). Various studies have shown that population sizes of bears, wolves, moose, and mountain lions are inversely proportional to road density (e.g. 22).

In short, forestry as commonly practiced places a radically new selective filter on the landscape, structurally much simplified compared to the natural, but also with new elements. Some species benefit, others are endangered, some may adapt. But none has evolved with that filter. The most vulnerable habitats in an intensive forestry regime are associated with forests older than harvest age (20 to 100 years depending on forest type and product), hardwoods (as most forestry focuses on conifers), riparian zones, wetlands, and streams, all of which are unique and biologically rich. In some cases, conifers of one species are replaced by others with faster initial growth; such is the case with longleaf pine in the south, which is often converted to loblolly and slash pines (113).

There is no doubt that many species lose habitat in intensively managed forest landscapes, though it should be borne in mind that habitat has never been a priority on the majority of industrial lands (there are notable exceptions) and has only recently become so on public lands. To most foresters, trees past their peak of growth are like savings bonds that no longer earn adequate interest; not cashing in and replacing with higher-yielding young stands makes no economic sense. The biological trade-offs involved in maximizing economic efficiency are now known to be severe. Research beginning with the International Biological Program (IBP) in the 1960s showed that forests past the peak of growth enter their biological prime, a period characterized by uniquely rich habitat (7, 47, 98). Riparian zones and wetlands are also biologically rich and functionally important (56, 90, 162), and road building, logging, and grazing within riparian zones have been major factors in the widespread decline of both migratory and resident fish in the northwest (100). In some areas, such as the southern United States, wetlands may be drained to plant commercial forests. A number of states and provinces now restrict logging and roading in riparian zones and near wetlands.

The standard approach to conserving diversity in forests has been to establish reserves, resulting in landscapes parceled out among two very different uses: intensive forestry and no forestry. The importance of reserves to conservation is clear. However, in most forested regions, the existing reserve network falls far short of that needed to adequately protect regional diversity (113), and many scientists agree successful conservation will require viewing landscapes for what they are—functional totalities in which both reserves and managed lands play a role (45, 59, 65, 140). The implications of that view have been profound for forestry, as maintaining habitat becomes a central issue in management rather than something to be dealt with "elsewhere." This strategy is not without controversy. Most conservation ecologists agree that the amount of land likely to be set aside will not, by itself, protect sensitive species. However, there are concerns that unproven and possibly ineffectual changes in forestry practices will be used to justify reduced commitment to reserves. The major scientific challenges involve three levels of understanding: (a) the relationship between managed forest structure and ecological function at the stand scale; (b) spatial patterning of stand structures that meet diversity goals for a given bioregion; and (c) the temporal dynamics of stand and landscape structures resulting from natural disturbance, anthropogenic disturbance, and interactions between natural and anthropogenic disturbances (which, as I discuss later, can be significant).

Research on how forestry might be adapted to better protect biological diversity did not begin in earnest until the 1970s. Until that time, habitat concerns on public forest lands focused largely on game animals, which were believed to benefit from the open areas and edge created by patchwork clear-cutting. Eventually, concerns about the impacts of forestry on native diversity led to passage of the National Forest Management Act of 1976 (NFMA), which directed the USFS to seek scientific advice on how to maintain diversity of plants and animals on lands under their jurisdiction. By the time NFMA was passed, US Forest Service biologists had already begun to deal with the issue of nongame bird habitat by hosting a National Symposium (150) followed by four regional symposia. Similar symposia since have dealt with amphibians, reptiles, small mammals, marten, fisher, wolverine, and lynx. In 1979, a group of USFS and BLM wildlife biologists published what came to be known as the "Blue Mountain Book" (because it dealt with the Blue Mountains of eastern Oregon and Washington), a guide for maintaining terrestrial habitat in managed forests (162). The first of its kind, the Blue Mountain Book was in demand throughout the world, and similar guides for other regions soon followed. FEMAT, the plan to protect old-growth associates and fish in the range of the northern spotted owl, relied mostly on reserves (41) [wildlife biologists on the Forest Ecosystem Management Assessment Team (FEMAT) were very wary of unproven silvicultural techniques]. However, 10 large (37,000 to 140,000 ha) Adaptive Management Areas (AMAs) were set aside to experiment with innovative management techniques.

Evidence from the northwest indicates that many species associated with old growth also occur in younger stands that originate from natural disturbances, presumably because of the rich structural legacies left behind in the form of large dead wood, remnant green trees, and sprouting shrubs (58, 60). Large dead wood (snags and logs) provides habitat for numerous terrestrial species, a fact that almost certainly holds for all forest types (62). Logs are also critically important in creating fish habitat because they dam streams to create pools (99, 139).

With the growing awareness of the needs of cavity-users in the late 1970s, federal foresters began leaving scattered trees in clear-cuts (when permitted by state safety inspectors), though dead wood on the ground was still routinely cleared (often because of concerns about wildfire). In the mid-1980s, researchers and managers on the HJA and the Williamette National Forest installed several trials using silvicultural techniques intended to capture the diversity of naturally disturbed forests. Initially termed "new forestry," and now "variable retention harvest," the basic idea is to leave a certain number of large, green trees at harvest, either dispersed or aggregated, with the objective of providing at least three functions not found in intensively managed forests: habitat continuity for species requiring large, green trees; a diversity of age classes with concomitant vertical and/or horizontal heterogeneity; and future sources of large dead wood (46). At least two companies have since initiated operational retention harvests (46, 158). Several replicated experiments are now installed in the United States and Canada that manipulate retention levels and, in at least one case, the physical arrangement of retention (6, 46, 76). These incorporate a level of biological detail far beyond anything in previous silvicultural experiments, including measurements of small mammals, birds, invertebrates, plants, and microbes.

Results to date (including studies conducted in earlier trials) show that young stands with remnant old trees support significantly greater abundance of some old-growth associates than do young stands without old remnants, including some not found at all in young stands without old trees (25, 60, 117, 137). Simulation predicts that bird richness will remain higher in mixtures of old and young trees than in stands composed only of young trees for 140 years following harvest (61). However, not all birds associated with old growth benefit from leaving residual trees, and some that do benefit exhibit threshold responses in which relatively small changes in the density of older trees have large effects on abundance (60). Research on this issue is in its infancy, and it would be surprising if new patterns did not emerge with time.

Resistance to Disturbances

A vast experiment is underway. Its unplanned and unwitting design is changing the spatial and temporal structure of terrestrial ecosystem...   R. L. Burgess and D. M. Sharp (24)

Insights gained with the emergence of landscape ecology and conservation biology as disciplines over the past two decades leave little doubt that the "vast experiment" to which Burgess & Sharp refer to in the above quotation has significantly perturbed the fabric of species relationships and ecological processes within regions, with largely unpredictable consequences that may take decades or centuries to unfold (85, 123, 163). Infestations of trees by a number of native pathogens and defoliating insects are already outside of known historic ranges in regions where humans have significantly altered forested landscapes, including the western spruce budworm, Swiss needle cast, and several root rots in western North America; fusiform rust and southern pine beetle in the south; and eastern spruce budworm and ash yellows (a viral disease) in the northeast United States and eastern Canada (26, 123, 127, 136, 173). Widespread changes in stand structure have also increased susceptibility of forests throughout the west to high-intensity crown fires (2).

Consistent with the resource concentration hypothesis of herbivore/pathogen dynamics, epidemics of recent years are frequently related to the spread of host trees. The spread of hosts stems in turn from various factors that depend on locale but all involve changes in the historic disturbance regime. In the south, agriculture, forestry, and fire exclusion have combined to sharply reduce the extent of once widespread longleaf pine forests. Much of that area has been planted with slash and loblolly pines, which have faster early growth than longleaf (hence are favored by foresters) but are also more susceptible to fusiform rust and southern pine beetle (127, 136). In the interior western United States, southern British Columbia, northeast United States, and eastern Canada, logging old growth coupled with fire exclusion allowed the spread of *Abies* sp., principle hosts for spruce budworm (118, 173). An ongoing epidemic of Swiss needle cast in the coastal mountains of Oregon is believed to have been triggered by the large areas of young Douglas-fir plantations (G Filip, personal communication). (Note, these are all native insects and pathogens—exotics are another story.)

Depending on the particular situation, various other factors are believed to have contributed to pest outbreaks. Roads have clearly accelerated the spread of several diseases in the Pacific Northwest (123). Disruption of the natural enemy complex through habitat loss and (in some cases) pesticides may be involved in outbreaks of herbivorous insects (137, 165), though that is difficult to demonstrate conclusively. The incidence of fusiform rust on southern pines increases with management intensity, particularly weed control (181) and fertilization (127). Eliminating broadleaved species from plantations also results in greater incidence of root rots on Douglas-fir (146) and defoliators on white spruce (52), phenomena that could stem from greater concentration of food plants (conifers), disruption of natural enemies that depend on broadleaved plants for habitat, or both.

Genetic and ecologic strategies have been employed to deal with herbivorous insects and pathogens in forestry. Selection for resistance has significantly reduced incidence of fusiform rust in southern pines (74); whether this is an evolutionarily stable situation remains to be seen. Experiments on the stability of resistance in slash pine indicate a highly dynamic relationship between pathogen and host, with significant interactions between tree genotypes and both environment (including the pathogen) and time (RA Schmidt, personal communication). A single gene having major control of fusiform resistance has been identified in loblolly pine (175), a situation in which rapid evolution of the pathogen would be expected.

Ecologic strategies center on diversifying landscapes and promoting habitat for natural enemies. Evidence from deliberate introductions of biotic control agents strongly supports the importance of natural enemies as an effective, and perhaps evolutionarily stable, source of control (77), and a great deal of spatially explicit modeling points to the importance of landscape patterns in host-pest dynamics (85). Ecologic approaches have so far been utilized mostly in European forestry, though concepts of integrated pest management are gaining a foothold in North American forestry.

THE FUTURE

> There is one outstandingly important fact regarding Spaceship Earth, and that is that no instruction book came with it.   Buckminster Fuller (50)

The scientific challenges facing forestry are precisely those confronting basic ecology: better understanding the relationships among structure, function, and spatiotemporal dynamics of systems interconnected at many scales; and coping with the "certain uncertainty" inherent in complex systems. Forest management alters forest structure and thereby influences processes, some purposely, as with productivity, some inadvertantly, as with erosion, population dynamics, and susceptibility to disturbances. How much can the structure, hence processes, of complex ecological systems be altered without compromising long-term integrity, or, to turn that question on its head, how much structure must be retained to maintain integrity? Can technology successfully substitute for nature's evolved controls? Implicit within these questions are numerous others having to do with system states and measureable variables that tell something useful about future system states.

During the 1990s a set of philosophies and general principles emerged under the rubric of ecosystem management (EM). During its short life EM has generated numerous papers and symposia, most of which share common themes: sustainability as the overall guiding principle; explicit goals; recognizing interconnections; working across ownerships; taking a broad view spatially and temporally; recognizing that ecological systems are dynamic but that some changes have greater long-term consequences than others; accommodating human needs within the constraints of ecological objectives (29, 57). The real challenges lie in translating abstract principles to what is (or is not) done on the ground, and developing tools to assess whether goals are being achieved.

Several strategies have been or are being used to translate the general principles of EM to specific practices. In most cases these were developed simultaneously over the past decade and are largely complementary rather than competing. The approach exemplified by FEMAT (41) focuses largely on reserves and their interconnectance. In FEMAT, the spotted owl serves as an umbrella species whose protection is hypothesized to de facto protect other OG associates, and monitoring protocols are established to test that hypothesis. FEMAT required relatively few modifications of forest practices in the managed matrix (which under the plan composed a minority of the landscape in many areas), though it took the ground-breaking step of devoting large areas specifically to experimentation and learning, and it set the important precedent of assembling multidisciplinary teams, along with protocols for synthesizing sometimes widely divergent opinions about the probable outcomes of different management scenarios.

A second approach, whose roots lie in the IBP-funded studies of OG coniferous forests and the emergence of landscape ecology as a discipline, deals primarily with how lands outside of reserves are managed and uses historic patterns of disturbance and recovery as templates for management. At the stand level, this has led to techniques for protecting biological legacies, particularly the variable retention harvests discussed earlier, and stimulated movement toward longer rotations (35), the latter a strategy adopted by the Germans 30 years previously. At the landscape level, the concept of natural range of variability (NV) has received considerable attention. NV uses historic patterns of disturbance to guide harvesting patterns, as, for example, identifying areas best suited for longer rotations or small patch cuts and areas where shorter rotations or larger cuts would be appropriate (30, 91). A third approach also focuses on managed lands but uses functional models rather than historic. A good example is Hansen et al (59), who used life histories to predict landscape patterns required to maintain habitats for a suite of vertebrate species.

Each strategy has strengths and weaknesses. I discussed issues surrounding dependence on reserves earlier. NV has had the positive effect of making foresters think about history (reliably determining history is another matter); however, taken too literally the approach promotes the false belief that logging can fit within a "natural" range of variability. Logging of any kind—whether clear-cut or single-tree selection—is an unnatural event in the history of a forest, and the relevant scientific question is not whether a practice fits within NV, but how far management can depart from NV before compromising system integrity. Answering that question requires a functional approach that deals with the suite of processes underpinning integrity, which in turn requires a measureable definition of integrity, testable hypotheses concerning keystone processes and their links to system structure, and protocols for choosing among hypotheses when the systems of interest are highly dynamic in space and time and the information base is weak (e.g. 75). To date, ecological processes such as disturbance, hydrology, and nutrient cycling have seldom been integrated with habitat in EM models (the integrity of soils is virtually never mentioned in the EM literature), though that is beginning to change (115, 151).

The move toward ecosystem management by federal agencies may have little relevance for forestry as a whole in the United States, where by far the greatest amount of timber is produced on private lands. In 1991, for example, 32% of timber harvested in the United States was from industry lands and 51% from nonindustrial private lands (167). As in agriculture, maximizing profit in forestry is most often associated with simplifying systems rather than creating or maintaining diversity (the degree to which the green certification movement changes that equation remains to be seen). However, while diversity may not be an explicit goal on most industrial lands, sustaining the productive base is, which leads directly to questions concerning the functions of diversity. A growing number of ecologists believe that complexity and stability of ecosystems are positively linked (where stability refers to ecological functions and potentials rather than species composition on a given piece of ground) (28, 36, 125); however, the nature of the linkages is poorly understood and "very few models...incorporate biological complexity as a regulating component of ecosystem function" (138). Maintaining genetic diversity within expansive areas of high-yielding cultivars shifts what in the natural forest was likely to have been a complex set of genetic, species, and landscape controls (top down and bottom up) to a much greater reliance on genetics (bottom up), an interesting and important experiment, unfortunately uncontrolled. With the discovery of single gene control over susceptibility to fusiform rust in loblolly pine (and probably slash pine), the stage is set for a classic, some might say mythic, confrontation between the forces of nature and the ingenuity of humans. The rust is likely to evolve relatively quickly around a single gene. Can biotechnologists put enough copies in the field to confound that process?

The forestry of the future seems likely to blend aspects of intensive management and EM, either on the same piece of ground, across landscapes, or (more likely) some combination of the two. What Seymour & Hunter (140) termed the triad approach—a mix of reserves, intensively managed lands, and EM lands—provides a useful starting point. The central issue then becomes one of assessing probabilities of outcomes associated with differing spatial and temporal patterns of the three general land-use types. That requires, in turn, improved understanding of structure-process-function at the scale of landscapes (e.g. habitat, disturbance) and sites (e.g. resilience following harvest or natural disturbance).

Science is faced with the challenge of providing knowledge that helps society achieve sustainability (96). Numerous scientists have concluded this can only be accomplished if we begin to grapple successfully with complexity (e.g. 34, 95, 119), a strategy that radically departs from the industrial approach of simplifying systems to make them more predictable. But coming to grips with complex reality requires more than change in management philosophies; some fundamental approaches of science and the role of scientists in society must be reassessed as well. The idea that ecological systems are quintessential complex systems is old; however, much knowledge concerning the characteristics of complex systems is new and growing. Though much of this knowledge has come from physicochemical systems and their models, ecological systems share many characteristics with these: existence far from thermodynamic equilibrium, interconnections through positive and negative feedbacks, self-organizing dynamics, metastability and vulnerability to threshold transitions (124, 166). Adding the social and economic dimensions defines the stage on which forestry plays and greatly magnifies the complexity.

Assessing how much such systems can be altered without triggering unforeseen consequences is far from straightforward. Some changes are obvious, others subtle; some set processes in motion that may not manifest for decades or centuries. Some, perhaps many, changes in system structure alter probabilities rather than being distinctly causal, as when susceptibility to herbivores, pathogens, fire, or wind is increased or decreased. Whether stands so altered actually burn down or are eaten up depends on factors such as climate, the evolutionary interplay between hosts and pests, and diverse biotic controls with complex spatial and temporal dependence. The logic of cause and effect has limited utility in such a milieu.

Scientists have played a much expanded role in forestry in the 1990s, a trend that will almost certainly continue. The traditional routes of experiment and modeling remain vital, and, despite inadequate funding (at least in the United States; 114), there are presently far more experiments dealing with links between structure and process in managed forests of the United States and Canada than at any time in the past. Within the scope of traditional science, however, issues of experimental analysis and data interpretation need to be dealt with, especially the obsession with "significance questing", which, as Rothman pointed out with regard to the medical sciences, and which is equally true for natural resources sciences, "has become for some a clumsy substitute for thought" (134). When the quest is for significance at very high levels, as is the norm, the burden placed on establishing differences among management approaches effectively favors implementing those with the highest economic payoff. Various scientists have argued that the burden of proof needs to be shifted away from practices that would change the management status quo and placed instead on practices that alter systems most dramatically or have the greatest economic payoff (e.g. 34, 72).

Funtowicz & Ravetz (51) argue that situations in which both system uncertainty and decision stakes are high require a "post-normal" or second-order science, which Costanza (34) interprets as taking the scientific method into new territory. "The scientific method," Costanza writes, "does not, in its basic form, imply anything about the precision of results achieved. It does imply a forum of open and free inquiry without preconceived answers or agendas ...." Whether one accepts the new arena as science or not, the fact is that scientists are increasingly playing nontraditional roles in resource management. This has taken various forms, including adaptive management, which effectively blurs the boundary between management and experiment (172); various regional scientific assessments (41, 72, 145, 151); expert testimony in courts and before legislative and regulatory bodies; and small teams of scientists that assess the suitability of lands for green certification. "We need a new model," wrote Gordon & Lyons (55), "for linking science, management, and policy ...." The circumstances of the past few years have thrust forestry scientists into a leading role in developing that model.

ACKNOWLEDGMENTS

Tom Adams, Tim White, Sharon Freidman, Dan Binkley, Gordon Grant, Lee Allen, Bob Wagner, and RA Schmidt were all generous with reprints and other guidance. I'm grateful to NSF's Long-Term Ecological Research Program and my Department Head, Logan Norris, for support over the years, and to the fertile learning environment I was provided as a member of the HJ Andrews LTER Program and Oregon State University's Sustainable Forestry Partnership.

> **Visit the *Annual Reviews* home page at**
> **http://www.AnnualReviews.org**

Literature Cited

1. Aber JD, Melilo JM. 1982. FORTNITE: a computer model of organic matter and nitrogen dynamics in forest ecosystems. *Univ. Wisc. Res. Bull. R3130*. Madison, WI. 49 pp.
2. Agee JK. 1990. The historical role of fire in Pacific Northwest forests. In *Natural and Prescribed Fire in Pacific Northwest Forests*, ed. JD Walstad, SR Radosovich, DV Sandberg, pp. 25–38. Corvallis, OR: Oregon State Univ. Press
3. Albaugh TJ, Allen HL, Dougherty PM, Kress LW, King JS. 1998. Leaf area and above- and belowground growth responses of loblolly pine to nutrient and water additions. *For. Sci.* In press
4. Alig R, Adams D, Chemelik J, Bettinger P. 1998. Private forest investment and long-run sustainable harvest volumes. In National Research Council 1997. *Forested Landscapes in Perspective.* Washington, DC: National Acad. Press. In press
5. Amaranthus MA, Perry DA. 1994. The

functioning of ectomycorrhiza in the field: linkages in space and time. *Plant Soil* 159:133–40

6. Amaranthus MP, Darbyshire R, Bormann B. 1998. Long-term ecosystem productivity integrated research sites. *Agron. Soc. New Zealand.* In press

7. Amaranthus MP, Trappe JM, Bednar L, Arthur D. 1994. Hypogeous fungal production in mature Douglas-fir forest fragments and surrounding plantation and its relation to coarse woody debris and animal mycophagy. *Can. J. For. Res.* 24: 2157–65

8. Amaranthus MP, Li CY, Perry DA. 1990. Influence of vegetation type and madrone soil inoculum on associative nitrogen fixation in Douglas-fir rhizospheres. *Can. J. For. Res.* 20:368–71

9. Angermeier PL, Karr JR. 1994. Biological integrity versus biological diversity as policy directives. *BioScience* 44:690–97

10. Baker FS. 1950. *The Principles of Silviculture.* New York: McGraw-Hill

11. Ballard R, Gessel SP, eds. 1983. *IUFRO Symposium on Forest Site and Continuous Productivity. USDA For. Serv. Gen. Tech. Rep. PNW-163.* Portland, OR

12. Behan M. 1997. Scarcity, simplicity, separatism, science—and systems. See Ref. 88, pp. 411–18

13. Beuter JH, Johnson KN. 1989. Economic perspectives on maintaining the long-term productivity of forest ecosystems. See Ref. 120, pp. 221–29

14. Binkley D. 1992. Mixtures of nitrogen-fixing and non-nitrogen-fixing tree species. In *The Ecology of Mixed Species Stands of Trees*, ed. MGR Cannell, DC Malcome, PA Robertson, pp. 99–123. London: Blackwell Sci.

15. Binkley D. 1986. *Forest Nutrition.* Chicester, NY: Wiley & Sons. 290 pp.

16. Bormann FH, Likens GE, Sicama TG, Pierce RS, Eaton JS. 1974. The export of nutrients and recovery of stable conditions following deforestation at Hubbard Brook. *Ecol. Monogr.* 44:255–77

17. Bormann BT, Bormann FH, Bowden WB, Pierce RS, Hamburg SP, et al. 1993. Rapid N_2 fixation in pines, alder, and locust: evidence from the sandbox ecosystem study. *Ecology* 74:583–98

18. Bormann BT, Wand D, Bormann FH, Benoit G, April R, Snyder MC. 1998. Rapid, plant-induced weathering in an aggrading experimental ecosystem. *Biogeochemistry* Submitted

19. Bormann FH, Likens GE. 1979. *Pattern and Process in a Forested Ecosystem.* New York: Springer-Verlag. 253 pp.

20. Bosworth B, Studer D. 1991. Comparisons of tree height growth on broadcast burned, bulldozer-piled, and non-prepared sites 15 to 25 years after clear-cut logging. *Proceedings—Management and Productivity of Western Montane Soils. USFS Gen. Tech. Rep. INT-280*, ed. AE Harvey, LF Neunshwander, pp. 197–200. Washington, DC: USGPO

21. Boyle JR. 1975. Nutrients in relation to intensive culture of forest crops. *Iowa State J. Res.* 49:297–303

22. Brocke RH, O'Pezio JP, Gustafson KAA. 1989. A forest management scheme for mitigating impact of road networks on sensitive wildlife species. *Is Forest Fragmentation A Management Issue in the Northeast? USFS NE Forest Exp. Sta. Gen. Tech. Rep. NE-140*, pp. 7–12. Washington, DC: USGPO

23. Burger JA, Pritchett WL. 1988. Site preparation effects on soil moisture and available nutrients in a pine plantation in the Florida flatwoods. *For. Sci.* 34:77–87

24. Burgess RL, Sharp DM. 1981. *Forest Island Dynamics in Man-Dominated Landscapes.* New York: Springer-Verlag

25. Carey AB. 1995. Sciurids in Pacific Northwest managed and old-growth forests. *Ecol. Appl.* 5:648–61

26. Castello JD, Leopold DJ, Smallidge PJ. 1995. Pathogens, patterns, and processes in forest ecosystems. *BioScience* 45:16–24

27. Chang SX, Preston CM, Weetman GF. 1995. Soil microbial biomass and microbial and mineralizable N in a clear-cut chronosequence on northern Vancouver Island, British Columbia. *Can. J. For. Res.* 25:1595–607

28. Chapin FS III, Sala OE, Burke IC, Grime JP, Hooper DU, et al. 1998. Ecosystem consequences of changing biodiversity. *BioScience* 48:45–52

29. Christensen NL, Bartuska AM, Brown JH, Carpenter S, D'Antonio C, et al. 1996. The report of the Ecological Society of America committee on the scientific basis of ecosystem management. *Ecol. Appl.* 6:665–91

30. Cissel JH, Swanson FJ. 1997. The present landscape-scale application of the range-of-historical-variability concept in the Oregon Cascades. *Abstracts of the 1997 Meeting of the Ecological Society of America*, p. 9

31. Clayton JL. 1979. Nutrient supply to soil by rock weathering. See Ref. 136, pp. 75–96

32. Cole DW. 1995. Soil nutrient supply in natural and managed forests. *Plant Soil.* 168–69:43–53
33. Compton J, Cole DW. 1990. Impact of harvest intensity on growth and nutrition of second rotation Douglas-fir. See Ref. 59, pp. 151–62
34. Costanza R. 1993. Developing ecological research that is relevant for achieving sustainability. *Ecol. Appl.* 3:579–81
35. Curtis RO. 1997. The role of extended rotations. See Ref. 88, pp. 165–70
36. Ehrlich PC. 1994. Biodiversity and ecosystem function: need we know more? In *Biodiversity and Ecosystem Function,* ed. E-D Schulze, HA Mooney, pp. vii–xi. New York: Springer
37. El-Kassaby YA, Ritland K. 1996. Impact of selection and breeding on the genetic diversity in Douglas-fir. *Biodiversity Conservation* 5:795–813
38. El-Kassaby YA, Namkoong G. 1995. Genetic diversity of forest tree plantations: consequences of domestication. In *Consequences of changes in biodiversity,* IUFRO World Congress 2:218–28. Tampere, Finland
39. Eriksson G, Namkoong G, Roberds JH. 1993. Dynamic gene conservation for uncertain futures. *For. Ecol. Manage.* 62:15–37
40. Farnum P, Timmis R Jr, Kulp K. 1983. Biotechnology of forest yield. *Science* 219.694–702
41. FEMAT. 1993. Forest ecosystem management: an ecological, economic, and social assessment *Rep. For. Ecosystem Manage. Assessment Team (FEMAT). 1996-793-071.* Washington, DC: USGPO
42. Foster DR, Orwig DA, McLachlan JS. 1996. Ecological and conservation insights from reconstructive studies of temperate old-growth forests. *Trends Ecol. Evol. (TREE)* 11:419–24
43. Fox TR, Morris LA, Maimone RA. 1989. The impact of windrowing on the productivity of rotation age loblolly pine. *Proc. 5th Biennial Southern Silviculture Conf.* Ashville, NC: USDA For. Serv.
44. Franklin JF. 1993. The fundamentals of ecosystem management with applications in the Pacific Northwest. In *Defining Sustainable Forestry,* ed. GH Aplet, N Johnson, JT Olson, VA Sample, pp. 127–44. Washington, DC: Island Press
45. Franklin JF. 1993. Preserving biodiversity: species, ecosystems or landscapes. *Ecol. Appl.* 3:202–5
46. Franklin JF, Berg DR, Thornburg DA, Tappeiner JC. 1997. Alternative Silvicultural approaches to timber harvesting:

47. Franklin JF, Cromack K Jr, Denison W, McKee A, Maser C, et al. 1981. Ecological characteristics of old-growth Douglas-fir forests. *USDA For. Serv. Gen. Tech. Rep. PNW-118.* Portland, OR
48. Freidman ST, Foster GS. 1997. Forest genetics on federal lands in the United States: public concerns and policy responses. *Can. J. For. Res.* 27:401–8
49. Fried JS, Boyle JR, Tappeiner JC II, Cromack K Jr. 1990. Effects of bigleaf maple on soils in Douglas-fir forests. *Can. J. For. Res.* 20:259–66
50. Fuller B. Quoted in *Explorations.* Spring/Summer 1998, p. 6
51. Funtowicz SO, Ravetz JR. 1991. A new scientific methodology for global environmental problems. In *Ecological Economics: The Science and Management of Sustainability,* ed. R Costanza, pp. 137–52. New York: Columbia Univ. Press
52. Gagnon RP, Chabot M. 1991. *Prevention des pertes de bois attribuables a la torduese des bourgeons de l'epinette.* Governement du Quebec, Ministere des Forets, Montreal
53. Gessel SP, Lacate DS, Weetman GF, Powers RF, eds. 1990. *Sustained Productivity of Forest Soils. Proceedings of the 7th North American For. Soils Conf.;* 1988 Jul; University of British Columbia. Faculty of Forestry, Univ. Br. Columbia. Vancouver, Canada V6T 1W5
54. Glover GR, Creighton JL, Gjerstad DH. 1989. Herbaceous weed control increases loblolly pine growth for twelve years. *J. For.* 87:47–50
55. Gordon JC, Lyons J. 1997. The emerging role of science and scientists in ecosystem management. See Ref. 88, pp. 447–53
56. Gregory SV, Lamberti GA, Erman DC, Koski KV, Murphy ML, Sedell JR. 1987. Influence of forest practices on aquatic production. In *Streamside Management: Forestry and Fisheries Interactions,* ed. EO Salo, TW Cundy, pp. 233–55. *Contribution No. 57,* Inst. For. Resources, Univ. Washington, Seattle
57. Grumbine RE. 1994. What is ecosystem management? *Conserv. Bio.* 8:27–38
58. Hansen AJ, Spies TA, Swanson FJ, Ohmann JL. 1991. Conserving biological diversity in managed forests. *BioScience* 41(6):382–92
59. Hansen AJ, Garman SL, Marks B. 1993. An approach for managing vertebrate diversity across multiple-use landscapes. *Ecol. Appl.* 3:481–96
60. Hansen AJ, McComb C, Vega R, Raphael

MG, Hunter M. 1995a. Bird habitat relationships in natural and managed forests in the west Cascades of Oregon. *Ecol. Appl.* 5:555–69

61. Hansen AJ, Garman SL, Weigand JF, Urban DL, McComb WC, Raphael MG. 1995b. Alternative silvicultural regimes in the Pacific Northwest: simulations of ecological and economic effects. *Ecol. Appl.* 5:535–54

62. Harmon ME, Franklin JF, Swanson FJ, Sollins P, Gregory SV, ct al. 1986. Ecology of coarse woody debris in temperate ecosystems. *Adv. Ecol. Res.* 15:133–302

63. Harr RD, Harper WC, Krygier JT, Hsieh FS. 1975. Changes in storm hydrographs after road building and clear-cutting in Oregon Coast Range. *Water Resour. Res.* 11:436–44

64. Harr RD. 1976. Hydrology of small forest streams in western Oregon. *Gen. Tech. Rep. PNW-55.* USFS Pacific Northwest Exp. Station. Portland, OR

65. Harris LD. 1984. *The Fragmented Forest.* Chicago, IL: Univ. Chicago Press

66. Hart SC, Binkley D, Campbell RG. 1986. Predicting loblolly pine current growth and growth response to fertilization. *Soil Sci. Soc. Am. J.* 50:230–33

67. Hart SC, Binkley D, Perry DA. 1997. Influence of red alder on soil nitrogen transformations in two conifer forests of contrasting productivity. *Soil Biol. Biochem.* 29:1111–23

68. Harvey AE, Neuenschwander LF, eds. 1991. *Proceedings—Management and Productivity of Western-Montane Forest Soils. USFS Gen. Tech. Rep. INT-280.* 254 pp.

69. Harvey AE, Larsen MJ, Jurgensen MF. 1980. Clear-cut harvesting and ectomycorrhizae: survival of activity on residual roots and influence on a bordering stand in western Montana. *Can. J. For. Res.* 10: 300–3

70. Harvey AE, Meurisse RT, Geist JM, Jurgensen MF, McDonald GI, et al. 1989. Managing long-term productivity in northwest-mixed conifers and pines. See Ref. 120, pp. 164–84

71. Harvey AE, Jurgensen MF, Larsen MJ, Graham RT. 1987. Relationships among soil microsite, ectomycorrhizae, and natural conifer regeneration of old growth forests of western Montana. *Can. J. For. Res.* 17:58–62

72. Henjum MG, Karr JR, Bottom DL, Perry DA, Bednarz JC, et al. 1994. *Interim Protection for Late-Successional Forests, Fisheries, and Watersheds: National Forests East of the Cascade Crest, Oregon and Washington.* Bethesda, MD: The Wildlife Society

73. Hodge GR, White TL, Powell GL, de Souza SM. 1989. Predicted genetic gains from one generation of slash pine tree improvement. *So. J. Appl. For.* 13:51–56

74. Hodge GR, White TL, Schmidt RA, Allen JE. 1993. Stability of rust infection ratios for resistant and susceptible slash and loblolly pines across rust hazard levels. *So. J. Applied For.* 17:188–92

75. Holling CS. 1992. Cross-scale morphology, geometry, and dynamics of ecosystems. *Ecol. Monogr.* 62:447–502

76. Hollstedt C, Vyse A. 1997. *Sicamous Creek Silvicultural Systems Project. Br. Columbia Min. For. Res. Prog. Work. Pap. 24.* Victoria, BC

77. Holt RD, Hochberg ME. 1997. When is biological control evolutionarily stable (or is it)? *Ecology* 78:1673–83

78. Hornbeck JW, Swank WT. 1992. Watershed ecosystem analysis as a basis for multiple-use management of eastern forests. *Ecol. Appl.* 2:238–47

79. Hutten M. 1998. *Performance of ponderosa pine seedlings outplanted in forest, patch-cuts, and clear-cuts in central Oregon.* MS thesis. Oregon State Univ., Corvallis

80. Ingestad T, Aronsson A, Agren GI. 1981. Nutrient flux density model of mineral nutrition in conifer ecosystems. *Studia Forestalia Suecica* 160:61–71

81. Johnson DW. 1992. Effects of forest management on soil carbon storage. *Water, Air, Soil Pollution* 64:83–120

82. Jones JA, Grant GE. 1996. Peak flow responses to clear-cutting and roads in small and large basins, western Cascades. *Oregon Water Resour. Res.* 32:959–74

83. Jongmans AG, van Breemen N, Lundstrom U, van Hees PAW, Finlay RD, et al. 1997. Rock-eating fungi. *Nature* 389:683–84

84. Jurgensen MF, Larsen MJ, Graham RT, Harvey AE. 1987. Nitrogen fixation in woody residue of northern Rocky Mountain conifer forests. *Can. J. For. Res.* 17: 1283–88

85. Kareiva P, Wennergen U. 1995. Connecting landscape patterns to ecosystem and population processes. *Nature* 373:299–302

86. Feller MC, Kimmins JP, Scoullar KA. 1983. FORCYTE 10: calibration data and simulation of potential long-term effects of intensive forest management on site productivity, economic performance, and energy benefit/cost ratio. See Ref. 82, pp. 179–200

87. Kimmins JP. 1977. Evaluation of the consequences for future tree productivity of the loss of nutrients in whole-tree harvesting. *For. Ecol. Manage.* 1:169–83

88. Kohm KA, Franklin JF, eds. 1997. *Creating a Forestry for the 21st Century.* Washington, DC: Island Press

89. Korzukin MD, Ter-Mikaelian T, Wagner RG. 1996. Process versus empirical models: which approach for forest ecosystem management? *Can. J. For. Res.* 26:879–87

90. Kuenzler EJ. 1989. Value of forested wetlands as filters for sediments and nutrients. In *Proc. Symposium: The Forested Wetlands of the Southern United States,* ed. DD Hook and R Lea, pp. 85–96. *USDA For. Serv. Gen. Tech. Rep. SE-50.* Asheville, NC

91. Landres P, Morgan P, Swanson F. 1997. Evaluating the usefulness of natural range of variability in managing ecological systems. *Abstr. 1997 Meet. Ecol. Soc. Am.,* p. 20

92. Langston N. 1994. *Forest Dreams, Forest Nightmares.* Seattle: Univ. Wash. Press

93. Ledig FT. 1986. Heterozygosity, heterosis, and fitness in outbreeding plants. In *Conservation Biology: The Science of Scarcity and Diversity,* ed. ME Soule, pp. 77–104. Sunderland, MA: Sinauer

94. Li CY, Hung LL. 1987. Nitrogen-fixing (acetylene-reducing) bacteria associated with ectomycorrhizae of Douglas-fir. *Plant Soil* 98:425–28

95. Lubchenco J, Olson AM, Brubaker LB, Carpenter SR, Holland MM, et al. 1991. The sustainable biosphere initiative: an ecological research agenda. *Ecology* 72:371–412

96. Lubchenco J. 1998. Entering the century of the environment: the need for a new social contract for science. *Science* 279:491–97

97. Mann LK, Johnson DW, West DC, Cole DW, Hornbeck JW, et al. 1988. Effects of whole-tree and stem-only clear-cutting on post-harvest hydrologic losses, nutrient capital, and regrowth. *For. Sci.* 34:412–28

98. Marcot BG. 1997. Biodiversity of old forests of the west: a lesson from our elders. See Ref. 130, pp. 87–105

99. McArthur JV. 1989. Aquatic and terrestrial linkages: flood plain functions. In *Proc. Symposium: The Forested Wetlands of the Southern United States,* ed. DD Hook, R Lea, pp. 107–16. Asheville, NC: *USDA For. Serv. Gen. Tech. Rep. SE-50*

100. McIntosh BA, Sedell JR, Smith JE,

Wissmar RC, Clarke SE, et al. 1993. Management history of eastside ecosystems: changes in fish habitat over 50 years, 1935–1992. In *Eastside Forest Ecosystem Health Assessment,* ed. PF Hessburg, Vol. III. Washington, DC: USDA For. Serv.

101. McKeand SE, Crook RP, Allen HL. 1998. Genotypic stability effects on predicted family responses to silvicultural treatments in loblolly pine. *So. J. Appl. For.* In press

102. McKeand S, Svensson J. 1997. Loblolly pine: sustainable management of genetic resources. *J. For.* March:4–9

103. Miller JH, Zutter B, Zedaker SM, Cain M, Edwards MB, et al. 1986. A region-wide study of loblolly pine growth relative to four competition levels after four growing seasons. In *Proc. 4th Biennial Res. Conf.,* ed. DR Phillips, pp. 581–91. Asheville, NC: *USDA For. Serv. Gen. Tech. Rep. SE-42*

104. Miller RE, Barker PR, Peterson C, Webster SR. 1986. Using nitrogen fertilizers in management of coast Douglas-fir. In *Douglas-fir: Stand Management for the Future,* ed. CD Oliver, DP Hanley, JA Johnson, pp. 290–303. Inst For Res. Contribution No. 55. Seattle: Univ. Wash. Coll. For. Resources

105. Moldenke A. 1990. One hundred twenty thousand little legs. *Wings* Summer:11–14

106. Morris LA. 1989. Long-term productivity research in the US southeast: experience and future directions. In *Research Strategies For Long-term Site Productivity,* ed. WJ Dyck, CJ Mees, *FRI Bull.* 152, pp. 221–35. Rotorua, NZ: For. Res. Inst.

107. Morris LA, Pritchett WL. 1983. Effects of site preparation on *Pinus elliotii-Pinus palustris* flatwoods forest soil properties. See Ref. 17, pp. 243–51

108. Morris LA, Pritchett WL, Swindel BF. 1983. Displacement of nutrients into windrows during site preparation in a flatwood forest. *Soil Sci. Soc. Am. J.* 47:591–94

109. Morrison PH, Swanson FJ. 1990. *Fire History and Pattern in a Cascade Mountain Landscape. USDA For. Serv. Gen. Tech. Rep. PNW-GTR-254.* Portland, OR

110. Namkoong G. 1984. A control concept for gene conservation. *Silvae Genetica* 33:160–33

111. Namkoong G, Bishir. 1987. The frequency of lethal alleles in forest tree populations. *Evolution* 41:1123–27

112. NC State Forest Nutrition Cooperative.

1998. HTTP://WWW2.ncsu.edu/unity/lockers/project/ncsfnchpg

113. Noss RF, LaRoe ET III, Scott JM. 1995. *Endangered Ecosystems of the United States: a Preliminary Assessment of Loss and Degradation. Biol. Rep. 28.* Washington, DC: USDI Nat. Biol. Service. 58 pp.

114. National Research Council. 1990. *Forestry Research: A Mandate for Change.* Washington, DC. Nat. Acad. Press. 84 pp.

115. Pastor J. 1993. Modelling the effects of timber management on populations dynamics, diversity, and ecosystem processes. In *Modelling Sustainable Forest Ecosystems,* ed. DC LeMasetr. Washington, DC: Soc. Am. Foresters

116. Pastor J, Post WM. 1986. Influence of climate, soil moisture, and succession on forest carbon and nitrogen cycles. *Biogeochemistry* 2:3–27

117. Peck JE, McCune BM. 1997. Remnant trees and canopy lichen communities in western Oregon: a retrospective approach. *Ecol. Appl.* 7:1181–87

118. Perry DA. 1994. *Forest Ecosystems.* Baltimore, MD: Johns Hopkins Univ. Press

119. Perry DA, Maghembe J. 1989. Ecosystem concepts and current trends in forest management time for reappraisal. *For. Ecol. Manage.* 26:123–40

120. Perry DA, Meurisse R, Thomas B, Miller R, Boyle J, et al, eds. 1989. *Maintaining Long Term Productivity of Pacific Northwest Forests.* Portland, OR: Timber

121. Perry DA, Amaranthus MP, Borchers JG, Borchers SL, Brainerd RE. 1989. Bootstrapping in ecosystems. *BioScience* 39:230–37

122. Perry DA, Rose SL. 1983. Soil biology and forest productivity: opportunities and constraints. See Ref. 17, pp. 229–38

123. Perry DA. 1988. Landscape patterns and forest pests. *Northwest Environ. J.* 4:213–28

124. Perry DA. 1995. Self-organizing systems across scales. *TREE* 10:241–44

125. Perry DA, Amaranthus MP. 1997. Disturbance, recovery, and stability. See Ref. 88, pp. 31–56

126. Plochmann R. 1989. The forests of central Europe: a changing view. In *Oregon's Forestry Outlook: An Uncertain Future. The 1989 Starker Lectures,* pp. 1–9. Corvallis, OR: Coll. Forestry

127. Powers HR, Miller T, Belanger RP. 1993. Management strategies to reduce losses from fusiform rust. *So. J. Appl. For.* 17: 146–49

128. Powers RF. 1988. Unpublished data given in Ref. 120.

129. Powers RF, Alban DH, Miller RE, Tiarks AE, Wells CG, et al. 1990. Sustaining site productivity in North American forests: problems and prospects. See Ref. 53, pp. 49–79

130. Radosovich SR, Ghersa CM. 1992. Weeds, crops, and herbicides: a modern-day "neckriddle." *Weed Tech.* 6:788–95

131. Read DJ. 1993. Plant-microbe mutualisms and community structure. In *Biodiversity and Ecosystem Function,* ed. ED Schulze, HΛ Mooney, pp. 181–236. New York: Springer-Verlag

132. Rehfeld GE. 1992. Breeding strategies for *Larix occodentalis:* adaptations to the biotic and abiotic environment in relation to improving growth. *Can. J. For. Res.* 22:5–13

133. Reiter ML, Beschta RL. 1995. Effects of forest practices on water. In *Cumulative Effects of Forest Practices in Oregon.* Rep. for Oregon Dept. For. Ch. 7, RL Beschta, JR Boyle, CC Chambers, WP Gibson, SV Gregory, et al, Salem, OR

134. Rothman KJ. 1986. Significance questing. *Ann. Int. Med.* 105:445–47

135. Savolainen O, Karkkainen K. 1992. Effect of forest management on gene pools. *New For.* 6:329–45

136. Schowalter TD, Turchin P. 1993. Southern pine beetle infestation development: interaction between pine and hardwood basal areas. *For. Sci.* 39:201–10

137. Schowalter TD. 1995. Canopy arthropod community response to forest age and alternative harvest practices in western Oregon. *For. Ecol. Manage.* 78:115–25

138. Schulze E-D, Mooney HA. 1994. Ecosystem function of biodiversity: a summary. In *Biodiversity and Ecosystem Function,* ed. E-D Schulze, HA Mooney, pp. 497–510. New York: Springer-Verlag

139. Sedell JR, Bison PA, Swanson FJ, Gregory SV, eds. 1988. *What we know about large trees that fall into streams and rivers. Gen. Tech. Rep. PNW-GTR-229.* Portland, OR: USDA For. Serv.

140. Seymour R, Hunter ML Jr. 1992. *Principles and Applications of New Forestry in Spruce-Fir Forests of Eastern North America. Publication 716.* Orono, Maine Ag. Exp. Sta.

141. Shainsky LJ, Radosovich SR. 1992. Mechanisms of competition between Douglas-fir and red alder seedlings. *Ecology* 73:30–45

142. Shainsky LJ, Rose CL. 1995. Effects of competition on the foliar chemistry of young Douglas-fir in monoculture and mixed stands with red alder. *Can. J. For. Res.* 25:1969–77

143. Sharitz RR, Boring LR, Van Lear DH, Pinder JE III. 1992. Integrating ecological concepts with natural resource management of southern forests. *Ecol. Appl.* 2: 226–37

144. Shumway JS, Chappell HN. 1995. Preliminary DRIS norms for coastal Douglas-fir soils in Washington and Oregon. *Can. J. For. Res.* 25:208–14

145. SICB. 1996. *Status of the Interior Columbia Basin: Summary of Scientific Findings.* USFS Gen. Tech. Rep. PNW GTR-385. Portland, OR. 144 pp.

146. Simard S, Vyse A. 1994. Paper birch: weed or crop tree in the interior cedar-hemlock forests of south British Columbia. In *Interior Cedar-Hemlock-White Pine Forests: Ecology and Management.* pp. 311–16. Pullman, WA: Wash. State Univ. Dept. Nat. Resource Sci.

147. Simard SW, Perry DA, Jones MD, Myrold DD, Durall DM, Molina R. 1997. Net transfer of carbon between ectomycorrhizal tree species in the field. *Nature* 388:579–82

148. Small MF, Hunter ML. 1988. Forest fragmentation and avian nest predation in forested landscapes. *Oecologia* 76:62–64

149. Smith CT Jr, McCormack ML Jr, Hornbeck JW, Martin CW. 1986. Nutrient and biomass removals from a red spruce—balsam fir whole-tree harvest. *Can. J. For. Res.* 16:381–88

150. Smith DR (Technical Coordinator). 1975. *Proceedings of the Symposium on Management of Forest and Range Habitats for Nongame Birds.* USFS Gen. Tech. Rep. WO-1. Washington, DC: USGPO. 343 pp.

151. SNEP. 1996. *Status of the Sierra Nevada. Sierra Nevada Ecosystem Project, Final Report to Congress. Wildland Rd Res. Center Rep. No. 36.* Davis, CA: Univ. Calif. Centers for Water and Wildland Resources

152. Deleted in proof

153. South DB, Zwolinski JB, Allen HL. 1995. Economic returns from enhancing loblolly pine establishment on two upland sites: effects of seedling grade, fertilization, hexazinone, and intensive soil cultivation. *New For.* 10:239–56

154. Spies TA, Franklin JF. 1991. The structure of natural young, mature, and old-growth Douglas-fir stands in western Oregon and Washington. In *Wildlife and Vegetation of Unmanaged Douglas-Fir Forests,* ed LF Ruggerio, KB Aubrey, AB Carey, MH Huff. *USDA For. Serv. Gen. Tech. Rep. PNW-GTR-285.* Portland, OR

155. Stelzer HE, Goldfarb B. 1997. Implementing clonal forestry in the southeastern United States: SRIEG satellite workshop summary remarks. *Can. J. For. Res.* 27:442–46

156. Stochr MU, El-Kassaby YA. 1997. Levels of genetic diversity at different stages of the domestication cycle of interior spruce in British Columbia. *Theor. Appl. Genet.* 94:83–90

157. Stonecypher RW, Piesch RF, Helland GG, Chapman JG, Reno HJ. 1996. Results from genetic tests of selected parents of Douglas-fir (*Pseudotsuga meneziessi* Mirb. Franco) in an applied tree improvement program. *For. Sci. Monogr.* 32:1–35

158. Stuart-Smith K, Hebert D. 1996. Putting sustainable forestry into practice at Alberta-Pacific. *CPPA Woodlands Pap.* April/May:57–60

159. Swanson FJ, Clayton JL, Megahan WF, Bush G. 1989. Erosional processes and long-term site productivity. See Ref. 120, pp. 67–81

160. Tappenier JC, Huffman D, Marshall D, Spies TA, Bailey JD. 1997. Density, ages, and growth rates in old-growth and young-growth Douglas-fir in coastal Oregon. *Can. J. For. Res.* 27:638–48

161. Taylor SP, Alfaro RI, DeLong C, Rankin L. 1996. The effects of overstory shading on white pine weevil damage to white spruce and its effects on spruce growth rates. *Can. J. For. Res.* 26:306–12

162. Thomas JW, ed. 1979. *Wildlife Habitats in Managed Forests of the Blue Mountains of Oregon and Washington. Agriculture Handbook No. 553.* Washington, DC: USDA For. Serv.

163. Tilman D, May RM, Lehman CL, Nowak MA. 1994. Habitat destruction and the extinction debt. *Nature* 371:65–66

164. Todd D, Pait J, Hodges J. 1995. The impact and value of tree improvement in the south. *Proc. 23rd So. Forest Tree Improvement Conf.* 7–15

165. Torgerson TR, Mason RR, Campbell RW. 1990. Predation by birds and ants on two forest insect pests in the Pacific Northwest. *Stud. Avian Biol.* 13:14–19

166. Turner MG, Romme WH, Gardner RH, O'Neill RV, Kratz TK. 1993. A revised concept of landscape equilibrium: disturbance and stability on scaled landscapes. *Landscape Ecol.* 8:213–27

167. USDA Forest Service. 1995. *The 1993 RPA Timber Assessment Update. General Tech. Rep. RM-GTR-259.* Fort Collins, CO. 66 pp.

168. Velazquez-Martinez A, Perry DA, Bell TE. 1992. Response of aboveground biomass increment, growth efficiency, and

foliar nutrients to thinning, fertilization, and pruning in young Douglas-fir plantations in the central Oregon Cascades. *Can. J. For. Res.* 22:1278–89

169. Vitousek PM, Matson PA. 1985. Disturbance, nitrogen availability, and nitrogen losses in an intensively managed loblolly pine plantation. *Ecology* 66:1360–76

170. Wagner RG. 1993. Research directions to advance forest vegetation management in North America. *Can. J. For. Res.* 23: 2317–27

171. Walstad JD, Kuch PJ, ed. 1987. *Forest Vegetation Management for Conifer Production.* New York: Wiley & Sons

172. Walters C. 1986. *Adaptive Management of Renewable Resources.* New York: MacMillan

173. Wickman BE, Mason RP, Swetnam TW. 1993. Searching for long-term patterns of forest insect outbreaks. In *Proceedings: Individuals, Populations, and Patterns, September 7–10.* Norwich, England

174. Wiersum KF. 1995. 200 years of sustainability in forestry: lessons from history. *Environ. Manage.* 19:321–29

175. Wilcox PL. 1995. *Use of DNA markers for the genetic dissection of breeding of fusiform rust resistance in loblolly pine.*

PhD thesis, North Carolina State Univ. Raleigh, NC. 125 pp.

176. Williams CG, Hamrick JL, Lewis PO. 1995. Multiple population versus hierarchical breeding populations: a comparison of genetic diversity levels. *Theor. Appl. Genet.* 90:584–94

177. Williams CG, Hamrick JL. 1996. Elite populations for conifer breeding and gene conservation. *Can. J. For. Res.* 26:453–61

178. Yarie J. 1989. A comparison of the nutritional consequences of intensive forest harvesting in Alaskan taiga forests as predicted by FORCYTE-10 and LINKAGES 2. *IEA/BE Project A6 Rep. No. 1. Forest Soils Lab. Univ. Alaska, Fairbanks.* 59 pp.

179. Zeimer RR, Lewis J, Rice RM, Lisle TE. 1991. Modelling the cumulative watershed effects of forest management strategies. *J. Environ. Qual.* 20:36–42

180. Zeimer RR, Lewis J, Lisle TE, Rice RM. 1991. Long-term sedimentation effects of different patterns of timber harvesting. *Int. Assoc. Hydrol. Sci.* 203:143–50

181. Zutter BR, Gjerstad DH, Glover GR. 1987. Fusiform rust incidence and severity in loblolly pine plantations following herbaceous weed control. *For. Sci.* 33:790–800

Annu. Rev. Ecol. Syst. 1998. 29:467–501
Copyright © 1998 by Annual Reviews. All rights reserved

PATHWAYS, MECHANISMS, AND RATES OF POLYPLOID FORMATION IN FLOWERING PLANTS

Justin Ramsey and Douglas W. Schemske
Department of Botany, University of Washington, Seattle, Washington 98195-5325;
e-mail: jramsey@u.washington.edu; schem@u.washington.edu

KEY WORDS: polyploidy, autopolyploidy, allopolyploidy, hybridization, speciation

ABSTRACT

Polyploidy is widely acknowledged as a major mechanism of adaptation and speciation in plants. The stages in polyploid evolution include frequent fertility bottlenecks and infrequent events such as gametic nonreduction and interspecific hybridization, yet little is known about how these and other factors influence overall rates of polyploid formation. Here we review the literature regarding polyploid origins, and quantify parameter values for each of the steps involved in the principal pathways. In contrast to the common claim that triploids are sterile, our results indicate that the triploid bridge pathway can contribute significantly to autopolyploid formation regardless of the mating system, and to allopolyploid formation in outcrossing taxa. We estimate that the total rate of autotetraploid formation is of the same order as the genic mutation rate (10^{-5}), and that a high frequency of interspecific hybridization (0.2% for selfing taxa, 2.7% for outcrossing taxa) is required for the rate of tetraploid formation via allopolyploidy to equal that by autopolyploidy. We conclude that the rate of autopolyploid formation may often be higher than the rate of allopolyploid formation. Further progress toward understanding polyploid origins requires studies in natural populations that quantify: (*a*) the frequency of unreduced gametes, (*b*) the effectiveness of triploid bridge pathways, and (*c*) the rates of interspecific hybridization.

0066-4162/98/1120-0467$08.00

INTRODUCTION

Polyploidy, defined as the possession of three or more complete sets of chromosomes, is an important feature of chromosome evolution in many eukaryote taxa. Yeasts, insects, amphibians, reptiles, and fishes are known to contain polyploid forms (100), and recent evidence of extensive gene duplication suggests that the mammalian genome has a polyploid origin (112). In plants, polyploidy represents a major mechanism of adaptation and speciation (24, 56, 95, 104, 120, 157, 159). It is estimated that between 47% and 70% of angiosperm species are polyploid (56, 110). Differences in ploidy have been observed among related congeners and even within populations of taxonomic species (24, 34, 56, 100, 156), and there is evidence that individual polyploid taxa may have multiple origins (154). These observations suggest that polyploid evolution is an ongoing process and not a rare, macroevolutionary event. Research in agricultural and natural systems indicates that polyploids often possess novel physiological and life-history characteristics not present in the progenitor cytotype (95, 104). Some of these new attributes may be adaptive, allowing a plant to enter a new ecological niche. Because plants of different ploidies are often reproductively isolated by strong post-zygotic barriers, polyploidy is also one of the major mechanisms by which plants evolve reproductive isolation (34, 56).

In spite of the prevalence and importance of polyploidy, the factors contributing to polyploid evolution are not well understood (165). Two critical stages of polyploid evolution can be identified: formation and establishment. To understand the process of polyploid formation requires information on the pathways, cytological mechanisms, and rates of polyploid formation. To assess the likelihood that a new polyploid will successfully establish requires information on the viability and fertility of new cytotypes, the extent of assortative mating and reproductive isolation within and between different cytotypes, and the ecological niche of new polyploids. Here we review the literature concerning polyploid formation to answer the following questions: (a) What are the primary pathways and mechanisms of polyploid formation? (b) What are the parameters for each of the steps involved in polyploid formation? (c) What are the numerical values reported for these parameters? and (d) What is the estimated rate of polyploid formation by each pathway?

One major motivation for this review is to synthesize the diverse literature on polyploid origins and thereby provide a resource for the development of future empirical and theoretical studies of polyploid evolution. To this end, we have tabulated data from many studies and made this information available on the *Annual Reviews* web site (http://www.annualreviews.org; see *Supplementary Materials*). We summarize these data throughout the text and identify the location of each database on the web site.

By necessity, many of the plants considered in this review are agricultural or horticultural cultivars and their wild relatives, as well as taxa widely used in classical genetic studies (e.g. *Oenothera* and *Datura*). We believe that the studies reviewed here provide insights into the process of polyploid formation in natural populations, but caution that further research in natural populations is needed to test our findings. Our survey draws from a wide range of plant taxa, but because of the limited number of studies, we do not interpret our results in a phylogenetic context.

In this chapter, $2n$ refers to the somatic chromosome number and n to the gametic chromosome number regardless of the degree of polyploidy, while x is the most probable base number. This gives the following cytological designations: diploids ($2n = 2x$), triploids ($2n = 3x$), tetraploids ($2n = 4x$), etc. In describing crosses within and between cytotypes, the maternal parent is always listed first.

MECHANISMS OF POLYPLOID FORMATION

Several cytological mechanisms are known to induce polyploidy in plants. Somatic doubling in meristem tissue of juvenile or adult sporophytes has been observed to produce mixoploid chimeras (2, 66, 82, 128, 153). For example, *Primula kewensis*, one of the first described allopolyploids, originated from fertile tetraploid shoots on otherwise sterile diploid F_1s of *P. floribunda* × *P. verticellata* (127). Similarly, a tetraploid shoot was observed on a diploid F_1 hybrid between *Mimulus nelsoni* and *M. lewisii* (66), and in wounded ("decapitated") tomato plants (82). Somatic polyploidy is known to be common in many non-meristematic plant tissues (30, 31). For example, normal diploid *Vicia faba* contains tetraploid and octoploid cells in the cortex and pith of the stem (26). Such polyploid cells occasionally initiate new growth, especially in wounds or tumors, and are a potentially important source of new polyploid shoots (30, 31, 99). The frequency of endopolyploidy, and the relative likelihood of polyploid formation from different endopolyploid tissues, are not well known.

Somatic doubling can also occur in a zygote or young embryo, generating completely polyploid sporophytes. This phenomenon is best described from heat shock experiments in which young embryos are briefly exposed to high temperatures (43, 140). Corn plants exposed to 40°C temperatures approximately 24 h after pollination produced 1.8% tetraploid and 0.8% octoploid seedlings (140). Polyploid seedlings are also known to arise from polyembryonic ("twin") seeds at a high frequency (122, 176), but it is now believed that such polyploids are generally of meiotic rather than somatic origin (29). In general, little is known about the natural frequency of somatic doubling in plants nor of the effects of interspecific hybridization on its occurrence.

A second major route of polyploid formation involves gametic "nonreduction," or "meiotic nuclear restitution," during micro- and megasporogenesis. This process generates unreduced gametes, also referred to as "$2n$ gametes," which contain the full somatic chromosome number (see reviews in 19, 63). The union of reduced and unreduced gametes, or of two $2n$ gametes, can generate polyploid embryos. As will be described in detail below, $2n$ gametes have been identified in many plant taxa. Polyspermy, the fertilization of an egg by more than one sperm nucleus, is known in many plant species (172), and has been observed to induce polyploidy in some orchids (59). However, it is generally regarded as an uncommon mechanism of polyploid formation (56).

Distinguishing between somatic doubling and $2n$ gametes as mechanisms of polyploid formation requires a system of genetic markers and a detailed knowledge of the cytological mechanism of gametic nonreduction, which are seldom available. There is, however, strong circumstantial evidence that $2n$ gametes are often involved in polyploid formation. The parents of spontaneous polyploids have, upon cytological analysis, commonly been found to produce $2n$ gametes (18, 21, 24, 46, 52, 55, 83, 90, 94, 101, 166, 168, 173). Conversely, plants known to produce $2n$ gametes can be crossed to produce new polyploids (18, 37, 81, 136, 138). In many cases, spontaneous polyploids have cytotypes that appear to have been formed by the union of reduced and unreduced gametes (21, 67, 68, 84, 125) rather than by somatic mutation, which generally only doubles the base chromosome number (e.g. $4x$ to $8x$) (30, 66, 82, 127). For example, Navashin (125) found triploids and pentaploids in the progeny of open-pollinated diploid *Crepis capillaris*, and these appear to have been produced by the union of reduced (n) and unreduced ($2n$ and $4n$) gametes. Similarly, triploids generated by backcrossing diploid hybrid *Digitalis ambigua* × *purpurea* are thought to have arisen from unreduced gametes produced by this interspecific hybrid (21). Because nonreduction appears to be the major mechanism of polyploid formation (19, 63, 165), we focus on the role of $2n$ gametes in polyploid origins. It is clear, however, that much research remains to determine the relative roles of the various cytological mechanisms of polyploid formation in natural populations.

Auto- and Allopolyploidy: An Evolving Terminology

Kihara & Ono (86) first described two distinct types of polyploids: "autopolyploids," which arise within populations of individual species, and "allopolyploids," which are the product of interspecific hybridization. Because chromosome pairing behavior was believed to be a reliable indicator of chromosome homology, early workers emphasized the frequency of multivalent formation at synapsis as a criterion for distinguishing auto- and allopolyploidy (32, 120). It was subsequently recognized that some polyploids of known hybrid origin exhibit multivalent pairing, while bivalent formation is prevalent in some non-hybrid polyploids (24, 156, 157). The term "segmental allopolyploid" was thus coined to denote polyploids of hybrid origin that possess chromosome pairing characteristics of autopolyploids, while "amphiploid" was used to indicate all polyploids that combine the chromosome complements of distinct species (24, 157). The term "autopolyploid" was reserved for polyploids that arose within single populations or between ecotypes or races of a single species (24, 56). Although this terminology is recognized by many students of polyploidy (56, 165), several confusing aspects remain. For example, there is considerable variation in the criteria used to delimit related taxa as "species." Moreover, some authors reserve the term allopolyploidy for hybrid polyploid derivatives of species that are largely reproductively isolated by barriers of hybrid sterility, because such species are more likely to differ in chromosome structure and pairing and to generate polyploids that behave cytogenetically as "true" allo-ploids (24, 160). Also, it is to be expected that some interpopulation polyploids may represent a class of polyploidy intermediate between auto- and allopoly-ploidy. Because of these and other difficulties, several alternate terminologies have been suggested. Jackson (71) proposed that the terms auto- and allopoly-ploidy be used in their original, cytological meaning (32, 120)—that autopoly-ploids exhibit multivalent pairing while allopolyploids do not—and developed statistical criteria for distinguishing these types of polyploids (72). Lewis (99) used "intraspecific" and "interspecific" polyploidy to distinguish polyploids that are morphologically distinguishable from those that are not, and considered these terms to correspond roughly to "autopolyploidy" and "allopolyploidy," respectively.

We believe that the primary criterion for classifying a polyploid is its mode of origin. We use the term "autopolyploid" to denote a polyploid arising from crosses within or between populations of a single species, and "allopolyploid" to indicate polyploids derived from hybrids between species, where species are defined according to their degree of pre- and/or post-zygotic isolation (biological species concept). We consider polyploids arising from hybridization between species with minor aneuploid differences (dysploidy) to be allopolyploids, following Clausen et al (24). Considerable differences in the mechanisms and rates of polyploid formation within "types" of autopolyploid and allopolyploid systems may exist. In particular, the frequency of meiotic irregularity and spontaneous polyploid production may differ between hybrids of recently and anciently diverged taxa.

Pathways of Polyploid Formation

Several different pathways of both auto- and allopolyploid formation have been described. In this section, we identify the major routes to polyploid formation and highlight examples in which one or more steps in the pathway have been directly observed.

AUTOTETRAPLOID, TRIPLOID-BRIDGE   Triploids are formed within a diploid population, and backcrossing to diploids, or self-fertilization of the triploid, produces tetraploids. For example, 1% tetraploid progeny were obtained by backcrossing a spontaneous triploid clone of *Populus tremula* to a diploid (15). Similarly, a small number of tetraploid progeny were obtained from triploid apple varieties that had themselves originated as spontaneous polyploids (14). Although all the steps in this pathway have rarely been observed in their entirety, the individual mechanisms are well substantiated. Triploids have often been observed in diploid populations (41, 47, 69, 143, 168), and it is generally believed that these are produced by the union of reduced (n) and unreduced ($2n$) gametes. Studies of such spontaneous triploids, as well as triploids produced by crossing diploids and tetraploids, indicate that many of the gametes produced by autotriploids are not functional, because they possess aneuploid, unbalanced chromosome numbers. However, triploids generate small numbers of euploid (x, $2x$) gametes (12, 40, 44, 87, 91, 148, 149) and can also produce $3x$ gametes via nonreduction (12, 91, 117). Autotriploids can produce tetraploids by self-fertilization or backcrossing to diploids (40, 44, 74, 174, 179).

AUTOTETRAPLOID, ONE-STEP   Tetraploids are formed directly in a diploid population by the union of two unreduced ($2n$) gametes or by somatic doubling. For example, Einset (47) found a small fraction (4.39×10^{-4}) of tetraploid seedlings while cytotyping the progenies of open-pollinated diploid apple varieties. Tyagi (168) crossed clones of *Costus speciosus* that were observed to produced some $2n$ pollen, and recovered a small number of tetraploid seedlings. This process has been observed in several other taxa (18, 69, 81, 85, 89, 122).

ALLOTETRAPLOID, TRIPLOID-BRIDGE   Hybrid triploids are formed by diploids in the F_1 or F_2 generation of interspecific crosses, and self-fertilization or backcrossing to diploids produces allotetraploids. For example, Müntzing (119) crossed *Galeopsis pubescens* and *Galeopsis speciosa* to generate a highly sterile, diploid F_1 hybrid. One of the 200 F_2 progeny was found to be triploid, and a backcross of this plant to *G. pubescens* formed a single viable seed, which was tetraploid (119). Similarly, Skalińska (152) found a single triploid plant in the F_1 progeny of a cross between *Aquilegia chrysantha* and *Aquilegia flavellata*. Selfing this triploid produced a small number of F_2 progeny, two thirds of which were tetraploid. Allotriploids have commonly been observed in the F_2 generation produced by backcrossing or selfing interspecific F_1 hybrids (21, 24, 94, 101, 118, 137), and in the F_1 generation by the union of reduced and unreduced gametes from the parent genotypes (27, 60, 67,

152). Studies of such spontaneous allotriploids, and of allotriploids obtained by crossing different diploid and tetraploid species, indicate that the production of diploid gametes (8, 54, 152, 169) and nonreduction (45, 70, 90, 162) enable allotriploids to produce allotetraploids by selfing or backcrossing (67, 80, 118, 152, 167).

ALLOTETRAPLOID, ONE-STEP   Allotetraploids are formed directly from diploids in the F_1 or F_2 generation of interspecific crosses. For example, 90% of the F_2 progeny of *Digitalis ambigua* and *Digitalis purpurea* were tetraploid (21). Half of the F_2 progeny of *Allium cepa* and *Allium fistolium* were tetraploid or hypotetraploid (i.e. $4x - 1$) (94), and 2% of the F_1 progeny of *Manihot epruinosa* × *glaziovii* were tetraploid (60). There are many other examples of tetraploids being produced in one step by F_1 interspecific hybrids (1, 21, 24, 46, 55, 66, 83, 128, 137), or in the F_1 generation of an interspecific cross (60, 79, 92, 169). Other important pathways involve the evolution of ploidy levels above tetraploidy.

HIGHER PLOIDY, ONE-STEP   Within a polyploid population, the union of reduced and unreduced gametes generates a new cytotype of higher ploidy. For example, 2% hexaploid cytotypes were recovered from the progeny of open-pollinated autotetraploid *Beta vulgaris*, apparently from the union of reduced ($2x$) and unreduced ($4x$) gametes (68). Similarly, 1% of the progeny of tetraploid alfalfa were found to be hexaploid (16). There is circumstantial evidence of autohexaploid formation in tetraploid populations in several other systems (23, 42). New odd-ploidy cytotypes could also be produced by this mechanism. For example, it has been suggested that unreduced gamete production in hexaploid *Andropogon gerardii* generated a $9x$ cytotype, which is now widely distributed (130).

ALLOPOLYPLOIDY, VIA HYBRIDIZATION OF AUTOPOLYPLOIDS   Hybridization between distinct autopolyploids directly produces allopolyploids. For example, crosses between autotetraploid *Lycopersicon esculentum* and autotetraploid *Lycopersicon pimpinellifolium* produced a fertile allotetraploid *Lycopersicon* that was identical to the allotetraploid made by doubling the diploid F_1 hybrid (103). An allotetraploid *Tradescantia* was produced by crossing autotetraploid forms of *T. canaliculata* and *T. subaspera* (4). It has repeatedly been found that the post-zygotic barriers that isolate diploid taxa break down in autopolyploids, so that interspecific hybrids are formed easily (23, 65). Not surprisingly, there may be extensive intergradation among polyploids, while diploid taxa remain morphologically distinct (65, 98).

ALLOPOLYPLOIDY, VIA HYBRIDIZATION OF DIFFERENT CYTOTYPES   Hybridization between different cytotypes (which may be of auto- or allopolyploid origin) generates intermediates of odd-ploidy, which subsequently produce new even-ploidy cytotypes. For example, a high frequency ($\sim$81%) of allohexaploids was generated by crossing triploid hybrids of *Nicotiana paniculata* ($2x$) and *Nicotiana rustica* ($4x$) (90). A similar process has been observed in many other systems (24, 25, 90, 166, 173).

Second-Generation Polyploids

The production of later-generation polyploids can be achieved through a variety of pathways. For example, a new self-compatible tetraploid can self to produce tetraploid offspring. For outcrossing taxa, second-generation tetraploids can be produced by matings between independently produced tetraploids. Alternatively, backcrossing to the diploid progenitor can produce triploids (18, 73, 76), which contribute to further tetraploid formation by crossing to either diploids (44, 80, 152, 170, 179) or tetraploids (40, 139, 152, 171, 174). Later-generation tetraploids can also be produced by backcrosses of tetraploids to diploids that produce unreduced gametes (18, 51, 69, 178). Clearly, the frequency and cytotype composition of later-generation polyploids will depend on factors such as the mating system and the degree of pre- and post-zygotic reproductive isolation between cytotypes (50, 144). There is a need for further empirical and theoretical research on these and other issues related to polyploid establishment.

UNREDUCED GAMETES AND THE ORIGIN OF NEW POLYPLOIDS

Unreduced gametes are believed to be a major mechanism of polyploid formation (19, 64). Both $2n$ pollen and $2n$ eggs have been observed in hybrid and non-hybrid agricultural cultivars and natural plant species (11, 13, 24, 36, 37, 83, 101, 116, 136, 142, 147, 151, 163, 164, 177). Unreduced pollen grains can often be identified by size, as they typically have a diameter 30–40% larger than that of reduced pollen (78, 168, but see 101, 105), and the distribution of pollen size in plants known to produce $2n$ pollen is often bimodal (131, 168). Unreduced female gametophytes can sometimes be identified by size (161), but more often the frequency of $2n$ gametes is indirectly estimated using controlled, inter-ploidy crosses in plants with very strong interploidy crossing barriers. The progeny generated are usually the products of $2n$ gametes (3, 36; see below). Little correlation has been observed between the production of $2n$ pollen and $2n$ eggs (36, 134, 142, 164, but see 147).

Meiotic aberrations related to spindle formation, spindle function, and cytokinesis have been implicated as the cause of $2n$ gamete production in nonhybrid crop cultivars (19). For example, a parallel spindle orientation at anaphase II results in the reconstitution of diploid nuclei in microsporogenesis, and pre-mature cytokinesis that immediately follows the first meiotic division creates diploid nuclei that never undergo a second meiotic division (116). The cyto-logical causes of nonreduction in hybrids are less well studied. Often, poor chromosome pairing in F_1 hybrids leads to asynapsis at the first meiotic di-vision, and a single "restitution" nucleus containing the full somatic chromo-some number forms in the spore mother cell (55, 83, 166, 173). Other cytolog-ical mechanisms related to cytokinesis are also known to produce $2n$ gametes in interspecific hybrids (64). The timing and type of cytological anomaly producing $2n$ gametes affects both the level of genic heterozygosity and the yield of polyploid cultivars (114). Unfortunately, there are few data on the cytological origins of $2n$ gametes in natural systems. Identifying the origins of unreduced gametes is complicated, because different individuals in the same species often produce $2n$ gametes by different cytological mechanisms, and more than one mechanism may operate within an individual plant (134, 177).

What Is the Frequency of Unreduced Gametes?

The frequency of $2n$ gametes determines the rate of new polyploid formation, as well as the types of polyploids being produced, and is therefore critical for understanding polyploid formation. We summarized the observed frequencies of $2n$ pollen in hybrid and nonhybrid systems, excluding those selected for their tendency to produce $2n$ gametes.

The mean frequency of $2n$ gametes found in studies of hybrids (27.52%) was nearly 50-fold greater than that in nonhybrids (0.56%), and this difference was significant (Mann-Whitney U test, $P < 0.001$) (web Table 1; 8 web Tables are located at www.AnnualReviews.org). This result is consistent with the qualita-tive impressions of Harlan & de Wet (63). Because interspecific hybrids often experience severe meiotic irregularities involving poor chromosome pairing and non-disjunction, the "reduced" gametes produced by hybrids often possess unbalanced, aneuploid cytotypes, and are thus inviable (22, 52). This suggests that the effective frequency of $2n$ pollen in hybrid systems may be even higher than estimated here.

The few existing data for $2n$ eggs suggest that the natural frequency of non-reduction is similar in megasporogenesis and microsporogenesis. The mean frequency of $2n$ eggs in a sample of approximately 100 field-collected individ-uals of *Dactylis glomerata* was 0.49% (36), while the frequency of $2n$ pollen in a similar collection of individuals was found to be 0.98% (105). The mean frequency of $2n$ eggs was 0.06% in *Trifolium pratense* (136) and 0.09% in maize (3). We know of no published reports on the frequency of $2n$ eggs in interspecific hybrids.

Polyploid Formation Is Facilitated by a Breakdown in Self-Incompatibility

The "effective" frequency of $2n$ pollen in plants with gametophytic self-incompatability may be increased by the breakdown of incompatibility in diploid pollen produced by either reduction division in established tetraploids, or nonreduction in diploids. This tendency, which appears to be related to genic interactions in diploid pollen grains (20, 96, 97), may allow self-pollination by $2n$ pollen, leading to polyploid formation. For example, Marks (108) obtained some polyploids by selfing diploid self-incompatible *Solanum*. Lewis (96) found only triploids in the selfed progeny of self-incompatible strains of *Pyrus* producing $2n$ pollen, and described this phenomenon as the "incompatibility sieve" for polyploid formation. This mechanism could contribute to polyploid formation where abiotic or biotic factors increase the frequency of self pollen deposition.

Genetic Factors Influence the Rate of Unreduced Gamete Production

Bretagnolle & Thompson (19) provide an exhaustive review of the genetic basis of $2n$ gamete production in nonhybrid crop species, and we provide only a brief summary here. Plant populations often possess heritable genetic variation for the capacity to produce $2n$ gametes, as illustrated by a rapid response to selection for $2n$ gamete production in crop cultivars (135, 164). For example, the mean frequency of $2n$ pollen increased from 0.04% to 47% in three generations of selection on *Trifolium pratense*, giving a realized heritability of 0.50 (135). In *Medicago sativa*, selection experiments on $2n$ pollen and $2n$ egg production gave realized heritabilities of 0.39 and 0.60, respectively (164). Meiotic analysis of progeny derived from crosses between plants differing in their level of $2n$ gamete production indicate that this phenotype can be under strong genetic control and is often determined by a single locus (115, 142, 147).

Why Is There Genetic Variation for Unreduced Gamete Production?

Because different cytotypes are typically reproductively isolated, $2n$ gametes do not contribute to the gene pool of their progenitor cytotype. Thus, we expect strong selection against $2n$ gamete production, and it is perhaps surprising to sometimes find high heritabilities for this trait. As yet, there is insufficient information to determine if the frequency of genes influencing $2n$ gamete production is different from that expected by mutation-selection balance. Polyploidy often occurs in perennial taxa capable of vegetative reproduction (58, 155).

Characters related to sexual reproduction may be under relaxed selection in these systems, resulting in a potentially higher frequency of $2n$ and nonfunctional gametes. This hypothesis is supported by the observation that many of the taxa in which $2n$ gamete production has been documented are perennials with means of vegetative propagation (9, 105, 131). Another possible mechanism contributing to $2n$ gamete production is that the cytological abnormalities leading to non-reduction are the pleiotropic effect of genes with other, perhaps beneficial, effects.

Environmental Factors Can Affect the Frequency of Unreduced Gametes

Several researchers have found that $2n$ pollen production is stimulated by environmental factors such as temperature, herbivory, wounding, and water and nutrient stress. Temperature, and especially variation in temperature, have particularly large effects (11, 38, 96, 111, 151). Belling (11) observed a dramatic increase in $2n$ pollen production in field and greenhouse cultures of *Strizolo bium* sp., *Datura stramonium*, and *Uvularia grandiflora* following aberrant cold spells. Potato genotypes selected for the tendency to undergo gametic non-reduction had approximately twice the mean frequency of $2n$ pollen in a coastal field as in a greenhouse, an effect attributed to the temperature differences of the two environments (111). Ramsey & Schemske (unpublished data) found that the frequency of $2n$ pollen in randomly selected *Achillea millefolium* plants reared in a temperature-cycling growth chamber was approximately six times that in the natural population from which the study plants had been sampled.

Plant nutrition, herbivory, and disease may also affect $2n$ gamete production. Grant (55) found that the rate of polyploid production per flower in F_1 *Gilia* hybrids grown in low-nutrient conditions was almost 900 times greater than that of plants grown in high-nutrient conditions, a result attributed to poor pairing at meiosis in the former treatment. However, the higher level of polyploid production per flower in the low-nutrient treatment was partially offset by a much lower flower number, such that the number of polyploids produced per plant was only seven-fold greater. Kostoff (88) and Kostoff & Kendall (89) described an effect of gall mites and tobacco mosaic virus on $2n$ pollen formation.

Many of the environmental factors known to influence $2n$ gamete production are experienced by plants in their natural habitats. This suggests that natural environmental variation, as well as large-scale climate change, could substantially alter the dynamics of polyploid evolution. The high incidence of polyploidy at high latitudes, high altitudes, and recently glaciated areas may be related to the tendency of harsh environmental conditions to induce $2n$ gametes and polyploid formation (23, 151).

TRIPLOIDS: FORMATION, MEIOSIS, FERTILITY, AND PROGENY

The evolution of tetraploidy may proceed directly from diploids via the union of two $2n$ gametes, or in two steps via a triploid bridge (19). Because the probability of the union of two $2n$ gametes is expected to be very low, it has been hypothesized that triploids usually play a role in the evolution of tetraploids (39, 63). However, the low fertility of triploids, coupled with the existence of cytological barriers that may prevent or limit triploid formation by diploids, may restrict the role of triploids (19, 150). We review components of the triploid bridge pathways, including the likelihood of triploid formation via $2n$ gametes, triploid fertility and meiotic behavior, and the cytotype composition of the progeny produced by triploid parents.

How Effective Are Unreduced Gametes in Triploid Formation?

In most flowering plants, fertilization of the egg by a sperm nucleus is accompanied by fusion of the other sperm nucleus with two haploid polar nuclei in the female gametophyte to form the triploid endosperm that functions to nourish the $2x$ embryo. In polyploids, the process proceeds in an analogous fashion, but the ploidies of all tissues are proportionately increased. Crosses between diploid and tetraploid plants often fail because intercytotype hybrid seed development does not proceed normally, and nonviable seeds are produced. The difficulty of obtaining viable triploid seeds by diploid-tetraploid and tetraploid-diploid crosses has been termed the "triploid block" (108). Barriers to intercytotype hybridization have been observed at higher ploidy levels, but are not well described. Viable seeds produced by crosses between diploids and tetraploids are often tetraploid and result from unreduced gametes produced by the diploid parent.

Abnormalities in the growth and structure of the endosperm have often been implicated as the source of triploid block (28, 49, 113, 178; see reviews in 62, 175). The ratios between the embryo, endosperm and/or maternal tissue, as well as the maternal:paternal ploidy ratio in the endosperm, are all altered in $2x \times 4x$ and $4x \times 2x$ crosses, and it has been suggested that normal seed development depends on these ploidy ratios. Müntzing (123) hypothesized that proper function requires an embryo:endosperm:maternal tissue ratio of 2:3:2, while others have suggested that it is the 2:3 ratio of the embryo to endosperm that is critical for proper seed development (175). An alternate explanation is that the maternal:paternal ploidy ratio in the endosperm, irrespective of the ploidy of the embryo or maternal tissue, determines seed viability (129). This has been described as the imprinting hypothesis (61, 102), and supporting evidence is provided by several studies showing that a 2:1 ratio of the maternal:paternal genomes in the endosperm is required for normal seed development (77, 102). Johnston et al (77) proposed a modification of this hypothesis to account for the anomalous findings that viable seed is produced in some systems where the 2:1 ratio is violated, but not in others where the 2:1 ratio is met. Their endosperm balance number hypothesis suggests that seed development is affected by the effective maternal:paternal ploidy ratio of the endosperm, which may not always reflect the actual ploidy composition (77). Although the genetic mechanisms responsible for interploidy crossing barriers remain an open area of investigation, there is general agreement that the ploidies of an embyro and/or its associated endosperm are the critical factors influencing successful seed development.

In diploids, fertilization of a reduced egg by an unreduced sperm nucleus will generate the same embryo:endosperm ploidy ratio (3:4) as a $2x \times 4x$ cross, and the union of an unreduced egg and a reduced sperm nucleus will form the ploidy ratio (3:5) of a $4x \times 2x$ cross [most described mechanisms of unreduced egg formation involve nonreduction in the megaspore mother cell, thus producing gametophytes with unreduced polar and egg nuclei (142, 147, 177, but see 29)]. Similarly, seed from a $2x \times 4x$ cross contains endosperm with the same maternal:paternal ratio (2:2) of seeds produced by the union of reduced eggs and unreduced pollen, while seed from a $4x \times 2x$ cross contains endosperm with the same maternal:paternal ratio (4:1) of seeds produced by the union of unreduced eggs and reduced pollen. Note that crosses in both directions violate the normal ($2x \times 2x$) embryo:endosperm ratio of 2:3 and the endosperm maternal:paternal ratio of 2:1. Irrespective of the cytological cause of triploid block, the success of crosses involving diploid reduced gametes from tetraploids and haploid reduced gametes from diploids should parallel that of crosses involving diploid unreduced gametes from diploids and reduced gametes from diploids. Here we use this approach to evaluate the likelihood of triploid formation via $2n$ gametes. We surveyed the literature for data on crosses between autotetraploids and their progenitor diploids. We excluded studies of naturally occurring polyploids because genic differences arising after autopolyploid formation could contribute to crossing barriers. Our review focuses on studies of autopolyploids because there are too few data on crossing success in allopolyploids.

No viable triploid seed production was observed in 13 of the 19 studies of $2x \times 4x$ crosses, or in 7 of the 17 studies of $4x \times 2x$ crosses (web Table 2 located at www.AnnualReviews.org). These data indicate that complete triploid block is present in many taxa, and that the possibility of triploid formation may differ for $2n$ pollen and eggs. In the 11 studies reporting viable triploid production in $2x \times 4x$ and/or $4x \times 2x$ crosses, 10 found that more viable triploid seeds were generated by $4x \times 2x$ crosses, while in only a single study was higher triploid production from $2x \times 4x$ crosses observed. This difference is statistically significant (Paired sign test, $P < 0.05$).

We calculated an index of triploid block via $2n$ gametes using studies in which data on intracytotype crosses were available. For diploid pollen, this is

$$1 - \frac{k_{(2\times4).3}}{k_{(2\times2)}},$$

1.

where $k_{(2\times4).3}$ is the viable triploid seed production from $2x \times 4x$ crosses, and $k_{(2\times2)}$ is the viable seed production from $2x \times 2x$ crosses (assumed to include only diploid seed). The corresponding likelihood of triploid formation from $2n$ eggs is

$$1 - \frac{k_{(4\times2).3}}{k_{(2\times2)}},$$

2.

where $k_{(4\times2).3}$ is the viable triploid seed production from $4x \times 2x$ crosses, and $k_{(2\times2)}$ is defined as above. For both calculations, values greater than 0 indicate a triploid block, and a value of 1.0 indicates a complete block.

In the nine studies involving $2x \times 4x$ crosses, the mean block via diploid pollen was 0.952 (range 0.57 to 1.0), and of the eight studies of $4x \times 2x$ crosses, the mean block via diploid eggs was 0.801 (range 0.34 to 1.0), a difference that is marginally significant (Wilcoxon Sign Rank test, $P = 0.07$). Together, these data demonstrate there is generally a large barrier to triploid formation via unreduced gametes, but that the barrier is often not complete. A direct estimate of triploid block would compare observed and expected triploid production in crosses between diploids producing $2n$ gametes. In *Dactylis glomerata*, this approach gave an overall block of 0.98 (18), which is similar to the mean value we report here.

The observed reciprocal differences in intercytotype crossing success suggest that unreduced eggs are more effective in polyploid formation than are unreduced pollen. This is consistent with the finding that spontaneous triploids in well-studied genetic systems arise via nonreduction in female parents (17, 27). Reciprocal differences in interploidy crossing success may be a consequence of the embryo:endosperm and endosperm maternal:paternal ploidy ratios, which differ with the direction of the cross (62, 175). By cytotyping the embryos and endosperm of *Citrus* seeds resulting from intracytotype and intercytotype crosses of parents producing both reduced and unreduced gametes, Esen & Soost (49) demonstrated that seeds with embryo:endosperm ratios of 2:3, 3:5, 4:6, and 6:10 were viable, while those with 3:4, the expected ratio resulting from the union of unreduced pollen with reduced eggs, were not. Using a meiotic mutant that generated endosperm of varying ploidy, Lin (102) showed an effect of endosperm genome composition on seed viability in $2x \times 2x$ and $2x \times 4x$

crosses in maize. Another possible cause of reciprocal differences is that some mutations in megasporogenesis create embryo sacs containing nuclei of varying ploidy, and thus triploid embryos produced by these megagametophytes can be accompanied by normal, functional triploid endosperm (29). These mutations are a cause of polyembryony (29) and may be an important route of viable triploid formation. Regardless of its cause, the possibility that unreduced eggs are of primary importance in triploid formation seems significant in light of the fact that nearly all studies of gametic nonreduction have focused on unreduced pollen (24, 46, 101, 168, 173).

Several taxa with low triploid block have atypical endosperm characteristics (62, 178). In *Oenothera*, only a single polar nucleus is involved in the formation of a diploid endosperm, and viable triploid seeds are easily produced by $2x \times 4x$ crosses (62). In *Populus*, mature seeds have no endosperm, and seed development is very rapid; viable seed set in $2x \times 4x$ crosses is 80% of the $2x \times 2x$ yield (75). These observations suggest a possible relationship between endosperm characteristics, the strength of triploid block, and polyploid formation. Plant families such as the Asteraceae, Crassulaceae, Onagraceae, Rosaceae, and Salicaceae that lack endosperm in mature seeds have a high incidence of polyploidy.

Triploids Generate Some Euploid Gametes, and Are Often Semi-Fertile

The reduction division in triploids is expected to generate aneuploid gametes with half the triploid chromosome number, or $3x/2$. However, the possession of an unmatched complement of chromosomes leads to the formation of multivalents and univalents during pairing, and subsequent irregularities during disjunction can create varied chromosome assortments (33).

We surveyed the literature to examine the cytotype composition of pollen produced by triploids of both hybrid and nonhybrid origin, as indicated by examination of anaphase I, metaphase II, and anaphase II in pollen mother cells, or the first postmeiotic mitosis in maturing pollen. In the 26 studies examined, the most common modal pollen chromosome number was $3x/2$ (17 studies), followed by $3x/2 - 1$ (5 studies; web Table 3). The tendency to produce aneuploid pollen is similar in auto- and allotriploids, so in the remaining analyses we consider both triploid types together. Figure 1 shows the average frequency of haploid, diploid, triploid, and the most common aneuploid cytotypes in pollen produced by triploids ($n = 25$ studies; web Table 3 located at www.AnnualReviews.org). Low mean frequencies were found for haploid (3%) and diploid (2%) pollen relative to the frequency of the common aneuploid class $3x/2$ (34%). Rare euploid gametes are formed when the separation of multivalents and unpaired chromosomes in a spore mother cell is so unequal

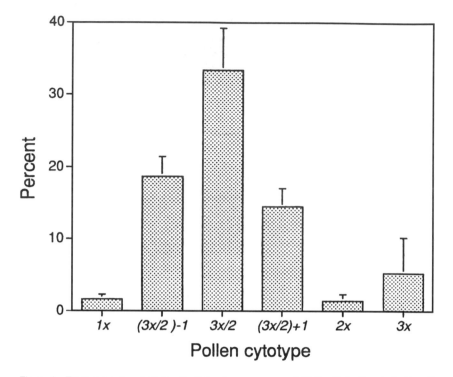

Figure 1   Frequency of euploid (1*x*, 2*x*, 3*x*) and common aneuploid (3*x*/2, 3*x*/2 − 1, 3*x*/2 + 1) cytotypes in pollen produced by hybrid and nonhybrid triploids, as determined by investigation of pollen mitoses as well as metaphase and anaphase of pollen mother cells. Data from web Table 1.

that haploid-diploid chromosome assortments are produced at the first meiotic division (12, 44, 149, 152). Triploid pollen, the result of gametic nonreduction (12, 45, 70, 91), was also observed at a low mean frequency (5.2%; $n = 9$ studies). Together, these analyses suggest that triploids produce mostly aneuploid classes of pollen. Unfortunately, there are relatively few studies of megasporogenesis to complement the data on microsporogenesis. Satina & Blakeslee (148) observed a modal egg chromosome number of $3x/2 − 2$ in triploid *Datura stramonium*, with 7% haploid and 1% diploid complements. These values are similar to those observed for pollen.

   Triploids are often expected to be sterile because of their meiotic irregularities and high frequency of aneuploid gametes. However, in a survey of the literature, we found a mean pollen fertility of 31.9% (range 0–97; web Table 4 located at www.AnnualReviews.org). The mean fertility of autotriploids (39.2%, $n = 23$ studies) was greater than that of allotriploids (23.7%, $n = 18$ studies), but this difference was not significant (Mann-Whitney U test, $P = 0.17$). For those studies with data on both pollen cytotype and fertility (web Table 4 located at www.AnnualReviews.org), there was a significant positive correlation between the frequency of euploid (x and $2x$) pollen and pollen fertility (Spearman Rank Correlation, $r = 0.63, P < 0.05, n = 11$), suggesting that euploid pollen contribute disproportionately to overall pollen fertility (the frequency of $3x$ pollen was not included in this analysis because most studies quantifying gametic nonreduction did not provide data on the frequency of $1x$ and $2x$ pollen). Further evidence that triploids are often semifertile comes from the few studies that have examined the relative fertility of crosses involving triploids. The available data suggest that some viable progeny are typically obtained from $2x \times 3x$, $3x \times 2x$, and $3x \times 3x$ crosses, and that crossing success may vary with the direction of the cross (51, 126, 152, 170).

New Polyploids Can Be Generated Through a Triploid Bridge

The cytotypes of the progeny derived from triploid crosses are often different from what might be expected from triploid meiotic behavior. Figure 2 illustrates this phenomenon in allotriploid *Aquilegia chrysantha* × *flavellata* ($3x = 21$) and autotriploid *Zea mays* ($3x = 30$). In both cases, pollen chromosome numbers had an approximately normal distribution (Figure 2a, b), with modes corresponding to the "expected" aneuploid value of $3x/2$. However, the cytotype distribution of the progeny differed significantly from the expected distributions calculated from the chromosome number distribution in microsporogenesis for self (Figure 2c, d) and backcross (not shown) progeny (Kolmogorov-Smirnov One-Sample test, $P < 0.01$). The selfed progeny in both studies had a bimodal cytotype distribution; in *Aquilegia*, most of the offspring were fully tetraploid (Figure 2c), while the modes in *Zea* were aneuploid (Figure 2d). These results suggest that gametes with cytotypes near the modal class of $3x/2$ do not function as well as other gametes, especially compared to those with euploid or near-euploid cytotypes.

We surveyed the literature to examine the frequency of polyploid cytotypes in the progeny of triploids. Auto- and allotriploids produced similar progeny cytotypes, so we combined them for the following analyses. Figure 3 illustrates the mean frequency of several euploid and aneuploid cytotypes resulting from $2x \times 3x$, $3x \times 2x$, $3x \times 3x$, $3x$ self, $3x \times 4x$, and $4x \times 3x$ crosses (see web Table 5 for the complete data set). We first investigated the two-step pathway of tetraploid formation, which proposes that triploids produced in diploid populations generate tetraploids via backcrossing to diploids, triploid selfing, or crossing among triploids. Tetraploid ($4x$) progeny were observed in four of 18 studies of $3x \times 2x$ crosses, with a mean frequency of 9.8% (range 0–85.7),

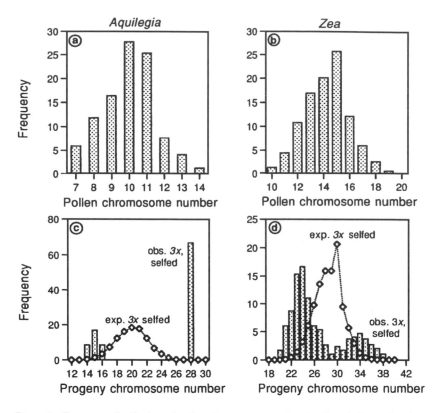

Figure 2 Frequency distribution of pollen chromosome number (*a, b*) and observed and expected selfed progeny cytotype (*c, d*) in allotriploid *Aquilegia chrysantha* × *flavellata* (*left*) and autotriploid *Zea mays* (*right*). Data from web Tables 3 and 5.

and two of ten studies of $2x \times 3x$ crosses, with a mean of 1.1% (range 0–7.4) (Figure 3*a, d*; web Table 5). Tetraploid ($4x$) progeny were observed in one of four studies of $3x \times 3x$ crosses (mean 0.2%, range 0.0–0.8), and in three of eight studies of $3x$ self crosses (mean 13.9%, range 0–66.7) (Figure 3*b, e*; web Table 5). Averaged across these four cross types, tetraploids constituted 6.3% of the progeny, suggesting that triploids can contribute to tetraploid formation.

Once some tetraploids have been produced, and mixed cytotype populations have been established, backcrossing between triploids and tetraploids may generate new tetraploids. The likelihood of this type of cross will depend on the relative frequency of the different cytotypes, and the extent of premating isolation between cytotypes. Our survey revealed that tetraploids were common in some $3x \times 4x$ and $4x \times 3x$ crosses, with mean frequencies of 31.6% (range

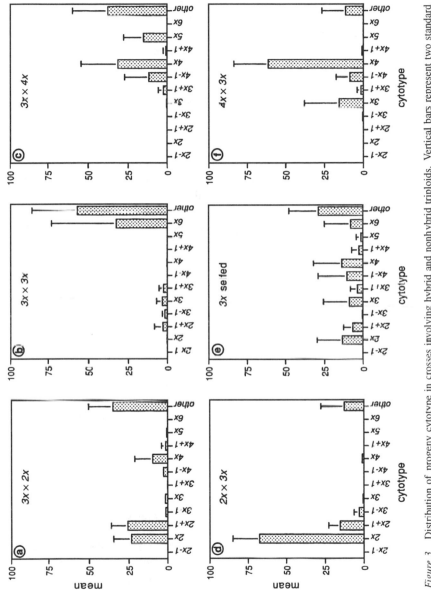

Figure 3 Distribution of progeny cytotype in crosses involving hybrid and nonhybrid triploids. Vertical bars represent two standard errors. Data from web Table 5.

0–100, $n = 10$ studies) and 61.1% (range 35.8–100, $n = 5$) studies, respectively (Figure 3c, f; web Table 5). Thus, tetraploid formation is enhanced under conditions favoring crossing between triploids and tetraploids, as might be expected from phenological differences between diploids and polyploids (99, 104).

These data also allowed us to examine the likelihood of hexaploid formation via a triploid intermediate. Hexaploids were observed in $3x \times 3x$ (mean 32.8%, range 0–81.1) and $3x$ self (mean 8.3%, range 0–66.7) crosses; they are presumably the result of the union of two unreduced ($2n = 3x$) gametes.

The mean frequency of triploids averaged across all cross types was 3.7% (range 0–66.7; web Table 5). Even in crosses between triploids (i.e. $3x \times 3x$ and $3x$ self), which would be expected to generate mostly triploid offspring, the mean frequency of triploids was less than 10% (Figure 3b, e). This suggests that triploids do not perpetuate themselves sexually at a high frequency, and that de novo production via $2n$ gametes in diploid populations, or crossing between tetraploids and diploids, is probably the primary means of triploid production. Tetraploid formation via triploids may thus be facilitated by the perennial habit and vegetative propagation of triploid clones, as has been observed in systems such as *Fritillaria lanceolata* (107) and *Calochortus longebarbatus* (10).

Diploids were recovered in each of the cross types that did not involve tetraploid parents. In crosses among triploids, diploids were observed at frequencies of less than 15% (Figure 3b, e). The frequency was higher in diploid backcrosses, with marked reciprocal differences. In $3x \times 2x$ crosses, the mean frequency of diploids was 23.4% (range 0–78.1), as compared to 67.4% (range 16.4–94.6) in $2x \times 3x$ crosses (Figure 3a, d; web Table 5). It is unclear if this and other apparent reciprocal differences (e.g. tetraploid production in $3x \times 4x$, and $4x \times 3x$ crosses) are related to the production or the viability of the various gamete cytotypes in eggs and pollen.

In general, aneuploids were common in the progeny of triploids. For all cross types combined, the mean frequency of all aneuploid offspring was 50.8%, varying from a mean of 23.2% in $4x \times 3x$ crosses to 64.3% in $3x \times 2x$ crosses (Figure 3; web Table 5). Polyploidy is not thought to evolve as a series of individual chromosome additions involving aneuploids as intermediate steps, because ploidy variation in natural systems generally occurs in complete, or nearly complete, steps (56, 159). Additionally, many aneuploids have low viability and fertility (15, 44, 121). However, the term aneuploid refers to a variety of cytotypes, from those which are very similar to euploids (the "hypo-" and "hyper-euploids," such as $2x + 1, 4x - 1$), to those with several to many chromosome additions or deletions (e.g. $2x + 4, 4x - 3$). The meiotic behavior and fertilities of hypo- and hyper-euploid cytotypes are often similar to those of true euploids (15, 121), and there is natural hypo- and hyper-euploid variation in many polyploid species (13, 16, 121). We have therefore distinguished near-euploids from all other aneuploid classes ("other" category in Figure 3). Hypo- and hyper-euploids represented, on average, 48.9% of the aneuploids, but the frequency of such near-euploid cytotypes ranged widely between cross types (Figure 3; web Table 5). The effective frequency of polyploid formation from triploids may be significantly influenced by the viability and fertility of hyper- and hypo-euploid offspring. For example, in $3x$ selfed crosses, if hypo- and hyper-tetraploids ($4x - 1, 4x + 1$) are equal in performance to true tetraploids, the mean production of "tetraploid" cytotypes would increase nearly twofold. Further research is necessary to determine the roles of different classes of aneuploids in polyploid formation. For the reasons discussed above, we consider $3x - 1$ and $3x + 1$ cytotypes as "triploids" and $4x - 1$ and $4x + 1$ cytotypes as "tetraploids" in the following sections.

THE FREQUENCY OF SPONTANEOUS POLYPLOIDS

Spontaneous polyploids have been observed in both hybrid and nonhybrid plant systems. These novel cytotypes are identified on the basis of morphological characteristics (41, 84, 143, 146), or in the course of cytological surveys (47, 68, 69, 125). Although the mechanisms responsible for the formation of these polyploids are often unknown, the frequency of appearance of spontaneous polyploids is informative as a direct estimate of the rate of polyploid formation.

Spontaneous Polyploids Are More Common in Hybrid than in Nonhybrid Systems

To assess the frequency of spontaneous polyploidy, we surveyed the literature for studies that screened for novel cytotypes in intraspecific crosses, in hybrid progeny generated by backcrossing, or in F_2 progeny from selfing or interhybrid crosses (web Table 6). We include hypo-euploid and hyper-euploid cytotypes (e.g. $2x + 1, 4x - 1$) in euploid categories. Here we focus on polyploids derived from diploids, but web Table 6 includes examples of higher-ploidy systems. The frequency and cytotype of spontaneous polyploids in diploid nonhybrid, backcross, and F_2 systems differ considerably (Figure 4). New polyploids in nonhybrid systems are very rare, and are typically triploid. In hybrid systems, polyploid progeny are often the primary, or only, cytotype produced. The frequency of tetraploids is much higher in progeny derived from F_1 outcrossed ($F_1 \times F_1$) and F_1 self crosses (63%) than from backcross progeny (2%), as might be expected from the higher frequency of $2n$ gametes observed in hybrids than in nonhybrids (Figure 4). However, the frequency of polyploids in hybrid crosses is higher than would be expected based on the frequency of $2n$ gametes we found previously. For example, on average, a diploid F_1 hybrid

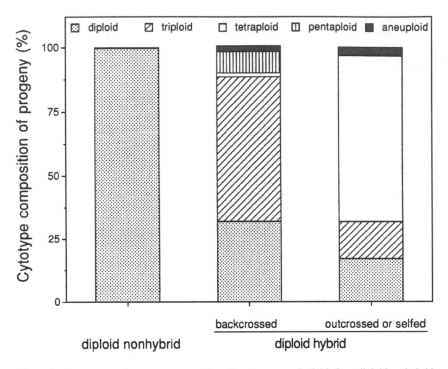

Figure 4   Frequency and cytotype composition of spontaneous polyploids from diploid nonhybrid and diploid hybrid systems. Frequencies of spontaneous polyploids in nonhybrids are 0.203% ($3x$), 0.013% ($4x$), and 0.067% ($5x$). Data from web Table 6.

produces 27.5% $2n$ gametes (web Table 1) and is therefore expected to produce only 7.6% tetraploid offspring on selfing or outcrossing to other F_1s. This difference in tetraploid production (63% vs 7.6%) may represent selection against gametes or progeny that are not polyploid.

Some homoploid interspecific crosses produce polyploids in the F_1 generation, occasionally in unexpectedly high frequencies (53, 60, 79, 93, 152, 163) (web Table 6). For example, crosses between the diploid *Brassica campestris* and *Brassica oleracea* produced four progeny, of which two were triploid and one was tetraploid (169). In Manihot, interspecific crosses produced 1.5% triploids and 2% tetraploids (60). These allopolyploid progeny were the result of $2n$ gametes produced by the nonhybrid parents. Given an average frequency of $2n$ pollen in nonhybrids of 0.05% (web Table 1), the expected frequency of spontaneous triploids and tetraploids is 0.05% and 0.0025%, respectively. The frequency of polyploids in F_1 progeny is commonly higher than these values, suggesting that $2n$ gametes may be at an advantage in some interspecific crosses.

Taken together, these data suggest that the formation of allopolyploids might be more common than that of autopolyploids. However, the rate of allopolyploid formation is a function of both the population-level frequency of hybridization and the rate of polyploid formation in interspecific hybrids. In many of the studies in which spontaneous allopolyploids were observed, it was clear that interspecific hybrids were secured only after considerable effort was made at crossing different species (1, 24, 128). Also, most F_1 hybrids observed to produce polyploids were highly sterile. The mean pollen viability of F_1 hybrids that produced polyploids (Figure 4) was 6.3% ($n = 8$ studies; web Table 7). Seed fertility, though rarely quantified, was often observed to be low (1, 21, 24, 35, 55). For example, Abdel-Hameed & Snow (1) made several hundred crosses between F_1 hybrids of *Clarkia amoena* × *lassenensis* and obtained only a single viable seed, which was tetraploid. After accounting for the ecological isolation that often separates natural species populations (56, 157), the crossing barriers between species, and the low fertility of interspecific hybrids, the overall rate of allopolyploid formation may be much lower than would be expected based only on the observed frequency of spontaneous polyploids from F_1 hybrids.

Allopolyploidy and Disturbed Habitats

The tendency of anthropogenic disturbance to encourage interspecific hybridization by breaking down ecological isolating barriers has long been noted by botanists (5, 56, 157). In the past century, several new allopolyploid species have evolved (106, 133, 145). In each case, the process involved nonnative plant taxa often invading disturbed habitats. For example, the appearance of two new allotetraploid species of *Tragopogon* followed the introduction of the diploid *T. dubius*, *T. porrifolius*, and *T. pratensis* into roadsides and waste areas of eastern Washington state (133). In Britain, the allohexaploid *Senecio cambrensis* was produced by hybridization between the native tetraploid *Senecio vulgaris* and the introduced diploid *Senecio squalidus* (7). Polyploid formation in interspecific hybrids can be essentially automatic (Figure 4), and it is possible that recent allopolyploid evolution is attributable to the high levels of hybridization found in habitats disturbed by human activities.

ESTIMATING THE RATE OF AUTO- AND ALLOPOLYPLOID FORMATION

While there is clear evidence that both auto- and allopolyploids exist in nature, there remains considerable uncertainty regarding the relative frequency of each, and of the factors influencing their abundance. Stebbins (157) and Grant (56) concluded that allopolyploids are much more frequent than autopolyploids (but see 34), and suggested that this was due in large part to heterosis and home-ostasis conferred by permanent hybridity in allopolyploids, which is lacking in autopolyploids. Another reason to expect a higher frequency of allo- than au-topolyploids is that autopolyploids often show reduced fertility due to meiotic irregularities (34, 159). These potential disadvantages of autopolyploidy are manifest only at the establishment phase, and it is therefore important to con-sider the likelihood of polyploid origins as well, and how the rate of polyploid formation may differ between autopolyploid and allopolyploid pathways. For example, interspecific hybridization is a potentially important rate-limiting step in allopolyploid formation (56). The frequency of $2n$ gametes may determine rates of autopolyploid formation, and a recent simulation model found that the rate of autopolyploid formation also influenced the likelihood of establishment (50).

Evaluating the mechanisms influencing the natural frequency of auto- and allopolyploids requires information on rates of polyploid formation by each pathway. Here we estimate the rate of tetraploid formation via the auto- and allopolyploidy pathways, based on the numerical values of parameters identi-fied in our review of the literature. Our objectives are to (*a*) estimate the total rate of autotetraploid formation, (*b*) determine the relative contribution of the triploid bridge to auto- and allotetraploid formation, (*c*) compare the rates of auto- and allotetraploid formation expected for selfing and outcrossing taxa, and (*d*) compare the rates of tetraploid formation by the two pathways. Be-cause there are few published reports of the frequency of F_1 hybrids in natural populations, this final objective is achieved by estimating the frequency of hy-bridization required to produce equal rates of tetraploid formation via auto- and allopolyploidy.

We believe these analyses are useful because they identify approaches that can be applied to individual natural systems, but we also emphasize several qualifications. The data used to estimate parameters may not be completely in-dicative of natural populations, given that they were taken from a relatively small number of systems, many of which are agricultural or horticultural. Although substantial variation between taxa was observed for many of the relevant parameters, it is likely that our analyses using the mean parameter values provide rough estimates of rates of polyploid formation.

The pathways of tetraploid formation, and the numerical values used to es-timate overall rates, are illustrated in Figure 5. For simplicity, we assume that unreduced gametes are the only cause of polyploid formation. The frequencies of autotriploids and autotetraploids were estimated from $2n$ gamete frequency and triploid block, rather than from estimates of spontaneous autopolyploid formation in diploids, because there are few studies of sufficient sample size to detect spontaneous autotetraploids (web Table 6). The results obtained using

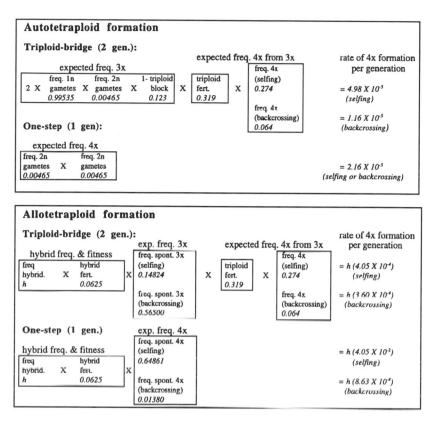

Figure 5 Estimated rates of auto- and allopolyploid formation via the triploid bridge and one-step pathways. Parameter values for each component were estimated from our review of published studies (see text).

data on spontaneous formation of auto-triploids and tetraploids were similar to those reported here. The frequency of $2n$ gametes for the autopolyploid pathway was estimated by taking the overall mean of the frequencies observed for $2n$ pollen and $2n$ eggs (web Table 1), and triploid block was calculated as the mean of that observed in $2x \times 4x$ and $4x \times 2x$ crosses (web Table 2). The expression for estimating the frequency of autotriploids includes a factor of two to account for allotriploid formation by both unreduced eggs and unreduced pollen (Figure 5). For the allotetraploid pathway, we used the expected frequency of spontaneous allotriploids and allotetraploids appearing in the F_2 progeny produced by selfing or backcrossing F_1 hybrids (web Table 6). This approach was considered more accurate than estimates based on $2n$ gametes. The low

Table 1 Summary of auto- and allotetraploid formation, based on the pathways shown in Figure 5

Type		Selfing	Backcrossing
Autotetraploid	Total rate of formation per generation	7.14×10^{-5}	3.32×10^{-5}
	% of total rate due to 3x bridge	69.7%	34.9%
Allotetraploid	Total rate of formation per generation	$h(4.09 \times 10^{-2})$	$h(1.22 \times 10^{-3})$
	% of total rate due to 3x bridge	1.0%	29.5%
Frequency of interspecific hybridization (h) required for the rate of autotetraploid formation to equal the rate of allotetraploid formation		0.0017	0.0272

viability of "reduced" gametes produced by the irregular meiotic divisions of interspecific hybrids makes the effective frequency of unreduced gametes much higher than would be expected based only on the frequency of gametic non-reduction. We estimated the fertility of F_1 hybrids from experimental studies of allopolyploid formation (web Table 7). For the triploid-bridge in both the auto- and allotetraploid pathways, triploid fertility was estimated as the mean pollen stainability from all studies (web Table 4). To estimate the expected frequency of tetraploids produced by triploids, we used the mean rate of tetraploid formation (defined as $4x - 1$, $4x$ and $4x + 1$) from both auto- and allotriploids resulting from either selfing or backcrossing (web Table 5). The sample sizes used to estimate these parameters are as indicated previously in the text.

The estimated total rate of autotetraploid formation (Table 1) is of the same order (10^{-5}) as estimates of the genic mutation rate obtained from studies in many organisms (57). The rate of tetraploid formation by selfing was greater than that by backcrossing for both autopolyploidy (1.7-fold greater) and allo-polyploidy (34-fold greater) (Table 1). The triploid bridge contributes 70% of the total rate of autotetraploid formation in selfing taxa, and 35% of the total in outcrossing taxa (Table 1). In contrast, the triploid bridge contributes only 1% of the total rate of allotetraploid formation in selfing taxa, as compared to 30% in outcrossers.

Most discussions of the role of mating systems in polyploid evolution have focused on establishment, including the contribution of self-pollination to the reproductive isolation of autopolyploids from their diploid progenitors (144, 159), and the selective value of outcrossing in maintaining heterozygosity in autopolyploids (158). We find that self-pollination may play an important role in tetraploid formation, particularly for allopolyploidy. This is consistent with

Grant's (56) conclusion that selfing can facilitate allopolyploid formation, and with the observation by Stebbins (158) that most self-fertilizing polyploids are allopolyploids. In our analysis, the high rate of polyploid formation in selfing allotetraploid systems is attributable to the high frequency of spontaneous allotetraploids in F_1 self crosses (Figure 5). Presumably, this results from self fertilization involving unreduced eggs and unreduced pollen, which may be produced at high frequency in F_1 hybrids. Because of the low frequency of $2n$ gametes in nonhybrid diploid progenitors, the frequency of spontaneous allotetraploids is much lower following backcrossing than following selfing (Figure 5).

Our results indicate that the triploid bridge contributes significantly to autotetraploid formation regardless of the mating system, but is important in allotetraploid formation only in outcrossing taxa. The striking effect of the mating system on the contribution of the triploid bridge to allotetraploid formation is attributed to the higher frequency of unreduced gametes in hybrids than in nonhybrids. Although tetraploid formation is higher in selfing than in outcrossing triploids (Figures 3, 5), this increase is greatly overshadowed by the high frequency of allotetraploids formed by selfing F_1 hybrids (Figures 4, 5). Thus the relative contribution of the triploid bridge is low in selfing allopolyploid systems. In contrast, unless there is a correlation between $2n$ pollen and $2n$ egg, selfing will not increase the frequency of autotetraploid formation via the one-step pathway. Hence, selfing increases the role of the triploid bridge in autopolyploid systems. Overall, the triploid bridge contributes equally to tetraploid formation in outcrossers for both pathways, but has a greater role in selfing auto- than allopolyploid systems.

The estimated frequency of hybridization required for equal rates of tetraploid formation via auto- and allopolyploidy is 0.002 for selfing, and 0.027 for backcrossing (Table 1). The very high frequency of F_1 hybrids required for equivalent rates of auto- and allotetraploid formation in outcrossing taxa seems unlikely in most natural populations. The level of hybridization required for equivalent rates of auto- and allotetraploid formation is much lower in selfing taxa, and this may provide a mechanistic explanation for the observation that self-pollination is the predominant mating system in allopolyploids (158). However, highly selfing taxa will likely have a low frequency of hybridization, so the maximum rate of allotetraploid formation may be expected in taxa with intermediate mating systems.

To estimate the absolute rates of allotetraploid formation requires information on the per generation rate of F_1 hybrid formation in natural plant populations. Surprisingly, despite the widely held view that interspecific hybridization is common in plants (6, 141, but see 48), there are very few estimates of the frequency of F_1 hybrids. The only data for polyploid systems are an estimate of 0.02% F_1 hybrids in mixed populations of *Senecio vulgaris* and *Senecio squalidus* (109), the progenitors of the allohexaploid *Senecio cambrensis* (145).

Although more data are clearly needed, our results suggest that the frequency of interspecific hybridization required for equivalent rates of auto- and allotetraploid formation is quite high, particularly for outcrossing taxa (Table 1). We conclude that the rate of autopolyploid formation may often be higher than the rate of allopolyploid formation, and that autopolyploidy represents a significant pathway of polyploid formation. The lower relative frequency of autopolyploids reported by many authors (24, 56, 157) may therefore be due more to constraints on the establishment of autopolyploids than to rates of formation (34). Alternatively, the difficulty of detecting autopolyploidy may have biased early estimates of the importance of this mode of formation.

CONCLUSIONS AND FUTURE DIRECTIONS

Although there is general agreement that unreduced gametes are the major mechanism of polyploid formation (19, 24, 63, 165), there is a need for detailed studies examining this and other mechanisms (somatic doubling, endopolyploidy, polyspermy) in natural populations. The cytological and embryological techniques required to examine these phenomena are well developed, but not commonly practiced by many of the plant evolutionary biologists of this generation.

Our review of the literature identified several factors that contribute to the overall rate of polyploid formation. Unreduced gametes are important in both auto- and allopolyploid formation, but their low frequency in nonhybrid systems suggests they have a greater influence on the rate of autopolyploid formation. Our finding of higher crossing success in $4x \times 2x$ than in $2x \times 4x$ crosses suggests that the likelihood of autotriploid formation is higher through $2n$ eggs than through $2n$ pollen. This warrants further study, given the emphasis in the literature on the investigation of $2n$ pollen (11, 24, 111, 132). In hybrid systems, the effective frequency of $2n$ gametes is often very high, and this results in a high rate of polyploid formation in F_1 hybrids. Further research is needed to provide estimates of (*a*) the frequency of $2n$ gametes in hybrid and nonhybrid systems, (*b*) the relative contribution of genetic and environmental factors to gametic nonreduction, and (*c*) the magnitude of spatial and temporal variation in the frequency of $2n$ gametes.

Although theoretical models of polyploid formation and establishment have generally assumed that triploids are either inviable or sterile (50, 144), our results indicate that triploids are often semifertile, and contribute to tetraploid formation. Surprisingly, the mean fertility of F_1 hybrids was lower than that of both auto- and allotriploids, suggesting that F_1 hybrid fertility is perhaps a greater barrier to polyploid formation. The triploid bridge contributes significantly to the rate of autopolyploid formation regardless of the mating system, and to allopolyploid formation in outcrossing taxa. Our summary of crossing studies revealed that triploid parents rarely produce triploid progeny, which suggests that polyploid formation via the triploid bridge will be favored in perennial taxa. Taken together, these findings demonstrate that future empirical and theoretical work on polyploid evolution should consider the impact of the triploid bridge on polyploid formation and establishment.

Aneuploids are commonly observed in cytological studies of polyploid systems (121, 124), but their role in polyploid formation is not well understood. Issues that need to be addressed include the effects of aneuploidy on viability and fertility, and the consequences of crosses between different euploid and aneuploid cytotypes for polyploid formation.

The estimated rate of autopolyploid formation in both selfing and outcrossing taxa is on the order of the genic mutation rate. New autopolyploids may possess novel physiological, ecological, or phenological characteristics (95) that allow them to colonize a new niche, and are often wholly or partly reproductively isolated from their diploid progenitors. Thus, autopolyploidy may represent a rapid means of adaptation and speciation. The frequency of interspecific hybridization required for equal rates of auto- and allotetraploid formation is so high for outcrossing taxa (2.7%) that this pathway of allopolyploidy is likely only after a breakdown of reproductive isolation, as might be observed in disturbed habitats, or following species introductions. The frequency of interspecific hybridization required for equal rates of auto- and allotetraploid formation (0.2%) is much lower in selfing than in outcrossing taxa, but is still sufficiently high that we suspect the frequency of hybridization is the major rate-limiting step in allopolyploid formation. We suggest that the rate of autopolyploid formation is high and that autopolyploidy is perhaps more common than previously thought, as recent investigations of polyploid taxa have suggested (154, 165).

For both auto- and allopolyploidy, we estimated that the rate of polyploid formation was greater in selfing than in outcrossing taxa. These observations, and the results of theoretical studies showing an advantage to selfing in polyploid establishment (144), demonstrate the importance of the mating system in polyploid evolution. Estimates of the mating system of polyploids and their progenitors are needed to determine the role of self-fertilization in polyploid formation. Phylogenetic studies are of particular interest in this regard to determine if polyploidy is more common in self-compatible lineages.

Despite a long history of cytological and biosystematic research documenting the importance of polyploidy in plant evolution, there are remarkably few empirical studies of polyploids in nature. Comprehensive examination of the pathways, mechanisms, and rates of polyploid formation in natural populations is a logical next step toward improving our understanding of plant speciation and adaptation. In addition, many questions remain regarding the establishment and persistence of new polyploids. To what extent are new polyploids adapted to novel ecological niches? How much reproductive isolation exists between new polyploids and their progenitors? Is the probability of establishment of a new polyploid related to its mode of origin? Despite nearly a century of research on polyploid evolution, these and other questions remain largely unresolved.

ACKNOWLEDGMENTS

We thank Toby Bradshaw, Luca Comai, Jerry "King" Coyne, Brian Husband, R. C. Jackson, Roselyn Lumaret, Doug Soltis, and Pam Soltis for comments on the manuscript, Barb Best for composing the web tables, and Chris Oakley and Kay Suiter for assistance with tabulating references. This material is based on work supported under a National Science Foundation Graduate Fellowship.

> Visit the *Annual Reviews home page* at
> http://www.AnnualReviews.org

Literature Cited

1. Abdel-Hameed F, Snow R. 1972. The origin of the allotetraploid *Clarkia gracilis*. *Evolution* 26:74–83
2. Ahloowalia BS, Garber FD. 1961. The genus *Collinsia*. XIII. Cytogenetic studies of interspecific hybrids involving species with pediceled flowers. *Bot. Gaz.* 122:219–28
3. Alexander DE, Beckett JB. 1963. Spontaneous triploidy and tetraploidy in maize. *J. Hered.* 54:103–6
4. Anderson E. 1936. A morphological comparison of triploid and tetraploid interspecific hybrids in *Tradescantia*. *Genetics* 21:61–65
5. Anderson E. 1949. *Introgressive Hybridization*. New York: Wiley
6. Arnold ML. 1997. *Natural Hybridization and Evolution*. New York: Oxford Univ. Press
7. Ashton PA, Abbott RJ. 1992. Multiple origins and genetic diversity in the newly arisen allopolyploid species, *Senecio cambrensis* Rosier (Compositae). *Heredity* 68:25–32
8. Avers CJ. 1954. Chromosome behavior in fertile triploid Aster hybrids. *Genetics* 39:117–26
9. Ballington JR, Galletta GJ. 1976. Potential fertility levels in four diploid

Vaccinium species. *J. Am. Soc. Hortic. Sci.* 101:507–9
10. Beal JM, Ownbey M. 1943. Cytological studies in relation to the classification of the genus *Calochortus*. III. *Bot. Gaz.* 104:553–62
11. Belling J. 1925. The origin of chromosomal mutations in *Uvularia*. *J. Genet.* 15:245–66
12. Belling J, Blakeslee AF. 1922. The assortment of chromosomes in triploid daturas. *Am. Nat.* 56:339–46
13. Belling J. Blakeslee AF. 1924. The distribution of chromosomes in tetraploid Daturas. *Am. Nat.* 58:60–70
14. Bergstrom I. 1938. Tetraploid apple seedlings obtained from the progeny of triploid varieties. *Hereditas* 21:383–93
15. Bergstrom I. 1940. On the progeny of diploid × triploid *Populus tremula* with special reference to the occurrence of tetraploidy. *Hereditas* 26:191–201
16. Bingham ET. 1968. Aneuploids in seedling populations of tetraploid alfalfa, *Medicago sativa* L. *Crop Sci.* 8:571–74
17. Bradshaw HD Jr, Stettler FR. 1993. Molecular genetics of growth and development in *Populus*. I. Triploidy in hybrid poplars. *Theor. Appl. Genet.* 86:301–7
18. Bretagnolle F, Lumaret R. 1995. Bilateral

polyploidization in *Dactylis glomerata* L. subsp. *lusitanica*: occurrence, morphological and genetic characteristics of first polyploids. *Euphytica* 84:197–207

19. Bretagnolle F, Thompson JD. 1995. Tansley review no. 78. Gametes with the somatic chromosome number: mechanisms of their formation and role in the evolution of atotpolyploid plants. *New Phytol.* 129:1–22

20. Brewbaker JL. 1958. Self-compatibility in tetraploid strains of *Trifolium hybridum*. *Hereditas* 44:547–53

21. Buxton BH, Newton WCF. 1928. Hybrids of *Digitalis ambigua* and *Digitalis purpurea*, their fertility and cytology. *J. Genet.* 19:1269–79

22. Chambers K. 1955. A biosystematic study of the annual species of *Microseris*. *Contrib. Dudley Herb.* 4:207–312

23. Clausen J, Keck DD, Hiesey WM. 1940. Experimental studies on the nature of species. I. Effect of varied environments on western North American plants. *Carnegie Inst. Wash. Publ. 520*

24. Clausen J, Keck DD, Hiesey WM. 1945. Experimental studies on the nature of species. II. Plant evolution through amphiploidy and autoploidy, with examples from the Madiinae. *Carnegie Inst. Wash. Publ. 564*

25. Clausen RE, Goodspeed TH. 1925. Interspecific hybridization in *Nicotiana*. II. A tetraploid *glutinosa-Tabacum* hybrid, an experimental verification of Winge's hypothesis. *Genetics* 10:278–84

26. Coleman L. 1950. Nuclear conditions in normal stem tissue of *Vicia faba*. *Can. J. Res.* 28:382–91

27. Collins JL. 1933. Morphological and cytological characteristics of triploid pineapples. *Cytologia* 4:248–56

28. Cooper DC. 1951. Caryopsis development following matings between diploid and tetraploid strains of *Zea mays*. *Am. J. Bot.* 38:702–8

29. Cutter GL, Bingham ET. 1977. Effect of soybean male-sterile gene msl on organization and function of the female gametophyte. *Crop Sci.* 17:760–64

30. D'Amato F. 1952. Polyploidy in the differentiation and function of tissues and cells in plants. *Caryologia* 4:311–58

31. D'Amato F. 1964. Endopolyploidy as a factor in plant tissue development. *Caryologia* 17:41–52

32. Darlington CD. 1932. *Recent Advances in Cytology*. London: Churchill.

33. Darlington CD. 1937. *Recent Advances in Cytology*. London: Churchill. 2nd ed.

34. Darlington CD. 1963. *Chromosome Botany and the Origins of Cultivated Plants*. New York: Hafner

35. Day A. 1965. The evolution of a pair of sibling allotetraploid species of Cobwebby Gilias (Polemoniaceae). *Aliso* 6:25–75

36. De Haan A, Maceira NO, Lumaret R, Delay J. 1992. Production of 2n gametes in diploid subspecies of *Dactylis glomerata* L. 2. Occurrence and frequency of 2n eggs. *Ann. Bot.* 69:345–50

37. De Mol WE. 1923. Duplication of generative nuclei by means of physiological stimuli and its significance. *Genetica* 5:225–72

38. De Mol WE. 1929. The originating of diploid and tetraploid pollen-grains in duc van thol-tulips (*Tulipa suaveolens*) dependent on the method of culture applied. *Genetica* 11:119–212

39. de Wet JMJ. 1980. Origins of polyploids. In *Polyploidy–Biological Relevance*, ed. W. H. Lewis, pp. 3–15. New York: Plenum

40. Dermen H. 1931. Polyploidy in petunia. *Am. J. Bot.* 18:250–61

41. DeVries H. 1915. The coefficient of mutation in *Oenothera biennis* L. *Bot. Gaz.* 59:169–96

42. Dewey DR, Asay KH. 1975. The crested wheatgrasses of Iran. *Crop Sci.* 15:844–49

43. Dorsey E. 1936. Induced polyploidy in wheat and rye. Chromosome doubling in *Triticum, Secale* and *Triticum-Secale* hybrids produced by temperature changes. *J. Hered.* 27.155–60

44. Dujardin M, Hanna WW. 1988. Cytology and breeding behavior of a partially fertile triploid pearl millet. *J. Hered.* 79:216-18

45. Dweikat IM, Lyrene PM. 1988. Production and viability of unreduced gametes in triploid interspecific blueberry hybrids. *Theor. Appl. Genet.* 76:555–59

46. Eckenwalder JE, Brown BP. 1986. Polyploid speciation in hybrid morning glories of *Ipomoea* L. sect. *Quamoclit* Griseb. *Can. J. Genet. Cytol.* 28:17–20

47. Einset J. 1952. Spontaneous polyploidy in cultivated apples. *Am. Soc. Hort. Sci.* 59:291–302

48. Ellstrand NC, Whitkus R, Rieseberg LH. 1996. Distribution of spontaneous plant hybrids. *Proc. Nat. Acad. Sci. USA* 93: 5090–93

49. Esen A, Soost RK. 1973. Seed development in *Citrus* with special reference to $2x \times 4x$ crosses. *Am. J. Bot.* 60:448–62

50. Felber F. 1991. Establishment of a tetraploid cytotype in a diploid population: effect of relative fitness of the cytotypes. *J. Evol. Biol.* 4:195–207

51. Gairdner AE, Darlington CD. 1931. Ring-formation in diploid and polyploid *Campanula persicifolia*. *Genetica* 13:113–50

52. Gajewski W. 1953. A fertile amphipolyploid hybrid of *Geum rivale* with *G. macrophyllum*. *Acta Soc. Bot. Pol.* 22:411–39

53. Gajewski W. 1954. An amphiploid hybrid of *Geum urbanum* L. and *G. molle* Vis. et Panc. *Acta Soc. Bot. Pol.* 23:259–78

54. Giles N. 1941. Chromosome behavior at meiosis in triploid *Tradescantia* hybrids. *Bull. Torrey Bot. Club* 68:207–21

55. Grant V. 1952. Cytogenetics of the hybrid *Gilia millefoliata* × *achilleaefolia*. I. Variations in meiosis and polyploidy rate as affected by nutritional and genetic conditions. *Chromosoma* 5:372–90

56. Grant V. 1981. *Plant Speciation*. New York: Columbia Univ. Press. 2nd ed.

57. Griffiths AJF, Miller JH, Suzuki DT, Lewontin RC, Gelbart WM. 1996. *An Introduction to Genetic Analysis*. New York: Freeman. 6th ed.

58. Gustafsson Å. 1948. Polyploidy, life-form, and vegetative reproduction. *Hereditas* 34:1–22

59. Hagerup O. 1947. The spontaneous formation of haploid, polyploid and aneuploid embryos in some orchids. *Kongel. Danske Videnskab. Selskab Biol. Meddelelser* 20:1–22

60. Hahn SK, Bai KV, Asiedu R. 1990. Tetraploids, triploids, and 2n pollen from diploid interspecific crosses with cassava. *Theor. Appl. Genet.* 79:433–39

61. Haig D, Westoby M. 1991. Genomic imprinting in endosperm: its effect on seed development in crosses between species, and between different ploidies of the same species, and its implications for the evolution of apomixis. *Philos. Trans. R. Soc. London Ser. B* 333:1–13

62. Håkansson A, Ellerstrom S. 1950. Seed development after reciprocal crosses between diploid and autotetraploid rye. *Hereditas* 36:256–96

63. Harlan JR, deWet JMJ. 1975. On Ö. Win ge and a prayer: the origins of polyploidy. *Bot. Rev.* 41:361–90

64. Deleted in proof

65. Heckard LR. 1960. Taxonomic studies in the *Phacelia magellanica* polyploid complex. *Univ. Calif. Publ. Bot.* 32:1–126

66. Hiesey WM, Nobs MA, Björkman O. 1971. Experimental studies on the nature of species. V. Biosystematics, genetics, and physiological ecology of the *Erythranthe* section of *Mimulus*. *Carnegie Inst. Wash. Publ.* 628:1–213

67. Hollingshead L. 1930. Cytological investigations of hybrids and hybrid derivatives of *Crepis capillaris* and *Crepis tectorum*. *Univ. Calif. Publ. Agric. Sci.* 6:55–94

68. Hornsey KG. 1973. The occurrence of hexaploid plants among autotetraploid populations of sugar beet (*Beta vulgaris* L.), and the production of tetraploid progeny using a diploid pollinator. *Caryologia* 26:225–28

69. Howard HW. 1939. The size of seeds in diploid and autotetraploid *Brassica oleracea* L. *J. Genet.* 38:325–40

70. Hunziker JH. 1962. The origin of the hybrid triploid willows cultivated in Argentina. *Silvae Genet.* 11:151–53

71. Jackson RC. 1982. Polyploidy and diploidy: new perspectives on chromosome pairing and its evolutionary implications. *Am. J. Bot.* 69:1512–23

72. Jackson RC, Casey J. 1982. Cytogenetic analysis of autopolyploids: models and methods for triploids to octoploids. *Am. J. Bot.* 69:487–501

73. Janick J, Stevenson EC. 1955. The effects of polyploidy on sex expression in spinach. *J. Hered.* 46:151–56

74. Johnsson H. 1940. Cytological studies of diploid and triploid *Populus tremula* and of crosses between them. *Hereditas* 26:321–51

75. Johnsson H. 1945. Chromosome numbers of the progeny from the cross triploid × tetraploid *Populus tremula*. *Hereditas* 31:500–1

76. Johnsson H. 1945. The triploid progeny of the cross diploid × tetraploid *Populus tremula*. *Hereditas* 31:411–40

77. Johnston SA, Nijs TPM, Peloquin SJ, Hanneman RE Jr. 1980. The significance of genic balance to endosperm development in interspecific crosses. *Theor. Appl. Genet.* 57:5–9

78. Jones A. 1990. Unreduced pollen in a wild tetraploid relative of sweetpotato. *J. Am. Soc. Hortic. Sci.* 115:512–16

79. Jones HA, Clarke AF. 1942. A natural amphidiploid from an onion species hybrid. *J. Hered.* 33:25–32

80. Jones RE, Bamford R. 1942. Chromosome number in the progeny of triploid gladiolus with special reference to the contribution of the triploid. *Am. J. Bot.* 29:807–13

81. Jongedijk E, Ramanna MS, Sawor Z, Hermsen JGT. 1991. Formation of first division restitution (FDR) 2n-megaspores through pseudohomotypic division in ds-1 (desynapsis) mutants of diploid potato: routine production of tetraploid

progeny from 2xFDR × 2xFDR crosses. *Theor. Appl. Genet.* 82:645–56

82. Jørgensen CA. 1928. The experimental formation of heteroploid plants in the genus *Solanum. J. Genet.* 11:133–210

83. Karpechenko GD. 1927. The production of polyploid gametes in hybrids. *Hereditas* 9:349–68

84. Khoshoo TN. 1959. Polyploidy in gymnosperms. *Evolution* 13:24–39

85. Khoshoo TN, Sharma VB. 1959. Cytology of the autotriploid *Allium rubellum. Chromosoma* 10:136–43

86. Kihara H, Ono T. 1926. Chromosomenzahlen und systematische Gruppierung der Rumex-Arten. *Z. Zellforsch. Mikr. Anat.* 4:475–81

87. King E. 1933. Chromosome behavior in a triploid *Tradescantia. J. Hered.* 24:253–56

88. Kostoff D. 1933. A contribution to the sterility and irregularities in the meiotic processes caused by virus diseases. *Genetica* 15:103–14

89. Kostoff D, Kendall J. 1931. Studies of certain *Petunia* aberrants. *J. Genet.* 24:165–78

90. Lammerts WE. 1931. Interspecific hybridization in *Nicotiana.* XII. The amphidiploid *rustica-paniculata* hybrid: its origin and cytogenetic behavior. *Genetics* 16:191–211

91. Lange W, Wagenvoort M. 1973. Meiosis in triploid *Solanum tuberosum* L. *Euphytica* 22.8–18

92. Levan A. 1937. Cytological studies in the *Allium paniculatum* group. *Hereditas* 23:317–70

93. Levan A. 1937. Polyploidy and self-fertility in *Allium. Hereditas* 22:278–80

94. Levan A. 1941. The cytology of the species hybrid *Allium cepa × fistulosum* and its polyploid derivatives. *Hereditas* 27:253–72

95. Levin DA. 1983. Polyploidy and novelty in flowering plants. *Am. Nat.* 122:1–25

96. Lewis D. 1943. The incompatibility sieve for producing polyploids. *J. Genet.* 45:261–64

97. Lewis D. 1949. Incompatibility in flowering plants. *Biol. Rev.* 24:472–96

98. Lewis H, Szweykowski J. 1964. The genus *Gayophytum* (Onagraceae). *Brittonia* 16:343–91

99. Lewis WH. 1980. Polyploidy in species populations. In *Polyploidy: Biological Relevance*, ed. WH Lewis, pp. 103–44. New York: Plenum

100. Lewis WH, ed. 1980. *Polyploidy: Biological Relevance*. New York: Plenum

101. Li HW, Yang KKS, Ho K-C. 1964. Cyto-
genetical studies of *Oryza sativa* L. and its related species. *Bot. Bull. Acad. Sinica* 5:142–53

102. Lin B-Y. 1984. Ploidy barrier to endosperm development in Maize. *Genetics* 107:103–15

103. Lindstrom EW, Humphrey LM. 1933. Comparative cyto-genetic studies of tetraploid tomatoes from different origins. *Genetics* 18:193–209

104. Lumaret R. 1988. Adaptive strategies and ploidy levels. *Acta Oecol. Oecol. Plant.* 9:83–93

105. Maceira NO, De Haan AA, Lumaret R, Billon M, Delay J. 1992. Production of 2n gametes in diploid subspecies of *Dactylis glomerata* L. 1. Occurrence and frequency of 2n pollen. *Ann. Bot.* 69:335–43

106. Marchant CJ. 1963. Corrected chromosome numbers for *Spartina × townsendii* and its parent species. *Nature* 19:929

107. Marchant CJ, Macfarlane RM. 1980. Chromosome polymorphism in triploid populations of *Fritillaria lanceolata* Pursh (Liliaceae) in California. *Bot. J. Linn. Soc.* 81:135–54

108. Marks GE. 1966. The origin and significance of intraspecific polyploidy: experimental evidence from *Solanum chacoense. Evolution* 20:552–57

109. Marshall DF, Abbott RJ. 1980. On the frequency of introgression of the radiate (T,) allele from *Senecio squalidus* L. into *Senecio vulgaris* L. *Heredity* 45:133–35

110. Masterson J. 1994. Stomatal size in fossil plants: evidence for polyploidy in majority of angiosperms. *Science* 264:421–24

111. McHale NA. 1983. Environmental induction of high frequency 2n pollen formation in diploid *Solanum. Can. J. Genet. Cytol.* 25:609–15

112. Miklos GLG, Rubin GM. 1996. The role of the genome project in determining gene function: insights from model organisms. *Cell* 86:521–29

113. Milbocker D, Sink K. 1969. Embryology of diploid × diploid and diploid × tetraploid crosses in poinsettia. *Can. J. Genet. Cytol.* 11:598–601

114. Mok DWS, Peloquin SJ. 1975. Breeding value of 2n pollen (diplandroids) in tetraploid × diploid crosses in potatoes. *Theor. Appl. Genet.* 46

115. Mok DWS, Peloquin SJ. 1975. The inheritance of three mechanisms of diplandroid (2n pollen) formation in diploid potatoes. *Heredity* 35:295–302

116. Mok DWS, Peloquin SJ. 1975. Three mechanisms of 2n pollen formation in diploid potatoes. *Can. J. Genet. Cytol.* 17: 217–25

117. Mok DWS, Peloquin SJ, Tarn TR. 1975. Cytology of potato triploids producing 2n pollen. *Am. Potato J.* 52:171–74

118. Müntzing A. 1930. Outlines to a genetic monograph of the genus *Galeopsis* with special reference to the nature and inheritance of partial sterility. *Hereditas* 13:185–341

119. Müntzing A. 1932. Cyto-genetic investigations on synthetic *Galeopsis tetrahit*. *Hereditas* 16:105–54

120. Müntzing A. 1936. The evolutionary significance of autopolyploidy. *Hereditas* 21:263–378

121. Müntzing A. 1937. The effects of chromosomal variation in *Dactylis*. *Hereditas* 23:8–235

122. Müntzing A. 1938. Note on heteroploid twin plants from eleven genera. *Hereditas* 24:487–91

123. Müntzing A. 1930. Über Chromosomenvermehrung in Galeopsis-Kreuzungen und ihre phylogenetische Bedeutung. *Hereditas* 14:153–72

124. Myers WM, Hill HD. 1940. Studies of chromosomal association and behavior and occurrence of aneuploidy in autotetraploid grass species, orchard grass, tall oat grass, and crested wheatgrass. *Bot. Gaz.* 102:236–55

125. Navashin M. 1925. Polyploid mutations in *Crepis*. Triploid and pentaploid mutants of *Crepis capillaris*. *Genetics* 10:583–92

126. Needham DC, Erickson HT. 1992. Fecundity of tetraploid × diploid crosses and fertility of the resultant triploids in *Salpiglossis sinuata*. *HortScience* 27:835–37

127. Newton WCF, Darlington CD. 1929. Meiosis in polyploids. *J. Genet.* 21:1–56

128. Newton WCF, Pellew C. 1929. *Primula kewensis* and its derivatives. *J. Genet.* 20:405–67

129. Nishiyama I, Inomata N. 1966. Embryological studies on cross incompatibility between 2x and 4x in *Brassica*. *Jpn. J. Genet.* 41:27–42

130. Norrmann G, Quarín C, Keeler K. 1997. Evolutionary implications of meiotic chromosome behavior, reproductive biology, and hybridization in 6x and 9x cytotypes of *Andropogon gerardii* (Poaceae). *Am. J. Bot.* 84:201–8

131. Orjeda G, Freyre R, Iwanaga M. 1990. Production of 2n pollen in diploid *Ipomoea trifida*, a putative wild ancestor of sweet potato. *J. Hered.* 81:462–67

132. Ortiz R, Vorsa N, Bruederle LP, Laverty T. 1992. Occurrence of unreduced pollen in diploid blueberry species, *Vaccinium* sect. *cyanococcus*. *Theor. Appl. Genet.* 85:55–60

133. Ownbey M. 1950. Natural hybridization and amphiploidy in the genus *Tragopogon*. *Am. J. Bot.* 37:487–99

134. Parrott WA, Smith RR. 1984. Production of 2n pollen in red clover. *Crop Sci.* 24:469–72

135. Parrott WA, Smith RR. 1986. Recurrent selection for 2n pollen formation in red clover. *Crop Sci.* 26:1132–35

136. Parrott WA, Smith RR, Smith MM. 1985. Bilateral sexual tetraploidization in red clover. *Can. J. Genet. Cytol.* 27:64–68

137. Poole CF. 1931. The interspecific hybrid, *Crepis rubra* × *C. foetida*, and some of its derivatives. I. *Univ. Calif. Publ. Agric. Sci.* 6:169–200

138. Pratassenja GD. 1939. Production of polyploid plants. *Akad. Nauk (Dokiady) NS SSSR* 22:348–51

139. Punyasingh K. 1947. Chromosome numbers in crosses of diploid, triploid and tetraploid maize. *Genetics* 32:541–54

140. Randolph LF. 1932. Some effects of high temperature on polyploidy and other variations in maize. *Proc. Natl. Acad. Sci. USA* 18:222–29

141. Raven PH. 1976. Systematics and plant population biology. *Syst. Bot.* 1:281–316

142. Rhoades MM, Dempsey E. 1966. Induction of chromosome doubling at meiosis by the elongate gene in maize. *Genetics* 54:505–22

143. Rick CM. 1945. A survey of cytogenetic causes of unfruitfulness in the tomato. *Genetics* 30:347–62

144. Rodríguez DJ. 1996. A model for the establishment of polyploidy in plants. *Am. Nat.* 147:33–46

145. Rosser EM. 1955. A new British species of *Senecio*. *Watsonia* 3:228–32

146. Sandfaer J. 1973. Barley stripe mosaic virus and the frequency of triploids and aneuploids in barley. *Genetics* 73:597–603

147. Satina S, Blakeslee AF. 1935. Cytological effects of a gene in *Datura* which causes dyad formation in sporogenesis. *Bot. Gaz.* 96:521–32

148. Satina S, Blakeslee AF. 1937. Chromosome behavior in triploid *Datura*. II. The female gametophyte. *Am. J. Bot.* 24:621–27

149. Satina S, Blakeslee AF. 1937. Chromosome behavior in triploids of *Datura stramonium*. I. The male gametophyte. *Am. J. Bot.* 24:518–27

150. Sato T, Maciera M, Lumaret R, Jacquard P. 1993. Flowering characteristics and fertility of interploidy progeny from normal

and 2n gametes in *Dactylis glomerata* L. *New Phytol.* 124:309–19

151. Sax K. 1936. The experimental production of polyploidy. *J. Arnold Arbor.* 17: 153–59

152. Skalińska M. 1945. Cytogenetic studies in triploid hybrids of *Aquilegia*. *J. Genet.* 47:87–111

153. Skalinska M. 1946. Polyploidy in *Valeriana officinalis* Linn. in relation to its ecology and distribution. *J. Linn. Soc. London Bot.* 53:159–86

154. Soltis DE, Soltis PS. 1992. Molecular data and the dynamic nature of polyploidy. *Crit. Rev. Plant Sci.* 12:243–73

155. Stebbins GL. 1938. Cytological characteristics associated with the different growth habits in the dicotyledons. *Am. J. Bot.* 25:189–98

156. Stebbins GL. 1947. Types of polyploids: their classification and significance. *Adv. Genet.* 1:403–29

157. Stebbins GL. 1950. *Variation and Evolution in Plants*. New York: Columbia Univ. Press

158. Stebbins GL. 1957. Self fertilization and population variability in the higher plants. *Am. Nat.* 91:337–54

159. Stebbins GL. 1971. *Chromosomal Evolution in Higher Plants*. London: Addison-Wesley

160. Stebbins GL. 1980. Polyploidy in plants: unsolved problems and prospects. In *Polyploidy: Biological Relevance*, ed. WH Lewis. New York: Plenum

161. Stelly DM, Peloquin SJ. 1985. Screening for 2n female gametophytes, female fertility, and 2x × 4x crossability in potatoes (*Solanum* spp.). *Am. Potato J.* 62:519–29

162. Stephens SG. 1942. Colchicine-produced polyploids in *Gossypium*. 1. An autotetraploid Asiatic cotton and certain of its hybrids with wild diploid species. *J. Genet.* 44:272–95

163. Storey WB. 1956. Diploid and polyploid gamete formation in orchids. *Proc. Am. Soc. Hortic. Sci.* 68:491–502

164. Tavoletti S, Mariani A, Veronesi F. 1991. Phenotypic recurrent selection for 2n pollen and 2n egg production in diploid alfalfa. *Euphytica* 57:97–102

165. Thompson JD, Lumaret R. 1992. The evolutionary dynamics of polyploid plants: origins, establishment and persistence. *Trends Ecol. Evol.* 7:302–7

166. Thompson RC. 1942. An amphidiploid *Lactuca*. *J. Hered.* 33:253–64

167. Thompson WP. 1931. Cytology and genetics of crosses between fourteen- and seven-chromosome species of wheat. *Genetics* 16:309–24

168. Tyagi BR. 1988. The mechanism of 2n pollen formation in diploids of *Costus speciosus* (Koenig) J. E. Smith and role of sexual polyploidization in the origin of intraspecific chromosomal races. *Cytologia* 53:763–70

169. U N. 1935. Genome-analysis in *Brassica* with special reference to the experimental formation of *B. napus* and peculiar mode of fertilization. *Jpn. J. Bot.* 7:389–457

170. Upcott M, Philp J. 1939. The genetic structure of *Tulipa* IV. Balance, selection and fertility. *J. Genet.* 38:91–123

171. Vardi A, Zohary D. 1967. Introgression in wheat via triploid hybrids. *Heredity* 22: 541–60

172. Vigfússon E. 1970. On polyspermy in the sunflower. *Hereditas* 64:1–52

173. Wagenaar EB. 1968. Meiotic restitution and the origin of polyploidy I. Influence of genotype on polyploid seedset in a *Triticum crassum* × *T. turgidum* hybrid. *Can. J. Genet. Cytol.* 10:836–43

174. Warmke HE, Blakeslee AF. 1940. The establishment of a 4n dioecious race in *Melandrium*. *Am. J. Bot.* 27:751–62

175. Watkins AE. 1932. Hybrid sterility and incompatibility. *J. Genet.* 25:125–62

176. Webber JM. 1940. Polyembryony. *Bot. Rev.* 6:575–98

177. Werner JE, Peloquin SJ. 1991. Occurrence and mechanisms of 2n egg formation in 2x potato. *Genome* 34:975–82

178. Woodell SRJ, Valentine DH. 1961. Studies in British primulas. IX. Seed incompatibility in diploid-autotetraploid crosses. *New Phytol.* 60:282–94

179. Zohary D, Nur U. 1959. Natural triploids in the orchard grass, *Dactylis glomerata*, L., polyploid complex and their significance for gene flow from diploid to tetraploid levels. *Evolution* 13:311–17

Annu. Rev. Ecol. Syst. 1998. 29:503–41

BACTERIAL GROWTH EFFICIENCY IN NATURAL AQUATIC SYSTEMS

Paul A. del Giorgio[1] and Jonathan J. Cole[2]
[1]Horn Point Laboratory, University of Maryland Center for Environmental Science, P.O. Box 775, Cambridge, Maryland 21613 and [2]Institute of Ecosystem Studies, Box AB, Millbrook, New York 12545-0129

KEY WORDS: bacteria, plankton, respiration, production, organic carbon

ABSTRACT

Heterotrophic bacteria perform two major functions in the transformation of organic matter: They produce new bacterial biomass (bacterial secondary production [BP]), and they respire organic C to inorganic C (bacterial respiration [BR]). For planktonic bacteria, a great deal has been learned about BP and its regulation during the past several decades but far less has been learned about BR. Our lack of knowledge about BR limits our ability to understand the role of bacteria in the carbon cycle of aquatic ecosystems. Bacterial growth efficiency (BGE) is the amount of new bacterial biomass produced per unit of organic C substrate assimilated and is a way to relate BP and BR: BGE = (BP)/(BP + BR). Estimates of BGE for natural planktonic bacteria range from <0.05 to as high as 0.6, but little is known about what might regulate this enormous range. In this paper we review the physiological and ecological bases of the regulation of BGE. Further, we assemble the literature of the past 30 years for which both BP and BR were measured in natural planktonic ecosystems and explore the relationship between BGE and BP. Although the relationship is variable, BGE varies systematically with BP and the trophic richness of the ecosystem. In the most dilute, oligotrophic systems, BGE is as low as 0.01; in the most eutrophic systems, it plateaus near 0.5. Planktonic bacteria appear to maximize carbon utilization rather than BGE. A consequence of this strategy is that maintenance energy costs (and therefore maintenance respiration) seems to be highest in oligotrophic systems.

0066-4162/98/1120-0503$08.00

INTRODUCTION

Bacteria are the most abundant and most important biological component involved in the transformation and mineralization of organic matter in the biosphere (20, 109, 155). Heterotrophic bacteria contribute to the cycles of nutrients and carbon in two major ways: by the production of new bacterial biomass (secondary production) and by the remineralization of organic carbon and nutrients. Understanding this dual character of planktonic bacteria in aquatic ecosystems is a central paradigm of contemporary microbial ecology (12, 37, 109). Much of the primary production in aquatic ecosystems is ultimately processed by planktonic bacteria. Comparative studies of a wide range of natural aquatic systems show that planktonic bacterial production is correlated with and averages about 30% of net primary production (NPP) (26, 37). The real magnitude of organic carbon flow through bacterioplankton remains largely unknown, however, because measurements of bacterial production are seldom accompanied by measurements of bacterial respiration (BR) (64). The amount of organic C assimilated by bacteria (A) is the sum of bacterial secondary production (BP) and BR. In most studies of organic carbon flow in aquatic ecosystems, this respiration term is derived from measurements of BP and assumed values of bacterial growth efficiency (BGE). BGE is defined as the ratio of BP to A. Thus, BGE $= BP/(BP + BR) = BP/A$.

Assumed values of BGE are often based on early measurements made with simple radiolabeled organic compounds (29, 59), and these values are now widely regarded as overestimates of the real growth efficiency of natural bacterioplankton that utilize complex natural substrates (17, 24, 32, 64, 83). Relative to the large body of data that has been gathered on BP and other microbial processes in the last 20 years, surprisingly little is known about BR and BGE and their regulation in natural systems. Our lack of knowledge is due to two factors. First, BR is simply more difficult to measure accurately than is BP. Second, there has been a general belief that rates of catabolism and anabolism are tightly coupled and that maximum efficiency and economy are achieved during growth. One of the ideas that we develop here is that catabolism and anabolism are not well coupled. This uncoupling provides bacteria with the metabolic flexibility necessary to cope with the vicissitudes of a largely oligotrophic and ever-changing environment.

In this review, we attempt to synthesize the results of research on bacterioplankton growth efficiency done in the last 30 years, focusing on data from natural ecosystems. Because most research on bacterial energetics has been conducted on pure bacterial cultures growing on defined media, we also briefly review current paradigms in microbial energetics. We then assemble direct measurements of BP, BR, and BGE taken from the literature and try to synthesize the state of knowledge of BGE and the factors that regulate it in natural aquatic ecosystems. This review focuses on aerobic, planktonic, heterotrophic bacteria. These bacteria utilize organic compounds to derive both their energy and carbon requirements and are responsible for the bulk of microbial biomass and activity in the water column of lakes and oceans.

Conceptual Framework

By definition, growth efficiency (or yield—here used interchangeably) is the quantity of biomass synthesized per unit of substrate assimilated. In the process of growth, various compounds, elements, and minerals are converted into cell material at the expense of the energy source (Figure 1). An organic substrate

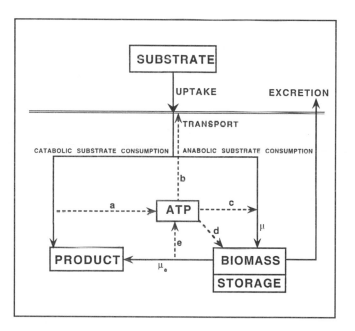

Figure 1 Simplified depiction of catabolic and anabolic pathways that influence BGE in aquatic bacteria. The oxidation of organic compounds contributes to the energy pool as ATP at a rate *a*. Active transport of substrates into the cell requires energy from this ATP pool at a rate *b*; anabolic reactions utilize ATP at a rate *c* and result in a growth rate μ. The anabolic pathways result not only in increases in biomass but also in storage products and organic compounds that may be excreted back to the medium. Maintenance expenditures consume ATP at rate *d*. In the absence of exogenous substrates, minimum maintenance energy requirements must be supported by degradation of biomass through endogenous metabolism (μe), which supplies ATP at a rate *e*. Endogenous metabolism is defined here as the state when no growth is possible, and by definition BGE is 0 under these conditions (116, 117). Adapted from Reference 6.

taken up by a bacterial cell will be partly used in catabolic reactions to generate ATP and partly used in anabolic reactions for biomass synthesis (31) (Figure 1). The purpose of this scheme is to emphasize that multiple processes determine growth efficiency. Each process may respond to a different set of controlling factors; the energy and carbon content of the organic substrate may determine whether it will be preferentially catabolized or incorporated into structural components; energy expenditures in active transport may be related to the concentration, variety, and nature of the exogenous substrates as well as to growth rate; excretion may be the result of energy-spilling reactions or may be an active process, as in the case of the production of exoenzymes; and storage may be a function of the physiological conditions of cells and also of the nature of organic substrate that is utilized. Thus, it is unlikely that growth efficiency in bacteria is regulated by a single factor.

Growth Efficiency in Bacterial Cultures with Defined Media

We cannot provide all the information on microbial bioenergetics in pure cultures, and so we refer the reader to several excellent reviews (100, 101, 117, 135, 140). In brief, there has been a continuous search for regular behavior in BGE by normalizing yield to substrate consumed (94) or to energy produced from the substrate (Y_{ATP} sensu Bauchop & Elsden [4]). The modern view, based on continuous-culture techniques, suggests that BGE is not constant, regardless of the parameter to which it is normalized (6, 116, 117, 133, 134). Continuous-culture techniques allow the growth rate to be varied over a wide range by varying the dilution rate. Whereas unconstrained growth of bacteria in batch culture often led to a rather constant yield for any given substrate, the constraining conditions of chemostat culture could provoke an enhanced rate of catabolism and variable and lower yields (134, 140). As experimental data accumulated, it become clear that Y_{ATP} was also not constant but varied at least sixfold around the presumably fixed value postulated by Bauchop & Elsden (4). Furthermore, theoretical calculations of the amount of energy (as ATP) that would be needed to synthesize bacterial biomass showed that measured bacterial yields in virtually all cases were much lower than those predicted from biochemical pathways (133). Subsequent research has confirmed that yields, whether based on substrate consumption, energy production, or thermodynamic efficiency, are usually at least 50% below expected values (57, 100–102), even in energy- and nutrient-sufficient cultures. Some of the variation and the generally low BGE can be explained by maintenance requirements.

MAINTENANCE ENERGY   Metabolic energy is distributed between two kinds of demands: the demands of biosynthetic processes that produce a net increase in biomass, and the demands of processes that do not (for example, regulation of internal pH and osmotic pressure, macromolecular turnover, membrane energization, and motility [102]). A common assumption is that the rates of energy demands of biosynthetic processes change in a continuous manner with specific growth rate while the energy demand of maintenance processes remains constant (116). Thus, maintenance energy becomes an increasingly greater fraction of the total energy flow in the cell at low growth rates, and growth efficiency declines. However, the concept of a constant maintenance energy requirement has been repeatedly challenged in recent years, because it has been experimentally shown that values vary by more than 30-fold (6, 135, 138–140). Variations in maintenance requirements have often been difficult to explain but are taken as evidence that bacteria often utilize large amounts of energy in reactions that are not directly related to growth, particularly when growth itself is constrained (117).

COUPLING BETWEEN CATABOLISM AND ANABOLISM   Results from experimental studies suggest that when growth is unconstrained, as in batch cultures, there is often a high degree of coupling between catabolism and anabolism. When growth is constrained by the supply of organic substrate or inorganic nutrients, as in most chemostat studies and certainly in most natural situations, different degrees of uncoupling are invariably observed. Washed suspensions of bacteria, for example, oxidize substrates such as glucose at a high rate under conditions at which cell synthesis is severely impeded (138). This uncoupling is manifested in various ways: high rates of oxygen and organic substrate consumption, metabolite overproduction and excretion, excess heat production, and energy-spilling pathways (117). All these processes result in reduced growth efficiency. Whereas growth is dependent on efficiency (i.e. cells must consume nutrients to grow), the reverse does not follow: Cells do not have to grow to consume carbon substrate. Anomalies in BGE are often found at low growth rates when growth is limited by some substrate other than the energy source. In general, catabolism appears to proceed at the maximum rate at which the organisms are capable under the conditions, irrespective of whether the energy so produced can be used for biosynthesis (116). Under conditions of severe constraints to growth, it has been suggested that maintaining the highest possible flow of energy would be advantageous (116, 117, 140). One of the potential advantages of a high energy flux in the cell may be to maintain the energization of cell membranes and the function of active transport systems, both of which are essential conditions for resumption of growth whenever environmental conditions change (31, 96). The conclusion that it is advantageous and even necessary for bacteria to maintain a high flow of energy is supported by thermodynamic analysis of microbial energetics, which suggests that microbial growth efficiency is usually low but is optimal for maximal growth (152).

Years of experimental work on microbial energetics have shown that even under the simplest culture conditions it is often difficult to predict bacterial growth efficiency. The difficulty in predicting BGE stems from the fact that bacteria can alter the coupling between catabolism and anabolism to maximize growth according to the conditions (117, 139). These considerations are particularly relevant in our interpretation of bacterioplankton energetics, because planktonic bacteria occupy an extremely dilute environment, in which energy, carbon, and other nutrients are often limiting and growth is usually slow. Therefore, it is expected that maintenance energy requirements, as well the nature of the organic substrates utilized by bacteria and the availability of nutrients, would play a significant role in determining BGE and that bacterioplankton should generally be in a region of low BGE. It is also expected that planktonic bacteria should exhibit a relatively large degree of uncoupling between catabolism and anabolism compared to their cultured counterparts. As discussed below, the data from natural aquatic systems generally support these expectations.

BGE IN NATURAL AQUATIC SYSTEMS

Measuring BGE

Measurement of BGE continues to be a challenge to microbial ecologists. Early studies monitored the uptake, incorporation, and respiration of simple radiolabeled compounds (59). The advantage of this approach is its high sensitivity, which allows rates of uptake and respiration to be measured in short incubations even in the most unproductive aquatic systems. The main disadvantage is that during these short incubations, the intracellular carbon pools may not attain equilibrium and so BGE can be grossly overestimated (14, 73). In addition, the single model compounds may not be representative of the range of substrates utilized by bacteria in nature. The use of single radiolabeled compounds has largely been replaced by techniques that attempt to measure the BGE of bacteria utilizing the in situ pool of organic matter. Two main approaches are used for this purpose.

1. The first is simultaneous measurements of BR and BP in relatively short (usually <36 h) incubations (11, 19, 24, 30, 49, 81, 107). There are two difficulties here. First, although BP can be measured in an incubation of <1 h, obtaining a measurable change in BR can take 24 h or more depending on the system. Second, bacteria must be physically separated from other planktonic components. This is usually attempted by filtration in the 2- to 0.6-μm range. Complete separation is seldom achieved, so a variable fraction of the measured BR is due to organisms other than bacteria. In addition, filtration disrupts the structure of the bacterial assemblage. Organic C consumption is

approximated as the sum of BP and BR. BR is generally measured as O_2 consumption (11, 19, 49) or, more rarely, as CO_2 production (54). BP is generally measured from the rate of protein or DNA synthesis, using radiolabeled leucine or thymidine, although in some studies the changes in bacterial abundance and size are monitored.

2. The second approach is dilution culture, in which filter-sterilized water is reinoculated with a small amount of the native bacterial assemblage and the subsequent growth of these bacteria is monitored, generally for days or weeks (14, 17, 80, 144, 156). In this type of long-term experiment, it is possible to monitor the changes in dissolved organic carbon (DOC) and particulate organic carbon (POC), and BGE is then calculated as $\Delta DOC/\Delta POC$. The obvious difficulty is the exceedingly long incubation and possible deviation from natural conditions.

There have been no explicit comparisons of BGE estimated from short- and long-term experiments, and both approaches have problems. Whichever approach is taken, bacteria are isolated from their natural sources of DOC, and separation of bacteria from microbial grazers also uncouples pathways of nutrient regeneration which may be important in maintaining higher BGE in natural systems. In long-term experiments, there may be increasing use of refractory DOC fractions and depletion of nutrients, and therefore the resulting estimate of BGE should be generally lower than in short-term incubations, in which presumably only the most labile fraction of DOC is utilized. Growth of heterotrophic nanoflagellates is almost inevitable in long-term incubations, and the resulting grazing may heavily affect the accumulation of bacterial biomass and the apparent BGE (43, 66, 83). Also, in long-term experiments the accumulation of toxic metabolic by-products may result in lower BGE (82). The actual consumption of DOC can seldom be directly measured in short-term experiments, and the assumption that BR + BP approximates C consumption does not always hold (18). Regardless of the time of incubation, there can be considerable variation in BR and BP rates (109, 129, 142), so the length of the incubation and the integration method for these rates become critical for the calculation of BGE (129). In general, it is thought that reducing the incubation times to hours results in ecologically more relevant data, but in many natural samples this is not possible with current methods.

All the methods used in determining BGE involve assumptions and the application of conversion factors, which contribute to the large variability observed in BGE. Some critical assumptions deal with the conversion of bacterial abundance to carbon, and a wide range of factors are used. Whereas some investigators measured the bacterial cells to estimate volume (119), others assumed a fixed cell size or carbon content per cell (8, 76, 118). Likewise, there is variability

510   DEL GIORGIO & COLE

in the conversion factors used for the calculation of BP (11, 19). Respiratory quotients (RQ) assumed by authors also vary (49, 91), although it is generally assumed that RQ = 1 and it is likely that RQ is a minor source of error compared to the problems discussed above.

Patterns in BGE in Natural Aquatic Systems

UPTAKE AND INCORPORATION OF SINGLE SUBSTRATES   Early research with radiolabeled single substrates demonstrated that growth efficiencies vary consistently among compounds or families of compounds. Amino acids are generally incorporated more efficiently (range, 40 to >80%) than sugars (<60%), for example; however, in general, simple substrates are incorporated with apparent efficiencies of more than 40% (29, 65). These BGE values were assumed to be representative of in situ bacterial processes, and for the next 20 years microbial ecologists applied a BGE range from 40 to 60% in ecological studies (26, 114). King & Berman (73) have shown that intracellular isotope dilution and non-steady-state conditions result in high apparent incorporation of radiolabeled compounds into biomass and therefore in an overestimation of BGE. Moreover, bacteria use many organic substrates simultaneously, and extrapolation from a single model compound may have led to significant overestimation of natural BGE (14, 49, 61, 64). The data on single compounds have been reviewed extensively by others (27, 64) and are not considered further in our review.

IN SITU MEASUREMENTS OF BGE   We surveyed the literature of the past 30 years for direct measurements of in situ bacterial growth efficiency in aquatic ecosystems. Data such as temperature, bacterial growth rate, DOC concentration, and substrate C:N were also recorded whenever possible. A total of 328 estimates of BGE were extracted from 39 published articles (Table 1). We pooled these data into four categories: marine, freshwater, estuarine, and riverine systems. Although there is a wide variation in measured BGE within each type of system (Figure 2), the data suggest consistent differences among systems with BGE increasing from marine areas to estuaries. We further explored the patterns in bacterioplankton metabolism by using a subset of the data (n = 237) made up of simultaneous measurements of bacterial production and respiration or DOC consumption in a variety of aquatic systems. In order to assess the possible effect of method on the estimate of BGE, we grouped the data according to the method: S for short-term metabolic measurements such as leucine uptake and oxygen consumption, and L for long-term experiments in which changes in POC and DOC were usually measured. We respected the assumptions and conversions used by the different authors and assumed RQ = 1 only when converting oxygen consumption rates to carbon units.

Table 1 List of published sources of direct measurements of in situ BGE that appear in Figures 2 and 3[a]

System	Method	BGE	Reference
Oceans			
Sargasso Sea	L	0.04–0.09	54
Coastal and shelf waters	S	0.08–0.69	24
Gulf of Mexico	S	0.02–0.23	107
North Pacific	L	0.01–0.33	18
Sargasso Sea	L	0.04–0.30	17
Coastal	L	0.31–0.64	68
Weddel Sea and Scotia Shelf	L	0.38–0.40	15
North Atlantic	L	0.04–0.06	76
Coastal and enclosures	S	0.07–0.46	30
Gulf of Mexico	L	0.26–0.61	80
Mississippi River plume	S	0.10–0.32	19
Peruvian upwelling	S	0.30–0.34	131
Louisiana shelf	S	0.18–0.55	11
Coastal waters	S	0.08–0.37	96
Coastal and shelf waters	S	0.01–0.25	49
Coastal waters	S	0.1–0.3	13
Baltic and Mediterranean Seas	L	0.21–0.29	156
Baltic Sea	S	0.25	103
Coastal waters	S	0.38–0.57	118
Estuaries			
Coastal Bay and salt marsh	S	0.11–0.61	24
Danish fjord	L	0.22–0.36	90
Hudson River	S	0.18–0.61	44
Danish fjord	L	0.19–0.23	14
Coastal bay	L	0.60–0.61	80
Brackish estuary	S	0.40	82
Lakes			
Swedish lakes	L	0.12–0.36	144
Latvian lakes	S	0.26	36
Cuban lakes	S	0.14–0.30	115
German lakes	L	0.17–0.22	147
Danish lakes	S	0.15–0.37	119
Temperate reservoir	S	0.14–0.66	71
Russian lake	S	0.04–0.24	38
Danish lakes	L	0.25–0.46	79
Lake Constance	S	0.16–0.35	53
Lake Constance	S	0.09–0.80	120
Danish lakes	L	0.34–0.43	92
Rivers			
River Meuse	L	0.30	121
River	L	0.32–0.36	80
Amazon River	S	0.03–0.46	9
Ogeechee River	L	0.31	89

[a]Data are grouped by system (marine, estuaries, lakes and rivers), and by method (S for short-term and L for long-term incubations).

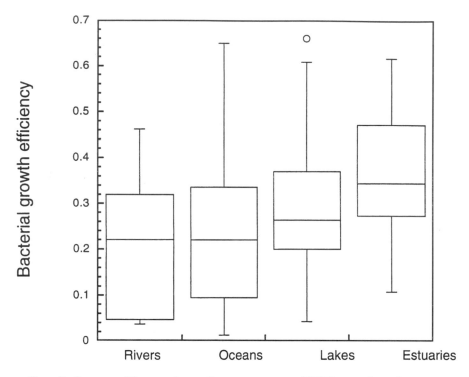

Figure 2 Summary of literature data on direct measurements of BGE in natural aquatic systems. Box-and-whisker plot shows median, and upper/lower quartiles (*box*), and range of values (*bars*). Extreme outliers are marked as *open circles*. The sources of the data are in Table 1.

There is a broad positive relationship between bacterial respiration and production (both in micrograms of C liter^{-1}h^{-1}, [Figure 3a]), with the following model I and II regression equations:

$$BR = 3.70 \times BP^{0.41}, \ r^2 = 0.46 \, (\text{model I}) \tag{1.}$$

$$BR = 3.42 \times BP^{0.61}, \ r^2 = 0.46 \, (\text{model II}) \tag{2.}$$

Equation 1 was obtained using ordinary least squares (OLS) regression, for comparison with previously published empirical models, and it provides the best predictive model of BR from measured BP. Because measurement error occurs in both BR and BP, the Model II regression equation, calculated following Ricker (110b), provides a better estimate of the true functional relationship between BR and BP. What these two models have in common is that the slopes of both regressions are significantly lower than 1. The ecological relevance of this low slope is that it determines a pattern of increasing BGE along a gradient of increasing BP (Figure 3*b*). The data from Griffith et al (50) consistently had one order of magnitude higher BR relative to BP than all the other studies and were excluded from the analyses. Had we included these data, the slope of the regression between BR and BP would have been even lower. While the asymptotic form of this relationship is in part a consequence of the form of the BGE equation, the magnitudes of both the slope and the asymptote are of interest. First, BGE approaches a maximal value (near 0.5) as BP reaches 5 μg C liter^{-1} h^{-1}. The mean value of BGE at BP > 5 μg C liter^{-1} h^{-1} is quite high, 0.46, a point at which BP and BR are nearly equal. Second, the lower values of BGE appear to be fixed above some minimum level. The following model best fits these data:

$$BGE = \frac{0.037 + 0.65BP}{1.8 + BP} \qquad\qquad 3.$$

F Roland & JJ Cole (submitted) found a similar relationship between BP and BGE for data from the Hudson River by using a consistent set of methods. We explored the possible effect of method on the estimates of BGE. Although the expectation is that long-term experiments (method L) should result in lower BGE (129), this effect was not evident in our data. An analysis of covariance showed no significant effect of the type of method (S or L) on the relationship between BP and BR or that between BP and BGE. When all the data are pooled by method, the average BGE values for the two groups are very similar (0.26 for S and 0.28 for L). It is evident that the combined effect of different conversion factors and assumptions applied by various authors add significant noise to the observed variability in the relationship between BP and BR. Of greater concern, perhaps, is whether this noise drives some of the patterns in BGE that we describe in this section, but at present we have no evidence that the various sources of error introduce a systematic bias to the BGE data.

BP is positively correlated with primary production in aquatic systems (26); therefore, the pattern described here represents a tendency for increasing BGE along broad gradients of primary production in aquatic systems. The systematic differences in BGE found among systems (Figure 2) reflect differences in average primary productivity, with marine systems being generally the least productive and estuaries being the most productive. Other than differences in productivity, there are no apparent systematic differences in BGE among systems, so that in all subsequent discussions, we refer to variation in BGE along trophic gradients rather than among specific systems. Several studies had already suggested that BGE increases systematically with primary productivity (11, 24, 49), and we show here that this pattern is general and extends from ultraoligotrophic oceans to highly productive lakes and estuaries.

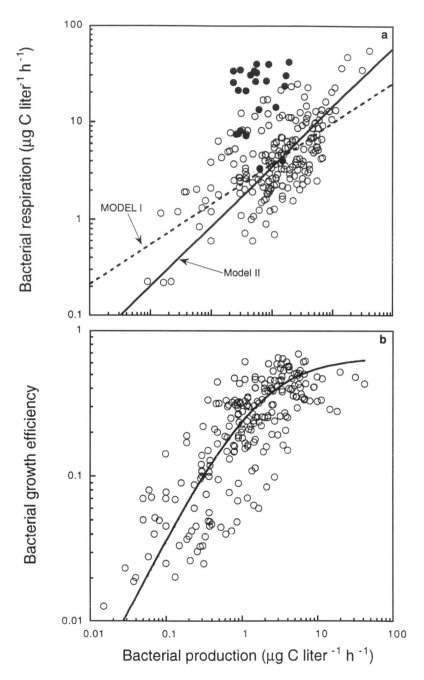

There is a large amount of variance in BGE for any given value of BP, which is the result of a large degree of uncoupling between bacterial production and respiration (Figure 3). Some of this variance may be simply the result of methodological artifacts. The relationship between BR and BP is also characterized by an intercept that is significantly different from zero, suggesting that at BP $= 0$ there would still be measurable rates of BR. This residual BR is not trivial, and since it corresponds to situations of virtually no bacterial growth, it must be somehow related to maintenance energy requirements of assemblages in ultraoligotrophic aquatic systems. It is clear from these patterns that maintenance energy requirements are a significant fraction of the energy flow in oligotrophic microbial assemblages.

TEMPORAL VARIATION IN BGE   There have been relatively few investigations of daily and detailed seasonal variation in natural bacterioplankton BGE. Coffin et al (24) reported a marked diel cycle, with BGE ranging from 37 to 72% and increasing during the day, presumably following inputs of alga derived organic substrates. Some of the daily variation in BGE has been linked to the differential effect of light on BP and BR (99a). Seasonal variation in BGE within systems is sometimes small; for example, Schwaerter et al (119) reported a range of 28–34% in one lake throughout the summer. However, generally there are large and often rapid variations in BGE (24, 44, 120; F Roland & JJ Cole, submitted). It appears that BGE responds quickly to subtle changes in the rate of supply and the quality of substrates and to any factor that alters BP.

Relationship Between Growth Rate and Growth Efficiency

The broad trends of BGE along gradients of productivity suggest a relationship between BGE and bacterial doubling time (growth rate), which, as pointed out above, is expected from theoretical considerations. Comparative analyses have shown that growth rate or cell-specific production tends to increase with NPP and chlorophyll concentration (26, 153), so over broad productivity gradients both bacterial growth rate and BGE should covary. We further explored this pattern with a subset of our data ($n = 52$) for which we had simultaneous measurements of BGE and bacterial generation times, mostly estimates of

←——

Figure 3   (*a*) BR a function of bacterial production in aquatic ecosystems. The data set collected from the literature is composed of 237 paired observations of BR and BP; the sources of these data appear in Table 1. *Lines* correspond to model I and II regressions fits to the data; the equations appear in the text. *Dark circles* are data from Griffith et al (50) that had significantly higher BR relative to BP and were excluded from the regression analyses. (*b*) BGE, calculated as BP/(BR + BP) by using the data in panel *a*, as a function of BP. The line is a rectilinear hyperbole with a fixed lower limit; the model is text Equation 3.

in situ rates. These data show no significant relationship between doubling time and BGE. Schweitzer & Simon (120) also found no relationship between growth rates and BGE in natural assemblages of bacterioplankton. In two studies, a positive relationship was found between growth rate and BGE in continuous cultures of natural bacterioplankton (79, 90), although Bjørnsen (14) found a negative relationship between BGE and growth rate. Søndergaard & Theil-Nielsen (129) found that the maximum BGE corresponded to the highest growth rates during batch incubations of bacterioplankton, but they found no consistent relationship between BGE and growth rates among samples. These data suggest that BGE may covary with growth rate in any given combination of temperature, organic substrates, nutrients, and other constraining factors but that the relationship between BGE and growth rate may be specific for each set of growth conditions.

One factor that uncouples growth rate from growth efficiency is the observation that bacteria may maximize growth at the expense of efficiency (116, 140, 151). This maximization is achieved with different energetic costs and different degrees of uncoupling between catabolism and anabolism. There are clear examples of this type of uncoupling for bacterioplankton. Addition of nutrients sometimes increases substrate consumption with no effect on net growth (8). Middelboe et al (93) found that viruses decreased BGE in bacterial cultures while increasing the growth rate of noninfected bacteria. They argued that lysed cells released P and N, which were used by uninfected cells for growth, at the expense of lowering BGE by the production of exoenzymes to hydrolyze polymeric P and N released by dead bacteria. Zweifel et al (156) observed a 70% increase in cell yield (number of cells) and a 20% decrease in BGE after phosphorus was added to the culture media. These authors suggested that P enhanced cell division while P-limited cells were able to store organic carbon without dividing and thus could maintain a higher carbon growth efficiency. Poindexter (104) showed that during P-limited growth in chemostats, the bacterial concentration but not the biomass was proportional to the substrate P content; conversely, during C-limited growth, the bacterial biomass but not the concentration was proportional to the substrate C content. Robinson et al (112) showed that addition of N did not increase BGE but sharply increased the rates of decomposition of detritus and the final yield of bacteria.

These examples, as well as the patterns in BGE discussed above, suggest that although the highest bacterial growth rates attained in natural aquatic systems correspond to conditions under which BGE is high, such as in estuaries and eutrophic lakes, it is not necessary to increase BGE in order to increase growth rates. This is important because total carbon consumption may be regulated by factors different from those that regulate growth or BP (75). Because most contemporary research in microbial ecology has focused on the regulation of bacterial growth and production, we know relatively little about what may regulate total bacterial carbon consumption in aquatic ecosystems.

REGULATION OF BACTERIOPLANKTON GROWTH EFFICIENCY

Research with cultured bacteria (117), as well as models of bacterial energetics and growth in aquatic systems (2, 12, 27, 61, 83, 151), suggests that the substrate supply and complexity and mineral nutrient availability are the most important variables controlling BGE. In addition, as described above, BGE is broadly correlated with BP. Hence, it is to be expected that the factors that influence BP in experiments would also affect BGE. However, the results of experiments in various systems have been inconsistent. In this section, we explore the empirical evidence on the regulation of BGE in natural bacterioplankton assemblages in an attempt to reconcile the pattern of increasing BGE along productivity gradients with observations and experimental results on carbon and nutrient quality and supply.

Effect of Temperature, Salinity, and Pressure

The growth rate declines as temperatures move away from the optima for each type of bacteria in laboratory studies (101). However, a strong positive relationship exists between growth rates and temperature in natural bacterioplankton assemblages (110, 153). If low temperatures result in lower growth rates, a positive relationship between temperature and growth efficiency would also be expected. Newell & Lucas (97) and F Roland & JJ Cole (submitted), for example, found higher BGE in summer than in winter. However, BGE tends to decline with increasing temperature in other systems (30, 50, 62), even though growth rates tend to increase. In all these cases, however, the effect of temperature was very weak. A subset of our data ($n = 151$) for which the BP, BR, BGE, and temperature are known shows no significant relationship between BGE and temperature or any significant effect of temperature on the relationship between BP and BR. However, this data set is biased towards higher temperature (mean $= 19°C$), and there are relatively few measurements in the low-temperature range ($<10°C$), so that we can conclude only that temperature is not an overriding regulating factor of BGE in natural systems.

The changes in BGE along gradients of salinity have been the subject of only a few studies. Griffiths et al (50) found a weak negative relationship between BGE on glucose or glutamate and salinity in a large-scale study but concluded that salinity had no direct effect on BGE. A gradient of increasing salinity may have corresponded to a gradient of declining productivity from coastal to open waters and, as shown above, a pattern of declining BGE would be expected.

The only study to our knowledge that has assessed the effect of pressure on BGE is that by Turley & Lochte (150), who concluded that deep-sea bacteria were able to mineralize more organic carbon at 450 atm than at 1 atm but that BGE was lower under high pressure.

Nutrient Limitation of BGE

The dependence of BGE on the relative availability of mineral nutrients and organic carbon was originally formulated by Fenchel & Blackburn (42) and later expanded by others (2, 12, 13, 45, 58, 151). The basic idea is that bacteria regulate the catabolism of organic substrates to attain the correct intracellular stoichiometry with respect to N (and other nutrients). Because the elemental composition of bacteria is relatively constant (45), BGE should be negatively related to the C:N ratio of the substrate, at least in the range of C:N where N, and not C, is limiting. This type of model is important not only to understand the regulation of BGE but also to assess the role of bacteria in nutrient cycling in aquatic environments (12, 45, 46).

Regulation of BGE by the availability of mineral nutrients implies that increases in the supply of nutrients should result in increased BGE. Billen (12) and Goldman et al (45) have unequivocally shown that the BGE of natural assemblages of marine bacteria grown on a range of substrates is inversely related to the C:N ratio of the substrate. However, the relationship between BGE and available C:N was considerably weakened when bacteria were exposed to multiple nitrogen and carbon sources (46), and under these circumstances the source of the nitrogen (i.e. NH_4 or amino acids) became important. Bacterial assemblages growing under ambient conditions are exposed to multiple sources of nutrients and organic matter, and so it is expected that the relationship between C:N and BGE may not always hold under natural conditions. Some experimental data support the role of N in regulating BGE; for example, Kroer (80) found that BGE increased with ammonium concentration and decreased with increasing C:N ratio of the bulk dissolved substrates in coastal areas, and Benner et al (8) found that BGE was limited by N in a salt marsh and by P in a freshwater marsh. However, most experimental additions of inorganic nutrients have had little or no effect in lakes and rivers (9, 144), coastal areas (30, 68, 112, 156), and open oceans (17, 18, 75). There is thus conflicting evidence about the role of N in regulating BGE in natural aquatic systems, and there is no clear pattern along trophic gradients.

That the C:N ratio may not be the key regulator of BGE in natural bacterial assemblages is also clear in a subset of our data, for which the C:N ratio of the presumed substrate was reported. These data show a very weak negative relationship between BGE and C:N. Figure 4 contains data from natural systems as well as experimental data from laboratory cultures (45, 46) to emphasize

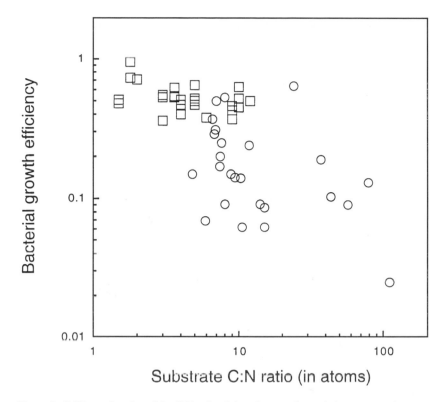

Figure 4 BGE as a function of the C:N ratio of the substrate. *Open circles* correspond to measurements of BGE on natural complex substrates. *Open squares* correspond to BGE in defined media (45, 46).

that growth in defined media creates conditions that are seldom experienced by natural bacterioplankton. From the weak relationship between substrate C:N and BGE, we draw three inferences. (*a*) Nutrients other than N may be limiting. There is increasing evidence that P may control BGE in both freshwater (9, 58) and marine (107, 156) systems and that iron may limit BGE in large areas of the oceans (143). (*b*) Bulk C:N may not be representative of the substrates actually available and taken up by bacteria (112). (*c*) Nutrients and organic carbon may colimit BGE (39, 40).

Energy and Organic Carbon Limitation of BGE

It has been repeatedly suggested that bacteria in oligotrophic systems are limited primarily by the supply of carbon and energy (18, 26, 37, 74, 75). The distinction between energy and carbon limitation is not always fully realized in microbial studies, however. From a bioenergetic point of view, energy limitation occurs when the ATP generated during the biological oxidation of a compound is insufficient to reduce all the available carbon in the molecule to the level of bacterial cell carbon (85). Growth on relatively oxidized substrates, such as acetate, glycolate, and even glucose, is usually energy limited from this perspective, and these compounds are incorporated into biomass with low efficiency, even if inorganic nutrients are in excess. Thus, it is the ratio between biologically available energy and carbon content of the organic molecules that determines the maximum BGE (27, 84, 85, 151).

In the context of this review, we should thus distinguish between control of BGE arising from the rate of supply of organic matter and control arising from the nature of the available organic matter. Although both may result in low BGE, they are ecologically distinct. If the supply of organic matter is low, whatever its nature, a large fraction of this substrate will be catabolized and used primarily for maintenance energy requirements rather than for growth, with a resulting low growth efficiency (55, 117, 135). Conversely, there might be a large supply of organic substrates which, because of their relative energy and carbon contents, are incorporated with low efficiency even under conditions of excess mineral nutrients. Distinguishing between these two types of limitation in natural situations is difficult, especially because of confounding by possible nutrient colimitation.

Regulation of growth efficiency by the supply of organic C implies that increases in the supply rate should result in increased BGE and production. Empirical and experimental results show that this is not always the case. For example, Kirchman (74) found that growth of planktonic bacteria in the oligotrophic subarctic Pacific was not stimulated by the addition of glucose, and others have found similar patterns in other areas (79, 107). Carlson & Ducklow (17) found that addition of glucose and amino acids resulted in a higher BGE but noted that with glucose addition, cells produced storage carbon and increased in mass rather than in abundance. A similar conclusion was reached by Cherrier et al (18). Perhaps the only common result of most addition experiments is that amino acids tend to enhance both BGE and bacterial growth (17, 18, 30, 68, 74). It has been suggested that it is energetically advantageous to use preformed compounds (74), but the energetic cost of transporting amino acids across the membranes greatly offsets this advantage (101, 133). It is more likely that because amino acids have relatively high energy and carbon contents and are also a source of N, they release bacteria from multiple limitation by carbon, energy, and N. Experimental evidence suggests that the quality of the organic C, rather than the rate of supply of organic matter, may regulate BGE in most natural aquatic systems (151). The reports of DOC accumulation in oligotrophic oceanic areas during the summer (42) provide further evidence that the supply of organic matter may not be the main factor regulating BGE.

SOURCES AND QUALITY OF ORGANIC SUBSTRATES   Qualitative aspects of natural DOC that are relevant to bacterial energetics are difficult to define (151). One approach has been to determine bacterial utilization of molecular size fractions of natural DOC. No clear patterns have emerged, because there are reports of highest BGE on either the low-molecular-weight (LMW) fractions (145, 148) or the high-molecular-weight (HMW) fractions (1, 89). Some of these differences can be explained by the C:N ratio of the weight fractions rather than by any qualitative characteristic of the organic carbon itself (89, 148). It is clear from these results that high bioavailability does not necessarily imply high BGE: Amon & Benner (1), for example, reported that HMW fractions were most bioreactive but LMW fractions were incorporated more efficiently into bacterial biomass. Both the absolute amount and the proportion of labile DOC, defined as DOC taken up by bacteria in batch incubations, tend to increase along trophic gradients (128), suggesting that a higher BGE in more productive systems may be the consequence of qualitative changes in the DOC pool. There is some experimental evidence for such a relationship (91).

Another approach has been to assess how different sources of organic matter affect BGE. We collected from the literature 85 direct measurements of BGE in five broad categories of organic matter depending on its source: organic matter excreted by phytoplankton (EOC), and organic detritus derived from phytoplankton, seaweeds, vascular vegetation, and animal feces (Table 2). These combined data show that the efficiency of conversion of detrital organic matter to bacterial biomass is generally low (<30%) for all categories except EOC, in which most values are above 50% (Figure 5). Algal EOC production and cycling are measured by monitoring the incorporation of ^{14}C into phytoplankton and its subsequent release and uptake by bacteria. The high values of BGE obtained for EOC most probably reflect the same type of problems that affect estimates of incorporation efficiency of single radiolabeled compounds, i.e. lack of isotopic equilibrium in the internal carbon pools of bacteria resulting in an overestimation of BGE (14, 73). Although organic carbon derived from vascular vegetation is usually considered of rather low quality and is a major component of the more refractory humic fraction of DOC, BGE measured on either vascular vegetation (Figure 5) or the humic DOC is well within the average values measured for bulk water and other organic components (3a, 8).

There are sources of organic matter in addition to the direct production of detritus from plants and animals. A potentially significant source in the open ocean is atmospheric deposition of volatile organic compounds. Heikes et al (56) estimated that formaldehyde is subject to atmospheric dry deposition rates of about 2.2 mg of C $m^{-2} day^{-1}$ in the Central Atlantic, and Nuncio et al (99) have shown that formaldehyde is rapidly utilized in the upper layer of the ocean. Another source of labile DOC is the photochemical oxidation of organic matter

Table 2 BGE on different types of detrital organic matter[a]

Source of detritus	System	BGE	Reference
Phytoplankton	Marine	0.13–0.22	10
Phytoplankton	Marine	0.17–0.27	5
Phytoplankton	Marine	0.07–0.13	98
Phytoplankton	Marine	0.09–0.24	83
Phytoplankton	Marine	0.50	150
Phytoplankton	Marine	0.18	130
EOC	Marine	0.71–0.81	154
EOC	Freshwater	0.57–0.75	7
EOC	Freshwater	0.50	28
EOC	Freshwater	0.31–0.56	23
EOC	Freshwater	0.77	62
Vascular plants	Marine	0.43	41
Vascular plants	Freshwater	0.09–0.11	89
Vascular plants	Freshwater	0.53	43
Vascular plants	Marine	0.025–0.10	83
Vascular plants	Freshwater/Marine	0.17–0.36	8
Vascular plants	Marine	0.11	16
Vascular plants	Freshwater	0.74–0.92	69
Vascular plants	Marine	0.04–0.17	111
Vascular plants	Freshwater	0.37–0.63	44
Vascular plants	Freshwater/Marine	0.27	95
Vascular plants	Marine	0.44	47
Vascular plants	Marine	0.19–0.64	52
Macroalgae	Marine	0.09	137
Macroalgae	Marine	0.06–0.07	86
Macroalgae	Marine	0.32	112
Macroalgae	Marine	0.25	78
Macroalgae	Marine	0.11	136
Macroalgae	Marine	0.07	97
Macroalgae	Marine	0.09–0.13	83
Tunicate feces	Marine	0.15	105
Bivalve feces	Marine	0.15	137
Bivalve feces	Marine	0.06–0.09	149

[a]Data were used in Figure 5 and are grouped according to the source of the detritus: phytoplankton detritus includes detritus from natural algal assemblages and from algal cultures; organic carbon excreted by phytoplankton (EOC), organic carbon derived from vascular vegetation, including seagrasses and other aquatic macrophytes and terrestrial vegetation; organic carbon derived from marine macroalgae; organic carbon derived from feces.

in the surface layers of lakes (110) and oceans (72, 95a). Kieber et al (72) estimated rates of pyruvate production in the Sargasso Sea of 1.6 mg m^{-2} day^{-1}, which would represent 1–4% of BR measured by Carlson & Ducklow (17). It is reasonable to expect that a wide variety of other LMW substrates, including acetate, acetaldehyde, formate, glucoxylate, and methanol, are formed together

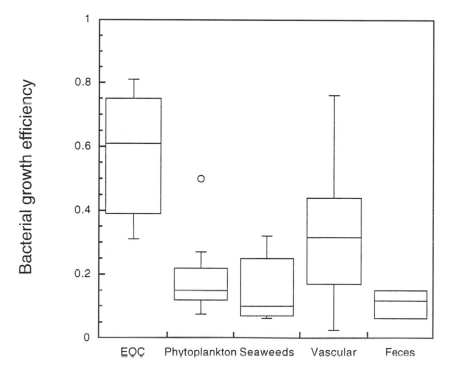

Figure 5 Summary of literature data on direct measurements of BGE for organic matter grouped according to source. Box-and-whisker plot shows median and upper/lower quartiles (*box*), and range of values (*bars*). Extreme outliers are marked as *open circles*. The sources of the data are in Table 2.

with pyruvate by the photochemical breakdown of DOC, and the summed production of these substrates may well be a significant source of C and energy for bacteria. Since many other simple organic compounds are deposited from the atmosphere, it is conceivable that these two pathways of carbon input into oceanic systems are significant to bacterial metabolism.

Atmospheric deposition and photochemical oxidation result in the production of LMW compounds that are characterized by a low heat of combustion and a high degree of oxidation relative to microbial biomass and are typically incorporated with low efficiencies (27, 85, 95a). Algal excretion is also dominated by small, low-energy organic molecules (23). In the large expanses of ultraoligotrophic ocean and even in ultraoligotrophic lakes, these compounds may form the bulk of biologically labile organic carbon. As systems become more productive, the relative importance of EOC as a bacterial substrate tends to decline (3) and the impact of atmospheric deposition and photolysis on pelagic metabolism will decline: These qualitative changes may positively influence BGE. However, because it is difficult to differentiate energy from carbon limitation, it is unclear whether the BGE of natural bacterioplankton assemblages is limited by the supply of organic matter, the chemical nature of the organic substrates present, or both.

GROWTH EFFICIENCY, ENERGY REQUIREMENTS, AND CELL ACTIVITY IN NATURAL BACTERIAL ASSEMBLAGES

We showed in the previous section that the supply and nature of the organic substrates, as well as the availability and sources of mineral nutrients, may influence BGE, although there are no clear patterns of resource regulation among systems. Research on bacterial bioenergetics has shown that when cultures are primarily energy limited as a result of the rate of supply of organic matter, bacteria tend to maximize the efficiency of utilization of the energy source through tight coupling between catabolism and anabolism and high BGE (117). The extremely low and often variable BGE values observed in most oligotrophic systems suggest a high degree of uncoupling between catabolism and anabolism and do not support the hypothesis that the rate of supply of energy is the main limiting factor for BGE. Moreover, cell-specific respiration rates are not consistently lower in oligotrophic areas (32), suggesting that the amount of energy available on a per-cell basis may be roughly similar among systems. Rather, cell-specific maintenance requirements appear to be higher in oligotrophic areas with extremely low concentrations of organic substrates and nutrients. Thus, per unit organic carbon input, more carbon is used for maintenance in oligotrophic areas than in eutrophic areas, and it is the interaction among the rate of supply of energy, the quality of the substrate, and the energetic demands of cells that determines BGE. These high apparent maintenance respiration rates may occur (a) when cells must transport solutes against a large concentration gradient, (b) when cells must produce extracellular substances in large amounts, (c) when cells must maintain a wide array of active transport systems and the corresponding arrays of catabolic enzymes, and (d) when a large fraction of the population is in a state of starvation survival, with only minimal metabolism. We briefly discuss these possibilities below.

Transport of Nutrients

Transport of nutrients and organic C could influence maintenance energy in two ways. First, as the concentrations become lower in dilute, oligotrophic environments, the energetic cost of active transport increases (87). Second, the appearance of nutrients and carbon sources for bacterial growth is transient, and the ability to respond to sudden increases in nutrient levels is an essential property for survival in dilute environments. It is also clear that bacteria under carbon limitation are able to simultaneously take up and catabolize a wide variety of substrates (39, 40), but there is a cost in maintaining the transport proteins, catabolic enzymes, and functional membrane systems needed to deal with both the variety and low concentrations of substrates (31, 96). The energetic cost of maintaining such a wide array of highly efficient transport systems has never been explicitly assessed in natural bacteria.

Metabolite Excretion

Most bacterial species excrete metabolites to the medium, even during aerobic growth. The causes and bioenergetic implications of excretion have been only minimally investigated in bacterioplankton, but the production of exoenzymes is perhaps the best-understood aspect of metabolite excretion in aquatic bacteria (60, 61a). A large fraction of DOC in natural aquatic systems is composed of polymeric substances that cannot be incorporated directly into bacteria. Large molecules and colloids present in the DOC pool must be acted upon by exoenzymes before they can be utilized by bacteria (61a, 125), and the hydrolysis of polymers has been suggested as the rate-limiting process to bacterial production in aquatic systems (22). The synthesis and excretion of enzymes must be coupled to active transport systems that can capture the products of extracellular hydrolysis and of enzymatic systems capable of catabolizing these substrates; it may thus represent a major energy expenditure of bacteria in natural aquatic systems. For example, Middelboe & Søndergaard (91) found an inverse relationship between lake BGE and β-glucosidase activity (Figure 6). Extracellular enzyme production increased toward the end of batch culture incubations of lake bacterioplankton, when most of the labile DOC had been consumed and the submicron and colloidal fractions were increasingly utilized (92). The need to perform extracellular hydrolysis of polymers thus makes a large energy demand on bacterial cells in natural environments. The increasing trend in BGE along productivity gradients suggests that bacteria may be deriving more of their C needs from exoenzymatic breakdown of polymeric substances in oligotrophic environments, with the consequent decline in BGE. This hypothesis remains to be empirically tested.

In addition to the excretion of enzymes, bacteria are capable of producing copious amounts of extracellular mucopolysaccharides (31a, 112), which form mucilaginous capsules around the cells (31a, 57a, 112) and also form loosely associated slimes and fibrils (82a, 112). The chemical nature of these extracellular compounds varies greatly, but uronic acids often form the bulk of these materials (71a). Metabolite excretion appears to be greater when the organic

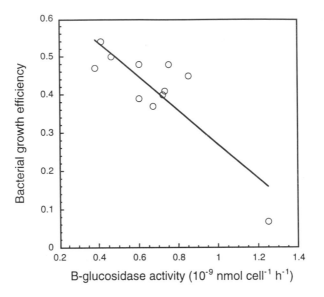

Figure 6 Relationship between BGE and β-glucosidase activity in lake bacterioplankton. Data from Middelboe & Søndergaard (91).

substrate is in excess of the growth requirement, and it also depends on the nature of the organic substrate (84, 139). It has been suggested that excretion of metabolites is a pathway of energy dissipation that may contribute to the maintenance of intracellular stoichiometry (31a, 84). However, excretion of organic metabolites, including polysaccharides, lipids, proteins, and humic-like substances, has also been found under conditions of carbon and nutrient limitation in aquatic bacteria (48, 63, 67, 146). Most excretion products are polymeric, and the biosynthesis of these substances typically exerts high energy requirements to the cell (117, 133). Not surprisingly, there is a general inverse relationship between the overproduction and excretion of metabolites and growth efficiency in bacterial cultures (84). In addition, natural bacterioplankton excrete both organic and inorganic N, even under nutrient and carbon limitation (21, 46, 141, 149). Current BP measurements, whether based on changes in bacterial biomass or on the incorporation of leucine or thymidine, are unlikely to include the production of exopolymers, and this will result in a more or less severe underestimation of BGE (31a). This underestimation of BP and BGE may not be trivial from the point of view of organic carbon flow in pelagic food webs, because there is evidence that a variety of grazers can effectively utilize bacterial exopolymers (31b, 78a).

Physiologic Condition of Cells

Although often treated as a black box in ecological studies, bacterioplankton assemblages display a large internal variation in cell size and morphology, taxonomic and functional characteristics, and physiological states (77, 88, 96). In any given bacterioplankton assemblage, there are cells in the entire range of physiological states, from extremely active to slowly growing, dormant, and even dead. Distinguishing cells in these various physiological states poses major technical and conceptual difficulties (88) and is rapidly becoming a major focus of research in aquatic microbial ecology. Stevenson (132) proposed that most bacterial cells present in aquatic systems are inactive, i.e. either dormant or dead. Subsequent research has demonstrated that under conditions of low organic substrate supply, heterotrophic marine bacteria can enter a phase of long-term survival; the literature on this subject has been extensively reviewed by Morita (96). Bacteria in starvation survival mode are not completely inactive but are able to take up substrate and engage in low but measurable rates of biosynthesis (77).

Several methods are currently used to determine single-cell activity in natural bacterioplankton assemblages (88). Since these methods are used and compared in various systems, it has become apparent that different approaches yield different estimates of the proportion of bacterial cells that are alive, viable, and/or metabolically active (70, 123). However, the growing consensus among microbiologists is that bacterioplankton assemblages are composed both of highly "active" and growing bacteria, which often comprise a small fraction of the total population, and cells that are dead, dormant, or slowly growing (96, 123). Most studies in aquatic microbial ecology focus on the average growth rates of bacterioplankton assemblages, by scaling the measured production rates to the total cell abundance. However, the BGE that is commonly measured in ecological studies is the average of the BGE values of these subpopulations of bacteria, which must vary within any given assemblage at least as much as growth rates. The assumption that all bacteria are growing with the same conversion efficiency is most probably wrong.

Figure 7 (left panel) shows the traditional approach to bacterioplankton assemblages, which assumes that all cells are participating equally in the metabolic processes. The right panel shows an alternative view, i.e. that the assemblage is composed of at least two fractions characterized by very different average growth rates and by different BGEs (the nongrowing fraction will still consume organic matter to fuel basic maintenance energy requirements). Overall BGE may vary as a result of changes in the relative size of the pools of active and inactive cells without any changes in the actual growth or metabolic rates of the bacteria in each pool. Likewise, measured values of adenylate

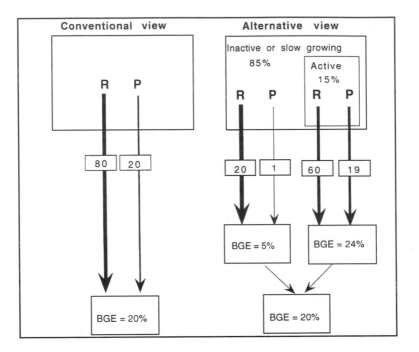

Figure 7 Two alternative depictions of the functions of bacterioplankton assemblages. The *left panel* assumes that production (P) and respiration (R) are homogeneous among all cells. BGE is related to the average growth rate. The *right panel* assumes that there are two distinct pools of cells, one highly active and the other relatively inactive. Each pool is characterized by distinct P, R, and BGE, so the resulting growth efficiency of the assemblage is dependent on the relative size of the active and the inactive pools. Numbers are for purposes of example only.

energy charge (EC_A) of marine bacterioplankton assemblages are often close to or below the range where growth and biosynthesis are theoretically possible (69a, 96a). These measurements are consistent with generally low BGE in unproductive marine waters, but EC_A, like BGE, is an average value for a mixed bacterial assemblage. A low community EC_A may indicate either a homogeneous population of bacteria severely limited by energy or the coexistence of actively growing cells (high EC_A) and dormant or inactive cells (low EC_A). We hypothesize that within bacterioplankton assemblages there is always a pool of highly active cells characterized by both high BGE and EC_A relative to the average values of the assemblage.

Previous comparative studies have found that the proportion of highly active bacteria increases along productivity gradients (34, 35), in much the same way we have shown here that BGE does. Regulation of the number and proportion of highly active and of less active or dormant cells in natural bacterioplankton assemblages is complex; there is evidence for both resource regulation and control through trophic interactions within microbial food webs (21a, 33). Whether there is a direct link between growth efficiency and the structure of bacterioplankton assemblages must be explicitly addressed in future studies. This link is ecologically important because it implies that processes which affect the proportion of different physiological subpopulations, such as selective grazing by protozoans, may have a bearing on the in situ BGE, regardless of the supply and nature of the inorganic and organic substrates.

TAXONOMIC COMPOSITION   Bacterial growth efficiency in laboratory studies is known to depend strongly on the type and supply of growth substrates, but for any given combination of growth parameters, different species of bacteria exhibit widely different patterns in growth efficiency (57). We have no information on how the taxonomic composition of the bacterioplankton assemblage may affect bacterial growth efficiency in natural aquatic systems. However, the advent of a new generation of molecular techniques is rapidly opening the genetic black box of planktonic bacteria, and soon we may be able to link broad taxonomic composition to aspects of microbial energetics and thus to explain some of the variance in BGE not accounted for by resource regulation.

ECOLOGICAL CONSEQUENCES OF PATTERNS IN BGE

Bacterial respiration is the major component of total respiration in most aquatic systems (61, 64, 126, 155), so that changes in bacterial respiration have profound effects on the overall carbon and gas balance in aquatic ecosystems (124). The magnitude and regulation of bacterial growth efficiency is of interest well beyond the realm of microbial ecology because the assumed value of BGE can greatly affect how one construes models of the C cycle in aquatic systems.

We have shown that BGE is postulated to be regulated by numerous factors. However, of all the possible factors, only BP gives a reasonable prediction of BR or BGE, and even this is associated with a high level of uncertainty. While this is perhaps intellectually not satisfying, it does allow us to estimate BGE and BR when BP is known. Since BP is much more commonly measured than is BR, this estimation is useful. We assert that it is more useful than using a generic value of BGE which is independent of measured BP. We propose a rectilinear hyperbole with a fixed lower limit as a predictive model of BGE from BP (Equation 3).

To infer the consequences of this model for BGE, we show in Figure 8 how the ratio of BR to net primary production (NPP) would vary across a gradient of NPP from ultraoligotrophic to eutrophic waters. This ratio is of interest because when it exceeds unity, the ecosystem is respiring more organic C than is fixed by photosynthesis and the system is net heterotrophic. In this exercise, we assumed that volumetric daily BP varied with NPP as specified by Cole et al (26):

$$\log(BP) = -0.483 + 0.814 \log(NPP)$$

and that hourly BP was constant over the day. At the oligotrophic end of the spectrum, BR would exceed NPP by sevenfold; at the eutrophic end of the spectrum, NPP would exceed BR. Compared with the assumption of a constant value of BGE across the gradient, the new model produces a much larger range in the ratio of BR/NPP. In the region for which NPP is 90–300 mg C m^{-3} day^{-1}, this model is in agreement with the traditionally assumed range of BGE of 0.25–0.45 (Figure 8).

We further examined the consequences of this new model of BGE on a large oceanic data set of simultaneous measurements of BP and NPP synthesized by

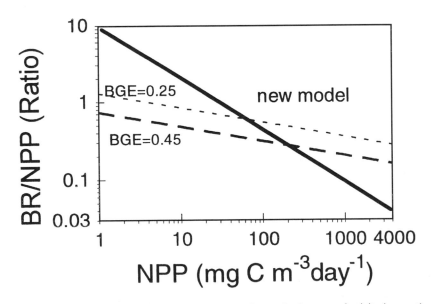

Figure 8   Model for the variation of BGE along gradients of primary productivity in aquatic ecosystems. BP was calculated along a simulated gradient of NPP by using the model in Reference 25, and BGE was then calculated from BP by using text Equation 3. The plot shows how the resulting ratio of BR to NPP would vary along a gradient of NPP and how this ratio would vary if we assumed that BGE is constant along this gradient. See text for a complete explanation.

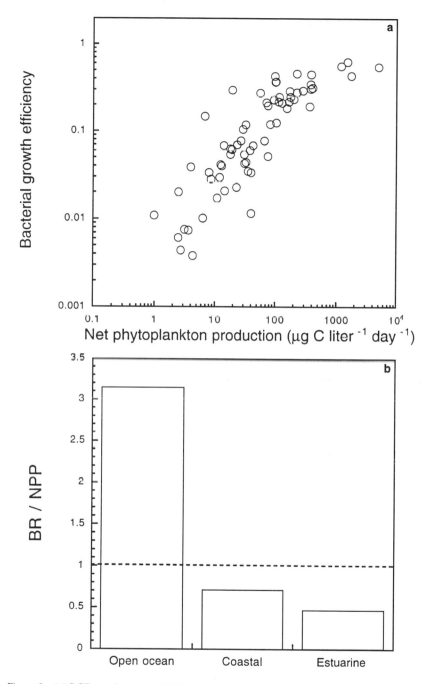

Figure 9 (*a*) BGE as a function of NPP in marine systems. The large data set on simultaneous measurements of BP and NPP in reference 37 was used, and BGE was calculated from BP by using text Equation 3. (*b*) Average ratio of BR to NPP for estuaries and coastal and open-ocean sites, calculated from the data in panel a.

Ducklow & Carlson (37). These values range in primary production from 5 to more than 5000 μg of C liter^{-1} day^{-1} and in BP from <0.01 to 11 μg of C liter^{-1} h^{-1}. There is a positive relationship between BP and NPP, not unlike previous reports (26). We used our model of BGE versus BP (Equation 3) to calculate BGE from these BP measurements, so that we could relate BGE to NPP. The estimated BGE increases with net phytoplankton production (Figure 9a), ranging from less than 0.1% in ultraoligotrophic ocean sites to slightly over 50% in the most productive estuarine and coastal sites. This pattern suggests virtually no bacterial net growth or production in the most unproductive aquatic ecosystems, as suggested previously (51, 126). The fact that BGE increases together with bacterial production along gradients of primary production also results in a pattern of relatively little change in bacterial respiration along this gradient. The relative invariance in microbial respiration, and even of plankton community respiration in general, has been noted before (31c, 32).

The covariation of BGE, BP, and NPP thus results in a pattern of BR being large relative to NPP in oligotrophic areas but small where NPP is high. The current paradigm generally assumes that coastal ecosystems, particularly estuaries, may be net heterotrophic (respiration exceeds primary production) because of the influence of allochthonous organic inputs (60, 127). In contrast, open-ocean systems are usually thought to be examples of autotrophic self-supporting systems because of their relative isolation from significant allochthonous sources of organic matter. However, when BGE is factored into the organic matter flow and the resulting BR is considered together with BP, the patterns in system functioning that emerge are strikingly different. Figure 9b shows the average balance between NPP and BR, calculated from BGE as described above by using the data of Ducklow & Carlson (37), grouped by system. There is a systematic increase in the BR:NPP ratio from estuaries, where BR:NPP is low on average, to open oceans, where the average BR:NPP ratio is well above unity (Figure 9b). This pattern agrees with actual measurements of bacterial metabolism relative to phytoplankton in a variety of marine systems (32, 110a).

Our model for BGE is derived from the empirical data culled from diverse ecosystems. We point out, however, that the data set is not very extensive and there is a great deal of variability in the plot of BGE versus BP. Thus, BGE may not be correctly estimated from Equation 3 in all environments. For example, in a detailed study of the organic C budget of oligotrophic Mirror Lake, Cole et al (25) concluded that BGE could not be below about 0.2 and have the C budget balance. Further, with the best estimates of all other variables in that budget, the best estimate of BGE would be near 0.4. The model derived here implies a BGE for Mirror Lake of about 0.1, which would imply a greater use of allochthonous DOC than is likely in that system. On the other hand, recent work on BGE in the Hudson River largely conforms to Equation 3: the relationship between BP and BGE was of the same form as equation 3, with slightly different constants (F Roland & JJ Cole, submitted).

CONCLUSIONS

There is a large range of variation in BGE in aquatic systems, but we have shown a consistent increase in BGE along gradients of productivity in aquatic systems. Growth seems to be energetically more costly in dilute systems, but at present we can only speculate on the causes of this. We have argued that maintenance of active transport systems and of basic metabolic machinery, and the production of extracellular enzymes, may exert disproportionately high energy demands on bacteria inhabiting oligotrophic aquatic systems and may result in low growth efficiency. We have also suggested that the cell-specific rates of organic matter utilization is similar in oligotrophic and eutrophic conditions, so that the rate of supply of organic matter may not be a factor limiting BGE. Rather, we suggest that a combination of the quality of this organic matter, nutrient availability, and the particular energetic demands in each type of system may regulate BGE.

We propose the following scenario. Bacterioplankton in oligotrophic lakes and oceans are exposed to generally low concentrations of dissolved substrates, including organic carbon and nutrients, and it is possible that cell growth is colimited by energy, carbon, and nutrients. The generally low concentrations of nutrients impose low growth rates on bacteria and place cells in the realm in which maintenance energy expenditures are high relative to the overall energy flux. Maintenance requirements are further enhanced by high costs of active transport, the need to maintain functional transport systems even when growth is impeded by lack of suitable substrates, and the production of exoenzymes needed to supply suitable substrates for growth. We have suggested that these relatively high-maintenance and other nongrowth energy requirements are met by the catabolism of relatively oxidized, LMW compounds which provide neither enough energy nor enough carbon to sustain growth. We suggest that these labile organic molecules, which support the bulk of BR in oligotrophic aquatic systems, particularly open oceans, originate from algal excretion, photooxidation of DOC, and atmospheric inputs of DOC. Thus, the low BGE values that characterize oligotrophic areas may be the result of relatively high maintenance energy requirements, the lack of mineral nutrients, and the predominance of low-energy compounds in the labile pool of DOC. As systems become enriched in nutrients and primary production increases, both the rate of supply and the quality of DOC increase, as does the availability of nutrients, with a general increase in BGE.

There is still much uncertainty surrounding the magnitude and variation of BGE in natural aquatic systems, and the present review has revealed significant gaps in our knowledge. We propose the following areas that should be given priority in future studies.

1. We need more and better estimates of BR of bacterioplankton in a wider variety of aquatic systems. Our current understanding of BGE is largely limited by the scarcity and uncertainty of BR measurements. We need more consistency in the methods to measure BGE.

2. We know relatively little about the regulation of BR in bacterioplankton, although this is possibly the largest single component of organic carbon flow in most aquatic systems. In particular, the concept of maintenance energy requirements has never been explicitly investigated for bacterioplankton.

3. The distinction between energy and organic carbon limitation of BGE should be further explored, at both the conceptual and experimental levels.

4. The BGE that we measure in natural assemblages is an integrated measure of the efficiency of utilization of a large number of organic compounds. Individual compounds may be incorporated with very different efficiencies, and perhaps the overall BGE that we measure is related to, and indicative of, the proportion of broad qualitative classes of organic compounds available to bacteria.

5. The BGE that we measure is the average of BGEs of different subpopulations of bacteria that coexist within the bacterioplankton assemblage. Understanding what controls the distribution of subpopulations of highly active versus dormant or slowly growing cells in bacterioplankton assemblages will no doubt advance our understanding of what controls BGE in natural aquatic systems.

6. The energetic costs of the production of exoenzymes, active solute transport, and excretion of a variety of polymers have seldom been investigated in natural bacterioplankton, but these processes may play an important role in determining differences in BGE among systems.

7. One fundamental question is whether the low BGE measured in most oceanic systems is only a reflection of external factors such as nutrient or carbon availability or whether it is genetically determined and an inherent characteristic of the dominant bacteria in these systems.

> Visit the *Annual Reviews home page* at
> http://www.AnnualReviews.org

Literature Cited

1. Amon RMW, Benner R. 1996. Bacterial utilization of different size classes of dissolved organic matter. *Limnol. Oceanogr.* 41:41–51
2. Anderson TR. 1992. Modeling the influence of food C:N ratio, and respiration on growth and nitrogen excretion in marine zooplankton and bacteria. *J. Plankton Res.* 14:1645–71
3. Baines SB, Pace ML. 1991. The production of dissolved organic matter by phytoplankton and its importance to bacte-

ria: patterns across marine and freshwater systems. *Limnol. Oceanogr.* 36: 1078–90

3a. Bano N, Moran MA, Hodson RE. 1997. Bacterial utilization of dissolved humic substances from a freshwater swamp. *Aquat. Microb. Ecol.* 12:233–38

4. Bauchop T, Elsden SR. 1960. The growth of microorganisms in relation to their energy supply. *J. Gen. Microbiol.* 23:457–69

5. Bauerfeind S. 1985. Degradation of phytoplankton detritus by bacteria. estimation of bacterial consumption and respiration in an oxygen chamber. *Mar. Ecol. Prog. Ser.* 21:27–36

6. Beeftink HH, van der Heijden RTJM, Heijnen JJ. 1990. Maintenance requirements: energy supply from simultaneous endogenous respiration and substrate consumption. *FEMS Microbiol. Ecol.* 73:203–10

7. Bell WH, Sakshaug E. 1980. Bacterial utilization of algal extracellular products. 2. A kinetic study of natural populations. *Limnol. Oceanogr.* 25:1021–33

8. Benner R, Lay J, K'nees E, Hodson RE. 1998. Carbon conversion efficiency for bacterial growth on lignocellulose: Implications for detritus-based food webs. *Limnol. Oceanogr.* 33:1514–26

9. Benner R, Opsahl S, Chin-Leo G, Richey JE, Forsberg BR. 1995. Bacterial carbon metabolism in the Amazon River system. *Limnol. Oceanogr.* 40:1262–70

10. Biddanda B. 1988. Microbial aggregation and degradation of phytoplankton-derived detritus in seawater. 2. Microbial metabolism. *Mar. Ecol. Prog. Ser.* 42:89–95

11. Biddanda B, Opsahl S, Benner R. 1994. Plankton respiration and carbon flux through bacterioplankton. *Limnol. Oceanogr.* 39:1259–75

12. Billen G. 1984. Heterotrophic utilization and regeneration of nitrogen. In *Heterotrophic Activity in the Sea*, ed. JE Hobbie, PJleB Williams, pp. 313–55. New York: Plenum

13. Billen G, Fontigny A. 1987. Dynamics of a *Phaeocystis*-dominated spring bloom in Belgian coastal waters. *Mar. Ecol. Prog. Ser.* 37:249–57

14. Bjørnsen PK. 1986. Bacterioplankton growth yield in continuous seawater cultures. *Mar. Ecol. Prog. Ser.* 30:191–96

15. Bjørnsen PK, Kuparinen J. 1991. Determination of bacterioplankton biomass, net production and growth efficiency in the Southern Ocean. *Mar Ecol. Prog. Ser.* 71:185–94

16. Blum K, Mills AL. 1991. Microbial growth and activity during the initial stages of seagrass decomposition. *Mar. Ecol. Prog. Ser.* 70:73–82

17. Carlson CA, Ducklow HW. 1996. Growth of bacterioplankton and consumption of dissolved organic carbon in the Sargasso Sea. *Aquat. Microb. Ecol.* 10:69–85

18. Cherrier J, Bauer JE, Druffel ERM. 1996. Utilization and turnover of labile dissolved organic matter by bacterial heterotrophs in eastern North Pacific surface waters. *Mar. Ecol. Prog. Ser.* 139:267–79

19. Chin-Leo G, Benner R. 1992. Enhanced bacterioplankton production and respiration at intermediate salinities in the Mississippi River plume. *Mar. Ecol. Prog. Ser.* 87:87–103

20. Cho BC, Azam F. 1988. Major role of bacteria in biogeochemical fluxes in the ocean's interior. *Nature* 332:441–43

21. Cho BC, Park MG, Shim JH, Azam F. 1996. Significance of bacteria in urea dynamics in coastal surface waters. *Mar. Ecol. Prog. Ser.* 142:19–26

21a. Choi JW, Sherr BF, Sherr EB. Dead or alive? A large fraction of ETS-inactive marine bacterioplankton cells, as assessed by the reduction of CTC, can become ETS-active with incubation and substrate addition. *Aquat. Microb. Ecol.* In press

22. Chróst RH. 1990. Microbial ectoenzymes in aquatic environments. In *Aquatic Microbial Ecology: Biochemical and Molecular Approaches*, ed. J Overbeck, RJ Chróst, pp. 47–78. Berlin: Springer-Verlag

23. Chróst RH, Faust MA. 1983. Organic carbon release by phytoplankton: its composition and utilization by bacterioplankton. *J. Plankton Res.* 5:477–93

24. Coffin RB, Connolly JP, Harris PS. 1993. Availability of dissolved organic carbon to bacterioplankton examined by oxygen utilization. *Mar. Ecol. Prog. Ser.* 101:9–22

25. Cole JJ, Caraco NF, Strayer DL, Ochs C, Nolan S. 1989. A detailed carbon budget as an ecosystem-level calibration of bacterial respiration in an oligotrophic lake during midsummer. *Limnol. Oceanogr.* 34:286–96

26. Cole JJ, Findlay S, Pace ML. 1988. Bacterial production in fresh and saltwater ecosystems: a cross-system overview. *Mar. Ecol. Prog. Ser.* 43:1–10

27. Connolly JP, Coffin RB, Landeck RE. 1992. Modeling carbon utilization by

bacteria in natural water systems. In *Modeling the Metabolic and Physiologic Activities of Microorganisms*, ed. C Hurst, pp. 249–76. New York: Wiley

28. Coveney MF, Wetzel RG. 1989. Bacterial metabolism of algal extracellular carbon. *Hydrobiologia* 173:141–49

29. Crawford CC, Hobbie JE, Webb KL. 1974. Utilization of dissolved free amino acids by estuarine microorganisms. *Ecology* 55:551–63

30. Daneri G, Riemann B, Williams PJL. 1994. In situ bacterial production and growth yield measured by thymidine, leucine and fractioned dark oxygen uptake. *J. Plankton Res.* 16:105–13

31. Dawes EA. 1985. Starvation, survival and energy reserves. In *Bacteria in Their Natural Environment*, ed. M Fletcher, GD Foodgate, pp. 43–79. New York: Academic

31a. Decho AW. 1990. Microbial exopolymer secretions in oceanic environments. *Oceanogr. Mar. Biol. Annu. Rev.* 28:73–153

31b. Decho AW, Moriarty DJ. 1990. Bacterial exopolymer utilization by a harpacticoid copepod: a methodology and results. *Limnol. Oceanogr.* 35:1039–49

31c. del Giorgio PA, Cole JJ, Caraco NF. 1998. Linking planktonic biomass structure to plankton metabolism and net gas fluxes in northern temperate lakes. *Ecology*. In press

32. del Giorgio PA, Cole JJ, Cimbleris A. 1997. Respiration rates in bacteria exceed phytoplankton production in unproductive aquatic systems. *Nature* 385:148–51

33. del Giorgio PA, Gasol JM, Vaque D, Mura P, Agusti S, et al. 1996. Bacterioplankton community structure: Protists control net production and the proportion of active bacteria in a coastal marine community. *Limnol. Oceanogr.* 41:1169–79

34. del Giorgio PA, Prairie YT, Bird DF. 1997. Coupling between rates of bacterial production and the number of metabolically active cells in lake bacterioplankton, measured using CTC reduction and flow cytometry. *Microb. Ecol.* 34:144–54

35. del Giorgio PA, Scarborough G. 1995. Increase in the proportion of metabolically active bacteria along gradients of enrichment in freshwater and marine plankton: implications for estimates of bacterial growth and production. *J. Plankton Res.* 17:1905–24

36. Drabkova VG. 1990. Bacterial production and respiration in the lakes of different types. *Arch. Hydrobiol. Ergeb. Limnol.* 34:209–14

37. Ducklow HW, Carlson CA. 1992. Oceanic bacterial production *Adv. Microb. Ecol.* 12:113–81

38. Dzyuban AN, Timakova TM. 1981. Microflora and decomposition of organic matter in the water and bottom sediments of lake Pert. *Hydrobiol. J.* 17:23–27

39. Egli T. 1991. On multiple-nutrient-limited growth of microorganisms, with special reference to dual limitation by carbon and nitrogen substrates. *Antonie Leeuwenhoek* 60:225–34

40. Egli T, Lendenmann U, Snozzi M. 1993. Kinetics of microbial growth with mixtures of carbon sources. *Antonie Leeuwenhoek* 63:289–98

41. Fallon RD, Pfaender FK. 1976. Carbon metabolism in model microbial systems from a temperate salt marsh. *Appl. Environ. Microbiol.* 31:959–68

42. Fenchel T, Blackburn TH. 1979. *Bacteria and Mineral Cycling*. New York: Academic

43. Findlay S, Carlough L, Crocker MT, Gill HK, Meyer JL, et al. 1986. Bacterial growth on macrophyte leachate and fate of bacterial production. *Limnol. Oceanogr.* 31:1335–41

44. Findlay S, Pace ML, Lints D, Howe K. 1992. Bacterial metabolism of organic carbon in the tidal freshwater Hudson Estuary. *Mar. Ecol. Prog. Ser.* 89:147–53

45. Goldman JC, Caron DA, Dennett MR. 1987. Regulation of gross growth efficiency and ammonium regeneration in bacteria by substrate C:N ratio. *Limnol. Oceanogr.* 32:1239–52

46. Goldman JC, Dennett MR. 1991. Ammonium regeneration and carbon utilization by marine bacteria grown on mixed substrates. *Mar. Biol.* 109:369–78

47. Gosselink JG, Kirby CJ. 1974. Decomposition of salt marsh grass, *Spartina alterniflora* Loisel. *Limnol. Oceanogr.* 19:825–32

48. Goutx M, Acquaviva M, Bertrand J-C. 1990. Cellular and extracellular carbohydrates and lipids from marine bacteria during growth on soluble substrates and hydrocarbons. *Mar. Ecol. Prog. Ser.* 61:291–96

49. Griffith PC, Douglas DJ, Wainright SC. 1990. Metabolic activity of size-fractioned microbial plankton in estuarine, nearshore, and continental shelf waters of Georgia. *Mar. Ecol. Prog. Ser.* 59:263–70

50. Griffiths RP, Caldwell BA, Morita RY. 1984. Observations on microbial percent respiration values in arctic and subarctic marine waters and sediments. *Microb. Ecol.* 10:151–64

51. Güde H, Jürgens K, Parth G, Walser R. 1991. Indications for low net productivity of pelagic bacterioplankton. *Kiel. Meeresforsch. Sonderh.* 8:309–16

52. Haines EB, Hanson RB. 1979. Experimental degradation of detritus made from the salt marsh plants *Spartina alterniflora* Loisel, *Salicornia virginica* L., and *Juncus roemerianus* Scheele. *J. Exp. Mar. Biol. Ecol.* 40:27–40

53. Hanisch K, Schweitzer B, Simon M. 1996. Use of dissolved carbohydrates by planktonic bacteria in a mesotrophic lake. *Microb. Ecol.* 31:41–55

54. Hansell DA, Bates NR, Gundersen K. 1995. Mineralization of dissolved organic carbon in the Sargasso Sea. *Mar. Chem.* 51:201–12

55. Harder J. 1997. Species-independent maintenance energy and natural populations sizes. *FEMS Microbiol. Ecol.* 23:39–44

56. Heikes B, McCully B, Zhou X, Lee YN, Mopper K, et al. 1996. Formaldehyde methods comparison in the remote lower troposphere during Manua Loa Photochemistry Experiment 2. *J. Geophys. Res.* 101:15741–55

57. Heijnen JJ, van Dijken JP. 1992. In search of a thermodynamic description of biomass yields for the chemotrophic growth of microorganisms. *Biotechnol. Bioeng.* 39:833–58

57a. Heissenberger A, Leppard GG, Herndl GJ. 1996. Relationship between the intracellular integrity and the morphology of the capsular envelope in attached and free-living marine bacteria. *Appl. Environ. Microbiol.* 62:4521–29

58. Hessen DO. 1992. Dissolved organic carbon in a humic lake: effects on bacterial production and respiration. *Hydrobiologia* 229:115–23

59. Hobbie JE, Crawford CC. 1969. Respiration corrections for bacterial uptake of dissolved organic compounds in natural waters. *Limnol. Oceanogr.* 14:528–32

60. Hopkinson CS Jr, Buffam I, Hobbie J, Vallino J, Perdue M, et al. 1998. Terrestrial inputs of organic matter to coastal ecosystems: an intercomparison of chemical characteristics and bioavailability. *Biogeochemistry*. In press

61. Hopkinson CS Jr, Sherr BF, Wiebe WJ. 1989. Size fractioned metabolism of coastal microbial plankton. *Mar. Ecol. Prog. Ser.* 51:155–66

61a. Hoppe HG. 1991. Microbial extracellular enzyme activity: a new key parameter in aquatic ecology. In *Microbial Enzymes in Aquatic Environments*, ed. RJ Chróst, pp. 60–83. New York: Springer-Verlag

62. Iturriaga R, Hoppe H-G. 1977. Observations of heterotrophic activity on photoassimilated matter. *Mar. Biol.* 40:101–8

63. Iturriaga R, Zsolnay A. 1981. Transformation of some dissolved organic compounds by a natural heterotrophic population. *Mar. Biol.* 62:125–29

64. Jahnke RA, Craven DB. 1995. Quantifying the role of heterotrophic bacteria in the carbon cycle: a need for respiration rate measurements. *Limnol. Oceanogr.* 40:436–41

65. Joint IR, Morris RJ. 1982. The role of bacteria in the turnover of organic matter in the sea. *Oceanogr. Mar. Biol. Annu. Rev.* 20:65–118

66. Johnson MD, Ward AK. 1997. Influence of phagotrophic protistan bacterivory in determining the fate of dissolved organic matter in a wetland microbial food web. *Microb. Ecol.* 33:149–62

67. Jørgensen NOG, Jensen RE. 1994. Microbial fluxes of free monosaccharides and total carbohydrates in freshwater determined by PAD-HPLC. *FEMS Microbiol. Ecol.* 14:79–94

68. Jørgensen NOG, Kroer N, Coffin RB, Yang X-H, Lee C. 1993. Dissolved free amino acids, combined amino acids, and DNA as sources of carbon and nitrogen to marine bacteria. *Mar. Ecol. Prog. Ser.* 98:135–48

69. Kaplan LA, Bott TL. 1983. Microbial heterotrophic utilization of dissolved organic matter in a piedmont stream. *Freshwater Biol.* 13:363–77

69a. Karl DM. 1980. Cellular nucleotide measurements and applications in microbial ecology. *Microbiol Rev.* 44:739–96

70. Karner M, Furhman JA. 1997. Determination of active marine bacterioplankton: a comparison of universal 16s rRNA probes, autoradiography, and nucleoid staining. *Appl. Environ. Microbiol.* 63:1208–13

71. Katretskiy YA. 1978. Oxygen consumption and efficiency with which the bacterioplankton in the Tsimlyansk Reservoir utilizes the energy contained in organic matter. *Hydrobiol J.* 15:16–19

71a. Kennedy AFD, Sutherland IW. 1987. Analysis of bacterial exopolysaccharides. *Biotechnol. Appl. Biochem.* 9:12–19

72. Kieber DJ, McDaniel, Mopper K. 1989. Photochemical source of biological substrates in seawater: implications for carbon cycling. *Nature* 341:637–39

73. King GM, Berman T. 1985. Potential effects of isotopic dilution on apparent respiration in ^{14}C heterotrophy experiments. *Mar. Ecol. Prog. Ser.* 19:175–80

74. Kirchman DL. 1990. Limitation of bacterial growth by dissolved organic matter in the subarctic Pacific. *Mar. Ecol. Prog. Ser.* 62:47–54

75. Kirchman DL, Rich JH. 1997. Regulation of bacterial growth rates by dissolved organic carbon and temperature in the equatonal pacific ocean. *Microb. Ecol.* 33:11–20

76. Kirchman DL, Suzuki Y, Garside C, Ducklow HW. 1991. High turnover rates of dissolved organic carbon during a spring phytoplankton bloom. *Nature* 352:612–14

77. Kjelleberg S, Flardh KBG, Nystrom T, Moriarty DJW. 1993. Growth limitation and starvation of bacteria. In *Aquatic Microbiology*, ed. TE Ford, pp. 289–320. Boston: Blackwell

78. Koop K, Newell RC, Lucas MI. 1982. Microbial regeneration of nutrients from the decomposition of macrophyte debris on the shore. *Mar. Ecol. Prog. Ser.* 9:91–96

78a. Korber DR, Lawrence JR, Lappin-Scott HM, Costerton W. 1995. Growth of microorganisms on surfaces. In *Microbial Biofilms*, ed. HM Lappin-Scott, JW Costerton, pp. 1–47. Cambridge: Cambridge Univ. Press

79. Kristiansen K, Nielsen H, Riemann B, Fuhrman JA. 1992. Growth efficiencies of freshwater bacterioplankton. *Microb. Ecol.* 24:145–60

80. Kroer N. 1993. Bacterial growth efficiency on natural dissolved organic matter. *Limnol. Oceanogr.* 38:1282–90

81. Laanbroek HJ, Verplanke JC. 1986. Tidal variation in bacterial biomass, productivity and oxygen uptake rates in a shallow channel in the Oosterschelde basin, The Netherlands. *Mar. Ecol. Prog. Ser.* 29:1–5

82. Landwell P, Holme T. 1979. Removal of inhibitors of bacterial growth by dialysis culture. *J. Gen. Microbiol.* 103:345–52

82a. Leppard GG. 1995. The characterization of algal and microbial mucilages and their aggregates in aquatic ecosystems. *Sci. Total Environ.* 165:103–31

83. Linley EAS, Newell RC. 1984. Estimates of bacterial growth yields based on plant detritus. *Bull. Mar. Sci.* 35:409–25

84. Linton JD. 1990. The relationship between metabolite production and the growth efficiency of the producing organism. *FEMS Microbiol. Rev.* 75:1–18

85. Linton JD, Stephenson RJ. 1978. A preliminary study on growth yields in relation to the carbon and energy content of various organic growth substances. *FEMS Microbiol. Lett.* 3:95–98

86. Lucas MI, Newell RC, Velimirov B. 1981. Heterotrophic utilization of kelp (*Ecklonia maxima* and *Laminaria pallida*). II. Differential utilization of dissolved organic components from kelp mucilage. *Mar. Ecol. Prog. Ser.* 4:43–55

87. Marden P, Nystrom T, Kjelleberg S. 1987. Uptake of leucine by a marine gram-negative heterotrophic bacterium during exposure to starvation conditions. *FEMS Microbiol. Ecol.* 45:233–41

88. McFeters GA, Yu FP, Pyle BH, Stewart PS. 1995. Physiological assessment of bacteria using fluorochromes. *J. Microbiol. Methods* 21:1–13

89. Meyer JL, Edwards RT, Risley R. 1987. Bacterial growth on dissolved organic carbon from a blackwater river. *Microb. Ecol.* 13:13–29

90. Middelboe M, Nielsen B, Søndergaard M. 1992. Bacterial utilization of dissolved organic carbon (DOC) in coastal waters—determination of growth yield. *Arch. Hydrobiol. Ergeb. Limnol.* 37:51–61

91. Middelboe M, Søndergaard M. 1993. Bacterioplankton growth yield: a close coupling to substrate lability and beta-glucosidase activity. *Appl. Environ. Microbiol.* 59:3916–21

92. Middelboe M, Søndergaard M. 1995. Concentration and bacterial utilization of sub-micron particles and dissolved organic carbon in lakes and a coastal area. *Arch. Hydrobiol.* 133:129–47

93. Middelboe MB, Jørgensen NOG, Kroer N. 1996. Effects of viruses on nutrient turnover and growth efficiency of noninfected marine bacterioplankton. *Appl. Environ. Microbiol.* 62:1991–97

94. Monod J. 1942. *Recherches sur la Criossance des Cultures Bactériennes.* Paris: Hermann

95. Moran MA, Hodson RE. 1989. Formation and bacterial utilization of dis-

solved organic carbon derived from detrital lignocellulose. *Limnol. Oceanogr.* 34:1034–47

95a. Moran MA, Zepp RG. 1997. Role of photoreactions in the formation of biologically labile compounds from dissolved organic matter. *Limnol. Oceanogr.* 42:1307–16

96. Morita RY. 1997. *Bacteria in Oligotrophic Environments.* New York: Chapman & Hall

96a. Nawrocki MP, Karl DM. 1989. Dissolved ATP turnover in the Bransfield Strait, Antarctica during a spring bloom. *Mar. Ecol. Prog. Ser.* 57:35–44

97. Newell RC, Lucas M. 1981. The quantitative significance of dissolved and particulate organic matter released during fragmentation of kelp in coastal waters. *Kiel. Meeresforsch.* 5:356–69

98. Newell RC, Lucas M, Linley EAS. 1981. Rate of degradation and efficiency of conversion of phytoplankton debris by marine microorganisms. *Mar. Ecol. Prog. Ser.* 6:123–36

99. Nuncio J, Seaton PJ, Kieber RJ. 1995. Biological production of formaldehyde in the marine environment. *Limnol. Oceanogr.* 40:521–27

99a. Pakulski JD, Aas P, Jeffrey W, Lyons M, Von Waasenbergen L, et al. 1998. Influence of light on bacterioplankton production and respiration in a subtropical coral reef. *Aquat. Microb. Ecol.* 14:137–48

100. Payne WJ. 1970. Energy yields and growth of heterotrophs. *Annu. Rev. Microbiol.* 24:17–52

101. Payne WJ, Wiebe WJ. 1978. Growth yield and efficiency in chemosynthetic microorganisms. *Annu. Rev. Microbiol.* 32:155–83

102. Pirt SJ. 1982. Maintenance energy: a general model for energy-limited and energy-sufficient growth. *Arch. Microbiol.* 133:300–2

103. Platpira VP, Filmanovicha RS. 1993. Respiration rate of bacterioplankton in the Baltic Sea. *Hydrobiol. J.* 29:87–94

104. Poindexter JS. 1987. Bacterial response to nutrient limitation. In *Ecology of Microbial Communities*, ed. M Fletcher, TRG Gray, JG Jones, pp. 283–317. Cambridge: Cambridge Univ. Press

105. Pomeroy LR, Hanson RB, McGillivary PA, Sherr BF, Kirchman D, et al. 1984. Microbiology and chemistry of fecal products of pelagic tunicates: rates and fates. *Bull. Mar. Sci.* 35:426–39

106. Pomeroy LR, Sheldon JE, Sheldon WM Jr. 1994. Changes in bacterial numbers and leucine assimilation during estimations of microbial respiratory rates in seawater by the precision Winkler method. *Appl. Environ. Microbiol.* 60:328–32

107. Pomeroy LR, Sheldon JE, Sheldon WM Jr, Peters F. 1995. Limits to growth and respiration of bacterioplankton in the Gulf of Mexico. *Mar. Ecol. Prog. Ser.* 117:259–68

108. Pomeroy LR, Wiebe WJ. 1993. Energy sources for microbial food webs. *Mar. Microb. Food Webs* 7:101–18

109. Pomeroy LR, Wiebe WJ, Deibel D, Thompson RJ, Rowe GT, et al. 1991. Bacterial responses to temperature and substrate concentration during the Newfoundland spring bloom. *Mar. Ecol. Prog. Ser.* 75:143–59

110. Reitner B, Herndl GJ, Herzig A. 1997. Role of ultraviolet-B radiation on photochemical and microbial oxygen consumption in a humic-rich shallow lake. *Limnol. Oceanogr.* 42:950–60

110a. Rich J, Gosselin M, Sherr E, Sherr B, Kirchman D. 1997. High bacterial production, uptake and concentrations of dissolved organic matter in the Central Arctic Ocean. *Deep-Sea Res.* 44:1645–63

110b. Ricker WE, 1973. Liner regression in fishery research. *J. Fish. Res. Bd. Can.* 30:409–34

111. Robertson ML, Mills AL, Zieman JC. 1982. Microbial synthesis of detritus-like particulates from dissolved organic carbon released by tropical seagrasses. *Mar. Ecol. Prog. Ser.* 7:279–85

112. Robinson JD, Mann KH, Novitsky JA. 1982. Conversion of the particulate fraction of seaweed detritus to bacterial biomass. *Limnol. Oceanogr.* 27:1072–79

113. Deleted in proof

114. Roman MR, Ducklow HW, Fuhrman JA, Garside C, Glibert PM, et al. 1988. Production, consumption, and nutrient recycling in a laboratory mesocosm. *Mar. Ecol. Prog. Ser.* 42:39–52

115. Romanenko VI, Perez Eiriz M, Kudryavtsev VM, Pubienes A. 1976. Microbiological processes in the cycle of organic matter in Anabanilja Reservoir, Cuba. *Hydrobiol. J.* 13:8–14

116. Russell JB. 1991. A re-assessment of bacterial growth efficiency: the heat production and membrane potential of *Streptococcus bovis* in batch and continuous culture. *Arch. Microbiol.* 155:559–65

117. Russell JB. Cook GM. 1995. Energetics of bacterial growth: balance of anabolic and catabolic reactions. *Microbiol. Rev.* 59:48–62

118. Sand-Jensen K, Jensen LM, Marcher S, Hansen M. 1990. Pelagic metabolism in eutrophic coastal waters during a late summer period. *Mar. Ecol. Prog. Ser.* 65:63–72

119. Schwaerter S, Søndergaard M, Riemann B, Jensen LM. 1988. Respiration in eutrophic lakes: the contribution of bacterioplankton and bacterial growth yield. *J. Plankton Res.* 3:515–31

120. Schweitzer B, Simon M. 1995. Growth limitation of planktonic bacteria in a large mesotrophic lake. *Microb. Ecol.* 30:89–104

121. Servais P. 1989. Bacterioplankton biomass and production in the river Meuse (Belgium). *Hydrobiologia* 174:99–110

122. Deleted in proof

123. Sherr BF, del Giorgio PA, Sherr EB. 1998. Estimating the abundance and single-cell characteristics of actively respiring bacteria via the redox dye, CTC. Aquat. Microb. Ecol. In press

124. Sherr EB, Sherr BF. 1996. Temporal offset in oceanic production and respiration process implied by seasonal changes in atmospheric oxygen: the role of heterotrophic microbes. *Aquat. Microb. Ecol.* 11:91–100

125. Sinsabaugh RL, Findlay S, Franchini P, Fischer D. 1997. Enzymatic analysis of riverine bacterioplankton production. *Limnol. Oceanogr.* 42:29–38

126. Smith REH, Harrison WG, Irwin B, Platt T. 1986. Metabolism and carbon exchange in microplankton of the Grand Banks (Newfoundland). *Mar. Ecol. Prog. Ser.* 34:171–83

127. Smith SV, Hollibaugh JT. 1993. Coastal metabolism and the oceanic organic carbon balance. *Rev. Geophys.* 31:75–89

128. Søndergaard M, Middelboe M. 1995. A cross-system analysis of labile dissolved organic carbon. *Mar. Ecol. Prog. Ser.* 118:283–94

129. Søndergaard M, Theil-Nielsen J. 1997. Bacterial growth efficiency in lakewater cultures. *Aquat. Microb. Ecol.* 12:115–22

130. Sorokin YI. 1971. On the role of bacteria in the productivity of tropical oceanic waters. *Int. Rev. Ges. Hydrobiol.* 56:1–48

131. Sorokin YI, Mameva TI. 1980. Rate and efficiency of the utilization of labile organic matter by planktonic microflora in

coastal Peruvian waters. *Pol. Arch. Hydrobiol.* 27:447–56

132. Stevenson LH. 1978. A case for bacterial dormancy in aquatic systems. *Microb. Ecol.* 4:127–33

133. Stouthamer AH. 1973. A theoretical study on the amount of ATP required for synthesis of microbial cell material. *Antonie Leeuwenhoek* 39:545–65

134. Stouthamer AH. 1979. The search for correlation between theoretical and experimental growth yields. *Int. Rev. Biochem. Microb. Biochem.* 21:1–47.

135. Stouthamer AH, Bettenhaussen C. 1973. Utilization of energy for growth and maintenance in continuous and batch cultures of microorganisms. *Biochim. Biophys. Acta* 301:53–70

136. Stuart V, Lucas MI, Newell RC. 1981. Heterotrophic utilization of particulate matter from the kelp *Laminaria pallida*. *Mar. Ecol. Prog. Ser.* 4:337–48

137. Stuart V, Newell RC, Lucas MI. 1982. Conversion of kelp debris and faecal material from the mussel *Aulacomya ater* by marine micro-organisms. *Mar. Ecol. Prog. Ser.* 7:47–57

138. Tempest DW. 1978. The biochemical significance of microbial growth yields: a reassessment. *Trends Biochem. Sci.* 3:180–84

139. Tempest DW, Neijssel OM. 1992. Physiological and energetic aspects of bacterial metabolite overproduction. *FEMS Microbiol. Lett.* 100:169–76

140. Tempest DW, Neijssel OM, Texeira de Mattos MJ. 1985. Regulation of carbon substrate metabolism in bacteria growing in chemostat culture. In *Environmental Regulation of Microbial Metabolism*, ed. IS Kulaev, EA Dawes, DW Tempest, pp. 53–68. New York: Academic

141. Therkildsen MS, Isaksen MF, Lomstein BA. 1997. Urea production by the marine bacteria *Delaya venuts* and *Pseudomonas stutzeri* grown in a minimal medium. *Aquat. Microb. Ecol.* 13:213–17

142. Thingstad TF, Hagström Å, Rassoulzadegan F. 1997. Accumulation of degradable DOC on surface waters: Is it caused by a malfunctioning microbial loop? *Limnol. Oceanogr.* 42:398–404

143. Tortell PD, Maldonado MT, Price NM. 1996. The role of heterotrophic bacteria in iron-limited ocean ecosystems. *Nature* 383:330–32

144. Tranvik LJ. 1988. Availability of dissolved organic carbon for planktonic bacteria in oligotrophic lakes of differing humic content. *Microb. Ecol.* 16:311–22

145. Tranvik LJ. 1990. Bacterioplankton growth on fractions of dissolved organic carbon of different molecular weights from humic and clear lakes. *Appl. Environ. Microbiol.* 56:1672–77

146. Tranvik LJ. 1992. Rapid microbial production and degradation of humic-like substances in lake water. *Arch. Hydrobiol. Beih.* 37:43–50

147. Tranvik LJ, Høffle MG. 1987. Bacterial growth in mixed cultures on dissolved organic carbon from humic and clear water lakes. *Appl. Environ. Microbiol.* 53:482–28

148. Tulonen T, Salonen K, Arvola L. 1992. Effects of different molecular weight fractions of dissolved organic matter on the growth of bacteria, algae and protozoa from a highly humic lake. *Hydrobiologia* 229:239–52

149. Tupas L, Koike I. 1990. Amino acid and ammonium utilization by heterotrophic marine bacteria grown in enriched seawater. *Limnol. Oceanogr.* 35:1145–55

150. Turley CM, Lochte K. 1990. Microbial response to the input of fresh detritus to the deep-sea bed. *Paleogeogr. Paleoclimatol. Paleoecol.* 89:3–23

151. Vallino JJ, Hopkinson CS, Hobbie JE. 1996. Modeling bacterial utilization of dissolved organic matter: optimization replaces Monod growth kinetics. *Limnol. Oceanogr.* 41:1591–1609

152. Westerhoff HV, Hellingwerf KJ, Van Dam K. 1983. Thermodynamic efficiency of microbial growth is low but optimal for maximal growth rate. *Proc. Natl. Acad. Sci. USA* 80:305–9

153. White PA, Kalff J, Rasmussen JB, Gasol JM. 1991. The effect of temperature and algal biomass on bacterial production and specific growth rate in freshwater and marine habitats. *Microb. Ecol.* 21:99–118

154. Wiebe WJ, Smith DF. 1977. Direct measurement of dissolved organic carbon release by phytoplankton and incorporation by microheterotrophs. *Mar. Biol.* 42:213–33

155. Williams PJleB. 1981. Microbial contribution to overall marine plankton metabolism: direct measurements of respiration. *Oceanol. Acta* 4:359–64

156. Zweifel UL, Riemann B, Hagström Å. 1993. Consumption of dissolved organic carbon by marine bacteria and demand for inorganic nutrients. *Mar. Ecol. Prog. Ser.* 101:23–32

Annu. Rev. Ecol. Syst. 1998. 29:543–66

THE CHEMICAL CYCLE AND BIOACCUMULATION OF MERCURY

François M. M. Morel
Department of Geosciences, Guyot Hall, Princeton University, New Jersey 08544;
e-mail: morel@geo.princeton.edu

Anne M. L. Kraepiel
Department of Chemistry, Frick Chemical Laboratory, Princeton University, Princeton, New Jersey 08544

Marc Amyot
Université du Québec, Institut National de la Recherche Scientifique, INRSEAU, C.P. 7500, Sainte-Foy, QC, G1V 4C7, Canada

KEY WORDS: methylation, biomagnification, speciation, solubility, microbial uptake

---

ABSTRACT

Because it is very toxic and accumulates in organisms, particularly in fish, mercury is an important pollutant and one of the most studied. Nonetheless we still have an incomplete understanding of the factors that control the bioconcentration of mercury. Elemental mercury is efficiently transported as a gas around the globe, and even remote areas show evidence of mercury pollution originating from industrial sources such as power plants. Besides elemental mercury, the major forms of mercury in water are ionic mercury (which is bound to chloride, sulfide, or organic acids) and organic mercury, particularly methylmercury. Methylmercury rather than inorganic mercury is bioconcentrated because it is better retained by organisms at various levels in the food chain. The key factor determining the concentration of mercury in the biota is the methylmercury concentration in water, which is controlled by the relative efficiency of the methylation and demethylation processes. Anoxic waters and sediments are an important source of methylmercury, apparently as the result of the methylating activity of sulfate-reducing bacteria. In surface waters, methylmercury may originate from anoxic

0066-4162/98/1120-0543$08.00

layers or be formed through poorly known biological or chemical processes. Demethylation is effected both photochemically and biologically.

INTRODUCTION

Mercury is a pervasive pollutant that accumulates in organisms and is highly toxic. As a result, it is probably the most studied of all trace elements in the environment. Known in mythology for its fleet-footedness, mercury rapidly spreads all over the earth from its natural and anthropogenic sources. It is also elusive, and the ways by which it is transformed in the environment and bioaccumulated remain perplexing.

In this review, after a brief overview of the global mercury cycle, we focus on the transformations of mercury in aquatic systems, for it is ultimately the accumulation of mercury in fish that is of concern to us. Even in remote regions, methylmercury in fish is often near to and sometimes exceeds the concentration deemed safe for human consumption (0.5–1 ppm). The question is straightforward, even if the answer is not: How do concentrations of parts per trillion of mercury in water yield concentrations of parts per million in fish? To begin to answer this question we do not provide an extensive review of the immense literature on mercury in the environment. Rather, we examine what is known of the chemical and biological mechanisms that effect the transformations of mercury in water and ultimately control its bioaccumulation in fish.

THE GLOBAL CYCLE OF MERCURY

The only metal to be liquid at room temperature, elemental mercury Hg^0 (1) is also a gas, Hg^0 (g), with little tendency to dissolve in water (57, 65). Natural waters are usually supersaturated in Hg^0 (aq) compared to the air above, and volatilization thus results in a flux of Hg^0 from the water into the atmosphere (see Figure 1) (28). This supersaturation is maximal during summer days, when photoreduction of Hg(II) in surficial waters is at its peak (5, 6, 8, 67, 83). In the atmosphere, where approximately 95% of total mercury is in the elemental state, Hg^0, it is slowly oxidized to the mercuric (+II) state, Hg(II). Most of this oxidation occurs at the solid-liquid interface in fog and cloud droplets. Ozone is probably the main oxidant in this process, with HClO, HSO_3^-, and $OH^{\bullet}$ being also significant (54–56). Gas-phase oxidation reactions of Hg^0 by O_3, Cl_2, and H_2O_2 may sometimes be important, although large uncertainties exist regarding their rates (70). Some of the Hg(II) produced in the atmosphere is re-reduced by mechanisms involving either SO_3^- as the reductant (56), or photoreduction of $Hg(OH)_2$ (84).

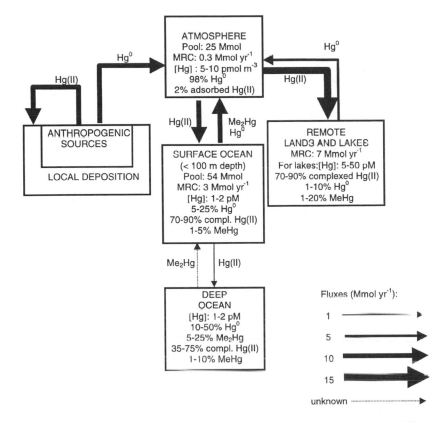

Figure 1   Global Hg cycle. The width of the *arrows* corresponds to the importance of the different fluxes, and are estimated from Mason et al (46). Main Hg species contributing to these fluxes are identified. MRC stands for modern rate of change for global reservoirs (35). Concentrations and species distributions are average from different sources, in particular Fitzgerald & Mason (27) and Meili (49). Caveat: The MRC estimates from Hudson et al (35) correspond to less evasion from land and reduced deposition compared to the fluxes from Mason et al (46).

The return of mercury from the atmosphere to the Earth's surface occurs chiefly via wet precipitation of dissolved Hg(II). Adsorption of mercury on aerosols such as soot also promotes its deposition, especially over land (46), where aerosols are abundant. Because Hg⁰ reoxidizes relatively slowly to the mercuric state Hg(II), its residence time in the atmosphere is on the order of a year (27) or perhaps less (SE Lindberg, personal communication). This is sufficient time for atmospheric mercury to be distributed over the entire planet before returning to the land, lakes, sea, and ice. As a result, while the principal emissions of mercury are from point sources concentrated in industrial regions, mercury pollution is truly global, affecting the most remote areas of the planet (see Figure 1). Historical records from lake sediments provide the most compelling evidence that remote areas receive significant inputs of anthropogenic Hg by long-range atmospheric transport (26).

Once oxidized, 60% of atmospheric mercury is deposited to land and 40% to water, even though land represents only 30% of the Earth's surface (46). The greater proportion of Hg deposition on land presumably reflects the proximity of its sources since water precipitation is three times less on land than on the oceans. In oceanic waters, after it undergoes a complex set of chemical and biological transformations, most of the Hg(II) is reduced to Hg^0 and returned to the atmosphere; only a small fraction is permanently exported to the sediments (46). Thus the mercury inventories in the atmosphere and surface seawater are tightly coupled by an effective precipitation/volatilization cycle driven by oxidation/reduction reactions. In lakes, the main loss mechanisms for mercury are sedimentation and gas evasion. The relative importance of each is still the subject of debate and seems to be a function of the concentration of reducible Hg in the epilimnion (28). Similar processes occur on land, resulting apparently in a smaller return of reduced mercury to the atmosphere and a greater permanent burial in soils. In the case of uncontaminated soils, net dry-weather Hg deposition (dry deposition > gas evasion) is sometimes observed, about three times less frequently than net emission (dry deposition < gas evasion) (42). Contaminated sites, however, consistently display important net emission fluxes (30, 42). Compared to its atmospheric flux, little mercury is transported by rivers.

Anthropogenic sources of mercury come from metal production, chlor-alkali, and pulp industries, waste handling and treatment, and coal, peat, and wood burning (43). Natural inputs to the atmosphere include degassing and wind entrainment of dust particles from land, notably from mercuriferous areas, volcanic eruptions, forest fires, biogenic emissions of volatile and particulate compounds, and degassing from water surfaces (63). Among those sources, degassing from natural mercury-rich geological formations may have been underestimated in the past (30, 31, 63). Based on lake sediment records (77), it is estimated that the atmospheric inputs of mercury have tripled over the past 150 years (46). This indicates that two thirds of the mercury now in the atmosphere, and hence in surface seawater, is of anthropogenic origin, and one third is from natural sources (see Figure 1).

THE CHEMISTRY OF MERCURY IN SURFACE WATERS

Chemical Speciation in Oxic Waters

In oxic surface freshwaters from uncontaminated sites, mercury at concentrations of 5–100 pM ($= 5$–100×10^{-12} mol/L $= 1$–20 parts per trillion) occurs in several physical and chemical forms (see Figures 1, 2, and 3). The

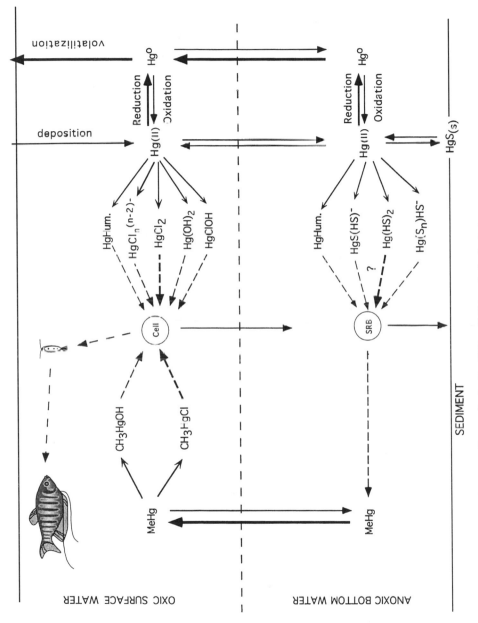

Figure 2   Aquatic cycle of mercury. See text for details. SRB = Sulfate-reducing bacteria.

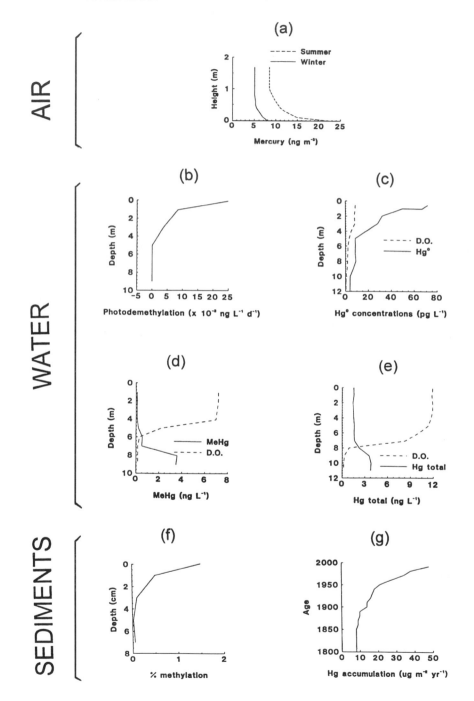

(a)

(b)

(c)

(d)

(e)

(f)

(g)

partitioning of Hg between the dissolved, colloidal and particulate phases varies widely spatially, seasonally and with depth in the water column. Some of this variation seems to be related to temporal changes in living particulate matter, mostly phytoplankton and bacteria (23, 36). The concentration of particulate Hg per unit particle weight is relatively constant reflecting perhaps a sorption equilibrium between dissolved and particulate phases (49). The exact chemical form of particulate mercury is unknown, although most of it is probably tightly bound in suspended organic matter. Adsorption of Hg to oxyhydroxides may also be important in lakes. In fact, the commonly observed enrichment of MeHg and Hg(II) in anoxic waters of lakes may result from the sedimentation of mercury-laden oxyhydroxides of iron and manganese from the epilimnion and their dissolution in the anoxic hypolimnion (49).

Dissolved Hg is distributed among several chemical forms: elemental mercury ($Hg^0_{(aq)}$), which is volatile but relatively unreactive, a number of mercuric species ($Hg(II)$), and organic mercury, chiefly methyl (MeHg), dimethyl (Me_2Hg), and some ethyl (EtHg) mercury.[1] In general, and particularly in stratified systems, concentrations of Hg^0 are higher near the air-water interface whereas levels of total Hg and MeHg are higher near the sediments (see Figure 3). An operationally defined fraction of total Hg (the sum of particulate and dissolved mercury), the so-called "reactive" Hg (measured after a $SnCl_2$ reduction step), is considered to be a good predictor of the naturally reducible Hg (28). It probably corresponds to the inorganically bound fraction of $Hg(II)$.

According to thermodynamic calculations (74), the divalent mercury in surface waters, $Hg(II)$, is not present as the free ion Hg^{2+} but should be complexed

---

[1]The simultaneous presence of Hg^0 and $Hg(II)$ in natural waters, both oxic and anoxic, brings up the question of the possible formation of $Hg(I)$, the mercurous form of Hg, which is only stable in water as the dimer Hg_2^{2+}. Simple calculations based on the constants in Table 1 show that a negligible fraction of either $Hg(II)$ or $Hg(0)$ may be present as $Hg(I)$ in natural waters when the concentrations are below 0.1 nM. Stabilization by an unknown ligand with much higher affinity for Hg_2^{2+} than for Hg^{2+} seems highly improbable. Note, however, that in many laboratory experiments performed with mercury concentrations in excess of 1 nM, and where Hg^0 may be present by design or as a contaminant, the formation of Hg_2^{2+} may be a complicating and easily overlooked factor.

Figure 3 Vertical profiles of mercury species concentrations and of transformation rates in air, water, and sediments. (*a*) Hg^0 height profile over the surface of a contaminated pond in summer and winter (10); (*b*)–(*e*) depth profiles of mercury photodemethylation (71), Hg^0 (8), MeHg (14), and total Hg (7) levels in different remote temperate forested lakes; (*f*) depth profile of mercury methylation in profundal lake sediments, expressed as percentage of total added mercury methylated after 24 h (40); (*g*) ^{210}Pd-dated depth profile of mercury accumulation rates in a western Minnesota lake (25).

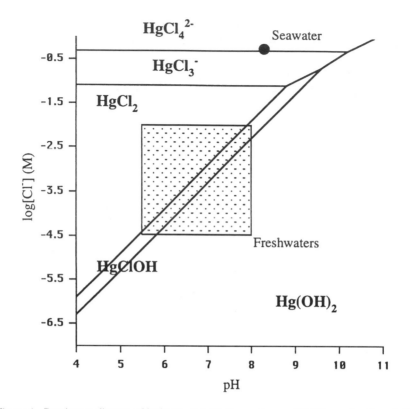

Figure 4 Dominance diagram of hydroxo- and chloro-complexes of Hg(II) as a function of pH and chloride concentrations (see Table 1). Ionic strength corrections were neglected. Seawater has a pH of 8.3 and a chloride concentration of 0.55 M. The pH and chloride concentration range of freshwaters was taken from Davies & DeWiest (24).

in variable amounts to hydroxide ($Hg(OH)^+$, $Hg(OH)_2$, $Hg(OH)_3^-$), and to chloride ($HgCl^+$, $HgClOH$, $HgCl_2$, $HgCl_3^-$, $HgCl_4^{2-}$) ions depending on the pH and the chloride concentration (see Figure 4 and Table 1). It is also possible that, even in oxic surface waters, some or much of Hg(II) might be bound to sulfides (S^{2-} and HS^-; see Table 1), which have been measured at nanomolar concentrations in surface seawater (45). In addition, an unknown fraction of Hg(II) is likely bound to humic acids, the assemblage of poorly defined organic compounds that constitute 50–90% of the dissolved organic carbon (DOC) in natural waters. According to Meili (49), nearly 95% of inorganic oxidized mercury in lakes is bound to dissolved organic matter. The nature of the chemical moieties responsible for the binding of Hg(II) and the thermodynamic

Table 1   Relevant acidity and thermodynamic constants for Hg_2^{2+}, Hg^{2+}, and CH_3Hg^+

Dissolution & Volatilization of Hg^0

$Hg^0(l) = Hg^0(aq)$	$K = 3.30\ 10^{-7}$ mol/L	[Clever et al (22)]
$Hg^0(g) = Hg^0(aq)$	$K = 2.56.10^{-3}$ mol.l^{-1}.atm^{-1}	[Sanemasa (65)]

Dismutation of Hg(I)

$Hg^{2+} + Hg^0(aq) = Hg_2^{2+}$	$K = 10^{8.46}$	[Hietanen & Sillen (33)]

Acidity constants

Acid/base couple	H_2S/HS^-	HS^-/S^{2-}	$Hg(SH)_2/Hg(SH)S^-$	$Hg(SH)S^-/HgS_2^{2-}$
pKa	7.02	14.6	6.33	8.12

Hg(I) complexes

Complex	Hg_2Cl_2
log K	12.4

Hg(II) complexes

Complex	$HgCl^+$	$HgCl_2$	$HgCl_3^-$	$HgCl_4^{2-}$	$HgOH^+$	$Hg(OH)_2$	$Hg(OH)_3^-$	HgClOH
log K	7.2	14	15.1	15.4	10.6	21.8	20.9	18.1

Complex	$Hg(SH)_2$	$Hg(SH)S^-$	HgS_2^{2-}	$Hg(S_n)HS$	HgS (red)	HgS (black)
log K	36.6	46.8	52.6	-3.7	53.3	52.7

MeHg complexes

Complex	CH_3HgCl	CH_3HgOH	CH_3HgS^-
log K	5.5	9.6	21.5

Constants are given as logarithms of the overall formation constants for complexes (e.g. for the reaction M + nL = ML_n, where M is Hg^{2+} or CH_3Hg^+ and L is a ligand). For solids ($HgS_{(s)}$), the constant corresponds to the precipitation reaction. For the mercury polysulfide complex, $Hg(S_n)HS^-$, the constant K is given for HgS (cinn) $SH^- + (n-1)\ S^0(rhom) = Hg(S_n)HS^-$. All constants are given at ionic strength I = 0 and are taken from Smith & Martell (72), except for $Hg(S_n)HS^-$ (58), Hg_2Cl_2 (59, 60), and those otherwise indicated in the table.

properties of the complexes have been little studied (34, 44). Through its binding to DOC, Hg can be mobilized from the drainage basin and transported to lakes (50, 51, 81). The reactions of ionic mercury are relatively fast, and it is thought that the various species of Hg(II), including those in the particulate phase, are at equilibrium with each other.

In the organometallic species of mercury, the carbon-to-metal bonds are stable in water because they are partly covalent and the hydrolysis reaction (see below), which is thermodynamically favorable (and makes the organometallic species of most others metals unstable), is kinetically hindered. As a result, the dimethyl mercury species, Me_2Hg ($= CH_3HgCH_3$), is unreactive. The monomethylmercury species, MeHg, is usually present as chloro- and hydroxo-complexes (CH_3HgCl and CH_3HgOH) in oxic waters (see Figure 5 and Table 1).

Reduction of Hg(II)

The processes that transform mercury between its elemental and ionic or organic forms determine how much mercury is in the elemental state, thus how

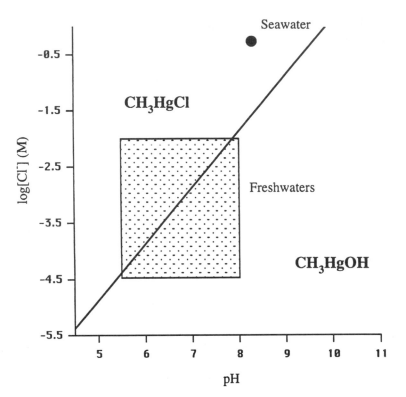

Figure 5 Dominance diagram of hydroxo- and chloro-complexes of methylmercury MeHg as a function of pH and chloride concentrations (see Table 1). Ionic strength corrections were neglected. Seawater has a pH of 8.3 and a chloride concentration of 0.55 M. The pH and chloride concentration range of freshwaters was taken from Davies & DeWiest (24).

quickly it volatilizes and, ultimately, how much total mercury remains in the water (see Figures 1 and 2). These processes are beginning to be understood. Reduction of ionic to elemental mercury may be effected by biological or chemical processes. Some published data show that most of the Hg(II) reduction in incubation bottles is linked to the presence of particles, implicating microorganisms (47). More recent data, however, show that, in many cases, photoreduction rather than microbial reduction is the principal mechanism (6, 8, 41). While it is likely that there are variations in time and space in the relative importance of these two processes, the explanation for this apparent contradiction may lie in the differences between experimental conditions. The experiments showing microbial Hg(II) reduction were conducted with additions of Hg(II) of 0.3–0.9 nM (47). These concentrations are above the threshold value of ca 50 pM, which is now known to induce the *mer*-operon[2] in bacteria (62; I Schaperdoth and FMM Morel, unpublished data). Microbial reduction via induction of the *mer*-reductase likely explains these data. In contrast, the experiments showing the dominance of photoreductive mechanisms were conducted at Hg(II) concentration of 3–20 pM, below the threshold for induction of the reductase. The efficiency of the photoreduction depends on levels of reducible Hg(II) complexes and radiation wavelength and intensity. When present at high concentrations, DOC seems to act as a competitive inhibitor for solar radiation, scavenging UV radiation before it can photoreduce Hg(II). As a result, higher photoreduction rates have been observed in clear, low-DOC lakes (6).

The mechanism for this photoreduction is still uncertain. Photoreduction of Fe, Mn, or humic acids may be implicated. The reduced metals [Fe(II), Mn(II)] or organic moieties (hydroquinones and semiquinones) formed photochemically could, in turn, reduce Hg(II) when they reoxidize, as they are known to do for other elements (39). Alternatively, direct photoreduction of $Hg(OH)_2$, $Hg(HS)_2$ (73), or DOC-bound mercury is possible (85). Part of the light dependence of the reduction may result from the activity of photosynthetic phytoplankton and cyanobacteria. Ben Bassat & Mayer (13) noted that reduction of Hg(II) to Hg^0 was accelerated by illumination of *Chlorella* cells. In their study, formation of Hg^0 decreased in concert with inhibition of photosynthesis. These authors suggested that light increased the amount of leakage from the cells of a metabolite capable of reducing Hg. Several studies have also shown that phytoplankton can externally reduce various species of Cu(II) and Fe(III) by cell-surface enzymatic processes that are inhibited by photosynthetic inhibitors (37, 38, 61). Such enzymatic processes also probably contribute to Hg reduction in the photic zone. Since photoreduction of Hg has been observed in uncontaminated environments under diverse conditions (pH: 4.5–8.3; DOC: 1–32 mg/L; total Hg: 2–20 pM; salinity: <1–30‰) and was induced by visible and UV radiation, it is likely that more than one of those processes are involved (5–8).

At the natural mercury concentrations in the low picomolar range, reduction thus seems to be effected chiefly by photochemical processes, whereas in polluted waters, when the mercury concentration exceeds 50 pM, microbial reduction via the MerA reductase likely becomes the predominant mechanism of Hg(II) reduction.

---

[2]The *mer*-operon, one of the best studied metal resistance mechanisms in bacteria, consists of a series of enzyme-encoding genes whose transcription is de-repressed by Hg(II). These enzymes include a MerT membrane protein that transports Hg(II) into the cell and a MerA reductase that reduces Hg(II) to Hg^0. Some *mer* also contain the gene for a MerB lyase that hydrolyzes organomercury compounds. The *mer*-operon is usually encoded on a plasmid and has been shown to be transferable among bacterial species (66, 76).

Oxidation of Elemental Mercury

Until very recently, it was thought that the oxidation of Hg^0 to $Hg(II)$ in natural waters was negligible or inexistent. However, recent data show that this may not be true in seawater (87). The presence of high chloride concentrations and of appropriate particle surfaces catalyze the oxidation of Hg^0 by oxygen, resulting in rates as high as 10% per hour in natural seawater (5). This oxidation may be more important in coastal areas, where particulate matter loadings are higher. One should note that an effective surface for the catalysis of Hg^0 oxidation is that of liquid mercury (M Amyot, unpublished). Thus, pools or droplets of liquid mercury that may be present in oxic seawater as a result of some human activity should be oxidized relatively efficiently.

Demethylation Reactions

As mentioned above, the hydrolysis reaction of MeHg,

$$CH_3\text{-}Hg^+ + H^+ \rightarrow CH_4 + Hg^{2+},$$

is thermodynamically favorable but kinetically hindered, and MeHg is thus stable in aqueous solution. However, the kinetic hindrance of this reaction can be overcome by enzymatic or photochemical mechanisms, and methylmercury has been shown to be degraded by some bacteria and by light.

Some *mer* operons (see above) carry a gene, MerB, for an organomercury lyase that confers bacterial resistance to organomercury compounds. The MerB enzyme catalyzes the hydrolysis reaction shown above, leading to the formation of $Hg(II)$. The $Hg(II)$ ion formed is then reduced to Hg^0 by the mercuric ion reductase MerA (53). There are no direct field data quantifying the importance of this mechanism in nature, but one may infer from the involvement of the MerA reductase that it may be induced in polluted water only when the Hg concentration exceeds 50 pM.

MeHg has been shown to be photodegraded in oxic waters in lakes and seawater (71, 75; see Figure 3). The reaction rate is first-order with respect to MeHg concentration and sunlight radiation, and is not associated with the particulate phase (71). Singlet oxygen generated by photochemical reactions is likely responsible for this degradation (75). Photodegradation is probably the main degradation pathway for methylmercury in oxic water bodies with low mercury concentrations (<50 pM).

Sources of Methylmercury in Surface Waters

Methylation is believed to occur mainly in anoxic waters and sediments; in most lakes, the MeHg at the surface originates from the anoxic water below, whence it is transported by diffusion and advection (see Figure 2). However, significant MeHg levels in the surface waters of the oceans and Great Lakes, for which transport of MeHg from deep waters is negligible, clearly indicate that there may be some MeHg production in oxic waters. The mechanism for the methylation is still uncertain, although most of the reaction is probably driven by microbial processes similar to those observed near the sediments (49). However, in lakes some of it may result from dark (82) or photochemical processes involving humic acids (see below). In the oceans, some MeHg could be formed by the partial demethylation of $(Me)_2Hg$ upwelled from deep waters, where it is itself formed by unknown biological mechanisms (27). In some rare cases, the atmosphere may be a significant source of MeHg, although most surface waters are a source rather than a sink for atmospheric organic mercury.

THE CHEMISTRY OF MERCURY IN ANOXIC WATERS AND SEDIMENTS

Chemical Speciation in Anoxic Waters

The mercuric ion exhibits extremely high affinity for sulfide. This property controls the chemistry of mercury in anoxic waters and sediments. The speciation of dissolved Hg(II) in sulfidic waters is completely dominated by sulfide and bisulfide complexes (HgS_2H_2, HgS_2H^- and HgS_2^{2-}), even at total sulfide, $S(-II)$, concentrations as low as 1 nM (see Figure 6 and Table 1). The only important sulfide complex of MeHg is CH_3HgS^- (see Figure 7). Two forms of solid mercuric sulfide, HgS(s), are known. the black form (metacinnabar) is metastable at room pressure and temperature, and in solution, it spontaneously evolves into the red form (cinnabar) over days. Both cinnabar and metacinnabar have a very low solubility product (see Table 1), and HgS(s) is thought to be the particulate mercury species that is buried in sediments and controls Hg(II) solubility in anoxic waters. It is difficult, however, to ascertain analytically the exact chemical nature of the traces of mercury present in natural sediments, and it is possible that, rather than being precipitated as HgS(s), sedimentary mercury is bound to particulate organic matter or even to inorganic particles such as iron oxides (78). Recently, authigenic submicron crystals of metacinnabar [black HgS(s)] have been identified in contaminated soils, using various electron microscopy techniques (11).

Although the solubility product of cinnabar is extremely low, its actual solubility increases at high $S(-II)$ concentrations, due to the formation of the dissolved sulfide and bisulfide mercuric complexes (see Figure 6). For example, at pH = 7, the dissolved mercury concentration of a water body at equilibrium with HgS(s) increases from 3 pM for $S(-II) = 1\ \mu M$ to 3 nM for $S(-II) = 1$ mM. This increasing solubility of mercury with sulfide concentration undoubtedly plays a role in the high dissolved mercury concentrations observed in many

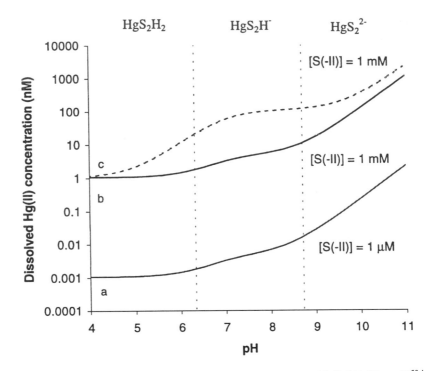

Figure 6 Calculated dissolved Hg(II) concentrations at equilibrium with HgS(s) [$K_s = 10^{52.1}$; $I = 0$; Schwarzenbach & Widmer (68)] in the presence of added sulfides (see Table 1); [Cl$^-$] = 1 mM. (*a*) and (*b*) (*solid lines*): no elemental sulfur is present. (*c*) (*dotted line*): the solution is at equilibrium with $S^0_{8(rhom.)}$; in that case, the dominant mercury complex is Hg(S$_n$)HS$^-$ for pH > 5. The *vertical lines* delimit the predominance regions of the sulfide and disulfide complexes.

anoxic waters. There is also recent evidence for the formation of polysulfide mercury complexes, Hg(S$_n$)SH$^-$ (n = 4–6) in the presence of elemental sulfur S(0) (58). Significant S(0) concentrations have often been measured in anoxic waters (58), and polysulfide complexes could in some cases dominate mercury speciation and increase its solubility even further (see Table 1, Figure 6).

In addition, we note that cinnabar, which is a semiconductor, can be dissolved by visible light. The dissolution rate increases at high sulfide concentrations and leads to the production of Hg^0 (AML Kraepiel and FMM Morel, unpublished data).

Reduction of Hg(II) in Anoxic Waters

As in oxic waters, Hg(II) can be reduced in anoxic waters by the activity of bacteria carrying the *mer*-operon, if the Hg levels are sufficiently high.

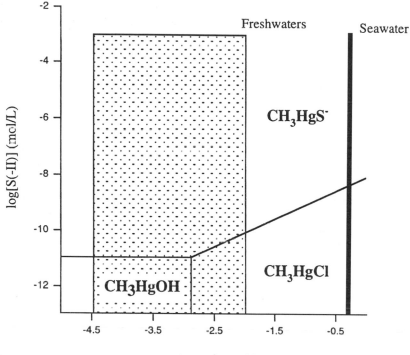

Figure 7 Dominance diagram of hydroxo-, chloro- and sulfide complexes of MeHg at pH = 7 as a function of chloride and total sulfide S(−II) concentrations (see Table 1). Ionic strength corrections were neglected. Seawater has a pH of 8.3 and a total sulfide concentration varying from 1 nM to 1 mM. The chloride concentration range of freshwaters was taken from Davies & DeWiest (24), and the total sulfide concentration was taken to vary from 1 nM to 1 mM.

Alternatively, abiotic reduction of mercury in the dark may be effected by humic substances (3, 9). According to Allard & Arsenie (4), this process is optimal in the absence of chloride and at pH circa 4.5. These authors have suggested that the formation of a complex between Hg and humic acids was necessary in order for the reduction to take place. A similar dark reduction process has been studied for other metals such as iron (79).

The Methylation of Mercury

Methylation of a metal—i.e. the transfer of a methyl group from an organic compound to the metal ion—is not a facile chemical reaction, at least in aqueous solution. It ultimately requires the transfer of an alkyl anion group (such as CH_3^-), a strong base highly unstable in water. Thus, methylation reactions either are the result of photochemical processes or need to be catalyzed by microorganisms. It is possible that photochemical reactions involving, for example, acetate or humic acids may lead to the formation of methylmercury in natural waters. Laboratory data have shown that Hg(II) is photomethylated in the presence of acetate (1,2), but there are no direct field data implicating photoproduction of MeHg. As discussed above, field studies show a net photochemical demethylation in oxic surface waters. In anoxic waters with sufficient light penetration (like those that support the growth of green and purple sulfur bacteria), it is conceivable that, in the absence of species such as singlet oxygen, net photomethylation would be observed. However, the methylmercuric sulfide ion (CH_3HgS^-), which is the dominant form of methylmercury in anoxic water (see Figure 7), has been shown to be readily decomposed by sunlight to CH_4 and HgS (12). Nonetheless, a balance of photochemically induced methylation and demethylation reactions may be important in maintaining low levels of MeHg in some natural waters (such as the surface of deep lakes and oceans).

There has long been massive circumstantial evidence that sulfate-reducing bacteria are responsible for the bulk of mercury methylation in natural waters (29): Sulfate-reducers in cultures are effective at methylating mercury; methylation rates are observed to correlate in time and space with the abundance and activity of sulfate-reducers; and the addition to natural samples of molybdate, a specific inhibitor of sulfate reduction, inhibits mercury methylation.

Recently, mechanistic evidence has been obtained to support the dominant role of sulfate-reducers in mercury methylation. In laboratory cultures with very elevated mercury concentrations (0.5 mM), the bacterium *Desulfovibrio desulfiricans* was shown to produce large amounts of MeHg (18, 19). The methylation of Hg(II) is enzymatically mediated in the presence of cobalamin (20). The higher methylation rates observed during fermentative growth compared to sulfate-reducing conditions may be due to the presence of pyruvate, which is necessary for the functioning of the enzyme. The nature of the enzyme has still to be investigated to resolve whether mercury methylation is the result of a specific process or of an aberrant side reaction of the enzyme at high mercury concentrations.

Although model sediment studies and pure culture studies are clearly showing the importance of sulfate-reducing bacteria in mercury methylation, its importance in the field, at natural concentrations, has yet to be demonstrated as convincingly. In particular, field observations and experiments with natural samples show that methylation increases with the sulfate concentration up to 200–500 μM and decreases at higher concentrations (29). Thus, sulfate concentrations in estuaries and seawater may be too high for methylation by sulfate-reducing bacteria to be efficient.

MICROBIAL UPTAKE OF MERCURY

To be methylated by sulfate-reducing bacteria or to enter the aquatic food chain via phytoplankton or bacteria, mercury must first be transported across the lipid membrane that surrounds unicellular organisms. The microbial uptake of mercury is thus a key step both in its methylation and its bioaccumulation.

Most metals enter cells via specialized transmembrane cation transporters, or they "leak" through the transporters of other metals. Indeed, at high concentrations, Hg(II) is transported into *mer*-carrying bacteria via a specialized MerT transport protein. At low concentrations, however, the cellular uptake of mercury, unlike that of other cationic metals, such as zinc or cadmium whose coordination properties are similar, appears to be effected chiefly by diffusion through the lipid membrane of lipid-soluble mercury complexes. The chemical bonding in the dichloro mercuric complex, $HgCl_2$, is largely covalent rather than ionic, such that the uncharged complex is relatively nonpolar and has fair lipid solubility. Lipid solubility is generally quantified by the "octanol-water partition coefficient," K_{ow}, which measures the relative solubilities of a compound in octanol and water and ranges from near zero for very hydrophilic molecules to 10^8 for very hydrophobic ones (69). The K_{ow} of $HgCl_2$ is 3.3, showing almost equal solubility in both solvents. Like other lipid-soluble species, this complex diffuses rapidly through lipid bilayers (32), leading to an efficient cellular uptake of mercury. This is, of course, not true of the charged chloride complexes such as $HgCl^+$ or $HgCl_3^-$. $Hg(OH)_2$, although uncharged, has a lower K_{ow} ($= 0.5$) than $HgCl_2$ and diffuses very slowly through membranes (32). The net result is that the chloride concentration and the pH (see Figure 3) greatly affect the cellular uptake of mercury in oxic waters, and all of its direct and indirect consequences such as toxicity or methylation.

While it seems clear that $HgCl_2$ is the key chemical species determining cellular uptake of inorganic mercury in oxic waters, the question remains of what species may play a similar role in anoxic waters, where most of the methylation occurs. A possible candidate is the uncharged di-bisulfide-mercury complex, $Hg(Hs)_2$, which dominates the speciation of Hg(II) at pH < 6.3 (see Figure 6). Except for the higher methylation rates observed at lower pHs (52, 64, 86), there are no reported experiments that directly or indirectly implicate $Hg(HS)_2$ in microbial uptake or methylation, however, and its K_{ow} is unknown. Perhaps the species of mercury that are important for bacterial uptake are the putative polysulfide complexes HgS_n, which carry no net charge. Some may have a low polarity and diffuse efficiently through cellular membranes. If this were the case, the presence of polysulfides might be an important factor determining the methylation rate in natural waters. One should note, however, that the only published study on mercury-polysulfide complexes (58) reports the existence of $Hg(S_n)HS^-$ complex but shows no evidence of an uncharged HgS_n species.

Like that of inorganic mercury, the microbial uptake of methylmercury is effected by diffusion of its uncharged chloride complex, CH_3HgCl. The lipid solubility of CH_3HgCl is similar to that of $HgCl_2$, and its permeability through cellular membrane is also similar ($K_{ow} = 1.7$). The accumulation of methylmercury in the food chain should thus be favored by conditions that maximize the formation of the CH_3HgCl species, namely low pH and high chloride concentration (see Figure 5). Field data generally support this conclusion (48).

Other nonpolar mercury species such as $(CH_3)_2Hg$ and Hg^0 also diffuse rapidly through lipid membranes. They are not bioaccumulated, however, as discussed below.

Biomagnification of Mercury in the Food Chain

To yield high concentrations in fish, mercury must not only be taken up efficiently by the microorganisms that are at the bottom of the food chain, it must also be retained by these organisms and passed on to their predators. Many trace metals are efficiently accumulated in planktonic bacteria and microalgae, but most are not biomagnified: Their concentrations in the biomass do not increase (they often decrease) at higher levels in the food chain. A key to understanding mercury bioaccumulation is provided by the contrast between Hg^0, $Hg(II)$, and Me_2Hg, which are not bioaccumulated, and MeHg, which is. Hg^0 and $(CH_3)_2Hg$ are not bioaccumulated, simply because they are not reactive and thus are not retained in phyto- or bacterio-pico-plankton: They diffuse out as readily as they diffuse in. (Note that intracellular oxidation of Hg^0 may be effected by catalase and hydrogen peroxide, as has been shown in red blood cells and brain cells; 21).

The difference between bioaccumulation of $Hg(II)$ and MeHg is more subtle. As we have seen, $HgCl_2$ and CH_3HgCl diffuse through membranes at about the same rate. Both are also reactive with cellular components and are efficiently retained by microorganisms. Laboratory experiments show, however, that the efficiency of transfer between a marine diatom and a copepod is four times greater for MeHg than for $Hg(II)$ (48). This is explained by the fact that $Hg(II)$ becomes bound chiefly to particulate cellular material (membranes) of the diatoms which are excreted rather than absorbed by the copepod. In contrast, MeHg is associated with the soluble fraction of the diatom cell and is efficiently assimilated by the copepod (see Figure 8; 48). Field data indicate that this difference in the efficiency of transfer between $Hg(II)$ and MeHg is applicable to other unicellular microorganisms and their predators (80).

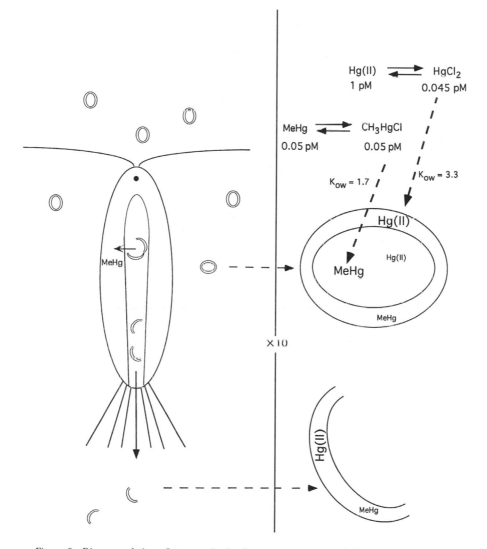

Figure 8   Bioaccumulation of mercury in the first steps of the food chain. Hg(II) and MeHg concentrations are estimates for average seawater (27); HgCl₂ and CH₃HgCl concentrations are calculated (see Table 1). See text for explanations.

To quantify the difference in the bioaccumulation of inorganic and organic mercury in the first steps of the food chain, we need to take into account three factors: the relative concentrations of Hg(II) and MeHg, the proportion of each that is in a lipid-permeable form, $HgCl_2$ and CH_3HgCl, and the relative efficiency of assimilation by grazers. As seen in Figure 8, even in seawater, where the excess of Hg(II) over MeHg is particularly large, organic mercury should be (and is) more bioaccumulated in grazers than is inorganic mercury.

Further efficient transfer of methylmercury through higher levels of the food chain seems to result from the lipid solubility of CH_3HgCl, which allows it to be partly retained in the fatty tissue of animals. In fish, however, MeHg burden in muscle tissue is more important than in lipids, clearly showing that bioaccumulation cannot be explained solely by MeHg liposolubility (15). In the case of fish, there seems to be a high specificity of the intestine wall toward MeHg absorption. In contrast, inorganic Hg is adsorbed at the microvilli interface, resulting in a very low uptake rate (15). As a result, the average proportion of MeHg over total Hg increases from about 10% in the water column to 15% in phytoplankton, 30% in zooplankton, and 95% in fish (80).

The accumulation of MeHg in higher organisms results mainly from the ingestion of MeHg-containing food rather than direct uptake of MeHg from the water. The structure of the foodweb determines the efficiency of transfer from algae to top predators. The number of trophic levels between predators and prey is critical, as shown by studies that correlate $\delta^{15}N$ (the normalized proportion of ^{15}N in biomass, a measure of trophic level) and Hg bioaccumulation (16, 17). In North American lakes, it has been observed (16, 17) that the presence of certain planktivores, such as lake herring, rainbow smelt, or mysids, which increases the number of trophic levels in the aquatic ecosystem, leads to higher mercury concentrations in top predators.

CONCLUSIONS

Over the past dozen years, much has been learned about the cycle of mercury in the environment. We now have good analytical data for the concentration of various mercury species in a number of environmental settings. We also have reasonable estimates for the various fluxes in the global mercury budget as well as in budgets for particular water bodies. The chemical and biological processes that control those fluxes are very difficult to ascertain and quantify, however, because of their complexities and the very low concentrations involved. Nonetheless, we are beginning to understand the redox mechanisms that control the exchange of mercury between natural waters and the atmosphere and the chemical/biological processes that control the bioaccumulation of mercury in the food chain. Less well understood are the mechanisms that control the removal of mercury from water to sediments. Most critical of all is the elucidation of the processes that determine the extent of mercury methylation in the environment, particularly the processes that control methylmercury concentrations in surface seawater and the nature of the chemical species that are available to the methylating bacteria in anoxic waters.

Visit the *Annual Reviews home page* at
http://www.AnnualReviews.org

Literature Cited

1. Akagi H, Fujita Y, Takabatake E. 1975. Photochemical methylation of inorganic mercury in the presence of mercuric sulfide. *Chem. Lett.* 171–76
2. Akagi H, Fujita Y, Takabatake E. 1977. Methylmercury:photochemical transformation of mercuric sulfide into methylmercury in aqueous solutions. *Photochem. Photobiol.* 26:363–70
3. Alberts JJ, Schindler JE, Miller RW. 1974. Elemental mercury evolution mediated by humic acid. *Science* 184:895–97
4. Allard B, Arsenie I. 1991. Abiotic reduction of mercury by humic substances in aquatic system—an important process for the mercury cycle. *Water Air Soil Pollut.* 56:457–64
5. Amyot M, Gill GA, Morel FMM. 1997a. Production and loss of dissolved gaseous mercury in the coastal waters of the Gulf of Mexico. *Environ. Sci. Technol.* 31:3606–11
6. Amyot M, Lean DRS, Mierle G. 1997b. Photochemical formation of volatile mercury in high arctic lakes. *Environ. Toxicol. Chem.* 16:2054–63
7. Amyot M, Mierle G, Lean DRS, McQueen DJ. 1994. Sunlight-induced formation of dissolved gaseous mercury. *Environ. Sci. Technol.* 28:2366–71
8. Amyot M, Mierle G, Lean DRS, McQueen DJ. 1997c. Effect of solar radiation on the formation of dissolved gaseous mercury in temperate lakes. *Geochim. Cosmochim. Acta* 61:975–88
9. Andersson A. 1979. Mercury in soils. In *Biogeochemistry of Mercury in the Environment,* ed. JO Nriagu, pp. 79–112, Amsterdam: Elsevier
10. Barkay T, Turner R, Saouter E, Horn J. 1992. Mercury biotransformations and their potential for remediation of mercury contamination. *Biodegradation* 3:147–59
11. Barnett MO, Harris LA, Turner RR, Stevenson RJ, Henson TJ, et al. 1997. Formation of mercuric sulfide in soil. *Environ. Sci. Technol.* 31:3037–43
12. Baughman GL, Gordon JA, Wolfe NL, Zepp RG. 1973. Chemistry of organomercurials in aquatic systems. *Ecol. Res. Ser.* U.S. Environmental Protection Agency
13. Ben-Bassat D, Mayer AM. 1978. Light-induced Hg volatilization and O_2 evolution in *Chlorella* and the effect of DCMU and methylamine. *Physiol. Plant* 42:33–38
14. Bloom NS, Watras CJ, Hurley JP. 1991. Impact of acidification on the methylmercury cycle of remote seepage lakes. *Water Air Soil Pollut.* 56:477–91
15. Boudou A, Ribeyre F. 1997. Mercury in the food web: accumulation and transfer mechanisms. *Metal Ions Biol. Syst.* 34:289–319
16. Cabana G, Rasmussen JB. 1994. Modelling food chain structure and contaminant bioaccumulation using stable nitrogen isotopes. *Nature* 372:255–57
17. Cabana G, Tremblay A, Kalff J, Rasmussen JB. 1994. Pelagic food chain structure in Ontario lakes: a determinant of mercury levels in lake trout (*Salvelinus namaycush*). *Can. J. Fish Aquat. Sci.* 51:381–89
18. Choi SC, Bartha R. 1993. Cobalamin-mediated mercury methylation by *Desulfovibrio desulfuricans* LS. *Appl. Environ. Microbiol.* 59:290–95
19. Choi SC, Chase T Jr, Bartha R. 1994a. Metabolic pathways leading to mercury methylation in *Desulfovibrio desulfuricans* LS. *Appl. Environ. Microbiol.* 60:4072–77
20. Choi SC, Chase T Jr, Bartha R. 1994b. Enzymatic catalysis of mercury methylation in *Desulfovibrio desulfiricans* LS. *Appl. Environ. Microbiol.* 60:1342–46
21. Clarckson TW. 1997. The toxicology of mercury. *Crit. Rev. Clin. Lab. Sci.* 34:369–403
22. Clever HL, Johnson SA, Derrick ME. 1985. The solubility of mercury and some sparingly soluble mercury salts in water

and aqueous electrolyte solutions. *J. Phys. Chem. Ref. Data Vol.* 14:3:632–80

23. Coquery M, Cossa D, Martin JM. 1989. The distribution of dissolved and particulate mercury in three Siberian estuaries and adjacent arctic coastal waters. *Water Air Soil Pollut.* 80:653–64

24. Davies SN, DeWiest RCM. 1966. *Hydrogeology* Chichester, UK: Wiley Interscience

25. Engstrom DR, Swain E. 1997. Recent declines in atmospheric mercury deposition in the upper midwest. *Environ. Sci. Technol.* 31:960–67

26. Fitzgerald WF, Engstrom DR, Mason RP, Nater EA. 1998. The case for atmospheric mercury contamination in remote areas. *Environ. Sci. Technol.* 32:1–12

27. Fitzgerald WF, Mason RP. 1997. Biogeochemical cycling of mercury in the marine environment. *Metal Ions Biol. Syst.* 34:53–111

28. Fitzgerald WF, Vandal GM, Mason RP, Dulac F. 1994. Air-water cycling of mercury in lakes. In *Mercury Pollution—Integration and Synthesis*, ed. CJ Watras, JW Huckabee, pp. 203–20, Boca Raton, FL: CRC

29. Gilmour CC, Henry EA. 1991. Mercury methylation in aquatic systems affected by acid deposition. *Environ. Pollut.* 71:131–69

30. Gustin MS, Taylor GE, Leonard TL, Keislar RE. 1996. Atmospheric mercury concentrations associated with both natural and anthropogenic enriched sites, central western Nevada. *Environ. Sci. Technol.* 30:2572–79

31. Gustin MS, Taylor GE, Maxey RA. 1997. Effect of temperature and air movement on the flux of elemental mercury from substrate to the atmosphere. *J. Geophys. Res.* 102(D3):3891–98

32. Gutknecht J, Tosteson DC. 1973. Diffusion of weak acids across lipid bilayer membranes: effects of chemical reactions in the unstirred layers. *Science* 182:1258–61

33. Hietanen S, Sillen LG. 1956. On the standard potentials of mercury, and the equilibrium $Hg^{2+} + Hg(I) = Hg_2^{2+}$ in nitrate and perchlorate solutions. *Arkiv For Kemi* 10:2:103–25

34. Hintelmann H, Welbourn PM, Evans RD. 1995. Binding of methylmercury compounds by humic and fulvic acids. *Water Air Soil Pollut.* 80:1031–34

35. Hudson RJM, Gherini SA, Fitzgerald WF, Porcella DB. 1995. Anthropogenic influences on the global mercury cycle: a model-based analysis. *Water Air Soil Pollut.* 80:265–72

36. Hurley JP, Watras CJ, Bloom NS. 1991. Mercury cycling in a northern seepage lake: the role of particulate matter in vertical transport. *Water Air Soil Pollut.* 56:543–51

37. Jones GJ, Palenik BP, Morel FMM. 1987. Trace metal reduction by phytoplankton: the role of plasmalemma redox enzymes. *J. Phycol.* 23:237

38. Jones GJ, Waite TD, Smith JD. 1985. Light-dependent reduction of copper (II) and its effect on cell-mediated, thiol-dependent superoxide production. *Biochem. Biophys. Res. Comm.* 128:1031–36

39. Kaczynski SE, Kieber RJ. 1993. Aqueous trivalent chromium photoproduction in natural waters. *Environ. Sci. Technol.* 27:1572–78

40. Korthals ET, Winfrey MR. 1987. Seasonal and spatial variations in mercury methylation and demethylation in an oligotrophic lake. *Appl. Environ. Microbiol.* 53:2397–404

41. Krabbenhoft DP, Hurley JP, Olson ML, Cleckner LB. 1998. Diel variability of mercury phase and species distributions in the Florida Everglades. *Biogeochemistry.* In press

42. Lindberg SE, Kim KH, Munthe J. 1995. The precise measurement of concentration gradients of mercury in air over soils: a review of past and recent measurements. *Water Air Soil Pollut.* 80:383–92

43. Lindqvist O, et al. 1991. Mercury in the Swedish environment. *Water Air Soil Pollut.* 55:23–30

44. Lövgren L, Sjöberg S. 1989. Equilibrium approaches to natural water systems—7. Complexation reactions of copper(II), cadmium(II) and mercury(II) with dissolved organic matter in a concentrated bog water. *Wat. Res.* 23:327–32

45. Luther III GW, Tsamakis E. 1989. Concentration and form of dissolved sulfide in the oxic water column of the ocean. *Mar. Chem.* 127:165–177

46. Mason RP, Fitzgerald WF, Morel FMM. 1994. The biogeochemical cycling of elemental mercury: anthropogenic influences. *Geochim. Cosmochim. Acta* 58:3191–98

47. Mason RP, Morel FMM, Hemond HF. 1995. The role of microorganisms in elemental mercury formation in natural waters. *Water Air Soil Pollut.* 80:775–87

48. Mason RP, Reinfelder JR, Morel FMM. 1996. Uptake, toxicity, and trophic transfer of mercury in a coastal diatom. *Environ. Sci. Technol.* 30:1835–45

49. Meili M. 1997. Mercury in lakes and rivers. *Metal Ions Biol. Syst.* 34:21–51

50. Mierle G. 1990. Aqueous inputs of mercury

from Precambrian Shield lakes in Ontario. *Environ. Toxicol. Chem.* 9:843–51

51. Mierle G, Ingram R. 1991. The role of humic substances in the mobilization of mercury from watersheds. *Water Air Soil Pollut.* 56:349–57

52. Miskimmin BM, Rudd JWM, Kelly CA. 1992. Influence of dissolved organic carbon, and microbial respiration rates on mercury methylation and demethylation in lake water. *Can. J. Fish. Aquat. Sci.* 49:17–22

53. Misra TK. 1992. Bacterial resistances to inorganic mercury salts and organomercurials. *Plasmid* 27:4–16

54. Munthe J. 1992. The aqueous oxidation of elemental mercury by ozone. *Atmos. Environ.* 26A:1461–68

55. Munthe J, McElroy WJ. 1992. Some aqueous reactions of potential importance in the atmospheric chemistry of mercury. *Atmos. Environ.* 26A:553–57

56. Munthe J, Xiao ZF, Lindqvist O. 1991. The aqueous reduction of divalent mercury by sulfite. *Water Air Soil Pollut.* 56:621–30

57. Onat E. 1974. Solubility studies of metallic mercury in pure water at various temperature. *J. Inorg. Nucl. Chem.* 36:2029–32

58. Paquett KE, Helz GR. 1997. Inorganic speciation of mercury in sulfidic waters: the importance of zero-valent sulfur. *Environ. Sci. Technol.* 31:2148–53

59. Pokrovskiy OS. 1996. Measurement of the stability constant of an Hg(I) chloride complex in aqueous solutions at 20–80°C. *Geochem. Int.* 33:83–97

60. Pokrovskiy OS, Savenko VS. 1995. A potentiometric study of the physicochemical state of mercury in sea water. *Geochem. Int.* 32:21–30

61. Price NM, Morel FMM. 1990. Role of extracellular enzymatic reactions in natural waters. In *Aquatic Chemical Kinetics*, ed. W Stumm, pp. 235–57. New York: Wiley

62. Rasmussen LD, Turner RR, Barkay T. 1997. Cell-density dependent sensitivity of a *mer-lux* bioassay. *Appl. Environ. Microbiol.* 63:3291–93

63. Rasmussen PE. 1994. Current methods of estimating atmospheric mercury fluxes in remote areas. *Environ. Sci. Technol.* 28:2233–41

64. Regnell O. 1994. The effect of pH and dissolved oxygen on methylation and partitioning of mercury in freshwater model systems. *Environ. Pollut.* 84:7–13

65. Sanemasa I. 1975. The solubility of elemental mercury vapor in water. *Bull. Chem. Soc. Jpn.* 48:1795–98

66. Schiering N, Kabsch W, Moore MJ, Distefano MD, Walsh CT, Pai EF. 1991. Structure of the detoxification catalyst mercuric ion reductase from *Bacillus* sp. strain RC607. *Nature* 352:168–172

67. Schroeder W, Lindqvist O, Munthe J, Xiao Z. 1992. Volatilization of mercury from lake surfaces. *Sci. Total Environ.* 125:47–66

68. Schwarzenbach G, Widmer M. 1963. Die löslichkeit von metallsulfiden I. schwarzes quecksilbersulfid. *Fasciculus VII*, 46(294):2613–28

69. Schwarzenbach RP, Gschwend PM, Imboden DM. 1993. *Environ. Organ. Chem.* Chichester, UK: Wiley-Intersci.

70. Seigneur C, Wrobel J, Constantinou E. 1994. A chemical kinetic mechanism for atmospheric inorganic mercury. *Environ. Sci. Technol.* 28:1589–97

71. Sellers P, Kelly CA, Rudd JWM, MacHutchon AR. 1996. Photodegradation of methylmercury in lakes. *Nature* 380:694–97

72. Smith RM, Martell AE. 1976. *Critical Stability Constants. Inorganic Complexes.* Vol. 4

73. Strömberg D, Strömberg A, Wahlgren U. 1991. Relative quantum calculations on some mercury sulfide molecules. *Water Air Soil Pollut.* 56:681–95

74. Stumm W, Morgan JJ. 1996. *Aquatic Chemistry. Chemical Equilibria and Rates in Natural Waters.* New York: Wiley-Interscience

75. Suda I, Suda M, Hirayama K. 1993. Degradation of methyl and ethylmercury by singlet oxygen generated from sea water exposed to sunlight or ultraviolet light. *Arch. Toxicol.* 67:365–68

76. Summers AO. 1986. Organization, expression, and evolution of genes for mercury resistance. *Annu. Rev. Microbiol.* 40:607–34

77. Swain EB, Engstrom DR, Brigham ME, Henning TA, Brezonik PL. 1992. Increasing rates of atmospheric mercury deposition in midcontinental North America. *Science* 257:784–87

78. Triffreau C, Lützenkirchen J, Behra P. 1995. Modeling the adsorption of mercury(II) on (hydr)oxides: I. Amorphous Iron Oxide and a-quartz. *J. Colloid Interf. Sci.* 172:82–93

79. Völker BM, Sulzberger B. 1996. Effects of fulvic acid on Fe(II) oxidation by hydrogen peroxide. *Environ. Sci. Technol.* 30:1106–14

80. Watras CJ, Bloom NS. 1992. Mercury and methylmercury in individual zooplankton: implication for bioaccumulation. *Limnol. Oceanogr.* 37:1313–18

81. Watras CJ, Morrisson KA, Host JS. 1995. Concentration of mercury species in relationship to other site-specific factors in the surface waters of northern Wisconsin lakes. *Limnol. Oceanogr.* 40:556–65

82. Weber JH. 1993. Review of possible paths for abiotic methylation of mercury(II) in the aquatic environment. *Chemosphere* 26:2063–77

83. Xiao ZF, Munthe J, Schroeder WH, Lindqvist O. 1991. Vertical fluxes of volatile mercury over forest soil and lake surfaces in Sweden. *Tellus* 43:267–79

84. Xiao ZF, Munthe J, Stromberg D, Lindqvist O. 1994. Photochemical behavior of inorganic mercury compounds in aqueous solution. In *Mercury Pollution—Integration and Synthesis*, ed. CJ Watras, JW Huckabee, pp. 581–592. Boca Raton, FL: CRC

85. Xiao ZF, Stromberg D, Lindqvist O. 1995. Influence of humic substances on photolysis of divalent mercury in aqueous solution. *Water Air Soil Pollut.* 80:789–98

86. Xun L, Campbell NER, Rudd JMW. 1987. Measurements of specific rates of net methyl mercury production in the water column and surface sediments of acidified and circumneutral lakes. *Can. J. Fish. Aquat. Sci.* 44:750–57

87. Yamamoto M. 1996. Stimulation of elemental mercury oxidation in the presence of chloride ion in aquatic environments. *Chemosphere* 32:1217–24

Annu. Rev. Ecol. Syst. 1998. 29:567–99

PHYLOGENY OF VASCULAR PLANTS

James A. Doyle
Section of Evolution and Ecology, University of California, Davis, California 95616;
e-mail: jadoyle@ucdavis.edu

KEY WORDS: tracheophytes, seed plants, angiosperms, cladistics, molecular systematics

ABSTRACT

Morphological and molecular analyses resolve many aspects of vascular plant phylogeny, though others remain uncertain. Vascular plants are nested within bryophytes; lycopsids and zosterophylls are one branch of crown-group vascular plants, and euphyllophytes (*Psilophyton*, sphenopsids, ferns, seed plants) are the other. In Filicales, Osmundaceae are basal; water ferns and Polypodiaceae sensu lato are both monophyletic. Seed plants are nested within progymnosperms, and coniferophytes are nested within platyspermic seed ferns. Morphology indicates that angiosperms and Gnetales are related, but detailed scenarios depend on uncertain relationships of fossils; molecular data are inconsistent but indicate that both groups are monophyletic. *Amborella*, Nymphaeales, *Austrobaileya*, and Illiciales appear basal in angiosperms. Groups with tricolpate pollen form a clade (eudicots), with ranunculids and lower hamamelids basal. Most eudicots belong to the rosid and asterid lines, with higher hamamelids in the rosid line and dilleniids scattered in both. Alismids, Arales, and *Acorus* are basal in monocots; palms are linked with Commelinidae.

INTRODUCTION

In the last two decades, unprecedented progress has been made in understanding the phylogeny of vascular plants, thanks to new methods and new data. First, cladistics (75), which uses shared derived character states (synapomorphies) to identify clades and the criterion of parsimony (minimizing character state changes) to decide among hypotheses, replaced subjective approaches to reconstructing phylogeny. Second, advances in molecular biology have allowed

567

0066-4162/98/1120-0567$08.00

mass use of molecular characters in phylogenetic analysis. To some (e.g. 84), it may appear that molecular systematics has replaced cladistics, but in fact it simply applies cladistic methods to a new kind of data.

The relative value of morphological and molecular data is a topic of debate. Clearly, molecular data are not infallible. Molecular data sets often disagree, which implies that—like any sort of data—they can be misleading. In addition, molecular data are especially susceptible to long-branch attraction (58): When branches are of different lengths (i.e. the number of character state changes differs) because of different rates of evolution or extinction of side branches, some changes on the long branches will be at the same sites; of these, a third will be to the same new base. As a result, cladistic analysis may indicate that terminal taxa on long branches are related to each other rather than to their true relatives. The same factors can cause incorrect rooting (where a clade is connected to its outgroups), because changes on a long branch may be reversals to the outgroup state. These effects are especially severe in radiations that occurred rapidly a long time ago (40, 41). In cases where stem lineages to modern groups split over a short amount of time, and when a long amount of time elapsed before radiation occurred of the crown groups that include living members, there may be few molecular synapomorphies on the short internodes between lines, and these may be erased by later changes at the same sites or overwhelmed by long-branch attraction.

An advantage of molecular over morphological systematics is that it involves fewer subjective decisions about whether features are similar or different enough to be treated as the same or different states. Except when alignment is uncertain, no one will argue whether a base at a particular site is an A or a G, as they might argue whether layers around the ovule are the same. However, cladistic analysis based on morphology is still the only method for determining relationships between fossils and living organisms. Fossils can alter inferred relationships among living organisms in morphological studies (40), especially when they are stem relatives that retain states lost in the crown group. Finally, even if molecular data give correct relationships among living taxa, fossil intermediates may be needed to infer homologies and the course of character evolution on the long branches leading to modern groups.

In this review, I summarize advances in understanding the phylogeny of vascular plants, considering both morphological and molecular evidence. In many cases, phylogenetic analyses have resolved long-standing controversies (e.g. whether angiosperms are monophyletic). Many hypotheses discussed at length in introductory texts can be considered obsolete and are mentioned there only for historical interest. However, other relationships seem just as uncertain as ever, if not more so (e.g. among seed plant groups). These are areas that need more investigation. In some cases, genes are unanimous but morphological analyses have given inconsistent results. This may seem to support the view that molecular evidence is more reliable, but in other cases the situation is reversed, apparently for the reasons just discussed. There is reason to hope that by combining different types of data (1, 48, 96, 117) such conflicts may be resolved: Even when the phylogenetic signal is overwhelmed by homoplasy in individual data sets, the patterns of homoplasy in different data sets should differ; the true signal, however, is presumably always the same.

The interest of phylogeny, especially for other areas of biology, lies not only in establishing branching relationships and thus providing the basis for a natural classification, but also in its implications for the evolution of structures and adaptations. Therefore, I note implications for character evolution in the origin of major groups and their relation to ecological factors. I discuss seed plants in the most detail; older groups are discussed by Bateman et al (9).

BASAL TRACHEOPHYTES

Morphological and molecular analyses indicate that vascular plants (tracheophytes) are a monophyletic subgroup of land plants (embryophytes), which in turn are nested within charophytes, fresh-water green algae that include Charales and *Coleochaete* (67). Within embryophytes, morphological analyses (95) indicate that bryophytes form a paraphyletic series of lines, with liverworts basal (the sister group of all other taxa) and anthocerotes and mosses closer to vascular plants. This implies that many key adaptations to the land arose in a stepwise fashion before vascular plants evolved: meiospores with a resistant exine, walled multicellular sporangia and gametangia (seen in liverworts); stomata (anthocerotes); and conducting tissue (mosses, with hydroids rather than tracheids). These results reaffirm the antithetic theory for origin of the land-plant life cycle: Land plants were derived from algae with a haploid life cycle (like charophytes), and the sporophyte was interpolated into the cycle by delay of meiosis and was progressively elaborated. The greatest advance of vascular plants themselves is the branched sporophyte. Studies of *rbc*L, 18S and 26S rDNA, and sperm morphology gave different results (64, 91, 96), but a combination of the last three data sets (96) reaffirmed the paraphyly of bryophytes and the monophyly of tracheophytes.

Because living "lower" vascular plants (pteridophytes) are separated by great morphological gaps, ideas about their early evolution have emphasized fossils, which include many remarkably preserved and reconstructed forms. The Early Devonian Rhynie Chert flora, dominated by rootless, leafless rhyniophytes with dichotomously branched stems and terminal sporangia, led to the telome theory of vascular plant morphology (141), which subsumed the classical theory that the sporophyte consists of stems, leaves, and roots by postulating that leaves originated by overtopping, planation, and webbing of dichotomous branches. In the 1960s, the view emerged that rhyniophytes gave rise to two lines (8): one included Devonian zosterophylls and lycopsids, with exarch xylem and lateral sporangia; the other included Devonian trimerophytes such as *Psilophyton*, with terminal sporangia but overtopped branches, which led to all other groups.

Although early morphological analyses (20, 99) placed Psilotaceae at the base of living vascular plants, because they lack roots, present evidence indicates that lycopsids are basal. This evidence includes the fact that lycopsids and bryophytes lack a chloroplast inversion found in other vascular plants (104) and a broad morphological analysis of fossil and living taxa by Kenrick & Crane (83) (Figure 1). However, studies of gene sequences (91, 92, 139) give inconsistent and anomalous results (e.g. with lycopsids polyphyletic). Fossils on the stem lineage leading to living groups show a stepwise transition from bryophytic to tracheophytic construction. *Horneophyton* still had a columella in the sporangium and moss-like hydroids (without secondary wall thickenings) rather than tracheids, but it also had a branched sporophyte; *Aglaophyton* (*Rhynia major*) had hydroids but no more columella; *Rhynia* had tracheids with secondary wall thickenings but of a primitive S-type. Kenrick & Crane (83) restricted the name tracheophytes to *Rhynia* and living groups and called the larger clade—including forms with a branched sporophyte but hydroids—the polysporangiophytes.

Lycopsids

Reniform sporangia with transverse dehiscence link lycopsids and zosterophylls with *Cooksonia* and *Renalia*, which had sporangia on lateral dichotomous branchlets (83). This suggests that lateral sporangia arose by overtopping and reduction. The fact that some zosterophylls (e.g. *Sawdonia*) had nonvascularized enations has been considered evidence that the one-veined leaves (microphylls) of lycopsids were derived from enations. However, Kenrick & Crane (83) found that taxa with enations are nested within smooth-stemmed zosterophylls, which led them to propose that microphylls are sterilized lateral sporangia. The basal lycopsid *Asteroxylon*, in which sporangia and leaves were intermixed, may support this scenario. Sporangia became associated with leaves (sporophylls) in the remaining groups. *Asteroxylon* also had leafless, dichotomous rhizomes that may be transitional to roots (not known in zosterophylls).

Within lycopsids, *Selaginella*, *Isoetes*, and Paleozoic arborescent forms (Lepidodendrales) are united by ligules and heterospory. The Devonian genus *Leclercqia* had a ligule but was homosporous, which implies that the ligule arose first. Lepidodendrales had secondary xylem (but no secondary phloem, reflecting the independent origin of secondary growth here and in seed plants),

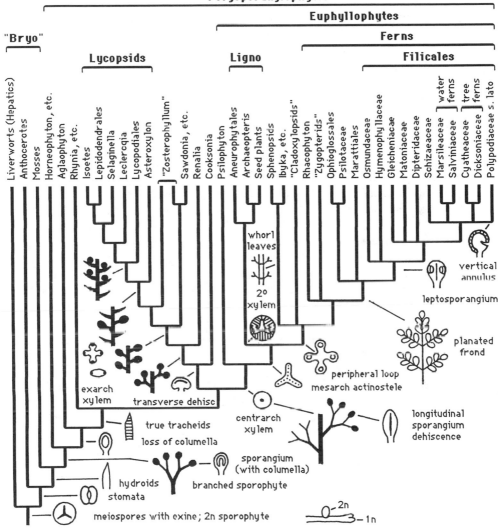

Figure 1   Synopsis of relationships among bryophytes and "lower" vascular plants found in morphological and molecular analyses [based primarily on the work of several others (83, 95, 102)], with sketches illustrating important synapomorphies in vegetative morphology, stem anatomy, and sporangia. "Bryo," bryophytes; Ligno, lignophytes.

periderm, and dichotomous "roots" (*Stigmaria*) covered with spirally arranged "rootlets;" developmentally and anatomically, the roots appear to be rhizomes and the rootlets modified leaves. The fact that *Isoetes* has anomalous cambial activity and hollow roots that resemble stigmarian rootlets, and the existence of the Triassic intermediate *Pleuromeia*, have suggested that *Isoetes* is reduced from Lepidodendrales. However, *Isoetes* and *Pleuromeia* appear to be related to smaller Carboniferous members of this clade (*Chaloneria*), not derived from large trees (11).

Euphyllophytes

Phylogenetic analyses also confirm the view that other vascular plants form a clade with trimerophytes, called euphyllophytes (83). Early Devonian members (*Psilophyton*) share such advances as longitudinal sporangial dehiscence, centrarch or mesarch xylem, and overtopped branches. Besides the chloroplast inversion mentioned (104), extant members are united by multiflagellate sperm (Lycopodiales and *Selaginella* are biflagellate, like bryophytes; the multiflagellate condition of *Isoetes* is presumably a convergence). Leaves in this group are often called megaphylls and are said to originate by overtopping, but this confuses different modes of origin and homologies of leaves. In some taxa (sphenopsids, progymnosperms), the leaf appears to represent a single overtopped branchlet; in others (ferns, seed plants), it is a branch system of several orders of overtopped axes that bear dichotomous branchlets. Roots apparently originated independently here and in lycopsids, where they are unusual in being dichotomous; roots are known in cladoxylopsids, *Rhacophyton*, and progymnosperms but not in *Psilophyton*.

Living euphyllophytes consist of three lines—sphenopsids, ferns, and seed plants—whose relationships are poorly resolved. Key fossil groups are the Devonian cladoxylopsids and progymnosperms. Both had lobed or dissected steles (actinosteles, polysteles), which may indicate they (and their derivatives) form a clade (83). In early progymnosperms (Aneurophytales), each lobe of the actinostele had several mesarch protoxylem points, whereas in cladoxylopsids, there was one parenchymatous protoxylem area near the tip of each lobe or stele segment, resulting in peripheral loops of metaxylem surrounding spongy protoxylem (83).

Sphenopsids

The oldest sphenopsids were long thought to be Devonian Hyeniales, which had highly overtopped branchlets that could be modified into the microphylls and sporangiophores of living *Equisetum*. This scenario is supported by the dichotomous leaves of *Sphenophyllum* and *Archaeocalamites*, the latter linked with *Equisetum* and Paleozoic trees (*Calamites*: Equisetales) by a hollow pith surrounded by vascular bundles (eustele). This view fell out of favor with evidence that Hyeniales cannot be separated from cladoxylopsids, then considered primitive ferns, and that the supposed jointing of their stems was a preservational artifact. However, it is possible that both ferns and sphenopsids are derived from cladoxylopsids (115, 122): The equisetalian eustele could be derived from a polystele or actinostele by differentiation of the central tissue into pith and transformation of the spongy protoxylem into protoxylem canals, connected with internode elongation. Furthermore, some cladoxylopsids were more like sphenopsids than had been formerly believed: Appendages were whorled in *Ibyka* and *Cladoxylon dawsonii* (29, 122).

Stewart (125) suggested that Sphenophyllales—which had whorled, wedge-shaped leaves—were related to lycopsids rather than sphenopsids because they had exarch actinosteles rather than eusteles. However, this characteristic is overwhelmed by features that link Sphenophyllales with other sphenopsids (83, 122). Their stele might be derived from the actinostele of primitive cladoxylopsids by loss of the outer metaxylem.

Ferns

Living ferns are traditionally divided into two categories: eusporangiate ferns (Ophioglossales, Marattiales), which retain thick-walled sporangia like other vascular plants; and leptosporangiate ferns or Filicales, with small, delicate sporangia. Paleozoic coenopterids were thought to show stages in evolution of the fern frond: Some had three-dimensional branch systems rather than leaves, but others (now considered primitive Filicales) had planated fronds. Ophioglossales have files of tracheids sometimes interpreted as secondary xylem; it has been suggested, therefore, that they are closer to progymnosperms than to ferns (16), but similar tissue occurred in supposed Paleozoic ferns (*Rhacophyton*, *Zygopteris*, etc).

Kenrick & Crane (83) did not analyze ferns in detail, but they did link modern ferns with sphenopsids, cladoxylopsids, and the Devonian genus *Rhacophyton*, based on one protoxylem area per stele arm (and peripheral loops). *Rhacophyton* and Carboniferous zygopterids had quadriseriate fronds, with pinnae in pairs perpendicular to the rachis; pinnae themselves bore dichotomous pinnules in two rows. *Rhacophyton* shows all degrees of reduction of one pinna of each pair, which suggests that planated fronds arose by an unexpected process of suppression of half of each pair. The Carboniferous genus *Ankyropteris*, which had zygopterid-like H-shaped petiole traces but planated fronds, might be a link between zygopterids and Filicales. However, Rothwell (110) separated living ferns from *Rhacophyton* and zygopterids, which were linked with lignophytes, based on secondary growth, which would otherwise be lost between zygopterids and Filicales. This would mean that no Devonian members of the line leading to modern ferns and no steps in origin of the fern frond are known.

Within extant ferns, analyses based on morphology (102), *rbc*L (73, 74), the two data sets combined (102), *cpITS* (98), and *atp*B (139) give more consistent results. They confirm that Filicales are a clade, united by leptosporangia and exposed antheridia. Marattiales and Ophioglossales are lower, but their exact relationships are poorly resolved.

Remarkably, four molecular data sets (74, 92, 98, 139) link Psilotaceae (*Psilotum*, *Tmesipteris*) with Ophioglossales. Because Psilotaceae have dichotomous stems and no roots, they have been considered living rhyniophytes, but they have also been linked with ferns based on their axial, subterranean gametophytes [as in Gleicheniaceae and Schizaeaceae (17)] and spore structure (89). They appear to be euphyllophytes because they have multiflagellate sperm and the chloroplast inversion cited above (83, 104). A relationship with Ophioglossales may seem preposterous because Psilotaceae have no roots, have highly branched stems, and have small appendages often considered enations rather than leaves, whereas Ophioglossales have thick roots, unbranched stems, and one large leaf at a time. However, they do share axial gametophytes and the unusual feature of sporangia borne on a structure attached to the adaxial side of an appendage. If this relationship is correct, the lack of roots in Psilotaceae is due to loss and is not primitive.

Within Filicales, Osmundaceae are basal in most trees, consistent with the fact that they have the least reduced leptosporangia (the exceptions are trees in which Hymenophyllaceae, which have extremely thin leaves, are one node lower). Subsequent branches, in uncertain order, are Gleicheniaceae, Dipteridaceae, and Matoniaceae, which have often been linked based on their round sori and seemingly dichotomous fronds, Schizaeaceae, and tree ferns (Cyatheaceae, Dicksoniaceae).

Marsileaceae and Salviniaceae, which are unusual in being both aquatic and heterosporous, have been thought to be derived from different terrestrial, homosporous ancestors (16, 19–19b), but both morphological and molecular analyses indicate they are a monophyletic group nested among Schizaeaceae, tree ferns, and more advanced ferns. Coincidentally, their monophyly has been confirmed by discovery of fossils intermediate between the two families (112).

Phylogenetic analyses have also refuted long-standing dogma concerning ferns with a vertical interrupted annulus—the majority of the group. These were once grouped as the huge family Polypodiaceae, but it has been argued (16, 19–19b) that they are polyphyletic—with different members related to Gleicheniaceae, Schizaeaceae, and tree ferns—based primarily on sporangial position and presence or absence of indusia rather than on structure of the sporangia themselves. However, molecular analyses indicate that they form a clade, nested among Schizaeaceae, tree ferns, and water ferns.

SEED PLANTS (LIGNOPHYTES)

Phylogenetic analyses have radically affected ideas on relationships of seed plants (gymnosperms plus angiosperms), which differ from "lower" groups not only in having seeds but also in vegetative advances such as secondary growth and axillary branching. Formerly, it was widely believed that seed plants are diphyletic (5, 23), consisting of two lines that differ in vegetative and seed characters. Cycadophytes (cycadopsids), including modern cycads and Paleozoic seed ferns or pteridosperms (which had seeds but fern-like fronds), have unbranched or sparsely branched stems with diffuse (manoxylic) wood, pinnately compound leaves, and radial seed symmetry (radiospermic). Coniferophytes (coniferopsids)—including conifers, *Ginkgo*, and Paleozoic Cordaitales—have highly branched stems with dense (pycnoxylic) wood, simple leaves with one vein or dichotomous venation, and bilateral (more precisely biradial) seed symmetry (platyspermic). Cycadophytes were assumed to be derived from ferns via seed ferns, coniferophytes from some Devonian group with dichotomous branches that might be transformed into simple leaves. This would imply that the seed originated twice; its different symmetry in the two lines was taken as support for this view. Discovery of Early Carboniferous seeds with the megasporangium (nucellus) surrounded by lobes showing all degrees of fusion into an integument confirmed suggestions that the seed itself was derived from a dichotomous fertile branch.

New insights came from recognition of the progymnosperms, Devonian trees with coniferophyte-like secondary xylem, phloem, and periderm that reproduced by spores, which implies that secondary growth arose before the seed (a prime example of mosaic evolution). Beck (12, 13) argued that both seed plant lines were derived from progymnosperms, but different subgroups. He linked coniferophytes with *Archaeopteris*, which had distichous branch systems with spiral, wedge-shaped leaves, recalling the branch systems of early conifers and the compound strobili of cordaites. The leaves would be barely modified in *Ginkgo* and cordaites and reduced to one-veined needles in conifers. He derived cycadophytes from the older Aneurophytales, which still had three-dimensional branch systems, because they and early seed ferns had actinosteles whereas *Archaeopteris* already had a eustele. The compound leaf would be derived by planation of a whole branch system, with leaves converted to leaflets.

This hypothesis was challenged by Rothwell (109), based on coniferophytic features in the Late Carboniferous seed fern *Callistophyton*: platyspermic seeds, saccate pollen, and frequent axillary branching. Deriving coniferophytes from a seed fern prototype would require a radical change in leaf morphology, which Rothwell proposed occurred not by reduction but by heterochrony. *Callistophyton* not only had fronds but also pointed cataphylls (bud scales), like seed plants in general; if fronds were suppressed and the plant continued to produce cataphylls its whole life, the result would be a coniferophytic morphology. Early conifers occurred in arid areas of the Euramerican tropical zone, which suggests that this shift adapted them to aridity.

Phylogenetic analyses largely confirm Rothwell's scenario. Although morphological analysis of modern plants (20) linked seed plants with ferns, they are nested within progymnosperms in analyses that include fossils (30, 46, 97, 111) (Figure 2). The combined clade has been called the lignophytes (46), because its most conspicuous synapomorphy is secondary wood. Several Late Devonian-Carboniferous seed ferns (lyginopterids, medullosans) form a basal paraphyletic series in seed plants, whereas coniferophytes are nested in a clade called platysperms (31) that includes *Callistophyton*, Permian-Mesozoic seed ferns (peltasperms, glossopterids, corystosperms, *Caytonia*), and some or all living taxa. This implies that the fronds of ferns and seed ferns arose from branch systems, but independently. Other seed plant innovations are a shift from pseudomonopodial to axillary branching, cataphylls (arrested frond primordia?), and fusion of microsporangia into synangia. Because it is heterosporous, *Archaeopteris* falls nearer seed plants than aneurophytes, but autapomorphies such as the eustele and grouped pitting imply that it is not ancestral. This shift to heterospory is the last of many iterative origins in vascular plants (10). The eustele apparently arose several times within seed plants, by formation of a pith at the center of an actinostele, leaving the lobes stranded as vascular bundles (13). The similarities between *Archaeopteris* and coniferophytes are convergences (eustele, branching pattern) or reversals (pycnoxylic wood, free microsporangia). Molecular data also support the monophyly of seed plants (1, 27, 66); this would not contradict belief that seed plants originate from two groups of progymnosperms, but it does rule out a separate relationship of cycadophytes and ferns.

These results clarify steps in the evolution of the seed (ovule) (Figure 2). Basal seed ferns had dichotomous cupules, shaped like two hands with seeds in the palms. These cupules may be derived from the fertile appendages of progymnosperms, the seeds from groups of sporangia, and the integument from sterilized outer sporangia (83). Successive taxa document steps from the original hydrasperman pollination mechanism (111), where pollen was captured by a drop secreted by the lagenostome (a nucellar beak with a pollen chamber formed by separation of the epidermis from a central column), to the modern condition, where a drop is exuded from the micropyle (made possible by fusion of the integument lobes—an example of transfer of function in the origin of

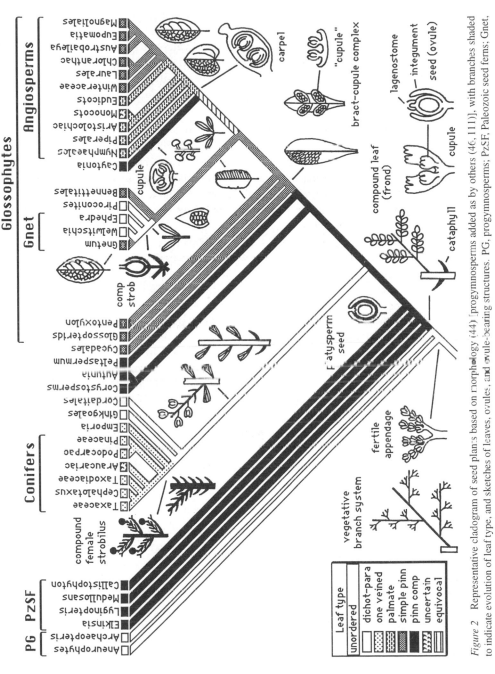

Figure 2 Representative cladogram of seed plants based on morphology (44) [progymnosperms added as by others (46, 111)], with branches shaded to indicate evolution of leaf type, and sketches of leaves, ovules, and ovule-bearing structures. PG, progymnosperms; PzSF, Paleozoic seed ferns; Gnet, Recent Gnetales.

new structures). Medullosans and primitive platysperms (including cycads and *Ginkgo*) still have a nucellar beak but have lost the central column. This is one synapomorphy of medullosans and platysperms; others are loss of the cupule, origin of a fleshy sarcotesta layer in the integument, and vascular tissue in the nucellus. It is possible that the cupule was not lost but is represented by the integument and that the original integument fused with the nucellus, leaving the nucellar vasculature as a vestige (136). Platysperms show additional steps in modernization of the seed: biradial symmetry and sealing of the micropyle after pollination.

Early seed plants, including not only lyginopterids and medullosans but also cordaites and Paleozoic conifers (101), had prepollen with a proximal tetrad scar, like the spores of lower vascular plants, rather than a distal sulcus, where the pollen tube germinates in modern groups. The pollen tube anchors the pollen and absorbs nutrients in cycads and *Ginkgo*, which still have motile sperm (zooidogamy), but it transfers nonmotile sperm to the egg in conifers, Gnetales, and angiosperms (siphonogamy). *Callistophyton* had a sulcus, and it is even known to have had a pollen tube, though the sperm type is unknown. Phylogenetic results indicate that the sulcus either originated more than once or was lost in cordaites and early conifers, which would be surprising considering that they are the oldest known platysperms. The first scenario would suggest that the pollen tube also arose several times, unless it originated before the exine was modified for its exit. Platysperms are united by a change from spongy-alveolar exine structure, as in progymnosperms, to honeycomb-alveolar structure, as in conifers and cycads. In some trees, saccate pollen originated at the same point, which implies that the nonsaccate monosulcate pollen of cycads, *Ginkgo*, and other groups arose later by loss of sacs. Other trees reverse this scenario; however, all known Carboniferous platysperms are saccate, and nonsaccate monosulcate pollen is rare until the Mesozoic. These uncertain relationships among platysperm lines are largely a function of incomplete information on Permian and Mesozoic fossils.

Cycads

Many early researchers derived cycads from medullosans, based on large ovules with a sarcotesta and nucellar vasculature and manoxylic stems with mucilage canals and numerous leaf traces, although cycads lack the internal secondary xylem of medullosans. In contrast, phylogenetic analyses that include fossils (30, 46, 47, 97, 111) place cycads either between medullosans and platysperms or nested in platysperms, with peltasperms and/or *Ginkgo* (a connection also found in molecular analyses, discussed below). This is because cycads and platysperms share many advances over medullosans, such as endarch primary xylem, abaxial microsporangia, linear megaspore tetrad, sulcate pollen,

honeycomb-alveolar exine, and sealed micropyle. Cycads are basal in morphological analyses of living seed plants (47, 87), which suggests that their lack
of axillary branching is primitive. However, in analyses that include fossils,
cycads are nested among taxa with axillary branching, which implies that this
feature was lost. Cycads do branch dichotomously (87), but this differs from
the progymnosperm state, which is pseudomonopodial (47).

It would be ironic if cycads, often used as the type example of radiospermic
seeds, were platysperms. *Cycas* has biradial ovules, however, and possible Permian cycad precursors with *Taeniopteris* leaves apparently did also (90). Present
trees confirm that the motile sperm of cycads is primitive, but it is uncertain
whether their monosulcate pollen is primitively or secondarily nonsaccate.

Relationships within cycads are better resolved. *Cycas* has been considered
primitive in having leaflike megasporophylls with several ovules along the
rachis, rather than peltate sporophylls with two ovules and likewise *Stangeria*
(87) in having fern-like leaflets with a midrib and secondary veins rather than
one vein (*Cycas*) or several parallel-dichotomous veins (*Zamia*, etc). Permian
plants with simple, pinnately veined *Taeniopteris* leaves and sporophylls with
numerous ovules support the primitive status of *Cycas* sporophylls but suggest
the unexpected possibility that *Cycas* and *Zamia* leaves were derived by dissection of a simple blade into segments of different widths (90). Morphological
and molecular analyses (25, 27, 123) indicate that *Cycas* is the sister group of
other genera, which confirms the standard scenario for sporophyll evolution
and is consistent with the *Taeniopteris* hypothesis for leaf evolution.

Conifers

Since the work of Florin on Paleozoic fossils (60), conifers have been associated
with Paleozoic cordaites, which had strap-shaped leaves, compound male and
female strobili, platyspermic seeds, and saccate pollen. Paleozoic conifers had
compound female strobili made up of bracts and axillary fertile short shoots,
apparently transformed into the cone scales of modern forms. However, there is
reason to doubt that conifers were directly derived from cordaites because they
have simple male strobili on normal branch systems, a condition that appears
more primitive. Thus, compound female strobili may have arisen independently by grouping of simple strobili. Some Podocarpaceae have grouped male
cones that have been compared with compound strobili of cordaites (138), but
Paleozoic conifers had solitary male cones.

In all morphological analyses, conifers and cordaites are nested among
platysperms. However, in some trees they are linked with each other (and ginkgos) whereas in others they are separate lines, which implies a parallel origin
of coniferophyte from cycadophyte features. Rothwell & Serbet (111) contradict Florin's scheme in separating modern conifers from Paleozoic conifers

(*Emporia*) and cordaites (which are still linked with each other) and placing them between cycads and anthophytes. This is because of characteristics in which modern conifers are more advanced than *Emporia* and cordaites, such as siphonogamy and loss of the lagenostome. However, it may also be a result of including only Pinaceae, Podocarpaceae, and Taxaceae, the last of which lack features—such as saccate pollen and compound strobili—that other conifers share with *Emporia* and cordaites. In analyses including all extant groups, modern conifers are linked with cordaites and ginkgos (44, 97) (Figure 2).

Because *Ginkgo*, cordaites, and early conifers have or are suspected to have had motile sperm (101), siphonogamy presumably originated between *Emporia* and modern conifers; this would coincide with loss of the lagenostome. An advance that occurred between cordaites or *Ginkgo* and *Emporia* is loss of the sarcotesta. Modern conifers and *Ginkgo* are derived over other seed plants (except Gnetales) in lacking scalariform pitting in the metaxylem (7, 16), but *Emporia* had scalariform metaxylem, which suggests that scalariform pitting was lost independently in the two groups.

Within conifers, morphological and molecular analyses (25, 27, 71, 121) have resolved several persistent problems. According to analyses of *rbc*L (25), conifers are not a clade but two adjacent lines (Pinaceae and other families). However, they are monophyletic in other studies. All analyses confirm the view (53) that Cupressaceae are a subgroup of Taxodiaceae that changed from helical to opposite phyllotaxy. These studies refute the concept (60) that Taxaceae are not conifers, based on the fact that they have simple female shoots with one terminal ovule rather than compound strobili with ovules on axillary cone scales. This view was questioned based on comparisons with *Cephalotaxus*, which is like Taxaceae in vegetative morphology and anatomy (tertiary spiral thickenings in the tracheids) and has cones that are highly reduced but still bear ovules on axillary structures; further reduction and a shift of the ovule to the apex could result in the taxaceous condition (70). All analyses link Taxaceae with *Cephalotaxus* and Taxodiaceae, confirming this scenario.

Morphological analyses give inconsistent basal relationships in living conifers, but molecular analyses agree that Pinaceae are the sister group of other families. Podocarpaceae and Araucariaceae form a clade; a possible synapomorphy is one ovule per cone scale, although the two ovules of supposed Triassic Podocarpaceae (131) pose a problem. Pinaceae and Podocarpaceae are the only living conifers with saccate pollen. Molecular trees are consistent with the view that this is a primitive feature retained from cordaites and Paleozoic conifers, and that the nonsaccate pollen of other conifers arose by loss of sacs and a shift from alveolar to granular exine structure—probably independently in Araucariaceae, which have large pollen, and the Taxodiaceae-Taxaceae clade, where the pollen is smaller.

Ginkgos

Ginkgo and its fossil relatives have been considered coniferophytes because they have simple, dichotomously veined leaves, simple strobili in the axils of leaves, and platyspermic seeds. Florin (59) proposed the Permian genus *Trichopitys* as a link because it had axillary female short shoots like those of Paleozoic conifers. *Karkenia* and *Baiera* had strobili bearing many ovules (4, 140), supporting the view that the biovulate stalk of *Ginkgo* is a reduced strobilus. However, this interpretation was questioned by Meyen (94), who argued that the short shoots of *Trichopitys* were actually pinnate structures. He associated ginkgos with Permian-Triassic peltasperms, which were also platyspermic, showed transitions from pinnate to dichotomous leaves, and sometimes had ginkgo-like nonsaccate monosulcate pollen. Mesozoic Czekanowskiales had ginkgo-like leaves and short shoots but female strobili of "capsules" composed of two valves with seeds on their facing surfaces; if these are related to ginkgos, they could be a link with peltasperms, which had seeds on paddle-like sporophylls. Other possible relatives are the Paleozoic Rufloriaceae of Siberia, which had cordaite-like strap-shaped leaves but simple male and female strobili (94).

Phylogenetic analyses have not resolved these issues. Some studies associate ginkgos with coniferophytes (30, 44, 46, 97), but others place them elsewhere in platysperms, sometimes with peltasperms and/or cycads (47, 111). Neither Czekanowskiales nor Rufloriaceae have been included in these studies because so many of their characteristics are unknown. Some molecular analyses link *Ginkgo* and cycads (27, 66); this could mean that both groups are peltasperm derivatives. Chaw et al (27) listed several morphological features—such as haustorial pollen tube, motile sperm, and free-nuclear megagametophyte and embryo development—as support for a *Ginkgo*-cycad relationship, but these are basic seed plant states that do not favor any relationship. However, nonsaccate monosulcate pollen could be a synapomorphy of *Ginkgo*, cycads, and some peltasperms.

Angiosperms and Gnetales

The most controversial topic in seed plant phylogeny is the relationship of angiosperms and Gnetales. The englerian theory (137) held that angiosperms were derived from Gnetales, homologizing the compound strobili of Gnetales, which consist of simple, unisexual, flower-like units in the axils of bracts, with the inflorescences of Amentiferae, which are also made up of simple flowers. This view lost favor with increasing evidence that Magnoliidae (which usually have complex, bisexual flowers) are primitive in angiosperms. Arber & Parkin (2, 3) also thought angiosperms and Gnetales were related, but they linked them with Mesozoic Bennettitales, which had large, often bisexual flowers, implying that the flowers of Gnetales and Amentiferae are reduced. Angiosperms have also been associated with Permian and Mesozoic seed ferns such as *Caytonia* (42, 65, 120) or glossopterids (106, 120). *Caytonia* had reflexed seed-bearing "cupules" borne in two rows on a rachis; glossopterids had one or more cupules or sporophylls on top of a leaflike bract (bract-sporophyll or bract-cupule complex). This would explain the morphology of most angiosperm ovules, which are bitegmic (with two integuments) and anatropous (reflexed): If the ovules in a *Caytonia* cupule were reduced to one, the cupule wall would provide a homolog for the outer integument, and the whole structure would be reflexed. The carpel might be derived by widening and folding of the *Caytonia* rachis, or by folding of the glossopterid bract. As for Gnetales, Eames (52) linked *Ephedra* with coniferophytes and *Welwitschia* and *Gnetum* with Bennettitales, whereas others (7, 16, 42) associated all three genera with coniferophytes, based on their compound stobili, lack of scalariform pitting in the primary xylem, and linear leaves (in *Ephedra* and *Welwitschia*).

Morphological analyses of seed plants have produced conflicting updated versions of these views, although they have resolved some issues. In particular, they all indicate that Gnetales are the closest living relatives of angiosperms, along with Bennettitales and the Cretaceous genus *Pentoxylon*. They also agree on the arrangement of the three genera of Gnetales, with *Welwitschia* and *Gnetum* linked by complex leaf venation, reduction of the male gametophyte, and tetrasporic development and incomplete cellularization of the female gametophyte, with free nuclei functioning as eggs. This implies that Gnetales originally had pollen of the striate ephedroid type seen in *Ephedra* and *Welwitschia*, which was reduced to spiny and inaperturate in *Gnetum*.

The first analyses (30, 46, 47) linked angiosperms, Gnetales, and Bennettitales, called anthophytes, with *Caytonia*, glossopterids, and Triassic corystosperms. This implies (*a*) that flowers (short axes with closely aggregated sporophylls) originated not in angiosperms but in their common ancestor with Gnetales and Bennettitales, before the ovules were enclosed in a carpel, and were reduced and grouped into compound strobili in Gnetales (a convergence with coniferophytes) and (*b*) that angiosperm ovules were derived from cupules of a *Caytonia* or glossopterid type. Crane (30) directly linked angiosperms with Gnetales, but Doyle & Donoghue (46, 47) placed angiosperms basal in anthophytes. Trees of the latter type imply that the vessels of angiosperms and Gnetales are convergent because Bennettitales and *Pentoxylon* were vesselless. Although both groups have two integuments, the outer integument of Gnetales corresponds to the perianth of the male flowers and of Bennettitales. The real synapomorphies of the two groups are more cryptic: a tunica in the apical meristem, lignin chemistry (Mäule reaction), loss of sacs and shift to granular exine in the pollen, and reduction of the megaspore wall (shared with Bennettitales and *Caytonia*). The most interesting synapomorphy is double fertilization, presumably of the type documented in Gnetales (22, 61), where both sperm produced by the male gametophyte fuse with nuclei in the female gametophyte but the second fusion produces an extra zygote rather than triploid endosperm. This implies that endosperm originated later on the angiosperm line, probably from the extra embryo (62).

These analyses treated angiosperms as a single taxon, which required questionable assumptions about basic states. One attempt to correct this problem (44) included 11 angiosperms and *Piroconites*, a Jurassic gnetalian relative with opposite, linear leaves and ephedroid pollen but glossopterid-like fertile structures (133). This gave generally similar trees, but with *Caytonia* on the line to angiosperms and glossopterids at or near the base (Figure 2). According to these trees, typical flowers arose independently in angiosperms, Bennettitales, and Gnetales. The whole clade originally had *Glossopteris*-like simple leaves with simple-reticulate venation and cupules attached to a bract, hence the name glossophytes. The bract-cupule complex was retained up to *Piroconites* on the line to Gnetales; cupules became anatropous on the *Caytonia*-angiosperm line only. In Gnetales, the bract-cupule complex was reduced to one ovule and shifted to the apex of an axillary branch; a possible intermediate is *Dechellyia* (6), a Triassic plant with opposite leaves and winged seeds (if the wing is derived from the bract). The original leaf became palmately compound in *Caytonia* (with four leaflets that resemble *Glossopteris* leaves), complex-reticulate by interpolation of finer veins in angiosperms, simple or pinnately dissected with secondarily open venation in Bennettitales and *Pentoxylon*, linear and opposite in *Piroconites* and Gnetales, and angiosperm-like in *Gnetum* (another convergence).

This scheme specifies a sequence of origin of angiosperm features: simple leaves, reticulate venation, loss of the lagenostome, and a carpel prototype below glossopterids; reduced megaspore wall, nonsaccate pollen, and granular exine above glossopterids; anatropous cupule/ovule and flat stomata below *Caytonia*; complex venation, flower, simple stamen with two pairs of microsporangia, and carpel closure in angiosperms. Features of angiosperms and Gnetales that are unknown in fossils—such as tunica, Mäule reaction, and double fertilization—could have arisen as far back as glossopterids. Unfossilizable features known only in angiosperms—such as companion cells in the phloem, three-nucleate male gametophyte, eight-nucleate female gametophyte, and endosperm—originated later, above or below *Caytonia*. By reducing the number of features that have no homologs in other groups, this scheme would reduce the mystery of the origin of angiosperms.

Other analyses have found very different trees, some of which are only slightly less parsimonious in terms of the data sets just discussed. Most notable are neo-englerian trees, in which anthophytes are linked with coniferophytes rather than *Caytonia* and glossopterids. This would mean that the simple flowers of Gnetales are primitive and derived from axillary short shoots of coniferophytes, whereas complex flowers in angiosperms originated by aggregation or elaboration of parts (43). The outer integument might be derived from the perianth of Gnetales and, ultimately, the sterile appendages of a short shoot; the carpel might be derived from the subtending bract. Somewhat similar trees were found by Rothwell & Serbet (111). Analyses of modern plants only (47, 87) also link angiosperms and Gnetales with conifers, although this would not exclude inserting glossopterids and *Caytonia* between them. A potential conifer-anthophyte synapomorphy is siphonogamy, which originates twice on previous trees; however, this is not so if *Emporia* is related to conifers and had no pollen tube (101). In trees of Nixon et al (97), who included 18 angiosperms, angiosperms are nested within Gnetales and linked with *Welwitschia* and *Gnetum* on leaf and embryological features, so that Gnetales are paraphyletic and angiosperms are derived from a gnetalian prototype.

Molecular studies appear to resolve some of these issues, but they introduce new conflicts (Figure 3). They do not directly address the position of fossils, which is essential in order to formulate and test hypotheses on the origin of flowers, carpels, and bitegmic ovules. However, except for early analyses with few taxa, they all indicate that both angiosperms and Gnetales are monophyletic, supported by high bootstrap values (1, 18, 27, 48, 66, 69, 72). They therefore refute trees in which Gnetales are paraphyletic (97), as does closer examination of the characters that link angiosperms, *Welwitschia*, and *Gnetum* (44). Most also confirm the basal position of *Ephedra* in Gnetales. However, in most molecular studies, angiosperms and Gnetales are not related at all. They are sister groups in some analyses of partial rRNA sequences (69), whereas cycads, *Ginkgo*, and conifers form a clade at the base of seed plants. However, with different taxon sampling or method of analysis, angiosperms are linked with cycads, *Ginkgo*, and conifers, and Gnetales are basal, a result also found with *rbc*L (1). With other *rbc*L data (72) and 28S rDNA (121), angiosperms are basal and Gnetales are linked with cycads, *Ginkgo*, and conifers. These variations are a function of rooting, which could be incorrect because living outgroups are so distant—angiosperms and Gnetales would be related if seed plants were rerooted among cycads, *Ginkgo*, and conifers, as in morphological trees. However, this is not true of trees based on *cpITS* (66), 18S (27), and *coxI* (18), in which angiosperms are basal but Gnetales are linked with conifers. If such trees are correct, we are farther from understanding the origin of angiosperms than we thought because there is no modern seed plant group that shares special homologies with angiosperms.

The lack of a molecular signal linking angiosperms and Gnetales may seem to be a major conflict between morphological and molecular data, but the

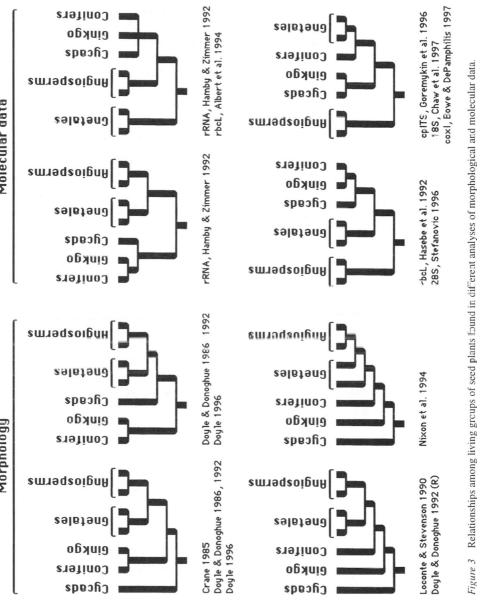

Figure 3   Relationships among living groups of seed plants found in different analyses of morphological and molecular data.

variations among molecular trees raise doubts about the ability of molecular data to resolve relationships at this level. In addition, the bootstrap support for arrangements that separate the two groups is low. The lack of resolution could be due to problems expected in rapid ancient radiations (41)—too few molecular synapomorphies and too much time for homoplasy on the long branches leading to modern taxa. Conifers, cycads, *Ginkgo*, and glossopterids diverged in the Late Carboniferous, all being known in that epoch or the Permian, and earlier seed plants being more primitive. Furthermore, the number of morphological characteristics that contradict the molecular arrangements is high. With the data set of Doyle (44), when angiosperms or Gnetales are forced to the base of living seed plants (45), the resulting trees are at least four to five steps longer than the shortest trees. They may be nine to ten steps longer because they assume that five states of angiosperms and Gnetales that are unknown in fossils (e.g. tunica, double fertilization) are primitive in seed plants whereas, other data suggest they are derived. These trees also conflict with the stratigraphic record (45): The oldest groups branch off first (Devonian-Carboniferous seed ferns), then the two youngest lines (angiosperms, Gnetales, and their fossil relatives), and finally taxa of intermediate age (cycads, ginkgos, conifers).

ANGIOSPERMS

Because of the vast diversity of angiosperms, there have been no morphological analyses of the whole group, although there have been studies of basal relationships and many subgroups. These have been overshadowed by molecular analyses of three genes in several hundred taxa: *rbc*L (25, 107), 18S (118), and *atp*B (114). These analyses give broadly consistent results (Figure 4), which in many cases conflict with earlier classifications but in hindsight make sense in terms of morphology. In particular, they break up many of the widely used subclasses of Takhtajan (127, 128) and Cronquist (34, 35). However, positions of many "floater" taxa vary among analyses. Combination of the three data sets promises to resolve many of the conflicts (117), but sampling of the same taxa is not yet complete.

Basal Relationships

Traditionally, much discussion focused on which living angiosperms are most primitive, a problem closely related to that of rooting (which taxa are basal). Most authors agreed that the most primitive angiosperms are in the subclass Magnoliidae (34, 35, 127, 128), a paraphyletic grade united by retention of monosulcate pollen and other gymnosperm features. However, this permits a wide range of hypotheses because magnoliids vary in habit from woody to herbaceous and in flowers from showy and bisexual to extremely simple.

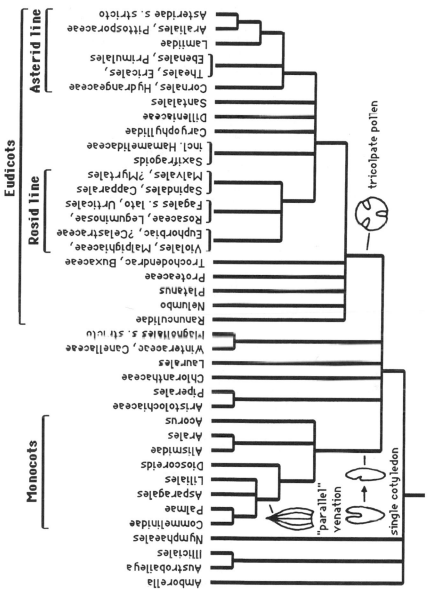

Figure 4 Synopsis of relationships within angiosperms based on analyses of *rbcL*, 18S rDNA, and *atp*B sequences (25, 107, 114, 118).

Morphological analyses have given several rootings, depending in part on choice of outgroups. Using Bennettitales, Gnetales, and *Caytonia* to polarize characters, Donoghue & Doyle (39) found that the basal branch was Magnoliales, which have granular monosulcate pollen (like Bennettitales and Gnetales), followed by other woody magnoliids (Laurales, Winteraceae, Illiciales). This implies that woody habit, pinnately veined leaves, showy bisexual flowers, and laminar stamens are primitive. Herbaceous magnoliids (Nymphaeales, Piperales, Aristolochiaceae) and monocots formed a clade called paleoherbs, showing a shift to herbaceous habit, palmately veined leaves, and stamens with a well-differentiated filament. These results generally agree with traditional ideas on polarity, except in implying that the lack of vessels in Winteraceae and other groups is due to loss because they are nested among taxa with vessels. However, trees rooted among paleoherbs were only one step longer, and these were favored in later seed plant analyses (44) (Figure 2). This implies that herbaceous habit, palmate venation, and differentiated stamens were primitive; flowers would still be bisexual, with a perianth, but trimerous. Loconte & Stevenson (88), who used Gnetales as the outgroup, found the basal line to be Calycanthales, a woody group usually placed in Laurales. Taylor & Hickey (129) and Nixon et al (97) nested angiosperms in Gnetales, and the basal line was Chloranthaceae (reduced Laurales in 39), which resemble Gnetales in having opposite leaves, simple flowers, and orthotropous ovules.

Analyses of partial rRNA sequences (48, 69) and *cpITS* (66) also rooted angiosperms among paleoherbs, with aquatic Nymphaeales basal. The first large analysis, of *rbcL* (25), placed the reduced aquatic genus *Ceratophyllum* at the base, whereas the remaining angiosperms split into (*a*) other magnoliids and monocots and (*b*) eudicots. However, judging from various experiments and other molecular results, this is probably a long-branch effect (38, 103). Both 18S (118) and *atp*B (114) place Nymphaeales near the base but variously associated with three woody taxa: *Amborella*, a vesselless shrub, and *Austrobaileya*, a liana, both often placed in Laurales, and Illiciales, which are small trees and lianas. When these data are combined with *rbcL*, the same result is found (117). Based on these trees, it cannot be inferred whether the first angiosperms were woody magnoliids or paleoherbs. However, the basal taxa do share one primitive-looking feature (56, 78): In most angiosperms the carpel margins are closed by postgenital fusion of the epidermal layers, but here they are sealed by a secretion.

Among magnoliids, molecular data imply that Magnoliales (purged of Winteraceae, *Austrobaileya*, and *Lactoris*) form a clade; apparently their granular exine structure is secondary, not primitive. Morphological analyses (39, 88) linked Winteraceae with Illiciales, but molecular data associate them with Canellaceae (the next-closest group in some morphological trees), which have fused stamens and carpels but winteraceous leaves. Although molecular data move some former Laurales to the base of the angiosperms, the rest (Caly-canthaceae, Monimiaceae s. lat., Lauraceae) form a clade united by nodal anatomy, hypanthium, and reduction in ovule number (105). The position of Chloranthaceae—which have been (a) linked with Piperales because they have no perianth and orthotropous ovules, (b) considered basal, and (c) inter-preted as reduced Laurales—seems isolated and unresolved and may remain so until Trimeniaceae, which have similar but less reduced flowers (39, 55), are included. Among paleoherbs, molecular data confirm morphological indica-tions (39) that Piperales and Aristolochiales form a clade and the controversial genus *Lactoris* belongs with Aristolochiaceae.

Eudicots

Phylogenetic analyses indicate that all non-magnoliid dicots, which make up six to nine subclasses recognized by Takhtajan (127, 128) and Cronquist (34, 35), form a monophyletic group named eudicots (49). Their most obvious mor-phological synapomorphy is tricolpate (and derived) pollen; another is loss of the oil cells typical of most magnoliids. An apparent exception is Illiciales, described as having three fused colpi, but these colpi are oriented—not like those of eudicots—but like the arms of a branched sulcus (77). Takhtajan and Cronquist assumed that tricolpate pollen arose several times, but they gave little evidence for links between eudicots and different magnoliids. Morphological analyses (39, 88) unite herbaceous Ranunculidae (kept in Magnoliidae by Cron-quist) and "lower" Hamamelidae, wind-pollinated trees with reduced flowers, such as Hamamelidaceae, *Platanus*, and Trochodendraceae (*Trochodendron*, *Tetracentron*). Eudicots are monophyletic in *rbc*L and *atp*B trees; with 18S, a few eudicots are inserted among magnoliids and vice versa, but the iden-tity and position of these vary among shortest trees. In general, 18S shows more anomalous relationships than *rbc*L and *atp*B, an impression confirmed by tests of congruence among data sets (117). Outgroup relations of the eudicots are unresolved; in morphological analyses (39, 88) they are earlier linked with paleoherbs based on palmate leaf venation and differentiated stamens, but in molecular trees they occupy various positions among magnoliids.

Molecular data agree remarkably on which groups are near the base of the eudicots, although their exact order is unresolved. These are ranunculids (in-cluding Papaverales and, surprisingly, the woody genus *Euptelea*, considered a hamamelid), the aquatic *Nelumbo* [often placed in Nymphaeales, but anoma-lous in having tricolpate pollen and earlier linked with ranunculids (39)], sev-eral lower hamamelids (*Platanus*, Trochodendraceae), and some unexpected groups. One of the latter is Proteaceae, placed in Rosidae but suspected of being lower (80); consistent with this, Proteaceae have triporate pollen with simple pores, whereas most triporate rosids have compound pores. Another unexpected group is Buxaceae, often associated with Euphorbiaceae, which may be related to Trochodendraceae; both groups have striate-reticulate pollen. Many of these taxa have dimerous (or seemingly tetramerous) flowers, with parts in alternating pairs (Papaverales, Proteaceae, *Tetracentron*, Buxaceae). This suggests that a shift from trimerous or spiral-chaotic to dimerous floral phyllotaxy, later modified to pentamerous, is another synapomorphy of eudicots (51). Most Early Cretaceous eudicots are related to these lines [*Nelumbites*, platanoid leaves and inflorescences, buxaceous flowers (33, 50, 132)].

These results support a modified version of a scenario for floral and pollination evolution proposed by Walker & Walker (135) and Ehrendorfer (54): that the transition from magnoliids to lower hamamelids was associated with floral reduction and loss of petals, as a result of spread into cooler areas where reversion from insect to wind pollination was favored, followed in the evolution of rosids and dilleniids by reorigination of petals from stamens to attract more advanced insects. However, more of the intermediate lines (ranunculids, *Nelumbo*) have petals than was thought, and the hypothesis that petals originated from stamens needs closer scrutiny in a phylogenetic and developmental context.

Above the lower eudicots, all three molecular data sets show a split into two huge clades, called Rosidae and Asteridae (25, 114, 118) but representing these groups only in a highly modified sense. I call these the rosid line and the asterid line. These results break up the Dilleniidae, which were distinguished from Rosidae on the direction of stamen initiation when stamens are numerous: centripetal in Rosidae (the normal acropetal order of appendage formation), centrifugal in Dilleniidae, often grouped into fascicles derived from five primordia. The specific partitioning of dilleniids was anticipated by Hickey & Wolfe (76), based on leaf venation: Their palmate dilleniids belong to the rosid line, pinnate dilleniids to the asterid line. Centrifugal stamens apparently originated several times, probably as part of a widespread tendency for increase in stamen number (secondary polyandry) among more advanced insect-pollinated groups, which in other cases followed the centripetal route (54, 85). Both lines have basically tricolporate pollen (with compound apertures), an advance over the tricolpate type of lower eudicots.

Several smaller groups are not consistently associated with one line or the other. Many have primitive features such as tricolpate pollen, which suggests they belong below the rosid-asterid split. One is Caryophyllidae, known for their peculiar anatomy, embryology, and betalain pigments; it includes Polygonaceae, Plumbaginaceae, and more unexpectedly the insectivorous Droseraceae and Nepenthaceae. Contrary to some suggestions, these have no direct connection with Ranunculaceae. The saxifragoids (116) include woody

Hamamelidaceae and *Cercidiphyllum*, put in Hamamelidae because of their reduced flowers, and mostly herbaceous Saxifragaceae, Crassulaceae (Rosidae), and *Paeonia* (Dilleniidae). This suggests a trend from wind-pollinated trees to insect-pollinated herbs. These groups often have palmately lobed leaves, two carpels, and a hypanthium. Many Saxifragaceae in the older sense, including most woody members, are scattered among rosids and asterids (25, 116). Three other lines are Dilleniaceae, which show no molecular relationship to other dilleniids and have more primitive tricolpate pollen and free carpels; Vitaceae, which like saxifragoids have two carpels and palmately lobed leaves; and Santalales, which show trends for parasitism and ovule reduction.

Molecular analyses have consistently revealed three major clades in the rosid line, plus several lines of more uncertain position.

One clade includes Rosaceae, Leguminosae, Rhamnaceae (Rosidae), and Cucurbitales (Dilleniidae). Most important, it also includes the wind-pollinated Amentiferae, or "higher" Hamamelidae, thought to represent the culmination of a trend to wind pollination begun in lower Hamamelidae. These consist of two separate lines, Urticales and Fagales s. lat. (93), including Fagaceae, Betulaceae, Juglandaceae, and Myricaceae. These results continue the dismemberment of the Hamamelidae. The idea that lower and higher hamamelids are unrelated was anticipated by palynology (134): Lower hamamelids have tricolpate pollen, whereas Urticales and most Fagales (except Fagaceae) have triporate pollen with compound pores, apparently derived from rosid-type tricolporate pollen via the triangular Normapolles of the Late Cretaceous (63). These groups thus represent independent later reversions from insect to wind pollination. Some data link Fagales with Cucurbitales; this may seem absurd, but both groups do have basically three carpels and an inferior ovary. Urticales are consistently linked with Rhamnaceae; this suggests that their four to five tepals, usually assumed to be sepals, are really petals because they are opposite the stamens, like the petals in Rhamnaceae. Legumes are always linked with Polygalaceae—an irony because polygalaceous flowers are superficially legume-like, but the similar-looking parts are not morphologically equivalent. A possible morphological synapomorphy of this whole clade is loss of endosperm in the seed.

The second clade includes Sapindales, many with pinnately compound leaves, and the dilleniid orders Capparales and Malvales (including Theales such as Dipterocarpaceae). Capparales are greatly expanded from older concepts (centered on Cruciferae and Capparaceae) to include almost all groups (except *Drypetes*) with glucosinolates (mustard oils), such as Caricaceae, Tropaeolaceae, and Limnanthaceae, formerly scattered among several orders (108). In this case, molecular data show that secondary compounds are of greater systematic value than was recognized. The connection of Capparales and Sapindales is not surprising because several glucosinolate groups were formerly placed in Sapindales. Two lines more equivocally linked with this clade are Myrtales [including Vochysiaceae, which have only one stamen but myrtalian anatomy (28)] and Geraniales in a narrow sense.

The third clade includes Euphorbiaceae, Malpighiaceae, Linales, and many Violales. Interestingly, it has been debated whether Euphorbiaceae are rosids or dilleniids. These results imply that stamen fascicles are less significant than was thought, but the fact that many of these taxa have three carpels may be more significant. Two less securely associated lines are Celastrales (in a restricted sense) and Oxalidales, including Cunoniaceae, often considered a link between lower hamamelids and rosids.

The asterid line contains Asteridae in the original sense, with a sympetalous corolla and epipetalous stamens, plus Cornales (Rosidae) and the remaining Dilleniidae, including Ericales, some Theales, Primulales, and Ebenales. Many of the latter groups were once associated with asterids as Sympetalae, whereas Takhtajan (127, 128) considered Cornales a link between Rosidae and Asteridae. Philipson (100) anticipated the whole group based on unitegmic-tenuinucellate ovules and chemistry (iridoid substances), another example of the value of chemical characters, although exceptions to the ovule characters in lower groups imply that they originated in parallel or were easily reversed. Hydrangeaceae, considered woody Saxifragaceae, are strongly linked with Cornales (116). These results imply that the numerous stamens of Theaceae are an example of secondary increase, not primitive. The dilleniid groups vary between free and fused petals; better resolution is needed to judge whether sympetaly is homologous here and in higher asterids but was often reversed or whether it arose many times.

Molecular analyses confirm Takhtajan's (128) separation of Lamiidae from Asteridae. The two groups are also supported by a shift from trilacunar to unilacunar nodes in Lamiidae and from superior to inferior ovary in Asteridae. Small groups near the base of the two lines include Aquifoliaceae, *Escallonia* (formerly Saxifragaceae), and a clade consisting of *Garrya*, *Aucuba* (Cornales), and *Eucommia* (misplaced in Hamamelidae because of its reduced flowers). Within Lamiidae, Labiatae and Verbenaceae are not linked with Boraginaceae (34, 35) but instead are nested within Scrophulariales, which implies that the gynobasic styles and four nutlets of Boraginaceae and Labiatae are purely convergent.

Molecular data link two former rosid groups, Araliales (Apiales) and Pittosporaceae, with each other and with Asteridae, a relationship supported by chemical features such as polyacetylenes. Araliales differ from typical asterids in having free petals, but in ontogeny the petals arise from a ring-like primordium. This suggests that they had sympetalous ancestors (57). Molecular data refute the hypothesis (34, 35) that Compositae are related to Rubiaceae

(Gentianales, Lamiidae); instead, they are nested in Campanulales, which like Compositae have inulin as a storage product and a "plunger" pollen presentation mechanism. Within Compositae, both a chloroplast inversion and restriction site data support the basal position of Barnadesiinae, with bilabiate florets, and the derived status of Asteroideae, with disk and ray florets (79).

Monocots

Monocots are clearly nested among magnoliids, consistent with the fact that they retain monosulcate pollen, but their closest outgroups are still uncertain. Some morphological analyses (39, 48, 88) support the classic concept that monocots are related to Nymphacales, based on loss of secondary growth and aquatic habit (as in alismids). This suggests that monocot advances such as linear leaves, scattered bundles, and one cotyledon arose in aquatic habitats. Other analyses of morphology (44) and partial rRNA and 18S sequences (15, 48) link them with Aristolochiaceae, the only outgroup with monocot-like PII sieve tube plastids. However, rbcL, the large 18S data set, and atpB (25, 114, 118) place them in inconsistent positions among paleoherbs and woody magnoliids.

Within monocots, molecular data confirm many suggestions of Dahlgren et al (36) and more recent morphological analyses (15, 124). Alismidae have often been considered primitive, based on free carpels, aquatic habit, and laminar placentation (as in Nymphaeales), but they have several advances over other monocots, such as trinucleate pollen and loss of endosperm. Instead, Dahlgren et al (36) linked alismids with Arales [placed in Arecidae by others (34, 35, 127)], based on extrorse anthers and amoeboid tapetum, and argued that the most primitive monocots are dioscoreids (Liliidae), which are lianas with palmately veined, dicot-like leaves, recalling Aristolochiaceae. Dioscoreids form a basal paraphyletic grade in morphological analyses (15, 124), but most molecular analyses (15, 25, 26, 37) place alismids and aroids together just below dioscoreids. An unexpectedly important taxon is *Acorus*, which was placed in Araceae because it has a spadix but differs in having equitant, unifacial leaves and magnoliid-like oil cells (68). *Acorus* is the sister group of all other monocots (25), linked with *Ceratophyllum* (114) or the primitive aroid *Gymnostachys* (37) in the same position, or dissociated from monocots entirely, near Piperales (15, 118). *rbcL* and fossil intermediates (126) confirm that the highly reduced, aquatic Lemnaceae are nested in Araceae, with the floating aquatic *Pistia*.

These results have important implications for monocot leaf evolution. In dicots, the blade develops from the upper zone of the leaf primordium, whereas in monocots, it usually develops from the lower leaf zone while the upper leaf zone becomes the apical "Vorläuferspitze" (14, 81). This led to suggestions that monocots went through a stage in which the blade was lost. This might seem less likely if the dicot-like leaves of dioscoreids were primitive. Although molecular trees indicate that dioscoreids are not basal, they still challenge the leaf reduction scenario because the blade develops from the upper leaf zone in *Acorus*, alismids, and aroids. This implies that the typical monocot pattern arose within the group (14). Contrary to usual assumptions, molecular data also imply that free carpels in monocots are secondarily derived from fused carpels (as in Aristolochiaceae) because apocarps (alismids, Melanthiaceae, some palms) are nested among syncarps.

As proposed by Dahlgren et al (36), *rbc*L data (24–26) group most Liliidae into two lines, Liliales and Asparagales. Derived features of Liliales include bulbs, spotted tepals, and tepalar nectaries. Asparagales, which include several secondarily woody lines, have capsules containing seeds with a black phytomelan crust or berries. Dahlgren et al (36) linked Iridaceae and Orchidaceae with Liliales, but *rbc*L places both taxa near Asparagales, with orchids linked with Hypoxidaceae.

Continuing the breakup of Arecidae, *rbc*L and restriction site data (25, 26, 37, 86) confirm the view of Dahlgren et al (36) that palms are the sister group of Commelinidae, based on fluorescent epidermal cell walls (cell-wall bound ferulic acids). Two other arecid groups, Cyclanthaceae and Pandanaceae, belong near Liliales. Commelinids themselves appear to be monophyletic, united by a shift to starchy endosperm (as in cereal grasses). Dahlgren et al (36) admitted one exception to the cell wall character in commelinids, Velloziaceae, but *rbc*L places this family near Liliales (26).

Petaloid taxa (Zingiberales, Bromeliaceae, etc) occupy poorly resolved positions near the base of commelinids, whereas Gramineae and Cyperaceae are the culmination of two lines that show floral reduction for wind pollination. Molecular data (25, 37, 86) confirm that Juncaceae and Cyperaceae are related, as inferred from their diffuse centromeres, the tetrad pollen of Juncaceae, and the pseudomonads (tetrads in which three nuclei abort) of Cyperaceae. Typhales belong near here rather than being wind-pollinated aroids (130). Gramineae are linked with Restionaceae and, more closely, with *Joinvillea* (37, 82), which resembles grasses in having long and short epidermal cells.

CONCLUSIONS

Judging from increasing congruence of results based on different kinds of data, phylogenetic analyses are approaching definitive answers to many of the most vexing questions of vascular plant phylogeny (relationships at the base of tracheophytes, within ferns, between cycadophytes and coniferophytes, and within eudicots and monocots). Other problems (rooting of angiosperms) may be resolved by addition of taxa and/or a combination of more molecular data sets.

However, some problems (relationship of angiosperms and Gnetales) may remain controversial for some time, requiring massive combination of molecular data, advances in analytical methods, or discovery of fossils on the stem lineages leading to modern groups. Even if molecular data resolve the relationships of modern groups, fossil taxa, integrated into phylogenies via morphology, will be needed to understand the evolution of many characters of modern plants.

ACKNOWLEDGMENTS

I wish to thank Vincent Savolainen, Doug Soltis, and Sasha Stefanovic for unpublished results, and Brent Mishler for comments on the manuscript.

Visit the *Annual Reviews* home page at
http://www.AnnualReviews.org

Literature Cited

1. Albert VA, Backlund A, Bremer K, Chase MW, Manhart JR, et al. 1994. Functional constraints and *rbc*L evidence for land plant phylogeny. *Ann. Mo. Bot. Gard.* 81: 534–67
2. Arber EAN, Parkin J. 1907. On the origin of angiosperms. *J. Linn. Soc. Bot.* 38:29–80
3. Arber EAN, Parkin J. 1908. Studies on the evolution of the angiosperms. The relationship of the angiosperms to the Gnetales. *Ann. Bot.* 22:489–515
4. Archangelsky S. 1965. Fossil Ginkgoales from the Tico Flora, Santa Cruz Province, Argentina. *Bull. Br. Mus. Nat. Hist. Geol.* 10:121–37
5. Arnold CA. 1948. Classification of gymnosperms from the viewpoint of paleobotany. *Bot. Gaz.* 110:2–12
6. Ash SR. 1972. Late Triassic plants from the Chinle Formation in northeastern Arizona. *Palaeontology* 15:598–618
7. Bailey IW. 1944. The development of vessels in angiosperms and its significance in morphological research. *Am. J. Bot.* 31: 421–28
8. Banks HP. 1968. The early history of land plants. In *Evolution and Environment*, ed. ET Drake, pp. 73–107. New Haven: Yale Univ. Press
9. Bateman RM, Crane PR, DiMichele WA, Kenrick PR, Rowe NP, et al. 1998. Early evolution of land plants: phylogeny, physiology and ecology of the primary terrestrial radiation. *Annu. Rev. Ecol. Syst.* 29:263–92
10. Bateman RM, DiMichele WA. 1994. Het-

erospory: the most iterative key innovation in the evolutionary history of the plant kingdom. *Biol. Rev.* 69:345–417
11. Bateman RM, DiMichele WA, Willard DA. 1992. Experimental cladistic analyses of anatomically preserved arborescent lycopsids from the Carboniferous of Euramerica: an essay in paleobotanical phylogenetics. *Ann. Mo. Bot. Gard.* 79:500–59
12. Beck CB. 1966. On the origin of gymnosperms. *Taxon* 15:337–39
13. Beck CB. 1970. The appearance of gymnospermous structure. *Biol. Rev.* 45:379–400
14. Bharathan G. 1996. Does the monocot mode of leaf development characterize all monocots? *Aliso* 14:271–79
15. Bharathan G, Zimmer EA. 1995. Early branching events in monocotyledons. Partial 18S ribosomal DNA sequence analysis. See Ref. 113, 1.81–107
16. Bierhorst DW. 1971. *Morphology of Vascular Plants*. New York: Macmillan
17. Bierhorst DW. 1977. The systematic position of *Psilotum* and *Tmesipteris*. *Brittonia* 29:3–13
18. Bowe LM, DePamphilis CW. 1997. Conflict and congruence among three plant genetic compartments: phylogenetic analysis of seed plants from gene sequences of *coxI*, *rbcL* and 18S rDNA. *Am. J. Bot.* 84(Suppl.):178–79
19. Bower FO. 1923. *The Ferns (Filicales)*, Vol. 1. Cambridge: Cambridge Univ. Press
19a. Bower FO. 1926. *The Ferns (Filicales)*,

Vol. 2. Cambridge: Cambridge Univ. Press

19b. Bower FO. 1928. *The Ferns (Filicales)*, Vol. 3, Cambridge: Cambridge Univ. Press

20. Bremer K. 1985. Summary of green plant phylogeny and classification. *Cladistics* 1:369–85

21. Camus JM, Gibby M, Johns RJ, eds. 1996. *Pteridology in Perspective*. Kew, UK: R. Bot. Gard.

22. Carmichael JS, Friedman WE. 1996. Double fertilization in *Gnetum gnemon* (Gnetaceae): its bearing on the evolution of sexual reproduction within the Gnetales and the anthophyte clade. *Am. J. Bot.* 83:767–80

23. Chamberlain CJ. 1935. *Gymnosperms: Structure and Evolution*. Chicago: Univ. Chicago Press

24. Chase MW, Duvall MR, Hills HG, Conran JG, Cox AV, et al. 1995. Molecular phylogenetics of Lilianae. See Ref. 113, 1:109–37

25. Chase MW, Soltis DE, Olmstead RG, Morgan D, Les DH, et al. 1993. Phylogenetics of seed plants: an analysis of nucleotide sequences from the plastid gene *rbc*L. *Ann. Mo. Bot. Gard.* 80:526–80

26. Chase MW, Stevenson DW, Wilkin P, Rudall PJ. 1995. Monocot systematics: a combined analysis. See Ref. 113, 2:685–730

27. Chaw SM, Zharkikh A, Sung HM, Lau TC, Li WH. 1997. Molecular phylogeny of extant gymnosperms and seed plant evolution: analysis of nuclear 18S rRNA sequences. *Mol. Biol. Evol.* 14:56–68

28. Conti E, Litt A, Sytsma KJ. 1996. Circumscription of Myrtales and their relationships to other rosids: evidence from *rbc*L data. *Am. J. Bot.* 83:221–33

29. Cordi J, Stein WE. 1997. A reinvestigation of *Cladoxylon dawsonii* Read: New data question taxonomic and phylogenetic distinctions between the Devonian Cladoxylopsida and Iridopteridales. *Am. J. Bot.* 84(Suppl.):132

30. Crane PR. 1985. Phylogenetic analysis of seed plants and the origin of angiosperms. *Ann. Mo. Bot. Gard.* 72:716–93

31. Crane PR. 1985. Phylogenetic relationships in seed plants. *Cladistics* 1:329–48

32. Crane PR, Blackmore S, eds. 1989. *Evolution, Systematics, and Fossil History of the Hamamelidae*. Oxford, UK: Clarendon

33. Crane PR, Pedersen KR, Friis EM, Drinnan AN. 1993. Early Cretaceous (early to middle Albian) platanoid inflorescences

associated with *Sapindopsis* leaves from the Potomac Group of eastern North America. *Syst. Bot.* 18:328–44

34. Cronquist A. 1968. *The Evolution and Classification of Flowering Plants*. Boston, MA: Houghton Mifflin

35. Cronquist A. 1981. *An Integrated System of Classification of Flowering Plants*. New York: Columbia Univ. Press

36. Dahlgren RMT, Clifford HT, Yeo PF. 1985. *The Families of the Monocotyledons. Structure, Evolution, and Taxonomy*. Berlin: Springer

37. Davis JL. 1995. A phylogenetic structure for the monocotyledons, as inferred from chloroplast DNA restriction site variation, and a comparison of measures of clade support. *Syst. Bot.* 20:503–27

38. Donoghue MJ. 1994. Progress and prospects in reconstructing plant phylogeny. *Ann. Mo. Bot. Gard.* 81:405–18

39. Donoghue MJ, Doyle JA. 1989. Phylogenetic analysis of angiosperms and the relationships of Hamamelidae. See Ref. 32, 1:17–45

40. Donoghue MJ, Doyle JA, Gauthier J, Kluge AG, Rowe T. 1989. The importance of fossils in phylogeny reconstruction. *Annu. Rev. Ecol. Syst.* 20:431–60

41. Donoghue MJ, Sanderson MJ. 1992. The suitability of molecular and morphological evidence in reconstructing plant phylogeny. See Ref. 119, pp. 340–68

42. Doyle JA. 1978. Origin of angiosperms. *Annu. Rev. Ecol. Syst.* 9:365–92

43. Doyle JA. 1994. Origin of the angiosperm flower: a phylogenetic perspective. *Plant Syst. Evol. Suppl.* 8:7–29

44. Doyle JA. 1996. Seed plant phylogeny and the relationships of Gnetales. *Int. J. Plant Sci.* 157(Suppl.):S3–39

45. Doyle JA. 1998. Molecules, morphology, fossils, and the relationship of angiosperms and Gnetales. *Mol. Phylog. Evol.*

46. Doyle JA, Donoghue MJ. 1986. Seed plant phylogeny and the origin of angiosperms: an experimental cladistic approach. *Bot. Rev.* 52:321–431

47. Doyle JA, Donoghue MJ. 1992. Fossils and seed plant phylogeny reanalyzed. *Brittonia* 44:89–106

48. Doyle JA, Donoghue MJ, Zimmer EA. 1994. Integration of morphological and ribosomal RNA data on the origin of angiosperms. *Ann. Mo. Bot. Gard.* 81:419–50

49. Doyle JA, Hotton CL. 1991. Diversification of early angiosperm pollen in a cladistic context. In *Pollen and Spores: Patterns of Diversification*, ed. S Black-

more, SH Barnes, pp. 169–195. Oxford, UK: Clarendon
50. Drinnan AN, Crane PR, Friis EM, Pedersen KR. 1991. Angiosperm flowers and tricolpate pollen of buxaceous affinity from the Potomac Group (mid-Cretaceous) of eastern North America. *Am. J. Bot.* 78:153–76
51. Drinnan AN, Crane PR, Hoot SB. 1994. Patterns of floral evolution in the early diversification of non-magnoliid dicotyledons (eudicots). *Plant Syst. Evol. Suppl.* 8:93–122
52. Eames AJ. 1952. Relationships of the Ephedrales. *Phytomorphology* 2:79–100
53. Eckenwalder JE. 1976. Re-evaluation of Cupressaceae and Taxodiaceae: a proposed merger. *Madroño* 23:237–56
54. Ehrendorfer F. 1989. The phylogenetic position of the Hamamelidae. See Ref. 32, 1:1–7
55. Endress PK. 1987. The Chloranthaceae: reproductive structures and phylogenetic position. *Bot. Jahrb. Syst.* 109:153–226
56. Endress PK, Igersheim A. 1997. Gynoecium diversity and systematics of Laurales. *Bot. J. Linn. Soc.* 125:93–168
57. Erbar C. 1991. Sympetaly—a systematic character? *Bot. Jahrb. Syst.* 112:417–51
58. Felsenstein J. 1978. Cases in which parsimony or compatibility methods will be positively misleading. *Syst. Zool.* 27:401–10
59. Florin R. 1949. The morphology of *Trichopitys heteromorpha* Saporta, a seedplant of Palaeozoic age, and the evolution of the female flowers in the Ginkgoinae. *Acta Horti Berg.* 15:79–109
60. Florin R. 1951. Evolution in cordaites and conifers. *Acta Horti Berg.* 15:285–388
61. Friedman WE. 1990. Sexual reproduction in *Ephedra nevadensis* (Ephedraceae): further evidence of double fertilization in a nonflowering seed plant. *Am. J. Bot.* 77:1582–98
62. Friedman WE. 1994. The evolution of embryogeny in seed plants and the developmental origin and early history of endosperm. *Am. J. Bot.* 81:1468–86
63. Friis EM. 1983. Upper Cretaceous (Senonian) floral structures of juglandalean affinity containing Normapolles pollen. *Rev. Palaeobot. Palynol.* 39:161–88
64. Garbary DJ, Renzaglia KS, Duckett JG. 1993. The phylogeny of land plants: a cladistic analysis based on male gametogenesis. *Plant Syst. Evol.* 188:237–69
65. Gaussen H. 1946. Les Gymnospermes, actuelles et fossiles. *Trav. Lab. For. Toulouse Tome II Etud. Dendrol.* 1(Sect. 1, Chap. 5):1–26

66. Goremykin V, Bobrova V, Pahnke J, Troitsky A, Antonov A, Martin W. 1996. Noncoding sequences from the slowly evolving chloroplast inverted repeat in addition to *rbcL* data do not support gnetalean affinities of angiosperms. *Mol. Biol. Evol.* 13:383–96
67. Graham LE. 1993. *Origin of Land Plants.* New York: Wiley
68. Grayum MH. 1987. A summary of evidence and arguments supporting the removal of *Acorus* from the Araceae. *Taxon* 36:723–29
69. Hamby RK, Zimmer EA. 1992. Ribosomal RNA as a phylogenetic tool in plant systematics. See Ref. 119, pp. 50–91
70. Harris TM. 1976. The Mesozoic gymnosperms. *Rev. Palaeobot. Palynol.* 21:119–34
71. Hart JA. 1987. A cladistic analysis of conifers: preliminary results. *J. Arnold Arbor. Harv. Univ.* 68:269–307
72. Hasebe M, Kofuji R, Ito M, Kato M, Iwatsuki K, Ueda K. 1992. Phylogeny of gymnosperms inferred from *rbcL* gene sequences. *Bot. Mag. Tokyo* 105:673–79
73. Hasebe M, Omori T, Nakazawa M, Sano T, Kato M, Iwatsuki K. 1994. *rbcL* gene sequences provide evidence for the evolutionary lineages of leptosporangiate ferns. *Proc. Natl. Acad. Sci. USA* 91:5730–34
74. Hasebe M, Wolf PG, Pryer KM, Ueda K, Ito M, et al. 1995. Fern phylogeny based on *rbcL* nucleotide sequences. *Am. Fern J.* 85:134–81
75. Hennig W. 1966. *Phylogenetic Systematics.* Urbana, IL: Univ. Ill. Press
76. Hickey LJ, Wolfe JA. 1975. The bases of angiosperm phylogeny: vegetative morphology. *Ann. Mo. Bot. Gard.* 62:538–89
77. Huynh KL. 1976. L'arrangement du pollen du genre *Schisandra* (Schisandraceae) et sa signification phylogénique chez les Angiospermes. *Beitr. Biol. Pflanzen.* 52:227–53
78. Igersheim A, Endress PK. 1997. Gynoecium diversity and systematics of the Magnoliales and winteroids. *Bot. J. Linn. Soc.* 124:213–71
79. Jansen RK, Michaels HJ, Palmer JD. 1991. Phylogeny and character evolution in the Asteraceae based on chloroplast DNA restriction site mapping. *Syst. Bot.* 16:98–115
80. Johnson LAS, Briggs BG. 1975. On the Proteaceae—the evolution and classification of a southern family. *Bot. J. Linn. Soc.* 70:83–182
81. Kaplan DR. 1973. The problem of leaf morphology and evolution in the monocotyledons. *Q. Rev. Biol.* 48:437–57

82. Kellogg EA, Linder HP. 1995. Phylogeny of Poales. See Ref. 113, 2:511–42

83. Kenrick P, Crane PR. 1997. *The Origin and Early Diversification of Land Plants: a Cladistic Study.* Washington, DC: Smithsonian

84. Kruckeberg AR. 1997. Essay: Whither plant taxonomy in the 21st century? *Syst. Bot.* 22:181–82

85. Kubitzki K. 1972. Probleme der Grosssystematik der Blütenpflanzen. *Ber. Deut. Bot. Ges.* 85:259–77

86. Linder HP, Kellogg EA. 1995. Phylogenetic patterns in the commelinid clade. See Ref. 113, 2:473–96

87. Loconte H, Stevenson DW. 1990. Cladistics of the Spermatophyta. *Brittonia* 42: 197–211

88. Loconte H, Stevenson DW. 1991. Cladistics of the Magnoliidae. *Cladistics* 7:267–96

89. Lugardon B. 1976. Sur la structure fine de l'exospore dans les divers groupes de Ptéridophytes actuelles (microspores et isospores). In *The Evolutionary Significance of the Exine*, ed IK Ferguson, J Muller, pp. 231–50. London: Academic

90. Mamay SH. 1976. Paleozoic origin of the cycads. *US Geol. Surv. Prof. Pap.* 934:1–48

91. Manhart JR. 1994. Phylogenetic analysis of green plant *rbc*L sequences. *Mol. Phylog. Evol.* 3:114–27

92. Manhart JR. 1995. Chloroplast 16S rDNA sequences and phylogenetic relationships of fern allies and ferns. *Am. Fern J.* 85: 182–92

93. Manos PS, Steele KP. 1997. Phylogenetic analyses of "higher" Hamamelididae based on plastid sequence data. *Am. J. Bot.* 84:1407–19

94. Meyen SV. 1984. Basic features of gymnosperm systematics and phylogeny as evidenced by the fossil record. *Bot. Rev.* 50:1–112

95. Mishler BD, Churchill SP. 1984. A cladistic approach to the phylogeny of the "bryophytes." *Brittonia* 36:406–24

96. Mishler BD, Lewis LA, Buchheim MA, Renzaglia KS, Garbary DJ, et al. 1994. Phylogenetic relationships of the "green algae" and "bryophytes." Ann. *Mo. Bot. Gard.* 81:451–83

97. Nixon KC, Crepet WL, Stevenson D, Friis EM. 1994. A reevaluation of seed plant phylogeny. *Ann. Mo. Bot. Gard.* 81:484–533

98. Pahnke J, Goremykin V, Bobrova V, Troitsky A, Antonov A, Martin W. 1996. Utility of rDNA internal transcribed spacer sequences from the inverted repeat of chloroplast DNA in pteridophyte molecular phylogenetics. See Ref. 21, pp. 217–230

99. Parenti LR. 1980. A phylogenetic analysis of the land plants. *Biol. J. Linn. Soc.* 13:225–42

100. Philipson WR. 1974. Ovular morphology and the major classification of the dicotyledons. *Bot. J. Linn. Soc.* 68:89–108

101. Poort RJ, Visscher H, Dilcher DL. 1996. Zoidogamy in fossil gymnosperms: the centenary of a concept, with special reference to prepollen of late Paleozoic conifers. *Proc. Natl. Acad. Sci. USA* 93: 11713–17

102. Pryer KM, Smith AR, Skog JE. 1995. Phylogenetic relationships of extant ferns based on evidence from morphology and *rbc*L sequences. *Am. Fern J.* 85:205–82

103. Qiu YL, Chase MW, Les DH, Parks CR. 1993. Molecular phylogenetics of the Magnoliidae: cladistic analyses of nucleotide sequences of the plastid gene *rbc*L. *Ann. Mo. Bot. Gard.* 80:587–606

104. Raubeson LA, Jansen RK. 1992. Chloroplast DNA evidence on the ancient evolutionary split in vascular land plants. *Science* 255:1697–99

105. Renner SS, Schwarzbach AE, Lohmann L. 1997. Phylogenetic position and floral function of *Siparuna* (Siparunaceae: Laurales). *Int. J. Plant Sci.* 158(Suppl.):S89–98

106. Retallack G, Dilcher DL. 1981. Arguments for a glossopterid ancestry of angiosperms. *Paleobiology* 7:54–67

107. Rice KA, Donoghue MJ, Olmstead RG. 1997. Analyzing large data sets: *rbc*L 500 revisited. *Syst. Biol.* 46:554–63

108. Rodman JE, Karol KG, Price RA, Sytsma KJ. 1996. Molecules, morphology, and Dahlgren's expanded order Capparales. *Syst. Bot.* 21:289–307

109. Rothwell GW. 1982. New interpretations of the earliest conifers. *Rev. Palaeobot. Palynol.* 37:7–28

110. Rothwell GR. 1996. Phylogenetic relationships of ferns: a palaeobotanical perspective. See Ref. 21, pp. 395–404

111. Rothwell GR, Serbet R. 1994. Lignophyte phylogeny and the evolution of spermatophytes: a numerical cladistic analysis. *Syst. Bot.* 19:443–82

112. Rothwell GR, Stockey RA. 1994. The role of *Hydropteris pinnata* gen. et sp. nov. in reconstructing the cladistics of heterosporous ferns. *Am. J. Bot.* 81:479–92

113. Rudall PJ, Cribb PJ, Cutler DF, Humphries CJ, ed. 1995. *Monocotyledons: Systematics and Evolution.* Kew, UK: R. Bot. Gard. Vols. 1, 2

114. Savolainen V, Chase MW, Morton CM, Hoot SB, Soltis DE, et al. Phylogenetics of flowering plants based upon a combined analysis of plastid *atp*B and *rbc*L gene sequences. *Syst. Biol.* In press
115. Skog JE, Banks HP. 1973. *Ibyka amphikoma*, gen. et sp. n., a new protoarticulate precursor from the late Middle Devonian of New York State. *Am. J. Bot.* 60: 366–80
116. Soltis DE, Soltis PS. 1997. Phylogenetic relationships in Saxifragaceae sensu lato: a comparison of topologies based on 18S rDNA and *rbc*L sequences. *Am. J. Bot.* 84: 504–22
117. Soltis DE, Soltis PS, Mort ME, Chase MW, Savolainen V, et al. 1998. Inferring complex phylogenies using parsimony: an empirical approach using three large DNA data sets for angiosperms. *Syst. Biol.* 47:32–42
118. Soltis DE, Soltis PS, Nickrent DL, Johnson LA, Hahn WJ, et al. 1997. Angiosperm phylogeny inferred from 18S ribosomal DNA sequences. *Ann. Mo. Bot Gard.* 84:1–49
119. Soltis PS, Soltis DE, Doyle JJ, eds. 1992. *Molecular Systematics of Plants.* New York: Chapman & Hall
120. Stebbins GL. 1974. *Flowering Plants: Evolution Above the Species Level.* Cambridge, MA: Harv. Univ. Press
121. Stefanovic S, Jager M, Deutsch J, Broutin J, Masselot M. 1998. Phylogenetic relationships of conifers inferred from partial 28S rRNA gene sequences. *Am. J. Bot.* 85:688–97
122. Stein WE, Wight DC, Beck CB. 1984. Possible alternatives for the origin of Sphenopsida. *Syst. Bot.* 9:102–18
123. Stevenson DW. 1990. Morphology and systematics of the Cycadales. *Mem. New York Bot. Gard.* 57:8–55
124. Stevenson DW, Loconte H. 1995. Cladistic analysis of monocot families. See Ref. 113, 2:543–78
125. Stewart WN. 1983. *Paleobotany and the Evolution of Plants.* Cambridge, UK: Cambridge Univ. Press
126. Stockey RA, Hoffman GL, Rothwell GW. 1997. The fossil monocot *Limnobiophyllum scutatum*: resolving the phylogeny of Lemnaceae. *Am. J. Bot.* 84:355–68
127. Takhtajan AL. 1966. *Sistema i Filogeniya Tsvetkovykh Rasteniy.* Moscow: Nauka
128. Takhtajan AL. 1997. *Diversity and Classification of Flowering Plants.* New York: Columbia Univ. Press
129. Taylor DW, Hickey LJ. 1992. Phylogenetic evidence for the herbaceous origin of angiosperms. *Plant Syst. Evol.* 180: 137–56
130. Thorne RF. 1976. A phylogenetic classification of the Angiospermae. *Evol. Biol.* 9:35–106
131. Townrow JA. 1967. On *Rissikia* and *Mataia*, podocarpaceous conifers from the lower Mesozoic of southern lands. *Pap. Proc. R. Soc. Tasmania* 101:103–36
132. Upchurch GR, Crane PR, Drinnan AN. 1994. The megaflora from the Quantico locality (upper Albian), Lower Cretaceous Potomac Group of Virginia. *Va. Mus. Nat. Hist. Mem.* 4:1–57
133. van Konijnenburg-van Cittert JHA. 1992. An enigmatic Liassic microsporophyll, yielding *Ephedripites* pollen. *Rev. Palaeobot. Palynol.* 71:239–54
134. Walker JW, Doyle JA. 1975. The bases of angiosperm phylogeny: palynology. *Ann. Mo. Bot. Gard.* 62:664–723
135. Walker JW, Walker AG. 1984. Ultrastructure of Lower Cretaceous angiosperm pollen and the origin and early evolution of flowering plants. *Ann. Mo. Bot. Gard.* 71:464–521
136. Walton J. 1953. The evolution of the ovule in the pteridosperms. *Adv. Sci.* 10:223–30
137. Wettstein RR. 1907. *Handbuch der Systematischen Botanik.* Leipzig, Germany: Deuticke. Vol. 2
138. Wilde MH. 1944. A new interpretation of coniferous cones: I. Podocarpaceae (*Podocarpus*). *Ann. Bot.* 8:1–41
139. Wolf PG. 1997. Evaluation of *atp*B nucleotide sequences for phylogenetic studies of ferns and other pteridophytes. *Am. J. Bot.* 84:1429–40
140. Zhou Z, Zhang B. 1992. *Baiera hallei* Sze and associated ovule-bearing organs from the Middle Jurassic of Henan, China. *Palaeontogr. Abt. B* 224:151–69
141. Zimmermann W. 1930. *Die Phylogenie der Pflanzen.* Jena, Germany: Fischer

SUBJECT INDEX

A

ABC model of organ identity,
    352–53
Abdel-Hameed, F, 489
Abies, 454
Acanthisitta chloris, 143
Acarapsis woodi, 92
Accretion rates, 185
Accumulation-attrition model for
    sex determination, 245
Achillea millefolium, 477
Acid lysis
    digestion by, 386
Aconitum napellus, 362
Acrocephalus luscinia, 120
A. melanopogon, 144
A. sechellensis, 150
A. vaughani taiti, 160
Actinomorphy, 348–49, 351
Activity patterns of pollinators,
    364–67
Adams, C, 24
Adaptedness trends, 304–6
Adaptive explanations for help,
    145, 147–62
    access to mates, 154–59
    acquisition of skills, 161–62
    enhanced production of
        nondescendant kin, 147,
        149–52
    formation of alliances, 160–61
    improvement of local
        conditions, 159–60
    payment of rent, 152–54
Adaptive Management Areas
    (AMAs)
    setting aside, 452
Adaptive suites and floral
    symmetry, 354
Aerodramus vanikorensis bartschi,
    114
Agents of selection, 16
Aglaophyton, 280
Agricultural practices
    changes in, 97–98
    effects on wild pollinators,
        89–90
Aizen, MA, 88
Alatoconchids, 196
Alces, 223
Algae
    benthic, 67
    coralline, 199–200
    dietary, 395

putative, 188
reef-associated, 200
upwelling water promoting
    activity in, 67
Allee effect, 87
Alleles
    under balancing selection, 4
    under directional selection, 4
    gene pool frequency, 9
    sex-linked, 166
    similar, 1
Allelic lineages
    agents of selection, 16
    age of, 10–12
    concept of, 9–10
    diversification of, 14–16
    HLA-DRB1, 5, 11, 15–16
    Mhc-DRB1, 17
    number of, 10, 12–14
    oldest, 12, 16
    trimmed form of, 14
Alliances
    facilitating formation of, 160
    improving prospect of
        reproduction, 160–61
A. fistolium, 473
Allium cepa, 473
Allopolyploid formation, 467,
    470–71
    and disturbed habitats, 489
    estimating the rate of, 489–94
    via hybridization of
        autopolyploids, 473
    via hybridization of different
        cytotypes, 473–74
Allotetraploid formation
    one-step, 473
    triploid-bridge, 472–73
Allozyme diversity, 25, 440–41
Allozyme heterozygosity
    and polymorphism, 88
Alluvial aquifers
    interacting with hyporheic zone,
        61
Altenberg, L, 302
AMAs
    See Adaptive Management
        Areas
Amia calva, 52
Ammonification, 63
Amon, RMW, 521
Amylases in herbivorous fish,
    388
Anabolism coupled with
    catabolism, 507–8

Anatomical phase of plant
    evolution, 265–66
Ancestral polymorphism, 2
Ancestral polymorphism vs
    convergence, 17
Ancient reef ecosystems
    complexity of, 184–88
    and photosymbiosis, 195–97
Aneuploids
    common, 486
    differences among, 471
    role in polyploid formation, 495
Angermeier, PL, 444
Angiosperms, 586–94
    eudicots, 589–93
    monocots, 593–94
    outcrossing in, 6
    pollination of, 84
    relationships among, 586–89
    storing oceanic buried carbon,
        413
Animal behavior in
    plant-pollinator systems,
        345–69
Animal movement patterns
    adjacent to roads, 211–12,
        225
Animal pollinators
    ecology of, 101
    interactions with plant
        pollinators, 102
Ankyropteris, 573
Anolis carolinensis, 114
    irruption on Guam, 124
Anoxic waters and sediment
    mercury in, 555–59
Anthocerotes, 569
Anthropogenic activities
    effect on coastal zones, 408
    sources of mercury, 546
Antirrhinum majus, 352–53
Aphelocoma coerulescens, 149
Apis mellifera, 92–93, 98
Aplodactylus punctatus, 389
Aplonis opaca, 114
Apocynaceae, 350
Aquatic ecosystems
    bacterial growth efficiency in,
        503–34
    carbon and carbonate
        metabolism in, 405–27
    distribution of macrophytes in,
        68
A. flavellata, 472, 483

601

CUMULATIVE INDEXES

CONTRIBUTING AUTHORS, VOLUMES 25–29

CHAPTER TITLES, VOLUMES 25–29

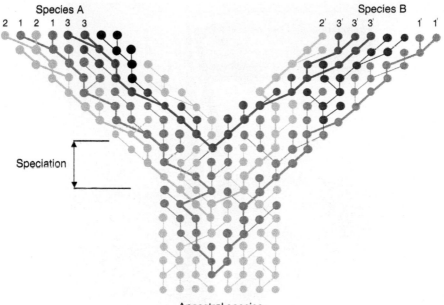

Species A
2 1 2 1 3 3

Species B
2' 3' 3' 3'   1' 1'

Speciation

Ancestral species

Figure 1 The concept of trans-species polymorphism.   Each *row of circles* symbolizes a gene pool of one generation, each *circle* a gene, different colors indicating different alleles.   *Connecting lines* indicate antecedent-descendent relationships, *thick lines* follow coalescence of alleles found in the extant generation. The *light blue, dark blue*, and *orange thick interconnecting lines* indicate allelic lineages 1, 1', 2, 2', and 3, 3' which arose in the ancestral species and were passed on to the two descendent species A and B.

Figure 1 Road corridor showing road surface, maintained open roadsides, and roadside natural strips.   Strips of relatively natural vegetation are especially characteristic of road corridors (know as road reserves) in Australia.   Wheatbelt of Western Australia. Photo courtesy of BMJ Hussey.